Alien Plant Invasions in Native Ecosystems of Hawai‘i

Alien Plant Invasions in Native Ecosystems of Hawai‘i: Management and Research

Edited by

Charles P. Stone,
Clifford W. Smith, and J. Timothy Tunison

University of Hawaii Cooperative National Park Resources Studies Unit
3190 Maile Way · Honolulu, Hawai‘i

Support for publication was provided by
the National Park Service and
Hawaii Natural History Association

Cover illustration by Dietrich Varez; used with permission
Camera-ready copy was prepared by the editors

Manufacture of this book was through
the production services of
University of Hawaii Press

The paper used in this publication meets the minimum
requirements of American National Standard for Information
Sciences/-permanence of Paper for Printed Library Materials

Library of Congress Cataloging-in-Publication Data

Alien plant invasions in native ecosystems of Hawai'i : management and
research / edited by Charles P. Stone, Clifford W. Smith, and J.
Timothy Tunison.
p. cm.
Papers from a symposium organized by the National Park Service and
the Cooperative National Resources Unit at the University of Hawaii,
which was held in conjunction with the Sixth Conference in Natural
Sciences at Hawaii Volcanoes National Park in 1986.
Includes bibliographical references and index.
ISBN 0-8248-1474-6
1. Alien plants--Hawaii--Congresses. 2. Alien plants--Control-
-Hawaii--Congresses. 3. Plant invasions--Hawaii--Congresses.
4. Alien plants--Hawaii--Ecology--Congresses. 5. Plant
introduction--Hawaii--Congresses. I. Stone, Charles P. II. Smith,
Clifford W., 1938- . III. Tunison, J. Timothy, 1948- .
IV. Cooperative National Park Resources Studies Unit, Hawaii.
SB612.H3A56 1993
639.9'9'09969--dc20 92-25041
CIP

Distributed by
University of Hawaii Press
Honolulu, Hawaii 96822

CONTENTS

Preface ix

Acknowledgments x

Explanatory Notes xi

Authors xiii

PERSPECTIVES 1

An Overview of Problems with Introduced Plant Species in National Parks and Biosphere Reserves of the United States 3
Lloyd L. Loope

Effects of Alien Plants on Native Ecosystems 29
Peter M. Vitousek

Characteristics of Invading Plant Species 42
Richard N. Mack

Alien Plant Management in Everglades National Park: An Historical Perspective 47
Anne Marie LaRosa, Robert F. Doren, and Lance Gunderson

Alien Plants at Channel Islands National Park 64
William L. Halvorson

STATUS AND DISTRIBUTION OF SOME IMPORTANT ALIEN PLANTS IN HAWAI'I 97

Origin and Distribution of Adventive Alien Flowering Plants in Hawai'i 99
Lyndon Wester

Distribution of Six Alien Plant Species in Upland Habitats on the Island of Hawai'i 155
James D. Jacobi and Fredrick R. Warshauer

Alien Plants on the Northwestern Hawaiian Islands 189
Derral R. Herbst and Warren L. Wagner

Firetree (*Myrica faya*) Distribution in Hawai'i 225
Louis D. Whiteaker and Donald E. Gardner

Distribution, Status, Phenology, Rate of Spread, and Management of Clidemia in Hawai'i 241
Clifford W. Smith

Mullein (*Verbascum thapsus*): the Spread and Adaptation of a Temperate Weed in the Montane Tropics 254
James O. Juvik and Sonia P. Juvik

The Status of Banana Poka in Hawai'i 271
Anne Marie LaRosa

Distribution and Spread of Alien Plants in Kīpahulu Valley, Haleakala National Park, above 2,300 ft Elevation 300
Stephen J. Anderson, Charles P. Stone, and Paul K. Higashino

PHYSICAL AND CHEMICAL CONTROL 339

Herbicidal Control of Selected Alien Plant Species in Hawaii Volcanoes National Park 341
Gregory L. Santos, Dina Kageler, Donald E. Gardner, Linda W. Cuddihy, and Charles P. Stone

Fountain Grass Control in Hawaii Volcanoes National Park: Management Considerations and Strategies 376
J. Timothy Tunison

Fire and Alien Plants in Hawai'i: Research and ManagementImplications for Native Ecosystems 394
Clifford W. Smith and J. Timothy Tunison

BIOLOGICAL CONTROL 409

Biological Control of Lantana, Prickly Pear, and Hāmākua Pāmakani in Hawai'i: a Review and Update 411
Clifton J. Davis, Ernest Yoshioka, and Dina Kageler

Plant Pathogens as Biocontrol Agents in Native Hawaiian Ecosystems 432
Donald E. Gardner

Review and Status of Biological Control of Clidemia in Hawai'i 452
Larry M. Nakahara, Robert M. Burkhart, and George Y. Funasaki

Status of Biological Control of Weeds in Hawai'i and Implications for Managing Native Ecosystems 466
George P. Markin, Po-Yung Lai, and George Y. Funasaki

AGENCY, ORGANIZATION, AND LANDOWNER APPROACHES 483

Alien Plant Control Strategies in Hawaii Volcanoes National Park 485
J. Timothy Tunison

Success in Controlling Localized Alien Plants in Hawaii Volcanoes National Park 506
J. Timothy Tunison and Nicholas G. Zimmer

Control of Alien Plants on Nature Conservancy Preserves 525
R. Alan Holt

Alien Plant Control on State Lands Including Natural Areas 536
Victor M. Tanimoto and Winona P. Char

Alien Plants in Haleakala National Park 551
Lloyd L. Loope, Ronald J. Nagata, and Arthur C. Medeiros

The U.S. Department of Agriculture's Rural Development Approach to Alien Plant Control in Hawai'i: a Case Study 577
Mike Tulang

Introduced Plant Control on Private Lands in Hawai'i 584
Carl "Soot" Bredhoff

Public/Private Sector Partnerships to Preserve Native Ecosystems: The Department of Hawaiian Home Lands Experience 589
Hardy Spoehr

INTRODUCED PLANTS IN THE ABSENCE OF UNGULATES 595

Relationships among Native and Alien Plants on Pacific Islands with and without Significant Human Disturbance and Feral Ungulates 597
Mark D. Merlin and James O. Juvik

Alien and Native Plant Response to Release from Feral Sheep Browsing on Mauna Kea 625
Paul G. Scowcroft and C. Eugene Conrad

Responses of Hawaiian Ecosystems to Removal of Feral Pigs and Goats 666
Charles P. Stone, Linda W. Cuddihy, and J. Timothy Tunison

CHOOSING WHICH PLANTS TO INTRODUCE 705

Plant Importation Rules and Regulations 707
Stan Higa

Activities and Needs of the Horticulture Industry in Relation to Alien Plant Problems in Hawai'i 712
Rylan S.N. Yee and Wayne C. Gagné

Plant Introduction Needs of the Hawaiian Sugar Industry 726
Robert V. Osgood and Robert D. Wiemer

Choosing Plant Introductions: Needs of Graziers 732
Gordon Cran and Ken Dillingham

CHOOSING WHICH PLANTS TO CONTROL IN NATIVE ECOSYSTEMS 737

Modelling to Predict the Fate of Invading Plants 739
Richard N. Mack and Michael E. Moody

Rule-Based Decision Making: a Way to Determine which Alien Species to Control 743
Kent W. Bridges

Controlling Weeds in Natural Areas in Hawai'i: a Manager's Perspective 752
Dan Taylor

Evaluating Proposed Biological Control Programs for Introduced Plants 757
George P. Markin and Ernest Yoshioka

ECOSYSTEM APPROACHES, EDUCATION, AND COMMUNITY INVOLVEMENT 779

Special Ecological Areas: An Approach to Alien Plant Control in Hawaii Volcanoes National Park 781
J. Timothy Tunison and Charles P. Stone

Environmental Education in Hawai'i: History and Overview 799
Charles P. Stone

Environmental Education in the Schools 807
John W. Hawkins

The Use of Volunteers for Alien Plant Control at Hawaii Volcanoes National Park and The Nature Conservancy's Kamakou Preserve 813
J. Timothy Tunison and Edwin T. Misaki

IN CONCLUSION . . . 819

ALIEN PLANT MANAGEMENT IN HAWAI'I: CONCLUSIONS 821
J. Timothy Tunison, Clifford W. Smith, and Charles P. Stone

INDEX 835

PREFACE

The problem of introduced or alien plants in Hawai'i's natural areas is as critical now as it ever has been. As the "crossroads of the Pacific," Hawai'i is subject to greater commerce and visitation with time. Alien plants are increasingly well distributed in other parts of the world for potential transport to Hawai'i. Growing numbers of residents in the State -- many from different parts of the world -- means more international and interisland travel, which also favors alien plant introduction and dispersal. Fragmentation of Hawai'i's few remaining natural areas provides more boundaries and corridors for alien plants to invade, and degradation of natural areas (for example, disturbance of soil by hoofed mammals and development for industry or housing) provides more sites for introduced plant establishment. Some alien plants can modify areas to favor their own survival and to discourage that of native plants through shading, chemical changes in soils and nutrient cycles, and changing fire regimes. Indeed, the vegetation of many large areas of Hawai'i is now almost entirely comprised of introduced plant species, and significant invasions of natural areas continue.

Most of the 44 papers in this volume began as presentations at a 1986 symposium entitled "Control of Introduced Plants in Hawai'i's Native Ecosystems." The Symposium, organized by the National Park Service and the Cooperative National Park Resources Studies Unit at the University of Hawaii, was held in conjunction with the Sixth Conference in Natural Sciences at Hawaii Volcanoes National Park in 1986. In the past few years, most of the papers have been updated considerably, and a few contributions, including a concluding chapter, have been added. One milestone publication in 1990, the *Manual of the Flowering Plants of Hawai'i*, caused the changing of plant taxonomy in the papers to conform with current usage. For some papers, this was a major task. Other important statewide and local progress in the past few years also necessitated significant revisions.

The purpose of the Symposium was to summarize and communicate research results, management implications, needs, and perspectives of special interest groups on the topic of introduced or alien plants in Hawai'i. Efforts at control of introduced plants in native Hawaiian ecosystems was the emphasis of the meeting and remains the primary focus of this book. We hope that those responsible for and those simply interested in alien plant problems will find something useful herein, whether in the Hawaiian Archipelago, other islands of the world, or in natural areas on the continents.

ACKNOWLEDGMENTS

Each manuscript in this volume received peer review by at least two individuals in addition to the three editors and the chairperson of one of the 10 sessions at which papers were presented. Session chairs (other than the editors) were R. Hobdy, D. Taylor, J.W. Beardsley, C.E. Conrad, J.D. Jacobi, L.L. Loope, and F.M. Pottenger. We thank these individuals and the anonymous reviewers of the manuscripts for their comments.

The one person most responsible for publication of this volume is D.B. Stone, Administrative Assistant for Research at Hawaii Volcanoes National Park. Danny spent endless hours with draft manuscripts, including tables, literature citations, and text. She helped edit all papers for grammar and content and prepared the index for typing. She communicated with the authors by phone and letter about many of the problems and questions that arose, including updating of materials. And she coordinated the progress throughout for the editors and kept us involved even when other duties called. Preparing copy for submission to U.H. Press was a major step, and Danny spent considerable time transforming the rough manuscripts to camera-ready format, using computer software that was by no means state-of-the-art by the time the volume was in its final stages of preparation. It is safe to say that without her watchful and focused eye, this publication would have been a mess and/or appeared sometime after the turn of the century, if at all!

Other folks involved in the production of the book include M.L. Fulton, who prepared many of the figures using "Freelance" software and assisted with editing the literature citations for consistency among papers; D.B. Barrère, who typed the index and checked it and the text for Hawaiian grammar and errors in usage; J. Heavenridge and L. Aono of University of Hawaii Press, who were as competent and pleasant to work with as ever; and Kathy English of the Hawaii Natural History Association, who facilitated partial funding for publication.

N. Sidaras provided the drawings of introduced plants that appear between the sections of the book and elsewhere (sketch identifications are listed below). L.W. Cuddihy helped us with numerous problems of botanical taxonomy and status at various times, and we appreciate her assistance. D. Varez and Bishop Museum Press allowed us to use the block print of Maui the demigod that appears on the cover (from *Maui: the Mischief Maker*, Bishop Museum Press, 1991).

Finally, we appreciate the patience and optimism of the authors and coauthors, who have worked with us so long on this project. Natural resource managers, educators, researchers, and private citizens, some of whom are no longer working on introduced plants and who have moved from previous jobs or locations, have maintained an interest in the "weed book" and seen it through to completion at last. Mahalo nui loa to all.

Quoth the raving, "Nevermore!"

C.P. Stone, C.W. Smith, and J.T. Tunison June 1992

EXPLANATORY NOTES

Scientific names of plants are italicized throughout the book, and only scientific names are used if common names or Hawaiian names do not exist. Hawaiian names and common names are not italicized, and scientific names appear with them on first usage in each paper. Diacritics are used in all words of Hawaiian origin where appropriate, except in names of organizations or institutions. Spelling of Hawaiian names is according to Pukui and Elbert (1981), and Hawaiian place names follow Pukui *et al.* (1986), except where common use differs. Scientific names of Hawaiian flowering plants are according to Wagner *et al.* (1990). Mentioning of products in this volume does not imply endorsement by authors or the agencies or organizations they represent.

Pen and ink drawings by Nanci Sidaras are as follows:

koa haole (*Leucaena leucocephala*) -- page 2
clidemia, or Koster's curse (*Clidemia hirta*) -- page 97
fountain grass (*Pennisetum setaceum*) -- page 339
pānini or prickly pear cactus (*Opuntia* sp.) -- page 409
Florida blackberry (*Rubus argutus*) -- page 483
yellow Himalayan raspberry (*Rubus ellipticus*) -- page 595
kāhili ginger (*Hedychium gardnerianum*) -- page 705
banana poka (*Passiflora mollissima*) -- page 737
firetree (*Myrica faya*) -- page 779
beardgrass (*Schizachyrium condensatum*) -- page 819
strawberry guava (*Psidium cattleianum*) -- page 833

Literature Cited

Pukui, M.K., and S.H. Elbert. 1981. *Hawaiian dictionary -- Hawaiian-English, English-Hawaiian*. Honolulu: Univ. Pr. of Hawaii.

Pukui, M.K., S.H. Elbert, and E.T. Mookini. 1986. *Place names of Hawai'i*, rev. and enlarged ed. Honolulu: Univ. Pr. of Hawaii.

Wagner, W.L., D.R. Herbst, and S.H. Sohmer. 1990. *Manual of the flowering plants of Hawai'i*. Bishop Mus. Spec. Pub. 83. Honolulu: Univ. Hawaii and Bishop Mus. Pr.

AUTHORS
and Affiliations during Preparation of Papers

Anderson, Stephen J.
Research Division,
Hawaii Volcanoes National Park
Hawai'i Island

Bredhoff, Carl "Soot"
Manager, Kahuku Ranch
Hawai'i Island

Bridges, Kent W.
Botany Department,
University of Hawaii - Manoa campus
Honolulu

Burkhart, Robert M.
Plant Pest Control Branch,
Hawaii Dept. of Agriculture, Honolulu

Char, Winona P.
Natural Area Reserves System,
Hawaii Dept. of Land & Natural Resources
Honolulu

Conrad, C. Eugene
Institute of Pacific Islands Forestry,
U.S. Forest Service, Honolulu

Cran, C. Gordon
Manager, Kapāpala Ranch
Hawai'i Island

Cuddihy, Linda W.
Research Division,
Hawaii Volcanoes National Park
Hawai'i Island

Davis, Clifton J.
Plant Pest Control Branch,
Hawaii Dept. of Agriculture (retired)
Hawai'i Island

Dillingham, Ken
Manager, Keauhou Ranch
Hawai'i Island

Doren, Robert F.
Resources Management Division,
Everglades National Park, Florida

Funasaki, George F.
Plant Pest Control Branch,
Hawaii Dept. of Agriculture, Honolulu

Gagné, Wayne C.*
Entomology Department,
B.P. Bishop Museum, Honolulu

Gardner, Donald E.
National Park Service
University of Hawaii - Manoa campus
Honolulu

Gunderson, Lance
Resources Management Division,
Everglades National Park, Florida

Halvorson, William L.
Research Division,
Channel Islands Natl. Park, California

Hawkins, John W.
Office of Instructional Services,
Hawaii Department of Education
Honolulu

Herbst, Derral R.
Office of Environmental Services,
U.S. Fish and Wildlife Service
Honolulu

Higa, Stanley J.
Quarantine Office,
Hawaii Dept. of Agriculture, Honolulu

Higashino, Paul K.
Research Division,
Hawaii Volcanoes National Park
Hawai'i Island

Holt, R. Alan
The Nature Conservancy of Hawaii
Honolulu

Jacobi, James D.
Hawaii Research Station,
U.S. Fish and Wildlife Service
Hawai'i Island

*deceased May 1988

Juvik, James O.
Geography Department,
University of Hawaii - Hilo campus
Hawai'i Island

Juvik, Sonia P.
Geography Department,
University of Hawaii - Hilo campus
Hawai'i Island

Kageler, Dina
Resources Management Divison,
Hawaii Volcanoes National Park
Hawai'i Island

Lai, Po-Yung
Plant Pest Control Branch,
Hawaii Department of Agriculture
Honolulu

LaRosa, Anne Marie
Resources Management Division,
Everglades National Park, Florida

Loope, Lloyd L.
Research Division,
Haleakala National Park, Maui

Mack, Richard N.
Botany Department,
Washington State University
Pullman, Washington

Markin, George P.
Institute of Pacific Islands Forestry,
U.S. Forest Service, Hawai'i Island

Medeiros, Arthur C.
Research Division,
Haleakala National Park, Maui

Merlin, Mark D.
General Science Department,
University of Hawaii - Manoa campus
Honolulu

Misaki, Edwin T.
Kamakou Preserve,
The Nature Conservancy of Hawaii
Moloka'i

Moody, Michael E.
Dept. of Pure and Applied Mathematics/
Program in Genetics and Cell Biology,
Washington State University
Pullman, Washington

Nagata, Ronald J.
Resources Management Division,
Haleakala National Park, Maui

Nakahara, Larry M.
Plant Pest Control Branch,
Hawaii Department of Agriculture
Honolulu

Osgood, Robert V.
Experiment Station,
Hawaiian Sugar Planters' Association
Aiea, Oahu

Santos, Gregory L.
Research Divison,
Hawaii Volcanoes National Park
Hawai'i Island

Scowcroft, Paul G.
Institute of Pacific Islands Forestry,
U.S. Forest Service, Honolulu

Smith, Clifford W.
Cooperative National Park
Resources Studies Unit, Univ. Hawaii/
Botany Dept., Univ. Hawaii - Manoa campus
Honolulu

Spoehr, Hardy
Department of Hawaiian Home Lands
Honolulu

Stone, Charles P.
Research Division,
Hawaii Volcanoes National Park
Hawai'i Island

Tanimoto, Victor M.
Divison of Forestry and Wildlife,
Hawaii Dept. of Land & Natural Resources
Honolulu

Taylor, Dan
Resources Management Divison,
Hawaii Volcanoes National Park
Hawai'i Island

Tulang, Mike
U.S. Department of Agriculture
Soil Conservation Service
Hilo and Honolulu

Tunison, J. Timothy
Resources Management Divison,
Hawaii Volcanoes National Park
Hawai'i Island

Vitousek, Peter M.
Department of Biological Sciences,
Stanford University, California

Wagner, Warren L.
Botany Department,
B.P. Bishop Museum, Honolulu

Warshauer, Fredrick R.
Hawaii Research Station,
U.S. Fish and Wildlife Service
Hawai'i Island

Wester, Lyndon
Geography Department,
University of Hawaii - Manoa campus
Honolulu

Whiteaker, Louis D.
Botany Department,
University of Hawaii - Manoa campus
Honolulu

Wiemer, Robert D.
Experiment Station,
Hawaiian Sugar Planters' Association
Aiea, O'ahu

Yee, Rylan S.N.
Botany Department, B.P. Bishop Museum
Honolulu

Yoshioka, Ernest
Plant Pest Control Branch,
Hawaii Department of Agriculture
Hilo, Hawai'i Island

Zimmer, Nicholas G. "Chris"
Resources Management Division,
Hawaii Volcanoes National Park
Hawai'i Island

PERSPECTIVES

AN OVERVIEW OF PROBLEMS WITH INTRODUCED PLANT SPECIES IN NATIONAL PARKS AND BIOSPHERE RESERVES OF THE UNITED STATES

Lloyd L. Loope

ABSTRACT

Although floristic lists for a large sample of U.S. reserves have 5-25% introduced species, most introductions are confined to drastically disturbed areas and pose little or no threat to native ecosystems. A survey of Biosphere Reserves of the United States suggests that many reserves have only very minor problems or no problems at all with introduced plant species. Problems with invasive introduced species are most severe on oceanic islands, but serious problems occur in some continental areas as well. In some U.S. parks and reserves, the primary concern for introduced plant species is that they pose economic threats to adjacent agricultural lands. In extreme situations, however, introduced plants may inhibit and entirely prevent reproduction of native plant species, degrade habitat for native animal species, and/or alter otherwise natural habitats through changing nutrient, fire, or water regimes. At the highly disruptive end of the spectrum, native ecosystems may become so totally transformed by invaders that they are scarcely recognizable, although this condition is rare within U.S. parks and reserves.

INTRODUCTION

U.S. national parks are multipurpose areas, mandated to "provide for the benefit and enjoyment of the people." A major objective involves preservation of ecosystems in as natural a state as possible (Ise 1961; Houston 1971). Many parks qualify as Biosphere Reserves, but the reserve system also contains units subjected to consumptive use by humans. Nonnative species are clearly undesirable in areas set aside for preservation of native ecosystems. National Park Service management policies address the issue directly and pragmatically (National Park Service 1988):

> Manipulation of population numbers of exotic plant and animal species, up to and including eradication, will be undertaken wherever such species threaten park resources or public health and when control is prudent and

> feasible. High priority will be given to the management of exotic species that have a substantial impact on park resources and that can be reasonably expected to be successfully controlled; lower priority will be given to exotic species that have almost no impact on park resources or that probably cannot be successfully controlled.

Floristic lists for a representative sampling of U.S. national parks (Table 1) include a large number of nonnative species, *ca* 5-25% of the total number of vascular plants listed. The majority of these nonnative species are "weeds" in the sense of Baker (1974), who stated that a plant is a weed "if, in any geographical area, its populations grow entirely or predominantly in situations markedly disturbed by man (without, of course, being deliberately cultivated plants)." National parks invariably have some habitats heavily disturbed by humans, such as roadsides, trailsides, campgrounds, garbage dumps, employee housing areas, visitor accommodations, historical sites, and water catchment areas. In some parks, domestic livestock grazing or other "traditional uses" continue, usually on a phaseout basis. The resulting disturbed habitats are usually colonized by weeds, most of which are nonnative. The majority of introduced species are present because of continuing human disturbance or because they persist from disturbance in the recent past. Weeds have high reproductive rates, excellent dispersing and colonizing abilities, wide environmental tolerances, and "general purpose genotypes" (Baker 1965); they thrive in disturbed sites.

By Baker's definition, weeds would be assumed to be replaced by native species when factors of human disturbance are removed. Native plants are, in theory at least, better adapted through the fine-tuning of evolution to local environmental conditions and will thrive if left alone. Nevertheless, significant numbers of nonnative species in parks and other natural areas of Hawai'i and elsewhere fit Baker's characterization of "weed" in most respects, except that they do not disappear when disturbance factors are removed. (An example is Hilo grass (*Paspalum conjugatum*) within an exclosure (Anderson *et al.*, this volume).) Many introduced species also continue to spread due to natural disturbance factors, *e.g.*, *Melaleuca quinquenervia* (LaRosa, this volume), and some may even invade undisturbed systems. In this paper, comparative information on alien plant problems in U.S. national parks and reserves is provided, with the hope of shedding light on Hawai'i's problems.

WHY ARE INTRODUCED PLANT SPECIES CONSIDERED "PROBLEMS" IN PARKS AND RESERVES?

Effects of introduced species in parks and reserves range along a gradient from negligible to highly disruptive. Even in Hawai'i, perhaps 90% of the alien species on floristic lists for a given reserve are of very little consequence. Although out of place and thus an aesthetic intrusion in an area devoted to preservation of native biota, such species seem to have little more negative effect than occupying ground that might have been available to native species. The threshold of concern is probably crossed

Table 1. Total number of vascular plant species and number and % of introduced species for selected areas managed by the National Park Service.[1]

Area	No. vascular plant species	No. introduced or (No. undetermined) vascular species	% introduced or introduced and undetermined species
Denali	615	27(14)	4-7
Glacier	1258	112(17)	9-10
Yellowstone	1101	120(12)	11-12
Rocky Mountain	967	67(15)	7-8
Sequoia-Kings Canyon	1331	82(38)	6-9
Pinnacles	410	51(9)	12-15
Redwood*	630	136(11)	22-23
Olympic	1344	260(22)	19-21
Mount Ranier	789	95(18)	12-14
Crater Lake*	588	20(9)	3-5
Death Valley	1049	84(20)	8-10
Grand Canyon	1473	118(21)	8-9
Organ Pipe Cactus*	517	32(9)	6-8
Big Bend	997	49(11)	5-6
Acadia	1042	222(55)	21-27
Isle Royale	689	95(18)	14-16
Indiana Dunes	1209	172(47)	14-18
Shenandoah	1138	221(55)	19-24
Great Smoky Mountains	1485	249(61)	17-21
Congaree Swamp*	327	23(7)	7-9
Big Thicket	1204	120(18)	10-11
Everglades	876	133(42)	15-20
Channel Islands*	383	62(10)	16-19
Haleakala	ca. 684	334	49
Hawaii Volcanoes	ca. 915	602	66

[1]Species lists and evaluation of introduced species (except for Hawaii Volcanoes and Haleakala) provided from current databases by Dr. Gary Waggoner, U.S. National Park Service, Denver, Colorado. Data for Hawaii Volcanoes based on Higashino *et al.* (1988). Data for Haleakala based on an unpublished species list by A.C. Medeiros and L.L. Loope. Parks marked with an asterisk (*) are those for which numbers are judged to be unusually low due to inadequate collections. A certain number of species for most parks fall in the category "undetermined," a status characterizing many widespread species of unknown original distribution (Baker 1972).

when the introduced species results in significant decline in populations of one or more native species, significantly alters ecosystem processes, causes aesthetic damage perceived to be unacceptable, or poses perceived threats to agriculture on adjacent lands. Effects of introduced plant species on native species and on ecosystem processes are not well understood and are poorly documented.

Inhibition of Prevention of Reproduction and/or Survival of Native Plant Species

I am unable to cite a single instance where an introduced plant species is believed to have contributed significantly to the extinction of a native species outside Hawai'i. However, Loope and Dunevitz (1981) considered shading effects of Brazilian peppertree, also known as Christmas berry (*Schinus terebinthifolius*), to threaten rare endemic herbaceous understory species of limestone rockland pine forests of the Miami rock ridge in southeastern Florida.

Smith (1985) has discussed the tendency of certain introduced plants in Hawai'i to form monotypic stands -- catastrophically reducing species diversity and threatening rare species (*e.g., Cyanea superba* on Ō'ahu). On East Maui, there can be little doubt that the alien kikuyu grass (*Pennisetum clandestinum*), velvet grass (*Holcus lanatus*), and Hilo grass have contributed to the decline or extirpation of many endemic plant species. For example, the greensword *Argyroxiphium virescens*, drastically reduced in distribution and numbers by foraging of domestic cattle, did not survive after 1959 in Haleakala National Park in spite of protection from cattle. A dense growth of velvet grass was probably a decisive factor in its demise. Loope, Nagata, and Medeiros (this volume) have addressed some of the specific effects of introduced plant species at Haleakala.

Degradation of Habitat for Native Animal Species

Tamarisk (*Tamarix* spp.), through lowering water tables and drying up water sources (Vitousek, this volume), degrades habitat for desert bighorn sheep (*Ovis canadensis nelsoni*) and various fishes. Scarcity of water sources is a major factor contributing to the decline of bighorn in Death Valley National Monument (National Park Service 1981). In Hawai'i, red mangroves (*Rhizophora mangle*) destroy habitat for endangered Hawaiian stilts (*Himantopus mexicanus knudseni*) in wetland areas (F. Kuailani, pers. comm.).

Alteration of Ecosystem Properties

Vitousek (this volume) has cited examples of introduced plant species that have lowered water tables, hastened aquatic succession, intensified fire regimes, and altered local biogeochemical cycles. Other examples of plants that modify ecosystems will be mentioned in subsequent sections of this paper.

Aesthetic Damage

Control of some introduced weeds in national parks and reserves is done primarily for aesthetic reasons. Thistles seem to fall in this category, although in some instances (especially in the case of the Canada thistle, *Cirsium arvense*) they form dense enough stands to crowd out native species. Thistles are probably controlled in parks more often than any other group of plants. For example, Pestana (1985) recorded the presence of thistles (Canada thistle, bull thistle (*Cirsium vulgare*), and/or musk thistle, *Carduus nutans*) in 15 National Park Service areas in the Midwest Region. Seven of the areas had thistle control programs.

Degradation of Agricultural Value of Adjacent Rangeland

Concern over introduced plant species is, in some cases, greater outside parks and reserves than within them. Park managers often receive requests from agricultural interests to eradicate certain "pest" species within park boundaries so that the park will not serve as a refuge from which the "pest" can invade or reinvade rangeland. "Weeds" on rangelands "interfere with grazing, lower the yield and quality of forage, increase costs of managing and producing livestock, slow livestock [weight] gains, and reduce quality of meat, milk, wool, and hides" (Klingman and Ashton 1982). Some weeds are poisonous to livestock. Many of these "pests" or "weeds" are introduced plants; many others are native. For example, two native species of tall larkspur (*Delphinium barbeyi* and *D. occidentale*), which are common in aspen (*Populus tremuloides*) and open subalpine habitats of the Intermountain West, are responsible for more cattle poisoning than any other range plant in that area (Williams and Cronin 1966). Locoweeds (*Astragalus* spp.), dozens of species of which are native to western U.S., are also notorious for their toxicity to livestock. Several "noxious" introduced species have been reduced in parks and reserves as well as on rangelands through biological control, invariably as a result of agricultural concerns (*e.g.*, Klamath weed (*Hypericum perforatum*), tansy ragwort (*Senecio jacobea*), spotted knapweed (*Centaurea maculosa*)). Mechanical and chemical control of introduced weeds has been conducted in many parks in response to external pressures.

A BIOGEOGRAPHIC OVERVIEW

An evaluation of the nature and severity of effects of introduced plants is attempted on a biogeographic basis in this section. Where generalizations are made below without citation of references, the Appendix is the source of supporting documentation.

Arctic and alpine vegetation is rarely invaded by introduced plant species, even if severely disturbed. Studies of revegetation of sites disturbed by oil and gas development in Alaskan Arctic tundra indicate that only native graminoids establish without fertilization. Attempts to seed such areas with introduced grasses have failed repeatedly (van Cleve 1977; Chapin and Chapin 1980). Willard and Marr (1971), working in the alpine tundra of Rocky Mountain National Park, found no tendency of introduced plant species to invade areas heavily trampled by visitors. Reinvasion of vegetation on the bare areas was slow but consisted entirely of native species. Likewise, Greller (1974) found no invasion of introduced plants on roadcuts in the alpine tundra of the same park.

Boreal and subalpine coniferous forests of North America experience little invasion by introduced plant species. More temperate coniferous forest ecosystems harbor numerous introduced species when disturbed (*e.g.*, high percentages of aliens in floras of Olympic and Acadia National Parks), but such aliens are seldom far from heavy disturbance. Tansy ragwort, spotted knapweed, white field daisy (*Chrysanthemum leucanthemum*), dandelion (*Taraxacum officinale*), smooth brome (*Bromus inermis*), and timothy (*Phleum pratense*) are the most serious invaders in northern Rocky Mountain and Pacific coniferous forests, but even these are generally associated with disturbed sites.

The native perennial grassland of California's Central Valley has been so altered by grazing, fire, and cultivation that its former nature is only sketchily known (details given by Heady 1977). Introduced species of Mediterranean origin (Jackson 1985) dominate it more than they do in any other potential natural vegetation type in the continental U.S. (A few relict areas with native valley grassland species remain (Heady 1977), with none in national park areas. Only in much of Hawai'i have the native plants and animals been obliterated to this extent.)

California coastal reserves (*e.g.*, Point Reyes National Seashore and Channel Islands National Park) appear to have more problems with invading plant species than other U.S. coastal areas. Invaders include sea-rocket (*Cakile edentula, C. maritima*), ice plant (*Mesembryanthemum chilense, M. edule*), and a number of annual grasses and forbs (Barbour and Johnson 1977). European beach grass (*Ammophila arenaria*) has been widely planted and is highly invasive today north of San Francisco. It substantially modifies dune topography, may harbor nitrogen-fixing bacteria, and forms a dense cover that appears to exclude many native taxa (Barbour and Johnson 1977). California chaparral and "coastal scrub" are largely free of shrub invaders (data in Hanes 1977 and Mooney 1977) but have an important introduced component (primarily annuals) in the understory (*e.g.*, Pinnacles National Monument; parts of Point Reyes National Seashore; and Sequoia and Kings Canyon National Parks).

Terrestrial ecosystems of the Channel Islands have been severely degraded by grazing animals and invasion of introduced plants. Philbrick and Haller (1977) commented: "In spite of grazing pressures, the four larger islands still contain fairly significant areas of at least depauperate native grasslands. Small areas of San Miguel and Anacapa also support native grasslands." There appear to be fairly good opportunities for eventual partial restoration of Channel Island ecosystems in the national park if grazing animals can be eliminated. Halvorson (this volume) treats the Channel Island situation in detail.

Cheatgrass (*Bromus tectorum*), Russian thistle (*Salsola kali*), bumble-mustard (*Sisymbrium altissimum*), and other annuals have changed succession throughout the Great Basin sagebrush desert wherever it has been degraded by severe overgrazing (Mack 1986). These annuals form such dense stands in wet years that they close the communities to recruitment by seedlings of native perennial grasses (Young *et al.* 1977). Halogeton (*Halogeton glomeratus*) is another introduced annual poisonous to livestock, which persists on overgrazed or otherwise disturbed lands in the Great Basin and Colorado Plateau. The importance of annual grasses and other weedy introduced species declines southward in these vegetation types. Humphrey and Mehrhoff (1958) reviewed data on marked historical vegetation changes involving shrub invasion in southern Arizona grasslands. All invading species mentioned were natives. Recent comprehensive vegetation surveys of the Grand Canyon (Warren *et al.* 1982) and the Santa

Catalina Mountains (Niering and Lowe 1984) in Arizona make scarce mention of introduced species.

By far the most serious introduced plant problem in the southwestern desert areas is tamarisk. Tamarisk, native to Eurasia, was introduced to the western U.S. in the late 1800s, arrived in southern Utah in 1925-1960 and in Death Valley in the late 1940s and 1950s (determined from historical, photographic and tree-ring evidence). It has spread and thrives along watercourses throughout the southwestern U.S. (Christensen 1962; Harris 1966; Robinson 1969; National Park Service 1981). The abundant seeds of this drought-resistant phreatophyte are mainly water-dispersed (also wind-dispersed), and spread is particularly rapid following infrequent flooding events. It grows rapidly (up to 12 in. or 30 cm/week) and attains heights of 10 ft (3 m) or more. Tamarisk is still spreading in Death Valley, where it is a potential invader of all streams, ponds, marshes and wet ground below 5,230 ft (1,585 m). It lowers the water table through evapotranspiration and displaces native species of plants and animals (including fish). It threatens many biologically important areas that have not yet been affected. About 198 a (80 ha) in Death Valley National Monument (National Park Service 1981) and 74 of 180 springs in Big Bend National Park are affected. Control is feasible but expensive, using mechanical cutting at ground level with chainsaws, brush-cutters, and other powered equipment, followed by application of herbicide (Tordon, a mix of picloram and 2,4-D) to the stump. Tamarisk occurs in dozens of areas managed by the National Park Service in the Southwest, including Zion, Canyonlands, Arches, Capitol Reef, and Grand Canyon National Parks.

Very little intact tallgrass prairie vegetation remains, but where it does remain (*e.g.*, Konza Prairie in Kansas), problems with invasive species are minimal. A number of introduced plants (*e.g.*, leafy spurge (*Euphorbia esula*), Canada thistle, *Centaurea* spp., and annual grasses) present problems for economic use of shortgrass prairie rangelands and are at least nuisances in parks and reserves in and adjacent to shortgrass prairie.

Introduced plant species present problems in parks and reserves in U.S. eastern deciduous forests and the related southeastern coniferous forest, but most are replaced in the course of succession and none invade undisturbed forests. Japanese honeysuckle (*Lonicera japonica*) and kudzu (*Pueraria lobata*) are among the most persistent in the southeastern U.S.

Several invasive aquatic weeds, including water hyacinth (*Eichhornia crassipes*), hydrilla (*Hydrilla verticillata*), and water milfoil (*Myriophyllum* spp.), pose problems in many wetland parks and reserves in Florida, though more often than not these are highly disturbed ecosystems. Biological control has proved at least partially successful for the first two species. Purple loosestrife (*Lythrum salicaria*) is a formidable invader of wetlands in the Midwest and Northeast (Wilcox and Seeling 1986).

After a century of repeated introductions of tropical species to southern Florida, some introduced tree species are beginning to attain local dominance, not only on highly disturbed sites, but in native plant communities of Everglades National Park. Establishment and spread of at least three major problem species -- melaleuca or cajeput, Brazilian peppertree, and casuarina (*Casuarina equisetifolia*) -- seem to guarantee that major vegetation changes will take place in Everglades National Park in the coming decades unless there is active intervention by managers (Olmsted and Loope 1984). Myers (1983) has proposed a "judiciously timed" (based on the plant's life cycle), integrated approach to control of melaleuca using chemicals, mechanical means, and fire. LaRosa (this volume) details these problems in southern Florida.

THE ROLE OF INTRODUCED UNGULATES AS VECTORS FOR ALIEN PLANT ESTABLISHMENT

Introduced ungulates have accentuated problems with introduced plant species both in the Hawaiian Islands and in continental situations. Mueller-Dombois *et al.* (1981), Loope and Scowcroft (1985), and others have stressed the fact that introduced plant species do not normally invade undisturbed Hawaiian ecosystems (exceptions do occur, however). Effects of overgrazing by domestic livestock in the western U.S., leading to range deterioration and establishment of nonnative plants, have also been well documented (*e.g.*, Mack 1981). Yet on the U.S. Mainland many situations of severe damage by ungulates apparently do not lead to invasion of introduced plant species.

In Olympic National Park, a population of about 1,000 individuals of the introduced mountain goat (*Oreamnos americanus*) has altered plant communities at high elevations and has affected the distribution, abundance, and morphology of endemic plant species (Schreiner 1984; D.B. Houston, pers. comm.). This introduced ungulate has apparently not exacerbated problems with introduced plants, however.

Feral burros (*Equus asinus*) may have a role in encouraging annual alien grasses in habitats where they thrive in the southwestern U.S., but such grasses are not considered a major problem at Death Valley, Grand Canyon, and other areas plagued by burros. The only relationship between burros and tamarisk is that use of springs and other water sources by burros intensifies negative effects of water table reduction by tamarisk on native wildlife, especially the desert bighorn.

Feral pigs (*Sus scrofa*) in the Great Smoky Mountains are highly destructive to the native early spring-blooming flora (Bratton 1974, 1975), but no significant invasion of introduced plants is occurring on the denuded areas. Baron (1982) found that feral pig rooting at Horn Island, Gulf Islands National Seashore, Mississippi, caused little alteration of native plant communities.

WHY ARE CERTAIN ECOSYSTEMS PARTICULARLY SUSCEPTIBLE TO BIOLOGICAL INVASION?

There is little question that islands or island-like situations invariably are more susceptible to invasion than comparable continental situations. The only areas identified in the Biosphere Reserve survey (Appendix) as having massive invasions of more than one introduced plant genus are Everglades,* Channel Islands, Haleakala, and Hawaii Volcanoes National Parks. The greater invasion of islands is not likely to surprise anyone. Darwin (1859) stated over one hundred years ago in *The Origin of Species*:

> He who admits the doctrine of the creation of each separate species, will have to admit that a sufficient number of the best adapted plants and animals were not created for oceanic islands; for man has unintentionally stocked them far more fully and perfectly than did nature.

Loope and Mueller-Dombois (1989) have reviewed possible reasons why island biotas in general, and the Hawaiian biota in particular, are highly vulnerable to invasions. These reasons can be summarized as follows:

1. **Evolution of oceanic island organisms in isolation from many of the powerful selective forces that continually shape most continental systems.** -- The Hawaiian biota and its ecosystems have evolved with disturbance of volcanism, windthrow, and landslides, but without grazing and trampling of ungulates and with a reduction in frequency and intensity of fire relative to most continental systems. Disturbance has been such an important evolutionary force over time in continental situations that opportunistic species have evolved that are adapted to persistence, dispersal, and colonization of unoccupied areas. The Hawaiian biota is well adapted to local disturbances related to volcanism, but fire does not appear to have played an important role in most native ecosystems of Hawai'i; few native plants of the Hawaiian Islands possess adaptations to it. Fires in modern Hawai'i, carried mainly by introduced grasses, are generally highly destructive to native plant species (Smith and Tunison, this volume). Opportunistic invasive plant species, on the other hand, spread rapidly following fire or other disturbance, notably on mineral soil exposed by pig digging. Island biotas have proved vulnerable to many other types of changes in conditions after the arrival of man. Unquestionably, the lack of such ecologically important native groups as ants, rodents, mammalian carnivores, and herbivorous ungulates in the Hawaiian Islands and other isolated island groups has resulted in vulnerability of endemics when members of these groups are introduced.

---

*The island-like nature of the Florida Everglades, a subtropical island at the end of a temperate peninsula, has been noted by Myers (1983). Its recent emergence from the sea and isolation from neotropical biota may result in lack of what Myers calls a complete species complement.

2. **The high degree of modification of island environments by humans.** -- The Hawaiian Islands were heavily exploited by the colonizing Polynesians, and much more so by continental man after 1778. Such exploitation is typical of islands in general, but the number of alien plants introduced to Hawai'i has been extraordinary.

3. **Increased invasions due to a small number of species and taxonomic disharmony of island biotas.** -- The total number of species per unit area is lower on islands than in continental situations. Island biotas also generally have low species numbers in certain taxonomic groups in relation to their relative proportions in continental areas of the world. "Disharmony" is a term often used in the context of island biology to denote a taxonomic balance that differs from continental norms. Low species numbers and disharmony appear to contribute in at least a minor way to vulnerability of islands to invasions. A dearth of species in certain functional groups or guilds, such as climax species and gap replacement species, probably also increases vulnerability.

4. **Reduced aggressiveness and vulnerability to extinction of island biotas.** -- Numerous workers have noted that native island species have reduced "aggressiveness" or increased vulnerability to extinction even under optimal environmental conditions. A theory first proposed by E.O. Wilson (1961) in describing this phenomenon for the Melanesian ant fauna is the concept of the taxon cycle. According to this theory, a continental taxon invading an archipelago undergoes increasing habitat specialization and fragmentation of populations accompanied by increasing vulnerability to extinction over evolutionary time. This theory has been neither fully supported nor refuted.

The genetic basis of the phenomenon of reduced aggressiveness with progressive island evolution is almost unexplored, but the extensive work with Hawaiian drosophilids by H.L. Carson and colleagues provides a basis for educated speculation. Carson (1981) stated that the levels of genetic variation found within a series of endemic and introduced species of *Drosophila* are basically similar to their continental counterparts. But continental species have many local populations, whereas island populations have few. Total genetic variance carried in a continental species should be far greater than that found in insular ones. A continental deme will be able to draw variability from adjacent demes, given a capacity for gene flow between demes. The isolated nature of most island demes may be conducive to the evolution of "restrictive specializations," whereas continental conditions are capable of giving rise to "general purpose genotypes."

One genetic factor, other than small population size, that could lead to increased vulnerability of island species through evolutionary time is the repetition (and compounding) of founder events in the genetic history of many island species. The rigor of natural selection in such an evolving insular system may be relaxed by the large number of genetic bottlenecks (founder events) that some groups have undergone in island-hopping.

CONTROL OF INTRODUCED SPECIES

The control of invasive introduced plant species in parks and reserves is too complex to be more than touched upon here. Control methods can be manual, mechanical, chemical, or biological. Manual control is rarely used in modern agriculture in the United States but is commonly feasible in management of natural areas. It is generally effective primarily for small areas but has been used with some success over large areas as well. Control of woody plants often involves a combination of manual, mechanical and chemical methods; small trees are uprooted, and larger trees are cut with chain saws and herbicides applied to stumps to prevent resprouting. Most government agencies, including the National Park Service, are committed to "Integrated Pest Management," which means the use of two or more control methods with appropriate integration to control a given pest with the least long-term environmental damage (Klingman and Ashton 1982). For Park Service management and many others, Integrated Pest Management implies control using minimal pesticides at critical times and stages. Important also is continual evolution of treatment efficiency and determination of pest impact levels.

Biological control has rarely been used for control of invaders in natural areas, undoubtedly because of the substantial investment in time and money involved in research and monitoring. The recent successful control of tansy ragwort in Redwood National Park and other reserves of Oregon and California (McEvoy 1985; Mastrogiuseppe *et al.*, n.d.; Holden, n.d.) stands out, but the momentum for the control of this weed came from the agricultural community. The first significant initiatives by the Park Service toward biological control are those against alien *Rubus* spp., firetree (*Myrica faya*), and banana poka (*Passiflora mollissima*), through participation in a cooperative program in Hawai'i (Markin *et al.*, this volume; Markin and Yoshioka, this volume). Much opportunity seems to exist in Hawai'i (Gardner and Davis 1982; Markin 1989) and in California (Pemberton 1985) for controlling some of the more serious invaders. Although biological control has some fine success stories in agriculture, it has its critics as well (*e.g.*, Krebs 1972; Howarth 1983). If done properly, with painstaking experimentation to ensure specificity for the target host, the only serious drawback to attempting it seems to be its cost and the time it takes for effectiveness (typically, 5-10 years for testing and another 5-10 years after release for population buildup, spread, and impact on host, according to Markin 1989). It succeeds about 50% of the time (Krebs 1972; Pemberton 1985; Markin 1989).

That there is much current interest in the topic of control of invasive introduced species in nature reserves is suggested by information disseminated in the *Natural Areas Journal* (*e.g.*, Evans 1983, 1984a, 1984b), *Restoration and Management Notes* (*e.g.*, Lawrence 1981; Galitz 1984), and *Fremontia* (*e.g.*, Boyd *et al.* 1984; Pemberton 1985). Some excellent pertinent reviews are available in more traditional journals (*e.g.*, Swan 1982; Trumble and Kok 1982; Lym and Messersmith 1985). Regional offices of The Nature Conservancy offer to provide computer printouts ("Stewardship Abstracts") regarding management of certain introduced plant species (*Fremontia* 13(2), p. 30, 1985). The Park Service is producing an "Integrated Pest Management Information Manual" (*e.g.*, National Park Service 1984). Some Park Service regional offices are trying to compile information on introduced species problems (Pestana 1985).

THE FUTURE

The severity of problems with introduced plant species in national parks and reserves ranges from overwhelming (Everglades, Hawaii Volcanoes) to negligible (Alaska parks, Isle Royale). In some instances (Haleakala, Channel Islands, Hawaii Volcanoes), feral ungulate damage has been the primary cause of massive invasions by introduced plant species. In other instances (Great Smoky Mountains, Death Valley, Olympic), serious damage to native vegetation by introduced ungulates has not appreciably exacerbated alien plant problems. Problems with alien organisms are formidable and will require larger political and financial commitments than currently exist to adequately cope with them. Because of far-reaching, ecosystem-level effects on water tables, the tamarisk problem in riparian areas of the desert southwest stands out as one of the worst nationwide and one of the most promising targets for biocontrol. Otherwise, alien plant problems in the Everglades, the Hawaiian parks, and perhaps the Channel Islands appear greater by an order of magnitude than those of other U.S. parks and reserves. Our knowledge of exactly what effects alien plants are having on our natural heritage is, however, quite meager and needs to be strengthened, if the needed stronger support for effective control measures is to be obtained.

Alien plants are apparently damaging native species and ecosystem processes less than alien animals (especially pigs, burros, goats, and insects) in U.S. parks and reserves. However, plants are persistent and more difficult to control than most vertebrate animals; their long-term effects are often difficult to predict and need to be explored further. The real cause for alarm in the continental United States is the progressive fragmentation and "insularization" of natural areas (Wilcove *et al.* 1986), combined with increasing establishment of alien species in the expanding matrix of disturbed habitats. Mooney *et al.* (1986) call attention to the rapid increase in invasive alien plants in California as suggested graphically by Frankel (1977). Even parks and reserves with minimal problems now may have serious problems in the future.

ACKNOWLEDGMENTS

Many thanks are due to Biosphere Reserve managers and staff members who responded to my inquiries concerning the nature of local introduced species problems (Appendix). I owe a special debt of gratitude to Gary Waggoner and his Park Service colleagues in Denver, Colorado, for making available computerized lists of vascular plant species of selected national parks. W.L. Loope, T. Weaver, and P.G. Sanchez provided special information on request. Useful suggestions for evaluation of introduced species problems were provided by J.S. Baron, J.J. Ewel, W. Gregg, D.B. Houston, R.N. Mack, C.J. Martinka, A.C. Medeiros, P.G. Sanchez, K. Schlom, and C.W. Smith.

APPENDIX: A COMPARATIVE SURVEY OF SEVERITY OF INTRODUCED PLANT PROBLEMS IN SELECTED RESERVES, BASED ON MANAGERS' PERCEPTIONS

METHODS: Essentially identical letters were sent to each U.S. Biosphere Reserve, requesting information on the severity of introduced species problems in the reserve. The rationale for requesting the information from Biosphere Reserves was that 1) these areas have been selected by the U.S. Man and the Biosphere Committee to provide as satisfactory a coverage as possible of the major "biogeographic provinces" of the U.S. (Risser and Cornelison 1979); 2) these areas have agreed to participate in a worldwide reserve network, cooperating in conservation, research, and education; part of such cooperation involves exchange of information (Batisse 1982); and 3) Biosphere Reserve areas were selected in part because of their relatively long history of ecological information-gathering. Addresses of reserve managers were obtained from a UNESCO document received accompanying a U.S. Department of State Man and the Biosphere Newsletter in December 1984. Letters were sent to 40 reserves. Responses were received from 70% of the reserves. Responses dealing with introduced plants, grouped by ecosystem location, are evaluated below. Supplementary evaluation based on the literature and personal knowledge has been added as deemed appropriate, especially to fill major information gaps.

Arctic

Managers of Denali National Park (J. Dalle-Molle) and Noatak National Preserve (G.R. Bane) replied that their reserves have essentially no problems with introduced species.

Rocky Mountains

The project leader of Fraser Experimental Forest in Colorado reported no problems with biological invasions (R.R. Alexander).

For Rocky Mountain National Park, Colorado, (H.E. McCutchen), the severity of impact from invading species was regarded as "low." Introduced plants obtained a foothold through settlement of private lands in this park prior to its establishment in 1915. An aggressive land acquisition program has allowed restoration of most disturbed sites, followed by natural plant succession. Three invading plants (*Cirsium arvense, Euphorbia esula*, and *Verbascum thapsus*) are controlled (mechanical and chemical methods) and monitored. At present, 19 areas totalling 62 ha (in a park of 110,400 ha) have been identified as needing treatment. [I must note that the park's new Chief of Resource Management feels that alien plants present a highly significant threat to the resources of Rocky Mountain National Park (C.C. Axtell, pers. comm. July 1986), illustrating that perceptions differ among knowledgeable individuals.]

Yellowstone National Park has minimal problems with introduced species. Houston (1982) has noted the presence of nonnative grasses such as *Agropyron cristatum* and *Bromus tectorum*, and several forbs in the heavily grazed northern Yellowstone elk winter range especially in the "boundary line area." *Cirsium arvense* invades thermal areas of sparse vegetation around hot springs in Yellowstone.

For Glacier National Park, Geographer C.H. Key stated that a number of introduced vascular plants (*Centaurea maculosa*, *Euphorbia esula*, *Bromus inermis*, *Chrysanthemum leucanthemum*, *Trifolium* spp., *Taraxacum officinale*, and *Phleum pratensis*) persist in, and sometimes invade, natural sites in the absence of disturbance. However, management attention is concentrated on the first two listed -- largely because neighboring agricultural interests are concerned about their economic impact. A project is under way by R.W. Tyser (1985) to assess success of biological control of *Centaurea maculosa*. Several biocontrol insects have been introduced in adjacent Canada to combat this species and have spread to western Montana (Maddox 1982; Story and Nowierski 1984). Weaver and Woods (1985) have assessed the overall impact of Glacier's introduced flora. These problems involving introduced plants seem to warrant concern and surveillance but appear to be modest in scale.

Northwestern Conifer Forests/Cascades

The responses from reserves in maritime northern California and Oregon indicated a definite awareness of and interest in problems with biological invasions and a competence and confidence in dealing with them.

California Coast Ranges Biosphere Reserve (3 units)

Landels Hill-Big Creek Preserve (Univ. Calif., Santa Cruz): Manager D.J. Usner reported the following introduced plant species as warranting special concern: *Cytisus monspessulanus*, *Cortaderia jubata*, *Silybum marianum*, *Foeniculum vulgare*, and *Conium maculatum*. *Cytisus* and *Cortaderia* are being removed by hand. Prescribed burning is planned to encourage an increase of native grasses at the expense of numerous introduced grasses.

Northern California Coast Range Preserve (The Nature Conservancy and Bureau of Land Management, Branscomb, California): Preserve Managers C. and K. Barrows reported concern with *Phalaris tuberosa*, which invades areas disturbed by feral pigs and outcompetes native bunchgrasses. Pigs are shot as the opportunity arises, and their impact remains modest. *Phalaris* is being controlled successfully by mowing during the period of active growth. *Cytisus scoparius* and *Verbascum thapsus*, plant invaders which tend to increase unless managed, are successfully controlled manually through assistance from volunteers.

Redwood National Park: Superintendent D. Warnock listed introduced plant species in order of decreasing concern: *Senecio jacobea, Cytisus scoparius, Pinus radiata, Cortaderia jubata*, and *Ammophila arenaria*. Warnock feels the severity of impact by introduced plants is "low to medium." *Senecio* is of special concern "not so much for its impact to park values, but rather for its toxicity to livestock outside the park." After years of "pulling, mowing, plowing, and burning," biological control is now proving effective on *Senecio*.

Cascade Head Experimental Forest and Scenic Research Area, Oregon (U.S. Forest Service): Scientist Sarah Greene stated that the major introduced pest at this Biosphere Reserve is tansy ragwort and that it has been partially controlled by biological control.

H.J. Andrews Experimental Forest, Oregon (U.S. Forest Service): Site Manager A. McKee stated that *Senecio sylvaticus* and *Cytisus scoparius* are invaders of minor concern.

Olympic National Park, Washington, has a high percentage (over 19%) of introduced taxa in its flora (see Table 1). Franklin (1985) stated:

Most introduced species are from relatively moist regions of Europe and Asia. In some areas, exotic plants are important components of the vegetation on extensive park areas. For example, *Agrostis stolonifera* and *Ranunculus repens* are dominants in the herbaceous meadow communities of lowland river valleys [D.B. Houston stated that these areas were formerly grazed by livestock and influenced by old homesteads.] Ruderal weeds such as *Senecio jacobea* or *Cirsium arvense* are ubiquitous in the park in areas disturbed by park facilities or operations. Other species (for example, *Cytisus scoparius*) are severe problems on lands adjacent to the park and have become established because of nearby land use patterns. Several species of east-Asian origin are found in undisturbed habitats of lowland areas of the park. Present management attention to exotic plant infestations is limited to occasional, qualitative observations of problem areas. Minor control efforts have been undertaken. The Park Resources Management Plan expresses concern about exotic plant control management, but it is not given a high priority for funding.

Sierra Nevada

Sequoia and Kings Canyon National Parks (B. Evison and D. Graber) characterized the severity of impact of known alien species as "minimal and localized, with the exception of the long-established Mediterranean annual grass flora below 2,000 m." Above 2,000 m, introduced plants are restricted to a few small developed areas.

California Chaparral, Grasslands, and Coastal Habitats

D.A. Duncan of San Joaquin Experimental Range replied that in the California "annual plant" (described by Talbot *et al.* 1939) and "oak-woodland" types, the great majority of the herbaceous plants are exotics, and have been for a long time. Most of the natives were replaced over 100 years ago, soon after the introduction of livestock. The substantial range livestock industry in the California annual plant rangelands is based on managing and utilizing these "invaders," and the experimental range has been mainly involved in assisting this type of management. San Joaquin Experimental Range's Research Natural Area, protected from grazing and burning for 50 years, has had a great increase in native woody plants, but the herbaceous vegetation is still primarily introduced annuals.

Southwestern Deserts and Mountains

At Death Valley National Monument, part of the Mojave and Colorado Deserts Biosphere Reserve, the two primary problems with introduced species are feral burros (*Equus asinus*) and *Tamarix* spp. Other problems exist but in comparison to these are quite minor. The area's resources management plan (National Park Service 1981) mentions need for localized control of introduced *Salsola kali*, *Phoenix* spp. and *Washingtonia filifera*, and *Ceratophyllum demersum*, an aquatic weed which crowds out native aquatic plants in one of the area's springs.

Organ Pipe Cactus National Monument: Several introduced plant species with histories of severe invasiveness elsewhere are or have been present in Organ Pipe Cactus National Monument without, as yet, serious consequences. *Tamarix ramosissima* and *T. aphylla* are present in the area but are not frequently encountered in the reserve. Whenever tamarisk establishes in intermittent watercourses from seed washed in from outside the reserve, it is removed by hand pulling. *Nicotiana glauca* has been recorded within the reserve but has been pulled whenever seen and is not known to currently persist within the reserve. *Pennisetum setaceum* is sufficiently widespread as to be regarded as uncontrollable (H.J. Smith and R.L. Anderson).

Big Bend National Park (C.R. LaFrance, Acting Superintendent) considered the potential impacts of *Tamarix* spp. very great, particularly when associated with backcountry water sources such as springs and seeps. Tamarisk has already invaded 74 of approximately 180 springs in the park and is considered to threaten the existence of available water for wildlife. Frequent use of springs by trespass cattle intensifies the problem. The current management strategy at Big Bend is to control tamarisk at springs where invasion is only beginning.

Great Basin Desert/Colorado Plateau

Information on Arches and Canyonlands National Parks is being used here in lieu of a natural area Biosphere Reserve in the area. More accessible areas of these parks underwent severe range deterioration through overgrazing by cattle prior to park establishment (maximum grazing pressure in the year 1900), but many portions were relatively or entirely inaccessible due to highly dissected topography (W.L. Loope 1976). Virginia Park, in Canyonlands, is believed to be a pristine grassland which has never been grazed. Kleiner and Harper (1971) found only subtle differences between *Hilaria jamesii*-dominated grassland vegetation of Virginia Park and of grazed Chesler Park. Introduced plant species were absent in both areas. Only four introduced species (*Bromus tectorum*, *Descurainia sophia*, *Lappula redowskii*, and *Salsola kali*) vs. 39 native species were encountered in vegetation sampling of over 200 sites in Canyonlands National Park by W.L. Loope (1976). Aliens (primarily *Bromus tectorum*) comprised an average of 50% of the plant cover in one physiographic situation, 22% in another, but were absent or had negligible cover in the four remaining physiographic situations. Riparian areas (not sampled by Loope 1976) of Canyonlands and Arches National Parks have been occupied by tamarisk.

Grasslands

Grasslands of the central U.S. have been drastically modified through cultivation, removal of the bison (*Bison bison*), and alteration of the fire regime, but a few nearly intact prairie areas remain. The reply from L.C. Hulbert, Director of Konza Prairie Research Natural Area in Kansas, indicated a noteworthy optimism regarding problems of biological invasions in this tallgrass prairie site. He responded:

> Konza Prairie Research Natural Area is a tallgrass prairie area of 3,487 ha. About 4% was plowed in the past when it was a cattle ranch, about 6% is forest, and the rest is unplowed native prairie, mostly in good to fine condition. Fire was a common natural part of the ecosystem, and the ranchers have commonly burned since settlement.
>
> We have some exotic plants on Konza Prairie, but they are of no serious trouble. Two are *Bromus japonicus* and *B. tectorum*, but they are present only where the land has been seriously disturbed, such as along trails or around salt boxes. Regular burning keeps them from invading the prairie. *Poa pratensis* is similar, although a perennial. It is native to the country, probably, but was not a normal part of the prairie. It takes over on unburned, ungrazed prairie, but is kept out by burning. We have it

> only where there has been appreciable grazing and little or no burning, so it is not a problem.
>
> *Carduus nutans* was present in colonies in heavily grazed areas when The Nature Conservancy acquired the area. Because it is a legally noxious weed, and owners are liable to eliminate it on their land, we have worked to remove it on the perimeter areas where it would likely invade adjoining private land. On the first 371 hectares acquired by The Nature Conservancy in 1971, I cut off each plant in the dozen or so colonies with a hand pick, doing so every week or 10 days for the first half of the growing season. The first year there were about 4,000, about half that number the next, and it was eliminated in 5 years. When a much larger addition was purchased by The Nature Conservancy in 1977 (3,116 ha) there were many more plants present than on the original acquisition. We spend a number of days cutting off below the crown the plants on areas near the perimeter, but have not done so on colonies farther from the perimeter. One of the entomologists at Kansas State has been using these colonies for research on biological control with insects from Europe from which the plant came. However, under good management these biennial plants are disappearing without control. We have removed them by hand only to avoid problems with the authorities and neighbors.
>
> *Melilotus* spp. and a few other species invade mowed areas but do not succeed where we burn. This seems to be the same for most all alien species. They are opportunists on disturbed and unburned areas. Therefore, except for the musk thistle we have spent no time on eradication and are not having problems with them.

M.C. Shoop, Range Scientist at Central Plains Experimental Range in Colorado, characterized the severity of biological invasions in the experimental range and surrounding rangelands as "moderate," overall. However, he expressed great concern over the potential invader *Euphorbia esula*:

> This weed could reduce the grazing capacity of our grasslands by an estimated 40 to 70%, if it were to invade. It is now spreading rapidly on the irrigated area 25 miles away. Our first approach to control would probably be through spot herbicidal control with Tordon of the initially invading plants. If the invasion becomes severe, then broadcast treatments would probably be undertaken. Tordon is moderately effective for controlling this weed, but very expensive.

Until now, the most invasive weed at Central Plains has been *Cirsium arvense*, which if not controlled (through annual spot applications of Tordon), "could reduce the grazing value of our lands by 10 to 20%, because it would become a severe competitor on the mesic sites." Two *Centaurea* species (*C. repens* and an unidentified species) are also noteworthy invaders.

Northern Coniferous-Deciduous Forest/Great Lakes

Acting Superintendent S.L. Croll of Isle Royale National Park, Michigan, responded that the park "has been fortunate thus far in its relative isolation from biological invaders." No alien plant problems were mentioned.

Eastern Deciduous Forest

Assistant Superintendent R. Wauer and Scientist P. White of Great Smoky Mountains National Park enumerated an abundance of concerns with introduced species, with the European wild hog (*Sus scrofa*), the rainbow trout (*Salmo gairdneri*), and the balsam woolly aphid (*Adelges piceae*) topping the list. The following introduced plants are considered to warrant concern (in order of decreasing concern): "moderate severity" -- *Pueraria lobata, Lonicera japonica, Microstegium vimineum*; "low severity" -- *Albizzia julibrissin, Paulownia tomentosa, Populus alba, Ailanthus altissima, Sorghum halpense, Rosa multiflora, Rosa wichuriana, Celastrus orbiculatus, Cirsium arvense, Ligustrum vulgare, Hedera helix, Vinca minor, Berberis thunbergii, Holcus lanatus, Lysimachia nummularia, Mentha spicata, M. piperata,* and *Nasturtium officinale*.

Most of the above-mentioned species are associated with old homesites, abandoned when the park was established in the 1930s. Kudzu (*Pueraria lobata*), a smothering vine (Leguminosae) causing the greatest concern, occurs in some 80 local patches (mostly along roads at the edge of the park or at old homesites) and is said to be capable of replacing native vegetation through rapid vegetative expansion. Fortunately, the species apparently sets seed seldom, if ever. Although the threat from kudzu is substantial, core resources (virgin forest) seem to be secure. Small-scale herbicidal control is being undertaken to hold populations in check. A control feasibility study has been carried out by Rosen (1982).

Superintendent R.S. McDaniel provided a preliminary list of introduced plant species of concern in Congaree National Monument, a newly acquired (1978) National Park Service area: *Albizzia julibrissin, Melia azedarach, Lonicera japonica, Ligustrum sinense, Nelumbo lutea*, and *Myriophyllum* spp. The problems appear to him significant but manageable.

Manager W. Miley stated that the Central Gulf Coastal Plain Biosphere Reserve, near Apalachicola in northern Florida, escapes the severity of biological invasions that characterize the central and southern portion of the state. The reserve has two problem plant species -- water hyacinths (*Eichhornia crassipes*) and Eurasian water milfoil (*Myriophyllum spicatum*) -- and feral hogs. *Eichhornia* control is carried out by the Northwest Florida Water Management District.

Former resource manager R.J. Krumenaker made the following observations regarding Big Thicket National Preserve:

> Chinese tallow [*Sapium sebiferum* -- Euphorbiaceae] . . . is an introduced hardwood tree species, extremely abundant, very prolific, and currently not subject to any serious control afforts. Since Preserve boundaries are so irregular and the areas protected are small and discontiguous, it is unlikely any efforts could be successful. Another exotic plant of interest . . . is slash pine (*Pinus elliottii*), which has been planted throughout the area as a commercial timber tree.

T.D. Moore, Executive Director of the New Jersey Pinelands Commission, stated that the impacts of biological invasions on the Pinelands National Reserve, "are not too obvious." Cosmopolitan vascular plant species have been introduced, but their effect in undeveloped portions of the Pinelands is usually limited to roadsides.

R.V. Kopple, Resident Biologist, states that threats from introduced species seem to be lacking in both terrestrial and aquatic communities at The University of Michigan

Biological Station. Plant communities there are still slowly recovering from large-scale logging and burning during 1880-1920, but exotics play only a very minor role.

Florida Everglades and Vicinity

Superintendent J. Morehead and his staff of Everglades National Park confirmed the severity of introduced species problems in southern Florida. They stated:

> Exotic plants are probably the greatest threat to the South Florida environment. They have demonstrated ability to displace entire communities of native plants and animals. The full extent of the problem is not really known as no assessment has been made of populations. However, the impacts to South Florida are already severe, and without action soon, they will be irreversible, if not already so.

Opportunistic control measures are taken locally against *Schinus terebinthifolius*, but no control actions are reported planned for *Melaleuca quinquenervia* and *Casuarina equisetifolia*.

Islands

Aleutian Islands National Wildlife Refuge: Although introduced animals (primarily the Arctic fox (*Alopex lagopus*), but also numerous herbivores--Arctic ground squirrel (*Citellus undulatus*), tundra vole (*Microtus oeconomus*), Greenland collared lemming (*Dicrostonyx groenlandicus*), reindeer (*Rangifer tarandus*), caribou (*Rangifer arcticus*), and domestic cattle) have had "a tremendous and lasting impact on the natural fauna and flora of the Aleutian Islands" (Refuge manager C.F. Zeillemaker), introduced plant species were not mentioned as a problem. These oceanic islands of volcanic origin were originally lacking in land mammals, so that the vegetation underwent several million years of evolution without the selective force of mammalian herbivory. Perusal of Hulten (1968) confirms the absence of high numbers of introduced plant species in the Aleutian Islands, however.

Channel Islands National Park: Superintendent W. Ehorn stated that all the islands in Channel Islands National Park have had major, ecosystem-changing, impacts from alien species. These impacts have been primarily due to land use practices before the Islands came under the administration of the National Park Service. The main impact has come from grazing and browsing organisms: cattle, sheep (*Ovis aries*), pigs, horses (*Equus caballus*), rabbits (*Oryctolagus cuniculus*), and donkeys (*Equus asinus*). Introduced annual grasses (*Avena* spp., *Hordeum* spp., etc.) have replaced the native perennial grasses in grasslands of all Islands. Many other introduced plant species have become established and are thriving under continued grazing disturbance. They dominate the vegetation over large areas. Among the more troublesome invaders are the introduced annual ice plants (*Mesembryanthemum crystallinum* and *M. nodiflorum*). These species accumulate salt and exclude other vegetation by increasing the salt content in the soil above the tolerance of potential competitors (Vivrette and Muller 1977).

Biological invasions into the waters of Channel Islands National Park/ Marine Sanctuary/ Biosphere Reserve have occurred but have had much less of an impact than terrestrial invasions. The alga *Sargasso muticum* is believed to have been introduced on the bottom of a freighter coming into Los Angeles harbor. It is well established around Anacapa Island and is moving westward.

Recent acquisition of Santa Rosa Island by the National Park Service and most of Santa Cruz Island by The Nature Conservancy provides opportunities for major conservation efforts. Ungulate control is already underway on Santa Cruz and the national park is planning a management and research program for Santa Rosa.

Luquillo Forest Reserve: Project Leader A. Lugo stated that although some introduced animals pose a problem to wildlife, "plants are not much of a problem except along highways."

Haleakala National Park: Detailed information is given in Loope *et al.* (this volume) and Anderson *et al.*, (this volume). Problems with introduced plants at Haleakala appear more manageable that those at Hawaii Volcanoes and Everglades but more difficult than other U.S. Reserves.

Hawaii Volcanoes National Park: Detailed information is given in chapters of this volume by Tunison and Zimmer, Whiteaker and Gardner, Stone *et al.*, and others. Introduced plant species have been controlled with mixed success. Many moderately aggressive species have been brought under control. The current approach emphasizes control of the more widespread, community-disruptive plant species only within designated Special Ecological Areas, biologically rich, relatively intact areas (see Tunison and Stone, this volume).

Literature Cited

Anderson, S.J., C.P. Stone, and P.K. Higashino. [this volume] Distribution and spread of alien plants in Kipahulu Valley, Haleakala National Park, above 2,300 ft elevation.

Baker, H.G. 1965. Characteristics and modes of origin of weeds. In *The genetics of colonizing species*, ed. H.G. Baker and G.L. Stebbins, 147-172. New York: Acad. Pr., Inc.

Baker, H.G. 1972. Migrations of weeds. In *Taxonomy, phytogeography, and evolution*, ed. D.H. Valentine, 327-347. London: Academic Pr., Inc.

Baker, H.G. 1974. The evolution of weeds. *Ann. Rev. Ecol. Syst.* 5:1-24.

Barbour, M.G., and F. Johnson. 1977. Beach and dune. In *Terrestrial vegetation of California*, ed. M.G. Barbour and J. Major, 223-261. New York: John Wiley & Sons.

Baron, J.S. 1982. Effects of feral hogs on the vegetation of Horn Island, Mississippi. *Amer. Midl. Nat.* 107(1):202-205.

Batisse, M. 1982. The biosphere reserve: a tool for environmental conservation and management. *Environ. Conserv.* 9(2):101-111.

Boyd, D., E. McClintock, J.H. Kerbauaz, S.S. van Hook, D. Hillyard, and W.M. Neill. 1984. Status reports on invasive weeds. *Fremontia* 12(4):16-23.

Bratton, S.P. 1974. The effect of the European wild boar (*Sus scrofa*) on high-elevation vernal flora in the Great Smoky Mountains National Park. *Bull. Torrey Bot. Club* 101:198-206.

Bratton, S.P. 1975. The effect of the European wild boar (*Sus scrofa*) on grey beech forest in the Great Smoky Mountains. *Ecology* 56:1356-1366.

Carson, H. 1981. Microevolution in insular ecosystems. In *Island ecosystems: biological organization in selected Hawaiian communities*, ed. D. Mueller-Dombois, K.W. Bridges, and H.L. Carson, 471-482. Stroudsburg, Penn: Hutchinson-Ross Pub. Co.

Chapin, F.S., and M.C. Chapin. 1980. Revegetation of an arctic disturbed site by native tundra species. *J. Appl. Ecol.* 17:449-456.

Christensen, E.M. 1962. The rate of naturalization of *Tamarix* in Utah. *Amer. Midl. Nat.* 68:51-57.

Darwin, C. 1859. *The origin of species*. Repub. in Vol. 49, *Great books of the western world*. Encyclopedia Britannica, Inc.

Evans, J.E. 1983. A literature review of management practices for multiflora rose. *Nat. Areas J.* 3(1):6-15.

Evans, J.E. 1984a. Japanese honeysuckle (*Lonicera japonica*): a literature review of management practices. *Nat. Areas J.* 4(2):4-10.

Evans, J.E. 1984b. Canada thistle (*Cirsium arvense*): a literature review of management practices. *Nat. Areas J.* 4(2):11-21.

Frankel, R.E. 1977. Ruderal vegetation along some California roadsides. *Univ. Calif. Publ. Geogr.* 20:1-163.

Franklin, J.F., ed. 1985. *The scientific information base and infrastructure at Olympic National Park and Biosphere Reserve*. Rep. Pac. Northwest Region, Natl. Park Serv. Pac. Northwest For. Range Exp. Sta., USDA For. Serv.

Galitz, D.S. 1984. Leafy spurge symposium. *Restoration and management notes* 2(2):86-87. Madison, Wisconsin: Univ. Wisconsin-Madison Herbarium.

Gardner, D.E., and C.J. Davis. 1982. *The prospects for biological control of nonnative plants in the Hawaiian national parks.* Tech. Rep. 45, Univ. Hawaii Coop. Natl. Park Resour. Stud. Unit. Honolulu.

Greller, A.M. 1974. Vegetation of roadcut slopes in the tundra of Rocky Mountain National Park, Colorado. *Biol. Conserv.* 6(2):84-90.

Halvorson, W.L. [this volume] Alien plants at Channel Islands National Park.

Hanes, T.L. 1977. Chaparral. In *Terrestrial vegetation of California*, ed. M.G. Barbour and J. Major, 417-470. New York: John Wiley & Sons.

Harris, D.R. 1966. Recent plant invasions in the arid and semi-arid southwest. *Amer. Geogr.* 56:408-422.

Heady, H.F. 1977. Valley grassland. In *Terrestrial vegetation of California*, ed. M.G. Barbour and J. Major, 491-514. New York: John Wiley & Sons.

Higashino, P.K., L.W. Cuddihy, S.J. Anderson, and C.P. Stone. 1988. *Checklist of vascular plants of Hawaii Volcanoes National Park.* Tech. Rep. 64, Univ. Hawaii Coop. Natl. Park Resour. Stud. Unit. Honolulu.

Holden, L.J. n.d. Biological control of tansy ragwort (*Senecio jacobaea*) in Redwood National Park. Draft Rep. Redwood Natl. Park.

Houston, D.B. 1971. Ecosystems of national parks. *Science* 172:648-651.

Houston, D.B. 1982. *The Northern Yellowstone elk: ecology and management.* New York: Macmillan Pub. Co.

Howarth, F.G. 1983. Classical biocontrol: panacea or Pandora's box. *Proc. Hawn. Entomol. Soc.* 24(2,3):239-244.

Hulten, E. 1968. *Flora of Alaska and neighboring territories: a manual of the vascular plants*. Stanford, Calif.: Stanford Univ. Pr.

Humphrey, R.R., and L.A. Mehrhoff. 1958. Vegetation changes on a southern Arizona grassland range. *Ecology* 39(4):720-726.

Ise, J. 1961. *Our national park policy: a critical history*. Baltimore, Maryland: Johns Hopkins Pr.

Jackson, L.E. 1985. Ecological origins of California's Mediterranean grasses. *J. Biogeogr.* 12:349-361.

Kleiner, E.F., and K.T. Harper. 1971. Environment and community organization in grasslands of Canyonlands National Park. *Ecology* 53:299-309.

Klingman, G.C., and F.M. Ashton. 1982. *Weed science: principles and practices*, 2nd ed. New York: John Wiley & Sons.

Krebs, C.J. 1972. *Ecology: the experimental analysis of distribution and abundance*. New York: Harper & Row.

LaRosa, A.M. [this volume] The status of banana poka in Hawai'i.

Lawrence, D.B. 1981. Control of common mullein. *Restoration and Management Notes* 1(1):19. Madison, Wisconsin: Univ. Wisconsin-Madison Herbarium.

Loope, L.L., and V.L. Dunevitz. 1981. *Impact of fire exclusion and invasion of* Schinus terebinthifolius *on limestone rockland pine forests of southeastern Florida*. Tech. Rep. T-645, Natl. Park Serv. So. Fla. Res. Cent. Homestead, Fla.

Loope, L.L., and D. Mueller-Dombois. 1989. Characteristics of invaded islands. In *Biological invasions: a global perspective*, ed. J. Drake, F. di Castri, R. Goves, F. Kruger, H. Mooney, M. Rejmanek, and M. Williamson, 257-280. United Kingdom: John Wiley & Sons.

Loope, L.L., and P.G. Scowcroft. 1985. Vegetation response within exclosures in Hawai'i: a review. In *Hawai'i's terrestrial ecosystems: preservation and management*, ed. C.P. Stone and J.M. Scott, 377-402. Univ. Hawaii Coop. Natl. Park Resour. Stud. Unit. Honolulu: Univ. Hawaii Pr.

Loope, L.L., R.J. Nagata, and A.C. Medeiros. [this volume] Alien plants in Haleakala National Park.

Loope, W.L. 1976. Relationships of vegetation to environment, Canyonlands National Park, Utah. Ph.D. diss., Utah State Univ., Logan, Utah.

Lym, R.G., and C.G. Messersmith. 1985. Leafy spurge control with herbicides in North Dakota: a 20-year summary. *J. Range Manage.* 38(2):149-154.

McEvoy, P.B. 1985. Depression in ragwort *Senecio jacobaea* abundance following introduction of *Tyria jacobaeae* and *Longitarsus jacobaeae* on the central coast of Oregon. *Proc. VI Internatl. Symp. Biol. Control Weeds*, 57-64. Vancouver.

Mack, R.N. 1981. Invasion of *Bromus tectorum* L. into western North America: An ecological chronicle. *Agro-Ecosys.* 7:145-165.

Mack, R.N. 1986. Alien plant invasion into the intermountain west: a case history. In *Ecology of biological invasions of North America and Hawaii*, ed. H.A. Mooney and J.A. Drake, 191-213. New York: Springer-Verlag.

Maddox, D.M. 1982. Biological control of diffuse knapweed (*Centaurea diffusa*) and spotted knapweed (*C. maculosa*). *Weed Sci.* 30:76-82.

Markin, G.P. 1989. Alien plant management by biological control. In *Conservation biology in Hawai'i*, ed. C.P. Stone and D.B. Stone, 70-73. Univ. Hawaii Coop. Natl. Park Resour. Stud. Unit. Honolulu: Univ. Hawaii Pr.

Markin, G.P., and E. Yoshioka. [this volume] Evaluating proposed biological control programs for introduced plants.

Markin, G.P., P.-Y. Lai, and G.Y. Funasaki. [this volume] Status of biological control of weeds in Hawai'i and implications for managing native ecosystems.

Mastrogiuseppe, R.J., N.T. Blair, and D.J. Vezie. n.d. Artificial and biological control of tansy ragwort, *Senecio jacobea* L., in Redwood National Park, California. Unpub. rep.

Mooney, H.A. 1977. Southern coastal scrub. In *Terrestrial vegetation of California*, ed. M.G. Barbour and J. Major, 471-490. New York: John Wiley & Sons.

Mooney, H.A., S.B. Hamburg, and J.A. Drake. 1986. The invasions of plants and animals into California. In *Ecology of biological invasions of North America and Hawaii*, ed. H.A. Mooney and J.A. Drake, 251-272. New York: Springer-Verlag.

Mueller-Dombois, D., K.W. Bridges, and H.L. Carson, eds. 1981. *Island ecosystems: biological organization in selected Hawaiian communities*. Stroudsburg, Penn.: Hutchinson Ross Pub. Co.

Myers, R.L. 1983. Site susceptibility to invasion by the exotic tree *Melaleuca quinquenervia* in southern Florida. *J. Appl. Ecol.* 20:645-658.

National Park Service. 1981. Proposed natural and cultural resources management plan and draft environmental impact statement, Death Valley National Monument, California-Nevada. National Park Serv. San Francisco, Calif.: Natl. Park Serv.

National Park Service. 1984. Thistles. *Natl. Park Serv. Integrated Pest Manage. Info. Manual* 36:1-16. Washington, D.C.: Natl. Park Serv.

National Park Service. 1988. Management policies, National Park Service. Wash., D.C.

Niering, W.A., and C.H. Lowe. 1984. Vegetation of the Santa Catalina Mountains: community types and dynamics. *Vegetatio* 58:3-28.

Olmsted, I.C., and L.L. Loope. 1984. Plant communities of Everglades National Park. In *Environments of South Florida: present and past*, ed. P.J. Gleason, 167-184. Miami Geol. Soc. Mem. 2. Miami, Fla.: Miami Geol. Soc.

Pemberton, R.W. 1985. Naturalized weeds and the prospects for their biological control in California. *Fremontia* 13(2):3-9.

Pestana, K.E. 1985. *Problem exotic plants in selected parks of the Midwest Region and a bibliography of their management.* Midwest Region, Natl. Park Serv. Res./Resour. Manage. Rep. MWR-6. Omaha, Nebraska.

Philbrick, R.N., and J.R. Haller. 1977. The southern California islands. In *Terrestrial vegetation of California*, ed. M.G. Barbour and J. Major, 893-906. New York: John Wiley & Sons.

Risser, P.G., and K.D. Cornelison, eds. 1979. *Man and the biosphere, U.S. information synthesis project, MAB-8 Biosphere Reserves*. Norman, Okla.: Okla. Biol. Survey.

Robinson, T.W. 1969. Introduction, spread, and areal extent of saltcedar (*Tamarix*) in the western states. U.S. Geol. Surv. Prof. Paper 491-A. Washington, D.C.

Rosen, A. 1982. Feasibility study: eradication of kudzu with herbicides and revegetation with native tree species in two national parks. Natl. Park Serv. Res./Resour. Manage. Rep. SER-59. Atlanta, Georgia.

Schreiner, E. 1984. Vegetation Studies. In *Mountain goat ecology and management investigations, Olympic National Park*, ed. D. Houston, B. Moorehead, E. Schreiner, and R. Olson, 29-78. Natl. Park Serv. Rep. Olympic Natl. Park, Port Angeles, Washington.

Smith, C.W. 1985. Impact of alien plants on Hawai'i's native biota. In *Hawai'i's terrestrial ecosystems: preservation and management*, ed. C.P. Stone and J.M. Scott, 180-250. Univ. Hawaii Coop. Natl. Park Resour. Stud. Unit. Honolulu: Univ. Hawaii Pr.

Smith, C.W., and J.T. Tunison. [this volume] Fire and alien plants in Hawai'i: research and management implications for native ecosystems.

Stone, C.P., L.W. Cuddihy, and J.T. Tunison. [this volume] Responses of Hawaiian ecosystems to removal of feral pigs and goats.

Story, J.M., and R.M. Nowierski. 1984. Increase and dispersal of *Urophora affinis* (Diptera: Tephritidae) on spotted knapweed in western Montana. *Environ. Entomol.* 13:1151-1156.

Swan, D.G. 1982. Long-term field bindweed (*Convolvulus arvensis*) control in two cropping systems. *Weed Sci.* 30:476-480.

Talbot, M.W., H.H. Biswell, and A.L. Hormay. 1939. Fluctuations in the annual plant vegetation of California. *Ecology* 20(3):394-402.

Trumble, J.T., and C.T. Kok. 1982. Integrated pest management techniques in thistle suppression in pastures of North America. *Weed Res.* 22:345-359.

Tunison, J.T., and C.P. Stone. [this volume] Special Ecological Areas: an approach to alien plant control in Hawaii Volcanoes National Park.

Tunison, J.T., and N.G. Zimmer. [this volume] Success in controlling localized alien plants in Hawaii Volcanoes National Park.

Tyser, R.W. 1985. Spotted knapweed in Glacier National Park: an ecological assessment. Research proposal to Glacier National Park. Biol. Dep., Univ. Wisconsin, La Crosse, Wisconsin.

van Cleve, K. 1977. Recovery of disturbed tundra and taiga surfaces in Alaska. In *Recovery and restoration of damaged ecosystems*, ed. J. Cairns, K.L. Dickson, and E.E. Henricks, 422-455. Charlottesville, Virginia: Univ. Pr. Virginia.

Vitousek, P.M. [this volume] Effects of alien plants on native ecosystems.

Vivrette, N.J., and C.H. Muller. 1977. Mechanism of invasion and dominance of coastal grassland by *Mesembryanthemum crystallinum*. *Ecol. Monogr.* 47:301-318.

Warren, P.L., K.L. Reichhardt, D.A. Mouat, B.T. Brown, and R.R. Johnson. 1982. *Vegetation of Grand Canyon National Park*. Univ. Ariz. Coop. Natl. Park Resour. Stud. Unit. Tucson, Ariz.

Weaver, T., and B. Woods. 1985. *The exotic flora of Glacier National Park: a preliminary atlas*. Bozeman, Montana: Montana State Univ. Biol. Rep. 37.

Whiteaker, L.D., and D.E. Gardner. [this volume] Firetree (*Myrica faya*) distribution in Hawai'i.

Wilcove, D.S., C.H. McClellan, and A.P. Dobson. 1986. Habitat fragmentation in the temperate zone. In *Conservation biology: the science of scarcity and diversity*, ed. M. Soulé, 237-256. Sunderland, Mass.: Sinauer Assoc.

Wilcox, D.A., and M.K. Seeling. 1986. Ecology and management potential for purple loosestrife (*Lythrum salicaria*). In *Conf. Sci. Natl. Parks, Program and Abstr.*, July 13-16, 1986, 197. Colorado St. Univ., Ft. Collins, Colorado.

Willard, B.E., and J.W. Marr. 1971. Recovery of alpine tundra under protection after damage by human activities in the Rocky Mountains of Colorado. *Biol. Conserv.* 3(3):181-190.

Williams, M.C., and E.H. Cronin. 1966. Five poisonous range weeds -- when and why they are dangerous. *J. Range Manage.* 19:274-279.

Wilson, E.O. 1961. The nature of the taxon cycle in the Melanesian ant fauna. *Amer. Nat.* 95:169-193.

Young, J.A., R.A. Evans, and J. Major. 1977. Sagebrush steppe. In *Terrestrial vegetation of California*, ed. M.G. Barbour and J. Major, 763-796. New York: John Wiley & Sons.

EFFECTS OF ALIEN PLANTS ON NATIVE ECOSYSTEMS*

Peter M. Vitousek

ABSTRACT

Invasions by alien plants can change the large-scale functioning of native ecosystems and alter the population dynamics and community structure of native species. Such ecosystem-level changes affect the ways local systems interact with surrounding ecosystems, with downstream ecosystems and groundwater, and with the atmosphere. Large-scale changes in ecosystems caused by biological invasions represent a specific and convincing demonstration that the characteristics of individual species are important at the ecosystem level, and studies of biological invasions thus offer the possibility of integrating the approaches employed by physiological, population, community, and ecosystem-level ecology. The evidence that alien animals alter native ecosystems is strong, and such effects are widespread. A few examples of the effects of alien plants are equally strong, but much of the evidence for ecosystem-level effects of plants is anecdotal. Hard evidence for the system-level effects of biological invasions may help to build support for control of invasions, target resources to the most important species, and address concerns as far afield as the ecological consequences of releasing recombinant ("genetically engineered") organisms.

INTRODUCTION

Can alien plants alter the ecosystem-level properties of the areas they invade? Ecosystem-level properties include coarse-scale processes such as primary production, consumption, decomposition, water balance, nutrient cycling and loss, soil fertility, erosion, and disturbance frequency. Many of these properties are directly relevant to human health and well-being

---

*Based on a chapter by the author in *Biological invasions of North America and Hawaii*, ed. H.A. Mooney and J. Drake, 1986.

(Ehrlich and Mooney 1983), and their alteration can have significant consequences beyond the altered site (Bormann and Likens 1979).

Ecosystem-level characterisitics are clearly influenced by the biota; in many cases they can fairly be regarded as wholly biotic properties. Nonetheless, it has been suggested that ecosystem-level properties are relatively little influenced by the identity or characteristics of any single species within an ecosystem (O'Neill *et al.* 1977). This point of view suggests that as long as some minimum diversity is present, most of the available resources will be fully utilized (Mooney and Gulmon 1983), and ecosystem characteristics will be independent of exactly which species are represented. This argument can be developed most easily for plants, because all plants require carbon dioxide, water, and the same mineral nutrients, and all these resources are obtained and used in similar ways. If this is true, then resource availability and environmental properties control both the biotic composition (Tilman 1982) and ecosystem-level properties (Rosenzweig 1968) of an area.

In contrast, a number of investigators have evaluated the significance of particular species or species groups in controlling ecosystem properties. The potential importance of understory trees as nutrient pumps (Thomas 1969); of early successional species in reducing run-off of surface water, erosion, and nutrient losses (Marks and Bormann 1972; Marks 1974; Foster *et al.* 1980; Boring *et al.* 1981); and of spring ephemeral species as "vernal dams" (Muller and Bormann 1976; Blank *et al.* 1980; Peterson and Rolfe 1982) has been suggested. Most of these studies demonstrated that a particular species takes up energy, nutrients, or water out of proportion to its biomass or apparent abundance in a community; but most studies to date have failed to show that ecosystem properties would be different in the absence of that species.

An examination of biological invasions of alien plants can provide unambiguous evidence for or against the ecosystem-level importance of individual species. We know that the characteristics of individual species are important on lower levels of ecological organization, and we know that biological invasions can alter native populations and communities (Elton 1958; Jarvis 1979; Clark *et al.* 1984). If biological invasions of alien species can alter ecosystem properties or processes, then the characteristics of individual species must be important on the ecosystem level as well.

There is substantial evidence that alien species can alter the characteristics of native ecosystems (Vitousek 1986). This evidence is much stronger for animal invaders than it is for plants (Singer *et al.* 1984; Stone and Scott 1985), in part because the effects of animals probably are greater than those of plants (on the average) (Vitousek 1986), and in part because too many studies of plants have been anecdotal rather than quantitative. Nonetheless, there is sufficient evidence suggesting that alien plants alter ecosystem properties. A systematic, quantitative study of the effects of alien plants is likely to prove rewarding, both practically and scientifically.

INVASIONS BY ALIEN PLANTS

Given the similarity in resource requirements among plants, it would seem that an alien that altered the properties of the natural ecosystem it invaded would have to either 1) gain access to resources at times that the natives are inactive or from locations that the natives cannot reach; or 2) be able to utilize those resources more efficiently than the natives. Different species of plants differ in resource use in a few striking and many subtle ways (Denslow 1980; Tilman 1982). For example, evergreens have access to light during "good" days within cold or dry seasons and therefore can obtain more energy than deciduous plants (Chabot and Hicks 1982). In contrast, deciduous plants may use water more efficiently than evergreens by dropping leaves during a dry season. One difficulty with studying the effects of invasions of alien plants is that many aliens tend to dominate disturbed sites (Allan 1936; Egler 1942; Smith 1985; Mooney and Drake 1986). It can, therefore, be difficult to separate the effects of alien species *per se* from the effects of the disturbance that allows them to establish. For example, Ewel (1986) showed that South Florida ecosystems dominated by the alien Brazilian peppertree, also known as Christmas berry (*Schinus terebinthifolius*), were very different from native ecosystems, but that an entirely novel anthropogenic disturbance (rock plowing) allowed *Schinus* to invade and succeed. Separating the effects of rock plowing and of Brazilian peppertree is difficult.

In this paper, I will concentrate on cases where aliens invade undisturbed native ecosystems, or where they clearly change the course of succession once they invade disturbed areas (Vitousek 1986). I will examine major ways in which plants can differ in resource acquisition or utilization, and cite examples in which biological invaders have been demonstrated to alter ecosystem properties because of differences in resource acquisition or use. I will then discuss interactions between alien plants and agents of disturbance such as fire and animals.

Differences in Life Form

In natural succession, increased productivity and nutrient use often occur when a new life form (e.g., perennial versus annual, shrub versus herb) is added (Odum 1960; Woodwell 1974). Perennials and woody plants can acquire more resources by maintaining a perennial root system that goes deeper into the soil, by maintaining evergreen leaves that are active year-round, or by regenerating deciduous leaves rapidly from storage (rather than *de novo* from current photosynthate). More efficient resource use could also result from the maintenance of internal storage pools of nutrients (Chapin 1980; Vitousek 1982; Fife and Nambiar 1984; Birk and Vitousek 1986).

It follows that alien plants could alter ecosystem properties when they add a life form that is not represented in the native flora. Perhaps the best example of this sort of change is provided by tamarisk or salt cedar (*Tamarix* spp.), a genus of riparian-zone invaders in the arid southwestern United States (Robinson 1969). Salt cedar requires wet areas for germination and early growth, but it is able to maintain itself on water from deep in the soil once it is established. Further, it has an extraordinarily rapid rate of transpiration, which represents a significant pathway of water loss from reservoirs in the arid and semi-arid Southwest (Horton 1977). Salt cedar also invades natural springs and water courses in desert regions. For example, Eagle Borax Spring in Death Valley was invaded by salt cedar in the 1930s or 40s. By the late 1960s, rapid transpiration by salt cedar had caused the complete disappearence of surface water from what had been a large marsh (Neill 1983; P.G. Sanchez, pers. comm.). When managers removed salt cedar from the site, surface water reappeared, and Eagle Borax Spring and its associated flora and fauna recovered (Neill 1983). It would be difficult to picture a more convincing illustration of the potential for an alien plant to alter ecosystem properties.

Floating aquatic weeds represent another life form that can greatly alter ecosystem properties. The water fern *Salvinia molesta* is a morphologically diverse floating weed that has been widely introduced in tropical river systems; its effects have been thoroughly studied in Africa (Mitchell 1970 in Mitchell *et al.* 1980), India (*cf.* Thomas 1981) and Papua New Guinea (Mitchell *et al.* 1980). It was introduced to the Sepik River in Papua New Guinea in the early 1970s. Within ten years, it had measurably altered productivity in floodplain lakes, reduced yields from fisheries, and changed surface water chemistry (Mitchell *et al.* 1980). In Kerala, India, floating islands of *Salvinia* 3-10 ft (1-3 m) thick were reported (Thomas 1981); paddy cultivation was under way on one of them! Such alien aquatic weeds represent an economic problem in many areas, and their ecosystem-level effects have been studied more thoroughly than those of most terrestrial invaders.

A number of additional examples of invasions by new life forms have been documented, among them Brazilian peppertree and *Melaleuca quinquenervia* in the Everglades (Ewel *et al.* 1981; Ewel 1986; LaRosa *et al.*, this volume). Melaleuca in particular is able to invade otherwise unaltered stands of dwarf cypress (*Taxodium distichum*) and to alter ecosystem-level water balance in the process (Ewel 1986). Similarly, North American and other treeline species grow well above local treeline in New Zealand (Wardle 1971) and Hawai'i. Wherever biological invasions add a major new life form to an area, ecosystem-level changes can be anticipated.

Alien plants with life forms that differ from natives in subtler ways could also have ecosystem-level effects. A large number of researchers have evaluated the effects of forest plantations, often composed of alien pines, on productivity, nutrient cycling, and soil properties (Nihlgard 1972; Egunjobi and Onweluzo 1979; Perala and Alban 1982; Feller 1983/84). The establishment of forest plantations generally requires substantial human intervention, so studies in plantations cannot fairly represent the influence of an alien species under natural conditions. Nonetheless, large effects related to the biology of the planted species are often observed. Pine plantations generally yield less water to runoff than the broad-leaf forests they replace (Swank and Douglass 1974; Feller 1981); this effect is due more to increased interception and evaporation in the pine canopy than it is to a longer season for transpiration. Additionally, increased litter production and decreased litter decomposition can lead to the formation of a massive forest floor (one with a thick organic layer) (Egunjobi and Onweluzo 1979).

I do not know if such effects can be documented outside of plantation conditions. Monterey pine (*Pinus radiata*) is spreading from plantation areas into adjacent dry sclerophyll forests in Australia, where it is outgrowing the native eucalypts by a wide margin (Burdon and Chilvers 1977; Chilvers and Burdon 1983).

Resource Acquisition and Utilization

One major way in which plants differ in resource requirements is the presence of nitrogen-fixing symbioses in a few groups of higher plants. An alien plant with the capacity to fix atmospheric nitrogen could clearly be at a competitive advantage when invading a nitrogen-limited community. More importantly, nitrogen fixed by the invader would ultimately (often rapidly) be released in the soil and become available to other organisms. Primary production in the majority of temperate, boreal, and montane tropical ecosystems is limited by nitrogen availability; fixation of nitrogen by alien plants can therefore increase soil fertility, the amount of nitrogen in circulation, and net primary productivity.

Nitrogen-fixing aliens occur in many regions, including *Ulex* and *Leucaena* on many Pacific islands (Allan 1936; Egler 1942), Scotch broom (*Cytisus scoparius*) in the Pacific Northwest, and *Casuarina* in Florida and Hawai‘i. Most often, relatively little is known of the nitrogen status of the area invaded. An exception is the recent invasion by firetree (*Myrica faya*) in young volcanic regions of Hawaii Volcanoes National Park on the island of Hawai‘i. The young ash deposits invaded are extremely nitrogen deficient (Balakrishnan and Mueller-Dombois 1983; Vitousek *et al.* 1983), and they contain no native symbiotic nitrogen fixers. Firetree can form nearly monospecific stands, and its actinorrhizal root nodules fix nitrogen rapidly (Vitousek *et al.* 1987). Its invasion clearly increases soil nitrogen availability in young sites, and this increased availability alters productivity, nutrient cycling, and possibly even the rate and direction of primary succession. Firetree adds three times more nitrogen to open-canopied ‘ōhi‘a *(Metrosideros polymorpha)* forests than do all natural sources combined (Vitousek *et al.* 1987). (Natural sources include rainfall, native lichens with blue-green alga symbionts such as *Stereocaulon vulcani*, blue-green algae growing on the lava or in association with cryptogams, particularly liverworts, and bacteria associated with decaying wood.) Moreover, biological invasions by a wide variety of alien plant species in Hawai‘i are most successful on more fertile sites (*cf.* Gerrish and Mueller-Dombois 1980), so the invasion of firetree may increase the probability of further invasions. A similar relationship between soil fertility and the success of alien species has been observed in Western Australia (Bridgewater and Backshall 1981).

Several other nitrogen-fixing aliens are established and aggressive in Hawai‘i (Smith 1985), although except for *Casuarina* they rarely dominate early in primary succession. All these aliens might have the potential to facilitate further invasions, and many would be worth intensive study.

Another way in which alien plants can alter soil characteristics is by concentrating salts. The ice plant *Mesembryanthemum crystallinum* is an aggressive invader in California (Vivrette and Muller 1977) and Australia (Kloot 1983). It usually requires human intervention to become established; it is most abundant in degraded pastures. Once established, it concentrates salt from throughout the rooting zone onto the soil surface, thereby altering soil physical and chemical properties and interfering with the growth of potential competitors.

It has been suggested that particular species could affect ecosystem properties by bringing nutrients to the surface from deep in the soil, thus serving as "nutrient pumps," which keep high levels of essential nutrients in circulation (Thomas 1969). I know of no direct evidence that alien plants have altered ecosystem properties in this way, but I also know of no serious attempts to look for such effects.

Photosynthetic Pathways

Three major photosynthetic pathways occur in higher plants. These are termed C_3 (the most common pathway), C_4 (tolerant of high temperatures and relatively water efficient), and CAM (a rather stable but extremely water-efficient pathway found in succulent plants). Alien plants with the C_4 and CAM photosynthetic pathways are highly aggressive invaders in many areas otherwise dominated by plants with the more common C_3 pathway (Rundel 1980). Many grasses (*Andropogon, Melinis, Pennisetum*, and many others) are successful alien C_4 species in Hawai'i, while the cactus *Opuntia* is a common CAM invader. Ecosystem-level effects of the alternative photosynthetic pathways are likely but not yet established.

Phenology

The effects of phenological differences between aliens and natives are somewhat better documented. For example, Mueller-Dombois (1973) described a situation in which an alien bunchgrass (*Andropogon*) in Hawai'i maintained its life cycle responses to photoperiodic cues from its range in the southeastern United States. Consequently, in Hawai'i it grows during the dry season and is inactive during the rainy season. Transpiration is much reduced with *Andropogon* relative to native rain forest, and boggy conditions develop in formerly mesic or dry situations (Mueller-Dombois 1973).

Successional Position

Invading plant species could also alter ecosystem properties if they differed from natives in their position in succession (*cf.* Pickett 1976). Individual species are most often argued to be important on the ecosystem level when they invade immediately after disturbance, grow rapidly, and take up nutrients that might otherwise be lost from the site as a whole (Marks and Bormann 1972; Foster *et al.* 1980; but see Vitousek 1986). The European alien plant wood groundsel (*Senecio sylvaticus*) apparently functions in this way in unburned clearcuts in the Oregon Cascades (Gholz *et al.* 1985), taking up a substantial fraction of the nitrogen and phosphorus cycled by the intact 450-yr-old Douglas-fir forest within two to three years following clearcutting. It is tempting to speculate that *Senecio* has entered a hitherto unfilled niche in these forest ecosystems, but it must be remembered that it colonizes following a novel disturbance. The functional equivalent of unburned clearcuts (large-scale disturbances that remove the canopy over a wide area but leave the forest floor intact) were uncommon in the Pacific Northwest prior to widespread logging. It is interesting that a native species, fireweed (*Epilobium angustifolium*), colonizes after both wildfires and slash burning in the Oregon Cascades (Gholz *et al.* 1985).

Disturbance Frequency and Intensity

Alterations in disturbance regimes represent another possible way in which invasions by alien species could alter ecosystem-level properties. Any subsequent ecosystem-level changes would result not from differences in resource acquisition or the efficiency of resource utilization, but rather from interactions between alien plants and agents of disturbance such as fire, animals, or diseases.

Alterations in fire frequency as a result of biological invasions have been clearly identified, although detailed work on their consequences remains to be done. Smith (1985) identified eight species of alien weeds in Hawai'i as "fire-enhancers" -- species that increase the areal extent or intensity of fires in areas where they are established. The most spectacular example involved broomsedge (*Andropogon glomeratus*) and bush beardgrass (*Schizachyrium condensatum*), both of which form dense, nearly complete canopies in seasonally dry areas after the removal of feral goats (*Capra hircus*). (*Andropogon* also spread and intensified in the presence of goats to some degree (J.T. Tunison, pers. comm.).) The areal extent of fires increased 1,800-fold after goats were removed, and the alien bunchgrasses were able to reestablish on the burned sites better than natives. The native shrub *pūkiawe (Styphelia tameiameiae)*, which successfully withstood centuries of feral goat browsing, was rapidly reduced by the fires (Smith *et al.* 1980). Species that alter hydrologic cycles (*e.g.*, the *Andropogon* species described by Mueller-Dombois 1973, above) could also alter disturbance frequency.

Animals are also agents of disturbance to vegetation as well as potential vectors for pollination and dispersal. Interactions between alien plants and alien animals can clearly cause ecosystem-level changes. Caldwell *et al.* (1981) demonstrated that *Agropyron desertorum*, an introduced bunchgrass in the Intermountain West, produces three to five times more tillers following clipping or defoliation than the otherwise very similar native bunchgrass *Agropyron spicatum*. Grazing pressure from cattle *Bos taurus*) gives *Agropyron desertorum* part of its competitive advantage. In Hawai'i, Smith (1985) stated that feral pigs (*Sus scrofa*) and a number of the most important and aggressive weedy plants, including strawberry guava (*Psidium cattleianum*) and a number of passionflower (*Passiflora*) species, reinforce each other's ability to invade native ecosystems. Pigs consume and distribute seeds of the invaders, and they also disturb areas of soil and create excellent conditions for seed germination and seedling growth. Soil characteristics and a community structure distinctly different from those in native forests can be maintained by this interaction.

Ecosystem-level effects of interactions between alien plants and pathogens are less well documented. Alien plants are known to have been involved in the spread of alien pathogens to native species (Gardner and Hodges 1983), but the consequences of such pathogens for native ecosystems have not been studied.

CONCLUSIONS

In general, the best-documented illustrations of the importance of invading plants in altering ecosystem properties involve effects on the hydrologic cycle, while effects on productivity or nutrient cycling are usually more speculative. Information on water yield from natural ecosystems has been collected much more precisely and over a longer period of time than information on most other ecosystem properties. Perhaps if information of similar quality were available for other properties, we could detect many more subtle but pervasive effects of biological invaders.

Studies of the ecosystem-level effects of biological invasions are likely to become more widespread and visible in the next few years. For example, the international Scientific Committee on Problems of the Environment (SCOPE) recently sponsored an international program to examine biological invasions. Moreover, biological invasions are being used as one of the very few sources of information that can help us to identify the consequences of releasing organisms with altered DNA ("genetic engineering") (Halvorson *et al.* 1985). Biological invasions by alien species are not identical to the sorts of environmental releases now planned, but they may represent a model for the consequences of releasing more highly modified organisms in the future. Hawai'i is a nearly ideal place to study the effects of biological invasions, and perhaps by studying those invasions that we cannot now control, we will be able to provide economic resources to Hawai'i and an additional impetus towards controlling alien plants in the future.

Literature Cited

Allan, H.H. 1936. Indigene versus alien in the New Zealand plant world. *Ecology* 17:187-193.

Balakrishnan, N., and D. Mueller-Dombois. 1983. Nutrient studies in relation to habitat types and canopy dieback in the montane rain forest ecosystem, island of Hawaii. *Pac. Sci.* 37:339-359.

Birk, E.M., and P.M. Vitousek. 1986. Nitrogen availability and nitrogen use efficiency in loblolly pine stands on the Savannah River plant, South Carolina. *Ecology* 67:69-79.

Blank, J.L., R.K. Olson, and P.M. Vitousek. 1980. Nutrient uptake by a diverse spring ephemeral community. *Oecologia* 47:96-98.

Boring, L.R., C.D. Monk, and W.T. Swank. 1981. Early regeneration of a clearcut southern Appalachian forest. *Ecology* 62:1244-1253.

Bormann, F.H., and G.E. Likens. 1979. *Pattern and process in a forested ecosystem*. New York: Springer-Verlag.

Bridgewater, P.B., and D.J. Backshall. 1981. Dynamics of some Western Australian ligneous formations with special reference to the invasion of exotic species. *Vegetatio* 46:141-148.

Burdon, J.J., and G.A. Chilvers. 1977. Preliminary studies on a native Australian eucalypt forest invaded by exotic pines. *Oecologia* 31:1-12.

Caldwell, M.M., J.H. Richards, D.A. Johnson, R.S. Nowack, and R.S. Dzurec. 1981. Coping with herbivory: photosynthetic capacity and resource allocation in two semiarid *Andropogon* bunchgrasses. *Oecologia* 50:14-24.

Chabot, B.F., and D.J. Hicks. 1982. The ecology of leaf life spans. *Ann. Rev. Ecol. Syst.* 13:229-259.

Chapin, F.S., III. 1980. The mineral nutrition of wild plants. *Ann. Rev. Ecol. Syst.* 11:233-260.

Chilvers, G.A., and J.J. Burdon. 1983. Further studies on a native Australian eucalypt forest invaded by exotic pines. *Oecologia* 59:239-245.

Clark, B., J. Murray, and M.S. Johnson. 1984. The extinction of endemic species by a program of biological control. *Pac. Sci.* 38:97-104.

Denslow, J.S. 1980. Gap partitioning among tropical forest trees. *Biotropica* 12 (Suppl.) :47-55.

Egler, F.E. 1942. Indigene versus alien in the development of arid Hawaiian vegetation. *Ecology* 23:14-23.

Egunjobi, J.K., and B.S. Onweluzo. 1979. Litterfall, mineral turnover, and litter accumulation in *Pinus caribea* L. stands at Ibadan, Nigeria. *Biotropica* 11:251-255.

Ehrlich, P.R., and H.A. Mooney. 1983. Extinction, substitution, and ecosystem services. *Bioscience* 33:248-253.

Elton, C.S. 1958. *The ecology of invasions by animals and plants*. London: Methuen and Co.

Ewel, J.J. 1986. Invasibility: lessons from South Florida. In *Biological invasions of North America and Hawaii*, ed. H.A. Mooney and J. Drake, 214-230. New York: Springer-Verlag.

Ewel, J.J., D. Ojima, and W. DeBusk. 1981. Ecology of a successful exotic tree in the Everglades. *Proc. 2nd Conf. Sci. Res. Natl. Parks* 8:419-422. Washington, D.C.

Feller, M.C. 1981. Water balances in *Eucalyptus regnans*, *Eucalyptus obliqua*, and *Pinus radiata* ecosystems in Victoria. *Austr. For. Res.* 44:153-161.

Feller, M.C. 1983/84. Effects of an exotic conifer (*Pinus radiata*) plantation on forest nutrient cycling in southeastern Australia. *For. Ecol. Manage.* 7:77-102.

Fife, D.N., and E.K.S. Nambiar. 1984. Movement of nutrients in radiata pine needles in relation to the growth of shoots. *Ann. Bot.* 54:303-314.

Foster, M.M., P.M. Vitousek, and P.A. Randolph. 1980. The effect of *Ambrosia artemisiifolia* on nutrient cycling in a first-year old-field. *Amer. Midl. Nat.* 103:106-113.

Gardner, D.E., and C.S. Hodges. 1983. Leaf rust caused by *Kuehneola uridinis* on native and non-native *Rubus* species in Hawaii. *Plant Diseases.* 67:962-963.

Gerrish, G., and D. Mueller-Dombois. 1980. Behavior of native and non-native plants in two tropical rain forests on Oahu, Hawaiian Islands. *Phytocoenologia* 8(2):237-295.

Gholz, H.L., G.M. Hawk, A. Campbell, K. Cromack, and A.T. Brown. 1985. Early vegetation recovery and element cycles on a clearcut watershed in western Oregon. *Can. J. For. Res.* 15:400-409.

Halvorson, H.O., D. Pramer, and M. Rogul, eds. 1985. *Engineered organisms in the environment: scientific issues*. Washington, D.C.: Amer. Soc. Microbiol.

Horton, J.S. 1977. The development and perpetuation of the permanent tamarisk type in the phreatophyte zone of the southwest. In *Importance, preservation, and management of riparian habitat: a symposium*, ed. R.R. Johnson and D.A. Jones, 124-127. U.S. For. Serv. Gen. Tech. Rep. RM-43. Washington, D.C.: US Dep. Agric.

Jarvis, P.J. 1979. The ecology of plant and animal introductions. *Progr. Phys. Geogr.* 3:187-214.

Kloot, P.M. 1983. The role of common iceplant (*Mesembryanthemum crystallinum*) in the deterioration of medic pastures. *Austr. J. Ecol.* 8:301-306.

LaRosa, A.M., R.F. Doren, and L. Gunderson. [this volume] Alien plant management in Everglades National Park: an historical perspective.

Marks, P.L. 1974. The role of pin cherry (*Prunus pensylvanica* L.) in the maintenance of stability in northern hardwood ecosystems. *Ecol. Monogr.* 44:73-88.

Marks, P.L., and F.H. Bormann. 1972. Revegetation following forest cutting: mechanisms for return to steady state nutrient cycling. *Science* 176:914-915.

Mitchell, D.S. 1970. Autecological studies of *Salvinia auriculata* Aubl. Ph.D. thesis, Univ. London, London.

Mitchell, D.S., T. Petr, and A.B. Viner. 1980. The water-fern *Salvinia molesta* in the Sepik River, Papua New Guinea. *Environ. Conserv.* 7:115-122.

Mooney, H.A., and J. Drake, eds. 1986. *Biological invasions of North America and Hawaii*. New York: Springer-Verlag.

Mooney, H.A., and S.L. Gulmon. 1983. The determinants of plant productivity -- natural versus man-modified communities. In *Disturbance and ecosystems: components of response*, ed. H.A. Mooney and M. Godron, 146-158. Berlin: Springer-Verlag.

Mueller-Dombois, D. 1973. A non-adapted vegetation interferes with water removal in a tropical rain forest area in Hawaii. *Trop. Ecol.* 14:1-18.

Muller, R.N., and F.H. Bormann. 1976. Role of *Erythronium americanum* Ker. in energy flow and nutrient dynamics of a northern hardwood forest ecosystem. *Science* 193:1126-1128.

Neill, W.M. 1983. The tamarisk invasion of desert riparian areas. *Educ. Bull.* 83-4, Desert Protect. Coun., Spring Valley, Calif.

Nihlgard, B. 1972. Plant biomass, primary production and distribution of chemical elements in a beech and a planted spruce forest in South Sweden. *Oikos* 23:69-81.

Odum, E.P. 1960. Organic production and turnover in old field succession. *Ecology* 41:34-49.

O'Neill, R.V., B.S. Ausmus, D.R. Jackson, R.I. van Hook, P. van Voris, C. Washburne, and A.P. Watson. 1977. Monitoring terrestrial ecosystems by analysis of nutrient export. *Water Air Soil Pollut.* 8:271-277.

Perala, D.A., and D.H. Alban. 1982. Biomass, nutrient distribution and litterfall in *Populus*, *Pinus* and *Picea* stands on two different soils in Minnesota. *Plant and Soil* 64:177-192.

Peterson, D.L., and G.L. Rolfe. 1982. Nutrient dynamics of herbaceous vegetation in upland and floodplain forest communities. *Amer. Midl. Nat.* 107:325-339.

Pickett, S.T.A. 1976. Succession: an evolutionary interpretation. *Amer. Nat.* 110:107-119.

Robinson, T.W. 1969. Introduction, spread, and areal extent of saltcedar (*Tamarix*) in the western states. U.S. Geol. Surv. Prof. Pap. 491-A.

Rosenzweig, M.L. 1968. Net primary productivity of terrestrial communities: prediction from climatological data. *Amer. Nat.* 102:67-74.

Rundel, P.W. 1980. The ecological distribution of C_4 and C_3 grasses in the Hawaiian Islands. *Oecologia* 45:354-359.

Singer, F.J., W.T. Swank, and E.E.C. Clebsch. 1984. Effects of wild pig rooting in a deciduous forest. *J. Wildl. Manage.* 48:464-473.

Smith, C.W. 1985. Impact of alien plants on Hawai'i's native biota. In *Hawai'i's terrestrial ecosystems: preservation and management*, ed. C.P. Stone and J.M. Scott, 180-250. Univ. Hawaii Coop. Natl. Park Resour. Stud. Unit. Honolulu: Univ. Hawaii Pr.

Smith, C.W., T. Parman, and K. Wampler. 1980. Impact of fire in a tropical submontane seasonal forest. *Proc. 2nd Conf. Sci. Res. Natl. Parks,* Vol. 10, *Fire ecology,* 313-324. San Francisco, Nov. 26-30, 1979. Natl. Park Serv.

Stone, C.P., and J.M. Scott, eds. 1985. *Hawai'i's terrestrial ecosystems: preservation and management*. Univ. Hawaii Coop. Natl. Park Resour. Stud. Unit. Honolulu: Univ. Hawaii Pr.

Swank, W.T., and J.E. Douglass. 1974. Streamflow greatly reduced by converting deciduous hardwood stands to pine. *Science* 185:857-859.

Thomas, K.J. 1981. The role of aquatic weeds in changing the pattern of ecosystems in Kerala. *Environ. Conserv.* 8:63-66.

Thomas, W.A. 1969. Accumulation and cycling of calcium by dogwood trees. *Ecol. Monogr.* 39:101-120.

Tilman, D. 1982. *Resource competition and community structure*. New Jersey: Princeton Univ. Pr.

Vitousek, P.M. 1982. Nutrient cycling and nutrient use efficiency. *Amer. Nat.* 119:553-572.

Vitousek, P.M. 1986. Biological invasions and ecosystem processes: can species make a difference? In *Biological invasions of North America and Hawaii*, ed. H.A. Mooney and J. Drake, 163-176. New York: Springer-Verlag.

Vitousek, P.M., K. van Cleve, N. Balakrishnan, and D. Mueller-Dombois. 1983. Soil development and nitrogen turnover in montane rainforest soils on Hawaii. *Biotropica* 15:268-274.

Vitousek, P.M., L.R. Walker, L.D. Whiteaker, D. Mueller-Dombois, and P.A. Matson. 1987. Biological invasion by *Myrica faya* alters ecosystem development in Hawaii. *Science* 238:802-804.

Vivrette, N.J., and C.H. Muller. 1977. Mechanism of invasion and dominance of coastal grassland by *Mesembryanthemum crystallinum*. *Ecol. Monogr.* 47:301-318.

Wardle, P. 1971. An explanation for alpine timberline. *New Zealand J. Bot.* 9:371-402.

Woodwell, G.M. 1974. Success, succession, and Adam Smith. *Bioscience* 24:81-87.

CHARACTERISTICS OF INVADING PLANT SPECIES

Richard N. Mack

ABSTRACT

We have little synthetic understanding of the characteristics of successful plant invaders, *i.e.*, those that become naturalized. I will make, however, three observations about the characteristics of these species: 1) If considered collectively, their morphologic, genetic, and ecological characteristics in many regions (including Hawai'i) are as diverse as the characteristics of the native vascular plants. 2) Despite this extreme diversity, many local floras worldwide share the same naturalized species. Consequently, successful invaders may indeed share traits, although not necessarily the traits often suggested. 3) The opportunity exists through comparative studies of congeners to determine the characteristics that allow successful invasions.

INTRODUCTION

Naturalized plants in Hawai'i aptly illustrate diversity in the characteristics of successful invaders. Prominent or conspicuous invaders in Hawai'i range from annual grasses (Glenwood grass, *Sacciolepis indica*) and herbs (Beggar's tick, *Bidens pilosa*) to tussock grasses (fountain grass, *Pennisetum setaceum*) and turf grasses (Hilo grass, *Paspalum conjugatum*) to shrubs (lantana, *Lantana camara*) and vines (banana poka, *Passiflora mollissima*). Even ferns (*e.g.*, blechnum fern, *Blechnum occidentale*) have become naturalized. Alien trees are particularly prominent and diverse in life form, including pines (Caribbean pine, *Pinus caribaea* and cluster pine, *Pinus pinaster*) and many angiosperms, for example, *koa haole (Leucaena leucocephala)*, firetree (*Myrica faya*), and strawberry guava (*Psidium cattleianum*) (Smith 1985). Vines and aquatic free-floating plants may have the fewest biological restrictions to their invasion, and some of the worst future invaders may be expected to come from these groups.

Genetic Characteristics of Invaders

Genetic attributes and breeding systems among invaders worldwide are also diverse. Examples of aggressive invaders with little genetic variation are at least as common as those with high variability (Barrett and Richardson 1986). Genetic variability is probably not of paramount importance even for invaders with such variability in their natural ranges; little of this variation would usually be expressed in a small founder population. Whether this population bottleneck in natural selection commonly hampers an invasion is unknown.

Among both natives and aliens, generalists usually have more genetic variation than specialists. But many successful invaders are cleistogamous, *i.e.*, the flowers are self-fertilized in the bud, and these plants show little or no isozyme variation. Several invaders, such as cocklebur (*Xanthium strumarium*) in Australia and cheatgrass (*Bromus tectorum*) in the U.S., show clear heritable variation among populations in ecologically important traits, yet they display little or no electrophoretic variation (Barrett and Richardson 1986; Novak and Mack, unpub. data). Isozyme variation is, however, only a crude gauge of overall genetic variation. Two genetic "main options" are most often invoked as the explanations for persistence of an invader: polymorphisms and phenotypic plasticity. Some invaders, such as cheatgrass, apparently evolved with both options.

Sexual recombination is not a prerequisite for a successful invader. Some of the most successful invaders either display sexual reproduction infrequently (*e.g.*, water hyacinth, *Eichhornia crassipes*) or are sterile (*Salvinia molesta*); their populations may possess just a few genotypes that persist through vegetative propagation (Barrett and Richardson 1986). Other widely successful invaders may display vivipary, such as Mauritius hemp (*Furcraea foetida*) in Hawai'i (Smith 1985). Dioecism (having male and female organs in separate plants), as seen in Brazilian peppertree or Christmas berry (*Schinus terebinthifolius*), also does not restrict the spread of some invaders (Loope and Dunevitz 1981).

Invaders are often polyploids, but this condition is probably more an expression of the high percentage of polyploidy among vascular plants than a trait advantageous in colonizing new habitats (Stebbins 1985; Barrett and Richardson 1986; Gray 1986). Furthermore, interpretation is clouded because polyploidy includes a wide range of conditions, from the simple doubling of the same chromosomes (autopolyploidy) to the possession of sets of chromosomes from genetically differentiated populations (allopolyploidy). There may be little or no advantage in the simple genomic multiplication of autopolyploidy (Gray 1986). Perhaps the only conclusion that can be drawn currently about the genetic traits of successful invaders concerns the lack of requisite traits. Successful invaders do not necessarily share high genetic variability, high polyploidy, or high levels of heterozygosity, and they may have wide variation in breeding system.

Demography of Invaders

The demographies of few invaders have been evaluated anywhere. The demography of cheatgrass in the interior Pacific Northwest is one of the few documented records. This grass, a native of arid Eurasia, is now prominent in western North America and temperate South America. It displays weak dormancy (Hulbert 1955); seeds can germinate within two months after leaving the parent in late spring. In the northern hemisphere its seedlings can emerge after rain any time from late August until early May. Even though mortality may be extensive over winter, some plants from these multiple cohorts always survive and produce seeds in late June (Mack and Pyke 1983). Furthermore, these seeds are viable even before seed filling is complete (Hulbert 1955).

Characteristics of Successful Invaders

Statements about diversity in the characteristics of plant invaders conflict, however, with an observation. Many of the same species recur among the lists of naturalized species in local floras, as reflected in the list of the 20 worst weeds worldwide (Holm *et al.* 1977). Do these species have extraordinary opportunities for dispersal, superior traits for invading, or both? The opportunity provided by their introduction leads to the first of the three categories of characteristics I consider important for a successful invader (Table 1).

Several questions arise from such a compilation of characteristics. Are the traits held in common by successful invaders associated more with enhancing their dispersal than with their ecological tolerance of the new range? Have the features that aid in dispersal become more varied through time? For a plant to make a transoceanic immigration without human intervention, possession of a floating seed is important, if not essential. In the age of sailing ships, successful dispersal was dependent on the immigrant's ability to survive as a seed or dormant stem or root (or a potted plant) during a long voyage. Today the chief criterion in dispersability may be only the plant's commercial (including ornamental) value; hardiness in transit is no longer essential. But once the alien arrives, how important to a successful invasion is the escape of the species from its predators, etc. in the natural range?

These questions and observations suggest a tool little employed in the search for common characteristics among invaders: detailed, simultaneous examinations of closely-related congeners in the same range. Such investigations could compare the performance of either several alien congeners in the same new range, or one alien with one native congener. The examination should ideally begin with a determination of the genetic variation in the species soon after alien entry. The demography of the species should be closely monitored for consecutive generations. Comparisons should also be made of the autecology of the species and their biotic interaction (predators, parasites, pollinators, competitors). All the invasions followed in this manner may fail. But tracking the epidemiology of aliens that never become prominent may nevertheless be instructive in deriving an understanding of why some invasions fail while others succeed. The Hawaiian flora offers many opportunities for such study, as in the alien congeneric pairing *Schinus molle* and *S.*

terebinthifolius, and the pairing *Myrsine salicina* (alien) and *M. lessertiana* (native). Other opportunities occur in the alien genera *Hedychium* (gingers), *Psidium* (guavas), *Rubus* (berries), and *Senecio* (groundsel). Only after these and other examples worldwide have been thoroughly investigated will predictions of future plant invaders and the sites of their invasions reach a level meaningful for control.

Table 1. Characteristics of successful invaders.

I. Possess characteristic(s) allowing the opportunity for dispersal (at any spatial scale).

Deliberate: the species is considered useful as:*
- forage
- food
- fiber
- medicine
- ornamental

Accidental: the species is dispersed:
- as a seed contaminant
- with animals (attached to or within)
- in contaminated goods (e.g. ballast)

*Plants not considered useful by humans or not associated with useful plants, and confined to the centers of continents, have little opportunity for dispersal. The importance of the species' other features then becomes moot.

II. Dispersal coincident with leaving predators, competitors, parasites, and pathogens behind in the home range.**

**Seeds usually leave behind all but surface fungi and viruses. Escape from predators, etc. is presumed to be essential but needs to be verified on a quantitative basis.

III. Ability to tolerate the new range beyond the point of entry.***

***For example: the alien may never leave a ballast heap if it requires well-drained soil but arrived in a region with only clayey soil.

The expression of morphologic, genetic, and most ecological characteristics of the invader arises under III. But the importance of these features is moot unless I and probably II are satisfied.

Literature Cited

Barrett, S.C.H., and B.J. Richardson. 1986. Genetic attributes of invading species. In *Ecology of biological invasions*, ed. R.H. Groves and J.J. Burdon, 21-33. Cambridge, England: Cambridge Univ. Pr.

Gray, A.J. 1986. Do invading species have definable genetic characteristics? *Phil. Trans. Royal Soc. London*, Ser. B, 314:655-672.

Holm, L.G., D.L. Plucknett, J.V. Pancho, and J.P. Herberger. 1977. *The world's worst weeds, distribution and biology*. East-West Center. Honolulu: Univ. Pr. Hawaii.

Hulbert, L.C. 1955. Ecological studies of *Bromus tectorum* and other annual bromegrasses. *Ecol. Monogr.* 25:181-213.

Loope, L.L., and V.L. Dunevitz. 1981. *Impact of fire exclusion and invasion of* Schinus terebinthifolius *on limestone rockland pine forests of southeastern Florida.* Tech. Rep. T-645, Natl. Park Serv. South Florida Res. Cent. Homestead, Fla.

Mack, R.N., and D.A. Pyke. 1983. The demography of *Bromus tectorum*: variation in time and space. *J. Ecol.* 71:69-93.

Smith, C.W. 1985. Impact of alien plants on Hawai'i's native biota. In *Hawai'i's terrestrial ecosystems: preservation and management*, ed. C.P. Stone and J.M. Scott, 180-250. Univ. Hawaii Coop. Natl. Park Resour. Stud. Unit. Honolulu: Univ. Hawaii Pr.

Stebbins, G.L. 1985. Polyploidy, hybridization, and the invasion of new habitats. *Annals Missouri Bot. Gard.* 72:824-832.

ALIEN PLANT MANAGEMENT IN EVERGLADES NATIONAL PARK: AN HISTORICAL PERSPECTIVE

Anne Marie LaRosa,
Robert F. Doren, and Lance Gunderson

ABSTRACT

Alien plants have been actively managed for the past 25 years at Everglades National Park, with primary emphasis on Australian pine (*Casuarina* spp.), cajeput (*Melaleuca quinquenervia*), and Brazilian peppertree (*Schinus terebinthifolius*). Cajeput has the highest potential for displacing native vegetation in South Florida, but the other two taxa have wide ecological tolerances. Australian pines have been controlled by chemicals (TORDON, 2,4,5-T and 2,4-D) and prescribed fire in the past, but the species continues to increase in the Park. Cajeput has been treated with the same two herbicides plus VELPAR and GARLON, but large, dense stands of the species adjacent to the Park still threaten. Brazilian peppertree has been subjected to mechanical treatment, removing altered substrate, thus increasing hydroperiod and eliminating a favorable substrate. Restoration of wetlands has proven more successful in reducing this species. GARLON 3A as a basal bark treatment is the most effective herbicide treatment. The first alien plant management plan was written in 1969 and has been revised many times. A comprehensive plan for control of some alien plant species has been prepared for Everglades, and parkwide mapping for major species is under way. The South Florida Exotic Plant Pest Council, comprised of over 50 groups, is taking a regional approach to the problem and has resulted in increased awareness and shared information. Long-term control of the target species can be achieved by continued planning and consistent and adequate funding.

INTRODUCTION

Everglades National Park is located in southern Florida at the southern terminus of the vast wetland complex known as the Everglades (Fig. 1). The park was established in 1947 to preserve the biological resources of the area and contains a variety of habitats within its 1,373,320 a (556,000 ha). The principal ecosystem types within the Park include shallow water marine habitats (558,220 a or 226,000 ha), saltwater wetland forests and marshes (449,540 a or 182,000 ha), freshwater marshes and

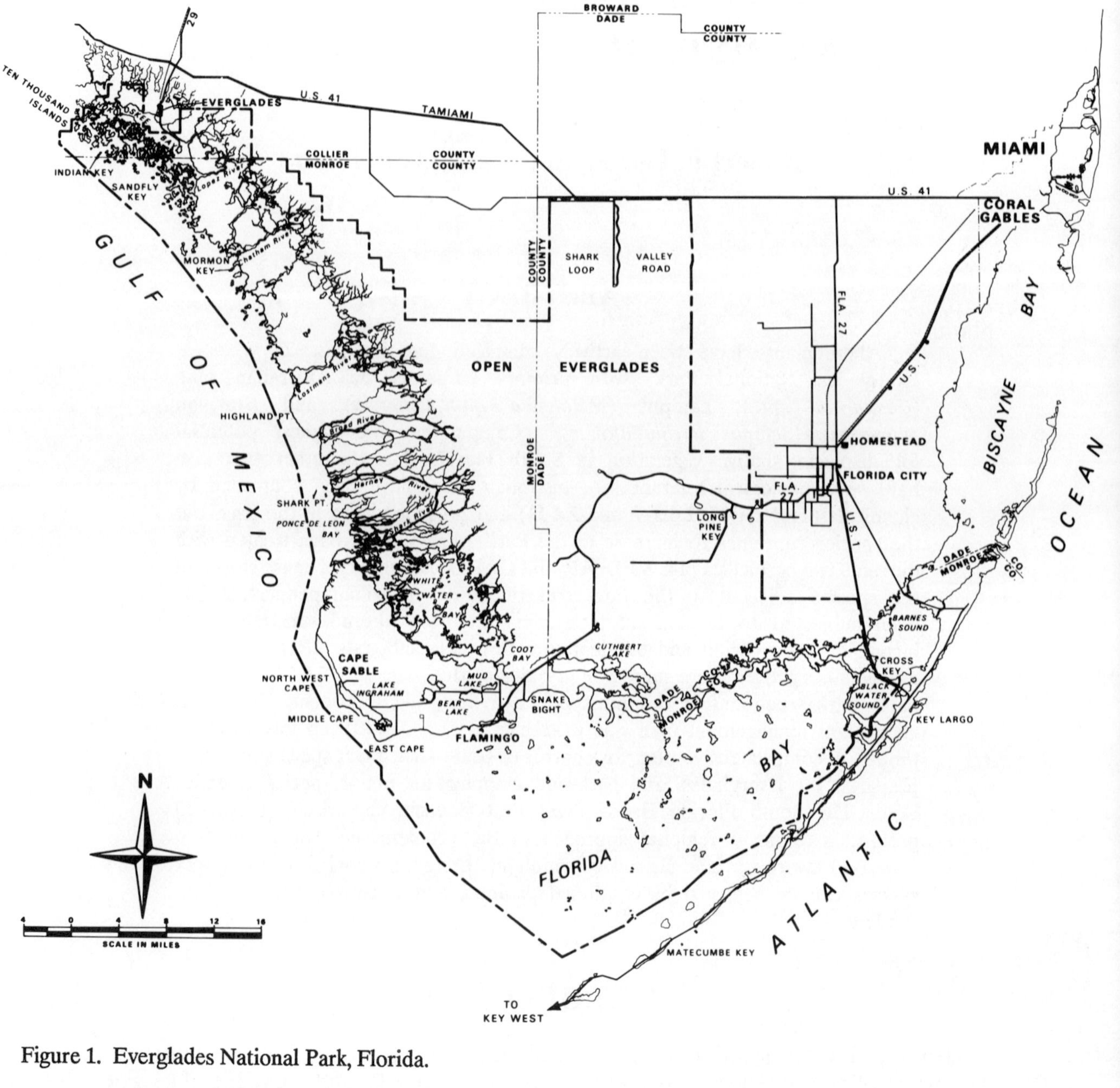

Figure 1. Everglades National Park, Florida.

prairie (370,500 a or 150,000 ha), and an upland complex of pine and hardwood forests (19,760 a or 8,000 ha).

One of the major factors controlling the distribution of the vegetation within the Everglades is the hydrologic regime, defined by the depth and duration of inundation as well as the quality and salinity of the source water. The flat topography, temporal distribution of rainfall, and proximity to the coast all interact to determine the hydrologic regime at a site. Surficial geology and overlying soil type also influence plant species composition and abundance. Natural disturbances that influence the vegetation are fires, hurricanes, freezes, and sea level changes (Craighead 1971; Alexander and Crook 1974; Olmsted and Loope 1984).

Weedy Plants in Southern Florida

The success of weedy plant species in southern Florida is due to a number of factors. Humans have accelerated the rate of species introduction, resulting in the transplanting of hundreds of thousands of plants for use as landscape ornamentals, food sources, and medicines. Southern Florida is essentially a subtropical island, surrounded on three sides by water and to the north by temperate ecosystems, which limits the natural rate of species arrival. Southern Florida is also geologically young, with the current plant communities in existence for only the past 5,000 years (Long 1974; Watts 1975). The natural disturbances that are part of the South Florida environment allow opportunities for weedy species to become established (Myers 1975; Wade *et al.* 1980; Ewel *et al.* 1982) and have been amplified by human activities. The most successful alien species (Myers 1983) are so well adapted to an altered niche that they outcompete native species (Meador 1977; Ewel *et al.* 1982).

The Exotic Pest Plant Council reported that over 400 introduced plant species have naturalized in southern Florida. In Everglades National Park, 840 plant species have been identified, of which 217 (25.8%) are nonnative in origin (Whiteaker and Doren 1989). Many of the aliens within the Park were brought in and planted by early settlers prior to park establishment. To date three taxa have been identified as the major pest aliens within the Park and are the primary objects of management actions. The targeted species are Australian pine (*Casuarina* spp.), cajeput (*Melaleuca quinquenervia*), and Brazilian peppertree or Christmas berry (*Schinus terebinthifolius*).

BIOLOGY AND DISTRIBUTION OF THE THREE WORST EXOTIC PLANTS IN EVERGLADES NATIONAL PARK

Australian Pine, *Casuarina* spp.

Two of the eight species of Australian Pine (*Casuarina*) introduced into Florida (Morton 1980) have been found in the Park: *Casuarina equisetifolia* and *C. glauca*. *Casuarina equisetifolia* was introduced into Florida in the late 1800s and was widely planted to stabilize banks, drain wetlands, and provide shade, as well as for timber and landscaping purposes. *Casuarina equisetifolia* is currently distributed from Key West north to Tampa (Geary 1981).

Casuarina invades many of the vegetation types in South Florida but is apparently restricted by extended periods of inundation. Survival is best on well-drained sites. *Casuarina equisetifolia* is tolerant of brackish soils and sea spray and readily colonizes open sand and shell beaches and coastal prairies (Egler 1952). Scattered individuals of this species are found in higher-elevation mangrove stands and the interior of keys in Florida Bay. *Casuarina glauca* can invade sawgrass marshes and slash pine forests but is most common along roadsides, berms, and in burned tropical hardwood forests (locally referred to as hammocks).

Dense stands of *Casuarina* are generally detrimental to wildlife, especially threatened and endangered species. Nesting loggerhead and green sea turtles (*Caretta caretta caretta* and *Chelonia mydas mydas*) require gently sloping beaches with soft sand for successful nesting (Klukas 1969; Conover and McElwee 1971). Beaches dominated by *Casuarina* are rarely used by these turtles, as the trees hinder nesting. Also, trees displace native, beach-stabilizing plant species and may alter the morphology of the beach to a steeper embankment due to waves undercutting the sand adjacent to roots. Sandy beaches used as nest sites for the American crocodile (*Crocodylus acutus*) are also colonized by *Casuarina*. Mazzotti *et al.* (1981) found little or no use of *Casuarina* stands by rodents. An exception to the presumptive negative effect on wildlife is the use of *Casuarina* as migration rest sites for the endangered peregrine falcon (*Falco peregrinus*) at Fort Jefferson National Monument, Dry Tortugas.

Cajeput, *Melaleuca quinquenervia*

The potential for displacement of native vegetation by cajeput is greater than for any other introduced plant species in Florida. Introduced from Australia in 1906 as an ornamental, it has since spread rapidly into a range of habitats, assisted by intentional plantings. Cost and Craver (1981) reported that cajeput ranges from Orlando south to the tip of the Florida peninsula, occupying nearly 459,420 a (186,000 ha), of which approximately 40,260 a (16,300 ha) of this total are dense, monotypic stands. Recent studies indicated that in some areas, especially following wildfires, *Melaleuca* is able to increase numbers of trees each year by a factor of 10 (R.F. Doren *et al.*, unpub. data).

Melaleuca has many adaptations that account for its success in South Florida. This plant can flower up to five times a year (Meskimen 1962). The small seeds produced are stored in capsules along the stem and remain viable up to 10 years (Conde *et al.* 1981). Estimates of seed storage range from 2 to 20 million seeds per tree (Meskimen 1962; Alexander and Hofstetter 1975). The seeds are released following a disruption of the stem transpiration stream either slowly by natural mortality, or *en masse* following a stress such as drought, fire, or frost (Woodall 1983). The seeds germinate in moist soil, even under standing water (Meskimen 1962; Myers 1975). Young seedlings are vulnerable to drought and fire, but once established they can withstand these perturbations as well as extended periods of inundation. *Melaleuca* is well adapted to frequent, severe fires; older individuals are well protected by thick, papery bark. A fire will release seeds onto a cleared seedbed, thereby increasing chances of regeneration (Wade 1981).

These adaptations allow cajeput to compete in native habitats but make it particularly successful in ecotones (vegetation junctions) and disturbed areas. In Australia, *Melaleuca* is found in wet, nutrient-poor savannahs with frequent fires (Coaldrake 1961). Myers (1975, 1983) has shown that the cypress prairie and the pine-cypress ecotone are similar habitats in South Florida and are the most vulnerable to cajeput invasion. Other habitats at risk include sawgrass marshes, muhly prairies, slash pine forests, hardwood hammocks, and the buttonwood-mangrove association. Large-scale alterations to the hydrologic regimes of south Florida and concomitant changes to the fire regime have also increased the distribution of this plant. In addition to the effects on native vegetation, *Melaleuca* provides poor habitat for native fauna (Schortemeyer *et al.* 1981; Mazzotti *et al.* 1981; Ostienka and Mazzotti 1988).

Brazilian Peppertree, *Schinus terebinthifolius*

Brazilian peppertree was first collected in Florida in the 1840s (Barkley 1944) but did not become a conspicuous component of natural systems until the 1960s (Alexander and Crook 1974; Craighead 1971). It was widely planted as an ornamental throughout central and southern Florida and currently ranges as far north as Jacksonville.

The success of Brazilian peppertree is due to a wide ecological tolerance, as shown by an ability to rapidly colonize disturbed areas and persist through later successional stages (Ewel *et al.* 1982; Krauss 1987; Doren and Whiteaker 1990a; Doren *et al.* 1990). As a pioneer species it grows rapidly, produces many seeds, and sprouts readily. Flowering occurs in a compressed synchronous period each October, and pollination of this dioecious species is primarily by insects. The bright red drupes mature in December and January, hence one of the common names, Christmas berry. The drupes are consumed by animals, especially robins (*Turdus migratorius*), which are the primary dispersal vectors. Seedlings can establish and survive in open, light areas as well as under dense canopies (Ewel *et al.* 1982).

Brazilian peppertree has invaded many habitats in Everglades National Park, including sawgrass marshes, muhly prairies, tropical hardwood hammocks, coastal hardwood hammocks, rockland pine forests, and at the saltmarsh/mangrove ecotone. *Schinus* rarely grows on sites flooded longer than three to six months and is not found in the deeper wetland communities. The largest populations occur on disturbed lands such as roadsides, canal banks, and abandoned farmland. The farming practices in South Florida, especially rockplowing (a method of crushing native limestone substrate), alter the substrate and allow for greater occurrence of mycorrhizae, a condition that allows *Schinus* to out-compete native species on these sites (Meador 1977) and alter successional vegetation patterns (Hilsenbeck 1976; Loope and Dunevitz 1981a; Krauss 1983, 1987; Doren and Whiteaker 1990b; Doren *et al.* 1990).

PAST ALIEN PLANT MANAGEMENT IN EVERGLADES NATIONAL PARK

The first efforts at plant eradication in Everglades National Park began prior to park establishment and were directed at a native species. In 1932, the U.S. Department of Agriculture began eradicating wild cotton, *Gossypium hirsutum*, to prevent the spread of the cotton boll weevil (*Anthonomus grandis*). The program was halted in 1972 due to unwarranted damage to park resources and lack of effects in boll weevil control.

In the early days of the Park, little attention was given to introduced species or their effects on the environment. Many plants were associated with abandoned homesites or dwellings of both native Indians and later Europeans. A memorandum by W.B. Robertson in 1956 raised the first alarm, noting that *Casuarina* spp. were found scattered throughout the Park and represented a future threat. The earliest control work focused on aquatic weeds in the Royal Palm Pond, a developed interpretive area, where some of the targeted species were actually native plants yet were being removed to improve wildlife viewing. Other control efforts were aimed at the alien water hyacinth, *Eichhornia crassipes*. Following Hurricane Donna in 1960, Park staff noticed an increase in populations of Australian pine. Initially, alien plants were located, then in 1963 the first control was attempted. This effort was a small-scale mechanical control aimed at *Casuarina* along beaches. *Casuarina* continued to expand after Hurricane Betsy in 1965.

In 1969 the first alien plant management plan was written, primarily targeting Australian pine (Klukas 1969). This report mentioned other problem plant species, including Brazilian peppertree, Australian pine, cajeput, common guava (*Psidium guajava*), and shoebutton ardisia (*Ardisia solanacea*). The first formal, comprehensive alien plant management plan for the Park was written in 1973 (Bancroft 1973) and contained information on policy, plant distributions, specific control objectives and associated costs, and research needs. Six management units were established within the Park, with specific treatment locations identified within each unit. The highest-priority actions were to control Australian pine and eradicate cajeput. Lather leaf (*Colubrina asiatica*) was added to the above-mentioned species as a problem alien plant. The potential for spread and known distribution were also included in the plan for an additional 100 plant species. The 1973 management plan was updated in 1977 (Bancroft 1977) and again targeted control efforts at *Casuarina* and *Melaleuca*. The 1977 plan also added another aquatic exotic, milfoil (*Hydrilla verticillata*), to the list of problem species. The management plan was updated in 1983 (Doren and Rochefort 1983) and again in 1988 (Whiteaker and Doren 1989). Specific programs for *Casuarina*, *Melaleuca*, and *Schinus*, which have received most of the control effort to date, will be discussed.

Australian Pine

The earliest alien plant control efforts within Everglades National Park emphasized Australian pine. Robertson (1956) reported scattered individuals within the Park in the following locales: the southeastern corner near Card Sound, along the Ingraham Highway to the mangrove zone, on several keys in Florida Bay, and along the Gulf Coast beaches. He reported that some of the populations were spreading slowly and recommended the initiation of experimental control studies. No action was taken until 1963, when post-hurricane surveys revealed rapidly increasing populations in Cape Sable and Highland Beaches. Several small-scale programs were initiated between 1963 and 1970 and were marginally successful. These programs involved mechanical control methods -- cutting down and uprooting individuals. Chemical control began in 1966; the herbicides 2,4,5-T and 2,4-D effectively killed *Casuarina* but were banned from use in 1969. Other herbicides have since been used, with limited success. Prescribed fire effectively killed scattered trees within prairies but did not prove effective in dense stands, because the fuel structure would not carry a fire except under extremely dry conditions.

In spite of these control projects, *Casuarina* continued to spread and by 1970 covered an estimated several thousand acres. Ogden (1970) warned of a four-fold increase in area within the habitat of the endangered American crocodile. This prompted another assessment of the problem (Wodehouse 1972) and a subsequent control program, which ran from 1971 through 1978. During this period, 86,300 trees were treated in the park interior, Highland Beach, and the northern shore of Florida Bay. Due to budget reduction, only 12,000 trees were treated from 1979 through 1985. A survey in 1983 estimated that dense stands covered 18,030 a (7,300 ha) in the southeastern corner of the park, a number at or above the 1970 estimate of coverage, but the Cape Sable and Highland Beach areas remain free of *Casuarina*. Minimal monitoring for seedling control is required.

Cajeput

Cajeput was first reported in Everglades National Park in 1967, and only a dozen more had been discovered by 1969. These isolated trees were found near Park headquarters and along the eastern and northern boundaries. Treatment prior to the mid-1970s consisted of felling the trees and applying herbicide (2,4,5-T and 2,4-D) to the remnant stump. Between 1979 and 1984 8,300 individuals were treated. Most of these were seedlings, which were pulled up by hand. Larger individuals were girdled and frilled, with herbicides (VELPAR or GARLON) applied in the frill. Large, dense stands of *Melaleuca* outside and adjacent to the Park, and in areas recently proposed for acquisition and addition to the Park, still pose the greatest threat.

Brazilian Peppertree

Previous efforts at controlling *Schinus* were concentrated in the "Hole-in-the-Donut" area of the Park and adjacent pinelands. Although Craighead noted invasion of *Schinus* into the mangrove areas around Everglades City in 1961, control efforts have yet to be mounted in these areas. The "Hole-in-the-Donut" is 9,880 a or 4,000 ha of abandoned farmland located in the middle of the Park. Rockplowing, invented in the 1950s, altered the substrate in about half of this area and resulted in conditions that favored the establishment of *Schinus*.

Attempts at restoring the "Hole-in-the-Donut" were initiated in 1972. A variety of techniques was employed to arrest succession and restore native communities (Koepp 1979). Various combinations of mowing, disking, burning, rolling, chopping, and bulldozing were used to remove standing *Schinus*. Subsequent studies indicated that post-treatment succession was toward *Schinus*-dominated vegetation because of the substrate alteration. In conjunction with some of these clearing efforts, pine seedlings were planted. Very few of the pines survived; this is considered by some to be due to poor planting practices and competition from *Schinus*.

One partially successful treatment, initiated in 1976, was the creation of artificial wetlands and uplands in the "Hole-in-the-Donut." Soil was removed from one area and mounded in an adjacent area. The higher, drier area was planted with native hardwood species to simulate a tropical hammock. The lower wetlands were not planted. Recent qualitative inspections of the area revealed that native upland hardwoods are being invaded by Brazilian peppertree, with numerous seedlings in the understory. Native wetland species have colonized the lower (less than 3.5 ft mean sea level) area, and Brazilian peppertree has still not established on this site. The intermediate elevations between the two extremes, however, are dominated by peppertree.

Another management strategy in the "Hole-in-the-Donut," initiated in 1980, was the planting of saplings of native hardwood species in cleared areas. The objective was to provide a seed source to create a forest of mixed species composition, including *Schinus*. The areas around the plantings were mowed for a year to promote growth of the natives. Qualitative observations in 1987 indicate that some of the planted hardwoods have survived, but peppertree has reestablished and is the dominant regrowth.

Brazilian peppertree in the "Hole-in-the-Donut" has been used to experiment with different techniques of herbicide application. Ewel *et al.* (1982) found that basal bark spraying of *Schinus* with GARLON 3A proved to be the most effective means of killing the unwanted plant (Doren *et al.* 1991) with minimum effect on the surrounding vegetation. Ewel *et al.* (1982) also recommended that the female trees in the population be targeted for herbicide treatment, a form of matricide. The objective was to gradually remove the reproductive females from the population, with the remnant males providing a dense canopy to shade and inhibit seedling growth. The matricide approach is difficult to carry out, as the multi-stem growth form of peppertree in the forest makes it difficult to identify stems as part of a male or a female plant. Matricide was especially uneconomical with *Schinus* seedling density in the understory (Doren and Whiteaker 1990b).

The native pine forests of Everglades National Park, dominated by South Florida slash pine (*Pinus elliottii* var. *densa*), are susceptible to invasion by *Schinus* (Ewel *et al.* 1982). Many of the remnant stands of this vegetation type outside the Park are dominated by *Schinus*, but prescribed burning at 3- to 7-year intervals has severely restricted its establishment within the Park (Loope and Dunevitz 1981b).

CURRENT ALIEN PLANT MANAGEMENT

Alien plant management is a program developed in response to several laws, general directives, and policies (National Park Service 1978; Everglades National Park 1982, 1983). National Park Service policy states that introduced species will be controlled or eradicated if park resources (native species, natural communities, and ecological processes) are threatened. This is always subject to funding availability. Management of alien plants was given a high priority in the Park's Resources Management Plan (Everglades National Park 1979, 1982). The management plan is developed by the Park Resources Management staff of the Ranger Division and articulated in the Everglades Exotic Plant Control Handbook (Doren and Rochefort 1983; Whiteaker and Doren 1989). The handbook establishes priorities, control methods, and guidelines.

Much alien plant control work is carried out by rangers in the various districts of the Park. They are guided by an annual work schedule outlined in the control handbook. Species and known locations are prioritized within each district for control. The work effort varies among districts and years, due to other work assignments and funding. For example, 35,000 trees were treated in 1984 and only 1,300 in 1985. Recent work efforts have concentrated on individuals and outliers rather than on dense stands, and primarily on *Casuarina* on the Florida Bay keys and *Melaleuca* throughout the Park.

Introduced plants are placed in five categories within the handbook for management purposes; these groupings aid in assigning priority and subsequent management action (Whiteaker and Doren 1989):

Category 1

In the first category are species that are widespread in the Park or South Florida and that have an established, documented potential to invade undisturbed native plant communities. In most cases, these species are too numerous and widespread for parkwide control with mechanical or chemical methods, given current and projected funding levels. Therefore, these species can only be controlled locally in high-value resource areas or when they occur as outlying populations or individuals that will potentially increase the distribution of the species. Examples of high-value resource areas include critical habitat for endangered animal species, such as nesting beaches for sea turtles and American crocodiles. These areas have high priority for control of *Casuarina*, which can cause severe beach erosion due to their fibrous root systems, thereby eliminating nesting habitat. *Casuarina* found on previously uncolonized keys or beaches would also have high priority for treatment. Control of outliers such as these has been statistically demonstrated to be most effective in controlling the spread of an invasive alien plant (Browder and Schroeder 1981; Moody and Mack 1988).

Category 2

The second category contains species that have localized populations in the Park but the distribution is expanding rapidly and/or the potential to invade native vegetation in other tropical or subtropical areas has been documented in the literature. These species have a high priority for treatment because they can be controlled parkwide before they become a widespread problem similar to the species in Category 1. For example, the Palay rubber vine (*Cryptostegia grandiflora*), native to Madagascar, is recorded from only one isolated site in the Park. However, in tropical Queensland, Australia, this species has invaded 11,580 mi^2 (30,000 km^2), establishing along river edges and moving into adjacent forest, covering trees up to 90 ft (30 m) tall and choking out native vegetation (McFayden 1988). Therefore, this species has a high priority for documentation of its current distribution in the Park and formulation and scheduling of control measures.

Category 3

In the third category are species able to naturalize and spread locally into undisturbed, native vegetation and form dense stands once they have been introduced to a site. Typically, these are large-fruited species that lack a long-distance dispersal mechanism. However, the distributions of these species need to be documented then monitored for rapid increases or large jumps in distribution. These species are prioritized for control relative to the value of the resources that they threaten, and they are treated as time and funding allow. The occurrence of these species is usually associated with abandoned homesites or other developed areas, where introduced landscape plants have persisted at the site and successfully reproduced, with the population gradually expanding radially from the original plant into the surrounding vegetation.

Category 4

The fourth category consists of species that are widespread in the Park and/or South Florida and form dense, monotypic populations, but primarily on disturbed sites such as roadsides, canal embankments, and agricultural lands. These species are usually identified as agricultural weeds in the literature and are a formidable problem on lands that are, or have been, intensively managed. However, large-scale invasions of undisturbed, native vegetation have never been documented for these species. Therefore, at this time these species are not considered to be threats to native plant communities. The best method for control of these species is by prevention or mitigation of disturbances that are not natural components of the Everglades environment. Included in this category are many weedy introduced grasses that form impressively dense, monotypic stands on disturbed areas just outside the Park, but have not successfully invaded adjacent native vegetation on Park land in spite of the production of copious wind-dispersed seeds.

Category 5

The fifth and largest (155 species) category contains all other nonnative species that have been listed, or observed, as occurring in Everglades National Park. These are species that persist from cultivation and landscape species that have not naturalized in the Park, or species that are ephemeral (often annual) weeds restricted to disturbed sites and not forming dense populations even on those sites. No management action is planned or needed for these species, because they have no potential to invade undisturbed native vegetation and/or disrupt natural processes.

In late 1987, work was initiated to conduct a systematic inventory of alien plant distribution (for cajeput, Australian pine, Brazilian pepper, and lather leaf) within Everglades National Park. The work effort will produce a parkwide map of these alien plants in order to guide control work. The map will be generated from interpretation of low-level false-color infra-red aerial photographs. The map will also form a baseline inventory from which future control work can be quantitatively evaluated by following changes in spatial distribution of these plants.

Interagency Efforts

Alien plant management in southern Florida is a regional problem that transcends political boundaries. Alien plants range throughout the area, and even if complete eradication is accomplished within natural, protected areas, external populations will still pose a threat. Governmental agencies, conservation groups, and concerned individuals have held informal meetings to exchange information on aliens since the early 1970s. Following several such meetings, the Exotic Plant Pest Council was formed in 1984 at Everglades National Park. The council is a multi-member task force to meet common objectives regarding management and control of alien plants. To date, over 50 groups, including local, state, and federal agencies, conservation groups, local native plant nurseries, and universities hold active memberships. The specific functions of the Council are:

- To provide a focus for the issues and concerns regarding introduced pest plants and promote understanding of problems and possible solutions.

- To facilitate communication and exchange of information on alien plant control and management and to disseminate this information.

- To serve as an advisory panel for various interests concerned with introduced pest plants, suggest management actions and coordinate the acquisition and dispensation of funds towards mutually beneficial programs.

The Council has a number of ongoing projects, including:

- Control of *Melaleuca* and *Casuarina* on lands adjacent to the Park that are slated for acquisition. The state legislature provided $20,000 for this project in 1986 to be administered by the Council. The State and Dade County Department of Environmental Resources Management have provided an additional $180,000 for the 1987-1989 budget.

- Active support of a proposal for interagency cooperation and funding for a biological control program for *Melaleuca*. The U.S. Army Corps of Engineers allocated $160,000, and the National Park Service allocated $70,000 in 1987 and $79,000 in 1989 to this program.

- Development of a draft alien plant ordinance for counties and municipalities, which would standardize ordinances throughout South Florida and help stem the spread of unwanted alien species.

- Mitigate 60 acres (24 ha) in a pilot study for control of peppertree in the "Hole-in-the-Donut" by removing disturbed substrate and increasing hydroperiod. This has been done through an offsite mitigation program consisting of $500,000 in outside funds.

- Develop a comprehensive list of known alien species, locations, and potential threats to South Florida ecosystems.

The regional, interagency approach has greatly improved awareness of the alien plant problem in South Florida. It has brought all agencies to a common level of control knowledge, thereby saving time and money associated with development of individual programs.

FUTURE EMPHASIS AND NEEDS

Complete eradication of cajeput is the number one alien plant control priority within the Park. The currently small number of individuals can be eradicated at a relatively low cost. Eradication of *Casuarina* will require a large commitment of both human and fiscal resources in order to eliminate the large, dense stands in the southeastern corner of the Park. *Schinus* control in the mangrove areas needs to be addressed, as well as an evaluation of options for future management of the "Hole-in-the-Donut" and surrounding areas; all these programs will also require large resources. All control programs should be supplemented with ongoing monitoring for evaluation of the programs.

Initial work should be continued on biological control options for these three pest species. This type of control is expensive and will take a long time but could result in eventual elimination or significant reduction of continued manual and chemical control in the region and the Park.

Other species, primarily shoebutton ardisia and lather leaf, should be examined for their potential for invasion and techniques for control. *Ardisia* has invaded areas in and around Royal Palm Hammock and the "Hole-in-the-Donut" and appears to be spreading. *Colubrina* has invaded the mangrove areas of the park, but the current and potential distributions are unknown.

The Park should continue participation in the Exotic Plant Pest Council. The Council and its activities further interagency cooperation in solving regional control problems. Participation should also prevent duplication of control efforts and research, thereby saving each member organization time and money. Control of populations outside the Park will eventually assist control within the Park. Councils should be considered for other areas.

The development of detailed distribution maps of both problem and potential pest species should aid in future management and control. Automation of these data on a Geographical Information System would provide this information in a usable format, easily retrievable and edited. Control plans and program evaluations could effectively employ such a system.

CONCLUSIONS

Alien plants have been recognized for 30 years and actively managed for the past 25 years at Everglades National Park. Primary control efforts have been aimed at Australian pine, Brazilian peppertree, and cajeput. The program over the years has been successful in some areas but has suffered some setbacks. Current management reflects a development of effective control strategies, such as attacking outliers of dense stands of Australian pine, cajeput, and Brazilian peppertree, and using new, cooperative strategies in controlling peppertree. The Park has developed safe, effective herbicide control techniques that kill targeted species and, when properly used, are not harmful to human users or surrounding nontarget vegetation.

Many of South Florida's natural ecosystems are subjected to recurring natural disturbances that allow establishment of alien species, especially in the context of large seed sources external to the park. Solution of the alien plant problem will require a long-term commitment of fiscal and human resources in order to meet the distant goal of an alien-free Everglades National Park. The ultimate cause of future problems will be insufficient and inconsistent funds. The State of Florida successfully controls aquatic weed species in waterways by spending $17 million annually. Long-term control can be achieved by consistent and adequate funding and planning.

ACKNOWLEDGMENTS

We wish to thank Amy Daraghy and Meg Heim for their assistance with many of the projects mentioned and cited in this paper. We also wish to thank Lou Whiteaker for supporting information.

Literature Cited

Alexander, T.R., and A.G. Crook. 1974. Recent vegetational changes in southern Florida. In *Environments of South Forida: present and past*, ed. P.J. Gleason, 61-72. Miami Geol. Soc. Mem. 2. Miami, Fla.: Miami Geol. Soc.

Alexander, T.R., and R.H. Hofstetter. 1975. Some current ecological aspects of *Melaleuca quinquenervia* (Cav.) Blake in southern Florida. Pres. Fla. Acad. Sci. 39th Meeting, Lakeland, Fla. Unpub. mimeo. Library, Everglades Natl. Park.

Bancroft, L. 1973. Exotic plant control plan 1974-1979. Everglades Natl. Park, Homestead, Fla. Mimeo.

Bancroft, L. 1977. Exotic plant control plan (update). Everglades Natl. Park, Homestead, Fla. Mimeo.

Barkley, F.A. 1944. *Schinus* L. *Brittonia* 5:160-198.

Browder, J.A., and P.B. Schroeder. 1981. Melaleuca seed dispersal and perspectives on control. In *Proc. Melaleuca Symp.* Sept. 23-24, ed. R.K. Geiger, 17-21. Tallahassee, Fla.: Fla. Div. For.

Coaldrake, J.E. 1961. *The ecosystem of the coastal lowlands ("wallum") of Southern Australia, Queensland*. CSIRO Bull. 283. Melbourne, Australia.

Conde, L.F., D.L. Rockwood, and R.F. Fisher. 1981. Growth studies on *Melaleuca*. *Proc. Melaleuca Symp.*, ed. R.K. Geiger, 23-28. Tallahassee, Fla.: Fla. Div. For.

Conover, C.A., and E.W. McElwee. 1971. *Selected trees for Florida homes*. Univ. Florida Coop. Ext. Serv. Bull. 182. Gainesville, Fla.

Cost, N.D., and G. C. Craver. 1981. Distribution of *Melaleuca* in south Florida measured from the air. *Proc. Melaleuca Symp.* Sept. 23-24, ed. R.K. Geiger, 1-8. Tallahassee, Fla.: Fla. Div. For.

Craighead, F.C. 1971. *The trees of South Florida*. Vol. 1: *The natural environment and their succession*. Coral Gables, Fla.: Univ. Miami Pr.

Doren, R.F., and G. Rochefort. 1983. Exotic plant control handbook (update). Everglades Natl. Park, Homestead, Fla. Mimeo.

Doren, R.F., and L.D. Whiteaker. 1990a. Comparison of economic feasibility of chemical control strategies using basal bark treatments with triclopyr on differing age and density classes of *Schinus terebinthifolius*. *Nat. Areas J.* 10(1):28-34.

Doren, R.F., and L.D. Whiteaker. 1990b. Effects of fire on different age/size individuals of *Schinus terebinthifolius*: an evaluation of the effectiveness of fire in exotic plant management. *Nat. Areas J.* 10(3):107-113.

Doren, R.F., L.D. Whiteaker, and A.M. LaRosa. 1990. Evaluation of fire as a management tool for controlling *Schinus terebinthifolius* as secondary successional growth on abandoned agricultural land. *Environ. Manage.* 15(1):121-129.

Doren, R.F., L.D. Whiteaker, and R. Rochefort. 1991. Seasonal effects on herbicide basal bark treatment of *Schinus terebinthifolius. Proc. Symp. Exotic Pest Plants*, Miami, Fla., Nov. 1988, ed. T.D. Center, R.F. Doren, R.L. Hofstetter, R.L. Myers, and L.D. Whiteaker, 115-130. Washington, D.C.: U.S. Dep. Int. Natl. Park Serv.

Egler, F.E. 1952. Southeast saline Everglades vegetation, Florida, and its management. *Veg. Act Geobot.* 3:213-265.

Everglades National Park. 1979. *General Management Plan.* U.S. Dep. Interior, Natl. Park Serv. Washington, D.C.: U.S. Dep. Interior.

Everglades National Park. 1982. *Everglades National Park Resource Management Plan and Environmental Assessment.* U.S. Dep. Interior, Natl. Park Serv.

Everglades National Park. 1983. Vegetation management policy. Everglades Natl. Park Memo. 5287-1. Everglades Natl. Park, Homestead, Fla.

Ewel, J.J., D.S. Ojima, D.A. Karl, and W.F. DeBusk. 1982. Schinus *in successional ecosystems of Everglades National Park.* Tech. Rep. T-676, Natl. Park Serv. So. Fla. Res. Cent., Everglades Natl. Park. Homestead, Fla.

Geary, T.F. 1981. Casuarinas in Florida (USA) and some Caribbean Islands. Pres. Internatl. Casuarina Workshop, 17-21 August 1981. Canberra, Australia.

Hilsenbeck, C. 1976. An investigation of old field succession in Everglades National Park, first interim report. Everglades Natl. Park, Homestead, Fla. Mimeo.

Klukas, R.W. 1969. The Australia pine problem in Everglades National Park. Management Rep., Everglades Natl. Park, Homestead, Fla.

Koepp, W.P. 1979. The status of *Schinus* manipulation in Everglades National Park. *Tech. Proc., Techniques for Control of* Schinus *in South Florida*, 45-47. Sanibel, Fla.: The Sanibel-Captive Conserv. Found. Inc.

Krauss, P. 1983. Old succession in Everglades National Park. M.S. thesis, Florida Atlantic Univ., Boca Raton, Fla.

Krauss, P. 1987. *Old field succession in Everglades National Park.* Tech. Rep. SFRC - 87/03. Natl. Park Serv. So. Fla. Res. Cent. Homestead, Fla.

Long, R.W. 1974. Origin of the vascular flora of Florida. In *Environments of south Florida: present and past*, ed. P.J. Gleason, 28-34. Miami Geol. Soc. Mem. 2. Miami, Fla.: Miami Geol. Soc.

Loope, L.L., and V.L. Dunevitz. 1981a. *Investigations of early plant succession on abandoned farmland in Everglades National Park.* Tech. Rep. T-644, Natl. Park Serv. So. Fla. Res. Cent. Homestead, Fla.

Loope, L.L. and V.L. Dunevitz. 1981b. *Impact of fire exclusion and invasion of* Schinus terebinthifolius *on limestone rockland pine forests of southeastern Florida*. Tech. Rep. T-645, Natl. Park Serv. South Florida Res. Cent. Homestead, Fla.

Mazzotti, F.J., W. Ostrenko, and A.T. Smith. 1981. Effects of the exotic plants *Melaleuca quinquenervia* and *Casuarina equisetifolia* on small mammal populations in the eastern Florida everglades. *Fla. Sci.* 44:65-71.

McFayden, R. 1988. Rubber vine, *Cryptostegia grandiflora*, a major threat to natural ecosystems in Northern Australia. *Proc. Exotic Pest Plant Symp.*, Atlanta, Georgia: Natl. Park Ser.

Meador, R.E. 1977. The role of mycorrhizae in influencing succession on abandoned Everglades farmland. M.S. thesis, Univ. Florida, Gainesville.

Meskimen, G.F. 1962. A silvical study of the melaleuca tree in south Florida. M.S. thesis, Univ. Florida, Gainesville.

Moody, M.E., and R.N. Mack. 1988. Controlling the spread of plant invasions: the importance of nascent fosci. *J. Applied Ecol.* 25:1009-1021.

Morton, J.F. 1980. The Australian pine or beefwood (*Casuarina equisetifolia* L.), an invasive "weed" tree in Florida. *Proc. Florida State Hort. Soc.* 93:87-95.

Myers, R.L. 1975. The relationship of site conditions to the invading capability of *Melaleuca quinquenervia* in southwest Florida. M.S. thesis, Univ. Florida, Gainesville.

Myers, R.L. 1983. Site susceptibility to invasion by the exotic tree *Melaleuca quinquenervia* in southern Florida. *J. Appl. Ecol.* 20:645-658.

National Park Service. 1978. *Management policies*. Washington, D.C.: U.S. Dep. Interior.

Ogden, J.C. 1970. Survival of the American crocodile in Florida. Everglades Natl. Park, Homestead, Fla. Mimeo.

Olmsted, I.C., and L.L. Loope. 1984. Plant communities of Everglades National Park. In *Environments of South Florida: present and past*, ed. P.J. Gleason, 167-184. Miami, Geol. Soc. Mem. 2. Miami, Fla.: Miami Geol. Soc.

Ostienko, W., and F. Mazzotti. 1981. Small mammal populations in *Melaleuca quinquinervia* communities in the Eastern Florida Everglades. *Proc. Melaleuca Symp.*, ed. R.K. Geiger, 91-98. Tallahassee, Fla.: Fla. Div. For.

Robertson, W.B. 1956. *Casuarina* in Everglades National Park. Memo to Chief Ranger Campbell. Everglades Memo Y1815, Everglades Natl. Park. Homestead, Fla.

Schortemeyer, J.L., R.E. Johnson, and J.D. West. 1981. A preliminary report on wildlife occurrence in *Melaleuca* heads in the Everglades wildlife management area. *Proc. Melaleuca Symp.*, 81-90. Tallahassee, Fla.: Fla. Div. For.

Wade, D. 1981. Some *Melaleuca*-fire relationships including recommendations for homesite protection. *Proc. Melaleuca Symp.*, ed. R.K. Geiger, 29-35. Tallahassee, Fla.: Fla. Div. For.

Wade, D., J. Ewel, and R. Hofstetter. 1980. *Fire in south Florida ecosystems.* Gen. Tech. Rep. SE-17, Southeastern For. Range Exp. Sta., USDA For. Serv.

Watts, W.A. 1975. A late Quaternary record of vegetation from Lake Annie, south-central Florida. *Geology* 3:344-346.

Whiteaker, L.D., and R.F. Doren. 1989. *Exotic plant species management strategies and list of exotic species in prioritized categories for Everglades National Park.* Natl. Park Ser. Res/Resour Manage. Rep. Ser.-89/04. Atlanta, Georgia: Natl. Park Ser.

Wodehouse, A. 1972. *Casuarina* in Everglades National Park. Everglades Natl. Park, Homestead, Fla. Unpub. rep.

Woodall, S.L. 1983. Establishment of *Melaleuca quinquenervia* in the pine-cypress ecotone of southwest Florida. *Florida Sci.* 46:65-72.

ALIEN PLANTS AT CHANNEL ISLANDS NATIONAL PARK

William L. Halvorson

ABSTRACT

Alien plant species are an important part of the biota of Channel Islands National Park. The Park plant list consists of 334 species; of these, 23.6% are considered to be alien to the southern California flora. However, alien species are much more important than floristics indicate; the vegetation of the islands is dominated by alien species. The primary cause is anthropogenic disturbance, not simply the introduction of species. In most cases, the original vegetation of the Channel Islands had a large component of shrubs and trees, whereas the alien plant communities on the Islands today are primarily herbaceous. In recovery from disturbance, native shrubs often return to dominate the landscape; the alien species, though not completely eliminated, are reduced in numbers of individuals and to subordinance rather than dominance. A few alien plant species on the Islands are quite aggressive. Without active control and management, these species can outcompete the natives and gain control. These are the species of most interest, given limited Park natural resources support.

INTRODUCTION

Alien plants (see terminology of Smith 1985) have been important in the flora of the Channel Islands (Fig. 1) since the early influence of European man in the late 1700s. A recent compilation of all available information on the flora of the Channel Islands (Wallace 1985) shows a total of 848 plant species present on nine islands (the eight California Channel Islands and Guadalupe Island, Mexico). Of the 848 species, 621 (73.2%) are considered native, 227 (26.8%) alien. Endemics constitute 16.2% of the total flora and 22.1% of the native flora. On the five islands (Anacapa (West, Middle, and East), Santa Barbara, and San Miguel) managed by Channel Islands National Park in 1986, there are 334 species, of which 79 (23.6%) are alien and 33 (9.9%) endemic (Appendix A). Management of the resources of Channel Islands National Park includes both the control of alien species and the encouragement of rare and endemic species. Management of the area

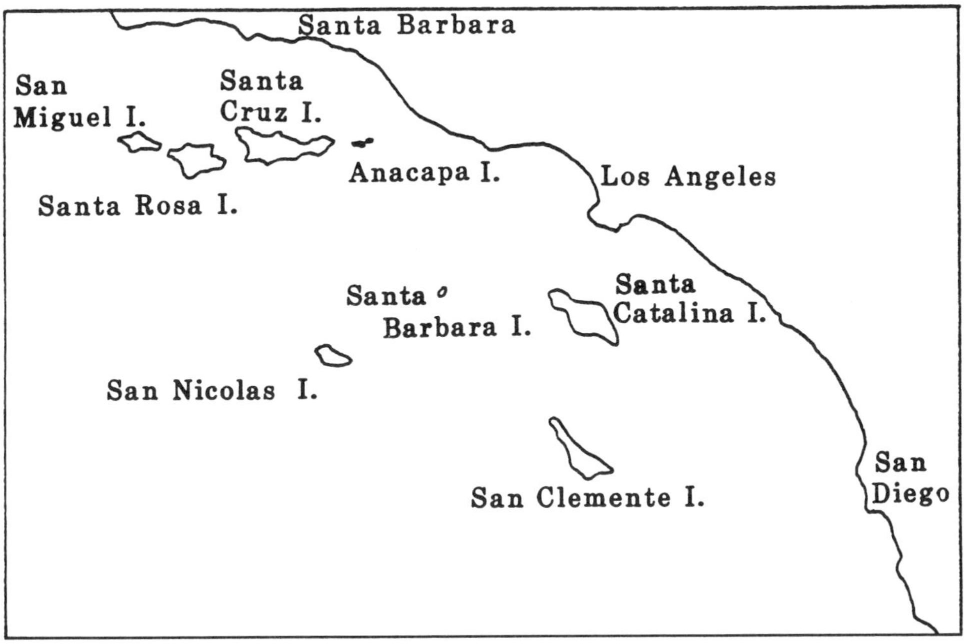

Figure 1. Channel Islands National Park as of 1986 (Anacapa West, Middle, and East; Santa Barbara; and San Miguel).

takes into consideration community integrity rather than simply the preservation of species populations.*

PHYSICAL FEATURES

With an area of approximately 640 a (260 ha), Santa Barbara Island is the smallest of the Channel Islands. It is located 38 mi (61 km) southwest of Los Angeles and 24 mi (39 km) west of Santa Catalina Island. Anacapa (690 a or 280 ha) is located about 12 mi (19 km) south of the mainland at Ventura. These islands are primarily of Miocene volcanic origin, with some small areas of Miocene sedimentary material (Kemnitzer 1933; Emery 1960; Vedder and Howell 1976). Quaternary sedimentary deposits of both marine and nonmarine origin cover the volcanic bedrock over most of Santa Barbara Island (Lipps *et al.* 1968; Johnson, 1979). Various joints, faults, and fractures are seen in sea cliffs about Anacapa, and where these intersect the sea, caves have formed (Scholl 1960; Valentine and Lipps 1963; Lipps 1964; Johnson 1979). Both islands are characterized by steep, rocky sea slopes.

*A recent (1988) list includes information for the five islands within Park boundaries. This list is available from the author and includes 802 species, of which 195 (24.3%) are alien and 85 (10.6%) are endemic.

San Miguel Island encompasses an area of approximately 9,885 a or 4,000 ha and is the westernmost of the northern Channel Islands. It lies about 28 mi (45 km) south of Point Conception and 62 mi (100 km) west-northwest of Ventura. This island is composed primarily of Cretaceous and early- to mid-Tertiary conglomerates, sandstones, siltstones, shales, and volcanics. For the most part, bedrock is covered by sand. Structurally, the island represents the northern flank of a folded and faulted anticline, whose axis trends northwest-southeast (Weaver and Doerner 1969; Weaver *et al.* 1969; Johnson 1979). Topographically, San Miguel is dominated by two rounded peaks with relatively gentle slopes leading to the sea. Prominent marine terraces are present, as are a number of large canyons and extensive beaches.

A reconnaissance soil survey of the Islands was conducted by Johnson (1979). He found three types of soils: 1) Entisol soils have formed in recent calcareous sands and in the steep, highly erodible areas around the perimeter of each island. These soils show very little profile development. 2) Vertisol soils, characterized by a high level of expandable clays that swell when wet and shrink when dry, creating deep cracks in the soil surface, occur in areas with deep deposits. Over time, this shrink-swell motion churns the soil into a fairly uniform, nonhorizonated profile with similar properties throughout. 3) Mollisol soils, characterized by a high clay content, a high base concentration, and a dark surface horizon containing a modest to high organic matter content, are usually found in the central portions of each island. These soils also show shrink-swell properties and form cracks when dry.

Specific weather/climate characteristics of the Islands are relatively undefined due to the lack of adequate records. The National Park Service has had automated, real-time reporting weather stations in operation since November 1985. From these data, together with information from other islands, nearby mainland stations, and a few records available from earlier periods, it is possible to gain a general picture of the weather. The Channel Islands lie in the Dry-summer Subtropical climatic region, commonly called the Mediterranean (Trewartha 1954). Rainfall generally comes only in the winter months, leaving the months of April through October rain free. Annual rainfall appears to be in the range of 50-65 in. (204-254 mm) for Santa Barbara, 40-50 in. (152-203 mm) for Anacapa, and 76-90 in. (300-355 mm) for San Miguel. In the scale reported by Dunkle (1950), this puts Santa Barbara Island in the arid maritime climatic zone, to the north of the desert maritime; San Miguel in the semi-humid maritime to the north and west; and Anacapa on the boundary between the two. Island temperatures are generally mild year-round due to the marine influence, and they show small diurnal and seasonal fluctuations. Mean annual temperatures are 60.8 F (16.0 C) for Santa Barbara and 56.6 F (13.7 C) for San Miguel, compared to 61.5 F (16.4 C) for Long Beach on the mainland.

Santa Barbara Island is far less windy than either San Miguel or Anacapa because it is protected from the strong northwesterly winds, which sweep down past Point Conception, and the Santa Ana winds, which come off the mainland. The winds on Santa Barbara and Anacapa are usually of low velocity except for those periods when the jet stream shifts south, causing strong westerly winds. San Miguel Island experiences wind nearly year-round, with brief lulls coming on the heels of passing storm fronts. Winds commonly blow out of the northwest at 12-19 mi/hr (20-30 km/hr), with storm periods bringing winds of 25-28 mi/hr (40-45 km/hr).

Cloudiness is also seasonal, with the winter season having cloudy, stormy weather interspersed with generally clear skies. The summer is characterized by a layer of fog (marine layer) that hangs offshore and generally covers the Islands, except for a variable amount of time at midday, when it burns off above land surfaces. San Miguel Island has the most consistently foggy weather (Dunkle 1950; Weissman and Rentz 1977; National Park Service, unpub. data).

VEGETATION COMMUNITIES OF CHANNEL ISLANDS NATIONAL PARK

The vegetation of the Channel Islands is Mediterranean in nature, ranging from grassland and scrubland through chaparral to woodland communities. The islands currently managed by the National Park Service -- Anacapa, Santa Barbara, and San Miguel -- all have low community diversity. Santa Barbara has no tree species, and Anacapa and San Miguel have a few trees in deeper canyons. Sea cliff and island grassland communities cover the largest areas of the islands, with perennial iceplant, coastal sage scrub, and coreopsis scrub communities forming patches of various sizes within the other two. Alien species dominate both the island grassland and perennial iceplant communities.

Native Shrublands

The native shrublands of the islands are varied and found to be more important on islands that have had the most amount of time to recover from ranching disturbance (Figs. 2-4). The more important shrub communities are described below.

Sea Cliffs. Due to the form of Santa Barbara and Anacapa Islands, with steep slopes from the sea to 245 ft (75 m), the sea cliff community is very important on these islands. The community is actually dominated by bare rock and soil, but it supports a number of species that are more or less restricted to cliffs. This is in part due to the fact that the slopes have been inaccessible to humans and domestic and feral herbivores; thus they have been refugia during periods of intense use. The widely scattered shrubs and succulents are all low growing (less than 1.6 ft or 0.5 m). Some of the rarer species found here are: *Eriogonum giganteum* var. *compactum*, *Eriophyllum nevinii*, and *Dudleya traskiae*, which are all island endemics. *Dudleya traskiae*, in fact, is restricted to Santa Barbara and is on the Federal list of endangered species. This is by far the least impacted of all communities on the Islands.

The sea cliff community on Anacapa Island is almost entirely restricted to the north-facing slopes, which catch most of the prevailing northwest winds. This community is dominated by bare rock and, as at Santa Barbara, is composed of a rather large number of species, none of which dominate the

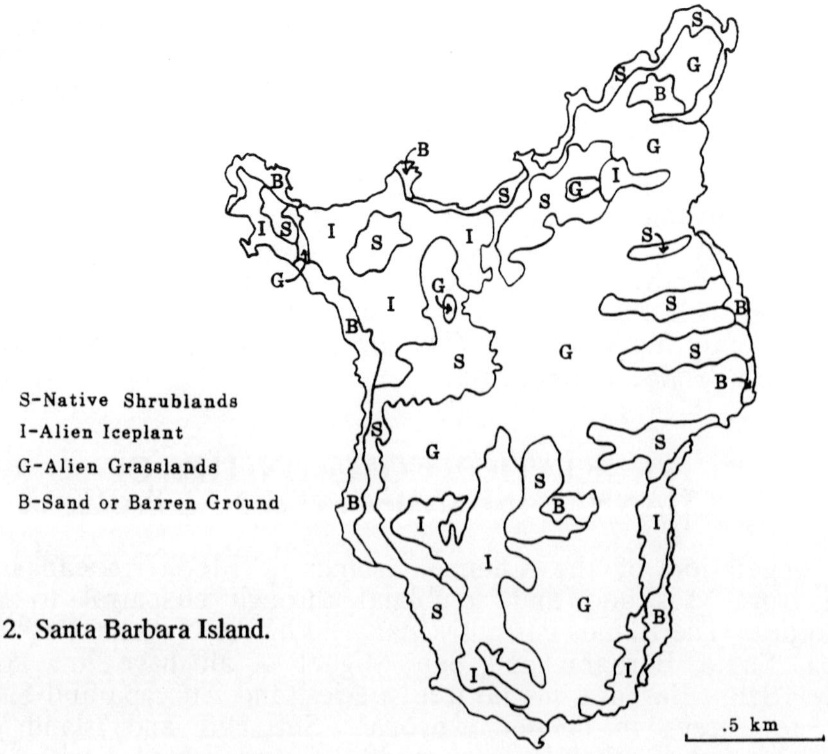

Figure 2. Santa Barbara Island.

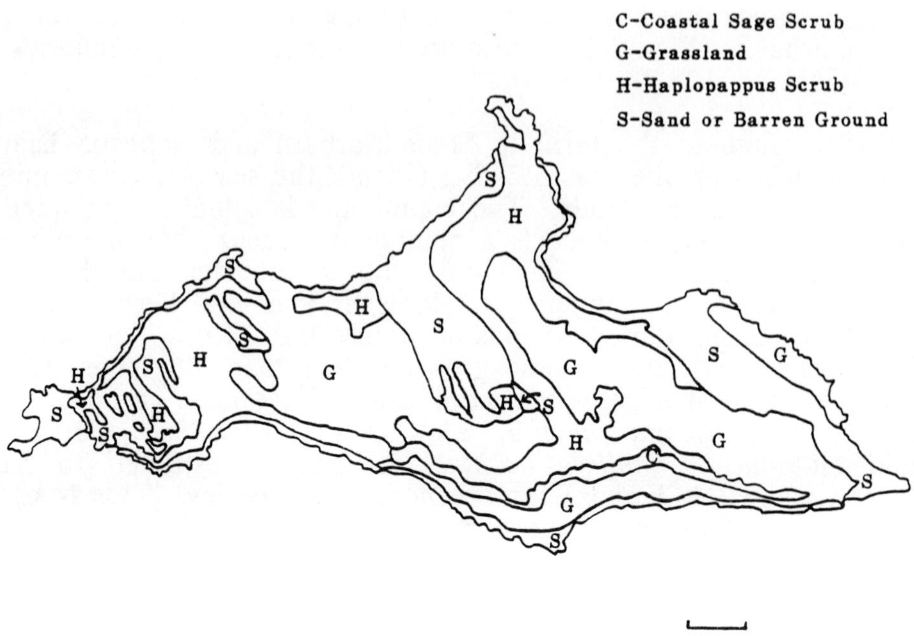

Figure 3. San Miguel Island.

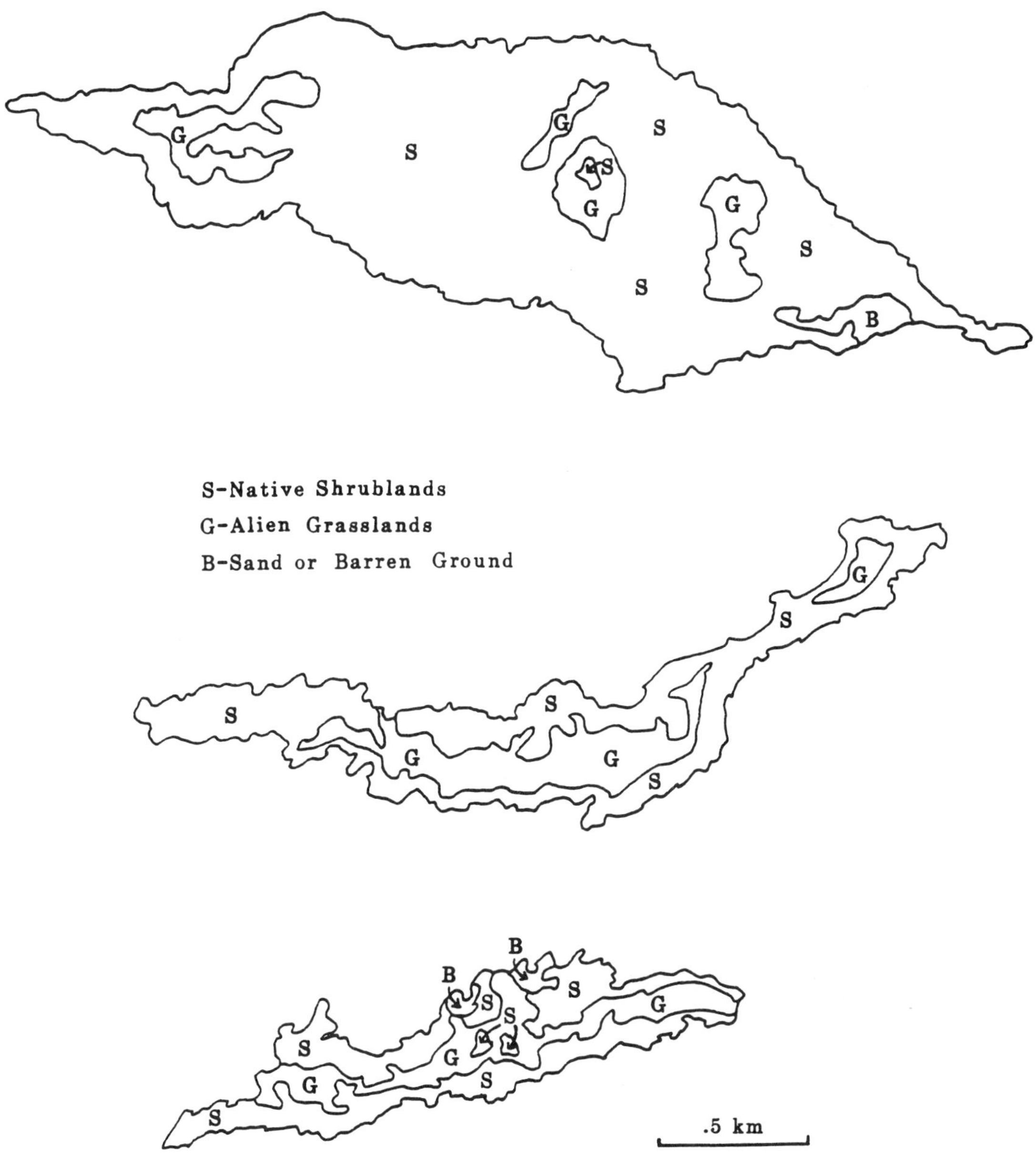

Figure 4. Anacapa Island (East, Middle, and West -- top to bottom).

landscape and most of which are also found on the island itself in other communities. Representative species include: *Dudleya caespitosa, Eriogonum grande* var. *grande, Malacothrix implicata, Eriophyllum* sp., *Haplopappus* sp., and various annual grasses.

On San Miguel Island, sandy beaches alternate with rocky headlands with their sea cliffs, which, though not as imposing as at Anacapa, are important to the island vegetation. As on the other islands, this community is widely scattered and dominated by species that are also found in other communities. Important species are: *Achillea millefolium, Artemisia californica, Coreopsis gigantea, Erigeron glaucus, Eriogonum grande* ssp. *rubescens, Galium buxifolium, Lotus dendroideus* var. *veatchii, Malacothrix saxatilis* var. *implicata*, and *Marah macrocarpus*.

Coreopsis Scrub. Giant coreopsis (*Coreopsis gigantea*) is one of the most picturesque species on the islands. It is a semi-succulent shrub that can attain heights of 8.2 ft (2.5 m). It is drought deciduous, with the dried leaves being persistent over the summer. Bright sunflower-like flowers follow the new leaves during the rainy period. *Coreopsis* once covered large portions of gentle island slopes, but because of its semi-succulent nature, it was reduced by farming, grazing, and fire to a few refugia on steep slopes and in canyons. These populations are now expanding to once again cover a major portion of each island. The physiognomy of this scrub community is of a tall shrub canopy, often closed enough to prohibit any understory. Newly developing stands have an understory of the grassland into which the species is expanding. There are no important associated shrub species. Herbaceous associates are: *Calystegia macrostegia* ssp. *amplissima, Marah macrocarpus, Pterostegia drymarioides*, and other shade-tolerant species.

Coastal Sage Scrub. This community is restricted in distribution to the southern and western slopes of the Islands and, except for San Miguel (see Fig. 4), to relatively small scattered patches. It is characterized by coastal sage, *Artemisia californica* var. *insularis*. Important associates are *Dudleya caespitosa, Encelia californica, Eriogonum arborescens, Opuntia oricola, O. prolifera, Haplopappus detonsus, Salvia mellifera, Lotus dendroideus* var. *veatchii, Bromus rubens, Calystegia macrostegia* var. *macrostegia, Haplopappus venetus*, and *Malacothrix saxatilis* var. *implicata*.

Maritime Cactus Scrub. Occurring only on Santa Barbara Island, this community is dominated by prickly pear, *Opuntia littoralis*, and cholla, *Opuntia prolifera*. Other important associates are *Amblyopappus pusillus, Marah macrocarpus, Perityle emoryi, Encelia californica*, and *Dudleya traskiae*.

Sea Blight Scrub. Also occurring only on Santa Barbara, this community is dominated by the shrub *Suaeda californica*. This is generally an open community with large bare or grass-covered spaces between the scattered shrubs. The alien grass *Hordeum glaucum* is the most common herbaceous species in the community. Other associated species are

Frankenia grandifolia, Spergularia macrotheca, and the alien iceplants *Mesembryanthemum crystallinum* and *M. nodiflorum*.

Boxthorn Scrub. Another community only associated with Santa Barbara, this is scattered in warmer microhabitats around the islands such as on south-facing slopes and is often associated with the maritime cactus scrub community. It is dominated by boxthorn, *Lycium californicum*. Where these half-meter tall shrubs are widely scattered, the spaces are filled with grasses from the grassland community. Often, however, they group together to form a dense canopy, which excludes most other species. The Federally listed threatened island night lizard, *Xantusia riversiana*, is frequently found inhabiting this community.

Haplopappus Scrub. The second-most abundant plant community on San Miguel Island (Fig. 4) is a low (<1.5 ft or <0.5 m) shrub community dominated by *Haplopappus venetus*. This community occurs on poorly developed soils that are either thin and rocky or sandy. It is unclear whether this is a climax community on these poorer soils, or a long-term seral stage which will give way to either a grassland community or a chaparral community with improving microenvironmental conditions. It is believed that this is such a widespread community on the island because of the vast disturbance the island has undergone. There is no comparable community on any other of the Channel Islands. Important associates are *Achillea millefolium, Astragalus miguelensis, Atriplex californica, Baccharis pilularis, Carpobrotus aequilaterus, Dudleya greenei, Eriogonum grande* ssp. *rubescens, Erysimum insulare*, and *Malacothrix incana*.

Island Chaparral. On West Anacapa Island in the larger canyons on the northern side of the island, there exist a few scattered stands of what appear to be relicts of a once more widely distributed chaparral community. These are stands with one or more of the following: toyon, *Heteromeles arbutifolia*; island bigpod ceanothus, *Ceanothus megacarpus* ssp. *insularis*; and Catalina cherry, *Prunus ilicifolia* ssp. *lyonii*. These stands are only a shadow of what one sees in the chaparral of the larger islands and the mainland. On San Miguel Island, this community was extirpated by grazing and by ranchers using the wood for fires. Occasionally one can find a large piece of wood lying on the surface as a reminder of the past, woodier, island. The largest patches are of shrub lupine, *Lupinus albifrons* and *L. arboreus*. These usually exist as an introduction to the grassland and therefore have a typical grassland assemblage of understory species. The other important species, lemonade berry (*Rhus intergrifolia*), is primarily found scattered along the sea slopes on the northern side of the island and has not yet begun to form a distinct community.

Native Woodlands

Island Woodland. This community type is represented by two small stands of island oak, *Quercus tomentella*, in Oak Canyon, on the northern side of West Anacapa Island. The understory of the community is rather sparse and made up mostly of native perennials such as *Penstemon cordifolia, Silene laciniata*, and *Lilium humboldtii*, rather than grasses.

Riparian

San Miguel Island has running water in a number of the larger canyons, and with such has developed a recognizable riparian community at some locations. Species that are important here are willows (*Salix lasiolepis*), *Coreopsis gigantea*, *Baccharis pilularis*, and *Typha domingensis*. These stands are widely scattered and only a minor part of the vegetation. For instance, willows occur in only four isolated locations.

Alien Communities

Perennial Iceplant. A community dominated by perennial, sea fig type iceplants of the genera *Carpobrotus* and *Malephora* covers extensive areas on the southern side of East Anacapa Island and on the upland, unstable sand areas of San Miguel. This community is a primary stabilizer and as such plays an important role in the restoration of island vegetation. Once established, the perennial iceplant community provides a more protected microhabitat into which other species colonize. This iceplant forms a low (<12 in. or <30 cm), tangled mat of vines. Associated species are *Astragalus miguelensis, Malacothrix incana, M. succulenta, Poa douglasii*, and *Distichlis spicata*.

Island Grassland. The most abundant vegetation type on most of the Channel Islands is grassland dominated by wild oats, *Avena fatua*. This community has an overstory (at about 3 ft or 1 m) of wild oats with a diverse understory of a number of alien brome grasses (*Bromus* spp.), other grass species, and a number of forbs such as *Erodium, Amsinckia*, and *Dichelostemma*. In response to protection, shrubs are rapidly invading this community. They include *Coreopsis gigantea, Atriplex semibaccata, Lycium californicum, Artemisia californica*, and *Baccharis pilularis*. Intermixed in this oat field are occasional patches dominated by the native bunchgrass, *Stipa pulchra*, and in depressions or wetter areas, patches dominated by saltgrass, *Distichlis spicata*, or by *Juncus* spp.

The Anacapa Island grassland is richer in native perennial grasses and forbs than that on Santa Barbara and San Miguel Islands, indicating that there has been less disturbance. It is also structurally distinct in that native perennials are also present, particularly *Dudleya caespitosa* and *Grindelia latifolia*. Common associates in this community are *Achillea millefolium, Calystegia macrostegia* ssp. *macrostegia, Castilleja affinis, Dichelostemma pulchellum, Lasthenia chrysostoma*, and *Marah macrocarpus*. Common alien species include *Erodium cicutarium, Sonchus oleraceus, Medicago polymorpha*, and *Atriplex semibaccata*.

A second type of alien grassland is a barley-dominated grassland. This is characterized by *Hordeum glaucum* as a canopy at about 1 ft (0.3 m), with many fewer associated herbaceous species than in the wild oats grassland. This is only widely scattered on the Islands, and the area covered is not extensive.

Annual Iceplant. On Santa Barbara Island (Fig. 2), crystalline iceplant, *Mesembryanthemum crystallinum*, forms large, dense patches and accumulates salt to the extent that all other species are generally excluded. A common associate is another annual iceplant, *Mesembryanthemum nodiflorum*. This is a serious management problem because only periodically are salt-tolerant shrubs such as *Suaeda californica* able to establish, if they can overtop the iceplant rapidly enough to overcome the shading effects. Otherwise, the iceplant leads to the extirpation of other plants from the area and then to erosion problems.

On Anacapa and San Miguel Islands, the annual iceplants form only a few large patches associated with erosion surfaces. Otherwise the species are restricted to narrow bands around small eroded areas and to a narrow strip along the top of sea cliffs.

ALIEN SPECIES

The influx of alien species to California started with the arrival of Europeans. In southern California, this was during the Mission Period (1769-1848) when exploration gave way to settlement. Adobe bricks used to build the earliest Spanish missions have been found to contain seeds of *Erodium cicutarium, Hordeum leporinum, Lolium perenne*, and *Sonchus asper* (Hendry 1931). Plants brought in for use as livestock forage included *Avena fatua*, *A. barbata*, and *Medicago polymorpha*. One story has it that Father Serra and other priests scattered seeds of wild mustard (*Brassica rapa* and *B. nigra*) along El Camino Real so that a trail of gold would guide travelers from mission to mission (Hochberg *et al.* 1979).

Following an initial period of alien plant increase, the native grasslands underwent major changes (beginning in the 1850s) as the cattle industry began to build up in California (Burcham 1956; Robbins 1970). The native grassland, originally dominated by perennial bunch grasses such as *Stipa pulchra*, came to be dominated by alien annual grasses. Adding to the problems of alien plants dominating native vegetation, especially in grassland areas, was the flood of miners coming into the State, bringing all sorts of seeds, bulbs, and cuttings; an eleven-year drought; and the development of farming starting in the 1880s (Hochberg *et al.* 1979).

In general, established weed populations on the mainland were the stock from which the islands were invaded later. Propagules were transported both on the winds and by explorers and ranchers. Human activities on the Islands became intense enough to start causing dramatic changes to the vegetation in the late 1800s and early 1900s, as reflected in the increase of weedy species on the Channel Islands at this time (Table 1). Sheep ranching began on San Miguel Island in 1850 and continued until 1948; sheep ranching on Anacapa did not begin until 1902 and stopped in 1937; goats (*Capra hircus*) were reported on Santa Barbara in 1849, but active farming and ranching occurred only between 1915 and 1922 (Farnham 1849; Hochberg *et al.* 1979). In addition, the U.S. military was active on the islands during and after WW II. A history of alien species on Santa Barbara is presented in Appendix B as an indication of site qualities and time scale of introductions on the Channel Islands.

Table 1. Approximate date of first report for selected weedy species on the California mainland and Santa Barbara, Anacapa, and San Miguel Islands.*

Species	MAIN	SBI	AI	SMI
Atriplex semibaccata	1901	1931	1932	1932
Avena fatua	1800	1939	1939	1887
Brassica nigra	1800	1939	----	----
Brassica rapa	1800	----	----	1887
Bromus diandrus	1862	1939	1939	1930
Centaurea melitensis	1797	1963	----	1887
Chenopodium murale	1797	1901	1932	1887
Conyza canadensis	1876	1978	1978	1973
Erechtites glomerata	1970	----	----	1984
Erodium cicutarium	1769	1939	1890	1887
Erodium moschatum	1856	1931	1962	1887
Larmarckia aurea	1875	1939	1959	1932
Lolium perenne	1775	1978	----	----
Malva parviflora	1791	1897	1962	1887
Medicago polymorpha	1800	1931	1932	1932
Mesembryanthemum crystallinum	1900	1897	1930	1887
Mesembryanthemum nodiflorum	1900	1932	1942	1932
Polypogon monspeliensis	1870	1939	1971	1890
Silene gallica	1837	1939	1940	1887
Silybum marianum	1854	1968	----	----
Sonchus oleraceus	1870	1931	1929	1927
Stellaria media	1870	----	----	1932
Torilis nodosa	1800	----	----	1930
Xanthium spinosum	1870	1931	----	1930

*based on Hochberg *et al.* 1979.

Main = Mainland, SBI = Santa Barbara Island, AI = Anacapa Island, SMI = San Miguel Island.

Species introductions have slowed since the National Park Service started protecting these islands, but they have not stopped. We continue to experience invasions, both from seeds blown across the Santa Barbara Channel and from seeds introduced with human activities (e.g., on planes and helicopters, with supplies, or on clothing and boots of visitors). At present, alien species range from a high of 29.9% on Santa Barbara to a low of 19.8% of the flora on Anacapa (Table 2). These figures are reflective of the amount of damage that has been caused by past overgrazing and the length of time that has transpired since grazing was last permitted.

Table 2. Percentage of native and alien species at Channel Islands National Park.

Island	Total (n)	Natives (n)	(%)	Endemics* (n)	(%)	Aliens (n)	(%)
Anacapa	222	178	80.2	18	8.1	44	19.8
Santa Barbara	107	75	70.1	13	12.1	32	29.9
San Miguel	233	174	75.1	13	5.6	58	24.9
All Three Islands	334	255	76.4	33	9.9	79	23.6

* = endemic to one or more of the Channel Islands.

Almost all alien species on the Islands are herbaceous (74 species); only two alien species are trees (see Appendix A). Four individuals from the southwestern U.S. desert, fan palm, *Washingtonia filifera*, were planted on the beach at Cuyler Harbor on San Miguel, the only location of this species in the Park; no reproduction is evident. Middle Anacapa Island has a grove of blue gum, *Eucalyptus globulus*, planted by sheep ranchers in the early 1920s. This native of Australia is commonly planted in southern California for shade and protection from wind. The eucalyptus has created an alien microhabitat and with it a somewhat unusual community of insects and birds, not as yet well studied. The species is not spreading from its planted location. The only other alien woody species on the Islands are the subshrubs Australian saltbush (*Atriplex semibaccata*) and nightshade (*Solanum nodiflorum*), which will be discussed later, and the suffrutescent perennial horehound (*Marrubium vulgare*).

Alien Plants of African Origin

Alien plants in Channel Islands National Park come from three regions of the world: Africa, Australia, and Europe. The African contingent is represented by a group of five species.

Authorities disagree about when and how the sea fig *Carpobrotus aequilaterus*, a perennial species, arrived on the shores of southern California. Since no clear data are available as to when it arrived, there is no agreement as to whether or not it should be considered native. This species is a vine and one of the most important dune-stabilizing plants on San Miguel Island, where it covers vast areas of sand dune and disturbed soil. It is also an important member of the shrub community inhabiting the tops of sea cliffs on East Anacapa Island.

Another perennial sea fig vine, *Malephora crocea*, is found only on East Anacapa. It was brought to the Island by U.S. Coast Guard personnel manning a lighthouse station there in the 1940s. The plant was used for erosion control and has not since spread to other islands.

The Hottentot fig (*Carpobrotus edulis*) is also only found on East Anacapa at this time. It is not known how this perennial vine was introduced to the Island; it only occurs as a few isolated patches among other perennial iceplants and appears to be slowly spreading at this time.

Crystalline iceplant (*Mesembryanthemum crystallinum*) covers extensive areas on Santa Barbara Island and scattered smaller areas on Anacapa and San Miguel. This annual species readily enters disturbed areas and became a dominant feature of the vegetation of Santa Barbara Island in the 1950s, when European rabbits (*Oryctolagus cuniculus*) and fire destroyed the native shrub cover. Crystalline iceplant forms large dense patches that both shade out other species and accumulate salt, which enters the soil profile upon the death of the plants (Vivrette and Muller 1977). Salt accumulation is a major resource problem on Santa Barbara Island. Another crystalline iceplant, *Mesembryanthemum nodiflorum*, similar to *M. crystallinum* but with smaller leaves, is found in the same situations but does not have as intense an impact because of smaller leaf area and generally lower density.

Alien Plants of Australian and New Zealand Origin

Alien species from Australia and New Zealand include one tree species, which has been planted, one shrub, and two herbs, one of which is particularly aggressive.

Blue gum (*Eucalyptus globulus*), a large tree, was planted in a grove on Middle Anacapa to provide protection for sheep ranchers. It is likewise planted on many of the larger Channel Islands. To date it has not been a problem for the Islands in that it has not spread from these groves.

Australian saltbush (*Atriplex semibaccata*) is a prostrate shrub, which is found in many habitats on the islands, especially disturbed areas and grasslands. In the grasslands it tends to produce a microhabitat that is favorable for grass species, causing a more lush growth around the perimeter of the shrub patch. This may be due to an amelioration of light and surface water conditions. The shrubs may be actually improving the moisture and nutrient content by bringing water and chemicals from deep in the soil profile. In any case, the species does have a dramatic affect on island grassland communities.

The herbaceous New Zealand spinach (*Tetragonia expansa*) occurs widely scattered along coastal areas of the islands and in some disturbed sites. It is not considered a problem species.

Australasian fireweed (*Erechtites glomerata*) is an aggressive composite that found its way to San Miguel Island in 1984 after having spread down the coast from the San Francisco area, where it was first reported in 1970 (Robbins 1970). It is invasive in established fields, does not need disturbance to become established, and very quickly dominates grasslands and fields. Fireweed is considered one of the most serious plant pests in Channel Islands National Park, and Park personnel are working quickly to keep it in check.

Alien Plants of European Origin

The great majority (87%) of the alien species in the Park (69 of the 79) are introductions from Europe. These are primarily herbaceous (21 species of grasses and 48 forbs). Many of the forbs are cosmopolitan garden and field weeds which thrive on some sort of disturbance. Those species that have the ability to invade and dominate for varying amounts of time and which are important to the ecology of Channel Islands National Park are listed below.

Grasses. 1) Wild oats (*Avena fatua*) and 2) slender oats (*A. barbata*), both annual species, were introduced to California at the time the cattle industry began (see Table 1). They now dominate much of the California grassland and, in that sense, the island grasslands merely reflect the general pattern throughout the State. Both are allelopathic, restricting the germination and growth of forbs in the vicinity (Tinnin and Muller 1971, 1972). By this and other means, oats maintain dominance over the grasslands for many years. 3) Soft chess (*Bromus mollis*), an annual grass, inhabits disturbed areas and also tends to dominate the understory in the wild oat grassland. 4) Red brome (*Bromus rubens*) is also widely distributed in the western United States and tends to replace natives and dominate drier slopes. 5) Foxtail (*Hordeum glaucum*), another annual, was noted as a nuisance in California as early as 1890 (Robbins 1970). On the Channel Islands it is a common associate of sea blight (*Suaeda californica*) in the native shrubland community, often being a co-dominant with this salt-tolerant shrub.

Forbs. 1) Yellow star thistle, or tocalote (*Centaurea melitensis*, an annual weed, inhabits areas of disturbance and has the ability to persist in shrublands and some grassland habitats. It is not a major problem at this time in the Channel Islands but has been slowly spreading over the past few years. 2) Pigweed (*Chenopodium murale*), a common annual weed, is often found scattered in disturbed areas, such as around campgrounds and along trails. It also forms large patches where there has been widespread destruction of vegetative cover through fire, overgrazing, etc. These large patches persist for many years and may even be the center of further dissemination of the population. 3) Red-stemmed filaree (*Erodium cicutarium*) is the most common and widespread of the filarees. Fremont found it already very dense in some areas on his 1844 exploration to southern California (Robbins 1970). On the Channel Islands (Table 1) it is found as an understory plant in the grassland and in scattered patches in disturbed areas. 4) Another filaree, *Erodium moschatum*, is also an important understory species in the island grasslands and is occasionally found as scattered patches in other habitats. 5) Cheese weed (*Malva parviflora*) is very similar to pigweed in its impact on the vegetation of the islands. 6) Bur clover (*Medicago polymorpha*) and 7) yellow sweet clover (*Melilotus indica*) were introduced to California during the Mission Period (1769-1848) and have been encouraged by ranchers ever since due to their forage value and the fact that they are nitrogen fixers. These annuals are found scattered throughout grasslands as understory plants and in disturbed areas. 8) Annual sowthistle (*Sonchus oleraceus*) is found scattered in areas of more disturbance. While individuals may be found throughout many plant communities, it is rare to find this annual weed clumped in large patches away from areas that have not been recently disturbed.

MANAGEMENT OF ALIEN SPECIES

Trees

The management strategy for the few groves of alien trees on the Channel Islands (fan palm and eucalyptus) has been determined from their importance in the history of the Islands. As important cultural features, they are managed for preservation. Since they are not extending their ranges, this appears to have only a minimal impact on native plant communities.

Shrubs

The three alien shrub species are all restricted in distribution. The control of these aliens is of low priority compared to the more important herbaceous species, and they are managed only on an *ad hoc* basis. In fact, as mentioned earlier, Australian saltbush has a somewhat beneficial impact, in that it tends to increase the growth of grasses when growing in association with them.

Herbaceous Plants

Perennial Iceplants. Perennial iceplants are not the subject of a control program at this time, because of their importance in erosion control and the fact that they tend to act as pioneer species that decline in importance with time and community development. Research is being conducted on sand dune stabilization on San Miguel Island to determine if any action can be taken which would speed the reduction of the dominance of *Carpobrotus aequilaterus* and improve stabilization of the dunes by a native plant community.

Annual Iceplants. These are a matter of great concern for Park managers. At present, research is being conducted to determine the best management strategy for the removal of the *Mesembryanthemum crystallinum* mat and replacement with species from the perennial native shrub community, such as *Suaeda californica*. Techniques will be designed to reduce the barren ground that is available for the crystalline iceplant to invade and to stop the spread of any existing patches. To date, mechanical methods of removal and revegetation have proven to be slow going and only minimally effective.

Annual Grasses and Forbs. In general, the native vegetation of Channel Islands National Park is dominated by perennial shrub communities. The major management strategy for the Park, therefore, is to protect and encourage native species and species assemblages. The islands in the Park have all experienced drastic native species reduction due to human misuse. This has led to the situation where the species list shows 23% alien species, but the vegetation shows anywhere from 25-70% alien species-dominated communities (primarily annual grasslands). Beginning with the time that each island came under the protection of the National Park Service and all grazers were removed, the vegetation has been recovering. Each island recovers at its own pace depending upon the length and intensity of past land use. With recovery, those species that are found principally in disturbance situations begin to diminish in importance.

A few species, however, are aggressive and require active management. Park personnel are currently working at removal of these species by mechanical means. We do not use fire or biological control because of a decision to not introduce anything new to the island ecosystems, whether it be fire, which has not had a role in vegetation development, or biological control agents, which were never part of these systems. The Park staff is also wary of the use of chemicals and has thus far chosen not to implement any herbicide programs, although a study was initiated in 1987 to test the feasibility of herbicide use on certain species. It has been necessary to use a rather large cadre of volunteers in order to accomplish control. Because of this we have chosen to work on one or two species per island at a time, for instance, *Erechtites glomerata* on San Miguel (Halvorson and Koske 1987) and *Centaurea melitensis* on Santa Barbara Island. Our technique is to hit any extant population hard with a large crew, after which resource management specialists and rangers periodically watch the area and remove any recruits. This technique has been successful in dealing with *Erechtites glomerata*, a species with a fairly restrictive distribution. The record of dealing with this species after its discovery in the spring of 1984 was as follows: 1) The species first discovered in Spring 1984. 2) Reconnaissance and mapping were completed in August 1984. 3) Park staff decided in October 1984 to carry out the eradication. 4) 200 worker days were required to remove the weed from 170 a (69 ha) in 1985. 5) 120 worker days were required to remove the weed from 123 a (50 ha) in 1986. 6) 19 worker days were required to remove the weed from 123 a (50 ha) in 1987. Park staff now spend only a few days during the spring removing any new *Erechtites* individuals originating from the remaining seed bank. For a number of years we will only be able to say that we have this one "under control," as we do not know how long the seed bank will remain viable.

CONCLUSIONS

Channel Islands National Park has a serious problem with alien plant species. Twenty-three percent of the plants on the Park list are introduced, but this is not a good indication of the impact of these species. In the Park, 25-70% of the vegetation cover is composed of plant communities that are dominated by alien plants. The most important of these communities in terms of area is a grassland dominated by alien annual grasses (primarily wild oats). Alien plant communities were largely caused by clearing, farming, fire, and grazing. Park personnel have begun to remove these disturbances from Santa Barbara, Anacapa, and San Miguel Islands. Native plant communities are now recovering, extending their ranges and thereby decreasing the coverage of the alien species. Management of introduced species seeks first to protect and encourage native species and second to mechanically remove alien species, population by population. Alien species that are aggressive and have restricted distributions are eliminated first. In time, it is planned to concentrate on less aggressive species with wider distributions.

ACKNOWLEDGMENTS

Many people have contributed to the information on which this report is based. Santa Barbara Botanical Garden continues to help the Park with species collections and identifications; in particular I would like to thank Steve Junak and Ralph Philbrick of that organization. I would also like to thank Charles Drost and Steve Veirs for information from their field work, and Ronie Clark for her field assistance and review of early drafts of the report.

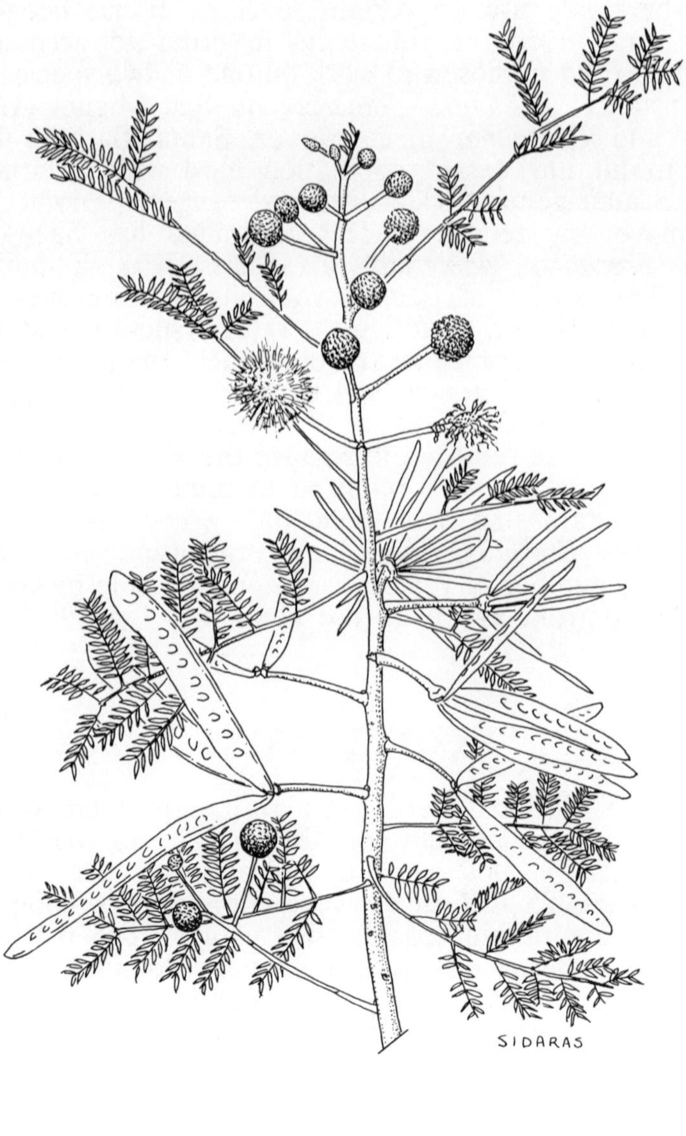

Appendix A. Status and distribution of plants of Channel Islands National Park.*

Taxon	Status	Island
AIZOACEAE		
Carpobrotus aequilaterus (Haw.) N.E.Br.	I	AM
Carpobrotus edulis (L.) Bolus	I	A
Malephora crocea (Jacq.) Schwant.	I	A
Mesembryanthemum crystallinum L.	I	ABM
Mesembryanthemum nodiflorum L.	I	ABM
Tetragonia tetragonioides (Pall.) O. Kuntze	I	AM
AMARYLLIDACEAE		
Allium praecox Bdg.	N	M
Brodiaea jolonensis Eastw.	N	M
Dichelostemma pulchellum (Salisb.) Heller	N	ABM
ANACARDIACEAE		
Toxicodendron diversilobum Greene	N	AM
Rhus integrifolia (Nutt.) Benth. & Hook.	N	AM
APIACEAE		
Apiastrum angustifolium Nutt. in T. & G.	N	ABM
Berula erecta (Huds.) Cov.	N	AM
Daucus pusillus Michx.	N	ABM
Lomatium caruifolium (H. & A.) Coult. & Rose	N	M
Sanicula arguta Greene ex Coult. & Rose	N	AM
Torilis nodosa (L.) Gaertn.	I	M
ARACEAE		
Zantedeschia aethiopica (L.) Spreng.	I	M
ARECACEAE		
Washingtonia sp.	I	M
ASPIDIACEAE		
Dryopteris arguta (Kaulf.) Watt	N	A
ASTERACEAE		
Achillea millefolium L.	N	ABM
Agroseris grandiflora (Nutt.) Greene	N	M
Amblyopappus pusillus H. & A.	N	ABM
Ambrosia chamissonis bipinnatisecta (Less.) Wiggins & Stockw.	N	AM
Ambrosia chamissonis chamissonis (Less.) Greene	N	M

*(Status: I = Alien, E = Insular Endemic, N = Native, R = Extirpated. Islands: A = Anacapa, B = Santa Barbara, M = San Miguel).

Appendix A, continued.

Appendix A, continued.

Taxon	Status	Island
ASTERACEAE (Cont)		
Artemisia californica californica Less.	N	AM
Artemisia californica insularis (Rydb.) Munz	N,E	B
Baccharis douglasii DC.	N	AM
Baccharis glutinosa Pers.	N	A
Baccharis pilularis DC. *consanguinea* (DC.) C.B. Wolf	N	ABM
Brickellia californica (T. & G.) Gray	N	A
Centaurea melitensis L.	I	BM
Cirsium occidentale (Nutt.) Jeps.	N	M
Conyza bonariensis (L.) Cronq.	I	AB
Conyza canadensis (L.) Cronq.	I	ABM
Coreopsis gigantea (Kell.) Hall	N	ABM
Corethrogyne filaginifolia (H. & A.) Nutt.	N	AM
Cotula coronopifolia L.	I	AM
Erechtites glomerata (Poir.) DC.	I	M
Erigeron foliosus Nutt. *foliosus* Nutt.	N	M
Erigeron foliosus Nutt. *stenophyllus* (Nutt.) Gray	N	AM
Erigeron glaucus Ker.	N	AM
Erigeron sanctarum Wats.	N	A
Eriophyllum confertiflorum (DC.) Gray	N	AM
Eriophyllum nevinii Gray	N,E	B
Eriophyllum staechadifolium Lag.	N	AM
Filago californica Nutt.	N	ABM
Gnaphalium beneolens A. Davids.	N	A
Gnaphalium bicolor Bioletti.	N	AB
Gnaphalium californicum DC.	N	AB
Gnaphalium chilense Spreng.	N	AM
Gnaphalium luteo-album L.	I	M
Gnaphalium microcephalum Nutt.	N	A
Gnaphalium purpureum L.	N	A
Grindelia latifolia Kell.	N	AM
Haplopappus canus (Gray) Blake	N,E	A
Haplopappus detonsus (Greene) Raven	N,E	A
Haplopappus ericoides (Less.) H. & A.	N	M
Haplopappus squarrosus H. & A.	N	A
Haplopappus venetus (HBK) Blake *sedoides* (Greene) Munz	N	AM
Haplopappus venetus (HBK.) Blake *vernonioides* (Nutt.) Hall	N	AM
Hemizonia clementina Bdg.	N,E	AB

*(Status: I = Alien, E = Insular Endemic, N = Native, R = Extirpated.
Islands: A = Anacapa, B = Santa Barbara, M = San Miguel).

Taxon	Status	Island
ASTERACEAE (Cont)		
Hemizonia fasciculata (DC.) T. & G.	N	BM
Hypochoeris sp.	I	A
Jaumea carnosa (Less.) Gray	N	M
Lactuca serriola L. *integrata* Gren. & Godr.	I	M
Lasthenia chrysostoma (F. & M.) Greene	N	ABM
Layia platyglossa (F. & M.) Gray	N	M
Malacothrix foliosa Gray	N,E	ABM
Malacothrix incana (Nutt.) T. & G.	N	M
Malacothrix indecora Greene	N,E	M
Malacothrix saxatilis implicata (Eastw.) Hall	N,E	AM
Malacothrix saxatilis tenuifolia (Nutt.) Gray	N	A
Malacothrix squalida Greene	N	A
Malacothrix succulenta Elmer	N	M
Microseris douglasii (DC.) Sch. Bip. *tenella* (Gray) Chamb.	N	M
Microseris elegans Greene ex Gray	N	M
Microseris linearifolia (Nutt.) Sch. Bip.	N	AB
Perityle emoryi Torr. in Emory	N	AB
Rafinesquia californica Nutt.	N	AB
Senecio vulgaris L.	I	AM
Silybum marianum (L.) Gaertn.	I,R	B
Solidago sp.	N	A
Sonchus asper (L.) Hill	I	AM
Sonchus oleraceus L.	I	ABM
Sonchus tenerrimus L.	I	AB
Stephanomeria virgata Benth.	N	M
Xanthium spinosum L.	I	BM
BERBERIDACEAE		
Berberis pinnata Lag. *insularis* Munz	N,E	A
BORAGINACEAE		
Amsinckia intermedia F. & M.	N	ABM
Amsinckia spectabilis F. & M.	N	ABM
Cryptantha clevelandii Greene	N	ABM
Cryptantha maritima (Greene) Greene	N	B
Cryptantha muricata jonesii (Gray) Jtn.	N	A
Heliotropium curassavicum L. *oculatum* (Heller) Thorne	I	AM
Plagiobothrys californicus californicus (Gray) Greene	N	AM

Appendix A, continued.

Appendix A, continued.

Taxon	Status	Island
Plagiobothrys californicus (Gray) Greene *gracilis* Jtn.	N	A
BRASSICACEAE		
Brassica geniculata (Desf.) J. Ball	I	M
Brassica nigra (L.) Koch in Rohling	I,R	B
Brassica rapa L. *sylvestris* (L.) Janchen	I	M
Cakile edentula californica (Heller) Hult.	N	M
Cakile maritima Scop.	I	AM
Capsella bursa-pastoris (L.) Medic.	I	M
Cardamine californica (Nutt.) Greene	N	AM
Descurainia pinnata menziesii (DC.) Detl.	N	A
Dithyrea californica maritima (A. Davids.) A. Davids. ex Rob.	N	M
Erysimum insulare Greene	N	AM
Hutchinsia procumbens (L.) Desv.	N	ABM
Lepidium nitidum Nutt. *nitidum*	N	AB
Lepidium oblongum Small *insulare*	I	AM
Lepidium strictum (Wats.) Rattan	I	A
Nasturtium officinale R. Br.	I	M
Thelypodium lasiophyllum (H. & A.) Greene	N	ABM
CACTACEAE		
Opuntia ficus-indica (L.) Mill.	I	A
Opuntia littoralis (Engelm.) Ckll.	N	ABM
Opuntia oricola Philbrick	N	ABM
Opuntia prolifera Engelm.	N	ABM
CARYOPHYLLACEAE		
Cardionema ramosissimum (Weinm.) Nels. & Macbr.	N	M
Cersatium glomeratum Thuill.	I	M
Sagina decumbens (Ell.) T&G *occidentalis* Wats. Crow	N	M
Silene antirrhina L.	N	M
Silene gallica L.	I	ABM
Silene laciniata Cav. *major* Hitchc. & Maguire	N	AM
Silene multinervia Wats.	N	A
Spergularia macrotheca (Hornem.) Heynh.	N	ABM
Stellaria media (L.) Vill.	I	AM
CHENOPODIACEAE		
Aphanisma blitoides Nutt. ex Moq. in DC.	N	B
Atriplex californica Moq. in DC.	N	ABM

---

*(Status: I = Alien, E = Insular Endemic, N = Native, R = Extirpated. Islands: A = Anacapa, B = Santa Barbara, M = San Miguel).

Taxon	Status	Island
Atriplex coulteri (Moq.) D. Dietr.	N	AM
Atriplex lentiformis breweri (Wats.) Hall. & Clem.	N	A
Atriplex leucophylla (Moq. in DC.) D. Dietr.	N	M
Atriplex pacifica Nels.	N	AB
Atriplex patula L. *hastata* (L.) Hall & Clem.	I	M
Atriplex semibaccata R. Br.	I	ABM
Beta vulgaris L.	I,R	M
Chenopodium californicum (Wats.) Wats.	N	ABM
Chenopodium murale L.	I	ABM
Monolepis nuttalliana (Schult.) Greene	N	BM
Salicornia subterminalis Parish	N	M
Salicornia virginica L.	N	M
Suaeda californica Wats.	N	ABM
CISTACEAE		
Helianthemum greenei Rob.	N,E	AM
CONVOLVULACEAE		
Calystegia macrostegia amplissima Brummitt	N,E	B
Calystegia macrostegia macrostegia (Greene) Brummitt	N,E	AM
Calystegia soldanella (L.) R. Br.	N	M
Cressa truxillensis HBK. *vallicola* (Heller) Munz	N	AM
Dichondra occidentalis House	N	M
CRASSULACEAE		
Crassula erecta (H. & A.) Berger	N	ABM
Dudleya caespitosa (Haw.) Britt. & Rose	N	A
Dudleya greenei Rose	N,E	M
Dudleya traskiae (Rose) Moran	N,E	B
CUCURBITACEAE		
Marah macrocarpus (Greene) Greene	N	ABM
CUSCUTACEAE		
Cuscuta salina Engelm. *salina*	N	AM
CYPERACEAE		
Scirpus cernuus Vahl *californicus* (Torr.) Thorne	N	M
ERICACEAE		
Comarostaphylos diversifolia Greene *planifolia* Jeps.	N	A

Appendix A, continued.

Appendix A, continued.

Taxon	Status	Island
FABACEAE		
Astragalus curtipes Gray	N	M
Astragalus didymocarpus H. & A.	N	AM
Astragalus miguelensis Greene	N,E	ABM
Astragalus traskiae Eastw.	N,E	B
Astragalus trichopodus Gray *leucopis* (T. & G.) Thorne	N	A
Lathyrus laetiflorus Greene *barbarae* (White) C.L. Hitchc.	N	A
Lotus argophyllus ornithopus (Greene) Raven	N,E	B
Lotus dendroideus (Greene) Greene *dendroideus*	N	A
Lotus dendroideus veatchii (Greene) Isley	N	M
Lotus salsuginosus Greene	N	AM
Lotus strigosus (Nutt. in T. & G.) Greene	N	A
Lotus subpinnatus Lag.	N	A
Lupinus albifrons Benth.	N	AM
Lupinus arboreus Sims	N	M
Lupinus bicolor Lindl.	N	AM
Lupinus bicolor Lindl. *umbellatus* (Greene) D. Dunn	N	A
Lupinus chamissonis Eschs.	N	M
Lupinus succulentus Dougl. ex Koch	N	AM
Lupinus truncatus Nutt. ex H. & A.	N	A
Medicago polymorpha L. *brevispina* (Benth.) Heyn.	I	M
Medicago polymorpha L. *polymorpha* L.	I	ABM
Medicago sativa L.	I	M
Melilotus indica (L.) All.	I	ABM
Trifolium amplectens T. & G.	N	AM
Trifolium barbigerum Torr.	N	M
Trifolium fucatum Lindl.	N	M
Trifolium gracilentum T. & G.	N	AM
Trifolium microcephalum Pursh	N	M
Trifolium palmeri Wats.	N,E	B
Trifolium tridentatum Lindl.	N	ABM
Vicia americana Muhl. ex Willd.	N	M
Vicia exigua Nutt. in T. & G.	N	AM
FAGACEAE		
Quercus tomentella Engelm.	N,E	A

*(Status: I = Alien, E = Insular Endemic, N = Native, R = Extirpated.
Islands: A = Anacapa, B = Santa Barbara, M = San Miguel).

Taxon	Status	Island
FRANKENIACEAE		
Frankenia grandifolia Cham. & Schlecht.	N	AM
GERANIACEAE		
Erodium cicutarium (L.) L'Her.	I	ABM
Erodium moschatum (L.) L'Her.	I	ABM
HYDROPHYLLACEAE		
Eucrypta chrysanthemifolia (Benth.) Greene	N	ABM
Nemophila pedunculata Dougl. ex Benth.	N	M
Phacelia cicutaria hispida (Gray) J.T.Howell	N	A
Phacelia distans Benth.	N	ABM
Phacelia insularis Munz *insularis*	N,E	M
Phacelia viscida (Benth.) Torr.	N	AM
Pholistoma auritum (Lindl.) Lilja ex Lind.	N	B
Pholistoma racemosum (Nutt.) Const.	N	B
IRIDACEAE		
Chasmanthe aethiopica (L.) N.E.Br.	I	M
Sisyrinchium bellum Wats.	N	M
JUNCACEAE		
Juncus balticus Willd.	N	M
Juncus bufonius L.	N	AM
Juncus mexicanus Willd.	N	M
LAMIACEAE		
Marrubium vulgare L.	I	M
LILIACEAE		
Lilium humboldtii Roezl & Leichtl.	N	A
Zigadenus fremontii Torr.	N	A
MALVACEAE		
Lavatera assurgentiflora Kell.	N,E	AM
Lavatera cretica L.	I	A
Malva parviflora L.	I	ABM
Sidalcea malvaeflora (DC.) Gray ex Benth.	N	M
MORACEAE		
Ficus carica L.	I,R	M

Appendix A, continued.

Appendix A, continued.

Taxon	Status	Island
MYRTACEAE		
Eucalyptus globulus Labill.	I	A
NYCTAGINACEAE		
Abronia latifolia Eschs.	N	M
Abronia maritima Nutt. ex Wats.	N	AM
Abronia umbellata Lam.	N	M
Mirabilis californica Gray *californica* Gray	N	AB
ONAGRACEAE		
Camissonia cheiranthifolia cheiranthifolia Raven	N	BM
Camissonia micrantha (Hornem. ex Spreng.) Raven	N	M
Clarkia epilobioides (Nutt.) Nels. & Macbr.	N	A
Epilobium adenocaulon Hausskn.	N	M
Zauschneria californica Presl	N	AM
Zauschneria cana Greene	N	A
OROBANCHACEAE		
Orobanche fasciculata Nutt.	N	M
Orobanche parishii (Jeps.) Heckard *brachyloba* Heck.	N	M
PAPAVERACEAE		
Eschscholzia californica Cham.	N	M
Eschscholzia ramosa (Greene) Greene	N,E	AB
Platystemon californicus Benth.	N	AM
Platystemon californicus Benth. *ciliatus* Dunkle	N,E	B
Stylomecon heterophylla (Benth.) G. Tayl.	N	ABM
PLANTAGINACEAE		
Plantago bigelovii Gray	N	M
Plantago erecta erecta Morris	N	A
Plantago hirtella HBK.	N	M
Plantago insularis Eastw.	N	A
Plantago ovata Forsk.	I	B
POACEAE		
Agropyron repens (L.) Beauv.	I	M
Agrostis semiverticillata (Forsk.) C. Chr.	I	A
Avena barbata Brot.	I	ABM

* (Status: I = Alien, E = Insular Endemic, N = Native, R = Extirpated.
Islands: A = Anacapa, B = Santa Barbara, M = San Miguel).

Taxon	Status	Island
Avena fatua L.	I	ABM
Bromus arizonicus (Shear) Steb.	N	B
Bromus carinatus H. & A.	N	A
Bromus diandrus Roth	I	ABM
Bromus madritensis L.	I	A
Bromus maritimus (Piper) Hitchc.	N	AM
Bromus mollis L.	I	ABM
Bromus pseudolaevipes Wagnon	N	A
Bromus rubens L.	I	ABM
Bromus trinii E. Desv. in Gray	N	ABM
Cortaderia selloana (Schult.) Asch. & Graebn.	I	M
Cynodon dactylon (L.) Pers.	I	A
Distichlis spicata (L.) Greene *stolonifera* Beetle	N	AM
Elymus condensatus Presl	N	AM
Elymus pacificus Gould	N	M
Hordeum californicum Covas & Steb.	N	AM
Hordeum glaucum Steud.	I	AB
Hordeum pusillum Nutt.	N	AB
Hordeum leporinum Link.	I	AM
Lamarckia aurea (L.) Moench	I	ABM
Lolium perenne L.	I,R	B
Melica imperfecta Trin.	N	ABM
Muhlenbergia microsperma (DC.) Kunth	N	AB
Parapholis incurva (L.) C.E.Hubb.	I	BM
Pennisetum clandestinum Hochst. ex Chiov.	I	M
Phalaris minor Retz.	I	ABM
Poa annua L.	I	M
Poa douglasii Nees	N	M
Poa scabrella (Thurb.) Benth. ex Vasey	N	A
Polypogon interruptus HBK.	I	M
Polypogon monspeliensis (L.) Desf.	I	BM
Stipa lepida Hitchc.	N	AM
Stipa pulchra Hitchc.	N	ABM
Vulpia dertonensis Volk.	I	AM
Vulpia megalura Rydb.	N	ABM
Vulpia octoflora (Walt.) Rydb.	N	ABM
Vulpia pacifica Rydb.	N	A
POLEMONIACEAE		
Gilia angelensis V. Grant	N	A

Appendix A, continued.

Appendix A, continued.

Taxon	Status	Island
POLEMONIACEAE, (Cont)		
Gilia clivorum (Jeps.) V. Grant	N	AM
Gilia nevinii Gray	N,E	A
Linanthus androcaceus micranthus (Steud.) Mason	N	M
POLYGONACEAE		
Eriogonum arborescens Greene	N,E	A
Eriogonum giganteum Wats. *compactum* (Dunkle) Munz	N,E	B
Eriogonum grande Greene *grande*	N	A
Eriogonum grande Greene *rubescens* (Greene) Munz	N,E	AM
Pterostegia drymarioides F. & M.	N	ABM
Rumex crispus L.	I	AM
Rumex fueginus Phil.	N	M
Rumex salicifolius Weinm.	N	M
POLYPODIACEAE		
Polypodium californicum Kaulf.	N	AB
PORTULACACEAE		
Calandrinia ciliata (R. & P.) DC. *menziesii* (Hook.) Macbr.	N	ABM
Calandrinia maritima Nutt.	N	AB
Claytonia perfoliata Donn	N	ABM
PRIMULACEAE		
Anagallis arvensis L.	I	AM
Dodecatheon clevelandii Greene *insulare* H.J.Thomps.	N	A
PTERIDACEAE		
Adiantum jordanii C. Muell.	N	A
Pellaea andromedaefolia (Kaulf.) Fee	N	A
Pellaea mucronata (D.C.Eat.) D.C.Eat.	N	A
Pityrogramma triangularis (Kaulf.) Maxon	N	A
RANUNCULACEAE		
Delphinium parryi Gray	N	AM
Ranunculus californicus Benth.	N	M

---

*(Status: I = Alien, E = Insular Endemic, N = Native, R = Extirpated.
Islands: A = Anacapa, B = Santa Barbara, M = San Miguel).

Taxon	Status	Island
RESEDACEAE		
Oligomeris linifolia (Vahl) Macbr.	N	ABM
RHAMNACEAE		
Ceanothus megacarpus Nutt.		
insularis (Eastw.) Raven	N,E,R	AM
Rhamnus pirifolia Greene	N,E	M
ROSACEAE		
Heteromeles arbutifolia (Ait.) M. Roem.	N	AM
Potentilla agedii Wormski		
grandis (Rydb.) J.T.Howell	N	M
Prunus lyonii (Eastw.) Sarg.	N	A
Rubus ursinus Cham. & Schlecht.	N	AM
RUBIACEAE		
Galium angustifolium Nutt.		
foliosum Hilend & Howell	N,E	A
Galium aparine L.	I	ABM
Galium buxifolium Greene	N,E	M
Galium californicum H. & A.		
miguelense (Greene) Jeps.	N,E	M
SALICACEAE		
Salix lasiolepis Benth.	N	M
SAXIFRAGACEAE		
Heuchera maxima Greene	N,E	A
Ribes malvaceum Sm.	N	A
SCROPHULARIACEAE		
Antirrhinum nuttallianum Benth. in DC.	N	AM
Castilleja affinis H. & A.	N	AM
Castilleja hololeuca Greene	N,E	AM
Linaria texana Scheele	N	AM
Mimulus flemingii Munz	N	A
Orthocarpus densiflorus Benth.	N	M
Orthocarpus purpurascens Benth.	N	M
Keckiella cordifolia (Benth.) Straw.	N	A
Scrophularia villosa Penn. in Millsp. & Nutt.	N	A

Appendix A, continued.

Appendix A, continued.

Taxon	Status	Island
SELAGINELLACEAE		
Selaginella bigelovii Underw.	N	A
SOLANACEAE		
Lycium californicum Nutt.	N	ABM
Lycopersicum esculentum Mill.	I,R	BM
Solanum douglasii Dunal in DC.	N	AM
TYPHACEAE		
Typha domingensis Pers.	N	M
URTICACEAE		
Hesperocnide tenella Torr.	N	B
Parietaria hespera Hinton	N	AB
Urtica dioica L. *holosericea* (Nutt.) Thorne	N	A
Urtica urens L.	I	M
VERBENACEAE		
Verbena lasiostachys Link.	N	M
ZOSTERACEAE		
Phyllospadix scouleri Hook.	N	AB
Phyllospadix torreyi Wats.	N	ABM
Zostera marina L.	N	A

*(Status: I = Alien, E = Insular Endemic, N = Native, R = Extirpated.
Islands: A = Anacapa, B = Santa Barbara, M = San Miguel).

Appendix B.

History of alien species on Santa Barbara Island.

To help in understanding how alien species came to gain such control in the vegetation of the Channel Islands, the following brief history of Santa Barbara Island is presented. The information, unless otherwise noted, is from Philbrick (1972).

1848 The island became part of the United States in the Treaty of Guadalupe Hildago (Weinman 1978) and was said to be densely populated with goats (Farnham 1849).

1852 The island was first surveyed and the report stated that "the whole surface is covered with a deep and apparently good soil, containing much lime, very light and ashy, and averaging four or five feet deep. It seems difficult to account for its uniformity of composition and depth, except on the supposition that it was to a great degree, formed from the decomposition of animal remains" (National Park Service, unpub. data).

1889 A Coast and Geodetic Survey report states that "there is not a drop of water on the island; no grass, but plenty of prickly pear and shrubs" (Weinman 1978).

1896 Feral house cats were introduced by a passing freighter. Several ground-nesting bird populations were subsequently destroyed (Weinman 1978).

1915 Alvin Hyder and his family moved to the island. The Hyders enlisted the aid of as many as 15 other people to assist in the homesteading pursuits. Eleven structures and three small reservoirs were built. Tillable portions of the island were burned each year and then plowed with a mold board plow. *Coreopsis* was pulled by hand and crystalline ice plant (*Mesembryanthemum*) was cut, dried and burned. Five acres (2 hectares) of potatoes were planted on the western slopes of the island and 150 acres (60.7 ha) of barley were cultivated on the eastern slope.

1918 The entire island was burned by the Hyders. Sometime during the early part of their stay they released 1,200 Belgian hares to raise for meat and pelts. The hares were eventually killed by the resident cats and eagles. Four hogs brought to the island died from eating poisoned rabbits or cats.

1922 The Hyders left the island with 12 goats, 300 sheep, four horses, and the family dogs. The island was described as covered with foxtail, iceplant, and jungles of coreopsis. Feral house cats were abundant (Weinman 1978).

1928 Hyder used the island to fatten 250 sheep for market after failing to get his lease renewed (Weinman 1978).

1938 F.D. Roosevelt designated Channel Island National Monument, which included Santa Barbara Island.

1942 New Zealand red rabbits were introduced by the Navy for meat and sport of the men stationed on coast watch duty.

Appendix B, continued.

Appendix B, continued.

History of alien species on Santa Barbara Island.

1949 Field notes of P.C. Orr stated that the whole eastern side of the island was covered with a dense forest of coreopsis (National Park Service, unpub. data).

1950 Lowell Sumner reported that a "jungle" of *Coreopsis* existed that was head high. He also expressed a concern that the National Park Service was actively exterminating the cats on the island because of their effect on the populations of ground nesting seabirds. He felt that the cats were also keeping rabbit numbers in check.

1958 Lowell Sumner reported that with a rise in the rabbit population, the coreopsis jungle had a stricken aspect. Many were girdled and felled by rabbits and "bare ground showed everywhere through a shriveled carpet of vegetation" (Sumner 1958).

1959 Lowell Sumner reported that acres of boxthorn and island sagebrush had been exterminated and a reduction of the coreopsis stands resulted in the extinction of the Santa Barbara song sparrow (Sumner 1959).

1959 The eastern two thirds of the island was accidentally burned.

1979 The National Park Service decided on an intensive rabbit removal program (National Park Service, unpub. data).

1981 The last of the rabbits was exterminated on the island (National Park Service, unpub. data).

1985 Halvorson, Drost, and Fowler reported on the recovery of a number of shrub species including coreopsis, island sagebrush, boxthorn, and seablight (National Park Service, unpub. data).

This pattern of alternating periods of heavy impact and recovery is evident on Anacapa and San Miguel Islands as well. On all islands, the number of reports is scanty and the changes in species composition and importance are not adequately documented. Further, grazing started on the islands before any biological inventories were accomplished; therefore it is impossible to reconstruct any sort of pristine condition. As far as our understanding of the biology of the islands is concerned, we have always had an impacted system.

Literature Cited

Burcham, L.T. 1956. Historical backgrounds of range land use in California. *J. Range Manage.* 9:81-86.

Dunkle, M.B. 1950. Plant ecology of the Channel Islands of California. *Allen Hancock Pac. Exped.* 13:247-386.

Emery, K.O. 1960. *The sea off Southern California, a modern habitat of petroleum.* New York: John Wiley & Sons.

Farnham, T.J. 1849. *Life, adventures, and travels in California, to which are added the conquest of California, travels in Oregon, and history of the old regions.* New York: Nafis and Cornish.

Halvorson, W.L., and R.E. Koske. 1987. Mycorrhizae associated with an invasion of *Erechtites glomerata* (Asteraceae) on San Miguel Island, California. *Madroño* 34:260-268.

Hendry, G.W. 1931. The adobe brick as a historical source. *Agric. Hist.* 5:110-127.

Hochberg, M.L., S. Junak, R. Philbrick, and S. Timbrook. 1979. Botany. In *Natural resources study of the Channel Islands National Monument, California*, ed. D.M. Power, 5.1-5.91. Santa Barbara, Calif.: Santa Barbara Mus. Nat. Hist.

Johnson, O.L. 1979. Geology, soils, and erosion. In *Natural resources study of the Channel Islands National Monument, California*, ed. D.M. Power, 3.1-3.12. Santa Barbara, Calif.: Santa Barbara Mus. Nat. Hist.

Kemnitzer, L.B. 1933. Geology of San Nicolas and Santa Barbara Islands, California. Master's thesis, Calif. Inst. Technology, Pasadena.

Lipps, J.H. 1964. Late Pleistocene history of West Anacapa Island, California. *Bull. Geol. Soc. Amer.* 25:1169-1176.

Lipps, J.H., H.W. Valentine, and E. Mitchell. 1968. Pleistocene paleontology and biostratigraphy, Santa Barbara Island, California. *J. Paleontol.* 42(2):291-307.

Philbrick, R.N. 1972. The plants of Santa Barbara Island, California. *Madroño* 21:329-393.

Robbins, W.W. 1970. Alien plants growing without cultivation in California. *Bull. Calif. Agric. Exp. Sta.* 637:1-128.

Scholl, D.W. 1960. Relationship of the insular shelf sediments to the sedimentary environments and geology of Anacapa Island, California. *J. Sedimentary Petrol.* 30:123-139.

Smith, C.W. 1985. Impact of alien plants on Hawai'i's native biota. In *Hawai'i's terrestrial ecosystems: preservation and management*, ed. C.P. Stone and J.M. Scott, 180-250. Univ. Hawaii Coop. Natl. Park Resour. Stud. Unit. Honolulu: Univ. Hawaii Pr.

Sumner, L. 1958. The rabbits of Santa Barbara Island, a progress report and summary. Rep. for Natl. Park Serv. Files, Channel Islands Natl. Park, Ventura, Calif.

Sumner, L. 1959. The battle for Santa Barbara. *Outdoor Calif.* 20:4-7.

Tinnin, R.O., and C.H. Muller. 1971. The allelopathic potential of *Avena fatua*: influence on herb distribution. *Bull. Torrey Bot. Club* 98:243-250.

Tinnin, R.O., and C.H. Muller. 1972. The allelopathic influence of *Avena fatua*: the allelopathic mechanism. *Bull. Torrey Bot. Club* 99:287-292.

Trewartha, G.T. 1954. *An introduction to climate*. New York: McGraw-Hill Book Co.

Valentine, J.W., and J.H. Lipps. 1963. Late Cenozoic rocky shore assemblages from Anacapa Island, California. *J. Paleont.* 37:1292-1302.

Vedder, J.G., and D.G. Howell. 1976. Neogene strata of the southern group of Channel Islands, California. In *Aspects of the geologic history of the California borderland*, ed. D.G. Howell, 80-106. Pac. Sec. Amer. Assn. Petrol. Engineers Misc. Pub. 24.

Vivrette, N.J., and C.H. Muller. 1977. Mechanism of invasion and dominance of coastal grassland by *Mesembryanthemum crystallinum*. *Ecol. Monogr.* 47:301-318.

Wallace, G.D. 1985. *Vascular plants of the Channel Islands of southern California and Guadalupe Island, Baja California, Mexico*. Nat. Hist. Mus. Los Angeles County Contr. Science 365.

Weaver, D.W., and D.P. Doerner. 1969. Lower Tertiary stratigraphy, San Miguel and Santa Rosa Islands. In *Geology of the northern Channel Islands and southern California borderland*, ed. D.W. Weaver, D.P. Doerne, and B. Nolf, 30-46. Spec. Pub., Pac. Sec. Amer. Assn. Petrol. Geol. and Soc. Econ. Paleont. and Mineralogy.

Weaver, D.W., D.P. Doerner, and B. Nolf, eds. 1969. *Geology of the northern Channel Islands and southern California borderland*. Spec. Pub., Pac. Sect. Amer. Assn. Petrol. Geol. and Soc. Econ. Paleont. and Mineralogy.

Weinman, L.J. 1978. *Historic resources study: Channel Islands National Monument and San Miguel Island, California*. Rep. for Natl. Park Serv. Santa Barbara, Calif.: Chambers Consultants and Planners.

Weissman, D.B., and D.C. Rentz. 1977. *Rainfall data for the California Channel Islands and adjacent mainland*. Private pub.

STATUS AND DISTRIBUTION OF SOME IMPORTANT ALIEN PLANTS IN HAWAI'I

ORIGIN AND DISTRIBUTION OF ADVENTIVE ALIEN FLOWERING PLANTS IN HAWAI'I

Lyndon Wester

ABSTRACT

Hawai'i has more than 800 known species of alien plants that are reproducing without direct human assistance. During this century, new plant species have been arriving at the rate of about five per year. Patterns of plant collecting in time and space suggest that other species await discovery and that data on the ranges of known species are far from satisfactory. The New and Old World tropics have supplied most of the adventive alien plants in Hawai'i, although other places, particularly those with which Hawai'i has close economic and cultural ties, have contributed significant numbers in recent years.

INTRODUCTION

The nature and derivation of native floras have long attracted the attention of biologists, but the rearrangement and wholesale homogenization of the world's biota caused by humans in the last 10,000 years has aroused comparatively little interest until recently. Contributing reasons for this are that alien plant invasions are generally short-lived, and that populations depend upon continued disturbance, are unstable, and are determined to a large extent on chance dispersal. Furthermore, there is a built-in preference for what is natural and rare and a prejudice against the commonplace and alien among biologists, except perhaps where economic factors are involved.

To protect natural ecosystems, agriculture, and even urban areas from undesirable invasion from alien plants, reliable data about the alien species present are needed. Information about how plants were introduced is helpful to plan means to limit further additions to the alien flora. Understanding generalized patterns of establishment and spread is also useful in managing problems that already exist.

DEFINITIONS AND DATA SOURCES

Considerable confusion exists about terminology applied to introduced plants and weeds (Frenkel 1984; Smith 1985), so it is important from the outset to define, in exact terms, the subject of this study. The focus of the work here will be plants introduced to Hawai'i that have shown the ability to reproduce themselves, either sexually or vegetatively, and to spread without direct human assistance. Excluded, therefore, are the introduced plants cultivated as crops or ornamentals, although it must be recognized that new recruits for the naturalized alien flora often come from the plants cultivated intentionally by humans.

Weeds may be categorized according to the degree of human disturbance they require for survival (Harlan and de Wet 1965; Hart 1976). Slight disturbance, such as a small opening in the canopy of a forest, may be sufficient to ensure the establishment of some species. However, ruderals, the plants of waste places, or the weeds of crops depend upon continual disturbance. Some species may depend for their survival on humans to supply water, to act as local dispersal agents, to set or extinguish fires, or to eliminate competitors by mowing, trampling, or spraying. All alien plants that are reproducing and actively spreading, or that have done so in the past, were considered in this survey. The degree of direct or indirect dependence on humans for the survival of the populations was not taken into account. Excluded, however, were plants confined to cultivation or persisting without showing any tendency to spread after abandonment. Opinions as to whether species are disliked, welcomed as valued additions to the flora, or simply tolerated were not taken into consideration.

There is no infallible way to determine whether a plant is indigenous or alien by its external characteristics. However, if a plant was not recorded in early botanical surveys, has no closely related indigenous relatives, seems incapable of dispersing over great stretches of open ocean without human assistance, and favors disturbed habitats created by humans, it seems reasonably safe to assume the species is an alien. There is substantial agreement of opinion about the provenance of most species in Hawai'i; however, there is some disagreement about a small number of species that might be indigenous, Polynesian introductions, or very early European introductions. Because many of these species have spread well beyond cultivation, this question must be addressed in more detail.

Exact and reliable information on the time and circumstances of an introduction is rarely available. Furthermore, many adventive aliens were first introduced as crop or ornamental species and only later did they escape from cultivation. It is rare for escapes to be recorded. In fact, a plant may spread for many years and build up a large population before it is noticed. For example, *Oxyspora paniculata* was observed growing as a planted species in a forest reserve in 1958 when a collector commented, "this plant does not spread, does not seem to be a pest" (B.P. Bishop Museum herbarium specimen 58128). Only in 1984 was a serious infestation found in the same area (Obata 1985). In the course of the present study, the earliest record of each alien species was sought, regardless of the situation where it was first found. Sources for this information included published floras, checklists, other written accounts, accession records of arboreta or botanical gardens, and botanical specimens. The majority of first records (72%) are from herbarium specimens. The B.P. Bishop Museum herbarium was the most important source. However, other herbaria containing the collections of 19th century botanists were also searched for plants believed to be early introductions to Hawai'i. The collections consulted were U.S. National Herbarium, Gray Herbarium of Harvard University, British Museum of Natural History, Royal Botanical Gardens at Kew, Museum National d'Histoire Naturelle, Paris, Universitatis Florentinae, Instituto Botanico, Academy of Sciences of the U.S.S.R. in Leningrad, Harold L. Lyon Arboretum at University of Hawaii, and Department of Botany, University of Hawaii.

EARLY ALIEN INTRODUCTIONS

Most of the main food and fiber crops used by the Hawaiians lack mechanisms of dispersal that would enable them to reach Hawai'i by natural processes. Handy and Handy (1972) believed that only 14 species were actually introduced by the Polynesians (Table 1), whereas St. John (1973, 1978) and Nagata (1985) have a rather longer list of candidates. Included are weedy species, which may have been introduced accidentally, and strand trees of widespread distribution in the Indo-Pacific (Merrill 1946). The status of several other plants, conventionally considered aliens introduced by Europeans, must be reconsidered in the light of the rediscovery of collections made by the first botanists to visit Hawai'i. As there is considerable controversy surrounding these species, each will be discussed separately.

'Ahu'awa, (*Mariscus javanicus*)

The first record of 'ahu'awa in Hawai'i comes from the collection by Lay and Collie, botanists on the Beechey voyage, in 1827 (Hooker and Arnott 1965). The plant is commonly found on sandy beaches, in taro (*Colocasia esculenta*) patches, and in marshes and may be one of the indigenous, pantropical wetland species whose seeds were widely dispersed by waterbirds; or it could have been carried as an accidental passenger on taro corms.

Kili'o'opu (*Kyllinga nemoralis*)

Mann (1866) was apparently the first botanist to collect kili'o'opu, which was regarded as indigenous by Hillebrand (1965) and "long established" by St. John (1973). It was reported in Tahiti in 1769 by Banks and Solander (Merrill 1954) and, like *Mariscus javanicus*, may either be indigenous or one of those species spread accidentally by the Polynesians.

Kūkaepua'a (*Digitaria setigera*)

The botanist David Nelson, who accompanied Cook to Hawai'i, was the first to collect botanical specimens of this plant in 1779 (St. John 1978). Since it is frequently found growing around pig pens and its Hawaiian name, *kūkaepua'a*, implies an association with pigs, it may have been carried along accidentally when the Polynesians brought the pig to Hawai'i (Degener 1962).

Table 1. Plants associated with Hawaiian culture and land use.

	Handy & Handy 1972	St. John 1973	St. John 1978	Nagata 1985
MAIN ECONOMIC PLANTS				
Aleurites moluccana		*		*
Alocasia macrorrhiza		*		*
Artocarpus altilis	*	*		*
Broussonetia papyrifera	*	*		*
Cocos nucifera	*	?		*
Colocasia esculenta	*	*		*
Cordyline fruticosa	*	*		*
Curcuma longa	*	*		*
Dioscorea alata	*	*		*
Dioscorea bulbifera		*		*
Dioscorea pentaphylla		*		*
Ipomoea batatas	*	*		*
Lagenaria siceraria	*	*		*
Musa spp.	*	*		*
Nasturtium sarmentosum		*		*
Piper methysticum	*	*		*
Saccharum officinarum	*	*		*
Schizostachyum glaucifolium	*	*		*
Syzygium malaccense		*		*
Tacca leontopetaloides	*	*		*
Tephrosia purpurea		*		*
Zingiber zerumbet		*		*
STRAND TREES AND SHRUBS				
Calophyllum inophyllum		*		*
Cordia subcordata		*		*
Morinda citrifolia		*		*
Thespesia populnea		*		*
ACCIDENTAL INTRODUCTIONS				
Digitaria setigera			*	
Indigofera suffruticosa			*	*
Ludwigia octovalvis			*	*
Merremia aegyptia			*	*
Oxalis corniculata			*	*
Urena lobata			*	*
Waltheria indica			*	
OTHER POSSIBLE POLYNESIAN INTRODUCTIONS				
Kyllinga nemoralis				
Lycopersicon pimpinellifolium				
Lythrum maritimum				
Mariscus javanicus				
Solanum americanum				
Vigna adenantha				

'Inikō (*Indigofera suffruticosa*)

Hillebrand (1965) noted that 'inikō was one of the most common weeds in the Islands but believed that it had been introduced in 1836 for the production of indigo. Nelson, the botanist with Cook, listed "*Indigofera*, Indigo" in a report, which Beaglehole's informant assumed was *Indigofera suffruticosa* (St. John 1978; Beaglehole 1967). No herbarium specimen has been found to substantiate this observation. No record of the use of the plant by Hawaiians has been found, and it is hard to imagine that such a useful dye plant would have escaped their attention had it been present in prehistoric times. It therefore seems likely that the plant was not a Polynesian introduction and that Nelson was mistaken in including it on his list.

Vigna adenantha

This plant is probably a native of South America but was reported by early botanists in many Polynesian islands as well as in eastern Australia. Plant specimens from Hawai'i bearing this name, including those collected by Menzies in 1794 (British Museum of Natural History), Macrae in 1825 (British Museum of Natural History), Nuttall in 1835 (British Museum of Natural History), and Rémy sometime between 1851 and 1855 (Gray Herbarium), can be found in a number of herbaria. Mann (1866) believed the species to be a Polynesian introduction, and Hillebrand (1965) reported that the Wilkes Expedition, which visited Hawai'i in 1840, found it on "Diamond Hill." A search for this specimen in the U.S. National Herbarium, where the collections from the expedition are said to be preserved, did not locate it. No more recent reports based on voucher specimens have been found, so the extant specimens should perhaps be reexamined to ensure that the earlier botanists did not confuse this species with other taxa.

Yellow wood sorrel (*Oxalis corniculata*)

Yellow wood sorrel or 'ihi'ai is a pantropical weed collected by Banks and Solander in Tahiti and was observed but not collected by Nelson in Hawai'i in 1779 (St. John 1978). The fact that seeds have been found in prehistoric archaeological sites makes it clear that the plant was not introduced by Europeans (Kirch 1985) but was either indigenous or an introduction, probably accidental, by the Polynesians.

Hau (*Hibiscus tiliaceus*)

Hau was a valuable plant for the Hawaiians and was widely cultivated. For this reason, it has been considered a likely Polynesian introduction (Krauss 1974). However, it is a widespread strand species whose seeds are adapted to long-distance dispersal over the ocean (Merrill 1946). The plant is common along sheltered coasts and in wet upland gulches in Hawai'i today. It is quite likely that it reached Hawai'i without human assistance.

Milo (*Thespesia populnea*)

The Polynesians held this tree in high regard as a shade tree and also used it for tannin, dye, medicines, and wood (Krauss 1974). It was widely planted around houses and may be found persisting and occasionally establishing along the coast today. As a plant of the coastal strand, it is adapted to transoceanic dispersal and hence may be indigenous, although St. John (1973) and Nagata (1985) judge it to be a Polynesian introduction.

Aramina (*Urena lobata*)

Nelson observed "*Urena*, Indian Mallow" in Hawai'i in 1779 but did not take a specimen of the plant. This was assumed to be the native ōpuhe, *Urera sandwicensis*, in the family Urticaceae; however St. John (1978) argued that Nelson, a competent botanist, would not have confused plants in such distinctively different families. Merrill (1954) stated that Banks and Solander found *Urena* in Tahiti. Seeds of this plant could have been carried by birds to Hawai'i, or it might have been a Polynesian introduction.

Kamani (*Calophyllum inophyllum*)

This tree was cultivated by Hawaiians and had many uses. It is not found naturalized to any great extent in Hawai'i today but may have been so in the past. It is generally regarded as a Polynesian introduction, but since its natural habitat is the coastal strand and its seeds are adapted to ocean dispersal, the possibility that it is indigenous should not be entirely discounted.

Carrot (*Daucus pusillus*)

Hillebrand (1965) believed the American carrot was undoubtedly in the Islands before European discovery. A collection made by Menzies (British Museum of Natural History) provides support for this conclusion. It might possibly have been brought to the Islands by migrating birds and thus be an indigenous species.

'Uhaloa (*Waltheria indica*)

'Uhaloa has weedy properties and is considered by most to be indigenous. It was collected by Nelson (St. John 1978) and has small seeds that might adhere to birds; it may have been transported to Hawai'i by this means.

Pūkāmole (*Lythrum maritimum*)

Today pūkāmole has been generally regarded as a European introduction in Hawai'i. It was collected by many of the early botanists, including Menzies in 1794. This would suggest that it is probably an indigenous species.

Primrose willow or Kāmole (*Ludwigia octovalvis*)

Nelson observed kāmole in Hawai'i in 1779, and Menzies collected it in 1794 (St. John 1978). It is most abundant in wetland habitats, and its small seeds may have been transported to Hawai'i by waterbirds; or, as it is a frequent weed in taro patches, it might possibly have been introduced accidentally by the Polynesians with taro.

Kou (*Cordia subcordata*)

The wood of the tree kou was one of the most prized by the Hawaiians and was used in prehistoric times (Krauss 1974). It is a plant of the littoral zone, and its seeds are dispersed on the ocean, so it may be an indigenous plant. However, it is not found in the wild in Hawai'i, possibly indicating either that it was brought and cultivated by the Polynesians or that its original habitat has been destroyed.

Hairy merremia or Kuahulu (*Merremia aegyptia*)

Most botanists regarded kuahulu as a weed introduced in the 19th century until it was realized that Nelson had collected it in Hawai'i in 1779 (St. John 1978; British Museum of Natural History). Thus, it may be regarded as indigenous or introduced.

Currant tomato (*Lycopersicon pimpinellifolium*)

The tomato is found as a volunteer in waste places and has been regarded as an alien introduced by Europeans. The fact that there is a specimen labelled as a collection by Menzies from Hawai'i (British Museum of Natural History) now casts doubt on this conclusion. It is conceivable that seeds might have been transported to Hawai'i stuck to migrating birds; but, if it were indigenous, it is surprising that Hawaiians did not make more conspicuous use of it. Furthermore, the fact that the plant was not collected or mentioned by other botanists until Hillebrand suggests that the Menzies specimen may have been collected elsewhere and is mislabeled. This is believed to have occurred with some other Menzies specimens.

Glossy nightshade or Pōpolo (*Solanum americanum*)

Although pōpolo was not recorded in Hawai'i until Nuttall made a collection of it in 1835 (British Museum of Natural History), it was one of the most important medicinal plants for the Hawaiians (Gutmanis 1979). Furthermore, its Hawaiian name, *pōpolo*, comes from a proto-Polynesian root word that is widely applied, with small variations, to the same plant throughout Polynesia (Biggs 1979). It would appear that it is either indigenous or was introduced, perhaps on purpose, by Polynesians.

Noni (*Morinda citrifolia*)

The noni was cultivated by Hawaiians in prehistoric times and was a valuable medicinal and dye plant. It has a very wide distribution in the tropical Pacific, and its buoyant seeds disperse by flotation. Although generally regarded as a Polynesian introduction it may have reached Hawai'i by means of natural dispersal.

The status of the above species as indigenes or aliens could be a subject of endless debate. However, it is unlikely that any of the disputes can be satisfactorily solved unless pollen or macrofossil evidence is found in sediments that unmistakably date from a period before human settlement.

ALIEN INTRODUCTIONS AFTER EUROPEAN CONTACT

Rate of Introduction

In the last 200 years, 813 species of alien plants have naturalized in Hawai'i (Appendix). (The nomenclature used in this study follows that of the *Manual of the Flowering Plants of Hawai'i* (Wagner *et al.* 1990) except in a few cases, where taxa used by earlier authorities are retained in order to minimize information loss in the present analysis. Instances of retained older nomenclature are noted in the Appendix.) The rate of introduction increased by the end of the 19th century to the present average rate of about five species per year (Fig. 1). A similar pattern was observed by Healy (1947) in New Zealand, where, in the hundred years prior to 1946, alien plants arrived at a more or less constant rate of 10 per year.

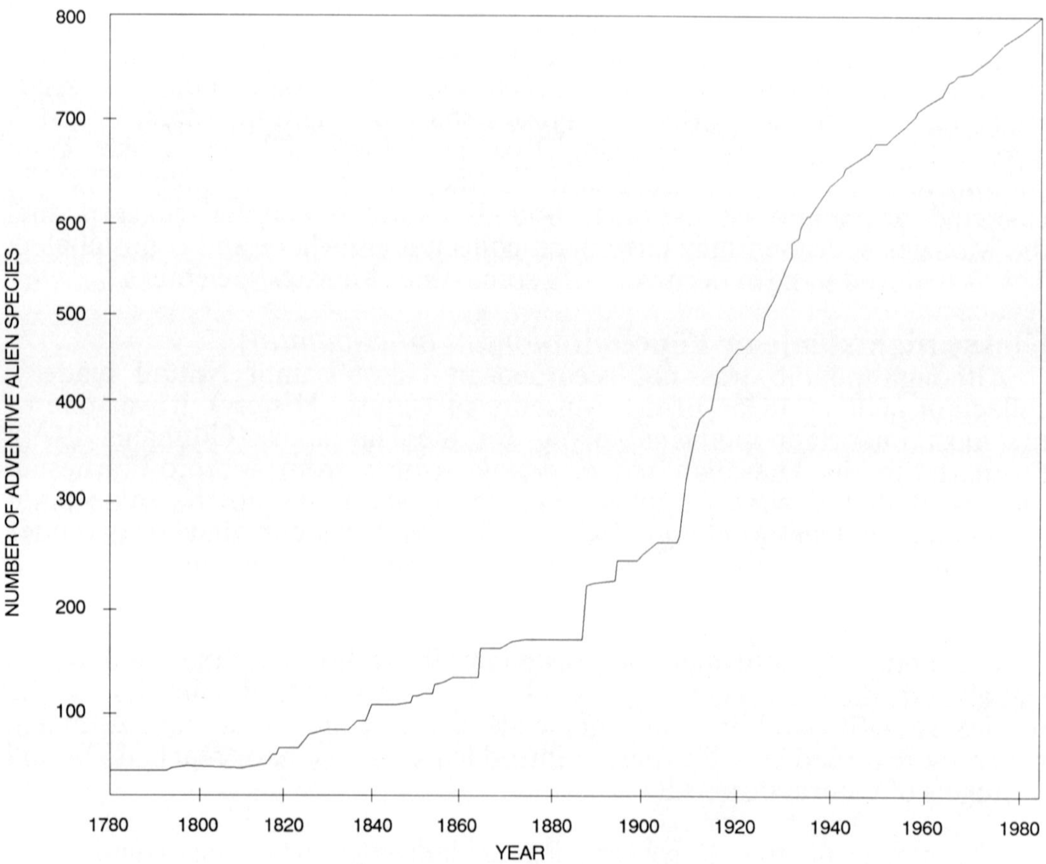

Figure 1. Accumulated number of adventive alien plant species in Hawai'i.

Bias in Herbarium Collections

Herbarium specimens may be an important source of information to establish the time of arrival of aliens in Hawai'i; thus, it is worthwhile to consider in detail the nature of the information preserved in collections. This approach is useful in interpreting the establishment and spread of colonizing alien plants and may also provide information useful in managing alien invasions. Extensive use has been made of herbarium material for mapping the ranges of taxa (Hastings *et al.* 1972) and for phenology (Holland 1975), as well as for reconstructing the introduction and colonization of species (Strother and Smith 1970). However, it is widely recognized that many biases exist in herbarium collections (Stearns 1951; Davis and Heywood 1965). For example, cultivated plants, as well as those difficult to preserve or store, tend to be poorly represented in collections. Furthermore, collecting activity tends to be concentrated in certain geographical areas and at certain times. This has special implications for alien species and will be discussed in more detail.

Temporal Biases. Sudden increases in accumulated species in Hawai'i herbaria during the 20th century reflect the intermittent nature of botanical surveys. In the 20th century, survey and systematic collecting of the flora were more frequent after the B.P. Bishop Museum and the University of Hawaii were founded and individuals with botanical training were employed permanently by private organizations or government agencies. As a result, newly introduced species were probably noticed and recorded on a more regular basis. Fluctuations in numbers of species accumulated in herbaria can be noted, for example, in the 1930s. This might be interpreted as a wave of invasion caused by the conscious attempt to revegetate degraded watershed with introduced plants. However, the historical records used to compile these data are not free from biases, and the observed fluctuations may well be the result of these instead.

The dates of collection of a randomized sample of approximately 2,000 sheets from the B.P. Bishop Museum herbarium reveal three main periods of intensive collecting activity in the 20th century (Fig. 2):

a. 1908-1920: This was the period soon after the founding of the B.P. Bishop Museum, when C.N. Forbes and others were building a representative collection.

b. 1930-1940: During this time several enthusiastic collectors worked in Hawai'i, including O. Degener, F.E. Egler, F.R. Fosberg, and H. St. John.

c. 1968-present: The last 18 years have been a period a renewed interest in ecological and taxonomic research.

Significant increases in the number of "first records" of alien plants correspond closely with these periods of intense collecting activity. Therefore, small fluctuations in the curve of accumulated numbers of species (Fig. 1) should not be given much significance. However, it is reasonable to assume that still more searching would reveal additional species, even though the period from 1968 to the present has been one of intensive collecting activity, because new aliens continue to colonize.

Geographic Biases. Considerable difference in intensity of botanical reconnaissance from place to place is obvious when examining the herbarium specimens preserved in the B.P. Bishop Museum herbarium. A randomized sample collected from O'ahu shows that many specimens were obtained from only a few favored areas such as the highest peaks of the Ko'olau and Wai'anae ranges, offshore islands, and in and around Honolulu (Fig. 3).

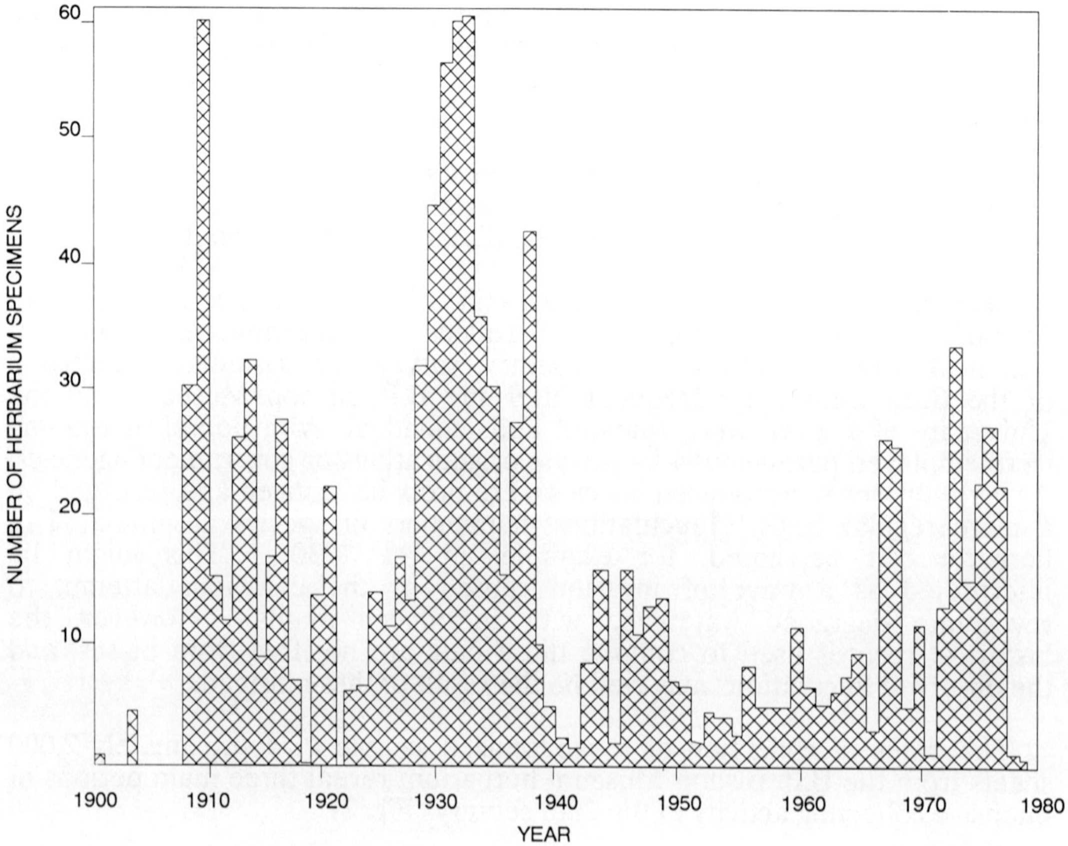

Figure 2. Number of herbarium specimens collected in Hawai'i that are preserved in the B.P. Bishop Museum herbarium.

The University of Hawaii campus at Mānoa and the B.P. Bishop Museum themselves figure prominently as sites of intense collecting activity. However, on the whole, most collecting, and presumably other botanical surveying, is concentrated in essentially natural areas where aliens are least common, not in the disturbed habitats where newly introduced plants are most likely to become established first. This is unfortunate because control or eradication of aliens is simplest when plants are recognized as soon as possible after arrival in an area, when populations are small and geographically confined.

If a policy of early detection and eradication were to be seriously adopted as a management technique, searches should be directed to sites other than those typically surveyed in the course of botanical research. For example, we might expect newly introduced aliens to establish near ports of entry such as airports, or perhaps more likely, near harbors where used automobiles are unloaded, as Wace (1979) demonstrated that cars carry a very large number of weed propagules. On the other hand, if many adventive plants in Hawai'i were first introduced as ornamental or crop

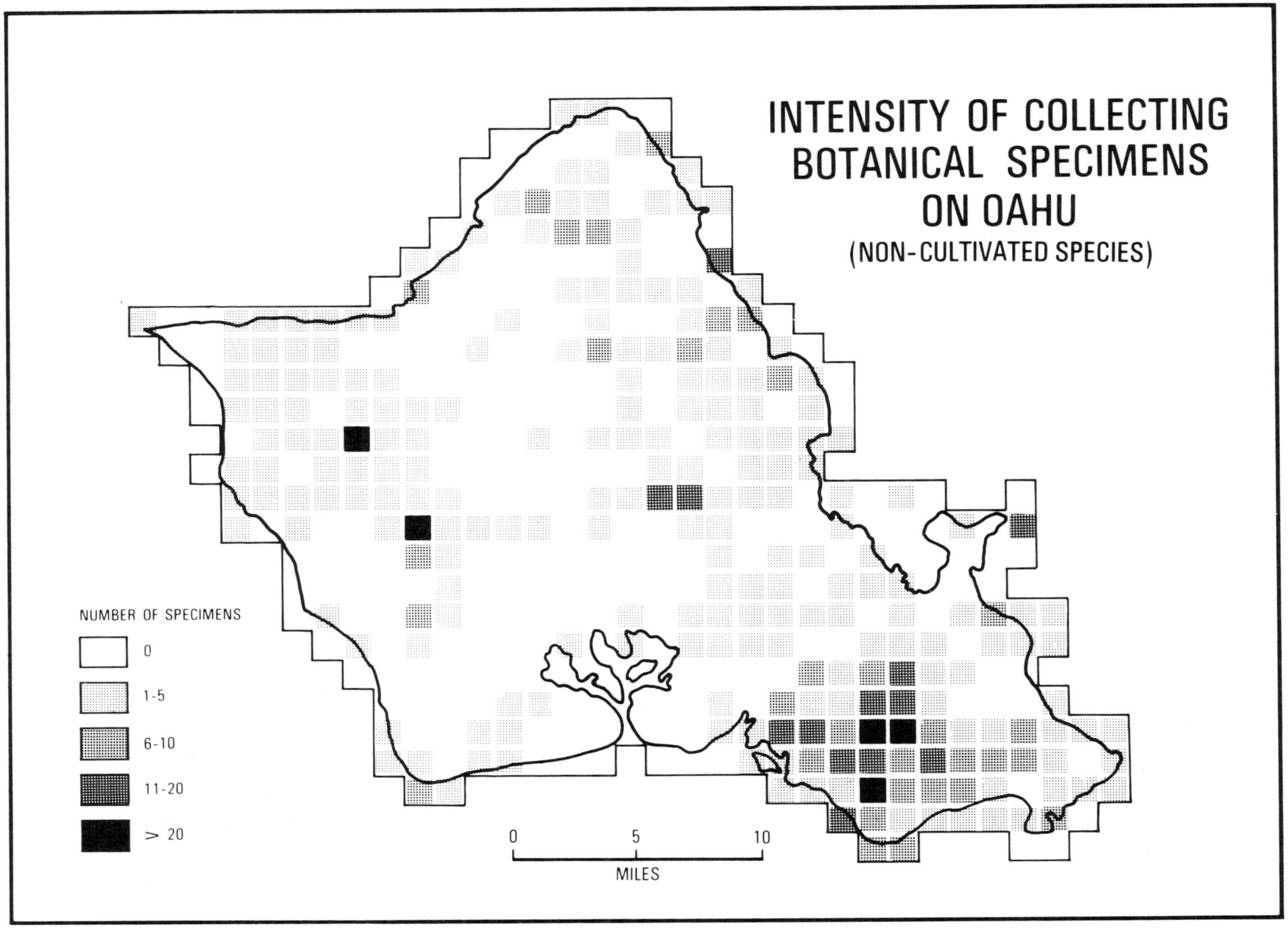

Figure 3. Number of herbarium specimens collected on O'ahu in the B.P. Bishop Museum Herbarium.

species, they may have first appeared in or near arboreta, botanical gardens, nurseries, or experimental plantings. In point of fact, many first records do come from these sites. Ornamental plants are also introduced and grown by hobbyists. Those that spread aggressively may colonize from these foci or be removed and dumped at official refuse sites, along roads, or in wasteland. Areas such as these should be searched for potential new invaders.

Roadsides not only provide a disturbed substrate for many alien species, but also provide corridors for dispersal through undisturbed landscapes otherwise inhospitable to aliens (Smith 1982; Wester and Juvik 1983). Roads allow aliens to achieve wide distribution rapidly.

Further sampling of B.P. Bishop Museum herbarium specimen sheets was done to determine relative numbers of collections from each island. The results show that the largest number of specimens come from O'ahu; with the exception of Ni'ihau, the small islands are represented by a large number of specimens in proportion to their area (Table 2). On the other hand, fewer specimens have been taken from Hawai'i Island than from O'ahu, which is only one-eighth its size.

A high correlation exists between the number of herbarium specimens taken from each island and the number of species known from that island, higher in fact than with other factors, which are often taken to determine the richness of an island flora (Table 3). It has been similarly noted in the Galapagos Archipelago that the number of botanical collecting trips to each of the islands is a better predictor of numbers of species than are area, elevation, or isolation (Conner and Simberloff 1978). Hence, it seems likely that our knowledge of the ranges of alien plants in the Hawaiian Islands, based on herbarium material, is inadequate; efforts can be most profitably be spent to correct this by concentrating on Hawai'i, Maui, Kaua'i, and Ni'ihau.

EXTINCTION OF ALIEN SPECIES

Invading species owe part of their success to an ability to increase populations rapidly. However, they are also subject to population crashes caused by environmental factors, such as drought or hurricanes, or to competition from other species. They often inbreed, forming homozygous populations vulnerable to attack by pests. As Baker (1965) pointed out, weeds are well adapted to rapid colonization and spread, but they have a limited future because they are often poorly equipped genetically to adjust to inevitable ecological and environmental changes. Although this does not seem to be the case in benign and isolated tropical areas, it is therefore of interest to consider whether abundant new arrivals are at least partly balanced by extinctions.

It is not possible to prove that a species has become extinct, but species once reported as aliens in Hawai'i that have not been collected at least since 1959 are listed in Table 4. The recent period of intensive collecting activity occurred after 1959, and it might usually be that a

Table 2. Alien species per unit area (mi^2) on major islands.

	Number of alien species	Area (mi^2)	Number of species per mi^2	Number of specimens per mi^2
NW Hawn. Ids.	76	5	14.62	20.77
Lāna'i	217	139	1.56	1.25
Kaho'olawe	65	45	1.44	1.56
Ni'ihau	77	73	1.05	0.62
O'ahu	571	608	0.94	2.60
Moloka'i	232	261	0.89	0.73
Kaua'i	345	553	0.62	0.58
Maui	379	729	0.52	0.45
Hawai'i	566	4,038	0.14	0.27
TOTAL	813	6,451	0.126	0.60

Table 3. Factors likely to influence the number of known adventive alien plant species on each island.

	Area (mi^2)	Maximum Elevation (feet)	Island Age (m. years)	Number of Alien Species	Number of Herbarium Specimens
Hawai'i	4,038	13,796	0.50	566	1,082
Maui	729	10,023	1.63	379	327
O'ahu	608	4,020	3.80	571	1,578
Kaua'i	553	5,243	5.72	345	319
Moloka'i	261	4,970	1.84	232	190
Lāna'i	139	3,370	1.46	217	174
Ni'ihau	73	1,281	n.a.	77	45
Kaho'olawe	45	1,477	1.03	65	70
NW Hawn. Ids.	5	874	11.30	76	108
Correlation coefficient	0.685*	0.757*	-0.312		0.888**

*$p < 0.05$

**$p < 0.01$

Table 4. Alien species not collected or observed since 1959.*

Species	Date of First Record	Date of Last Record	Number of Herbarium Specimens
Bidens laevis	1855	1912	2
Blumea laciniata	1923	1937	7
Blumea sessiliflora	1920	1937	1
Bothriochloa intermedia	1916	1940	2
*Digitaria longiflora***	1916	1938	5
*Echinochloa phyllopogon****	1928	1930	2
Geranium dissectum	1910	1948	4
Laportea interrupta	1819	1959	1
Potamogeton nodosus	1825	1920	5
Potamogeton pectinatus	1947	1949	2
Spergula arvensis	1888	1958	5
Stachytarpheta x intercedens	1939	1940	2
Stachytarpheta x trimeni	1928	1951	2
Typha domingensis#	1939	1942	2

*Nomenclature according to herbarium labels; a number of commonly cultivated species and those observed as naturalized at only one location on one occasion have not been included in this list.

**Not listed in Wagner *et al.* 1990; listed in St. John 1973.

***Not listed in Wagner *et al.* 1990 or St. John 1973.

#Sparingly naturalized, according to Wagner *et al.* 1990.

species would be collected if populations persisted. Among the species not collected and possibly extinct are a number that once had populations established on several islands and that were collected in considerable numbers. This would suggest that aliens may be suppressed, or possibly exterminated, without purposeful eradication programs, and that large populations do not always survive. The present regression of many koa haole *(Leucaena leucocephala)* populations in the Hawaiian Islands and elsewhere is a vivid illustration of the fact that, once introduced, alien plant populations may undergo dramatic fluctuation without direct human interference. This observation should not be a surprise: most aliens are opportunist species with general ecological roles. Their strategy for survival is effective dispersal and rapid population increase, rather than specialized adaptation to local environments. Successful competition against all other plants over an extended period of time is not favored by the generalist strategy. Disturbance, however, does favor such generalists.

The number of species that seem to disappear from herbarium records has steadily increased during the 20th century; during the decade of the 1950s, this disappearance occurred at an average rate of less than one per year. Preliminary evidence thus suggests that turnover in the weed flora may occur, but the number of arrivals of new alien species is far greater than the number being exterminated.

Place of Origin

The immediate origin of a new alien is usually not known because the colonizing propagule may not come directly from its native home. However, except in a few cases of pantropical species, it is possible to determine the provenance of aliens. It appears that the New World supplied a large proportion of aliens in the early European period of contact (Fig. 4). This is not surprising, since sailing ships coming to Hawai'i typically stopped in South America and hence had the opportunity to pick up tropical plants, on purpose or by accident, suited to the Hawaiian environment. In later times, the Old World made a larger contribution to the Hawaiian alien flora, perhaps as contacts with Asia became more frequent and as the circumtropical weed flora became more homogenized. A significant number of species arrived from North America in the 20th century, and these introductions reflect the close cultural and economic ties with that region.

Mode of Introduction

Direct evidence of the means of introduction of alien plants can only be found for a small proportion of the species. These are mostly plants purposefully introduced as crops, reforestation species, or ornamentals and observed to establish from the original plantings. If we assume that introduction was purposeful for species that are commonly crop or ornamental species, and that introduction was accidental for species with no redeeming social value, it appears that about half the introductions were on purpose and the other half accidental (Table 5).

CONCLUSIONS

The native flora of the Hawaiian Islands has been known in more detail, and for a longer period of time, than that of most other oceanic islands. However, the alien flora is not nearly so well known despite the fact that it represents a large component of the present-day flora; additionally, many aliens are now ecological dominants over much of the land area of Hawai'i. A clearer knowledge of the nature and development of the alien flora is desirable so that further introductions can be minimized and the spread of aliens already present can be monitored and controlled to a greater extent. Competition that aliens exert on the diminishing native ecosystems can thus be reduced.

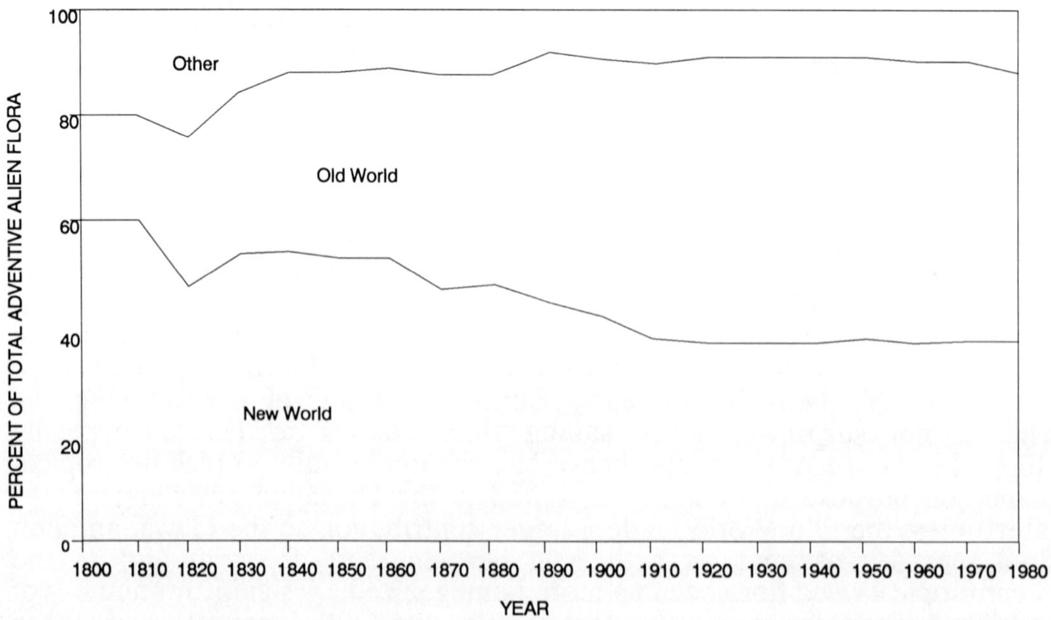

Figure 4. Region of origin of adventive alien plants in Hawai'i.

Table 5. Mode of introduction of alien species.

Mode of Introduction	Number of Species	Percent of Species
Purposeful		
Crop	199	21.8
Ornamental	121	35.9
Accidental	236	42.4

Appendix. A Preliminary List of Alien Adventive Seed Plants in the Hawaiian Islands.

The following abbreviations are used in this list:

Life form	Origin (Place of Origin)
B - tuber, bulb	AA - Asia, Australia
H - herb	AF - Africa
S - shrub	AS - Asia
T - tree	ASP - Asia, Pacific
U - succulent	AUP- Australia, Pacific
V - vine	AUS- Australia
	COS- cosmopolitan
	DIS - disjunct
	EAS - Eurasia
	EUR- Europe
	HOR- horticulture
	MED- Mediterranean
	NAM- North America
	NH - Northern Hemisphere
	NW - New World
	OW - Old World
	P - Pacific
	SAM- South America
	TAM- Tropical America
	TOW- Tropical Old World
	USM- United States Mainland

Mode of Introduction
- A - accidental
- C - crop
- O - ornamental

Record (date of reports of species in the Hawaiian Islands)
- first - earliest report
- recent - most recent record

Info. Source (information source for earliest report of species in Hawai‘i)
- P - published reference
- H - herbarium specimen

No. spms. (Number of herbarium specimens of species collected in Hawai‘i and preserved in B.P. Bishop Museum Herbarium)

Island (Name of island where presence of species can be verified by an herbarium specimen. Note: many species occur more widely than indicated here.)

- Hi - Hawai‘i
- Ma - Maui
- Mo - Moloka‘i
- Ln - Lāna‘i
- Kh - Kaho‘olawe
- Oh - O‘ahu
- Ka - Kaua‘i
- Ni - Ni‘ihau
- NW - Northwest Hawaiian Islands

Appendix, continued.

Appendix, continued.

Taxon	Life Form	Origin	Mode of Intro.	Record in Hi. First	Record in Hi. Recent	Info. Source	No. Spms.	Hi	Ma	Mo	Ln	Kh	Oh	Ka	Ni	NW
GYMNOSPERMS																
PINACEAE (Pine family)																
Pinus elliottii	T	USM	C	1976	1976	H	1							+		
Pinus patula	T	NAM	C	1944	1976	H	2		+							
Pinus pinaster	T	EUR	C	1976	1976	H	1		+							
ANGIOSPERMS																
MONOCOTYLEDONS																
ALISMATACEAE (Water plantain family)																
* *Echinodorus rostratus*	H	NW	A	1939	1939	H	1						+			
Sagittaria latifolia	H	EAS	O	1916	1986	H	13						+			
AGAVACEAE (Agave family)																
Agave sisalana	U	TAM	C	1893	1977	P	4	+			+		+			
Cordyline fruticosa	S	ASP	C	1000	1985	P	68	+	+	+	+		+	+	+	
Furcraea foetida	U	SAM	C	1870	1946	P	10	+		+			+			
Phormium tenax	S	P	O	1933	1949	H	4	+						+		
ARACEAE (Taro family)																
Alocasia macrorrhiza	B	ASP	C	1000	1972	P	6			+			+			
Colocasia esculenta	B	AS	C	1000	1986	P	4		+				+			
Pistia stratiotes	H	OW	O	1932	1985	P	3						+	+		
CANNACEAE (Canna family)																
Canna edulis	B	TAM	C	1898	1929	P	1						+			
Canna indica	B	TAM	C	1817	1986	H	21	+	+		+		+	+		
COMMELINACEAE (Spiderwort family)																
Callisia fragrans	H	TAM	O	1928	1975	H	8						+	+		
Commelina benghalensis	H	OW	A	1909	1982	H	16		+				+	+		
Commelina diffusa	H	AS	A	1835	1983	H	30	+	+	+	+		+	+		+

Taxon	Life Form	Origin	Mode of Intro.	Record in Hi. First	Record in Hi. Recent	Info. Source	No. Spms.	Hi	Ma	Mo	Ln	Kh	Oh	Ka	Ni	NW
Commelina erecta	H	NAM	O	1925	1963	P	0						+			
Tradescantia geniculata	H	TAM	O	1888	1973	P	0									
Tradescantia spathacea	H	TAM	O	1937	1976	H	6	+					+			
Tradescantia zebrina	H	TAM	O	1926	1976	H	9							+	+	
CYPERACEAE (Sedge family)																
Bulbostylis capillaris	H	NAM	A	1911	1985	H	31	+								
Cyperus alternifolius	H	AF	O	1921	1985	H	20	+	+				+	+		+
Cyperus compressus	H	COS	A	1935	1982	H	6	+								
Cyperus difformis	H	OW	A	1895	1984	H	15						+	+		
Cyperus esculentus	H	COS	C	1956	1956	H	2	+								
Cyperus gracilis	H	AUS	O	1893	1985	P	12	+					+	+		
Cyperus halpan	H	EAS	A	1957	1985	H	14	+	+					+		
Cyperus pilosus	H	AS	A	1916	1916	H	2							+		
* *Cyperus rivularis*	H	-	A	1949	1975	H	3	+								
Cyperus rotundus	H	COS	A	1850	1983	P	21	+			+		+	+	+	+
Cyperus trinervis	H	AUS	A	1926	1966	H	3	+								
Cyperus virens	H	TAM	A	1976	1978	H	4	+								
Eleocharis geniculata	H	TAM	A	1909	1984	H	13			+			+	+		
Eleocharis obtusa	H	NAM	A	1852	1985	H	64	+	+	+			+	+		
Eleocharis radicans	H	TAM	A	1917	1982	H	6		+				+			
Fimbristylis aestivalis	H	-	A	1977	1983	H	4							+		
Fimbristylis schoenoides	H	OW	A	1984	1985	H	2	+								
Kyllinga brevifolia	H	OW	A	1865	1984	H	66	+	+		+		+	+		
Kyllinga nemoralis	H	-	A	1929	1985	H	16						+	+		
Schoenoplectus californicus	H	NW	A	1912	1986	H	21	+	+	+			+	+	+	
Torulinium odoratum	H	AS	A	1825	1982	H	3		+							

Appendix, continued.

Appendix, continued.

Taxon	Life Form	Origin	Mode of Intro.	Record in Hi. First	Record in Hi. Recent	Info. Source	No. Spms.	Hi	Ma	Mo	Ln	Kh	Oh	Ka	Ni	NW
DIOSCOREACEAE (Yam family)																
Dioscorea bulbifera	B	AS	C	1000	1984	P	40	+	+	+			+	+		
Dioscorea pentaphylla	B	ASP	C	1000	1984	P	26			+			+	+		
HYDROCHARITACEAE (Frog's-bits family)																
Egeria densa	H	SAM	O	1937	1971	H	4	+					+			
IRIDACEAE (Iris family)																
Belamcanda chinensis	B	AS	O	1888	1976	P	7						+			
Crocosmia x crocosmiiflora	B	AF	O	1932	1985	H	9	+					+	+		
Sisyrinchium rosulatum	H	TAM	A	1944	1945	H	4	+								
Sisyrinchium strictum	H	USM	A	1938	1938	H	1	+								
Trimezia martinicensis	H	TAM	O	1943	1984	H	2						+			
JUNCACEAE (Rush family)																
Juncus acuminatus	H	NAM	A	1935	1935	H	1	+								
Juncus bufonius	H	COS	A	1910	1971	H	17	+	+	+						
Juncus effusus	H	COS	A	1900	1975	P	10	+		+			+			
Juncus ensifolius	H	--	A	1911	1980	H	4	+								
Juncus planifolius	H	AUP	A	1930	1984	H	14	+		+				+		
Juncus polyanthemos	H	AUS	A	1953	1967	H	2	+	+							
Juncus tenuis	H	NH	A	1915	1974	H	8	+		+						
LEMNACEAE (Duckweed family)																
Lemna perpusilla	H	COS	O	1895	1961	H	10		+	+			+	+		
Spirodela polyrhiza	H	COS	O	1918	1976	H	5							+		
LILIACEAE (Lily family)																
Hippeastrum puniceum	B	TAM	O	1928	1986	P	3						+	+		
LIMNOCHARITACEAE**																
** *Hydrocleys nymphoides*	H	SAM	O	1939	1986	H	4						+			

Taxon	Life Form	Origin	Mode of Intro.	Record in Hi. First	Record in Hi. Recent	Info. Source	No. Spms.	Island Hi	Ma	Mo	Ln	Kh	Oh	Ka	Ni	NW
MUSACEAE (Banana family)																
Musa troglodytarum	H	P	C	1960	1960	H	3						+			
Musa x paradisiaca	H	ASP	C	1000	1957	P	4			+			+			
ORCHIDACEAE (Orchid family)																
Arundina graminifolia	H	AS	O	1950	1983	H	22	+	+				+	+		
Epidendrum x obrianianum	H	TAM	O	1940	1984	H	5						+	+		
Phaius tankarvilleae	H	ASP	O	1931	1985	H	17	+					+	+		
Spathoglottis plicata	H	AS	O	1929	1975	H	34	+	+		+		+	+		
POACEAE (Grass family)																
Agrostis avenacea	H	AUP	A	1888	1982	H	63	+	+	+	+		+	+		
Agrostis semiverticillata	H	EUR	C	1864	1972	P	11	+						+		
Agrostis stolonifera	H	EUR	C	1912	1969	H	13	+	+	+			+	+		
Aira caryophyllea	H	EUR	A	1916	1985	H	10	+	+	+						
Ammophila arenaria	H	EUR	A	1913	1962	P	2									+
Andropogon virginicus	H	USM	C	1924	1973	H	32	+					+			
Anthoxanthum odoratum	H	EUR	C	1907	1984	H	33	+		+						
Arthraxon hispidus	H	AS	-	1972	1972	H	1	+								
Avena barbata	H	EUR	A	1916	1961	P	4				+		+			
Avena fatua	H	EUR	C	1888	1943	P	5	+					+	+		
Axonopus fissifolius	H	USM	C	1912	1984	P	16	+	+		+		+	+		
Bothriochloa barbinodis	H	TAM	C	1910	1967	H	9		+	+			+			
Bothriochloa intermedia	H	AUS	C	1916	1940	H	2			+			+			
Bothriochloa pertusa	H	TOW	C	1916	1980	P	17	+	+	+	+		+		+	
Brachiaria mutica	H	AF	C	1902	1984	P	27	+	+		+		+	+		+
Briza maxima	H	EUR	C	1917	1984	H	7	+	+					+		
Briza minor	H	EUR	C	1840	1984	P	30	+	+	+	+		+	+		

Appendix, continued.

Appendix, continued.

Taxon	Life Form	Origin	Mode of Intro.	Record in Hi. First	Record in Hi. Recent	Info. Source	No. Spms.	Hi	Ma	Mo	Ln	Kh	Oh	Ka	Ni	NW
Bromus mollis	H	EUR	C	1909	1978	H	10	+	+				+	+		
Bromus rigidus	H	EUR	C	1910	1984	H	19	+	+					+		
Bromus rubens	H	EUR	C	1920	1967	P	3	+		+						
Bromus tectorum	H	EUR	C	1888	1969	P	3		+	+						
Bromus willdenowii	H	-	-	1909	1982	H	30	+	+	+			+			+
Cenchrus ciliaris	H	AF	C	1932	1982	H	7	+			+		+			
Cenchrus echinatus	H	TAM	A	1867	1984	P	51	+	+	+		+	+	+	+	+
Chloris barbata	H	TAM	A	1902	1984	H	70	+	+	+	+	+	+	+	+	+
Chloris divaricata	H	AUS	C	1924	1980	H	12		+			+	+		+	
Chloris gayana	H	AF	C	1913	1985	H	19	+	+		+		+		+	
Chloris petraea	H	TAM	-	1968	1980	H	2									+
Chloris radiata	H	TAM	C	1851	1976	P	30	+	+		+		+	+		
Chloris truncata	H	AUS	C	1915	1980	P	2					+				
Chloris virgata	H	TAM	C	1903	1982	H	28		+	+	+	+	+	+		+
Chrysopogon aciculatus	H	TAM	A	1865	1972	H	34	+		+	+		+	+		
Coix lachryma-jobi	H	AS	C	1895	1984	P	25	+	+	+			+	+		
Cymbopogon refractus	H	AUS	A	1929	1982	H	10	+		+			+	+		
Cynodon dactylon	H	OW	C	1888	1984	H	93	+	+		+	+	+	+		+
Dactylis glomerata	H	EUR	C	1911	1984	H	51	+	+	+			+	+		
Dactyloctenium aegyptium	H	TOW	A	1909	1983	H	33	+	+	+		+	+			
Danthonia pilosa	H	AUS	C	1910	1985	P	6	+	+							
Danthonia semiannularis	H	AUS	C	1903	1975	P	2	+	+							
Dichanthium annulatum	H	AS	-	1927	1963	P	3	+					+			
Dichanthium aristatum	H	AS	-	1911	1976	H	18		+		+		+		+	
Dichanthium sericeum	H	AUS	-	1912	1980	P	11	+			+		+		+	
Digitaria ciliaris	H	-	C	1912	1982	H	44	+	+	+	+	+	+	+	+	

	Taxon	Life Form	Origin	Mode of Intro.	Record in Hi. First	Record in Hi. Recent	Info. Source	No. Spms.	Hi	Ma	Mo	Ln	Kh	Oh	Ka	Ni	NW
	Digitaria fuscescens	H	SAM	C	1852	1969	H	38	+	+	+	+		+	+	+	
	Digitaria insularis	H	COS	-	1948	1957	H	2			+			+			
*	*Digitaria longiflora*	H	AS	C	1916	1938	H	5	+					+			
	Digitaria pentzii	H	AF	C	1950	1978	P	2	+				+				
	Digitaria sanguinalis	H	EUR	A	1864	1984	P	75	+	+	+	+	+	+	+		+
	Digitaria setigera	H	AS	A	1826	1983	P	64	+	+	+	+		+	+	+	
	Digitaria violascens	H	AS	A	1916	1985	H	10	+	+				+			
	Echinochloa colona	H	AS	A	1864	1982	P	54	+	+	+	+	+	+	+	+	
	Echinochloa crus-galli	H	AS	C	1846	1983	H	62	+	+	+	+		+	+		
*	*Echinochloa cruspavonis*	H	SAM	C	1916	1966	H	2	+					+			
	Echinochloa glabrescens	H	AS	-	1895	1966	H	13	+	+		+		+			
**	*Echinochloa phyllopogon*	H	-	-	1928	1930	H	2						+			
	Ehrharta stipoides	H	AUP	A	1916	1983	H	16	+	+				+			
	Eleusine indica	H	AS	A	1940	1984	P	62	+	+	+	+		+	+		+
	Eragrostis brownei	H	AUS	C	1916	1973	H	11	+	+	+						
	Eragrostis cilianensis	H	EUR	C	1864	1985	P	36	+		+		+	+	+		
	Eragrostis ciliaris	H	COS	-	1976	1976	H	1						+			
	Eragrostis elongata	H	AS	C	1949	1983	P	4	+						+		
	Eragrostis pectinacea	H	NAM	C	1916	1977	H	16	+	+				+			
	Eragrostis pilosa	H	EUR	C	1913	1983	P	1							+		
	Eragrostis tenella	H	TOW	A	1895	1984	H	40	+	+			+	+		+	+
	Eragrostis unioloides	H	AS	C	1958	1958	H	1	+								
	Eustachys petraea	H	TAM	-	1978	1978	H	1									+
	Festuca rubra	H	EUR	C	1912	1982	P	5	+	+							
	Garnotia acutigluma	H	-	-	1888	1972	H	9	+	+	+						
	Gastridium ventricosum	H	EUR	C	1909	1982	H	13	+	+					+		

Appendix, continued.

Appendix, continued.

	Taxon	Life Form	Origin	Mode of Intro.	Record in Hi. First	Record in Hi. Recent	Info. Source	No. Spms.	Hi	Ma	Mo	Ln	Kh	Oh	Ka	Ni	NW
	Holcus lanatus	H	EUR	A	1909	1985	H	35	+	+	+			+	+		
	Hordeum leporinum	H	MED	-	1909	1938	H	12	+	+		+					+
	Hordeum vulgare	H	OW	C	1935	1984	H	6	+	+				+	+		
	Hyparrhenia hirta	H	OW	C	1962	1981	H	4		+		+					
	Hyparrhenia rufa	H	TOW	C	1933	1982	P	16	+	+	+	+		+	+		
	Lolium multiflorum	H	EUR	C	1910	1980	H	18	+	+	+			+	+		
	Melinis minutiflora	H	EUR	C	1913	1983	P	35	+	+	+	+	+	+	+		
	Muhlenbergia microsperma	H	NW	C	1930	1974	H	11						+			
	Oplismenus compositus	H	AS	C	1966	1966	H	1						+			
	Oplismenus hirtellus	H	TAM	A	1841	1983	P	84	+	+	+	+		+	+		
	Panicum maximum	H	AF	C	1888	1984	H	14	+				+	+	+		
	Panicum miliaceum	H	EAS	C	1923	1981	P	7	+	+							
	Panicum repens	H	OW	C	1916	1979	H	12	+	+		+					
*	*Paspalidium radiatum*	H	AUS	C	1939	1939	H	1								+	
	Paspalum conjugatum	H	TAM	A	1840	1983	P	55	+	+	+	+		+	+	+	
	Paspalum dilatatum	H	SAM	C	1911	1984	H	41	+	+	+	+		+	+		
	Paspalum distichum	H	TAM	C	1916	1978	P	3						+	+		
	Paspalum fimbriatum	H	TAM	C	1916	1983	P	21						+	+		+
	Paspalum paniculatum	H	TAM	C	1980	1980	H	1	+								
	Paspalum scrobiculatum	H	NW	-	1852	1982	H	49	+	+	+			+	+		
	Paspalum urvillei	H	SAM	C	1914	1984	H	26	+	+	+	+		+	+		
	Paspalum vaginatum	H	TAM	C	1936	1974	H	8	+					+	+		
	Pennisetum clandestinum	H	AF	C	1924	1984	H	8	+	+		+		+			
	Pennisetum macrostachyum	H	AS	C	1916	1967	P	8	+					+	+		
	Pennisetum polystachion	H	TOW	-	1929	1985	H	30	+	+	+	+	+	+	+		
	Pennisetum purpureum	H	AF	C	1912	1976	P	18	+	+	+	+		+	+		

	Taxon	Life Form	Origin	Mode of Intro.	Record in Hi. First	Record in Hi. Recent	Info. Source	No. Spms.	Island Hi	Ma	Mo	Ln	Kh	Oh	Ka	Ni	NW
	Pennisetum setaceum	H	AF	C	1914	1983	H	27	+			+		+	+		
	Poa annua	H	EUR	C	1841	1983	P	35	+	+		+		+	+		+
	Poa pratensis	H	EUR	C	1911	1978	H	16	+	+				+			
	Polypogon interruptus	H	SAM	C	1871	1970	P	14	+	+				+	+		
	Polypogon monspeliensis	H	EUR	C	1909	1967	H	25	+	+				+			+
	Rhynchelytrum repens	H	AF	C	1895	1984	P	63	+	+	+	+	+	+	+	+	+
	Sacciolepis indica	H	AS	C	1908	1984	H	43	+	+	+			+	+		
	Schizachyrium condensatum	H	NAM	C	1961	1974	H	15	+								
	Schizostachyum glaucifolium	H	P	C	1000	1980	P	37	+	+	+			+	+		
	Setaria glauca	H	OW	C	1910	1983	P	16	+			+		+	+		
	Setaria gracilis	H	TAM	C	1855	1984	H	45	+	+	+	+		+	+		
	Setaria palmifolia	H	OW	C	1903	1983	H	30	+	+		+		+			
	Setaria verticillata	H	OW	A	1895	1984	H	97	+	+	+	+	+	+	+	+	+
	Sorghum halpense	H	MED	C	1909	1983	H	24	+	+				+	+		
*	*Sorghum sudanense*	H	AF	C	1914	1929	H	4				+		+			
	Sporobolus africanus	H	AS	C	1911	1984	H	11	+	+	+	+		+	+		+
	Sporobolus diander	H	AF	C	1911	1983	H	39	+					+			
	Sporobolus elongatus	H	AUS	C	1909	1936	H	17	+	+		+		+			
	Sporobolus indicus	H	TAM	C	1912	1983	P	15	+	+	+	+		+			
	Sporobolus pyramidatus	H	TAM	-	1978	1979	H	4						+			+
	Stenotaphrum secundatum	H	USM	C	1840	1984	P	36	+		+	+		+	+		+
	Stipa cernua	H	USM	C	1957	1982	H	8	+								
	Themeda villosa	H	AS	C	1924	1979	H	11	+					+	+		
	Tragus berteronianus	H	OW	A	1935	1984	P	8		+	+		+				
#	*Tricachne insularis*	H	TAM	C	1913	1984	H	43	+	+	+	+		+	+		+

Appendix, continued.

Appendix, continued.

Taxon	Life Form	Origin	Mode of Intro.	Record in Hi. First	Record in Hi. Recent	Info. Source	No. Spms.	Island Hi	Ma	Mo	Ln	Kh	Oh	Ka	Ni	NW
Vulpia bromoides	H	EUR	-	1901	1976	P	39	+	+	+	+		+	+		
Vulpia myuros	H	EUR	-	1911	1985	H	53	+	+		+		+	+		+
* *Vulpia octoflora*	H	USM	-	1966	1966	H	2	+								
PONTEDERIACEAE (Pickerel weed family)																
Eichhornia crassipes	H	TAM	O	1900	1976	P	5	+	+				+			
Monochoria vaginalis	H	AS	A	1935	1946	H	4							+		
POTAMOGETONACEAE (Pondweed family)																
Potamogeton foliosus	H	NAM	A	1817	1984	H	15	+	+				+	+		
Potamogeton nodosus	H	COS	A	1825	1920	H	5		+							
Potamogeton pectinatus	H	COS	A	1947	1949	H	2							+	+	
TACCACEAE (Tacca family)																
Tacca leontopetaloides	B	AS	C	1000	1977	P	40	+	+	+			+	+		
TYPHACEAE (Cattail family)																
Typha domingensis	H	OW	O	1939	1942	H	2						+			
Typha latifolia	H	COS	A	1977	1981	P	3						+			
XYRIDACEAE (Yellow-eyed grass family)																
Xyris complanata	H	AUS	A	1973	1985	P	2	+								
Xyris platylepis	H	USM	A	1951	1962	H	6	+								
ZINGIBERACEAE (Ginger family)																
Alpinia purpurata	B	P	O	1925	1979	H	16	+		+			+			
Alpinia zerumbet	B	AS	O	1888	1976	P	12	+					+			
Hedychium coronarium	B	AS	O	1888	1986	P	12	+	+	+	+		+			
Hedychium flavescens	B	AS	O	1913	1972	H	16	+	+	+	+		+	+		
Hedychium gardnerianum	B	AS	O	1954	1975	H	3	+					+			
Zingiber zerumbet	B	AS	C	1000	1983	P	35		+	+	+		+	+		

Taxon	Life Form	Origin	Mode of Intro.	Record in Hi. First	Record in Hi. Recent	Info. Source	No. Spms.	Hi	Ma	Mo	Ln	Kh	Oh	Ka	Ni	NW
DICOTYLEDONS																
ACANTHACEAE (Acanthus family)																
* *Aphelandra aurantiaca*	S	TAM	O	1958	1975	H	6						+	+		
Asystasia gangetica	H	TOW	O	1925	1976	H	18		+	+			+	+		
Barleria cristata	H	AS	O	1929	1974	H	8	+					+			
Dicliptera chinensis	H	AS	O	1942	1985	H	13						+			
Hemigraphis reptans	H	AS	O	1957	1979	H	5						+			
Justicia betonica	H	TOW	O	1943	1984	H	14	+		+	+		+	+		
Justicia spicigera	S	-	O	1934	1983	H	3			+			+			
** *Odontonema cuspidatum*	S	-	O	1937	1973	H	7	+					+			
Ruellia brittoniana	S	TAM	O	1930	1981	H	11						+			
Ruellia graecizans	S	SAM	O	1925	1893	H	10						+			
Ruellia prostrata	H	AS	O	1944	1980	H	5						+			
Thunbergia alata	V	AF	O	1865	1946	H	6			+			+			
Thunbergia fragrans	V	AS	O	1916	1984	H	12	+	+				+	+		
AIZOACEAE (Fig-marigold family)																
Tetragonia tetragonioides	H	DIS	C	1909	1976	H	10	+					+			+
AMARANTHACEAE (Amaranth family)																
Achyranthes aspera	H	AS	A	1855	1983	H	26	+			+		+			+
Alternanthera brasiliana	H	TAM	A	1973	1973	H	1			+						
Alternanthera caracasana	H	AF	A	1925	1985	H	14		+	+	+		+	+		
Alternanthera pungens	H	TAM	A	1959	1985	H	9			+			+			
Alternanthera sessilis	H	COS	A	1935	1983	H	8	+					+			
Alternanthera tenella	H	TAM	A	1936	1944	H	2						+			
Amaranthus dubius	H	TOW	A	1930	1977	P	4	+			+		+			
Amaranthus hybridus	H	NW	O	1930	1982	H	4		+				+			

Appendix, continued.

Appendix, continued.

Taxon	Life Form	Origin	Mode of Intro.	Record in Hi. First	Record in Hi. Recent	Info. Source	No. Spms.	Hi	Ma	Mo	Ln	Kh	Oh	Ka	Ni	NW
Amaranthus lividus	H	COS	A	1971	1985	H	3	+					+			
Amaranthus spinosus	H	TAM	A	1911	1984	P	12	+	+	+		+	+	+		+
Amaranthus viridis	H	COS	A	1819	1984	P	25	+	+		+	+	+			+
Gomphrena globosa	H	TAM	O	1929	1948	H	4						+			
ANACARDIACEAE (Mango family)																
Anacardium occidentale	T	TAM	C	1915	1983	H	9						+			
Schinus terebinthifolius	T	SAM	O	1909	1972	H	23	+	+	+	+		+			+
ANNONACEAE*																
* *Annona cherimola*	T	SAM	C	1791	1975	P	7	+			+					
APIACEAE (Parsley family)																
Anethum graveolens	H	EAS	O	1888	1953	P	3						+			
Centella asiatica	H	AS	A	1888	1972	P	23	+		+			+	+		
Ciclospermum leptophyllum	H	NW	A	1909	1983	H	26	+		+	+		+	+	+	
Coriandrum sativum	H	EAS	C	1926	1952	H	5	+					+			
Cryptotaenia canadensis	H	NAM	C	1925	1979	H	22	+					+	+		
Daucus pusillus	H	NAM	A	1794	1981	H	14	+		+	+		+			
Eryngium foetidum	H	TAM	A	1944	1976	H	2						+	+		
Foeniculum vulgare	H	EUR	C	1865	1984	P	15	+	+		+		+	+		
Hydrocotyle bowlesioides	H	-	A	1931	1984	H	6	+								
Hydrocotyle sibthorpioides	H	AS	A	1932	1983	H	5	+	+				+			
Hydrocotyle verticillata	H	OW	A	1817	1966	H	31	+	+	+			+	+	+	
Petroselinum crispum	H	MED	C	1910	1984	H	8	+	+				+			
Torilis nodosa	H	OW	A	1855	1911	H	3	+								
APOCYNACEAE (Dogbane family)																
Catharanthus roseus	H	TAM	O	1865	1983	H	22	+	+	+			+			+

Taxon	Life Form	Origin	Mode of Intro.	Record in Hi. First	Record in Hi. Recent	Info. Source	No. Spms.	Island Hi	Ma	Mo	Ln	Kh	Oh	Ka	Ni	NW
AQUIFOLIACEAE (Holly family)																
Ilex cassine	T	USM	O	1949	1985	H	15	+					+			
Ilex paraguariensis	T	SAM	C	1934	1984	H	7						+			
ARALIACEAE (Ginseng family)																
Schefflera actinophylla	T	AUS	O	1900	1976	P	6						+			
ASCLEPIADACEAE (Milkweed family)																
Asclepias curassavica	H	TAM	O	1850	1976	P	44	+		+	+	+	+	+		
Asclepias physocarpa	H	AF	O	1919	1974	H	22	+	+		+		+			
Stapelia gigantea	H	AF	O	1942	1984	H	5						+			
ASTERACEAE (Sunflower family)																
Acanthospermum australe	H	TAM	A	1840	1974	P	28	+		+	+	+	+	+		
Acanthospermum hispidum	H	TAM	A	1951	1966	H	2			+						
Achillea millefolium	H	NH	A	1911	1975	H	10	+					+			
Ageratina adenophora	H	TAM	A	1909	1967	H	63	+	+	+	+		+	+		
Ageratina riparia	H	TAM	A	1925	1976	P	13	+					+			
Ageratum conyzoides	H	TAM	O	1825	1974	P	52	+	+	+	+		+	+	+	
Ageratum houstonianum	H	TAM	O	1909	1977	H	15	+					+	+		
Ambrosia artemisiifolia	H	NAM	A	1840	1984	H	40	+	+	+	+		+			
Anthemis cotula	H	EUR	A	1909	1976	H	14	+			+			+		
Arctium lappa	H	EAS	O	1910	1960	H	7	+			+		+			
Artemisia pontica	S	EUR	A	1959	1959	H	1						+			
Artemisia vulgaris	H	EUR	A	1909	1959	H	14	+					+	+		
Aster subulatus	H	TAM	O	1826	1977	H	10	+	+	+			+	+	+	
Baltimora recta	H	TAM	A	1984	1984	P	0						+			
Bidens alba	H	-	A	1958	1983	H	5	+				+	+			+
Bidens cynapiifolia	H	TAM	A	1929	1976	H	20	+			+		+			

Appendix, continued.

Appendix, continued.

Taxon	Life Form	Origin	Mode of Intro.	Record in Hi. First	Record in Hi. Recent	Info. Source	No. Spms.	Hi	Ma	Mo	Ln	Kh	Oh	Ka	Ni	NW
Bidens gardneri	H	-	A	1983	1984	H	1			+						
Bidens laevis	H	NW	A	1855	1912	P	2			+			+			
##*Bidens lanceolata*	H	-	A	1963	1963	H	1	+								
Bidens pilosa	H	TAM	A	1819	1975	H	56	+	+		+		+			
Blumea laciniata	H	AS	A	1923	1937	H	7						+			
Blumea sessiliflora	H	AS	A	1920	1937	H	1		+							
Calyptocarpus vialis	H	TAM	A	1971	1985	H	11		+	+	+		+	+		
Centaurea maculosa	H	EUR	A	1965	1965	H	1	+								
Centaurea melitensis	H	EUR	A	1855	1976	H	31	+	+	+	+		+	+	+	
Chrysanthemum leucanthemum	H	EAS	A	1911	1975	H	4	+	+							
* *Chrysanthemum maximum*	H	EUR	O	1917	1959	H	4	+								
Cirsium vulgare	H	EUR	A	1909	1983	H	19	+	+	+	+	+	+	+		
Conyza bonariensis	H	SAM	A	1888	1985	P	53	+	+	+	+	+	+			
Conyza canadensis var. *canadensis*	H	NW	A	1911	1984	H	6	+	+							
Conyza canadensis var. *pusilla*	H	NW	A	1817	1984	H	57	+	+	+	+		+	+	+	
Coreopsis lanceolata	H	USM	O	1938	1946	H	4	+					+	+		
Cosmos caudatus	H	TAM	O	1865	1953	H	4	+					+			
Cotula australis	H	AUP	A	1943	1983	H	3	+	+				+			
Crassocephalum crepidioides	H	AF	A	1929	1984	H	16						+	+		
Crepis capillaris	H	EUR	A	1934	1975	H	8	+					+	+		
Eclipta alba	H	COS	A	1865	1974	P	16	+					+		+	
Elephantopus mollis	H	TAM	A	1926	1974	H	16		+	+				+	+	
Elephantopus spicatus	H	TAM	A	1935	1975	H	6	+					+	+		
Emilia coccinea	H	AS	A	1909	1963	H	11				+		+			

Taxon	Life Form	Origin	Mode of Intro.	Record in Hi. First	Record in Hi. Recent	Info. Source	No. Spms.	Island Hi	Ma	Mo	Ln	Kh	Oh	Ka	Ni	NW
Emilia fosbergii	H	-	A	1920	1984	H	92	+	+	+	+	+	+	+	+	+
Emilia sonchifolia var. *javanicus*	H	AS	A	1927	1976	H	4						+	+		
Emilia sonchifolia var. *sonchifolia*	H	AS	A	1931	1983	H	9						+	+		
Encelia farinosa	H	USM	A	1909	1983	H	10		+							
Erechtites hieracifolia	H	NW	A	1909	1984	H	15	+		+	+		+	+		
Erechtites valerianifolia	H	TAM	A	1916	1984	H	34	+	+	+	+		+			
Erigeron annuus	H	COS	A	1936	1936	H	1	+								
Erigeron bellioides	H	TAM	A	1984	1985	H	3						+			
Erigeron karvinskianus	H	TAM	O	1911	1983	H	17	+	+				+	+		
Filago gallica	H	EUR	A	1966	1967	H	3	+								
Flaveria trinervia	H	NAM	A	1937	1978	H	8						+			
Gaillardia pulchella	H	USM	O	1908	1984	H	12						+			+
Galinsoga parviflora	H	TAM	A	1908	1976	H	27	+					+	+		
Galinsoga quadriradiata	H	TAM	A	1964	1976	H	4	+		+			+			
Gnaphalium japonicum	H	AS	A	1909	1981	H	35	+	+	+	+		+			
Gnaphalium purpureum	H	NAM	A	1888	1984	P	69	+	+	+		+	+	+		
Helichrysum foetidum	H	NW	O	1949	1982	H	9	+								
Heterotheca grandiflora	H	USM	A	1909	1975	H	33	+	+		+	+	+	+		
Hypochoeris glabra	H	EUR	A	1909	1974	H	11	+			+		+			
Hypochoeris radicata	H	MED	A	1895	1975	P	39	+	+	+	+			+		
** *Kalimeris pinnatifida*	H	-	A	1983	1983	H	1							+		
Lactuca serriola	H	EUR	A	1928	1985	H	6	+	+							
Lapsana communis	H	EUR	A	1909	1984	H	24	+	+							
Madia sativa	H	SAM	A	1927	1937	H	4		+							

Appendix, continued.

Appendix, continued.

Taxon	Life Form	Origin	Mode of Intro.	Record in Hi. First	Record in Hi. Recent	Info. Source	No. Spms.	Hi	Ma	Mo	Ln	Kh	Oh	Ka	Ni	NW
Matricaria chamomilla	H	EAS	A	1909	1948	H	4	+								
Montanoa hibiscifolia	S	TAM	O	1919	1975	P	18	+	+		+		+	+		
Osteospermum calendulaceum	H	-	A	1983	1983	H	1	+								
Palafoxia callosa	H	USM	A	1971	1971	H	1			+						
Parthenium hysterophorus	H	USM	A	1960	1985	H	7	+	+	+				+		
Picris hieracioides	H	EAS	A	1930	1974	H	9	+			+					
Pluchea indica	S	AS	A	1909	1972	H	41		+				+	+	+	
Pluchea symphytifolia	S	TAM	A	1931	1974	H	44	+	+	+	+		+	+	+	
Pluchea x fosbergii	S	-	A	1934	1980	H	21		+	+			+	+		+
Reichardia picroides	H	OW	A	1909	1955	H	42	+					+			
Rudbeckia laciniata	H	NAM	O	1958	1958	H	1						+			
Senecio mikanioides	V	AF	O	1909	1975	H	14	+								
Senecio sylvaticus	H	EUR	A	1910	1985	H	24	+	+	+			+			
Senecio vulgaris	H	EUR	A	1932	1982	H	4	+	+							
Sigesbeckia orientalis	H	OW	A	1888	1976	P	16	+	+				+	+		
Solidago canadensis	H	USM	O	1926	1977	P	3	+					+	+		
Sonchus asper	H	COS	A	1975	1980	H	2	+								
Sonchus oleraceus	H	EUR	A	1817	1972	H	38	+	+		+		+	+	+	
Synedrella nodifolia	H	TAM	A	1926	1983	H	14	+	+	+		+	+	+		
Tagetes minuta	H	SAM	A	1932	1974	H	14	+					+			
Tanacetum vulgare	H	EUR	A	1959	1959	H	1						+			
Taraxacum laevigatum	H	-	A	1981	1981	H	1						+			
Taraxacum officinale	H	EUR	A	1909	1976	H	18	+	+		+		+			
Tithonia diversifolia	S	TAM	O	1917	1975	H	11	+	+				+	+		
Tragopogon porrifolius	H	EUR	O	1938	1982	H	3	+								
Tridax procumbens	H	TAM	A	1922	1976	H	26	+	+	+	+	+	+			

Taxon	Life Form	Origin	Mode of Intro.	Record in Hi. First	Record in Hi. Recent	Info. Source	No. Spms.	Island Hi	Ma	Mo	Ln	Kh	Oh	Ka	Ni	NW
Verbesina encelioides	H	NAM	A	1888	1976	H	26		+	+	+	+	+	+		
Vernonia cinerea	H	TOW	A	1870	1967	P	30	+			+		+	+	+	
Wedelia trilobata	H	TAM	O	1965	1975	P	13	+	+				+	+		
Xanthium strumarium	H	NAM	A	1840	1983	P	42	+	+	+	+	+	+	+	+	+
Youngia japonica	H	AS	A	1865	1976	H	44	+	+	+	+		+	+		
Zinnia palmeri	H	TAM	O	1980	1980	H	1						+			
Zinnia peruviana	H	TAM	O	1909	1983	H	25		+	+	+	+				
BALSAMINACEAE (Touch-me-not family)																
Impatiens oliveri	H	AF	O	1956	1968	H	2							+		
Impatiens wallerana	H	-	O	1939	1984	H	10	+					+			
BASELLACEAE (Basella family)																
Anredera cordifolia	V	SAM	O	1940	1973	H	8	+					+			
Basella alba	V	OW	C	1865	1945	H	3						+			
BATACEAE (Saltwort family)																
Batis maritima	H	TAM	A	1859	1985	P	47	+	+	+	+		+	+	+	
BEGONIACEAE (Begonia family)																
Begonia foliosa v miniata	H	SAM	O	1917	1937	H	3	+								
Begonia hirtella	H	SAM	O	1952	1984	H	6	+					+			
Begonia reniformis	H	-	O	1976	1985	H	2	+								
BIGNONIACEAE (Bignonia family)																
Spathodea campanulata	T	AF	O	1890	1986	H	22		+				+	+		
BIXACEAE (Arnotto family)																
Bixa orellana	T	TAM	C	1830	1986	P	29	+	+	+			+	+		
BORAGINACEAE (Borage family)																
Amsinckia intermedia	H	USM	A	1944	1944	H	1	+								
Bothriospermum tenellum	H	AS	A	1854	1975	H	7	+					+			

Appendix, continued.

Appendix, continued.

Taxon	Life Form	Origin	Mode of Intro.	Record in Hi. First	Record in Hi. Recent	Info. Source	No. Spms.	Island Hi	Ma	Mo	Ln	Kh	Oh	Ka	Ni	NW
Cordia subcordata	T	TOW	C	1000	1973	P	31	+	+	+	+		+	+	+	
Cynoglossum amabile	H	AS	O	1930	1976	H	11	+					+			
Heliotropium amplexicaule	H	SAM	O	1871	1983	P	20	+	+	+			+			
Heliotropium procumbens	H	TAM	A	1975	1983	H	10		+				+			+
Mytosis discolor	H	-	-	1981	1981	H	1	+								
Tournefortia argentea	T	AS	O	1888	1974	P	21		+		+		+	+	+	
BRASSICACEAE (Mustard family)																
Brassica campestris	H	EUR	A	1909	1980	H	13	+		+	+		+	+		
Brassica juncea	H	AS	C	1942	1970	H	4	+					+			
Brassica nigra	H	EUR	A	1888	1982	P	8	+	+				+			
Capsella rubella	H	EUR	A	1909	1985	H	20	+	+		+		+			
Cardamine flexuosa	H	EUR	A	1840	1984	H	43	+	+				+			
Cardamine hirsuta	H	EUR	C	1840	1979	H	3	+						+		
Coronopus didymus	H	OW	A	1840	1976	H	44	+	+	+	+		+	+	+	
Lepidium bonariense	H	SAM	A	1975	1975	H	1	+								
Lepidium densiflorum	H	NAM	A	1935	1935	H	1	+								
Lepidium hyssopifolium	H	-	A	1975	1975	H	1	+								
Lepidium oblongum	H	-	A	1914	1980	H	14			+	+	+	+		+	
Lepidium virginicum	H	NAM	A	1872	1985	H	37	+	+	+			+			+
Lobularia maritima	H	MED	O	1939	1980	H	15	+	+				+			+
Nasturtium microphyllum	H	EUR	C	1888	1932	P	7	+	+				+	+		
Nasturtium sarmentosum	H	P	C	1912	1984	H	20	+	+				+	+		
Raphanus raphanistrum	H	EUR	C	1963	1984	H	5	+					+	+		
Raphanus sativus	H	EUR	C	1888	1980	P	9	+		+		+	+	+		
Sisymbrium altissimum	H	EAS	A	1927	1981	H	7	+	+			+		+		
Sisymbrium irio	H	EUR	C	1963	1982	H	4	+								
Sisymbrium officinale	H	OW	A	1909	1984	H	25	+	+	+	+			+		

Taxon	Life Form	Origin	Mode of Intro.	Record in Hi. First	Record in Hi. Recent	Info. Source	No. Spms.	Hi	Ma	Mo	Ln	Kh	Oh	Ka	Ni	NW
BUDDLEIACEAE (Butterfly bush family)																
Buddleja asiatica	S	AS	O	1908	1984	H	58	+		+			+			
Buddleja madagascariensis	S	AF	O	1931	1984	H	7	+	+				+			
CACTACEAE (Cactus family)																
Cereus uruguayanus	U	SAM	O	1930	1959	H	4						+	+		
Harrisia martinii	U	SAM	O	1959	1959	P	0									
Hylocereus undatus	U	TAM	O	1830	1985	P	9						+			
Opuntia cochenillifera	U	TAM	O	1888	1985	P	1						+			
Opuntia ficus-indica	U	NW	O	1985	1985	H	4	+					+			
Opuntia vulgaris	U	SAM	O	1932	1944	P	1						+			
CAMPANULACEAE (Bellflower family)																
Hippobroma longiflora	H	TAM	A	1865	1983	H	24	+					+	+		
Lobelia erinus	H	AF	O	1931	1931	H	1	+								
Triodanis biflora	H	NW	A	1945	1945	H	2	+								
Wahlenbergia gracilis	H	COS	O	1928	1985	H	12	+		+						
CAPPARACEAE (Caper family)																
Cleome gynandra	H	AF	A	1857	1984	P	38	+	+	+	+		+	+		
CAPRIFOLIACEAE (Honeysuckle family)																
Lonicera japonica	V	AS	O	1882	1965	P	11	+					+			
Sambucus mexicana	T	TAM	O	1909	1974	P	12	+					+	+		
CARICACEAE (Papaya family)																
Carica papaya	T	SAM	C	1823	1984	P	13	+					+			
CARYOPHYLLACEAE (Pink family)																
Arenaria serpyllifolia	H	EUR	A	1935	1983	H	9	+	+							
Cerastium fontanum	H	EUR	A	1865	1985	P	32	+	+	+	+		+	+		
Dianthus armeria	H	EUR	A	1964	1982	H	2	+								

Appendix, continued.

Appendix, continued.

Taxon	Life Form	Origin	Mode of Intro.	Record in Hi. First	Record in Hi. Recent	Info. Source	No. Spms.	Hi	Ma	Mo	Ln	Kh	Oh	Ka	Ni	NW
Drymaria cordata	H	COS	A	1895	1984	H	35	+	+	+	+		+	+		
Petrorhagia velutina	H	MED	A	1938	1962	H	5	+								
Polycarpon tetraphyllum	H	EUR	A	1909	1982	H	22	+	+	+	+	+				
Sagina decumbens	H	USM	A	1865	1974	H	3	+		+						
Sagina japonica	H	AS	A	1985	1985	H	1						+			
Silene gallica	H	EUR	A	1865	1984	H	55	+	+	+	+		+	+		
Spergula arvensis	H	EUR	A	1888	1958	P	5	+					+			
Spergularia marina	H	NH	A	1909	1980	H	35		+	+			+	+		+
Stellaria media	H	EAS	A	1930	1982	H	17	+	+		+		+	+		
CASUARINACEAE (She-oak family)																
Casuarina equisetifolia	T	ASP	C	1895	1985	H	59	+	+		+	+	+	+		
Casuarina glauca	T	AUS	C	1916	1985	H	27			+	+	+	+	+		
CECROPIACEAE (Cecropia family)																
Cecropia obtusifolia	T	TAM	C	1926	1976	H	14	+					+	+		
CERATOPHYLLACEAE (Hornwort family)																
Ceratophyllum demersum	H	COS	O	1934	1985	H	2						+			
CHENOPODIACEAE (Goosefoot family)																
Atriplex eardleyae	H	AUS	A	1984	1984	H	1	+								
Atriplex lentiformis	S	USM	C	1915	1974	H	10			+						
Atriplex semibaccata	S	AUS	C	1895	1977	P	34	+	+	+	+	+	+	+		
Atriplex suberecta	H	AUS	C	1923	1984	H	41	+	+	+			+			+
Bassia hyssopifolia	H	EAS	A	1983	1983	H	1		+							
Chenopodium ambrosioides	H	NAM	A	1858	1975	H	35	+	+		+		+			
Chenopodium carinatum	H	AUS	A	1922	1984	H	24	+	+	+	+	+				
Chenopodium murale	H	EUR	A	1865	1984	H	43	+	+	+		+	+	+	+	+
** *Euchylaena tomentosa*	S	-	-	1985	1985	H	1		+							

Taxon	Life Form	Origin	Mode of Intro.	Record in Hi. First	Record in Hi. Recent	Info. Source	No. Spms.	Hi	Ma	Mo	Ln	Kh	Oh	Ka	Ni	NW
Salicornia virginica	H	NAM	A	1968	1968	H	1									+
Salsola kali	H	EUR	A	1947	1975	P	3	+								
CLUSIACEAE (Mangosteen family)																
* *Calophyllum antillanum*	T	TAM	C	1968	1968	H	1						+			
Calophyllum inophyllum	T	ASP	C	1000	1986	P	21	+	+	+			+	+		
Clusia rosea	T	TAM	O	1934	1985	H	11	+					+	+		
Hypericum canariense	S	-	A	1985	1985	H	1		+							
Hypericum gramineum	H	P	A	1911	1984	H	5	+								
Hypericum mutilum	H	NAM	A	1911	1983	P	8	+		+						
Hypericum parvulum	H	-	A	1932	1974	H	11	+								
Hypericum perforatum	S	EUR	A	1961	1964	H	2	+								
COMBRETACEAE (Indian almond family)																
Terminalia catappa	T	AS	O	1888	1986	P	16	+	+				+	+		+
CONVOLVULACEAE (Morning glory family)																
Convolvulus arvensis	V	EUR	A	1918	1974	H	5		+				+			
* *Dichondra repens*	H	P	O	1967	1967	H	1		+							
Ipomoea alba	V	COS	O	1819	1976	H	29	+	+	+	+		+	+		
Ipomoea aquatica	V	TOW	C	1888	1969	P	8		+				+			
Ipomoea batatas	V	TAM	C	1000	1984	P	106	+	+	+			+	+	+	
Ipomoea cairica																
* var. *hederacea*	V	COS	A	1840	1983	H	80	+		+	+	+	+	+	+	
Ipomoea hederifolia	V	NW	A	1915	1985	H	5	+			+					
Ipomoea obscura	V	AF	A	1926	1983	H	22	+					+	+		
Ipomoea ochracea	V	AF	O	1919	1985	H	8	+	+				+			
Ipomoea purpurea	V	TAM	O	1826	1961	P	2						+			+
Ipomoea triloba	V	TAM	A	1943	1985	H	18		+				+			+

Appendix, continued.

Appendix, continued.

Taxon	Life Form	Origin	Mode of Intro.	Record in Hi. First	Record in Hi. Recent	Info. Source	No. Spms.	Hi	Ma	Mo	Ln	Kh	Oh	Ka	Ni	NW
Ipomoea violacea	V	-	A	1910	1984	H	9	+	+	+			+			
* *Merremia peltata*	V	AS	O	1911	1911	H	2							+		
Merremia tuberosa	V	TAM	O	1932	1976	H	10	+					+	+		
Merremia umbellata	V	TAM	O	1911	1985	H	10	+					+			
Stictocardia tiliifolia	V	AS	O	1837	1955	H	24	+	+	+			+	+		
CORYNOCARPACEAE (Karakanut family)																
Corynocarpus laevigatus	T	P	C	1891	1976	P	27			+			+	+		
CRASSULACEAE (Orpine family)																
Kalanchoe pinnata	H	AF	O	1888	1984	P	18	+	+	+	+		+	+		
Kalanchoe tubiflora	H	AF	O	1930	1983	P	6	+	+		+		+			
CUCURBITACEAE (Gourd family)																
Coccinia grandis	V	OW	O	1969	1986	P	2						+			
Cucumis dipsaceus	V	AS	C	1903	1984	H	39	+	+		+		+	+	+	
Lagenaria siceraria	V	TOW	C	1000	1986	P	6	+					+			
Momordica charantia	V	OW	C	1909	1986	H	36	+	+	+	+		+	+	+	
Sechium edule	V	TAM	C	1965	1986	P	4						+	+		
CUSCUTACEAE (Dodder family)																
Cuscuta campestris	H	NAM	A	1955	1955	H	1	+								
ELAEAGNACEAE (Oleaster family)																
Elaeagnus umbellata	S	AS	O	1963	1984	H	5	+								
EUPHORBIACEAE (Spurge family)																
Acalypha indica	S	AS	O	1888	1888	P	0									
Aleurites moluccana	T	ASP	C	1000	1973	P	0									
Chamaesyce albomarginata	H	NAM	A	1947	1984	H	3						+	+		
Chamaesyce hirta	H	COS	A	1817	1983	H	53	+	+	+	+	+	+	+	+	+
Chamaesyce hypericifolia	H	TAM	A	1913	1983	H	16			+			+	+		+

Taxon	Life Form	Origin	Mode of Intro.	Record in Hi. First	Record in Hi. Recent	Info. Source	No. Spms.	Hi	Ma	Mo	Ln	Kh	Oh	Ka	Ni	NW
Chamaesyce hyssopifolia	H	NW	A	1959	1983	H	11	+					+	+		
Chamaesyce maculata	H	COS	A	1980	1980	H	1									+
Chamaesyce prostrata	H	NW	A	1909	1983	H	18						+			+
Chamaesyce thymifolia	H	OW	A	1888	1978	P	5	+					+			
* *Euphorbia cotinifolia*	S	TAM	O	1959	1985	H	3	+					+			
Euphorbia cyathophora	H	NW	O	1917	1980	H	13		+	+			+	+		+
Euphorbia graminea	H	-	A	1978	1978	H	1						+			
Euphorbia heterophylla	H	TAM	O	1895	1984	H	41	+	+		+		+	+	+	+
Euphorbia peplus	H	EAS	A	1855	1984	H	18	+	+				+			
Jatropha curcas	T	TAM	O	1937	1963	H	12	+	+				+		+	
Jatropha gossypiifolia	S	TAM	O	1950	1958	H	4						+			
Macaranga mappa	T	AS	C	1927	1971	H	10	+					+			
Phyllanthus debilis	H	AS	A	1865	1978	P	19	+					+	+		
Phyllanthus tenellus	H	AF	A	1963	1983	H	6						+	+		
Ricinis communis	S	AF	C	1819	1974	P	20	+	+		+	+	+	+	+	
FABACEAE (Pea family)																
Abrus precatorius	V	AS	O	1888	1986	P	8	+	+	+			+	+	+	
Acacia confusa	T	AS	C	1915	1986	P	29	+	+	+	+	+	+	+		
Acacia farnesiana	S	TAM	C	1865	1976	H	21		+	+		+	+			+
Acacia mearnsii	T	-	C	1911	1985	P	13	+	+		+		+	+		
Albizia lebbeck	T	AS	C	1865	1983	H	17									+
Alysicarpus vaginalis	H	AS	C	1909	1985	H	18		+				+	+		
Caesalpinia bonduc	S	COS	O	1794	1972	H	37	+	+				+	+	+	
Caesalpinia decapetala	S	AS	O	1888	1984	P	19	+		+			+	+		
Caesalpinia major	S	COS	A	1865	1960	H	24	+		+	+		+		+	
* *Caesalpinia pulcherrima*	S	TAM	O	1828	1985	P	20						+			

Appendix, continued.

Appendix, continued.

	Taxon	Life Form	Origin	Mode of Intro.	Record in Hi. First	Record in Hi. Recent	Info. Source	No. Spms.	Hi	Ma	Mo	Ln	Kh	Oh	Ka	Ni	NW
	Cajanus cajan	S	AS	C	1865	1986	H	15	+		+			+	+		
	Canavalia cathartica	V	OW	O	1908	1972	H	54		+	+			+	+		
	Canavalia sericea	V	P	O	1930	1985	H	12		+				+			
	Chamaecrista nictitans	S	COS	A	1870	1976	P	105	+	+	+	+	+	+	+	+	
	Clitoria ternata	S	COS	O	1865	1945	H	7						+			
	Crotalaria assamica	H	AS	C	1865	1974	H	13	+					+			
	Crotalaria berteroana	H	AS	C	1895	1985	H	3		+				+			
	Crotalaria incana	H	TAM	C	1895	1976	H	29	+	+		+		+	+		
	Crotalaria longirostrata	H	TAM	C	1865	1985	P	6		+				+			
	Crotalaria micans	H	SAM	O	1911	1983	H	9	+					+			
	Crotalaria pallida	H	OW	C	1965	1983	H	35	+	+		+		+			
	Crotalaria retusa	H	AS	A	1931	1975	H	3						+			
	Crotalaria spectabilis	S	AS	A	1865	1961	P	8			+			+			
*	*Crotalaria verrucosa*	H	AA	O	1926	1944	H	3			+			+			
*	*Cytisus monspessulanus*	S	MED	O	1956	1956	H	2							+		
	Cytisus palmensis	S	AF	O	1927	1985	H	9		+							
	Cytisus scoparius	S	EAS	O	1909	1909	H	3	+								
	Desmanthus virgatus	S	NAM	A	1900	1980	P	22	+	+			+	+	+	+	
	Desmodium cajanifolium	S	TAM	C	1933	1984	H	7	+					+			
	Desmodium heterocarpon	H	AS	C	1916	1917	H	3						+			
	Desmodium incanum	H	TAM	C	1916	1983	H	32	+	+				+	+		
	Desmodium intortum	H	TAM	C	1938	1982	H	3	+	+				+			
	Desmodium sandwicense	H	-	-	1847	1986	H	62	+	+		+	+	+	+	+	+
	Desmodium tortuosum	H	TAM	A	1913	1986	H	18	+	+			+	+	+		
	Desmodium triflorum	H	OW	A	1855	1985	P	34	+	+	+	+	+	+	+		
	Dioclea wilsonii	V	TAM	O	1825	1974	H	10	+						+		

Taxon	Life Form	Origin	Mode of Intro.	Record in Hi. First	Record in Hi. Recent	Info. Source	No. Spms.	Hi	Ma	Mo	Ln	Kh	Oh	Ka	Ni	NW
* *Flemingia strobilifera*	S	ASP	O	1910	1979	H	4		+				+			
Haematoxylum campechianum	T	TAM	C	1917	1962	H	7						+			
Indigofera spicata	H	TOW	C	1977	1977	H	1						+			
Indigofera suffruticosa	H	TAM	C	1836	1986	P	48	+	+		+	+	+	+		
Lablab purpureus	V	AS	O	1840	1971	P	25	+		+			+	+		
Lathyrus latifolius	H	EUR	O	1985	1985	H	1		+							
Lathyrus tingitanus	H	-	O	1985	1985	H	1		+							
Leucaena leucocephala	S	TAM	C	1837	1985	H	24			+	+	+	+			
Lotus subbiflorus	H	OW	O	1937	1983	H	4	+	+							
Macroptilium lathyroides	H	TAM	C	1865	1985	H	44	+	+		+	+	+	+	+	
* *Medicago intertexta*	H	EUR	C	1895	1979	P	6		+		+					
Medicago lupulina	H	OW	C	1888	1984	P	19	+	+							
Medicago polymorpha	H	OW	C	1882	1984	P	30	+	+		+		+	+		
Medicago sativa	H	EAS	C	1929	1982	H	11	+			+		+	+		+
Melilotus indica	H	EAS	C	1909	1985	H	8	+	+			+	+	+		+
Melilotus officinalis	H	EAS	C	1920	1984	P	10	+	+				+			+
Mimosa pudica	S	SAM	A	1855	1985	P	20	+			+		+	+		
Mucuna urens	V	SAM	O	1819	1961	H	8		+				+			
Paraserianthes falcataria	T	AS	C	1930	1976	H	11	+	+				+	+		
Parkinsonia aculeata	T	NW	O	1975	1976	H	2		+				+			
Pithecellobium dulce	T	TAM	O	1870	1985	P	20	+					+	+		
Prosopis pallida	T	SAM	C	1828	1974	P	29	+	+	+	+					
Rhynchosia minima	V	TAM	O	1965	1965	H	1							+		
Samanea saman	T	TAM	C	1919	1985	H	15	+					+			
Senna didymobotrya	S	AF	C	1930	1986	H	13		+		+		+	+		
Senna obtusifolia	S	-	A	1945	1960	H	3	+					+			

Appendix, continued.

Appendix, continued.

	Taxon	Life Form	Origin	Mode of Intro.	Record in Hi. First	Record in Hi. Recent	Info. Source	No. Spms.	Hi	Ma	Mo	Ln	Kh	Oh	Ka	Ni	NW
	Senna occidentalis	S	TAM	O	1836	1983	H	26	+	+				+	+	+	
	Senna pendula var. *advena*	S	-	O	1908	1979	H	20	+	+		+		+			
	Senna septemtrionalis	S	-	O	1913	1984	H	20	+	+		+		+			
	Senna surattensis	S	AA	O	1888	1985	P	24		+	+			+	+		
	Sesbania sesban	S	TOW	C	1909	1982	H	15						+	+		
	Spartium junceum	S	EUR	O	1956	1972	H	2							+		
*	*Tamarindus indica*	T	TOW	C	1797	1985	P	16	+	+	+			+	+		
	Tephrosia purpurea	H	ASP	C	1000	1984	P	66	+	+	+	+	+	+	+	+	
	Trifolium arvense	H	EUR	C	1932	1984	H	14	+	+							
	Trifolium dubium	H	EUR	C	1985	1985	H	1		+							
	Trifolium pratense	H	EUR	C	1932	1982	H	4	+	+							
	Trifolium procumbens	H	EUR	C	1966	1966	H	1	+								
	Trifolium repens	H	EAS	C	1909	1984	H	9	+	+					+		
	Ulex europaeus	S	EUR	C	1915	1983	H	10	+	+							
	Vicia angustifolia	H	-	-	1909	1985	H	13	+	+							
	FRANKENIACEAE (Frankenia family)																
	Frankenia salina	H	USM	A	1968	1968	H	1									+
	GENTIANACEAE (Gentian family)																
	Centaurium erythraea	H	OW	A	1909	1984	H	75	+	+	+	+		+			+
	GERANIACEAE (Geranium family)																
	Erodium cicutarium	H	MED	A	1888	1975	P	15	+	+	+	+		+			
	Geranium dissectum	H	AA	A	1910	1948	H	4	+	+							
	Geranium homeanum	H	-	A	1909	1984	H	25	+	+					+		
	Geranium pusillum	H	EUR	A	1909	1909	H	1	+								
	Geranium retrorsum	H	-	A	1888	1980	H	8	+	+		+					
	Pelargonium zonale	S	AF	O	1888	1940	P	1							+		

Taxon	Life Form	Origin	Mode of Intro.	Record in Hi. First	Record in Hi. Recent	Info. Source	No. Spms.	Hi	Ma	Mo	Ln	Kh	Oh	Ka	Ni	NW
HALORAGACEAE (Water milfoil family)																
Gonocarpus chinensis	H	-	-	1966	1983	H	2		+							
Myriophyllum aquaticum	H	SAM	O	1919	1982	H	7		+					+		
LAMIACEAE (Mint family)																
Hyptis pectinata	H	TAM	A	1930	1983	P	15		+	+				+		
Leonotis leonurus	H	AF	O	1929	1929	H	1						+			
Leonotis nepetifolia	H	AF	O	1938	1985	H	13		+					+	+	
Leonurus sibiricus	H	AS	O	1909	1971	H	19		+	+				+	+	
Marrubium vulgare	H	EAS	C	1913	1978	H	8		+			+				
Mentha pulegium	H	OW	C	1920	1948	H	2		+	+						
Mentha spicata	H	EUR	C	1936	1959	H	2		+							
Mentha x villosa	H	-	C	1953	1959	H	3		+					+		
Ocimum basilicum	H	TOW	C	1912	1984	H	11		+	+	+			+		+
Ocimum gratissimum	H	TOW	C	1924	1984	H	22		+	+	+			+		+
Plectranthus scutellarioides	H	-	O	1890	1972	P	13		+	+				+		
Prunella vulgaris	H	EAS	A	1909	1975	H	15		+	+	+					
Salvia coccinea	H	USM	O	1888	1975	P	31		+	+	+	+		+		
Salvia occidentalis	H	TAM	O	1835	1976	H	43		+	+		+		+	+	
Stachys arvensis	H	EUR	A	1865	1976	H	40		+	+	+	+		+	+	
LAURACEAE (Laurel family)																
Cinnamomum burmannii	T	AS	C	1934	1984	H	18							+		
Persea americana	T	TAM	C	1825	1954	P	7		+					+		
LINACEAE (Flax family)																
Linum bienne	H	-	-	1982	1982	H	1			+						
Linum trigynum	H	EUR	A	1912	1985	H	3							+		

Appendix, continued.

Appendix, continued.

Taxon	Life Form	Origin	Mode of Intro.	Record in Hi. First	Record in Hi. Recent	Info. Source	No. Spms.	Hi	Ma	Mo	Ln	Kh	Oh	Ka	Ni	NW
LYTHRACEAE (Loosestrife family)																
Ammannia auriculata	H	-	A	1936	1974	H	3		+				+			
Ammannia coccinea	H	-	-	1977	1977	H	1							+		
Cuphea carthagenensis	H	TAM	A	1855	1975	H	41	+	+	+			+	+		
Cuphea hyssopifolia	S	NAM	A	1909	1982	H	8	+					+			
Cuphea ignea	S	NAM	O	1915	1933	H	5	+					+			
Lythrum maritimum	S	SAM	A	1794	1975	H	58	+	+	+	+		+	+		
MALVACEAE (Mallow family)																
Abelmoschus moschatus	H	AS	C	1895	1941	H	2						+	+		
Abutilon grandifolium	S	COS	A	1903	1985	H	61	+	+	+	+	+	+	+	+	+
Gossypium barbardense	S	TAM	C	1817	1985	P	20	+			+		+		+	
Gossypium hirsutum	S	TAM	C	1914	1916	H	2				+		+			
Hibiscus macrophyllus	T	AS	O	1917	1936	P	7						+			
Malachra alceifolia	H	TAM	A	1940	1971	H	7						+	+		
Malva parviflora	H	EUR	A	1825	1982	P	32	+	+		+	+	+	+	+	+
Malvastrum coromandelianum	H	TAM	A	1840	1984	H	51	+	+	+	+	+	+	+	+	+
Modiola caroliniana	H	NW	A	1911	1985	H	30	+	+							
Sida acuta	S	TAM	A	1985	1985	H	10	+		+			+			
Sida rhombifolia	S	COS	A	1837	1985	H	43	+		+	+		+	+		
Sida spinosa	S	TAM	A	1888	1981	H	8	+					+			
Thespesia populnea	T	AS	C	1000	1984	P	26	+		+			+		+	
MELASTOMATACEAE (Melastome family)																
Arthrostema ciliatum	S	SAM	O	1939	1985	H	10	+					+			
Clidemia hirta	S	TAM	A	1941	1978	P	17						+			
Dissotis rotundifolia	H	TAM	O	1958	1975	H	4						+			
Heterocentron subtriplinervium	S	TAM	O	1926	1973	P	20	+	+				+	+		

Appendix, continued.

Taxon	Life Form	Origin	Mode of Intro.	Record in Hi. First	Record in Hi. Recent	Info. Source	No. Spms.	Hi	Ma	Mo	Ln	Kh	Oh	Ka	Ni	NW
Medinilla venosa	S	-	O	1957	1973	H	2		+							
Melastoma candidum	S	-	O	1916	1982	P	22	+						+		
Melastoma sanguineum	S	-	O	1957	1979	H	2	+					+			
Miconia calvescens	T	-	O	1985	1985	P	0									
Oxyspora paniculata	S	AS	O	1954	1974	H	5						+			
Pterolepis glomerata	S	TAM	A	1949	1983	H	14						+	+		
Tetrazygia bicolor	S	TAM	O	1932	1985	H	6	+					+			
Tibouchina herbacea	S	-	O	1979	1979	H	4	+		+						
Tibouchina urvilleana	S	SAM	O	1910	1983	P	29	+	+				+	+		
Trembleya phylogiformis	H	-	O	1947	1947	H	1							+		
MELIACEAE (Mahogany family)																
Melia azedarach	T	AS	O	1839	1985	P	30	+	+		+		+	+	+	
Toona ciliata	T	AA	C	1929	1983	H	19	+	+		+		+	+		
MOLLUGINACEAE (Carpetweed family)																
Mollugo cerviana	H	-	A	1975	1982	H	3	+								
MORACEAE (Mulberry family)																
Artocarpus altilis	T	P	C	1000	1976	H	16		+	+			+	+	+	
Broussonetia papyrifera	T	AS	C	1000	1984	P	71	+	+	+	+		+	+	+	
Ficus microcarpa	T	AS	O	1943	1976	H	4		+				+	+		
Morus alba	T	AS	C	1920	1984	H	13	+	+				+	+		
MYRICACEAE (Bayberry family)																
Myrica faya	T	EUR	C	1900	1976	P	38	+	+		+		+	+		
MYRSINACEAE (Myrsine family)																
Ardisia crenata	S	AS	O	1930	1985	H	16	+					+			
Ardisia elliptica	S	AS	O	1888	1986	P	30	+	+				+	+		

Appendix, continued.

Taxon	Life Form	Origin	Mode of Intro.	Record in Hi. First	Record in Hi. Recent	Info. Source	No. Spms.	Hi	Ma	Mo	Ln	Kh	Oh	Ka	Ni	NW
MYRTACEAE (Myrtle family)																
Eugenia uniflora	T	TAM	C	1888	1986	P	15	+		+			+	+		+
Leptospermum flavescens	S	AUS	C	1927	1983	H	8	+					+			
Leptospermum laevigatum	S	AUS	C	1930	1985	P	1				+					
Leptospermum scoparium	S	AUP	C	1927	1986	H	44				+		+	+		
Melaleuca quinquenervia	T	AUP	C	1920	1985	P	31	+	+	+			+	+		
Psidium cattleianum	T	TAM	C	1908	1983	H	49	+	+	+	+		+	+		
Psidium guajava	T	TAM	C	1825	1986	P	42	+	+	+	+		+	+		
Rhodomyrtus tomentosa	S	AS	O	1920	1986	P	19	+			+		+	+		
Syzygium cumini	T	AS	C	1888	1978	P	27	+	+	+			+	+		
Syzygium jambos	T	AS	C	1825	1972	P	16	+	+		+		+	+		
Syzygium malaccense	T	ASP	C	1000	1986	P	40	+	+	+			+	+		
NYCTAGINACEAE (Four-o'clock family)																
Boerhavia coccinea	H	-	A	1974	1985	H	10	+	+				+	+		
Mirabilis jalapa	H	TAM	O	1888	1985	P	18	+	+		+		+	+		+
OLEACEAE (Olive family)																
* *Chionanthus ligustrina*	T	TAM	A	1973	1973	P	1									
Fraxinus uhdei	T	NAM	C	1926	1985	H	14	+			+		+			
Olea europaea subsp. *africana*	T	AF	C	1932	1973	H	7	+					+			
Olea europaea subsp. *europaea*	T	MED	C	1910	1982	H	5	+					+			
ONAGRACEAE (Evening primrose family)																
Epilobium billardierianum	S	ASP	A	1909	1984	H	37	+	+							
Epilobium ciliatum	H	NAM	A	1929	1967	P	2	+								
Fuchsia magellanica	S	SAM	O	1922	1983	H	10	+	+				+			

Taxon	Life Form	Origin	Mode of Intro.	Record in Hi. First	Record in Hi. Recent	Info. Source	No. Spms.	Island Hi	Ma	Mo	Ln	Kh	Oh	Ka	Ni	NW
Fuchsia paniculata	S	TAM	O	1933	1980	H	3	+								
Ludwigia octovalvis	H	OW	A	1000	1975	P	40	+	+	+	+		+	+		
Ludwigia palustris	H	EUR	A	1934	1982	H	6	+					+	+		
Oenothera affinis	H	SAM	O	1915	1961	H	4	+								
Oenothera laciniata	H	NAM	O	1933	1982	H	14	+	+							+
Oenothera stricta	H	SAM	O	1911	1984	H	20	+	+							
OXALIDACEAE (Wood sorrel family)																
Oxalis corniculata	H	EUR	A	1000	1976	P	45	+	+	+	+	+	+	+	+	+
Oxalis corymbosa	H	SAM	O	1840	1985	P	15	+			+		+	+		
PAPAVERACEAE (Poppy family)																
Argemone mexicana	H	TAM	O	1934	1984	H	7		+				+	+		
Bocconia frutescens	S	TAM	O	1920	1983	H	16	+	+							
Hunnemannia fumariifolia	H	TAM	O	1920	1983	P	9		+				+			
PASSIFLORACEAE (Passion flower family)																
Passiflora x caerulea	V	SAM	O	1865	1960	H	6						+	+		
Passiflora edulis	V	SAM	C	1888	1985	P	35	+	+		+		+			
Passiflora foetida	V	SAM	O	1888	1984	P	59	+	+				+	+	+	
Passiflora laurifolia	V	TAM	C	1888	1985	P	14	+		+			+	+		
Passiflora ligularis	V	TAM	C	1910	1977	H	21	+			+		+	+		
Passiflora mollissima	V	TAM	O	1926	1983	H	14	+						+		
Passiflora pulchella	V	TAM	O	1924	1985	H	6	+					+			
Passiflora quadrangularis	V	TAM	C	1888	1945	P	4	+					+			
Passiflora suberosa	V	TAM	O	1888	1986	P	39	+	+				+			
Passiflora subpeltata	V	SAM	C	1914	1986	H	24		+				+	+		
PHYTOLACCACEAE (Pokeweed family)																
Phytolacca octandra	H	TAM	C	1910	1986	H	52			+	+		+	+		

Appendix, continued.

Appendix, continued.

Taxon	Life Form	Origin	Mode of Intro.	Record in Hi. First	Record in Hi. Recent	Info. Source	No. Spms.	Hi	Ma	Mo	Ln	Kh	Oh	Ka	Ni	NW
Rivina humilis	H	TAM	C	1888	1985	P	15	+					+	+		
PIPERACEAE (Pepper family)																
Peperomia pellucida	H	TAM	O	1918	1986	P	3						+			
Piper methysticum	S	P	C	1000	1984	P	48	+	+	+			+	+		
PITTOSPORACEAE (Pittosporum family)																
Pittosporum undulatum		T	AUS	O	1875	1982	P	11		+			+			
Pittosporum viridiflorum	T	AF	C	1954	1978	H	5	+			+		+			
PLANTAGINACEAE (Plantain family)																
Plantago aristata	H	NAM	A	1941	1985	H	3	+								
Plantago australis	H	-	A	1908	1985	H	24	+	+							
Plantago debilis	H	-	A	1934	1985	H	9						+			
Plantago lanceolata	H	EAS	A	1895	1985	H	58	+	+	+	+		+	+		+
Plantago major	H	EAS	A	1850	1985	H	45	+	+	+	+		+	+		+
POLEMONIACEAE (Phlox family)																
Gilia capitata	H	NAM	O	1929	1982	H	5		+							
POLYGALACEAE (Milkwort family)																
Polygala paniculata	H	SAM	A	1974	1984	H	6	+	+							
POLYGONACEAE (Buckwheat family)																
Antigonon leptopus	V	NAM	O	1909	1967	H	14				+		+			
Emex spinosa	H	MED	A	1928	1983	H	17	+	+	+			+			
* *Fagopyrum esculentum*	H	EAS	C	1925	1925	H	2				+					
Muehlenbeckia axillaris	S	P	A	1937	1963	H	3	+								
Polygonum argyrocoleon	H	AS	A	1937	1937	H	1		+							
Polygonum aviculare	H	EUR	A	1932	1982	H	2	+	+							
Polygonum capitatum	H	AS	A	1960	1983	H	15	+	+				+			
Polygonum chinense	H	AS	C	1966	1983	P	3	+								

Taxon	Life Form	Origin	Mode of Intro.	Record in Hi. First	Record in Hi. Recent	Info. Source	No. Spms.	Island Hi	Ma	Mo	Ln	Kh	Oh	Ka	Ni	NW
Polygonum convolvulus	H	COS	A	1981	1981	H	1		+							
Polygonum glabrum	H	AS	A	1819	1981	H	43	+	+	+			+	+		
Polygonum punctatum	H	NW	A	1909	1983	H	10	+			+					
Rumex acetosella	H	EAS	A	1895	1975	H	28	+	+					+		
Rumex brownei	H	AUS	A	1958	1978	H	2	+								
Rumex crispus	H	EAS	A	1910	1978	H	16	+	+		+		+	+		
Rumex obtusifolius	H	OW	A	1917	1984	H	16	+						+		
PORTULACACEAE (Purslane family)																
Portulaca oleracea	H	COS	A	1817	1985	H	104	+	+	+	+		+	+	+	+
Talinum paniculatum	H	NW	O	1931	1984	H	10	+					+			
Talinum triangulare	H	TAM	O	1979	1979	H	1	+								
PRIMULACEAE (Primrose family)																
Anagallis arvensis	H	EUR	A	1895	1985	P	61	+	+	+	+	+	+	+		+
PROTEACEAE (Protea family)																
Grevillea banksii	T	AUS	C	1909	1982	H	16	+	+				+	+	+	
Grevillea robusta	T	AUS	C	1909	1984	H	16	+					+	+	+	
RANUNCULACEAE (Buttercup family)																
Anenome hupehensis	H	AS	O	1942	1983	H	8	+								
Ranunculus muricatus	H	EUR	A	1939	1943	H	2	+								
Ranunculus parviflorus	H	MED	A	1911	1932	H	3	+								
Ranunculus plebius	H	AUS	A	1949	1970	P	0	+								
Ranunculus repens	H	EUR	A	1922	1951	P	4	+								
RHIZOPHORACEAE (Mangrove family)																
Bruguiera gymnorrhiza	T	AS	C	1922	1946	P	10						+			
Rhizophora mangle	T	TAM	C	1902	1984	P	41	+		+	+		+			
ROSACEAE (Rose family)																
Eriobotrya japonica	T	AS	C	1915	1967	H	6	+					+			

Appendix, continued.

Appendix, continued.

Taxon	Life Form	Origin	Mode of Intro.	Record in Hi. First	Record in Hi. Recent	Info. Source	No. Spms.	Island Hi	Ma	Mo	Ln	Kh	Oh	Ka	Ni	NW
Fragaria vesca	H	EUR	C	1829	1983	P	20	+						+		
* *Prunus x cerasifera*	T	AF	C	1933	1975	P	6	+	+					+		
Pyracantha angustifolia	S	AS	O	1937	1983	H	7	+					+	+		
* *Pyracantha crenati-serrata*	S	AS	O	1945	1984	H	4	+			+		+			
Rosa sp.	S	-	O	1910	1983	H	24	+	+	+	+					
Rubus argutus	S	USM	C	1894	1983	P	53	+	+				+	+		
Rubus discolor	S	-	C	1983	1983	H	2		+				+			
Rubus ellipticus	S	AS	C	1961	1984	H	5	+								
Rubus niveus	S	-	C	1965	1965	H	1	+								
Rubus rosifolius	S	AS	C	1880	1967	P	44	+	+				+			
Rubus sieboldii	S	ASP	C	1971	1976	H	3							+		
RUBIACEAE (Coffee family)																
Coffea arabica	S	AF	C	1823	1985	P	27	+	+	+	+		+	+		
Galium divaricatum	H	-	A	1981	1981	H	1		+							
Hedyotis corymbosa	H	SAM	A	1965	1986	H	8	+	+				+			
Mitracarpus hirtus	H	TAM	A	1982	1982	H	1	+								
Morinda citrifolia	S	ASP	C	1000	1985	P	60	+	+	+	+		+	+	+	
Paederia scandens	V	AS	A	1854	1985	P	39	+					+	+		
Richardia brasiliensis	H	TAM	A	1855	1983	P	36	+	+		+			+		
Sherardia arvensis	H	EUR	A	1909	1984	H	19	+	+							
Spermacoce assurgens	H	AS	A	1929	1986	H	23	+					+	+		
Spermacoce mauritiana	H	-	A	1973	1983	H	4	+								
SAPINDACEAE (Soapberry family)																
Cardiospermum grandiflorum	V	COS	O	1951	1985	H	5						+			
Cardiospermum halicacabum	V	COS	O	1817	1984	H	27	+	+	+			+	+	+	
SCROPHULARIACEAE (Figwort family)																
Antirrhinum orontium	H	MED	O	1950	1960	H	4						+			

Taxon	Life Form	Origin	Mode of Intro.	Record in Hi. First	Record in Hi. Recent	Info. Source	No. Spms.	Hi	Ma	Mo	Ln	Kh	Oh	Ka	Ni	NW
Castilleja arvensis	H	TAM	A	1968	1983	H	6	+	+							
Dopatrium junceum	H	-	A	1977	1977	H	1							+		
Linaria canadensis var. *texana*	H	NW	A	1904	1979	P	15	+					+			
Lindernia crustacea	H	TOW	A	1954	1984	P	7	+								
Lophospermum erubescens	H	NAM	O	1917	1982	H	14	+					+			
Maurandya antirrhiniflora	H	NAM	O	1943	1943	H	1						+			
Orthocarpus purpurascens	H	USM	A	1945	1945	H	1	+								
Parentucellia viscosa	H	EUR	-	1971	1971	H	1	+								
Torenia asiatica	H	AS	O	1975	1983	H	4	+								
Verbascum blattaria	H	EAS	A	1955	1955	H	1	+								
Verbascum thapsus	H	EAS	A	1932	1975	H	14	+								
Verbascum virgatum	H	EAS	A	1943	1982	H	2	+								
Veronica arvensis	H	EUR	A	1943	1985	H	10	+	+	+			+			
Veronica plebeia	H	AUS	A	1911	1974	H	31	+								
Veronica serpyllifolia	H	NH	A	1926	1983	H	20	+						+		
SOLANACEAE (Nightshade family)																
Brugmansia candida	S	TAM	O	1890	1976	H	23	+	+				+	+		
Capsicum annuum	H	TAM	C	1815	1985	P	16		+				+	+		
Cestrum diurnum	S	TAM	O	1888	1984	P	20			+			+	+		
Cestrum nocturnum	S	TAM	O	1927	1985	H	18	+			+		+	+		
Datura stramonium	H	AS	A	1837	1985	H	20	+	+	+			+			
Lycopersicon esculentum	H	SAM	C	1888	1983	P	1						+			
Lycopersicon pimpinellifolium	H	SAM	C	1895	1986	H	40	+	+		+		+	+	+	
Nicandra physalodes	H	SAM	O	1850	1985	P	24	+	+	+			+	+		

Appendix, continued.

Appendix, continued.

Taxon	Life Form	Origin	Mode of Intro.	Record in Hi. First	Record in Hi. Recent	Info. Source	No. Spms.	Hi	Ma	Mo	Ln	Kh	Oh	Ka	Ni	NW
Nicotiana glauca	S	SAM	A	1865	1982	H	31		+		+	+	+			
Nicotiana tabacum	H	SAM	C	1812	1986	P	42	+	+	+	+		+	+	+	+
Physalis angulata	H	TAM	-	1976	1983	H	2							+		
Physalis peruviana	H	SAM	O	1825	1985	P	61	+	+	+	+		+	+		
* *Salpichroa rhomboidea*	H	SAM	O	1931	1985	H	4						+	+		
Solanum aviculare	S	OW	A	1955	1964	H	5				+					
Solanum capsicoides	S	TAM	A	1837	1985	H	34	+	+	+	+		+	+		
Solanum elaeagnifolium	H	USM	A	1958	1979	P	3			+			+			
Solanum linnaeanum	H	MED	A	1895	1983	H	33	+	+	+	+	+	+	+		
Solanum mauritianum	H	SAM	O	1909	1985	H	29						+			
Solanum pseudocapsicum	S	OW	O	1888	1986	P	24	+	+	+			+			
Solanum robustum	S	-	A	1977	1977	H	1		+							
Solanum seaforthianum	V	TAM	O	1916	1985	H	14		+	+			+	+		
Solanum torvum	S	-	-	1954	1965	H	3		+							
Solanum tuberosum	B	SAM	C	1811	1919	P	2		+							
Solanum viride	H	-	-	1937	1937	H	1							+		
* *Streptosolen jamesoni*	H	TAM	O	1932	1974	H	7	+	+				+	+		
STERCULIACEAE (Cacao family)																
Melochia umbellata	T	AS	C	1929	1984	H	15	+	+		+		+			
THEACEAE (Tea family)																
Camellia sinensis	S	AS	O	1932	1977	H	5						+			
TILIACEAE (Linden family)																
Heliocarpus popayanensis	T	TAM	O	1941	1968	P	6	+					+			
Triumfetta rhomboidea	H	AF	A	1888	1976	H	27	+		+			+			
Triumfetta semitriloba	H	COS	A	1910	1983	H	20	+	+				+	+		

Taxon	Life Form	Origin	Mode of Intro.	Record in Hi. First	Record in Hi. Recent	Info. Source	No. Spms.	Island Hi	Ma	Mo	Ln	Kh	Oh	Ka	Ni	NW
TROPAEOLACEAE (Nasturtium family)																
Tropaeolum majus	V	SAM	O	1840	1975	P	4	+	+	+						
TURNERACEAE (Turnera family)																
Turnera ulmifolia	S	TAM	C	1888	1984	P	10			+			+	+		
ULMACEAE (Elm family)																
Trema orientalis	T	AS	C	1870	1986	H	50	+	+	+			+	+		
URTICACEAE (Nettle family)																
Boehmeria nivea	H	AS	O	1928	1976	H	12	+	+				+			
Laportea interrupta	H	ASP	A	1819	1959	P	1	+					+			
Pilea microphylla	H	TAM	O	1926	1985	H	2						+			
Urtica urens	H	OW	A	1909	1975	H	8	+								
VALERIANACEAE (Valerian family)																
Centranthus ruber	H	EAS	O	1965	1982	P	1		+							
VERBENACEAE (Verbena family)																
Citharexylum caudatum	T	TAM	O	1929	1983	P	17						+			
Citharexylum spinosum	T	TAM	O	1937	1985	H	8						+			
Clerodendrum philippinum	S	AS	O	1839	1978	H	17	+	+	+	+		+	+		
Lantana camara	S	TAM	O	1858	1976	P	50	+			+		+	+	+	
Stachytarpheta dichotoma	H	TAM	A	1910	1978	H	46	+	+		+		+	+		
Stachytarpheta jamaicensis	H	TAM	A	1895	1983	H	23	+	+		+		+			+
Stachytarpheta mutabilis	H	TAM	O	1888	1980	P	7	+					+	+		
Stachytarpheta urticifolia	H	COS	A	1895	1973	H	19	+	+	+			+	+		
Stachytarpheta x intercedens	H	AS	A	1939	1940	H	2						+			
Stachytarpheta x trimeni	H	HOR	A	1928	1951	H	2							+		
Verbena bonariensis	H	SAM	A	1914	1918	H	4				+					

Appendix, continued.

Appendix, continued.

Taxon	Life Form	Origin	Mode of Intro.	Record in Hi. First	Record in Hi. Recent	Info. Source	No. Spms.	Hi	Ma	Mo	Ln	Kh	Oh	Ka	Ni	NW
Verbena litoralis	H	TAM	A	1835	1975	H	61	+	+	+	+		+	+	+	
Verbena rigida	H	SAM	O	1932	1938	H	3	+								
VITACEAE (Grape family)																
Cissus nodosa	V	AS	O	1975	1975	H	1	+								
ZYGOPHYLLACEAE (Creosote bush family)																
Tribulus terrestris	V	OW	A	1947	1984	H	5	+						+		

* Taxon is not discussed in Wagner *et al.* 1990.
** Taxon is not discussed in Wagner *et al.* 1990 or St. John 1973.
\# *Tricachne insularis* = *Digitaria insularis*; taxon is retained to avoid loss of data.
\#\# *Bidens lanceolata* = *Coreopsis lanceolata*; taxon is retained to avoid loss of data.

Literature Cited

Baker, H.G. 1965. Characteristics and modes of origin of weeds. In *The genetics of colonizing species*, ed. H.G. Baker and G.L. Stebbins, 147-172. New York: Acad. Pr.

Beaglehole, J.C. 1967. *The journals of Captain James Cook. The voyage of Resolution and Discovery 1776-1780*. Vol. 3. Cambridge, England: Cambridge Univ. Pr.

Biggs, B. 1979. Pro-Polynesian word list II. *Working papers in anthropology, archaeology, linguistics and Maori studies*, No. 53. Anthropology Dep., Univ. Auckland, New Zealand.

Conner, E.F., and D. Simberloff. 1978. Species number and compositional similarity of the Galapagos flora and avifauna. *Ecol. Monogr.* 48:219-248.

Davis, P.H., and V.H. Heywood. 1965. *Principles of angiosperm taxonomy*. Princeton, New Jersey: Van Nostrand.

Degener, O. 1962. *Flora Hawaiiensis: Digitaria pruriens*. Privately published.

Frenkel, R.E. 1984. Weeds and aliens: a conceptual and terminological jungle. Ann. Meet., Assn. Amer. Geographers, Washington D.C. April 22-25. Mimeo.

Gutmanis, J. 1979. *Kahuna la'au lapa'au*. Honolulu: Island Heritage.

Handy, E.S.C., and E.G. Handy. 1972. *Native planters of old Hawaii: their life, lore, and environment*. B.P. Bishop Museum Bull. 233. Honolulu: Bishop Mus. Pr.

Harlan, J.R., and J.M.J. de Wet. 1965. Some thoughts about weeds. *Econ. Bot.* 19:16-24.

Hart, R. 1976. An index for comparing weediness in plants. *Taxon* 25:245-247.

Hastings, J.R., R.M. Turner, and D.K. Warren. 1972. *An atlas of some plant distributions in the Sonoran Desert*. Meteorology and Climatology Arid Regions Tech. Rep. 21. Tucson, Ariz.: Univ. Arizona Inst. Atmospheric Physics.

Healy, A.J. 1947. Evolution of the alien flora of New Zealand. *Proc. N.Z. Sci. Congress*, 160-162.

Hillebrand, W.F. [1888] 1965. *Flora of the Hawaiian Islands*. New York and London: Hafner.

Holland, P.G. 1975. The use of herbarium records in biogeography. *Professional Geographer* 27:475-479.

Hooker, W.J., and G.A. Arnott. [1830] 1965. *The botany of Captain Beechey's voyage: Sandwich Islands*. Weinheim: Historiae Naturalis Classica 39:78-111.

Kirch, P. 1985. *Feather gods and fishhooks*. Honolulu: Univ. Hawaii Pr.

Kornas, J. 1971. Changements récents de la flore polonaise. *Biol. Conserv.* 4:43-47.

Krauss, B.H., comp. 1974. *Ethnobotany of Hawai'i.* Botany Dep. Honolulu: Univ. Hawaii.

Mann, H. 1866. Flora of the Hawaiian Islands. *Communications Essex Institute* 5:113-248, 6:105-111.

Mann, H. 1868. Enumeration of Hawaiian plants. *Proc. Amer. Acad. Arts and Sci.* 7:143-235.

Merrill, E.D. 1946. Man's influence on the vegetation of Polynesia, with special reference to introduced species. *Chronica Botanica* 10:334-45.

Merrill, E.D. 1954. The botany of Cook's voyages. *Chronica Botanica*. 14:164-384.

Nagata, K.M. 1985. Early plant introductions to Hawai'i. *Hawn. J. History* 19:35-61.

Obata, J.K. 1985. Another noxious melastome? *Oxyspora paniculata*. *Newsletter, Hawn. Bot. Soc.* 24:25-26.

St. John, H. 1973. *List and summary of the flowering plants in the Hawaiian Islands*. Mem. 1. Lawai, Kauai, Hawaii: Pac. Trop. Botan. Gdn.

St. John, H. 1978. The first collection of Hawaiian plants by David Nelson in 1779. Hawaiian Plant Studies 55. *Pac. Sci.* 32:315-324.

Smith, C.W. 1985. Impact of alien plants on Hawai'i's native biota. In *Hawaii's terrestrial ecosystems: preservation and management*, ed. C.P. Stone and J.M. Scott, 180-250. Univ. Hawaii Coop. Natl. Park Resour. Stud. Unit. Honolulu: Univ. Hawaii Pr.

Smith, J.B.M. 1982. *Establishment of exotic woody plants along roadsides in New England (New South Wales)*. Research Ser. Applied Geogr., Geography Dep. Armidale, Australia: Univ. New England.

Stearns, W.T. 1951. Mapping the distribution of species. In *The study of the distribution of British plants*, ed. J.E. Lousey, 48-64. Oxford: Botan. Soc. British Isles.

Strother, J.L., and A.R. Smith. 1970. Chorology, collection dates and taxonomic responsibility. *Taxon* 19:871-874.

Wace, N.M. 1979. Assessment of dispersal of plant species -- The car-borne flora in Canberra. *Proc. Ecol. Soc. Australia* 10:167-186.

Wagner, W.L., D.R. Herbst, and S.H. Sohmer. 1990. *Manual of the flowering plants of Hawai'i*. Bishop Mus. Spec. Pub. 83. Honolulu: Univ. Hawaii and Bishop Mus. Pr.

Wester, L., and J.O. Juvik. 1983. Roadside plant communities on Mauna Loa, Hawai'i. *J. Biogeography* 10:307-316.

DISTRIBUTION OF SIX ALIEN PLANT SPECIES IN UPLAND HABITATS ON THE ISLAND OF HAWAI'I

James D. Jacobi and Fredrick R. Warshauer

ABSTRACT

The 1976-1981 distribution, elevation, and median annual rainfall of six alien plant species in upland habitats on the island of Hawai'i are described based on data collected at 7,864 sampling points (stations) along 117 transects within a 1,930-mi^2 (5,000-km^2) study area. Focal species were Malabar melastome (*Melastoma candidum*), banana poka (*Passiflora mollissima*), fountain grass (*Pennisetum setaceum*), strawberry guava (*Psidium cattleianum*), yellow Himalayan raspberry (*Rubus ellipticus*), and German ivy (*Senecio mikanioides*). Three species, strawberry guava, fountain grass, and banana poka, were widely distributed throughout the study area, while the remaining three species were found on less than 2% of the stations sampled. Relatively xeric habitats (median annual rainfall 49 in. or <1,250 mm) and low-elevation areas (1,640-4,265 ft or 500-1,300 m) were found to be consistently invaded by three or more of the species. Forty-six (72%) of the 64 vegetation types sampled contained at least one of the six alien plant species. Portions of the two largest vegetation units, 'ōhi'a *(Metrosideros polymorpha)* rain forest, and 'ōhi'a-koa *(Acacia koa)* rain forest, were colonized by five of the six species analyzed. Elevation and median annual rainfall ranges were used to predict the potential geographical distribution of each species in the study area, assuming no restriction in plant invasion over time. Four of the six species (banana poka, German ivy, strawberry guava, and yellow Himalayan raspberry) showed great potential for expansion of their ranges if they continue to invade areas with environmental conditions similar to those where they were found during this survey.

INTRODUCTION

Nearly 5,000 taxa of flowering plants have been purposely or accidentally introduced to Hawai'i during the past 200 years (St. John 1973). Fortunately, of this large number of alien species, fewer than 15% are established outside cultivation, and less than 2% are currently considered serious threats to native ecosystems (Smith 1985; Wester, this volume). However, the relatively small number of recognized noxious plants is expected to increase with time, both as new species are brought into the islands and as species that are already established expand in distribution and abundance.

The degree of threat posed by alien plants depends not only on site-related impacts where they have become established, but also on their potential for dispersal into new areas. For this reason, it is important to recognize the types of habitats the species are potentially able to colonize, and the possible extent of the distribution of the species if they are not effectively controlled.

In this paper we describe the current distributions and some of the habitat characteristics of six alien plants that are potential threats to the natural integrity of native ecosystems on the island of Hawai'i. This information is then used to develop a generalized model describing the potential ranges of these species in upland habitats, based on their current distribution in relation to two habitat variables: median annual rainfall and elevation. The prime objective of this analysis is to present a regional overview of the current and potential distribution of the six species. Of particular interest is the identification of general geographic and habitat zones that could be invaded if the species are allowed to continue to spread unabated.

The species discussed herein include one grass (fountain grass, *Pennisetum setaceum*), two vines (banana poka and German ivy, *Passiflora mollissima, Senecio mikanioides*), two shrubs (Malabar melastome and yellow Himalayan raspberry, *Melastoma candidum*, *Rubus ellipticus*), and a small tree (strawberry guava, *Psidium cattleianum*) (Table 1). Besides being recognized as threatening to the native ecosystems they presently occupy, these species were chosen as representatives of different life forms and different times of invasion into habitats they potentially affect.

Historical Background of the Six Focal Species

Malabar melastome is a moderate-sized shrub native to Southeast Asia. It was introduced as an ornamental to the island of Kaua'i from Florida in 1916 (Haselwood and Motter 1983). This species is currently considered a problem weed on the islands of Kaua'i and Hawai'i where it grows into dense thickets up to 6.5 ft (2 m) tall, displacing native vegetation. Dispersal is primarily by birds and other animals that feed on the plant's berry-like fruits. All species in the genus *Melastoma* are listed as noxious by the Hawaii Department of Agriculture (1978).

Banana poka is native to the Andes region of Peru, Bolivia, and Venezuela (Escobar 1980). It was brought to Hawai'i for its edible fruit in the early 1900s and was first collected in the wild near Pu'uwa'awa'a Ranch in North Kona in 1921 (LaRosa 1984). This species is now established on the islands of Maui, Hawai'i, and Kaua'i, with extremely heavy infestations on the latter two islands (Warshauer *et al.* 1983).

Table 1. Pertinent information on six alien plant species discussed in text. Data on origin, establishment date, and dispersal from Haselwood and Motter (1983); nomenclature follows Wagner *et al.* (1990).

Scientific Name [Common Name]	Family	Life[1] Form	Origin	First Collected	Dispersal (agent)	Considered[2] Noxious By
Melastoma candidum [Malabar melastome]	Melastomataceae	Ch	SE Asia	1916	Seed; (bird; mamm.)	A
Passiflora mollissima [Banana poka]	Passifloraceae	Vi	Tropical America	ca. 1900	Seed; (bird, mamm.)	C
Pennisetum setaceum [Fountaingrass]	Poaceae	HC	North Africa	1914	Seed; (wind; anim., vehic.)	A,C
Psidium cattleianum [Strawberry guava]	Myrtaceae	Ph	Brazil	1825	Seed; (bird, mamm.)	C
Rubus ellipticus [Yellow Himalayan raspberry]	Rosaceae	Ch	Himalayas	1961	Seed; (bird, mamm.)	B,C
Senecio mikanioides [German ivy]	Asteraceae	Vi	South Africa	1909	Seed; veg. (wind, anim.)	C

[1]Ch = Chamaephyte; HC = Hemicryptophyte; Ph = Phanerophyte; Vi = Vine.

[2]A = Hawaii Department of Agriculture (1978).
B = Proposed for inclusion on Hawai'i Dept. of Agriculture Noxious Weed List (M. Isherwood, pers. comm.).
C = National Park Service (1986).

Banana poka is a perennial vine that can climb up to 80 ft (25 m) into the tree canopy or densely cover the forest understory vegetation. The seeds are contained in a fleshy fruit that is eaten by many species of animals. Primary dispersal agents are believed to be several species of birds, rodents, feral pigs (*Sus scrofa*), and humans (Warshauer *et al.* 1983). Although banana poka is listed as noxious only by the National Park Service (1986), a cooperative research program is currently being conducted by the state of Hawai'i (Department of Land and Natural Resources), U.S. Forest Service, and the National Park Service to identify biological control agents for use against this species in Hawai'i (Markin, this volume).

Fountain grass is a perennial bunchgrass native to North Africa. It has been introduced to many areas around the world as an ornamental and was first collected by G. Munro on the island of Lāna'i in 1914. The species has since become established on most of the major Hawaiian Islands (Haselwood and Motter 1983). Although it is a dominant grass throughout most of the dry ranchland in the North Kona and South Kohala districts on the island of Hawai'i, it is not considered a particularly good pasture species (Whitney *et al.* 1939). The seeds of fountain grass are easily dispersed short distances by wind, but long-distance dispersal is probably the result of propagules being carried into new areas on vehicles or livestock from an infested area. Once established, this species can significantly increase the potential for fire and rapidly colonize an area that has been burned (Smith 1985). Fountain grass is recognized as a noxious weed by both the Hawaii Department of Agriculture (1978) and the National Park Service (1986).

A native of Brazil, strawberry guava was brought to Hawai'i as a fruit tree around 1825 (St. John 1973). This species is currently found on the six largest Hawaiian Islands. It is a small tree that can form extremely dense thickets up to 30 ft (10 m) tall. The seeds are widely dispersed by many species of birds and other animals, such as feral pigs, that feed on the fleshy fruit. Fruits of strawberry guava and other *Psidium* species are regularly picked by local residents to be either eaten directly or made into juice, jelly, or jam. Despite its local importance as a fruit tree, strawberry guava is recognized as a serious threat to native forest ecosystems below 4,900 ft (1,500 m) elevation (Smith 1985). The National Park Service (1986) considers it a species that needs to be controlled or eliminated within Park boundaries.

Yellow Himalayan raspberry was first collected in the town of Volcano on the island of Hawai'i in 1961, where it had been introduced for its edible fruit (Degener and Degener 1968). Native to tropical and subtropical regions of the Himalaya Mountains, this extremely thorny raspberry forms nearly impenetrable thickets where it has become established. Degener and Degener (1968) recognized the potential threat this species could pose to native ecosystems and recommended that it be controlled immediately to stop its spread. As with other *Rubus* species, the seeds of yellow Himalayan raspberry are readily dispersed by birds. Currently, the species is considered a noxious species by the National Park Service (1986) and has also been proposed for inclusion on the State of Hawai'i's noxious weed list (M. Isherwood, pers. comm.).

The history of introduction and current distribution of German ivy in Hawai'i is not well documented. Haselwood and Motter (1983) reported that this vine is native to South Africa and probably was introduced as an ornamental to the Kona side of the island of Hawai'i around 1909. German ivy apparently has become naturalized only on that island, with most populations concentrated in the North Kona and Mauna Kea-Mauna Loa saddle areas. This species is dispersed by either wind-blown seeds, or vegetatively as cuttings (Haselwood and Motter 1983). The National Park Service (1986) recognized German ivy as a species that may adversely impact native ecosystems in Hawaii Volcanoes National Park if it were to become established there.

METHODS

Field Sampling

Field work for this study was conducted between 1976 and 1981 during the U.S. Fish and Wildlife Service Hawai'i Forest Bird Survey (Scott *et al.* 1981). The Survey research areas covered approximately 47% (1,960 mi^2 or 4,952 km^2) of the island of Hawai'i, focusing primarily on forest bird habitats above 1,640 ft (500 m) elevation (Fig. 1). Field work was conducted along 117 transects, located 2 mi (3.2 km) apart and perpendicular to elevation contours, with a total combined length of 668 mi (1,078 km). Sampling points (stations) were established at 328 ft, 440 ft, or 492 ft (100, 134, or 150 m) intervals along the transects, depending on the study area (Scott *et al.* 1981, 1986). A total of 7,864 stations was sampled for the island of Hawai'i. Although in this paper we report primarily on data gathered on the distribution of six alien plant species, detailed information was also collected for several other introduced plants and animals, the distribution and abundance of native and nonnative forest birds, vegetation structure and composition, substrate characteristics, and phenology of selected plant species (Scott *et al.* 1981).

Data Analysis

Location coordinates used in plotting species presence for the stations sampled were generated by digitizing the start and end points for each transect into a computer mapping program, which then interpolated the relative locations of the stations based on transect length and station interval. For the analysis of habitat characteristics, elevation was taken from 1:24,000-scale U.S. Geological Survey quadrangle maps, vegetation types were identified using 1:24,000-scale maps prepared for the Hawai'i Forest Bird Survey study areas (Scott *et al.* 1981; Jacobi 1990), and median annual rainfall was determined from maps prepared by the Hawaii Division of Water and Land Development (Hawaii Department of Land and Natural Resources 1980). A summary of the distribution and frequency of all stations sampled on the island in relation to elevation and median annual rainfall classes is presented in Figure 2.

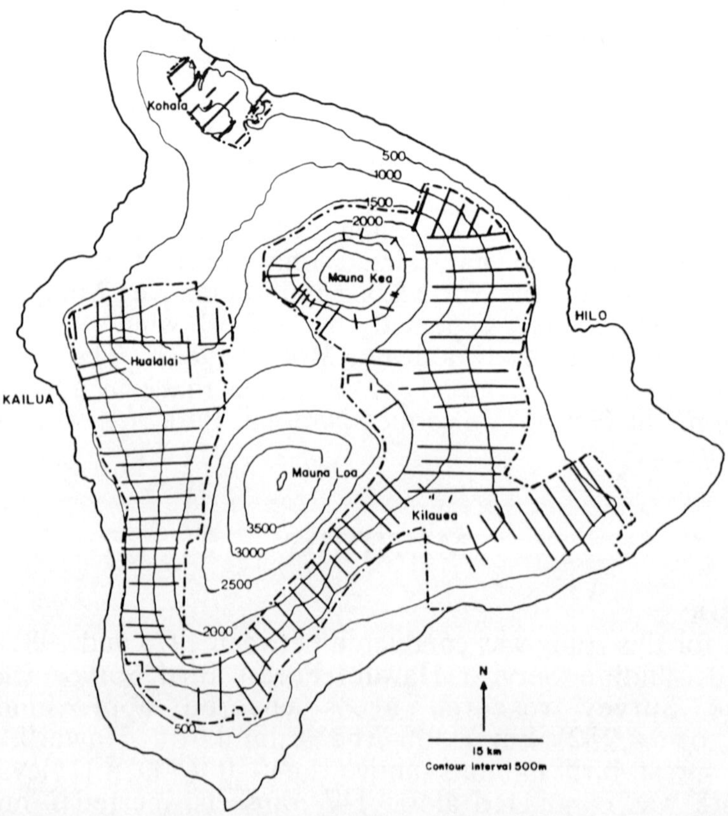

Figure 1. Location of study areas (dash-dot line) and transects (straight solid lines) established on the island of Hawai'i during the Hawai'i Forest Bird Survey.

MEDIAN ANNUAL RAINFALL (mm)

ELEVATION	0–1250	1250–2500	2500–3500	3500–5000	5000–7000	>7000	ELEV TOTALS No.Sta	%Tot
>3300 m							0	0.0
2900–3300 m	11						11	0.1
2500–2900 m	166						166	2.1
2100–2500 m	306	10					316	4.0
1700–2100 m	794	368	9				1171	14.8
1300–1700 m	1067	724	387	88			2266	28.8
900–1300 m	334	796	264	678	101		2173	27.6
500–900 m	194	408	217	362	334	58	1573	20.0
100–500 m	16	19	53	73	27		188	2.3
0–100 m							0	0.0
PPT. No.Sta	2888	2325	930	1201	462	58	7864	
%Tot	36.7	29.5	11.8	15.2	5.8	0.7		

Figure 2. Summary of all stations sampled during the Hawai'i Forest Bird Survey plotted by elevation and median annual rainfall. Heavy line indicates elevation-rainfall cells sampled during the Survey.

Predicting Potential Range of the Plant Species

Elevation and median annual rainfall ranges for the six alien species were used to develop a general model to predict the potential geographic range of each species in the Hawai'i Forest Bird Survey study area, assuming no other restrictions of plant invasion over time. The two-way tabulation of stations sampled by rainfall and elevation classes (Fig. 2) was used as the basis for the model. It was assumed that each species has the potential to become established in the proximity of all stations falling within elevation and rainfall class cells that the species were found to occupy during the Survey. For example, a species may have been found at stations located in dry habitats at high elevation. The model would predict the species could eventually be found only on those additional stations showing the same combined low rainfall and high elevation characteristics, but not at low-elevation dry or high-elevation wet sites.

Although the elevation-rainfall model used in this paper is rather simplistic, it allows for a general prediction of the potential ranges of the six species on the island of Hawai'i based on data gathered systematically in the field. Further refinement of this model may be possible with the inclusion of additional information on other habitat variables (*e.g.*, substrate type, soil moisture availability, seasonality of precipitation) for each documented occurrence. Unfortunately, data on additional habitat variables are not available for the study area. Despite the small number of model variables, we believe the potential range maps presented in this paper provide a general overview of the areas that may be colonized by these species if they are allowed to continue to spread.

Finally, it should be emphasized that the model used to predict potential distribution is based only on information relating to stations containing the species during the Hawai'i Forest Bird Survey. The model does not take into account the fact that these recorded occurrences may not reflect the full range of environmental conditions the species may ultimately tolerate with continued invasion. This may certainly be the case with species in an early stage of invasion that could become established in areas outside the range of environmental conditions in which they were found during the Survey. This recognized limitation of the model may result in a conservative prediction of potential distribution for such species. A more accurate prediction may be possible as more information is collected on additional occurrences of these invading species.

RESULTS

Distribution and Environmental Ranges of the Six Species Studied

Three basic patterns are apparent in the distribution maps for the six species studied (Figs. 3-14). Two species (Malabar melastome and yellow Himalayan raspberry) were concentrated on the eastern portion of the Island (Figs. 3 and 11), two species (fountain grass and German ivy) were generally found on the western side of the Island (Figs. 7 and 13), and the

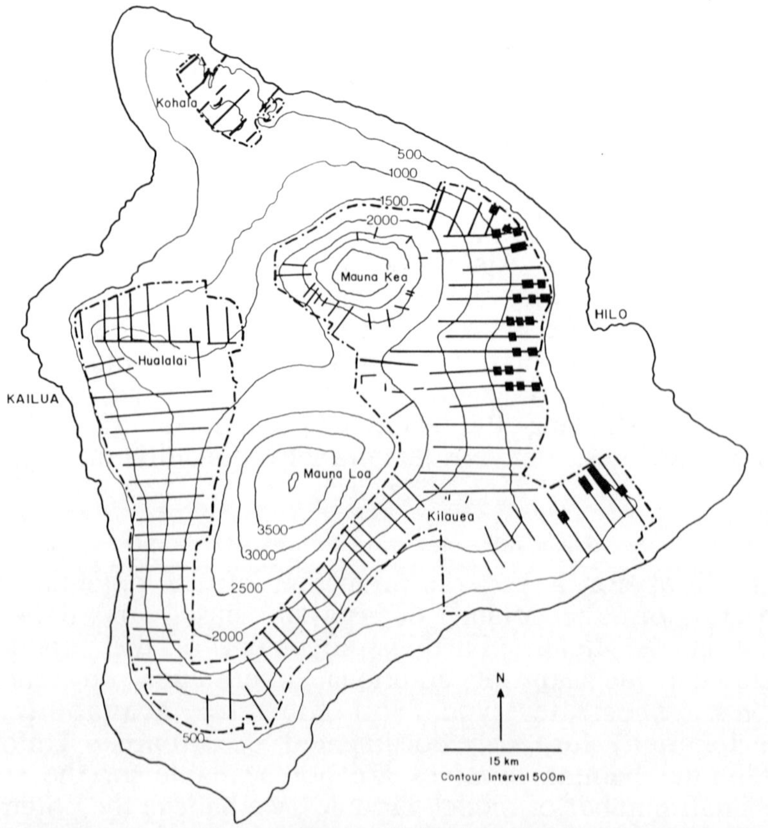

Figure 3. Occurrence of Malabar melastome during the Hawai'i Forest Bird Survey on the island of Hawai'i.

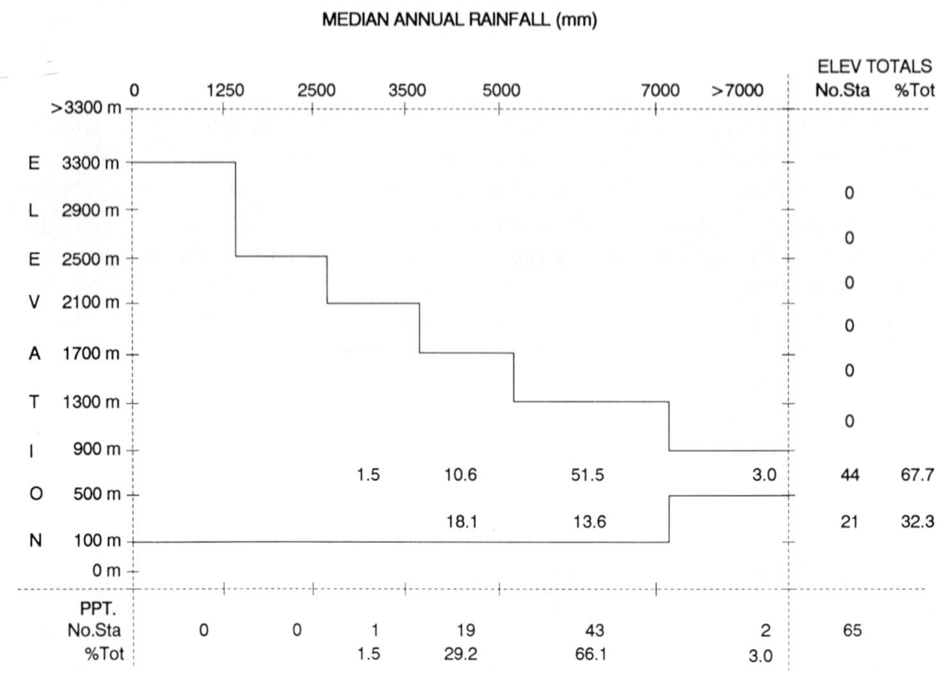

Figure 4. Plot of percent of total stations containing Malabar melastome in elevation and rainfall classes for the Hawai'i Forest Bird Survey study areas on the island of Hawai'i.

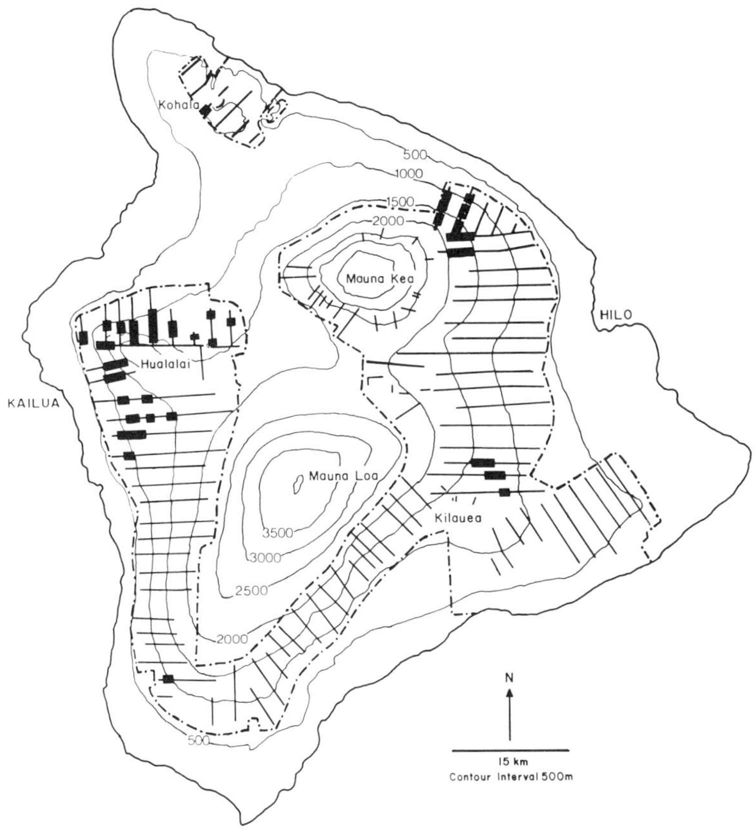

Figure 5. Occurrence of banana poka during the Hawai'i Forest Bird Survey on the island of Hawai'i.

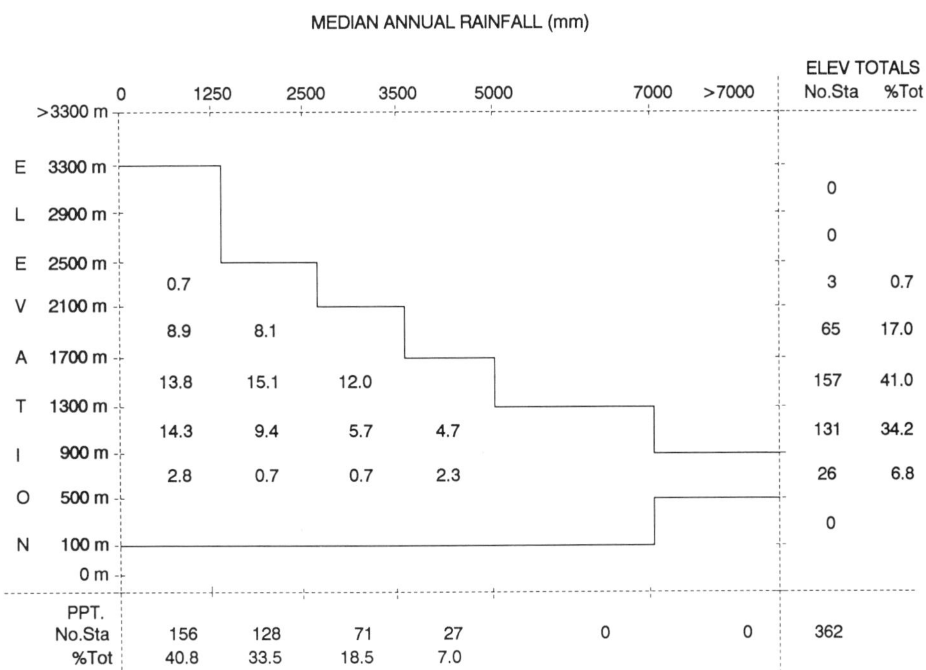

Figure 6. Plot of percent of total stations containing banana poka in elevation and rainfall classes for the Hawai'i Forest Bird Survey study areas on the island of Hawai'i.

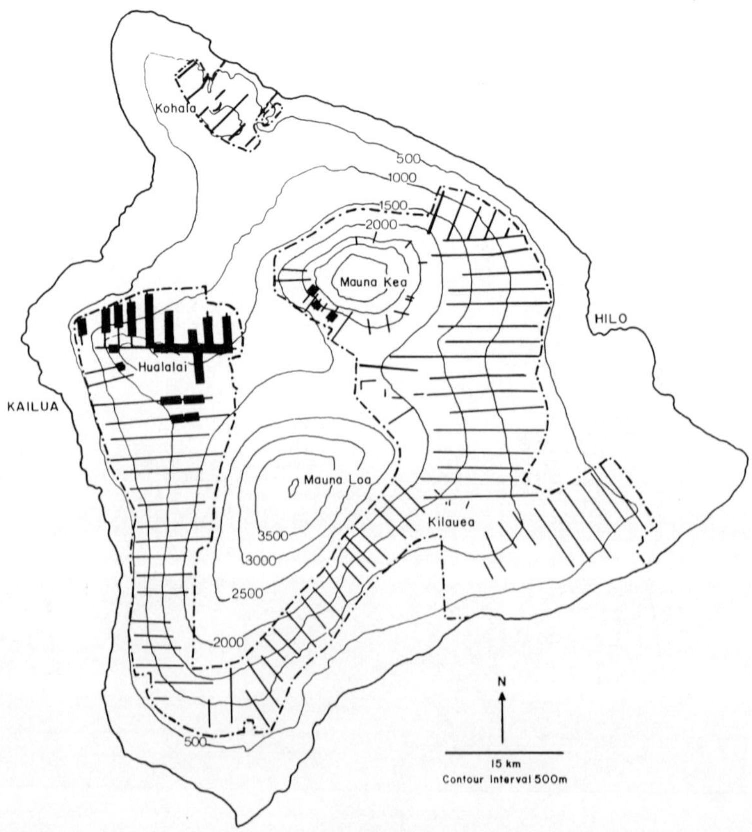

Figure 7. Occurrence of fountain grass during the Hawai'i Forest Bird Survey on the island of Hawai'i.

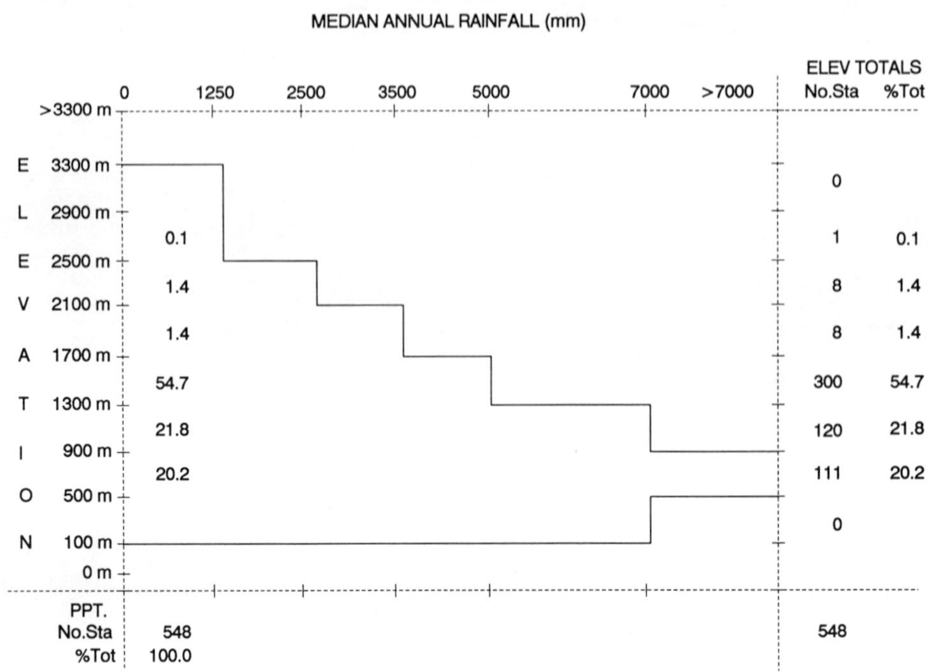

Figure 8. Plot of percent of total stations containing fountain grass in elevation and rainfall classes for the Hawai'i Forest Bird Survey study areas on the island of Hawai'i.

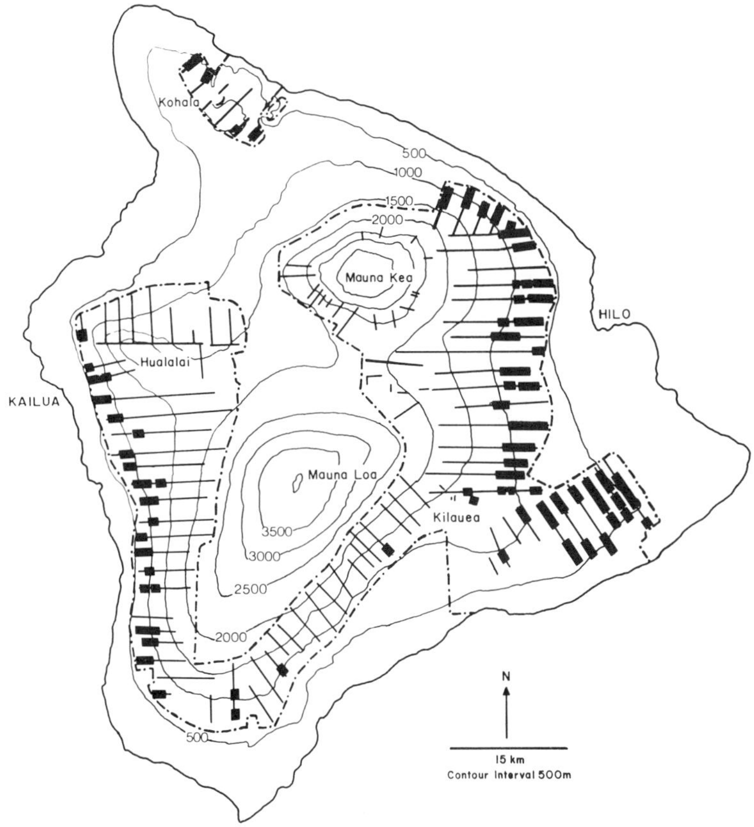

Figure 9. Occurrence of strawberry guava during the Hawai'i Forest Bird Survey on the island of Hawai'i.

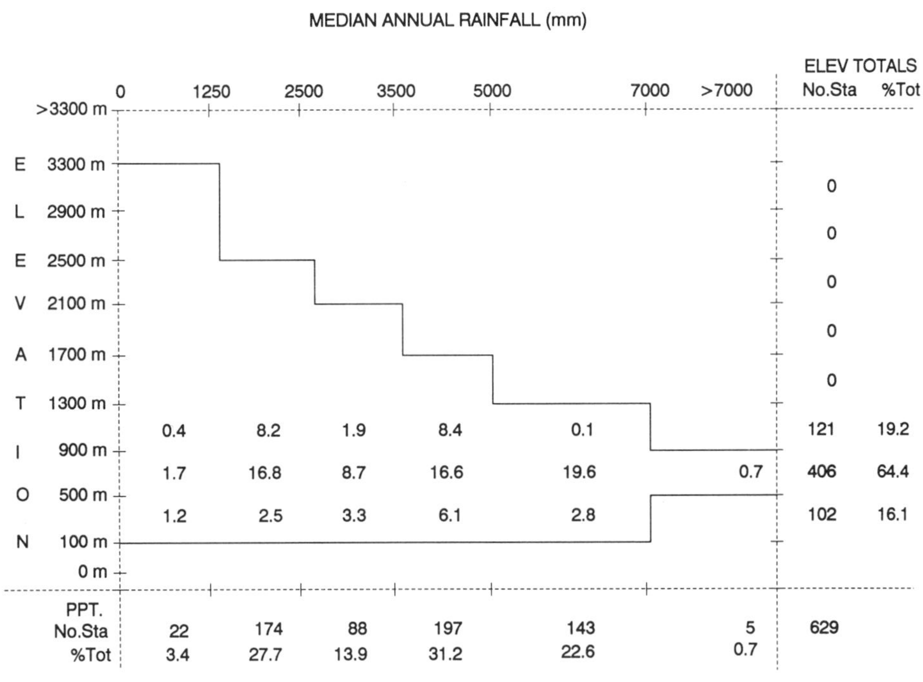

Figure 10. Plot of percent of total stations containing strawberry guava in elevation and rainfall classes for the Hawai'i Forest Bird Survey study areas on the island of Hawai'i.

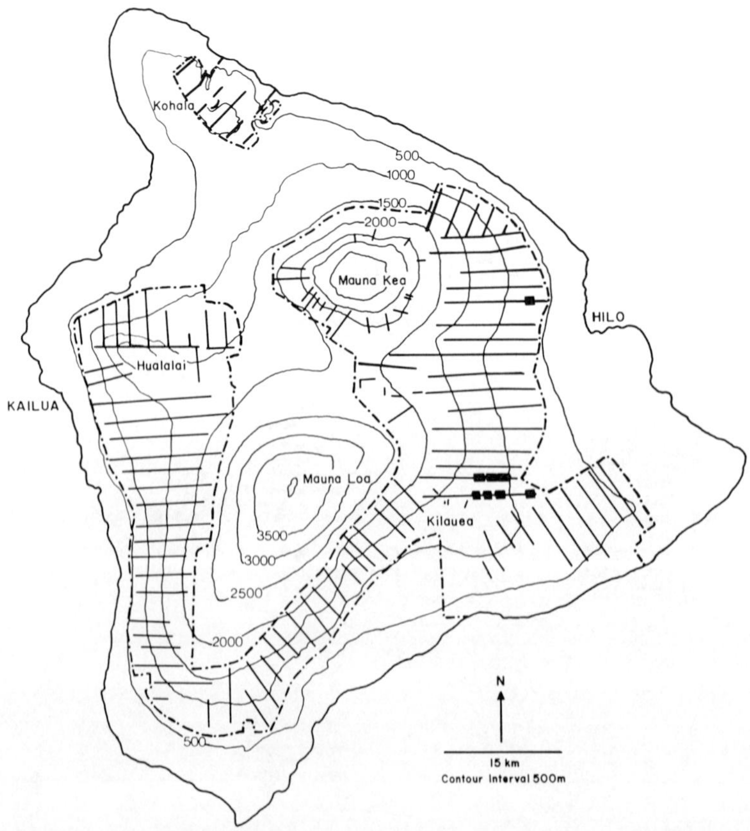

Figure 11. Occurrence of yellow Himalayan raspberry during the Hawai'i Forest Bird Survey on the island of Hawai'i.

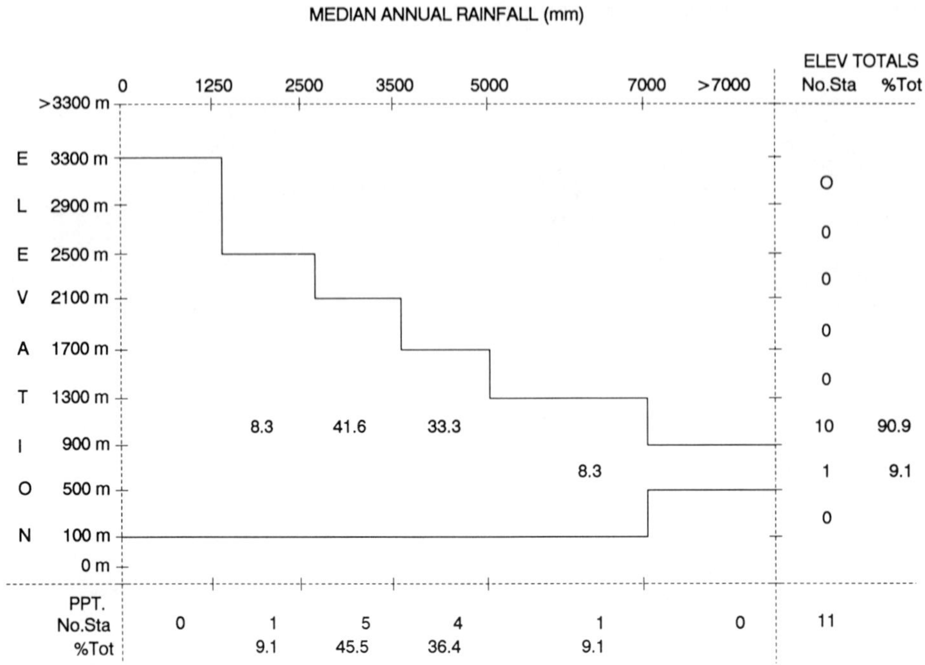

Figure 12. Plot of percent of total stations containing yellow Himalayan raspberry in elevation and rainfall classes for the Hawai'i Forest Bird Survey study areas on the island of Hawai'i.

Figure 13. Occurrence of German ivy during the Hawai'i Forest Bird Survey on the island of Hawai'i.

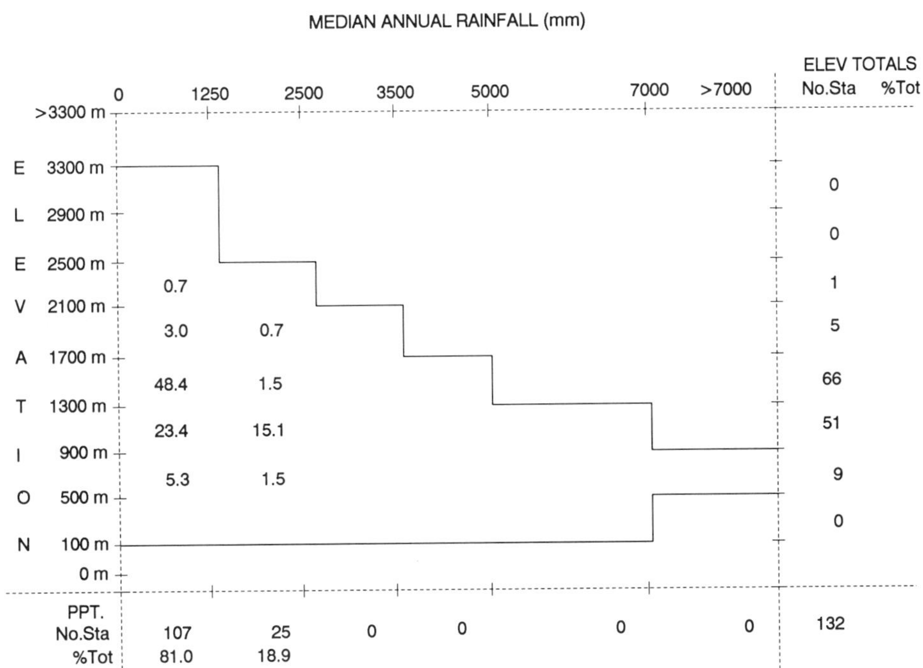

Figure 14. Plot of percent of total stations containing German ivy in elevation and rainfall classes for the Hawai'i Forest Bird Survey study areas on the island of Hawai'i.

remaining two species (banana poka and strawberry guava) were found in nearly all areas sampled (Figs. 5 and 9). Further separation in the distributions of these species is evident when elevation range, rainfall range, and vegetation types occupied are examined (Table 2; Appendix).

Malabar melastome was recorded at 65 stations, representing 0.8% of the total number sampled (Table 2). This species was found in two different areas below 2,950 ft (900 m) elevation on the eastern slopes of Mauna Kea, Mauna Loa, and Kīlauea volcanoes (Fig. 3). The majority of occurrences was in areas where the median annual rainfall is between 140 and 275 in. (3,500-7,000 mm), although it was also recorded at three stations just outside this range (Fig. 4). Malabar melastome was only found in wet habitats, ranging from open sedge-rush-native shrub communities, native koa (*Acacia koa*) and koa-ʻōhiʻa (*Metrosideros polymorpha*) forest, to highly disturbed communities dominated by many introduced plant species (Appendix). *Melastoma* was more common in wet ʻōhiʻa and ʻōhiʻa-koa forests than in the other six vegetation types occupied.

Banana poka was recorded at 382 stations, equivalent to 4.9% of the stations surveyed (Table 2). It was most abundant and widespread on the western side of the Island, concentrated on the slopes of Hualālai and Mauna Loa volcanoes (Fig. 5). Additionally, another relatively large population was found on the eastern flank of Mauna Kea, and a smaller one in the ʻŌlaʻa forest area northeast of Kīlauea Volcano. Two very small outlying populations were identified, one on the northern portion of the island in the Kohala region, and the other on the southwestern flank of Mauna Loa near Manukā State Park. Each of these outlying populations was characterized by a few immature plants recorded at a single station.

Banana poka was found between 1,640 and 8,200 ft (500-2,500 m) elevation, but most of the sightings were between 2,950 and 6,890 ft (900-2,100 m). Annual rainfall values for stations occupied by this species ranged from < 50 in. (<1,250 mm) to 200 in. (5,000 mm) (Fig. 6). The wide variation shown by banana poka with regard to both rainfall and elevation ranges is further reflected in the relatively large number of vegetation types (28) in which it was found (Appendix). Occupied vegetation units included dry native shrub communities, dry native forest, mesic forest and shrub units, to tall koa-ʻōhiʻa rain forest. Approximately 38% (147 stations) of the occurrences of banana poka were in the disturbed mesic koa-ʻōhiʻa forest type with native and alien shrub and grass understory. Other vegetation units with high incidence of this species were the less disturbed wet ʻōhiʻa forest and koa-ʻōhiʻa forest with tree fern (*Cibotium* spp.) and native shrub understory.

Fountain grass was found on the northern, eastern, and western slopes of Hualālai, and on the southwestern side of Mauna Kea (Fig. 7). The sampling grid did not cover other known populations on the southern flank of Kīlauea Volcano, in and adjacent to Hawaii Volcanoes National Park (Tunison, this volume). A total of 548 stations (7.0%) sampled was occupied by the species (Table 2). Fountain grass was found to have a relatively wide elevation range (1,640 to 9,515 ft or 500-2,900 m) but was limited to areas with median annual rainfall of less than 50 in. (1,250 mm) (Fig. 8). Only

Table 2. The six alien species, with a summary of elevation range, median annual rainfall range, and habitat types occupied for sightings made during the Hawai'i Forest Bird Survey (HFBS), station occurrence during the Survey for the island of Hawai'i, and predicted station occurrence in the Survey study area.

Species [Common Name]	Elevation Range (m)	Rainfall Range (mm)	Habitat Types Occupied	Station Occ. in HFBS (%)	Predicted Stations (%)
Melastoma candidum [Malabar melastome]	100 - 900	2,500 - >7,000	Hydric scrub and forest	65 (0.8)	1,071 (13.6)
Passiflora mollissima [banana poka]	500 - 2,500	<1,250 - 5,000	Xeric to mesic scrub and forest; hydric forest	382 (4.9)	6,899 (87.7)
Pennisetum setaceum [fountain grass]	500 - 2,900	<1,250	Xeric to mesic grassland, scrub, and open forest	548 (7.0)	2,861 (36.4)
Psidium cattleianum [strawberry guava]	100 - 1,300	<1,250 - >7,000	Mesic to hydric scrub and forest; xeric scrub and forest	629 (8.0)	3,934 (50.0)
Rubus ellipticus [yellow Himalayan raspberry]	500 - 1,300	1,250 - 7,000	Mesic to hydric forest	11 (0.1)	2,072 (26.3)
Senecio mikanioides [German ivy]	500 - 2,500	1,250 - 2,500	Xeric to mesic scrub and forest; hydric forest	132 (1.7)	4,991 (63.5)

dry and mesic habitats were colonized by this species (21 different vegetation types in all). The list includes grassland, shrub, and tree dominated communities (Appendix). Most of the records of fountain grass were from dry 'ōhi'a forest (149 stations) and the alien grassland-shrub community with scattered native trees (114 stations).

Strawberry guava had the most widespread distribution of the six species discussed in this paper. A total of 629 stations (8.0% of all stations sampled) was occupied by this species (Table 2). Plants were found in the lower portions of all study areas on Hawai'i, with greatest concentrations on the eastern side of the Island (Fig. 9). Strawberry guava was found only below 4,265 ft (1,300 m) elevation but occupied stations in all rainfall classes (50-275 in. or <1,250->7,000 mm) sampled (Fig. 10). It was established in 23 different vegetation types ranging from dry grass and shrub communities to tall native rain forest (Appendix). The two vegetation units occupied most often were the wet 'ōhi'a and tree fern forest (203 stations) and the wet koa-'ōhi'a and tree fern forest (159 stations).

Yellow Himalayan raspberry had the most limited distribution of the species analyzed. It was found at only 11 (0.1%) of the 7,864 stations sampled (Table 2), with the largest population concentrated in wet and mesic forest habitats immediately north of Kīlauea (Fig. 11). One additional sighting was made in the forest on the eastern slope of Mauna Kea, west of the city of Hilo. All but one of the stations with yellow Himalayan raspberry were in the 2,950 to 4,265 ft (900-1,300 m) elevation range, with rainfall distribution between 50 and 275 in. (1,250-7,000 mm) per year (Fig. 12). Plants were found in five different plant communities, including both mesic and hydric forest types (Appendix).

German ivy was found predominately on the western flanks of Mauna Kea and Mauna Loa, and on the northern and western slopes of Hualālai Volcano (Fig. 13). A single plant was also recorded on the eastern flank of Mauna Loa. This species occupied 132 (1.7%) of the Survey stations (Table 2). It was found only at stations with less than 100 in. (2,500 mm) of annual rainfall but encompassed an elevation range between 1,640 and 8,200 ft (500-2,500 m) (Fig. 14). German ivy was established in 19 different vegetation units ranging from dry shrub communities to rain forest (Appendix).

Predicted Potential Distribution

The predicted range model, based on median annual rainfall and elevation ranges, indicated a great potential for expansion of the geographical distribution of at least four of the six focal species if they continue to spread into areas with environmental conditions similar to habitats in which they were found during the Hawai'i Forest Bird Survey (Table 2). Banana poka showed the greatest potential range, with a predicted occupancy of 87.7% of the stations sampled (Table 2). The only major portions of the study area that this species does not appear to be able to invade are below 1,640 ft (500 m) elevation and where the median annual rainfall is greater than 200 in. (5,000 mm), both on the eastern side of the island. Additionally, banana poka has not become established above 8,200 ft (2,500 m) elevation, specifically in the upper portions of the Mauna Kea transects (Fig. 15).

A second species showing a large potential distribution is German ivy (Fig. 16). The model predicted that this species is capable of becoming established on 63.5% of the Hawai'i Forest Bird Survey stations (Table 2). The potential distribution map for German ivy suggests that all the sampled areas on the western and southwestern sides of the Island may be invaded, but only the upper portions of the eastern side are vulnerable. Based on the current environmental tolerance ranges of the species, high annual rainfall on the windward, eastern section of the Island could preclude the spread of German ivy into that area.

Strawberry guava is predicted to be able to occupy 50% of the stations sampled, limited primarily by elevation (Table 2; Fig. 17). Little difference between the distribution of this species on the map showing actual sightings during the Hawai'i Forest Bird Survey (Fig. 9) and the potential distribution map (Fig. 17) is evident. The relative evenness of its current distribution around the island of Hawai'i probably reflects the length of time it has been established there. If left uncontrolled, the major distributional change expected over time would be a filling-in of the area within the present geographical range of this species.

Fountain grass shows a relatively small potential distribution, primarily due to its observed restriction to dry habitats. Based on the predictive model, fountain grass has the potential to invade 36.4% of the Hawai'i Forest Bird Survey stations (Table 2), generally confined to the upper portions of the western side of the Island (Fig. 18). As mentioned earlier, however, the Survey did not sample a currently expanding population of this species on the southwestern flank of Kīlauea Volcano within Hawaii Volcanoes National Park and the adjacent Ka'ū and Puna districts.

Although the predictive model used data from only 11 stations for yellow Himalayan raspberry, this species appears to have a great potential for spread into most wet habitats at moderate to low elevations (26.3% of the stations sampled) on all portions of the island of Hawai'i (Table 2; Fig. 19). Other recent observations indicate yellow Himalayan raspberry has already invaded more dry habitats than those in which it was found during the Hawai'i Forest Bird Survey (J. Jacobi, unpub. data), suggesting the geographical distribution predicted from the Survey data set may be conservative.

The species that appears to be most limited in its potential spread is Malabar melastome (Fig. 20). The model suggests it could become established on just 13.6% of the total stations sampled (Table 2). The predicted range does not differ greatly from where the species was recorded during the Hawai'i Forest Bird Survey (Fig. 3), suggesting the current distributional area may just continue to fill in over time. Additional populations of the species appear to have the potential to become established on the lower southeastern flank on Mauna Loa and in the Kohala region.

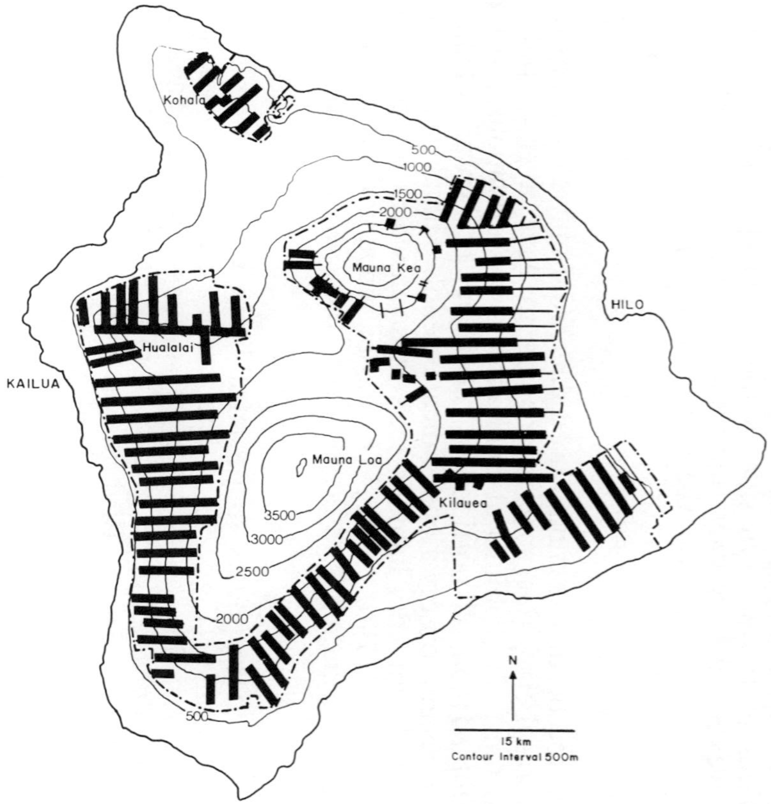

Figure 15. Predicted distribution of banana poka along the Hawai'i Forest Bird Survey transects on the island of Hawai'i, based on median annual rainfall and elevation of sites where the species was found in 1976-1981.

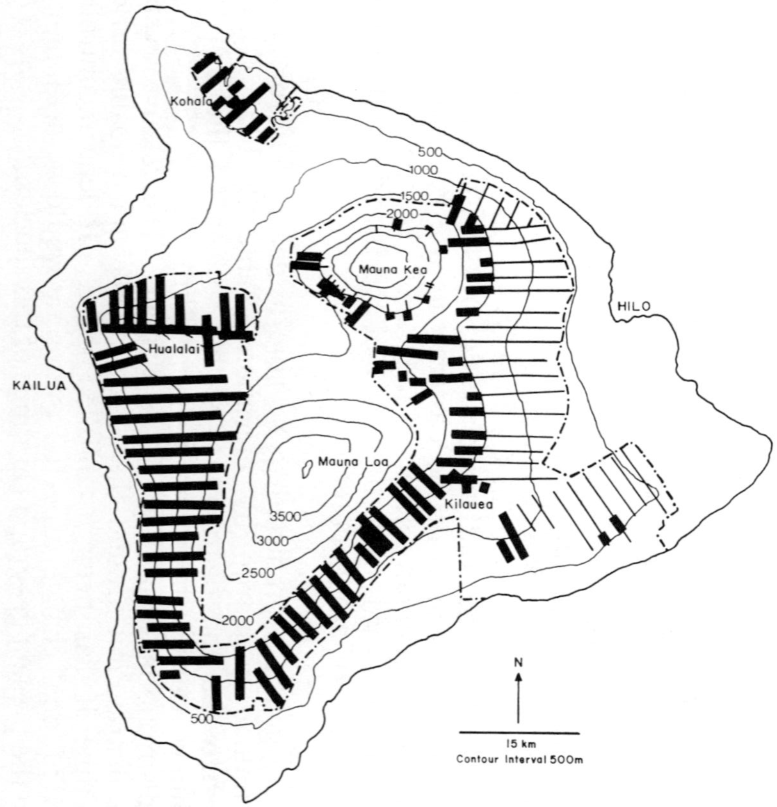

Figure 16. Predicted distribution of German ivy along the Hawai'i Forest Bird Survey transects on the island of Hawai'i, based on median annual rainfall and elevation of sites where the species was found in 1976-1981.

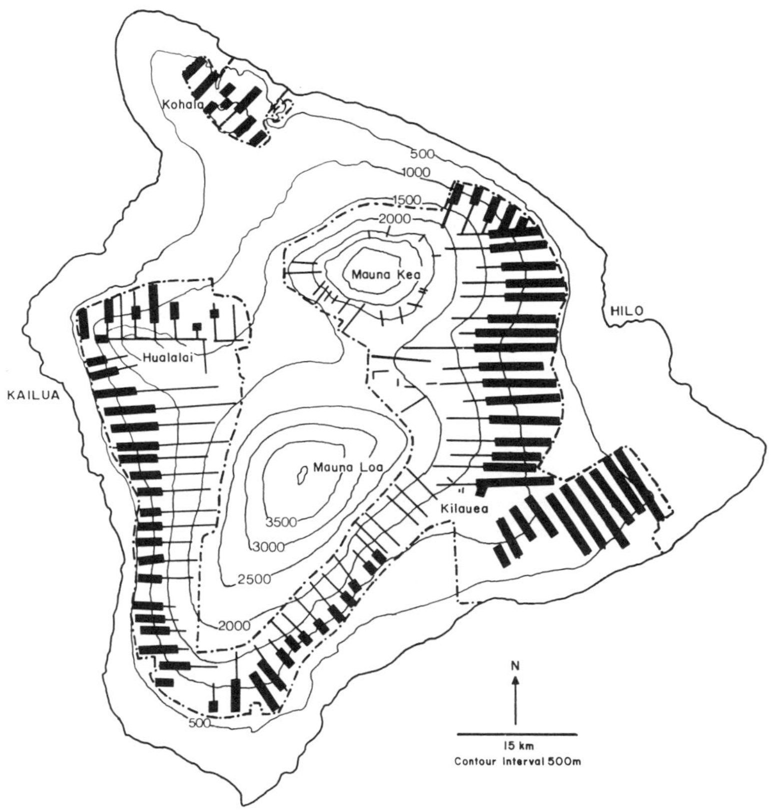

Figure 17. Predicted distribution of strawberry guava along the Hawai'i Forest Bird Survey transects on the island of Hawai'i, based on median annual rainfall and elevation of sites where the species was found in 1976-1981.

Figure 18. Predicted distribution of fountain grass along the Hawai'i Forest Bird Survey transects on the island of Hawai'i, based on median annual rainfall and elevation of sites where the species was found in 1976-1981.

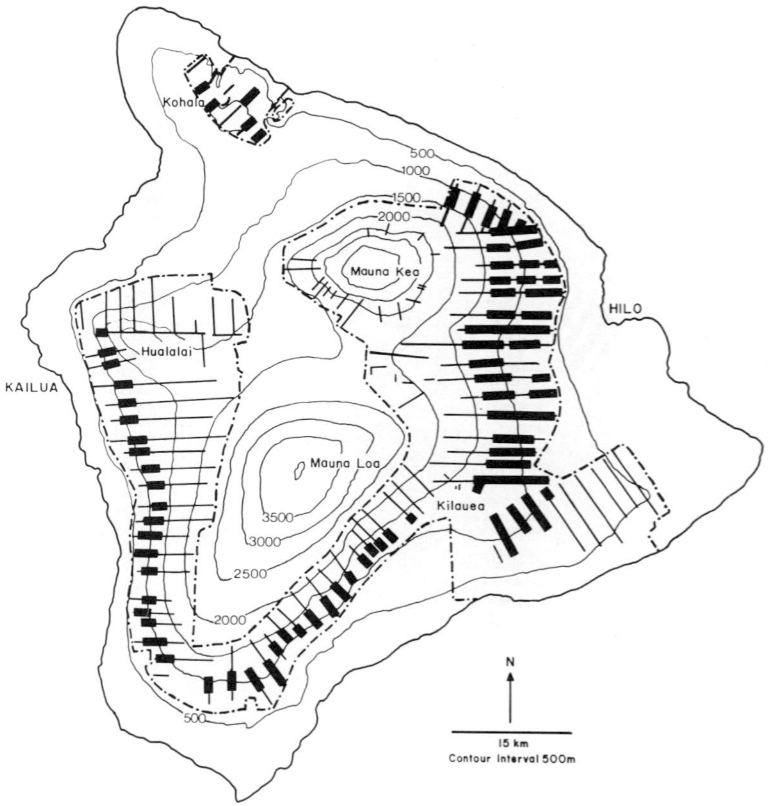

Figure 19. Predicted distribution of yellow Himalayan raspberry along the Hawai'i Forest Bird Survey transects on the island of Hawai'i, based on median annual rainfall and elevation of sites where the species was found in 1976-1981.

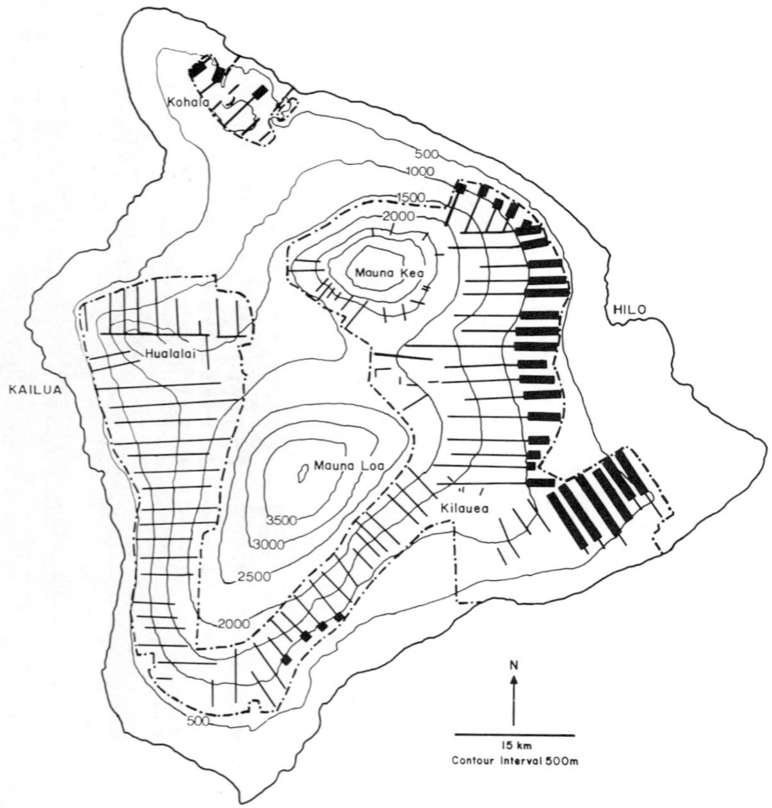

Figure 20. Predicted distribution of Malabar melastome along the Hawai'i Forest Bird Survey transects on the island of Hawai'i, based on median annual rainfall and elevation of sites where the species was found in 1976-1981.

DISCUSSION

Each of the six species examined shows a unique distributional pattern in terms of geographical areas, vegetation types, and habitat conditions (*i.e.*, median annual rainfall and elevation classes) occupied within the Hawai'i Forest Bird Survey study areas. The size of the area occupied (*i.e.*, number of stations) generally corresponds with the length of time a species has been established in the islands (Table 3). A noticeable exception is seen in the distribution of fountain grass, which was introduced at least by 1914 and was subsequently found on 548 stations during the Survey. The fact that this grass is easily dispersed by vehicles and the wind, and the considerable natural and unnatural disturbance in the habitats it occupies, may explain the large area it has invaded over such a relatively short period of time.

Table 3. Dates of introduction and number of stations occupied by the six alien species during the Hawai'i Forest Bird Survey, 1976-1981.

Species	Common Name	Date Introduced	No. stations Occupied
Psidium cattleianum	Strawberry guava	1825	629
Passiflora mollissima	Banana poka	1900	382
Senecio mikanioides	German ivy	1909	132
Melastoma candidum	Malabar melastome	1916	65
Pennisetum setaceum	Fountain grass	1914	548
Rubus ellipticus	Yellow Himalayan raspberry	1961	11

Combined Distributional Patterns

When the geographical ranges for all six species were plotted together (Fig. 21), the eastern and western sections of the Hawai'i Forest Bird Survey study areas were found to contain the broadest distribution of the six species. The ranges of two species (Malabar melastome and strawberry guava) overlapped on most of the lower eastern slopes of Mauna Kea, Mauna Loa, and Kīlauea volcanoes. In this area the alien species were generally confined to elevations below 3,280 ft (1,000 m). On the western side of the island, however, at least one of the six species was found throughout all areas sampled on Hualālai and up to 5,740 ft (1,750 m) on the eastern flank of Mauna Loa. The ranges of two of the species studied overlapped throughout a portion of this region, and three species were found to coexist in two smaller localities. The wide distribution of alien plants in this area was probably the result of a long history of disturbance to the native ecosystems by ranching and logging activities, and large populations of introduced ungulates.

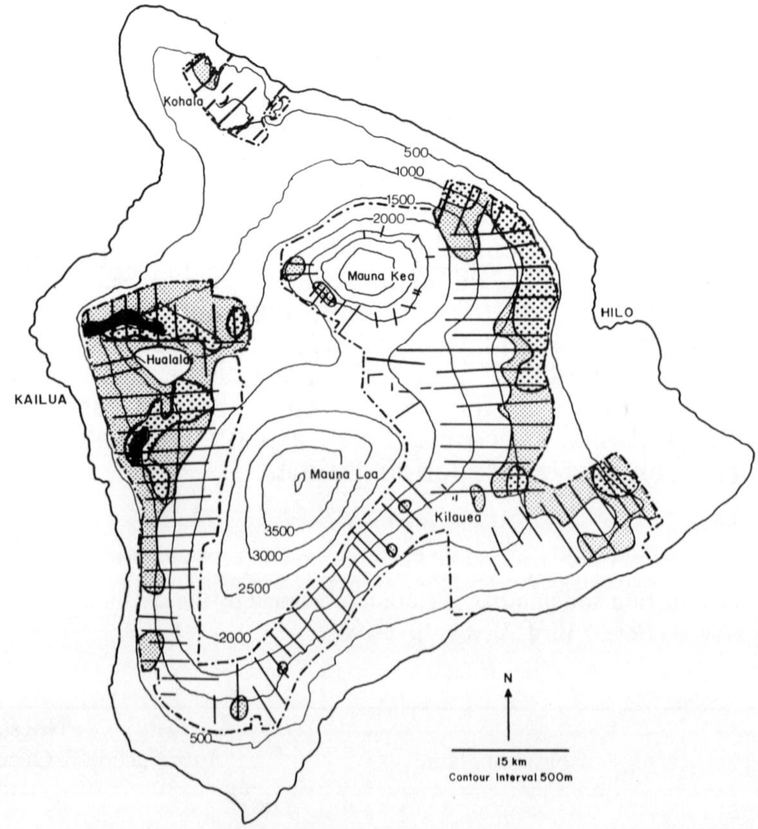

Figure 21. Overlap of the distributions of the six alien species within the Hawai'i Forest Bird Survey study areas on the island of Hawai'i, based on data collected during 1976-1981. Small dots = 1 species; large dots = 2 species overlap; black areas = 3 species overlap.

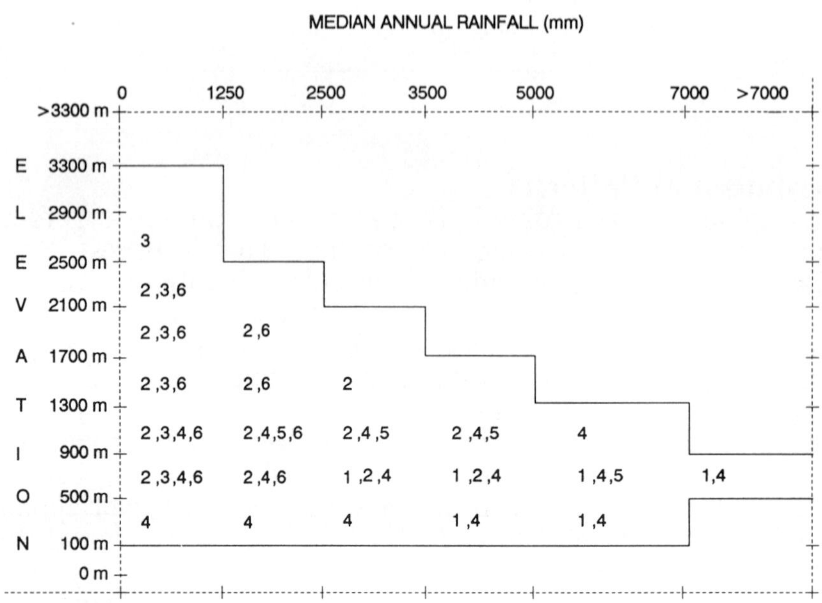

Figure 22. Summary of species recorded within each rainfall-elevation cell. 1 = Malabar melastome, 2 = banana poka, 3 = fountain grass, 4 = strawberry guava, 5 = yellow Himalayan raspberry, 6 = German ivy.

Most of the rainfall and elevation cells sampled were occupied by at least one of the alien plants (Fig. 22). Two habitat zones consistently invaded by three or more species were relatively dry sites (median annual rainfall <50 in. or <1,250 mm) between 1,640 and 8,200 ft (500-2,500 m) elevation), and low-elevation areas (1,640-4,265 ft or 500-1,300 m) with annual rainfall between <50 and 275 in. (<1,200-7,000 mm) (Fig. 22). The small number of alien species found established below 1,640 ft (500 m) in this study was probably a function of both the small sample size in this elevation band (188 stations) and the restricted number of species examined. Other studies have shown that low-elevation habitats in Hawai'i generally contain the greatest concentration of established alien plant species (Smith 1985).

Of the 64 plant communities sampled on the island of Hawai'i during the Hawai'i Forest Bird Survey, 46 were found to contain one or more of the six alien plant species (Appendix). The vegetation types invaded ranged from closed forest to treeless communities in dry, mesic, and wet habitats. Portions of the two largest communities, 'ōhi'a rain forest and koa-'ōhi'a rain forest, had been colonized by five of the six species analyzed.

Management Considerations

Developing an effective management program to restrict the spread and impacts of introduced species in native ecosystems requires an integrated approach. Mapping the current and potential range of the invading species should be an initial step in such a program. Information on distributional areas, species abundance, habitat requirements, and vegetation units currently occupied should be gathered as systematically as possible. However, care must be taken not to use so much time or other resources in this step that the invading population is allowed to expand significantly before actual control measures are initiated.

The analysis of current and potential ranges of the species discussed in this paper does not address population size or density. There is no doubt that different habitats will ultimately support different abundance levels of a particular species. Although the greatest direct effects on an ecosystem result from high abundance of the invading species, areas with low abundance in non-optimal habitats may serve as dispersal corridors for spread of the species into adjacent, more optimal areas. Management programs to reduce the overall effects of alien species must also focus control efforts on the smaller, and often discontinuous, populations in either an early stage of invasion or in sub-optimal habitats, to minimize their potential role in expansion of the distribution of a species. Warshauer *et al.* (1983) discussed the significance of such low-abundance areas in the continuing spread of banana poka on the island of Hawai'i. Mack (this volume) further stresses the importance of directing control efforts to small outlying populations of an invading species that have the greatest potential for rapid expansion in both range and abundance.

Finally, and most importantly, major emphasis should be placed on maintaining the natural structural and compositional integrity of the native ecosystems as a preventive measure to reduce the possibility of establishment of new alien plants in these communities. The relationship between animal damage and alien plant invasion has been frequently documented in the literature (*e.g.*, Harper 1965; Warshauer *et al.* 1983; Stone 1985). Hawaiian ecosystems with minimal animal or human disturbance appear to be most resistant to colonization by noxious introduced plants (Mueller-Dombois *et al.* 1981).

Outlook for Control

Most of the species discussed in this paper appear to be beyond the feasibility of mechanical or chemical control in areas where they have already become well established. This is especially true with fountain grass and German ivy on the western side of the island of Hawai'i, and with banana poka, Malabar melastome, and strawberry guava elsewhere. Yellow Himalayan raspberry is still in an early stage of invasion into wet and mesic habitats on the island. However, the rapid increase seen in its distribution in the Kīlauea region (J. Jacobi, unpub. data; T. Tunison, pers. comm.), even since the Hawai'i Forest Bird Survey, suggests mechanical control may already not be possible except in small areas managed intensively (Tunison and Stone, this volume).

The best potential for broad-scale control of species, such as those that are already well established, lies with the development of a successful biological control program. However, active steps can be taken to reduce the establishment of these species into new geographical areas. The potential range maps presented here may be useful in developing a surveillance program to watch for new infestations of these species. If small, localized infestations are detected early, conventional mechanical or chemical control techniques may be effective in eliminating a new population before it becomes well established. Specifically, the isolated patches of banana poka in the Kohala and Manukā regions, and German ivy on the southeastern flank of Mauna Loa, should be surveyed in greater detail and conventional control methods applied before they spread any further. Similarly, any new populations of yellow Himalayan raspberry outside its current range should be eliminated as quickly as possible.

The potential for range expansion of banana poka, German ivy, fountain grass, strawberry guava, and yellow Himalayan raspberry on the island of Hawai'i should be reason for great concern. Malabar melastome, on the other hand, appears to have the least potential for a significant increase in its distribution on the island of Hawai'i. Although additional information on the habitat requirements of these species may allow further refinement of their predicted ranges, we believe that the general picture of the possible expansion of the areas affected is not exaggerated.

ACKNOWLEDGMENTS

We thank Carmelle Crivellone, Paul Higashino, Holly McEldowney, Steve Montgomery, Rick Schuster, and Lani Stemmermann for their cheerful assistance with botanical field work during the Hawai'i Forest Bird Survey. Derral Herbst, Bob Hobdy, Jack Jeffrey, Jeff Kelly, Keith Miles-Iverson, Steve Mountainspring, Grey Pendleton, Sandra Pletschet, Mike

Scott, Cliff Smith, Don Sparling, Lani Stemmermann, Peter Stine, Chuck Stone, Robert Sugihara, Tim Tunison, and Julie Williams provided many useful comments, both during the study and at various stages in preparation of the manuscript. We also thank Robert Sugihara for his help with data management and computer programming during the analysis, and Sheila Doyle and Julie Williams for drafting some of the figures. Finally, we express our appreciation to the many landowners who allowed us access to conduct this field work on their property. Support for this project was provided by the U.S. Fish and Wildlife Service's Patuxent Wildlife Research Center (Region 8) and Pacific Islands Office (Region 1).

APPENDIX. List of vegetation types in which each of the focal species was found for stations sampled on the island of Hawai'i during the Hawai'i Forest Bird Survey.

VEGETATION TYPE	TOTAL NUMBER OF STATIONS	NUMBER OF STATIONS OCCUPIED (% OF STATIONS FOR VEG. TYPE)					
		M.c.[1]	P.m.	P.s.	P.c.	R.e.	S.m.
XERIC HABITAT							
Dry alien grassland with scattered koa[2], māmane[3] and other native trees.	79						3 (4)
Dry alien grassland with scattered koa, 'ōhi'a[4] and other native trees.	45		8 (18)	3 (7)	5 (11)		
Dry alien grassland with scattered koa and other native trees.	13		1 (8)				
Dry alien grassland with scattered māmane trees.	22			4 (18)			
Dry alien grassland with scattered native and alien trees.	14		1 (7)	12 (86)			
Dry alien grassland with some native shrubs, scattered 'ōhi'a and other native trees.	164		7 (4)	114 (70)	3 (2)		1 (1)
Dry alien grassland with native and alien shrubs and scattered native trees.	64		6 (9)	62 (97)			10 (16)
Dry mixed grass-native shrub community with scattered māmane trees.	77		12 (16)	2 (3)			

VEGETATION TYPE	TOTAL NUMBER OF STATIONS	NUMBER OF STATIONS OCCUPIED (% OF STATIONS FOR VEG. TYPE) M.c.[1]	P.m.	P.s.	P.c.	R.e.	S.m.
Dry mixed grass-native shrub community with scattered māmane and naio[5] trees.	22			(5)	1		
Dry native shrub with mixed grass community.	17		4 (24)				
Dry native shrub community with scattered ʻōhiʻa and other native trees.	323		4 (1)	26 (8)			1 (<1)
Dry alien tree community with a native shrub and mixed grass understory.	1		1 (100)				
Dry koa community with mixed native trees and a native shrub-alien grass and shrub understory.	16		1 (6)				
Dry koa-māmane community with an alien grass and scattered native shrub understory.	164						12 (8)
Dry koa-ʻōhiʻa community with a native shrub-alien grass understory.	3		3 (100)	2 (67)			3 (100)
Dry māmane-mixed native trees community with an alien grass-native shrub understory.	115		8 (7)	9 (8)			2 (2)

APPENDIX, continued.

APPENDIX, continued.

VEGETATION TYPE	TOTAL NUMBER OF STATIONS	NUMBER OF STATIONS OCCUPIED (% OF STATIONS FOR VEG. TYPE) M.c.[1]	P.m.	P.s.	P.c.	R.e.	S.m.
Dry māmane-naio scrub forest with a mixed grass and native shrub understory.	122			5 (4)			1 (1)
Dry mixed native tree community with an alien grass and native and alien shrub understory.	31			31 (100)			
Dry naio scrub forest with scattered māmane and an alien grass and native shrub understory.	22			2 (9)			2 (9)
Dry naio scrub forest with other native trees and native shrub-alien grass understory.	26		1 (4)	25 (96)			3 (11)
Dry ʻōhiʻa forest with mixed native trees and a native shrub and mixed grass understory.	726		10 (14)	149 (21)			16 (2)
Dry ʻōhiʻa forest with mixed native trees and a native shrub-alien grass-shrub understory.	161		9 (6)	76 (47)	1 (1)		1 (1)
MESIC HABITAT							
Mesic alien grassland with scattered koa, ʻōhiʻa and other native trees.	120		4 (3)		1 (1)		3 (3)

VEGETATION TYPE	TOTAL NUMBER OF STATIONS	NUMBER OF STATIONS OCCUPIED (% OF STATIONS FOR VEG. TYPE) M.c.[1]	P.m.	P.s.	P.c.	R.e.	S.m.
Mesic alien grass and native shrub community with scattered ʻōhiʻa and other native trees.	70		1 (1)	14 (20)	4 (6)		
Mesic alien shrub community with scattered ʻōhiʻa and alien trees.	3		2 (67)	1 (33)			
Mesic alien tree community with some native trees and native and alien grass-shrub understory.	84		12 (14)	3 (4)	18 (21)		
Mesic koa-māmane forest with other native trees and an alien grass, native shrub understory.	12		6 (50)				
Mesic koa-ʻōhiʻa forest with other native trees and a native-alien shrub and grass understory.	685		147 (22)	4 (1)	9 (1)		13 (2)
Mesic koa-ʻōhiʻa forest with other native trees and a native fern and shrub understory.	134		16 (12)		1 (1)		9 (7)
Mesic māmane-naio-alien tree community with an alien grass understory.	1		1 (100)				

APPENDIX, continued.

APPENDIX, continued.

VEGETATION TYPE	TOTAL NUMBER OF STATIONS	NUMBER OF STATIONS OCCUPIED (% OF STATIONS FOR VEG. TYPE) M.c.[1]	P.m.	P.s.	P.c.	R.e.	S.m.
Mesic native tree community with scattered alien trees and alien grass-shrub understory.	3		3 (100)				
Mesic ʻōhiʻa forest with other native trees and native shrub-alien grass and shrub understory.	259		13 (5)	65 (25)	3 (1)	27 (10)	
Mesic ʻōhiʻa forest with other native trees and a native shrub, matted-fern understory.	244			3 (1)	12 (5)		1 (<1)
HYDRIC HABITAT							
Wet matted-fern, mixed sedge-rush-native shrub community with scattered koa and ʻōhiʻa trees.	14	2 (14)			2 (14)		
Wet matted-fern and/or native shrub community with scattered ʻōhiʻa and other native trees.	143				6 (4)		
Wet native-alien shrub community with scattered ʻōhiʻa and other native trees.	4				4 (100)		
Wet tree fern and native shrub community with scattered ʻōhiʻa and other native trees.	186		9 (5)		15 (8)	4 (2)	

VEGETATION TYPE	TOTAL NUMBER OF STATIONS	NUMBER OF STATIONS OCCUPIED (% OF STATIONS FOR VEG. TYPE) M.c.[1]	P.m.	P.s.	P.c.	R.e.	S.m.
Wet alien tree plantation with some native trees and an alien shrub understory.	37	2 (5)			12 (32)		
Wet alien tree, ʻōhiʻa forest with a matted-fern and native shrub understory.	75				21 (28)		
Wet koa-ʻōhiʻa forest with other native trees and an alien and native shrub understory.	35	19 (54)			29 (83)		
Wet koa-ʻōhiʻa forest with other native trees and a native shrub, matted-fern and tree fern understory.	106	5 (5)			24 (23)		
Wet koa-ʻōhiʻa forest with other native trees and a tree fern, native shrub understory.		21 720	50 (3)	(7)	159	1 (22)	23 (<1) (3)
Wet pioneer ʻōhiʻa forest with a matted-fern and native shrub understory.	70				7 (10)		
Wet ʻōhiʻa forest with native shrubs/matted-fern and some tree ferns in the understory.		1 177	(1)		21	(12)	
Wet ʻōhiʻa forest with other native trees and a native shrub-alien grass and shrub under-story.	61				7 (12)	1 (2)	

APPENDIX, continued.

APPENDIX, continued.

VEGETATION TYPE	TOTAL NUMBER OF STATIONS	NUMBER OF STATIONS OCCUPIED (% OF STATIONS FOR VEG. TYPE) M.c.[1]	P.m.	P.s.	P.c.	R.e.	S.m.
Wet ʻōhiʻa forest with other native trees and a tree fern and native shrub understory.	1871	15 (1)	42 (2)		203 (11)	2 (<1)	1 (<1)
NUMBER OF STATIONS OCCUPIED:		65	382	548	629	11	132
VEGETATION TYPES OCCUPIED:		7	28	21	23	5	19

[1] M.c. = *Melastoma candidum*; P.m. = *Passiflora mollissima*; P.s. = *Pennisetum setaceum*; P.c. = *Psidium cattleianum*; R.e. = *Rubus ellipticus*; S.m. = *Senecio mikanioides*

[2] *Acacia koa*

[3] *Sophora chrysophylla*

[4] *Metrosideros polymorpha*

[5] *Myoporum sandwicense*

Literature Cited

Degener, O., and I. Degener. 1968. *Rubus ellipticus*, Family 167. *Flora Hawaiiensis*. Private pub.

Escobar, L.K. 1980. Interrelationships of the edible species of *Passiflora* centering around *Passiflora mollissima* (HBK.) Bailey subgenus *Tacsonia*. Ph.D. diss., Univ. Texas, Austin.

Harper, J.L. 1965. Establishment, aggression, and cohabitation in weedy species. In *The genetics of colonizing species*, ed. H.G. Baker and G.L. Stebbins, 243-265. New York: Academic Pr.

Haselwood, E.L., and G.G. Motter, eds. 1983. *Handbook of Hawaiian weeds*, 2nd ed., revised and expanded by R.T. Hirano. Harold L. Lyon Arboretum. Honolulu: Univ. Hawaii Pr.

Hawaii Department of Agriculture. 1978. Regulation NW 10. Noxious weeds. Div. Plant Indus., Hawaii Dep. Agric. Honolulu.

Hawaii Department of Land and Natural Resources. 1980. *Median annual rainfall, island of Hawai'i*. Div. Water Land Devel., Hawaii Dep. Land Nat. Resour. Honolulu.

Jacobi, J.D. 1990. Distribution maps, ecological relationships, and current status of native plant communities on the island of Hawai'i. Ph.D. diss., Univ. Hawaii, Honolulu.

LaRosa, A.M. 1984. *The biology and ecology of* Passiflora mollissima *in Hawai'i*. Tech. Rep. 50, Univ. Hawaii Coop. Natl. Park Resour. Stud. Unit. Honolulu.

Mack, R.N. [this volume] Characteristics of invading plant species.

Markin, G.P., P.-Y. Lai, and G.Y. Funasaki. [this volume] Status of biological control of weeds in Hawai'i and implications for managing native ecosystems.

Mueller-Dombois, D., K.W. Bridges, and H.L. Carson, eds. 1981. *Island ecosystems: biological organization in selected Hawaiian communities*. Stroudsburg, Penn.: Hutchinson Ross Pub. Co.

National Park Service. 1986. Natural Resources Management Plan and Environmental Assessment, revised January 1986. Hawaii Volcanoes National Park.

St. John, H. 1973. *List and summary of the flowering plants in the Hawaiian Islands*. Mem. 1. Lawai, Kauai, Hawaii: Pac. Trop. Botan. Gdn.

Scott, J.M., J.D. Jacobi, and F.L. Ramsey. 1981. Avian surveys of large geographical areas: a systematic approach. *Wildl. Soc. Bull.* 9(3):190-200.

Scott, J.M., S. Mountainspring, F.L. Ramsey, and C.B. Kepler. 1986. *Forest bird communities of the Hawaiian Islands: their dynamics, ecology, and conservation*. Studies in Avian Biology 9. Berkeley, Calif.: Cooper Ornithol. Soc.

Smith, C.W. 1985. Impact of alien plants on Hawai'i's native biota. In *Hawai'i's terrestrial ecosystems: preservation and management*, ed. C.P. Stone and J.M. Scott, 180-250. Univ. Hawaii Coop. Natl. Park Resour. Stud. Unit. Honolulu: Univ. Hawaii Pr.

Stone, C.P. 1985. Alien animals in Hawai'i's native ecosystems: toward controlling the adverse effects of introduced vertebrates. In *Hawai'i's terrestrial ecosystems: preservation and management*, ed. C.P. Stone and J.M. Scott, 251-297. Univ. Hawaii Coop. Natl. Park Resour. Stud. Unit. Honolulu: Univ. Hawaii Pr.

Stone, C.P., and J.M. Scott, eds. 1985. *Hawai'i's terrestrial ecosystems: preservation and management*. Univ. Hawaii Coop. Natl. Park Resour. Stud. Unit. Honolulu: Univ. Hawaii Pr.

Tunison, J.T. [this volume] Fountain grass control in Hawai'i Volcanoes National Park: management considerations and strategies.

Tunison, J.T., and C.P. Stone. [this volume] Special Ecological Areas: an approach to alien plant control in Hawaii Volcanoes National Park.

Wagner, W.L., D.R. Herbst, and S.H. Sohmer. 1990. *Manual of the flowering plants of Hawai'i*. Bishop Mus. Spec. Pub. 83. Honolulu: Univ. Hawaii and Bishop Mus. Pr.

Warshauer, F.R., J.D. Jacobi, A.M. LaRosa, J.M. Scott, and C.W. Smith. 1983. *The distribution, impact and potential management of the introduced vine* Passiflora mollissima *(Passifloraceae) in Hawai'i*. Tech. Rep. 48, Univ. Hawaii Coop. Natl. Park Resour. Stud. Unit. Honolulu.

Wester, L.L. [this volume] Origins and introductions of adventive alien flowering plants in Hawai'i.

Whitney, L.D., E.Y. Hosaka, and J.C. Ripperton. 1939. *Grasses of the Hawaiian ranges*. Hawaii Agric. Exp. Sta. Bull. 82. Honolulu: Univ. Hawaii.

ALIEN PLANTS ON THE NORTHWESTERN HAWAIIAN ISLANDS

Derral R. Herbst and Warren L. Wagner

ABSTRACT

The Northwestern Hawaiian Islands is a chain of approximately 40 emergent low islands extending in a west-northwesterly direction for a distance of nearly 750 mi (1,210 km). The islands are currently managed as a wildlife refuge for birds, sea turtles, and the Hawaiian monk seal (*Monachus schauinslandi*). The vegetation is composed primarily of Pacific Basin strand species, naturalized species, and, on Kure and Midway Atolls and Tern Island, cultivated plants. The present vascular flora consists of 317 taxa, of which 8 are endemic (2.5%), 42 are indigenous (13.2%), 114 are naturalized (36.0%), and 132 are cultivated (41.6%). The flora is summarized on an island by island basis in an appendix to this paper. Some alien species have been deliberate or unintentional introductions by humans; others have arrived by "natural" means. Ironwood (*Casuarina equisetifolia*), a species well adapted to dry conditions, is an example of an intentional introduction. It has been introduced on seven of the islands and has become naturalized on Tern Island and Midway Atoll; it is a management problem only on Midway. *Heliotropium procumbens*, introduced by military flights from O‘ahu to two of the three leeward islands with runways, is an example of an accidental introduction that has not become a management problem. Little has been done to manage the alien plants of the Northwestern Hawaiian Islands other than to take precautions to prevent new ones from being introduced. There has been little need for management. Proposed management needs include monitoring the vegetation of the islands, producing a manual of plants, and developing a contingency plan for the control of any alien species of plant or animal impacting the vegetation or wildlife.

INTRODUCTION

The Northwestern Hawaiian Islands, or Leeward Islands as they are commonly (and hereafter) called, are a chain of islands stretching in a west-northwesterly direction from Nihoa Island to Kure Atoll, a distance of nearly 750 mi (1,210 km). Politically, all except Midway Atoll are part of the City and County of Honolulu; Midway is property of the U.S. Navy and is not part of the State of Hawai'i. President Theodore Roosevelt signed Executive Order 1019 on February 3, 1909, setting them aside (except Midway) as a preserve and breeding ground for native birds. Today, all except Kure and Midway Atolls are in the Northwestern Hawaiian Islands National Wildlife Refuge and are managed by the U.S. Fish and Wildlife Service. President Franklin Roosevelt signed Executive Order 7299 on February 20, 1936, placing Kure Atoll under the jurisdiction of the Secretary of the Navy; however, there are no records of any activity on the Atoll during World War II. On November 17, 1952, President Harry Truman signed Executive Order 10413, giving Kure Atoll to the Territory of Hawai'i. Kure presently is maintained as a wildlife refuge by the State Division of Forestry and Wildlife, and the Coast Guard maintains a long-range navigation (LORAN) station on Green Island, the only permanent land mass in the Atoll.

The Leeward Islands (Fig. 1) comprise approximately 40 emergent islands; numbers vary as the small sand islets are built up, fuse, separate, or wash away over time. The islands total about 3.9 mi^2 (10.1 km^2) of dry land, an amount less than 1% of the total land area of the Hawaiian Islands. They vary in elevation from 10 to 910 ft (3-277 m) (Table 1). The highest is Nihoa Island, and the lowest are French Frigate Shoals and Pearl and Hermes Atoll. The eastern islands (Nihoa, Necker, La Pérouse Pinnacle of French Frigate Shoals, and Gardner Island) are eroded remnants of shield volcanoes, the last two being residual volcanic and limestone islands. The western islands (French Frigate Shoals except for La Pérouse Pinnacle, Laysan and Lisianski Islands, and Pearl and Hermes, Midway, and Kure Atolls) are limestone islands resting on the summits of submerged volcanoes. More than a dozen other reefs and shoals mark volcanoes completely submerged below the surface of the ocean.

Compared to the main Hawaiian Islands, the Leeward Islands are poorly known geologically. They are much older than the main islands; the ages of five of the islands are known and range from about 7 to 27 million years (Table 1).

VEGETATION

The vegetation of the Leeward Islands is composed chiefly of typical strand species and the naturalized alien plants that can tolerate strand conditions. Cultivated ornamental plants are maintained on the three inhabited islands, and some also persist around abandoned buildings. On Nihoa, the highest island, several genera usually considered to be elements of mid-elevation, diverse mesic forests can be found. Thus, hu'ahu'akō *(Rumex albescens), Schiedea verticillata,* and perhaps *Pritchardia remota* and 'akoko *(Chamaesyce celastroides)* may be remnants of an earlier vegetation before the island eroded to its present state. Several of the strand species, such as pōpolo *(Solanum nelsonii)*, 'ānaunau *(Lepidium bidentatum* var. *o-waihiense)*, and 'ohai *(Sesbania tomentosa)*, which are rare on the main Hawaiian Islands, are still common on several of the Leeward Islands. This may be due to the greater disturbance of strand areas on the main Hawaiian islands.

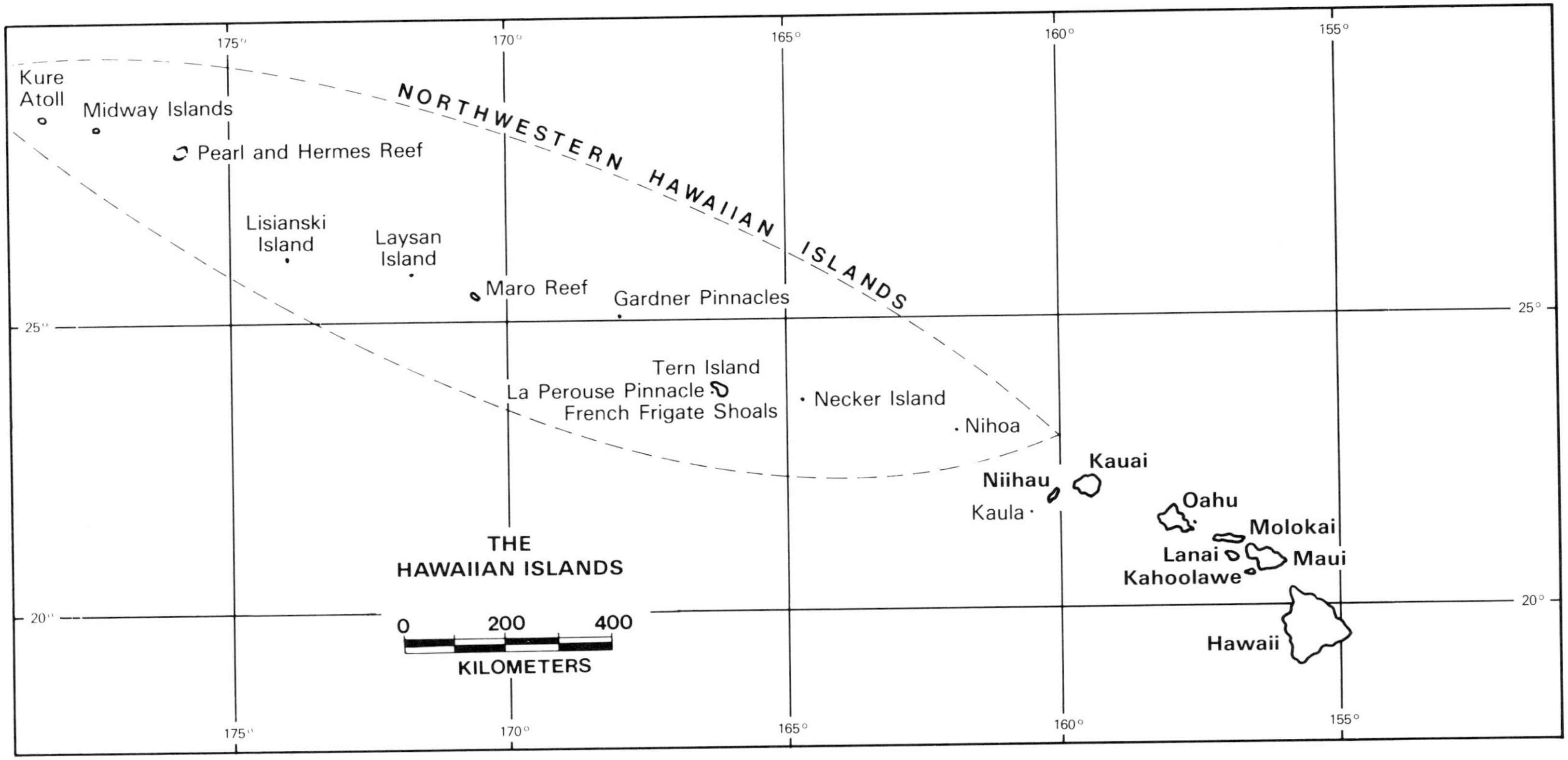

Figure 1. The Hawaiian Archipelago.

Table 1. Area, elevation, and age of the Leeward Hawaiian Islands.

Island Group	Area[a] (acres)	(hectares)	Maximum elevation[a] feet	(meters)	Age (in millions of years by Potassium-Argon)
Nihoa Island	191	(77)	910	(277)	7.0 + 0.3[c]
Necker Island	58	(24)	277	(84)	10.0 + 0.4[c]
French Frigate Shoals (13 islets)	56	(23)	135	(41)[b]	11.7[b,c]
Gardner Pinnacles	3	(1)	190	(58)	--
Laysan Island	981	(397)	35	(11)	--
Lisianski Island	432	(175)	20	(6)	--
Pearl and Hermes Atoll (7 islets)	78	(32)	10	(3)	20.1 + 0.5[d]
Midway Atoll (4 islets)	1,280	(518)	12	(4)	27.0 + 0.6[e]
Kure Atoll (2 islets)	237	(96)	20	(6)	--
Northwestern Hawaiian Islands (Total)	3,316	(1,343)			

[a]from Armstrong 1983.
[b]La Perouse Pinnacles only.
[c]Dalrymple *et al.* 1974; unreliable due to sampling problems.
[d]Clague *et al.* 1975; probably too young.
[e]Dalrymple *et al.* 1977; probably too young.

Table 2. Summary of status information for plant taxa on Leeward Islands.

Island Group	Status[a] E	I	X	C	S	P	?	Total Taxa
Nihoa Island	3	17	5	-	-	-	-	25
Necker Island	-	5	1	-	1	7	-	14
French Frigate Shoals	-	12	27	22	2	5	-	68
Gardner Pinnacles	-	1	-	-	2	-	-	3
Laysan Island	6	26	12	2	6	18	1	71
Lisianski Island	-	15	3	2	2	2	1	25
Pearl and Hermes Atoll	1	15	9	3	1	2	-	31
Midway Atoll	3	26	101	124	-	-	1	255
Kure Atoll	3	16	41	15	-	-	-	75
Totals[b]	8	42	114	132	8	11	2	317

[a]E = endemic; I = indigenous; X = naturalized alien species; C = cultivated alien species; S = seeds that arrived via natural dispersal; P = intentionally introduced propagules; ? = unknown.
[b]In calculating the totals, each species was counted only once, although it may have occurred on more than one island.

Of the 317 taxa of vascular plants recorded from the Leeward Islands, 8 (2.5%) are endemic and 42 (13.2%) are indigenous to these islands (although the indigenous species may be endemic to the Hawaiian Archipelago), while 114 (36.0%) are naturalized alien species and 132 (41.6%) are cultivated plants. The remaining 21 (6.6%) represent the seeds, cuttings, and seedlings deliberately introduced to the islands, or the seeds that floated to the islands but did not become established. The flora is summarized on an island by island basis in the Appendix and in Table 2 (page 192). A history of major disturbances on each of the Leeward Island groups is given in Table 3.

Plants have been introduced by two means: deliberate or unintentional introductions by humans, and introductions that did not involve the intervention of humans. Naturalized species and cultivated plants fall into the first category and native species into the last. For the purpose of this paper, alien species that moved from island to island by natural means (wind, ocean currents, or birds) are still considered alien species. Examples of different categories of naturalized alien species, based upon their means of introduction and their impacts, are given below.

CASE HISTORIES

Intentional Introduction of an Alien Plant: Ironwood

Casuarinaceae is a distinctive family of a single genus of shrubs and trees native primarily to Australia. A few species occur on Pacific Basin islands from Java to New Caledonia and in Asia as far north as Burma. The earliest documented record of the family in Hawai'i is a collection of ironwood (*Casuarina equisetifolia*) made by Heller on O'ahu in 1895 (Heller 1955: herbarium specimen, B.P. Bishop Museum). The species probably was introduced in 1882 by P. Isenberg, who planted trees at Kilohana Crater, Kaua'i. In 1890, A.S. Cleghorn, the father of Princess Ka'iulani, planted an avenue of ironwood trees on his estate in Waikīkī. This is the avenue of ironwoods still standing in Kapi'olani Park and probably was the source of Heller's collection.

Ironwood has several qualities that help it compete under adverse conditions (Apfelbaum *et al.* 1983). It is capable of fixing atmospheric nitrogen and therefore is able to colonize nutrient-poor soils, especially those lacking nitrogen. The root system takes up most of the nutrients from the soil, which, along with the thick mulch of fallen branchlets that accumulates under the trees, prevents nearly all other plants from growing near it (Neal 1965). Ironwoods can reproduce both sexually and asexually, and the last, by root suckers, appears to be more prominent in Hawai'i. The leaves are reduced to whorls of minute scales with stomata (minute openings) located in longitudinal grooves where they are protected by minute cilia, reducing transpiration, an important adaptation for dry conditions. Moreover, ironwood is a facultative halophyte (salt lover). These qualities make ironwood a desirable, hardy ornamental and shade tree on the inhabited Leeward Islands. These same qualities make this species a potential management problem.

Table 3. History of major disturbances in the Leeward Hawaiian Islands.

ISLAND GROUP	EVENT	YEAR
Nihoa Island[a]	* Prehistoric habitation; house foundations, terraces, shelters,and religious sites.	----
	* Fire apparently destroyed much vegetation.	1885
	* HIRAN[+] projects probably introduced *Cenchrus echinatus* and *Paspalum* sp. (since disappeared), and *Portulaca oleracea* (replacing, perhaps via hybridization, native *P. lutea*).	1961
Necker Island[b]	* Religious sites and terraces indicate much impact by prehistoric Hawaiians.	----
	* Some evidence of unofficial bombing.	1940s
French Frigate Shoals[c]	* Shipwreck crew of 27 on island about one month.	1867
	* Vegetation partly destroyed during military exercise on East Island.	1932-1940
	* Permanent military station on East Island.	1942-1945
	* Tern Island enlarged by dredging, naval base built, all vegetation destroyed.	1942-1946
	* LORAN[+] station on East Island.	1942-1952
	* East Island inundated by a tidal wave.	1946
	* LORAN Station on Tern Island.	1952-1979
	* Tern Island inundated during heavy seas, much vegetation damaged.	1969
	* U.S. Fish & Wildlife Service refuge.	1979 to present
Gardner Pinnacles[d]	* None known.	

ISLAND GROUP	EVENT	YEAR
Laysan Island[e]	* Guano mining.	1890-1910
	* Various invasions by bird poachers.	? -1915
	* Introduced rabbits virtually denuded the island.	1903-1923
	* Various plants reintroduced.	1923-1966
	* HIRAN operation may have accidentally introduced *Cenchrus echinatus* and *Conyza bonariensis*.	1961-1962
Lisianski Island[f]	* Crews from 3 shipwrecks on island for up to 5 months.	1844-1872
	* Bird poachers (at least 5 groups, largest of 77 people, shacks built).	1904-1910
	* Rabbits destroyed all vegetation, then starved to death.	after 1904-1915/16
Pearl and Hermes Atoll[g]	* Crews of 2 shipwrecks lived on island for 2 months.	1822
	* Rabbits damaged vegetation.	before 1916-1928
	* Hawaiian Sea Products Co. erected buildings and harvested oyster shell.	1929 - 1931
	* Temporary naval base on SE Island.	1930s
	* Abandoned Sea Products buildings bombed and strafed; landing party burned them during a war exercise.	1942
	* Navy constructed a 15 ft steel tower on SE Island.	1961
Midway Atoll	* Shipwrecked crew lived on islands for 14 months until rescued.[h]	1888-1889
	* Commercial Pacific Cable Co. installation built. About 9,000 tons of soil imported from Honolulu;	

Table 3, continued.

Table 3, continued.

ISLAND GROUP	EVENT	YEAR
	Windbreaks and sand-binding plants, vegetables, and ornamentals planted.[h]	1903-?
	* Pan American Airways built an airport, hotel, power plant, warehouses, and other buildings; planted over 200 species of ornamentals.[h,i]	1935-1941?
	* Military activities, especially during WWII, completely changed the ecology of the island.[i]	1941 to present
Kure Atoll[j]	* Crew of shipwreck on island.	1837-1838
	* Crew of shipwreck on island (killed and ate ca 7000 seabirds, 60 seals, and 1 dog).	1842-1843
	* Crew of shipwreck on island ca 9 weeks before rescued.	1870-1871
	* Emergency building and water tanks built, tree seeds sown (none grew).	1886
	* Navy built radar reflector.	1955
	* LORAN station and runway built and maintained by Coast Guard.	1960 to present

[a]Clapp *et al.* 1977.
[b]Clapp and Kridler 1977.
[c]Amerson 1971.
[d]Clapp 1972.
[e]Ely and Clapp 1973.
[f]Clapp and Wirtz 1975.
[g]Amerson *et al.* 1974.
[h]Bryan 1978.
[i]Neff and DuMont 1955.
[j]Woodward 1972.

[+]HIRAN: High-precision short-range navigation; LORAN: Long-range navigation.

Ironwood has been introduced to seven of the Leeward Islands: Necker Island, French Frigate Shoals, Laysan and Lisianski Islands, and Pearl and Hermes, Midway, and Kure Atolls. Seeds were sown on Tern, Necker, and Laysan Islands by C.S. Judd in 1923 during the Tanager Expedition (Christophersen and Caum 1931); if any of these seeds germinated they did not survive. Ironwood trees were planted on Pearl and Hermes by naval personnel in 1963 to make this low island more visible from a distance. The seedlings subsequently were removed in 1964 by the U.S. Fish and Wildlife Service, as it violates Service policies to introduce alien plants on the Refuge. Ironwood trees were introduced to Laysan at least three times (Ely and Clapp 1973). G.P. Wilder planted a box of ironwood seedlings on Laysan in September 1905 and additional trees in 1930, and Judd sowed seed in 1923 (Christophersen and Caum 1931). Although several of the trees grew, only one is presently alive; it is believed to be one of the 1905 plantings. Coast Guard personnel planted trees around their installations on Tern Island and Kure in the mid 1950s. The original plantings are still alive, but only those on Tern Island have reproduced and become sparingly naturalized. Trees were planted on Lisianski in 1931 (Clapp and Wirtz 1975). Of the original plantings, only about four or five are still alive.

Ironwood has become abundant enough only on Midway to cause a management problem. In 1903, the Commercial Pacific Cable Company took possession of Sand Island, Midway Atoll, and erected permanent concrete buildings. Over the ensuing years, more than 9,000 tons of top soil was brought from Honolulu, vegetable gardens were established, and ornamental trees, shrubs, and flowers were planted. The island changed from a barren sand waste to an oasis around the company installations (Bryan 1978). In 1907, Daniel Morrison, then Assistant Superintendent of the cable station, planted hundreds of ironwood trees to give protection against the wind and sun (Bryan 1978). Descendants of those trees now dominate the island's ecosystem and have impacted the use of the island by both wildlife and humans. Changes in bird populations have occurred with ironwood establishment, favoring birds associated with forested areas and selecting against the species of open habitats. Although black noddies (*Anous minutus melanogenys*) and white terns (*Gygis alba rothschildi*) benefit from the increase in their preferred nesting habitat, other species, such as the black-footed albatross (*Diomedea nigripes*), which nests on the ground in open areas, suffer habitat loss. Also, several of the larger seabird species need open areas because they must run in order to become airborne. Native vegetation also has been impacted by ironwood trees. For example, the beach naupaka (*Scaevola sericea*) is an important sand stabilizer on these islands. Since beach naupaka cannot tolerate heavy shade, the gradual encroachment and replacement of the naupaka by ironwood has not only changed the vegetation components of the island but has rendered the shoreline more susceptible to erosion.

The Federal Aviation Administration requires that the vegetation along the sides and ends of runways be kept below certain heights for safety. The runway aprons were paved with "soil cement," an asphalt that was sprayed on the ground to prevent the encroachment of the ironwoods and reduce maintenance. According to L. Pinter (pers. comm.), however, seeds became lodged in seams and cracks in the soil cement and germinated, their roots threatening to break up the aprons. Graders were used to remove the unwanted vegetation but broke through the soil cement; the ironwoods then invaded the area heavily. Tagged trees, measured at 6-month intervals by the Navy, grew as much as 40 ft (12 m) in 18 months. When the trees were cut down, they regrew from the stumps and roots. Herbicides were not used, although the species is very susceptible to them, because the runoff of the landing strips is the main source of water for the installation. It has been necessary to keep the runway area cleared mechanically, using manual labor to remove the seedlings. Backhoes, chain saws, and heavier equipment are used around the buildings, antenna fields, and other areas. Additionally, the older trees often are infested with ground termites (*Coptotermes formosanus*), which eat out the core of the plant. These trees break easily or lose large limbs during storms, damaging power lines and buildings. Ironwood control is, indeed, a major problem on Midway, but oddly, not on any of the other Leeward Islands.

Accidentally Introduced Alien Plants: *Heliotropium procumbens* and Golden Crown-beard

The heliotrope *Heliotropium procumbens* is a low, mat-forming, perennial herb. It superficially resembles, and shares the same habitat as, the indigenous nena (*Heliotropium curassavicum*) but can easily be distinguished from the latter by its nonsucculent, densely pubescent leaves. *Heliotropium procumbens* is native from the southern United States south to Central and South America and east to the West Indies. It also is found on Guam, where it is considered native (Stone 1970; Fosberg *et al.* 1979).

For about three centuries, Spanish galleons regularly laid over on Guam to load fresh supplies, drop off goods, and unload passengers during the annual Acapulco to Manila trip. Because the Guam populations of this heliotrope most closely resemble those of Mexico (Fosberg, pers. comm.), and because *Heliotropium procumbens* is otherwise native only in America, it seems likely that it represents an early, accidental introduction into Guam from Mexico.

H. procumbens was first discovered in 1975 in Hawai'i, at the Campbell Industrial Park at Barbers Point, O'ahu (Herbst and Ishikawa 5466: herbarium specimen, B.P. Bishop Museum), where it may have been brought accidentally by one of the frequent military flights between Hawai'i and Guam. On O'ahu, it was next seen at construction sites or other areas having a direct connection to Barbers Point. For example, it now grows at the Sand Island gas storage tanks that store gasoline refined at Barbers Point (Herbst and Ishikawa 5643: herbarium specimen, B.P. Bishop Museum). This heliotrope also occurs at the ready-mix concrete plant on Sand Island (D.R. Herbst observation), which uses raw materials brought from Barbers Point. *Heliotropium procumbens* has been spread from the concrete plant to various construction sites on O'ahu, such as the Entomology building at the University of Hawaii at Mānoa, (D.R. Herbst observation) and a Mariner's Ridge housing development (Nagata 1286: herbarium specimen, B.P. Bishop Museum). In recent years, it has gradually migrated to other disturbed sites on O'ahu and, more recently, has been collected on Maui

(Hobdy 1042: herbarium specimen, B.P. Bishop Museum). *H. procumbens* was collected on Tern Island in 1978 and Midway Atoll in 1980. The Midway and Tern Island plants probably were accidentally introduced by military flights from Barbers Point (Herbst 1980).

Although *H. procumbens* is not presently considered a management problem, it does represent a foreign element in the Leeward Islands and potentially could replace native vegetation. No management plan is being considered at present.

Golden crown-beard (*Verbesina encelioides*) has become a management problem. It was first reported on Kure Atoll by H.F. Clay in October 1959, growing near the radar reflector on Green Island. Bermuda grass (*Cynodon dactylon*), ironwood, and sourbush (*Pluchea symphytifolia*) were growing with it. Clay surmised that seeds of these plants were brought on equipment from Midway Atoll in 1955 when the reflector was built (Clay 1961). Golden crown-beard and sweet alyssum (*Lobularia maritima*) have become widespread on Green Island and are undergoing population explosions (Corn *et al.* 1981). Golden crown-beard forms a thicket that limits the space for ground-nesting seabirds, and it has become necessary to mow the central plain of Green Island to restore nesting areas for the birds.

Three taxa of plants endemic to the Leeward Islands are believed to have become extinct within the last 20 or 25 years. These are kāmanomano (*Cenchrus agrimonioides* var. *laysanensis*), *Achyranthes atollensis*, and *Phyllostegia variabilis*. On Green Island, all three grew only in the central plain area. Competition from the golden crown-beard and sweet alyssum undoubtedly had a major role in their extinction on that island.

The Leeward Islands have had a long history of human disturbance, dating back on Nihoa and Necker to prehistoric times (Table 3). Other plants and animals, such as *Pritchardia* sp., sandalwood (*Santalum ellipticum*), Laysan rail (*Porzana palmeri*), and Laysan honeycreeper (*Himatione sanguinea freethii*), on Laysan have become extinct as a direct result.

Naturally Introduced Alien Species: *Nephrolepis multiflora* and Tree Heliotrope

Nephrolepis multiflora, a sword fern, is known from two of the Leeward Islands, Midway (Herbst, unpub. data) and Nihoa (Conant and Herbst 1983; Conant 1985). On Midway, the fern may have been intentionally introduced as an ornamental, or it may have been accidentally brought in with soil or other ornamental plants. Today, it persists in abandoned cultivated areas and appears to have been deliberately cultivated at one time.

The fern was first collected on Nihoa by S. Conant in 1981. A colony was found on the western ridge of Miller Valley, a good distance away from the usual landing sites, and Conant believed that the spores arrived by wind dispersion rather than by accidental or intentional human intervention. In 1983, Conant discovered two additional small colonies that probably were established after her 1981 sighting. In 1983 the original colony had been reduced in size due to an invasion of its habitat by *Eragrostis variabilis*, a native grass. The fern does not appear to be aggressive enough to impact native vegetation, and no management plan is being considered for its control.

Tree heliotrope, *Tournefortia argentea*, naturalized on the main islands since the mid 1800s, is known from six of the Leeward Islands. The species was growing on Pearl and Hermes Atoll as early as 1923 (Amerson *et al.* 1974), where it probably was introduced as an ornamental. It likely was planted on Midway for the same purpose at an earlier date, but we have no record of its original introduction. The oldest trees on Tern Island are believed to have been planted in the mid 1950s by the Coast Guard when that agency was upgrading and landscaping the old naval facility on the island. In 1959, H. Clay found a single young tree growing on the windward shore of Green Island, Kure Atoll (Clay 1961). Two years later a single small tree was found growing on the northwestern beach of Laysan (Tsuda 1965). The fruits of tree heliotrope contain corky tissue and are well adapted for water dispersal; the Kure and Laysan plants probably were the results of natural dispersal by ocean currents.

Studies of mercury in Hawaiian seabird eggs indicated that Midway and Laysan are influenced by the eastward-flowing Kuroshio Extension and North Pacific Current, while the currents around the other sites sampled during the study were more affected by the westward-flowing California Current Extension and North Equatorial Current (Ohlendorf and Harrison 1968). If this is correct, it would seem most probable that Midway was the source of both the Kure and the Laysan introductions of this species.

MANAGEMENT NEEDS

Except for the examples cited above, little has been done to manage the alien plants in the Leeward Islands. This is partly due to the logistics involved, as the islands are remote and getting to them is expensive and time consuming. Moreover, it is difficult or impossible to land on most islands except during the few times of the year that the seas are calm. The main reason for inaction, however, is that there has been little need seen for managing alien plants, other than taking precautions that no new species are introduced. These precautions should include careful cleaning and inspection of all equipment and supplies used in temporary camps established to accommodate scientists studying the wildlife of the Islands.

Most of the naturalized species, except for those on the three islands inhabited by humans, appear to be in balance with the rest of the flora. Some even appear to be beneficial to some of the native birds because they provide additional nesting sites and food and nesting material. Sandbur (*Cenchrus echinatus*), for example, provides additional nesting sites for wedge-tailed shearwaters (*Puffinus pacificus chlororhynchus*) and other burrowing birds. The fibrous root system binds the loose sand, allowing the birds to construct nesting burrows beneath them, while the leaves and seeds are used for nesting material and food by the Laysan finch (*Telespyza cantans*). Tree heliotrope, ironwood, and the naturalized shrubs provide additional nesting sites for noddies and other birds that prefer to nest off the ground.

RECOMMENDATIONS

The vegetation dynamics in the Leeward Islands should be closely monitored to determine changes in composition and to detect new introductions of plants or animals that may impact the existing vegetation. Also, general vegetation maps of each island are needed. Vegetation mapping methods for the Leeward Islands, using aerial photography, have been developed. Detailed distributional maps of all alien species could be completed while ground-truthing the general vegetation maps for all Islands.

Most of our knowledge about the vegetation of the Leeward Islands is based upon casual observations made by ornithologists during the course of their primary research, or by botanists or other biologists during brief visits. The study of ironwood by Apfelbaum *et al.* (1983) on Midway, although limited, constitutes one of the few detailed studies of plants undertaken in the Leeward Islands. Permanent transects and photo stations and the collection of baseline data are necessary for the development of long-term management plans.

A well-illustrated, nontechnical manual of the plants of the Leeward Islands would aid in monitoring the vegetation. It could be used by non-botanists to detect new island introductions or populations as well as distributional changes of both native and alien plants and would be a valuable tool for individuals requiring plant identification as an aspect of their research (*e.g.*, the host plant of an insect or the nesting site of a bird). A complete species list of the plants of Midway Atoll does not yet exist, but when a floristic survey of that atoll is completed, sufficient information about the Leeward Island plants will be available to develop such a manual. The appendix provided here is intended as an interim aid.

It is equally important that the Islands be monitored for the introduction of alien animals, particularly rats (*Rattus* spp.). Shipwrecks provide the most likely means for introducing rats to the islands. Within recent years, fishing vessels have gone aground on Laysan Island and French Frigate Atoll, but fortunately no rats were aboard. Rats have been established on Midway Atoll since 1943 (Kramer 1971), where they have caused damage to both birds and vegetation, as they eat both the seeds and the vegetative parts of plants. Rat damage to beach naupaka is most conspicuous; they chew through the branches just below the tuft of leaves to get the nutrients and moisture in the stem's pith, and the remaining tissue cannot support the weight of the leaves, causing the branches to break.

Mosquitoes, which are naturalized on Midway (U.S. Fish and Wildlife Service, unpub. report), carry fowl pox, infecting young albatross chicks. During their inactive periods, mosquitoes rest in the shelter of the dense vegetation surrounding the nests. Mowing the nesting areas has greatly decreased the incidence of fowl pox in the albatross colony. The "Scaevola dieback" on Lisianski and Laysan Islands may have been caused by an introduced scale insect. The insect may have weakened the shrubs to the point where they could not withstand the stress of a prolonged period of drought followed by a severe winter storm.

Finally, it is important to develop a management plan for the control of alien species, both plant and animal, impacting the vegetation and wildlife of the islands. The plan should include a means for monitoring and detecting new introductions and for determining population and distribution changes of existing naturalized organisms, and it should discuss the various methods of containing or removing the alien species.

ACKNOWLEDGMENTS

The contribution by Warren L. Wagner was supported by a grant from the Irwin Charity Foundation of San Francisco to S.H. Sohmer. We thank S.W. Mill for assisting with the manuscript preparation and editing. We appreciate comments on an earlier draft of the paper by S. Conant and S. Fefer.

APPENDIX

Checklist of the Vascular Plants of the Leeward Hawaiian Islands

This list is based on the collections in the herbaria of the B.P. Bishop Museum and the University of Hawaii, literature citations, and personal collections and observations. It should be considered a preliminary report, as the identification of many of the specimens has not been verified. Each entry includes the scientific name, the common name, the distribution, and the status of the species within the Leeward Islands.

The symbols used for status are:

E - endemic to the Leeward Islands; *i.e.*, occurring naturally nowhere else in the world, including the main Hawaiian Islands.
I - indigenous; *i.e.*, native to the Leeward Islands but occurring naturally (without the aid of humans) elsewhere; however, the plant may be endemic to the Hawaiian archipelago.
X - non-cultivated alien species; *i.e.*, plants of accidental or deliberate introduction that have become naturalized on the islands.
C - cultivated species which have not become naturalized.
S - seeds that arrived through natural dispersal means. These are most commonly found in the debris line.
P - intentionally introduced propagules, as seeds or cuttings, that did not germinate or live.
? - status unknown.
\- an underlined symbol denotes a non-vouchered species reported in the literature as an observation.

In the case of multiple entries, only the symbol which best describes the present status of the species is used. For example, Tern Island, French Frigate Shoals, has both cultivated ironwood trees (C) and wild plants that have seeded from these planted specimens (X); only the "X" symbol is used in the list, as it gives the most dynamic statement of that plant's present status on that island.

Symbols for the island groups are: N = Nihoa; Ne = Necker; FF = French Frigate Shoals; GP = Gardner Pinnacles; La = Laysan Island; Li = Lisianski Island; PH = Pearl and Hermes Atoll; Mi = Midway Atoll; Ku = Kure Atoll.

Appendix, continued.

Appendix, continued.

Taxon	Common Name	N	Ne	FF	GP	La	Li	PH	Mi	Ku
					Island Group					
PTERIDOPHYTA										
DAVALLIACEAE (Sword fern family)										
Nephrolepis multiflora (Roxb.) Morton	sword fern	X							C	
DICKSONIACEAE (Dicksonia family)										
Cibotium sp.	Hawaiian tree fern								C	
POLYPODIACEAE (Common fern family)										
Phymatodes scolopendria (J. Burm.) Ching	laua‘e								C	
PSILOTACEAE (Psilotum family)										
Psilotum nudum (L.) Griseb.	moa								?	
GYMNOSPERMAE										
ARAUCARIACEAE (Araucaria family)										
Araucaria heterophylla (Salisb.) Franco	Norfolk Island pine								C	C
CUPRESSACEAE (Cypress family)										
Cupressus sp.									C	
CYCADACEAE (Cycas family)										
Cycas circinalis L.	sago palm								C	
Cycas revoluta Thunb.	sago palm								C	

Taxon	Common Name	N	Ne	FF	GP	La	Li	PH	Mi	Ku
					Island Group					
MONOCOTYLEDONAE										
AGAVACEAE (Agave family)										
Agave sisalana Perrine	sisal								C	
Cordyline fruticosa (L.) A. Chev.	ti								C	
Cordyline sp.									C	
Dracaena sp.									C	
Sanseviera sp.	bowstring hemp								C	
ALOEACEAE (Aloe family)										
Aloe sp.	aloe									C
ARACEAE (Taro family)										
Alocasia cucullata (Lour.) G. Don	Chinese taro								C	
Anthurium andraeanum Lindl.	anthurium								C	
Caladium bicolor (Aiton) Vent.	caladium								C	
Colocasia esculenta (L.) Schott	taro								C	
Dieffenbachia sp.	dumb cane								C	
Monstera deliciosa Liebm.	monstera								C	
Philodendron sp.	philodendron			C					C	
Rhaphidophora aurea (Andre) Birdsey	taro vine								C	
Syngonium podophyllum Schott	syngonium								C	
Xanthosoma sp.	‘ape								C	
ARECACEAE (Palm family)										
Cocos nucifera L.	coconut			C		X	C	C	C	C
Livistona australis (R. Br.) Mart.	cabbage palm		P	P						

Appendix, continued.

Appendix, continued.

Taxon	Common Name	N	Ne	FF	GP	La	Li	PH	Mi	Ku
					Island Group					
Phoenix dactylifera L.	date palm					P				
Phoenix sp.								C	C	
Pritchardia martii (Gaud.) H.A. Wendl.	loulu			P						
Pritchardia pacifica Seem. & H.A. Wendl.	vin		P	P		P		P		
Pritchardia remota (Kuntze) Becc.	loulu	E								
Pritchardia sp.			P			?			C	
Roystonea sp.									C	
CANNACEAE (Canna family)										
Canna indica L.	canna								C	
COMMELINACEAE (Spiderwort family)										
Commelina diffusa N.L. Burm.	honohono								X	
Dichorisandra thyrsiflora Mikan.	blue ginger								C	
Tradescantia spathacea Sw.	oyster plant								C	
Tradescantia zebrina Bosse	wandering Jew								C	
CYPERACEAE (Sedge family)										
Cyperus alternifolius subsp. *flabelliformis* (Rottb.) Kukenth.	umbrella plant								X	
Cyperus laevigatus L.	makaloa					I				
Cyperus papyrus L.	papyrus								C	
Cyperus rotundus L.	nutgrass			X					X	X
Fimbristylis cymosa subsp. *spathacea* (Roth) T. Koyama	button sedge			I		I			I	
Fimbristylis cymosa subsp. *umbellato-capitata* (Hillebr.) T. Koyama	button sedge					I			I	I

Taxon	Common Name	N	Ne	FF	GP	La	Li	PH	Mi	Ku
					Island Group					
Mariscus javanicus (Houtt.) Merr. & Metcalfe	‘ahu‘awa								I	
Mariscus pennatiformis subsp. *bryanii* (Kukenth.) T. Koyama						E				
Pycreus polystachyos (Rottb.) P. Beauv. subsp. *polystachyos*									I	
LILIACEAE (Lily family)										
Allium cepa L.	onion					X		X	C	
Allium fistulosum L.	bunch onion			C						
Asparagus setaceus (Kunth) Jessop	asparagus fern								C	
Chlorophytum capense (L.) Voss	bracket plant			C						
Crinum asiaticum L.	spider lily								C	C
Hemerocallis sp.	day lily								C	
Hippeastrum sp.									C	
Pancratium littorale Jacq.	spider lily								C	
MUSACEAE (Banana family)										
Heliconia psittacorum L. fil.									C	
Musa sp.	banana								C	
Strelitzia reginae Banks	bird of paradise								C	
ORCHIDACEAE (Orchid family)										
Vanda sp.									C	
Various orchids									C	
PANDANACEAE (Screwpine family)										
Pandanus tectorius S. Parkinson ex Z.	hala			C					C	C

Appendix, continued.

Appendix, continued.

Taxon	Common Name	N	Ne	FF	GP	La	Li	PH	Mi	Ku
					Island Group					
POACEAE (Grass family)										
Ammophila arenaria (L.) Link	European beachgrass								X	
Brachiaria mutica (Forssk.) Stapf	paragrass								X	
Bromus willdenowii Kunth	prairie grass								X	
Cenchrus agrimonioides var. *laysanensis* F. Br.						E			E	E
Cenchrus echinatus L.	sandbur	X		X		X	X		X	X
Chloris barbata (L.) Sw.	swollen fingergrass								X	X
Chloris virgata Sw.	feather fingergrass									X
Cymbopogon citratus (Nees) Stapf	lemon grass			<u>C</u>						
Cynodon dactylon (L.) Pers.	Bermuda grass			X		X		X	X	X
Digitaria ciliaris (Retz.) Koeler	kūkaepua‘a			X					X	X
Digitaria insularis (L.) Ekman	sourgrass								X	
Eleusine indica (L.) Gaertn.	goosegrass			X					X	X
Eragrostis tenella (L.) Roem. & Schult.	Japanese lovegrass								X	X
Eragrostis variabilis (Gaud.) Steud.	‘emoloa	I				I	I	I	I	I
Eragrostis paupera Jedwabn.				I				I	I	I
Eustachys petraea Desf.				X					X	
Hordeum leporinum Link									X	
Lepturus repens (G. Forst.) R. Br. var. *repens*				I					I	
Lepturus repens var. *cinereus* (Burch.) Fosb.										I
Lepturus repens var. *occidentalis* Fosb.								I		
Lepturus repens var. *subulatus* Fosb.				I		I	I	I	I	
Melinis minutiflora P. Beauv.	molasses grass					<u>P</u>				
Panicum torridum Gaud.	kākonakona	I	I							
Paspalum fimbriatum Kunth	Columbia grass									X
Paspalum urvillei Steud.	vasey grass								X	
Paspalum sp.			<u>X</u>							X

Taxon	Common Name	N	Ne	FF	GP	La	Li	PH	Mi	Ku
					Island Group					
Poa annua L.	annual bluegrass								X	X
Polypogon interruptus Kunth	ditch polypogon								X	X
Polypogon monspeliensis (L.) Desf.	rabbitfoot grass								X	
Rhynchelytrum repens (Willd.) Hubb.	Natal redtop								X	
Setaria verticillata (L.) P. Beauv.	bristly foxtail	X		X				X	X	X
Sporobolus africanus (Poir.) Robyns&Tournay	African dropseed								X	
Sporobolus indicus (L.) R. Br.	West Indian dropseed								X	
Sporobolus pyramidatus (Lam) Hitchc.				X					X	X
Sporobolus virginicus (L.) Kunth	beach dropseed					I			I	
Stenotaphrum secundatum (Walter) Kuntze	buffalo grass								X	
Vulpia myuros (L.) Gmelin	foxtail fescue								X	
Zea mays L.	corn			C						
ZINGIBERACEAE (Ginger family)										
Catimbium speciosum (Wendl.) Holttum	shell ginger								C	
Hedychium gardnerianum Ker-Gawl.	kāhili ginger								C	
DICOTYLEDONAE										
ACANTHACEAE (Acanthus family)										
Asystasia gangetica (L.) T. Anderson	asystasia								X	
Odontonema stictum (Nees) Kuntze	odontonema								C	
Ruellia brittoniana E. Leonard									X	
AIZOACEAE (Carpetweed family)										
Sesuvium portulacastrum (L.) L.	‘ākulikuli		I			I	I	I		
Tetragonia tetragonioides (Pall.) Kuntze	New Zealand spinach	X								

Appendix, continued.

Appendix, continued.

Taxon	Common Name	N	Ne	FF	GP	La	Li	PH	Mi	Ku
					Island Group					
AMARANTHACEAE (Amaranth family)										
Achyranthes aspera L. var. *aspera*						X				
Achyranthes atollensis St. John						E		E	E	E
Alternanthera tenella Colla	joyweed								<u>C</u>	
Amaranthus brownii Christoph. & Caum		E								
Amaranthus dubius Mart.	pakai								X	
Amaranthus hybridus L.	green amaranth								X	
Amaranthus spinosus L.	spiny pigweed								<u>X</u>	X
Amaranthus viridis L.	slender amaranth					X			X	X
ANACARDIACEAE (Mango family)										
Mangifera indica L.	mango								<u>C</u>	
Schinus terebinthifolius Raddi	Christmas berry								X	
APIACEAE (Carrot family)										
Ciclospermum leptophyllum (Pers.) Sprague	fine-leaved celery								X	X
APOCYNACEAE (Periwinkle family)										
Allamanda cathartica L.	allamanda								<u>C</u>	
Carissa macrocarpa (Eckl.) DC.	Natal plum								<u>C</u>	
Cascabela thevetia (L.) Lippold	be-still tree								C	
Catharanthus roseus (L.) G. Don	periwinkle								<u>C</u>	
Ervatamia sp.	crape jasmine								<u>C</u>	
Nerium oleander L.	oleander								<u>C</u>	<u>C</u>
Plumeria obtusa L.	Singapore plumeria			C					<u>C</u>	
Plumeria sp.	plumeria								<u>C</u>	

Taxon	Common Name	Island Group N	Ne	FF	GP	La	Li	PH	Mi	Ku
ARALIACEAE (Ginseng family)										
Polyscias guilfoylei (Bull) Bailey	panax								<u>C</u>	
Schefflera actinophylla (Endl.) Harms	octopus tree								<u>C</u>	
ASTERACEAE (Daisy family)										
Arctium lappa L.	gobo								C	
Bidens alba var. *radiata* (Schultz-Bip.) Melchert	Spanish needles								X	X
Bidens pilosa L.	Spanish needles								X	
Conyza bonariensis (L.) Cronq.	ilioha			X		X			X	X
Emilia fosbergii Nicolson	red pualele									X
Gaillardia pulchella Foug.	waikāhuli									X
Gnaphalium purpureum L.	cudweed								X	
Gnaphalium sandwicensium Gaud. var. *sandwicensium*	‘ena‘ena								I	I
Helianthus annuus var. *macrocarpus* (DC.) Cockerell	sunflower								<u>C</u>	C
Lactuca sativa L.	lettuce			<u>C</u>						
Lipochaeta integrifolia (Nutt.) A. Gray	nehe					I				I
Pluchea x *fosbergii* Cooperr. & Galang									X	
Pluchea indica (L.) Less.	Indian pluchea					X			X	
Pluchea symphytifolia (Mill.) Gillis	sour bush			X					X	X
Sonchus oleraceus L.	sow thistle							X	X	X
Tridax procumbens L.	coat buttons								X	
Verbesina encelioides (Cav.) Benth. & Hook.	golden crown-beard								X	X
Wedelia trilobata (L.) Hitchc.	wedelia								X	
Xanthium strumarium var. *canadense* (Mill.) Torr. & A. Gray	cocklebur								X	

Appendix, continued.

Appendix, continued.

Taxon	Common Name	N	Ne	FF	GP	La	Li	PH	Mi	Ku
					Island Group					
BIGNONIACEAE (Bignonia family)										
Spathodea campanulata P. Beauv.	African tulip tree								C	
Tabebuia sp.									C	
BORAGINACEAE (Heliotrope family)										
Cordia sebestena L.	kou-haole								C	
Cordia subcordata Lam.	kou					P				
Heliotropium curassavicum L.	nena	I		I		I				
Heliotropium procumbens var. *depressum* (Cham.) Fosb.				X					X	
Tournefortia argentea L. fil.	tree heliotrope			X		X	X	X	X	X
BRASSICACEAE (Mustard family)										
Brassica campestris L.	field mustard							X		
Brassica nigra (L.) W. Koch	black mustard								X	
Capsella rubella Reut.	shepherd's purse								X	
Coronopus didymus (L.) Sm.	swine cress			X				X	X	X
Lepidium bidentatum var. *o-waihiense* (Cham. & Schlecht.) Fosb.	ʻānaunau					I		I	I	I
Lepidium virginicum L.	pepper grass								X	X
Lobularia maritima (L.) Desv.	sweet alyssum								X	X
Raphanus sativus L.	radish								C	
CACTACEAE (Cactus family)										
Epiphyllum oxypetalum (DC.) Haw.	gooseneck cactus								X	
Hylocereus undatus (Haw.) Britton & Rose	night blooming cereus								C	
CAPPARACEAE (Caper family)										
Capparis sandwichiana DC.	puapilo					I		I	I	

Taxon	Common Name	N	Ne	FF	GP	La	Li	PH	Mi	Ku
		Island Group								
CARICACEAE (Papaya family)										
Carica papaya L.	papaya								<u>C</u>	
CARYOPHYLLACEAE (Pink family)										
Cerastium fontanum										
var. *triviale* (Link) Jalas	common mouse ear								X	
Schiedea verticillata F. Br.		E								
Spergularia marina (L.) Griseb.	mimi'ilio			X					X	X
Stellaria media (L.) Vill.	common chickweed								X	X
CASUARINACEAE (Casuarina family)										
Casuarina equisetifolia L.	ironwood		<u>P</u>	X		C	C	C	X	C
CHENOPODIACEAE (Goosefoot family)										
Atriplex suberecta Verd.	salt bush			X		<u>P</u>				
Beta vulgaris var. *cicla* L.	Swiss chard			<u>C</u>						
Chenopodium murale L.	nettle-leaved goosefoot			X					X	X
Chenopodium oahuense (Meyen) Aellen	'āweoweo	I	I	I		I	I			
Salicornia virginica L.	samphire			X						
CLUSIACEAE (Mangosteen family)										
Calophyllum inophyllum L.	kamani			S		<u>P</u>			C	
Clusia rosea Jacq.	copey								<u>C</u>	
COMBRETACEAE (Indian almond family)										
Conocarpus erectus L.	buttonwood			C		<u>P</u>			<u>C</u>	
Terminalia catappa L.	false kamani			C		<u>P</u>			C	<u>C</u>
Terminalia myriocarpa Heurck & Muell.-Arg.						<u>P</u>				

Appendix, continued.

Appendix, continued.

Taxon	Common Name	N	Ne	FF	GP	La	Li	PH	Mi	Ku
					Island Group					
CONVOLVULACEAE (Morning-glory family)										
Convolvulus sp.						<u>P</u>				
Ipomoea batatas (L.) Lam.	sweet potato								<u>C</u>	
Ipomoea indica f. *albiflora* Stone							I			
Ipomoea indica (J. Burm.) Merr. f. *indica*	koali'awahia	I				I	I		I	I
Ipomoea pes-caprae subsp. *brasiliensis* (L.) Ooststr.	beach morning glory	I		I		I	I		I	I
Ipomoea sp.							<u>S</u>		<u>C</u>	
Ipomoea triloba L.	little bell								X	
Merremia tuberosa (L.) Rendle	wood rose								<u>C</u>	
CRASSULACEAE (Stonecrop family)										
Crassula sp.									<u>C</u>	
Kalanchoe daigremontiana x *tubiflora*									<u>C</u>	
Kalanchoe pinnata (Lam.) Pers.	air plant								<u>X</u>	
Kalanchoe tubiflora (Harv.) Raym.-Hamet	chandelier plant								<u>C</u>	
CUCURBITACEAE (Gourd family)										
Cucumis sativus L.	cucumber			<u>C</u>						
Cucurbita pepo L.	squash, zucchini			<u>C</u>		<u>P</u>			<u>C</u>	
Sicyos maximowiczii Cogn.						I	I	I		I
Sicyos pachycarpus Hook. & Arnott		I				I				
Sicyos semitonsus St. John						E				
EUPHORBIACEAE (Spurge family)										
Acalypha wilkesiana Muell.-Arg.	beefsteak plant								C	
Aleurites moluccana (L.) Willd.	kukui		S	S		S	S	S		

Taxon	Common Name	N	Ne	FF	GP	La	Li	PH	Mi	Ku
		Island Group								
Breynia disticha var. *rosi-picta* Hort.	snow bush								C	
Chamaesyce celastroides (Boiss.) Croizat & Degener var. *celastroides*	‘akoko	I								
Chamaesyce hirta (L.) Millsp.	hairy spurge			X					X	X
Chamaesyce hypericifolia (L.) Millsp.				X					X	X
Chamaesyce maculata (L.) Small									X	X
Chamaesyce prostata (Aiton) Small	small ground fig			X					X	
Codiaeum variegatum var. *pictum* (Lodd.) Muell.-Arg.	croton								C	C
Euphorbia cyathopora Murray	wild poinsettia								X	
Euphorbia heterophylla L.									X	
Euphorbia peplus L.	petty spurge								X	X
Euphorbia pulcherrima Klotz.	poinsettia								C	
Pedilanthus tithymaloides (L.) Poit.	slipper flower								C	
Ricinus communis L.	castor bean								X	
FABACEAE (Pea family)										
Acacia farnesiana (L.) Willd.	klu								X	
Albizia lebbeck (L.) Benth.	siris tree								C	
Caesalpinia bonduc (L.) Roxb.	kākalaioa, kinikini					S				
Canavalia ensiformis (L.) DC.	jackbean					P				
Cassia sp.									C	
Crotalaria incana L.	fuzzy rattle-pod								X	
Crotalaria pallida Aiton	rattle-pod								X	
Delonix regia (Hook.) Raf.	royal poinciana								C	
Desmanthus virgatus (L.) Willd.	virgate mimosa								X	
Desmodium sandwicense E. Mey.	Spanish clover								X	

Appendix, continued.

Appendix, continued.

Taxon	Common Name	N	Ne	FF	GP	La	Li	PH	Mi	Ku
		Island Group								
Dioclea wilsonii Standl.	sea bean					S				
Entada scandens (Rosb.) Benth.						S				
Erythrina variegata var. *orientalis* (L.) Merr.	tiger's claw								C	
Erythrina sp.										<u>C</u>
Glycine soja (L.) Sieb. & Zucc.	soy bean								<u>C</u>	
Haematoxylum campechianum L.	logwood tree		<u>P</u>	<u>P</u>		<u>P</u>				
Leucaena leucocephala (Lam.) de Wit	koa haole			X		<u>P</u>			X	
Medicago lupulina L.	black medic								X	
Medicago polymorpha L.	bur clover									X
Medicago sativa L.	alfalfa								X	
Melilotus alba Medik.	white sweet clover								X	
Melilotus indica (L.) All.	sweet clover								X	
Mucuna gigantea (Willd.) Medik.	kā'e'e					S				
Mucuna urens (L.) Medik.	sea bean					S				
Mucuna sp.					S					
Phaseolus vulgaris L.	green bean			<u>C</u>					<u>C</u>	
Prosopis pallida (Willd.) Kunth	kiawe								C	
Samanea saman (Jacq.) Merr.	monkeypod								<u>C</u>	
Senna surattensis (N. L. Burm.) H. Irwin & Barneby	kolomona								C	
Sesbania tomentosa Hook & Arnott	'ohai	I	I							
Seedling legume							?			
FRANKENIACEAE (Frankenia family)										
Frankenia salina (Molina) I.M. Johnst.				X						

Taxon	Common Name	N	Ne	FF	GP	La	Li	PH	Mi	Ku
		Island Group								
GENTIANACEAE (Gentian family)										
Centaurium erythraea Raf. subsp. *erythraea*	bitter herb								X	
GERANIACEAE (Geranium family)										
Pelargonium hortorum Bailey	fish geranium								C	
GOODENIACEAE (Goodenia family)										
Scaevola sericea Vahl	naupaka kahakai			I		I	I	I	I	I
HYDROPHYLLACEAE (Water-leaf family)										
Nama sandwicensis A. Gray	Hawaiian nama					I	I			
LAMIACEAE (Mint family)										
Phyllostegia variabilis Bitter						E			E	E
Plectranthus scutellarioides (L.) R. Br.	coleus								C	
Stachys arvensis L.	staggerweed								X	
LECYTHIDACEAE (Barringtonia family)										
Barringtonia asiatica (L.) Kurz	barringtonia			C		P	P			
MALPIGHIACEAE (Malphigia family)										
Tristellateia australasiae Rich.	bagnit								C	
MALVACEAE (Mallow family)										
Abutilon grandifolium (Willd.) Sweet	hairy abutilon								X	
Hibiscus esculentus L.	okra								C	

Appendix, continued.

Appendix, continued.

Taxon	Common Name	N	Ne	FF	GP	La	Li	PH	Mi	Ku
					Island Group					
Hibiscus rosa-sinensis L.	red hibiscus								<u>C</u>	
Hibiscus tiliaceus L.	hau			C		<u>C</u>		<u>P</u>	C	
Hibiscus sp.									C	<u>C</u>
Malva parviflora L.	cheese weed			X					X	
Malvastrum coromandelianum (L.) Garcke subsp. *coromandelianum*	false mallow							X	X	
Malvaviscus arboreus Cav.	turks cap								C	
Sida fallax Walp.	ʻilima	I							I	
Thespesia populnea (L.) Correa	milo		<u>P</u>	<u>P</u>		<u>P</u>			C	<u>C</u>
MORACEAE (Mulberry family)										
Ficus benghalensis L.	Indian banyan								C	
Ficus elastica Hornem.	Indian rubber tree								<u>C</u>	
Ficus microcarpa L. fil.	Chinese banyan								C	
Ficus sp.				C					<u>C</u>	
Morus alba L.	white mulberry								C	
MYRTACEAE (Myrtle family)										
Eugenia uniflora L.	Surinam cherry								C	
Psidium guajava L.	guava								<u>C</u>	
NYCTAGINACEAE (Four o'clock family)										
Boerhavia herbstii Fosb.							I	I		
Boerhavia repens L.	alena	I		I		I	I	I	I	I
Bougainvillea spectabilis Willd.	bougainvillea								<u>C</u>	
Mirabilis jalapa L.	four-o'clock								X	
Pisonia grandis R. Br.							I			

Taxon	Common Name	N	Ne	FF	GP	La	Li	PH	Mi	Ku
					Island Group					
OLEACEAE (Olive family)										
Jasminum sambac (L.) Aiton	pikake								$\underline{C}$	
Noronhia emarginata (Lam.) Stadm.	Madagascar olive								$\underline{C}$	$\underline{C}$
ONAGRACEAE (Evening primrose family)										
Oenothera laciniata J. Hill									X	X
OXALIDACEAE (Wood sorrel family)										
Oxalis corniculata L.	yellow wood sorrel								X	X
Oxalis corymbosa DC.	pink wood sorrel								X	
PASSIFLORACEAE (Passion flower family)										
Passiflora edulis Sims	liliko'i								C	
PLANTAGINACEAE (Plantain family)										
Plantago lanceolata L.	English plantain			X					X	
Plantago major L.	common plantain								X	
PLUMBAGINACEAE (Leadwort family)										
Plumbago auriculata Lam.	plumbago								$\underline{C}$	
POLYGONACEAE (Buckwheat family)										
Coccoloba uvifera (L.) L.	sea grape			C		$\underline{P}$			C	$\underline{C}$
Rumex albescens Hillebr.		I								
PORTULACACEAE (Purslane family)										
Portulaca lutea G. Forster	'ihi	I	I	I	I	I	I	I	I	
Portulaca oleracea L.	common purslane	X		X		X		X	X	X

Appendix, continued.

Appendix, continued.

Taxon	Common Name	N	Ne	FF	GP	La	Li	PH	Mi	Ku
		Island Group								
Portulaca villosa Cham.		I								
Portulacaria afra Jacq.	jade tree			<u>C</u>					<u>C</u>	
PRIMULACEAE (Primrose family)										
Anagallis arvensis L.	scarlet pimpernel								X	X
ROSACEAE (Rose family)										
Rosa sp.	rose								<u>C</u>	
RUBIACEAE (Coffee family)										
Gardenia sp.									<u>C</u>	
RUTACEAE (Citrus family)										
Citrus sp.									<u>C</u>	
Murraya paniculata (L.) Jack	mock orange								C	
SANTALACEAE (Sandalwood family)										
Santalum ellipticum Gaud.	‘iliahialo‘e					I				
SCROPHULARIACEAE (Figwort family)										
Bacopa monnieri (L.) Wettst.	‘ae‘ae								I	
SOLANACEAE (Nightshade family)										
Capsicum annuum L.	nīoi								<u>C</u>	
Lycopersicon esculentum Mill.	tomato		<u>P</u>	<u>C</u>						
Nicotiana glauca R.C. Graham	tree tobacco					<u>P</u>				
Nicotiana tabacum L.	tobacco					X	<u>X</u>			

Taxon	Common Name	N	Ne	FF	GP	La	Li	PH	Mi	Ku
					Island Group					
Solanum americanum Mill.	pōpolo	I				I	I	I	I	I
Solanum melongena L.	eggplant			<u>C</u>						
Solanum nelsonii Dunal		I				I		I	I	I
Solanum tuberosum L.	potato					X				
STERCULIACEAE (Cocoa family)										
Waltheria indica L.	ʻuhaloa								I	
TROPAEOLACEAE (Nasturtium family)										
Tropaeolum majus L.	nasturtium								X	
URTICACEAE (Nettle family)										
Pilea microphylla (L.) Liebm.	artillery plant								X	
VERBENACEAE (Verbena family)										
Lantana camara L.	lantana								X	
Phyla nodiflora (L.) Greene									X	
Stachytarpheta dichotoma (Ruiz & Pav.) Vahl	oī								X	
Stachytarpheta jamaicensis (L.) Vahl	ōwī								X	
Verbena litoralis Kunth	haʻuōwī								X	
Vitex trifolia var. *subtrisecta* (Kuntze) Mold. f. *subtrisecta*	pōlinalina								C	
Vitex trifolia var. *subtrisecta* f. *variegata* Mold.	pōlinalina								C	
VIOLACEAE (Violet family)										
Viola odorata L.	violet								<u>C</u>	

Appendix, continued.

Appendix, continued.

Taxon	Common Name	Island Group								
		N	Ne	FF	GP	La	Li	PH	Mi	Ku
VITACEAE (Grape family)										
Vitis sp.	grape								C	
ZYGOPHYLLACEAE (Tribulus family)										
Tribulus cistoides L.	nohu	I		I	S	I	I	I	I	I

Literature Cited

Amerson, A.B., Jr. 1971. The natural history of French Frigate Shoals, Northwestern Hawaiian Islands. *Atoll Res. Bull.* 150.

Amerson, A.B., Jr., R.B. Clapp, and W.O. Wirtz II. 1974. The natural history of Pearl and Hermes Reef, Northwestern Hawaiian Islands. *Atoll Res. Bull.* 174.

Apfelbaum, S.I., J.P. Ludwig, and C.E. Ludwig. 1983. Ecological problems associated with disruption of dune vegetation dynamics by *Casuarina equisetifolia* L. at Sand Island, Midway Atoll. *Atoll Res. Bull.* 261.

Armstrong, R.W., ed. 1983. *Atlas of Hawai'i*, 2nd ed. Honolulu: Univ. Hawaii Pr.

Bryan, E.H., Jr. 1978. *The Northwestern Hawaiian Islands: an annotated bibliography*. Honolulu: U.S. Fish and Wildl. Serv.

Christophersen, E., and E.L. Caum. 1931. *Vascular plants of the Leeward Islands, Hawaii*. B.P. Bishop Mus. Bull. 81. Honolulu: Bishop Mus. Pr.

Clague, D., G.B. Dalrymple, and R. Moberly. 1975. Petrography and K-Ar ages of dredged volcanic rocks from the western Hawaiian Ridge and the southern Emperor Seamount Chain. *Bull. Geol. Soc. Amer.* 86: 991-998.

Clapp, R.B. 1972. The natural history of Gardner Pinnacles, Northwestern Hawaiian Islands. *Atoll Res. Bull.* 163.

Clapp, R.B., and E. Kridler. 1977. The natural history of Necker Island, Northwestern Hawaiian Islands. *Atoll Res. Bull.* 206.

Clapp, R.B., E. Kridler, and R.R. Fleet. 1977. The natural history of Nihoa Island, Northwestern Hawaiian Islands. *Atoll Res. Bull.* 207.

Clapp, R.B., and W.O. Wirtz II. 1975. The natural history of Lisianski Island, Northwestern Hawaiian Islands. *Atoll Res. Bull.* 186.

Clay, H.F. 1961. Narrative report of botanical field work on Kure Island. *Atoll Res. Bull.* 78.

Conant, S. 1985. Recent observations on the plants of Nihoa Island, Northwestern Hawaiian Islands. *Pac. Sci.* 39(2):135-149.

Conant, S., and D.R. Herbst. 1983. A record of *Nephrolepis multiflora* from Nihoa Island, Northwestern Hawaiian Islands. *Newsletter, Hawn. Bot. Soc.* 22:17-19.

Corn, C.A., D.R. Herbst, and C.H. Lamoureux. 1981. Floristic changes on Green Island, Kure Atoll. In *Abstracts, XIII Internatl. Bot. Congr.*, coll. D.J. Carr, 110.

Dalrymple, G.B., M.A. Lanphere, and E.D. Jackson. 1974. Contributions to the petrography and geochronology of volcanic rocks from the Leeward Hawaiian Islands. *Bull. Geol. Soc. Amer.* 85: 727-738.

Dalrymple, G.B., D.A. Clague, and M.A. Lanphere. 1977. Revised age for Midway volcano, Hawaiian volcanic chain. *Earth Planetary Sci. Letters* 37:107-116.

Ely, C.A., and R.B. Clapp. 1973. The natural history of Laysan Island, Northwestern Hawaiian Islands. *Atoll Res. Bull.* 171.

Fosberg, F.R, M.-H. Sachet, and R. Oliver. 1979. A geographical checklist of the Micronesian Dicotyledonae. *Micronesia* 15: 41-295.

Herbst, D.R. 1980. Miscellaneous notes on the Hawaiian flora I. *Phytologia* 45(1):67-81.

Kramer, R.J. 1971. *Hawaiian land mammals*. Tokyo: Charles E. Tuttle Co.

Neal, M.C. 1965. *In Gardens of Hawaii*. B.P. Bishop Mus. Spec. Pub. 50. Honolulu: Bishop Mus. Pr.

Neff, J.A., and P.A. DuMont. 1955. A partial list of the plants of the Midway Islands. *Atoll Res. Bull.* 45.

Ohlendorf, H.M., and C.S. Harrison. 1968. Mercury, selenium, cadmium and organochlorides in eggs of three Hawaiian seabird species. *Environmental Pollution* (Ser. B) 11: 169-191.

Stone, B.C. 1970. The flora of Guam. *Micronesia* 6:1-659.

Tsuda, R.T. 1965. Marine algae from Laysan Island with additional notes on the vascular flora. *Atoll Res. Bull.* 110.

Woodward, P.W. 1972. The natural history of Kure Atoll, Northwestern Hawaiian Islands. *Atoll Res. Bull.* 164.

FIRETREE (MYRICA FAYA) DISTRIBUTION IN HAWAI'I

Louis D. Whiteaker and Donald E. Gardner

ABSTRACT

The primary objective of this study was to map the distribution and abundance of firetree (*Myrica faya*) in the state of Hawai'i. Reconnaissance data were used to map 85,912 a (34,365 ha) of infestation throughout the State. Infestations of 72,265 a (28,906 ha) occur on the island of Hawai'i, 4,770 a (1,908 ha) on Maui, 2,518 a (1,007 ha) on Lāna'i, 435 a (174 ha) on O'ahu, and 5,925 a (2,370 ha) on Kaua'i. Distribution ranges in elevation from as low as 1,400 ft (425 m) on Lāna'i to as high as 6,400 ft (1,940 m) on the slopes of Haleakalā on Maui. Firetree occurs on recent, thin ash over pāhoehoe lava as well as on deep, well-developed silty clay loam soil. It is found in montane rain forest habitats and in dry scrub (marginal through submontane seasonal) forest. The distribution patterns of firetree suggest that this species has not yet reached the limits of its potential distribution in Hawai'i. Observations on the phenology of firetree were made at two-week intervals from March 1983 through March 1985. Data were taken at three sites and were related to the data from nearby weather stations. Information on flowering, fruiting (mature and immature), leaf flushing, leaf fall, and fruit drop was collected. Analysis of the phenologic patterns observed showed flowering, fruiting, and fruit drop to be endogenously controlled, while leaf flushing and leaf fall were influenced by environmental variables.

INTRODUCTION

Fire or faya tree (*Myrica faya*, Myricaceae) is a native of the Azores, Madeira, and the Canary Islands in the Atlantic Ocean. It was introduced to Hawai'i in the late 1800s by Portuguese immigrants, presumably as an ornamental (Fosberg 1937). Because it forms multiple branches near the base of the main stem (Smathers and Gardner 1979), firetree has often been described as a shrub or small tree reaching heights of only 13 to 20 ft (4-6 m) (Fosberg 1937; Neal 1965; Hasselwood and Motter 1983). However, in the Hāmākua region of the island of Hawai'i, firetree grows to over 50 ft (16 m) high, forming dense canopies with an understory devoid of other plant life (Smathers and Gardner 1979).

Firetree has narrow, pointed, smooth, shiny, dark green leaves with entire or toothed margins (Neal 1965; Hasselwood and Motter 1983). Although firetree has a strong tendency toward dioecism, a few staminate flowers are found on pistillate plants and a few pistillate flowers are found on staminate plants (Gardner 1985). Staminate flowers with four stamens each are borne on small catkins. Three pistillate flowers that may be joined and accompanied by one bract are also grouped in small catkins. Fruits are small, edible drupes that form dense clusters changing from green through red to purple when ripe (Fosberg 1937; Lawrence 1951; Neal 1965).

The purpose of this study was to map the distribution and abundance of firetree and to observe the phenological cycles of this species in the state of Hawai'i. Distribution studies will provide a database for location of all infestations, both for implementation of a biological control program, should an appropriate agent be found, and for future studies of the spread of this species in Hawai'i. The phenology data will help to determine the timing of control programs.

HISTORY OF SPREAD AND CONTROL EFFORTS

The Hawaiian Sugar Planters' Association obtained seeds of firetree from a Portuguese farmer on the island of Hawai'i for use in reforestation attempts (Fosberg 1937). Plantings were made on the islands of Kaua'i, O'ahu, and Hawai'i (Skolmen 1979), most of them in the 1920s. The aggressive, noxious character of this species was soon noted, along with its spread to Maui (Fosberg 1937). The continued spread of firetree led the Territorial Board of Agriculture and Forestry to begin attempts to eradicate it in 1944 (Neal 1965). Managers of Shipman Estate observed firetree spreading in the Volcano area on the island of Hawai'i in the mid-1940s. In the mid-1960s, Shipman attempted to control the species there (T. Lindsey, pers. comm.). It was declared noxious for state land leases because it spreads rapidly and forms a dense cover that crowds out desirable species (Hosaka 1945; Haselwood and Motter 1983). Firetree has continued to spread despite efforts by the State and the National Park Service, and it now occurs on all major Hawaiian Islands except Kaho'olawe, Moloka'i, and Ni'ihau. Estimates of the total infested area have increased over the years: 8,200 a (3,280 ha) (Yamayoshi 1954), 21,375 a (8,550 ha) (Anonymous 1962), 40,000 a (16,000 ha) (Walters and Null 1970), and the most recent estimate, showed 53,938 a (21,575 ha) to be supporting firetree populations of various densities (Watanabe 1982). Hawai'i contained 50,000 a (20,000 ha); Maui, 3,000 a (1,200 ha); O'ahu, 100 a (40 ha); Kaua'i, 325 a (130 ha); and Lāna'i, 512 a (205 ha).

Firetree was thought to occur where average annual rainfall is 35 in. (900 mm) or more in mesic low-elevation forest and wet low- and high-elevation forest [vegetation zones C_1, C_2, D_1, and D_2 of Ripperton and Hosaka (1942)] (Hosaka and Thistle 1954; Anonymous 1962).

Within these broad limits, however, the tree can adapt to a wide range of habitats. Clarke (1978) described the distribution of firetree in Hawaii Volcanoes National Park as a horizontal band between 2,200 and 4,000 ft (665-1,210 m) elevation, including 11 soil types and 15 of the vegetation units described by Mueller-Dombois and Fosberg (1974). Average annual rainfall within this distribution is from 50 in. (1,270 mm) to over 100 in. (2,540 mm), and average annual temperature ranges from 60 to 72 F (22 C) (Clarke 1978). The wide range of habitats where firetree has been found suggests that it has not yet occupied its full potential range.

An estimated 15,000 a (6,000 ha) of infestation along the Hāmākua coast on the island of Hawai'i has been classified as "heavy" (Watanabe 1982). In this region firetree grows over 50 ft (16 m) tall and forms dense, interlocking canopies with no understory (Smathers and Gardner 1979). The absence of other plant species under the canopy may be due partially to shading. However, allelopathic activity has been reported for a closely related species (*M. cerifera*) from the southeastern United States (Dunevitz and Ewel 1981). Thus, the lack of understory in firetree stands may be due not only to canopy shading, but also to allelopathic activity (Smith 1985).

Current evidence suggests that the seeds of firetree are dispersed primarily by birds (Smathers and Gardner 1979; LaRosa *et al.* 1985). The extensive, uniform distribution of the tree in remote areas and the close spatial association of this species with other tree species, especially 'ōhi'a (*Metrosideros polymorpha*), suggest dispersal of seeds by birds. Birds commonly associated with dispersal include the Japanese white-eye (*Zosterops japonica*), the common myna (*Acridotheres tristis*), the red-billed leiothrix (*Leiothrix lutea*), and the 'ōma'o (*Phaeornis obscurus*) (Clarke 1978; Smathers and Gardner 1979; Gardner and Davis 1982; LaRosa *et al.* 1985). Low germination rates of firetree seeds collected in the field, contrasted with copious seed production and rapid dispersal of the species, has led to the hypothesis that scarification from bird ingestion greatly improves germination rates (Clarke 1978). This idea was supported by experiments that showed increased amounts and rates of germination due to mechanical or chemical scarification of seeds of *M. cerifera* (Ewel *et al.* 1982). However, germination tests of firetree in Hawai'i showed no significant difference in amounts or rates of germination between seeds that passed through captive birds and apparently mature, viable seeds collected in the field from trees (LaRosa *et al.* 1985). Feral pigs (*Sus scrofa*) have also been identified as possible dispersal agents for seeds of firetree. Firetree seedlings have been observed growing directly from pig- rooted areas in Hawaii Volcanoes National Park (Clarke 1978). Stomach content analyses of feral pigs in the Park have shown significant percentages of firetree seeds during certain times of the year (Stone and Taylor 1984).

Firetree control efforts have been implemented primarily by the state of Hawai'i and the National Park Service. The effort by the State has been conducted over the last 25 years, but variations in availability of funding and manpower have resulted in fluctuations in intensity of this effort. Herbicides are the primary control agents. Of the various herbicides used by the State, Tordon 22K (picloram) has been the most effective, producing a complete canopy kill and a 99% control of resprouting (Walters and Null 1970; Smathers and Gardner 1979; R. Kami, pers. comm.). In Hawaii Volcanoes National Park, basal bark application of a 4% solution of KURON (silvex) in diesel oil was effective, and a method to introduce ROUNDUP (glyphosate) directly into the vascular tissue of firetree through a cut branch was suggested for remote areas and less-than-ideal weather conditions (Gardner and Kageler 1982). In some areas, pasturelands have been cleared of invading firetree with bulldozers by private landowners. However, follow-up monitoring and/or herbicide treatment of stumps and brush piles are needed with mechanical methods to control resprouting (R. Kami, pers. comm.).

F. Bianchi, an exploratory entomologist from Hawai'i, was sent by Hawaiian Sugar Planters' Association to the native habitats of firetree in 1955 to search for potential biological control agents. A fungus disease caused by *Dothiorella berengeriana* initially appeared promising but was later rejected because tests conducted in Portugal showed a lack of host specificity (Gardner and Davis 1982; Gardner 1984). Other disease-causing organisms were observed on this trip, but none have been sufficiently tested for use as control agents. Krauss (1964) summarized his and others' previous observations of insects associated with firetree and related species. However, insects collected during exploratory efforts either proved ineffective in tests of control or failed to propagate. Krauss (1964) also discussed other apparently pathogenic organisms on *Myrica* spp., but no attempt has been made to test these as control agents in Hawai'i (Gardner and Davis 1982). Observations from another exploratory trip indicated that firetree is generally abundant and healthy in its native range and that biological agents in these areas are not important factors in limiting the growth, reproduction, or distribution of this species. However, at least two diseases and two insects were found to merit further investigation; these should be the subjects of current research in a biological control program for firetree in Hawai'i (Gardner 1984; Hodges and Gardner 1985).

DATA COLLECTION

Maps of the distribution of firetree in Hawai'i presented herein (Figs. 1-5) are based on reconnaissance data. Reconnaissance was done in or near areas where firetree was reported by knowledgeable individuals. Final drafts of the distribution maps were plotted on U.S. Geological Survey topographic maps (Whiteaker and Gardner 1985). Phenology data (flowering, immature fruiting, mature fruiting, leaf flushing, leaf fall, and fruit drop) were collected at three sites on the island of Hawai'i (Whiteaker and Gardner 1987).

RESULTS

Distribution

Island of Hawai'i. Firetree was found between Laupāhoehoe and Honoka'a (Hāmākua District) from 2,000 to 4,400 ft (600-1,335 m) elevation, with 29,390 a (11,755 ha) infested (Whiteaker and Gardner 1985). Much of this area is ranchland where firetree has been cleared from pasture areas, but extensive areas are still covered with heavy or moderate density stands of large (65 ft or 20 m tall) trees.

Another major infestation on Hawai'i Island is in the vicinity of Volcano Village, including a large portion of Hawaii Volcanoes National Park. The distribution ranges between 1,800 and 4,000 ft (545-1,210 m) elevation and covers 41,500 a (16,600 ha). The Volcano area has a wide variety of habitats, including montane 'ōhi'a rain forest near Volcano Village and the Kīlauea summit area, submontane seasonal forest near 'Āinahou Ranch and Hilina Pali Road, montane seasonal forest converted to pasture on Keauhou Ranch adjacent to the Mauna Loa Strip Road area of the Park, and dry scrubland with scattered trees in the Ka'ū Desert. The vegetation, climate, and substrates of the wide variety of habitats in which firetree is found within the Park have been described in detail by Doty and Mueller-Dombois (1966) and Mueller-Dombois and Fosberg (1974).

An infestation of firetree occurs on the Kapāpala Ranch (Ka'ū District) adjacent to the Ainapō Cabin between 3,500 and 3,760 ft (1,060-1,140 m) elevation. The infestation occupies 80 a (32 ha). This ranchland consists of pasture mixed with stands of alien shrubs and trees on steeper slopes. Firetree has also been reported in the Kiolaka'a-Kea'ā Homesteads Addition portion of the Ka'ū Forest Reserve at 2,300 ft (720 m) elevation. A single plant was recorded on a Hawaii State Division of Forestry survey of this area (L.W. Cuddihy and S.J. Anderson, pers. comm.).

Firetree occurs on the slopes of Hualālai (North Kona District) between 4,850 and 6,040 ft (1,470-1,830 m) elevation. The infestation covers 1,300 a (520 ha). This population is in a remote area on privately owned ranchland (R. Kami, pers. comm.).

The total area of firetree distribution on the island of Hawai'i is 72,265 a (28,906 ha) (Fig. 1), including 30,500 a (12,200 ha) within Hawaii Volcanoes National Park.

Island of Maui. Firetree is found in the Kula area on the western slopes of Haleakalā between 3,200 and 6,400 ft (970-1,940 m) elevation (Fig. 2). The area consists of ranchland and small, private residential parcels of land. Total area of firetree distribution is 4,770 a (1,908 ha). The distribution may be limited by the efforts of Haleakalā Ranch, which controls weed species. Only one individual firetree along the Haleakalā Highway, at 4,250 ft (1,290 m) elevation, and a small stand in Hāpapa Gulch were noted within the boundaries of Haleakalā Ranch (Fig. 2). However, these individuals indicate that the ranch is suitable habitat for an infestation of this species. The Hawaii Department of Agriculture attempted to control firetree on Maui in the mid-1970s, but logistical

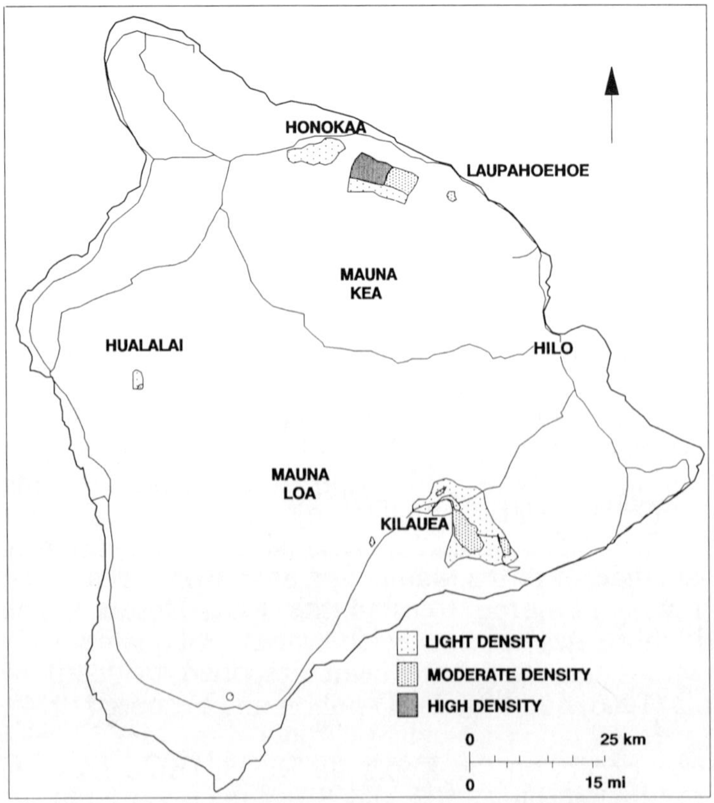

Figure 1. Distribution of firetree on the island of Hawai'i.

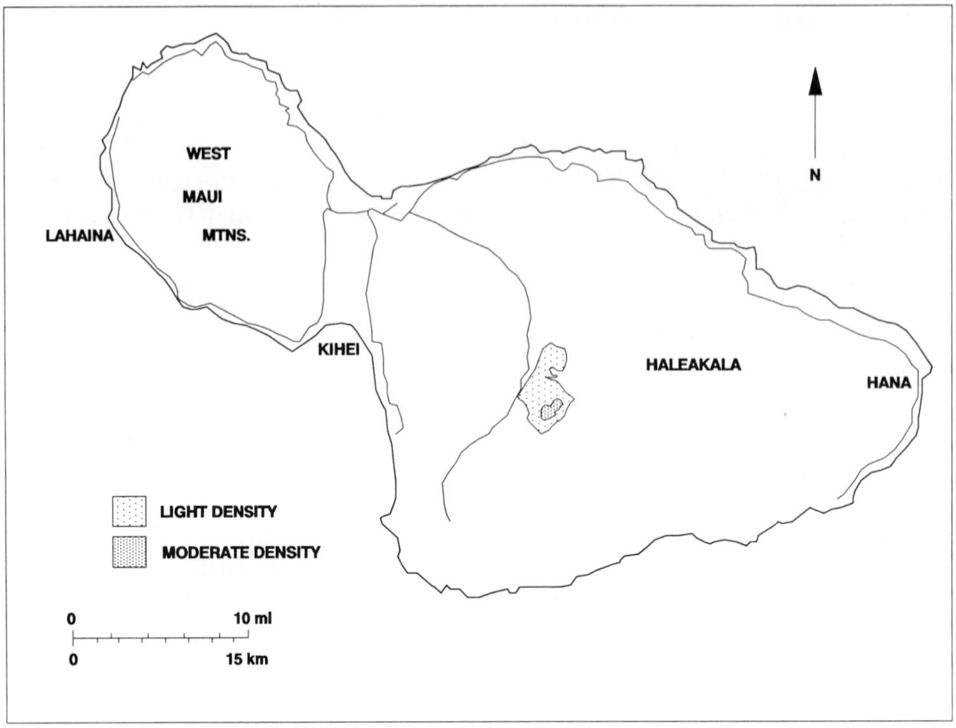

Figure 2. Distribution of firetree on the island of Maui.

difficulties related to the large number of small, privately owned parcels of land led to abandonment of the effort (E. Tamura, pers. comm.).

Island of Lāna'i. Firetree occurs in the mountainous portion, or Lāna'i Hale area, of the Island (Fig. 3). The distribution ranges from 1,400 ft (425 m) elevation in protected gullies to the summit of the Island at 3,370 ft (1,020 m). The total area occupied is 2,518 a (1,020 ha). This range of firetree on Lāna'i is rough, mountainous terrain consisting of narrow ridges and steep slopes.

Island of O'ahu. Firetree is found in the Wai'anae mountains between 2,000 and 3,127 ft (610-948 m) elevation (Fig. 4). The center of distribution is at the southern end of the Wai'anae mountains between Mauna Kapu and Pu'ukaua in the Honouliuli Forest Reserve, where it was planted by the Territory of Hawaii for reforestation (Skolmen 1979). A smaller, noncontiguous population occurs to the northwest at Pu'u Kawiwi, and to the north at Pu'u Hāpapa (Fig. 4). The total area occupied is 435 a (174 ha). As on the island of Lāna'i, the range is rough, mountainous terrain with narrow ridges and steep slopes.

Firetree was planted in several other areas on O'ahu, including sites in the Ko'olau mountains (Skolmen 1979). However, populations in these locations have not been observed by persons familiar with these areas (J. Obata, pers. comm.), and it is assumed that plantings were at elevations not suited for reproduction.

Island of Kaua'i. The major infestation of firetree is found just west and north of Waimea Canyon, mostly within Waimea Canyon and Kōke'e State Parks. A separate population just east of Waimea Canyon is centered around Wai'alae Cabin on Wai'alae Stream in the Nā Pali-Kona Forest Reserve and is partially within the Alaka'i Wilderness Preserve. Firetree is also found on the northwestern slopes of the Island below the mesic native forest and/or forestry plantings (Fig. 5). These populations occur between 1,800 ft (545 m) on the northwestern slopes and 4,200 ft (1,270 m) within Kōke'e State Park. The total area occupied is 5,925 a (2,370 ha). On Kaua'i, firetree occurs in a variety of habitats, from montane rain forest with little topographic relief to sparsely vegetated, dry, eroded, steep slopes.

All Islands. Total area of firetree distribution in Hawai'i is 85,912 a (34,365 ha). More detailed maps and descriptions of the distribution of firetree in Hawai'i are presented in a separate report (Whiteaker and Gardner 1985).

Phenology

Krauss (1964) reported that firetree in its native habitat showed abundant male flowers and a few small fruits in June; in July, most male flowers were dry, and many green (and a few purple) fruits were seen. Krauss observed all stages of fruiting in September, and by November many ripe and fallen fruits were reported. Gardner (1985) reported various

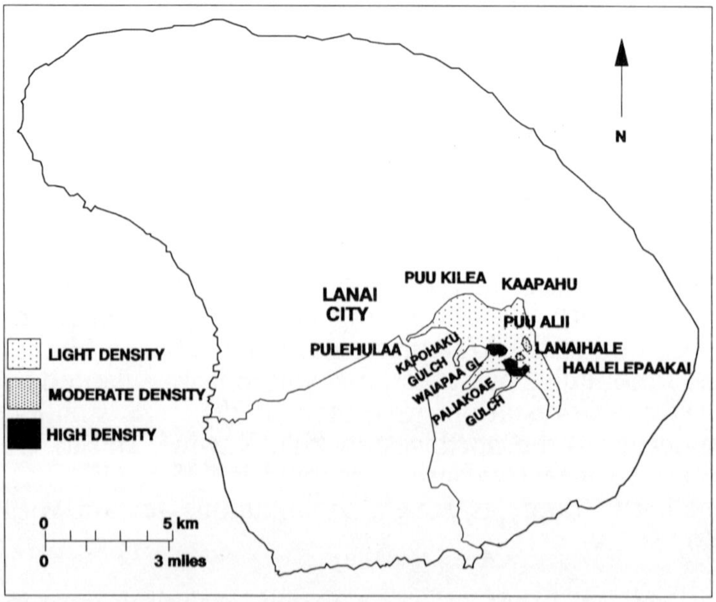

Figure 3. Distribution of firetree on the island of Lāna'i.

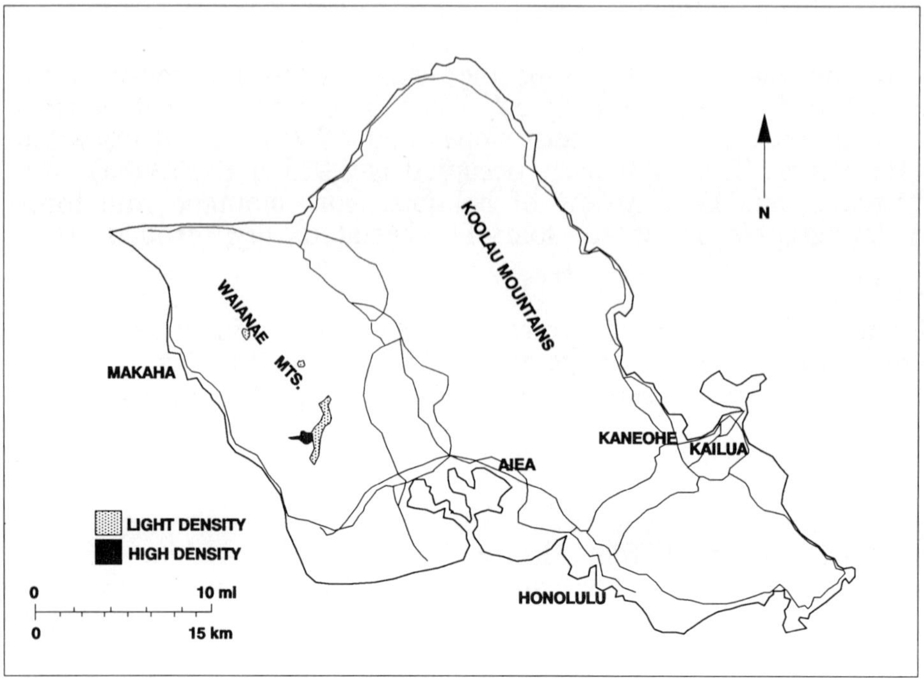

Figure 4. Distribution of firetree on the island of O'ahu.

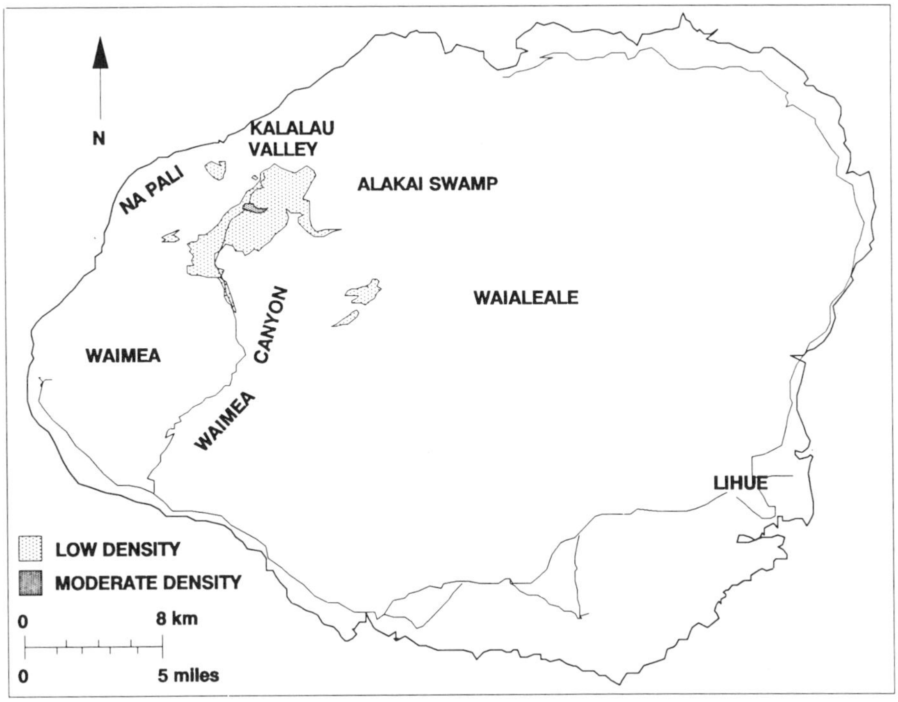

Figure 5. Distribution of firetree on the island of Kaua'i.

stages of flowering with some immature fruit development during a visit to the native habitats of firetree in April and May.

In this study, flowering, immature fruiting, mature fruiting, and fruit drop showed similar patterns at all three study sites on Hawai'i Island (Hilina Pali and Byron Ledge in Hawaii Volcanoes National Park, and the Eldon English property in Volcano). Flowering showed maximum activity in June and minimum activity in December and January in the two years observed. Peaks of immature fruiting followed in August and September, with minimum activity in February and March. Maximum mature fruiting was observed in November, and minimum mature fruiting was seen in May. Fruit drop followed a similar pattern to mature fruiting in the one year observed. Leaf flushing showed similar patterns at two of the sites, with maximum activity in June. At the third site, leaf flushing also had peaks in June, but the pattern over the entire observation period differed somewhat from the other two sites. Two sites also had similar patterns in the amounts of leaf fall, but the amounts of leaf fall at the third site showed a different pattern. A more extensive presentation and discussion of these data, including graphs, appears in a separate report (Whiteaker and Gardner 1987).

DISCUSSION

Distribution

The distribution of firetree described here indicates that this alien species is adapted to a wide variety of habitats in Hawai'i. It occurs on recent, thin ash over pāhoehoe lava on Hawai'i Island and on soil classified by Foote *et al.* (1972) as deep, well-developed silty clay loam on Kaua'i and Hawai'i. It occurs in montane rain forest habitats on Hawai'i and Kaua'i islands and in dry scrub marginal to submontane seasonal forest on Kaua'i and Hawai'i. It grows as low as 1,400 ft (425 m) elevation on Lāna'i and as high as 6,400 ft (1,940 m) on Maui. It is found on nearly vertical slopes on Lāna'i, in volcanic craters in Hawaii Volcanoes National Park, and on nearly level terrain or gentle slopes in Kōke'e State Park, Kaua'i. These extremes of several habitat factors in areas where firetree occurs suggest that the species may be able to occupy sites that are intermediate along these habitat factor gradients where it does not now occur. Competition from other plants may also be a factor on some sites, and some small, outlying populations may have been missed during the survey. The distribution of firetree has likely not reached its potential extent in Hawai'i.

Firetree infestations in leeward areas such as Kula on Maui and North Kona, Hawai'i, extend up to 6,400 ft (1,940 m) and 6,040 ft (1,830 m) elevation. These records imply that infestations on Kapāpala and Keauhou ranches and Hawaii Volcanoes National Park (in Ka'ū District), with upper limits of 3,760 ft (1,140 m) and 4,000 ft (1,210 m), have not yet reached their potential upper elevational extent. Similarly, firetree is found down to 2,000 ft (600 m) and lower in protected sites in Hāmākua and Hawaii Volcanoes National Park on Hawai'i Island, on Lāna'i, and on Kaua'i. Infestations along Highway 11 near Volcano Village and on Kapāpala Ranch may not have reached their potential lower elevation extent.

The rate at which firetree can spread into suitable habitats appears to be very rapid. A single individual was noted in Hawaii Volcanoes National Park between Kilauea Military Camp and Hawaiian Volcano Observatory by F.R. Fosberg in 1961 (Doty and Mueller-Dombois 1966). By 1978, 1,304 a (609 ha) in the Park were mapped as supporting firetree infestations of various densities (Clarke 1978). Data from the present study show 30,134 a (12,200 ha) infested with firetree in the Park, including a 704-a (285-ha) increase during the two-year course of the study. These records indicate a 20-fold increase in infested area within Hawaii Volcanoes in approximately eight years. Firetree has since spread into suitable habitats that were not occupied at the time of this distribution survey (1983-1985).

The distribution of firetree on Kaua'i (Fig. 5) includes infestations on the northwestern slopes of the Island that are noncontiguous with the center of distribution in the Kōke'e/Waimea Canyon area. Separating these areas of firetree invasion is an area of closed-canopy mesic native forest. This pattern of invasion may indicate that the tree is limited in its ability to invade intact closed-canopy native forests. If so, it is possible that firetree is somewhat shade intolerant, and that its spread may be partially controlled by excluding disturbances, such as feral animals and forest fragmentation, in areas that support a closed canopy forest.

More generally, the distribution of firetree throughout the State seems associated with disturbance. Major areas of infestation are pastures, roadsides, trails, secondary forest, and steep, unstable slopes. In fact, the aggressive nature of the species was first noted in pasturelands. Even the areas of its distribution that are within Hawaii Volcanoes National Park have been subjected to fire, volcanic activity, and foraging by feral goats (*Capra hircus*), pigs, and cattle (*Bos taurus*). Additionally, firetree has been identified as an actinorrhizal nitrogen-fixing species (Miguel and Rodriguez-Barrueco 1974; Turner and Vitousek 1988). Thus, the widespread invasions observed in Hawai'i may be due to a competitive advantage gained through nitrogen fixation on sites undergoing primary or secondary plant succession, especially on nitrogen-poor substrates such as those occurring in Hawaii Volcanoes National Park. This advantage may be lost on sites that have not been recently disturbed. However, in a landscape that is predominately influenced directly and/or indirectly by human activities, the further spread of firetree seems inevitable.

The complex of environmental factors in a particular area may affect the potential elevational range of firetree, even on a single island. In leeward areas such as North Kona, Hawai'i, and Kula, Maui, the distribution is between 3,200 and 6,400 ft (970-1,940 m). However, in windward areas such as Hāmākua, Hawai'i, the distribution is between 2,000 and 4,400 ft (600-1,335 m) elevation. This implies that the species may be adapted to a lower elevational range in windward (wetter) areas than in leeward (drier) areas in Hawai'i. In regions that are intermediate along rainfall gradients, firetree may have an intermediate potential elevational range.

Phenology

Flowering, immature fruiting, mature fruiting, and fruit drop seem to follow an annual cycle that is endogenously controlled and/or initiated by consistent annual environmental cycles, such as day length. The timing of these phases in Hawai'i seems to be similar to that observed in the native habitat of the species (Krauss 1964; Gardner 1984). Fruiting was heaviest at the warmest and driest site and lightest at the coolest and wettest site. This suggests that firetree may not be well adapted to reproducing in cold and/or extremely wet habitats. Also, our data (Whiteaker and Gardner 1987) show that while these phenophases have definite maximum and minimum activity periods, at no time during the course of this study were any of these phenophases competely absent.

Leaf flushing may also have a generally consistent annual cycle, as illustrated by the similar patterns at two of the study sites, and by the peak of activity occurring in June of each year at all three sites. However, when the slightly different overall pattern of leaf flushing at the driest site is compared with precipitation, maximum flushing seems to follow periods of high precipitation; minimum flushing seems to follow periods of low precipitation (Whiteaker and Gardner 1987). Thus, flushing of firetree may be sensitive to variations in precipitation where total annual precipitation is relatively low. Flushing was only absent at one site for one observation. Thus, although the amount of flushing may vary, this phenophase of firetree is also almost never absent.

The variation among sites in the pattern of the amount of leaf fall is shown by large peaks of leaf fall occurring at different times of the year at one study site relative to the other two sites. This may be due to the geographical position of the sites relative to the volcanically active Pu'u 'Ō'ō vent and the corresponding incidence of high sulfur dioxide levels in the atmosphere. Thus, the large peaks in the amount of leaf fall may reflect the response of firetree to high sulfur dioxide concentrations superimposed on the otherwise steady rate of leaf fall of an evergreen species.

MANAGEMENT CONSIDERATIONS

Most ecologists familiar with the problem think that, if no action is taken, the present trend of rapid spread of firetree into native habitats will continue, resulting eventually in formation of dense, closed-canopy stands, beneath which native plant life will be excluded. This, in turn, will result in the destruction of habitats for native birds, including endangered species. However, sufficient funds and manpower are not available, nor is it anticipated that they will be provided in the future, for large-scale control of firetree through the labor-intensive chemical or mechanical removal approaches used in the past. Such continued efforts would serve only to divert limited funding and manpower away from efforts to control other alien plant species for which these approaches may still be feasible.

Although chemical and/or mechanical control over large areas would be impractical, these approaches may be within the manpower and funding capability of management on comparatively small areas that contain particularly valuable resources, such as rare and endangered endemic species and the native plant communities that support them. In conjunction with this more focused approach to chemical and/or mechanical control, biological control is the only long-term hope for widespread control.

By promoting awareness of the threat of firetree to native ecosystems and ranchlands, the support of public land managers and ranchers adjacent to conservation areas can be obtained. This would aid in preventing firetree from being continually reintroduced into protected areas. It would also enable managers to consider direct control in larger areas with increased possibility of success.

The Hawaii Department of Agriculture and the Department of Land and Natural Resources should again declare firetree a noxious weed and increase efforts toward its control on a statewide basis on State as well as privately owned land. Support for biocontrol research for firetree should be encouraged by providing funds for foreign exploration and related work. Federal agencies should cooperate in any State-supported control program and could provide biological control quarantine facilities as well as services of federally funded personnel. A cooperative Statewide control program would enhance the ability of management to control firetree and to prevent this species from becoming established in protected areas such as Haleakala National Park. Public awareness would then be much more widespread than at present, and public attitudes toward permitting the occurrence of apparently innocuous infestations of firetree on private land could be altered, with increased understanding of the need for control. The public should be less tolerant of firetree infestations on State and federal public lands than it is.

ACKNOWLEDGMENTS

We would like to express our appreciation to S. Anderson, P. Connally, L. Cuddihy, R. Daehler, R. Hobdy, R. Kami, T. Lindsey, J. Obata, D. Sugawa, E. Tamura, and T. Watanabe for their knowledge, advice, and logistical support. Thanks to J. Dybdal and the Estate of James Campbell for access to the Palikea area on O'ahu. Gratitude is also due to A. de la Cruz and the Kōke'e Museum, and H. Voris for lodging and logistical support. Thanks also to Mr. and Mrs. Eldon English for their permission to monitor trees on their land and for use of the precipitation data that they maintain.

This study was supported by funds made available through the University of Hawaii Cooperative National Park Resources Studies Unit, by National Park Service Cooperative Agreement CA 8004-2-0001.

Literature Cited

Anonymous. 1962. Noxious weeds of Hawaii. Hawaii Dep. Agric. Div. Plant Industry. Mimeo.

Clarke, G. 1978. The distribution of *Myrica faya* and other selected problem exotics within Hawaii Volcanoes National Park. Pres. 2nd Conf. Nat. Sci., Hawaii Volcanoes Natl. Park, June 1-3, 1978. Abstract, *Proc. 2nd Conf. Nat. Sci., Hawaii Volcanoes Natl. Park*, 51. Honolulu: Univ. Hawaii Coop. Natl. Park Resour. Stud. Unit.

Doty, M.S., and D. Mueller-Dombois. 1966. *Atlas for bioecology studies in Hawaii Volcanoes National Park.* Hawaii Bot. Sci. Paper No. 2. Botany Dep. Honolulu: Univ. Hawaii.

Dunevitz, V.L., and J.J. Ewel. 1981. Allelopathy of wax myrtle (*Myrica cerifera*) on *Schinus terebinthifolius*. *Fla. Scientist* 44:1-13.

Ewel, J.J., D.S. Ojima, D.A. Karl, and W.F. DeBusk. 1982. Schinus *in successional ecosystems of Everglades National Park.* Tech. Rep. T-676, Natl. Park Serv. So. Fla. Res. Cent., Everglades Natl. Park. Homestead, Fla.

Foote, D.E., E.L. Hill, S. Nakamura, and F. Stephens. 1972. *Soil survey of the islands of Kauai, Oahu, Maui, Molokai, and Lanai, State of Hawaii.* Honolulu: U.S. Dep. Agric., Soil Conservation Service, in cooperation with Univ. Hawaii Agric. Exp. Sta.

Fosberg, F.R. 1937. Immigrant plants in the Hawaiian Islands. I. *Univ. Hawaii Occas. Paper* No. 32. Honolulu: Univ. Hawaii.

Gardner, D.E. 1984. Current biocontrol investigations: exploration for control agents in the native habitats of firetree, and evaluation of rust diseases of *Rubus* spp. in the southeastern U.S. *Proc. Fifth Conf. Nat. Sci., Hawaii Volcanoes Natl. Park*, 53-60. Honolulu: Univ. Hawaii Coop. Natl. Park Resour. Stud. Unit.

Gardner, D.E. 1985. Observations on some unusual flowering characteristics of *Myrica faya*. *Newsletter, Hawn. Bot. Soc.* 24:14-17.

Gardner, D.E., and C.J. Davis. 1982. *The prospects for biological control of nonnative plants in Hawaiian national parks.* Tech. Rep. 45, Univ. Hawaii Coop. Natl. Park Resour. Stud. Unit. Honolulu.

Gardner, D.E., and V.A.D. Kageler. 1982. *Herbicidal control of firetree in Hawaii Volcanoes National Park: a new approach.* Natl. Park Serv. Ecol. Serv. Bull. 7. Washington D.C.: U.S. Dep. Interior.

Haselwood, E.L., and G.G. Motter, eds. 1983. *Handbook of Hawaiian weeds*, 2nd ed., revised and expanded by R.T. Hirano. Harold L. Lyon Arboretum. Honolulu: Univ. Hawaii Pr.

Hodges, C.S., Jr., and D.E. Gardner. 1985. Myrica faya*: potential biological control agents*. Tech. Rep. 54, Univ. Hawaii Coop. Natl. Park Resour. Stud. Unit. Honolulu.

Hosaka, E.Y. 1945. *Noxious weeds of Hawaii.* Honolulu: Board of Commissioners of Agriculture and Forestry, in cooperation with Univ. Hawaii Agric. Exp. Sta.

Hosaka, E.Y., and A. Thistle. 1954. *Noxious plants of the Hawaiian ranges.* Univ. Hawaii Agric. Ext. Serv. Bull. 62. Honolulu: Univ. Hawaii.

Krauss, N.L.H. 1964. Insects associated with firebush (*Myrica faya* Aiton). *Proc. Hawn. Entomol. Soc.* 18(3):405-411.

LaRosa, A.M., C.W. Smith, and D.E. Gardner. 1985. Role of alien and native birds in the dissemination of firetree (*Myrica faya* Ait. - Myricaceae) and associated plants in Hawaii. *Pac. Sci.* 39:372-378.

Lawrence, G.H.M. 1951. *Taxonomy of vascular plants.* New York: The MacMillan Co.

Miguel, C., and C. Rodriguez-Barrueco. 1974. Acetylene-reducing activity of detached root nodules of *Myrica faya* Ait. *Plant and Soil* 41:521-526.

Mueller-Dombois, D., and F.R. Fosberg. 1974. *Vegetation map of Hawaii Volcanoes National Park.* Tech. Rep. 4, Univ. Hawaii Coop. Natl. Park Resour. Stud. Unit. Honolulu.

Neal, M.C. 1965. *In gardens of Hawaii.* B.P. Bishop Mus. Spec. Pub. 50. Honolulu: Bishop Mus. Pr.

Ripperton, J.C., and E.Y. Hosaka. 1942. *Vegetation zones of Hawaii.* Hawaii Agric. Exp. Sta. Bull. 89. Honolulu: Univ. Hawaii.

Skolmen, R.G. 1979. *Plantings on the forest reserves of Hawaii 1910-1960.* Inst. Pac. Islands For., Pac. SW For. Range Exp. Sta. USDA For. Serv., Honolulu.

Smathers, G.A., and D.E. Gardner. 1979. Stand analysis of an invading firetree (*Myrica faya* Aiton) population, Hawaii. *Pac. Sci.* 33(3):239-255.

Smith, C.W. 1985. Impact of alien plants on Hawai'i's native biota. In *Hawai'i's terrestrial ecosystems: preservation and management*, ed. C.P. Stone and J.M. Scott, 180-250. Univ. Hawaii Coop. Natl. Park Resour. Stud. Unit. Honolulu: Univ. Hawaii Pr.

Stone, C.P., and D.D. Taylor. 1984. Status of feral pig management and research in Hawaii Volcanoes National Park. *Proc. 5th Conf. Nat. Sci., Hawaii Volcanoes Natl. Park*, 106-117. Honolulu: Univ. Hawaii Coop. Natl. Park Resour. Stud. Unit.

Turner, D.R., and P.M. Vitousek. 1988. Nodule biomass of the nitrogen-fixing alien *Myrica faya* Ait. in Hawaii Volcanoes National Park. *Pac. Sci.* 41(1-4):186-190.

Walters, G.A., and W.S. Null. 1970. *Controlling firetree in Hawaii by injection of Tordon 22K.* Research Note PSW-217. Berkeley: USDA For. Serv.

Watanabe, T. 1982. Noxious weeds of Hawaii. Hawaii Dept. Agric. Div. Plant Industry. Unpub. report.

Whiteaker, L.D., and D.E. Gardner. 1985. *The distribution of* Myrica faya *Ait. in the state of Hawai'i.* Tech. Rep. 55, Univ. Hawaii Coop. Natl. Park Resour. Stud. Unit. Honolulu.

Whiteaker, L.D., and D.E. Gardner. 1987. *The phenology and stand structure of* Myrica faya *Ait. in Hawaii.* Tech. Rep. 62, Univ. Hawaii Coop. Natl. Park Resour. Stud. Unit. Honolulu.

Yamayoshi, H.T. 1954. *M. faya* survey of Paauilo and Kukaiau Ranch. Plant Quarantine Inspector, Territorial Agric. Dep. Rep. Unpub.

DISTRIBUTION, STATUS, PHENOLOGY, RATE OF SPREAD, AND MANAGEMENT OF CLIDEMIA IN HAWAI'I

Clifford W. Smith

ABSTRACT

Clidemia (*Clidemia hirta*, Melastomataceae) is a significant disruptive weed in many ecosystems in Oceania and the Southeast Asian and Indian subcontinents. In Hawai'i, it forms dense monotypic stands in mesic to wet environments between 33 and 4,920 ft (10-1,500 m) elevation. First reported in the Hawaiian Islands on O'ahu in 1941, it has since spread to Hawai'i Island (1972), Moloka'i (1973), Maui (1976), Kaua'i (1982) and Lāna'i (1988). Humans are long-distance dispersal agents, whereas short-distance dispersal is effected by birds, feral pigs (*Sus scrofa*), mongooses (*Herpestes auropunctatus*), humans, and other organisms. Wildfires, landslides, and windstorms accelerate dominance of this weed. All efforts to eradicate newly established populations failed when they were not initiated before first fruit set. The impacts of this weed on native species and ecosystems are devastating. Clidemia has the potential of driving a number of species to extinction. However, impacts on the water balance of entire ecosystems may be more devastating in the long run.

INTRODUCTION

Clidemia hirta (Melastomataceae), commonly known as clidemia or "Koster's curse," is a serious weed in many ecosystems in Oceania as well as on the Southeast Asia and Indian subcontinents. First reported in the Hawaiian Islands on O'ahu in 1941 (Anonymous 1954) and included by Hosaka and Thistle (1954) in their list of noxious plants because of its history in Fiji, it remained a relatively uncommon plant in the Islands into the early 1960s. Clidemia was declared noxious by the Hawaii Department of Agriculture in 1957 (Regulation 10). However, Plucknett and Stone (1961) stated that it was not common and "at present cannot be classed as dangerous." It was still confined to O'ahu at that time, but in the early 1970s it was spread to the other Islands. In 1975, the State Senate, in response to pressure from agriculturists, foresters, and conservationists and concerned about further spread to islands other than O'ahu, passed a resolution recommending action; in the past 10 years considerable management of, and research on, this weed has been conducted.

This paper documents the ecology, phenology, distribution and dissemination of *C. hirta* in the Islands. Its spread, potential range and probable impact are also considered. Biological control is discussed in Nakahara *et al.*, this volume.

ECOLOGY

Clidemia is found in secondary-succession tropical rain forests in Central and South America and the Caribbean Islands, where it is not a common plant (Cook 1929; Wester and Wood 1977). In its native habitat in the West Indies and Central America, it is an early colonizer of cleared ground, rapidly becoming dominant before being smothered by gap-replacement vines (Burkhart, pers. comm.). It is constantly present in open forested areas, plantations, and along roads and trails, but only as an occasional plant (Wester and Wood 1977). All plants show signs of heavy predation by insects and other organisms. Burkhart (pers. comm.) has noted that it also faces strong competition from other melastomes in its native habitat.

Clidemia invades undisturbed areas in Hawai'i, but population levels remain low. However, once an area is disturbed, colonization proceeds rapidly. Wester and Wood (1977) recorded the domination of an area on O'ahu by clidemia within two years after a fire. Similar responses have been seen elsewhere on O'ahu after fire. Hurricane Iwa, in 1981, disturbed much of the area along the Contour Trail in the Wai'anae Mountains on O'ahu. Two years later there was an almost impenetrable thicket of clidemia 10 ft (3 m) or more tall. Feral pigs (*Sus scrofa*) play a significant role in the intensification of infestations of this weed. Along the Aiea Loop Trail in Honolulu, areas dug up by pigs showed large numbers of seedlings within two months of disturbance. These later grew into dense thickets within six months. Pigs not only bring seeds into an area, but disturb the ground sufficiently to provide an excellent environment for germination and establishment.

In Hawai'i, clidemia is free of competition from other melastomes and gap succession vines. The only significant insect pest is a thrips, *Liothrips urichi* (Thysanoptera: Phlaeothripidae), an introduced biocontrol agent. Plants are now found in forested areas and open situations where rainfall is high and cloud cover frequent. In open areas with high insolation, *Liothrips* debilitates the weed sufficiently to effectively control it. In shaded areas, however, *Liothrips* is not as effective.

Clidemia occurs in dense, monotypic stands in mesic to wet environments from 36 ft (11 m) above sea level (never directly exposed to salt spray) to at least 4,000 ft (1,220 m) elevation. In exposed windy areas, it remains a scrambling shrub not more than 30 in. (75 cm) tall. In moist, shaded environments, the bush reaches 16 ft (5 m) into the subcanopy. Pendant plants up to 50 ft (15 m) long have been observed along precipitous streams in Wailau Valley, Moloka'i (A.C. Medeiros, pers. comm.).

The impact of heavy infestations of clidemia on associated vegetation is devastating. Most plants below the clidemia canopy disappear, including even mosses and liverworts, which are normally able to survive in shaded habitats. A few tussock mosses remain as spindly strands, but the soil-binding, mat-forming species, such as *Leucobryum gracile* and *Rhizogonium spiniforme*, disappear. Rare subcanopy species cannot compete against this weed and are rapidly displaced. Even common species are unable to reproduce. Their survival, dependent on their seed banks, declines as the infestation continues.

PHENOLOGY

Growth, flowering, and fruiting of clidemia have been observed throughout the year in areas where the rainfall exceeds 98 in./yr (2,500 mm/yr) with no dry season. However, in areas where rainfall is as low as 39 in./yr (1,000 mm/yr), and particularly where there is a dry season, the plant does not flower or fruit much after the dry season has begun. Leaves fall off, growth ceases, and some death of shoot tips occurs. As long as drought does not last more than about six months, plants recover and resume reproduction within a very short period.

DISTRIBUTION

Considerable inconsistency in estimates of the distribution of clidemia in the Islands exists. Some estimates refer only to known populations, which tends to minimize estimated area of occupation because of inadequate field surveys. Other estimates join isolated populations together and consider the whole area infested unless there is reasonable explanation for the disjunction. I have followed the latter approach because the ease of dispersal, rapid growth and maturation, and the abundance and diversity of dispersal agents indicate that on O'ahu, in particular, clidemia has saturated the environment.

O'ahu

How and when *Clidemia hirta* first arrived in the Islands is not known. However, Whitten (1977) reported that H.L. Lyon (who thought that it "was very promising because it won't be spread by birds"), grew it in the Wahiawa Botanic Garden. Clidemia was first reported from Pōamoho in 1941 (Hawaii Department of Agriculture 1975). Otto Degener, on a B.P. Bishop Museum herbarium label dated 1949, noted that clidemia was "escaping." In 1952, State foresters estimated that it was growing vigorously in an area of less than 250 a (100 ha) in the Ko'olau Mountains (Hawaii Department of Agriculture 1975). In retrospect, we can be almost certain that the plant occurred at low densities in an area three times the size of this estimate. By the early 1960s, clidemia had spread as far south as Mt. Tantalus and as far north as Kawailoa (Thistle 1962). In 1975, State Forester J. Pelly (1975) estimated that approximately 96,000 a

(38,400 ha) were infested, 77,000 a (30,800 ha) in the Ko'olau Range and 19,000 a (7,600 ha) in the Wai'anae Range. However, in 1977 Wester and Wood (1977) estimated that it occupied only 77,455 a (31,350 ha). Today, the infestation ranges as far south as Mariner's Ridge above Hawai'i Kai, and to Pūpūkea in the Ko'olau Range (Fig. 1).

Clidemia spread to the Wai'anae Range before 1970. The initial infestation was probably in the Schofield Barracks military reservation, transported there from the Kawailoa and Kahuku military training area. In 1973, it had reached the contour trail below Pu'u Hāpapa. The weed is now present throughout the Honouliuli Forest Reserve, and in the north it almost reaches the tracking station above Ka'ena Point. The rapid spread and increase in cover of clidemia in the Wai'anae Range was particularly evident after Hurricane Iwa in 1981. The only area in this mountain range that remains free of infestation is 'Ōhikilolo; this is probably due to the very dry conditions there, even in the forest, and the inability of pigs to climb the cliffs.

On O'ahu, clidemia is found from 35 ft (11 m) above sea level at Pōamoho Gulch, Waialua, to the summit of Mt. Ka'ala at 4,025 ft (1340 m) elevation, where several plants were removed from the Mt. Ka'ala Natural Area Reserve in 1986. However, its abundance on O'ahu is quite erratic. Some gullies are almost filled by this weed, whereas adjacent areas may be little affected. The heavy cover in most gullies and other protected areas is effected by a few large individuals. In contrast, in more open situations, such as on the windward summit slopes of the Ko'olau Range, there are solid stands of many short, stunted plants.

I estimate that the plant is now present in over 252,000 a (100,000 ha) on O'ahu. This figure is tentative because I have assumed that all areas somewhere between the 40-50 in. (1,016-1,270 mm) isohyet are habitat, and that if it is present in the bottom of a gulch it is on the adjacent ridge, and vice versa. The range extension from 1972 (estimated 77,435 a or 31,350 ha by Wester and Wood 1977) is dramatic, but it appears that this weed has now invaded all the territory on O'ahu that it can. It is reported as an occasional weed in gardens in Aiea Heights, Tantalus, Mānoa, and Pālolo.

Kaua'i

By 1982, clidemia was established on Kaua'i. The largest infestation is now in Limahuli and Wai'oli valleys, where it is fairly common (Fig. 2). The plant has also been seen where the Nā Pali trail crosses Hanakoa Stream and along the trail on the Kalalau side just beyond Hanakāpī'ai. The Hawaii Division of Forestry has been sending crews to eradicate clidemia on Mt. Kāhili, where it is fairly common, and a small population, recently found along the Nounou Forest Reserve tree plantation trail, with little success so far. There are also unconfirmed reports of the plant in the back of some of the valleys along the Nā Pali Coast and along Highway 581 between Wailua and Kapa'a.

The current infestation, consisting of five separate confirmed populations and a number of other unconfirmed locations, suggests that

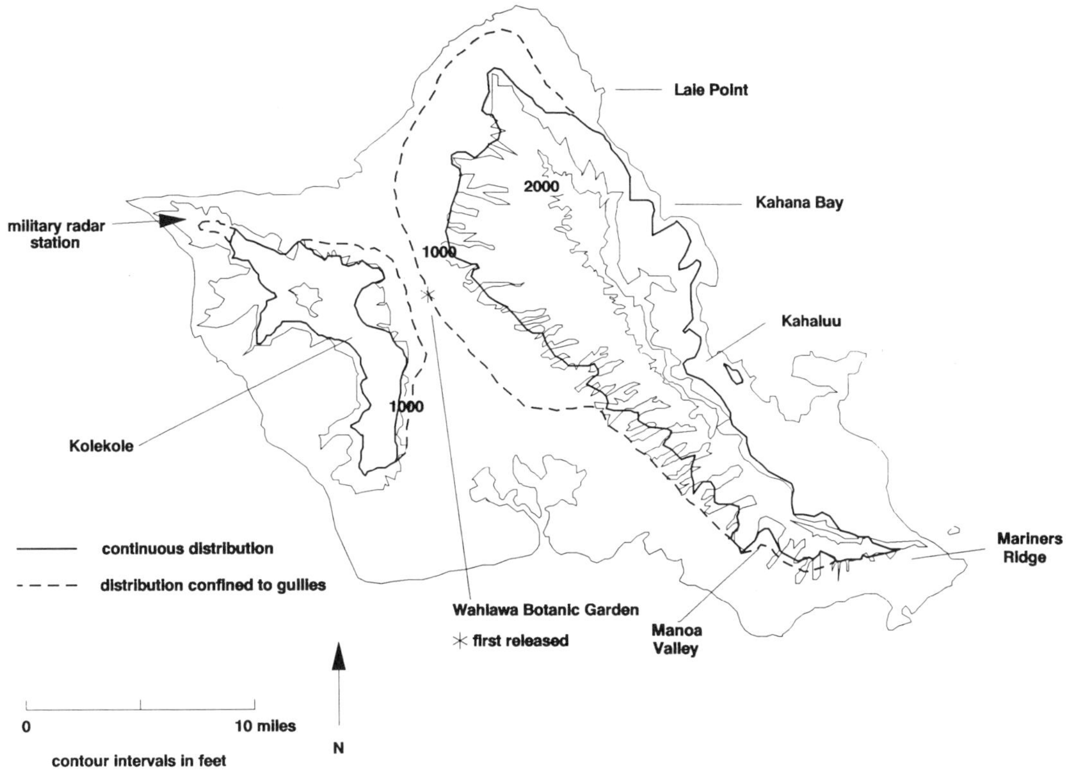

Figure 1. Distribution of *Clidemia hirta* on O'ahu, Hawaiian Islands.

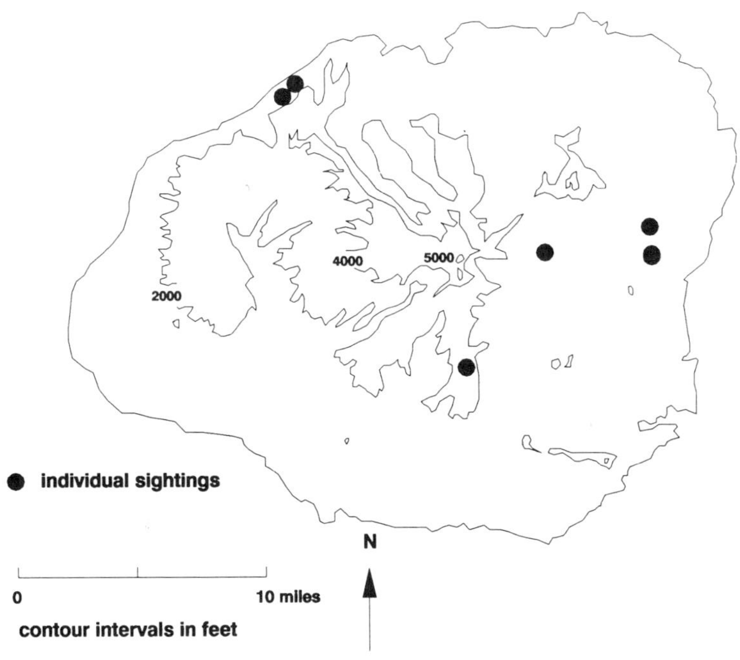

Figure 2. Distribution of *Clidemia hirta* on Kaua'i, Hawaiian Islands.

clidemia has become well established on Kaua'i. The fact that less than 99 a (40 ha) are occupied is no comfort. The prognosis is that these populations will expand very rapidly.

Maui

Clidemia was first reported on East Maui in 1976, along Makapipi Stream just below Pogues Tunnel. One and one-half years later it was quite common above Kailua as far as O'opuola Stream, where it was growing in deep shade under uluhe *(Dicranopteris linearis)* and even epiphytically. It has since spread to many of the adjacent ridges and along the Ko'olau Ditch banks and associated roads. Clidemia is now seen fairly frequently along the Hāna highway in the Kailua area (Fig. 3). Approximately 4,940 a (2,000 ha) are infested. Clidemia was first seen in Kīpahulu Valley in 1988.

On West Maui, it was first reported in 1983-1984 in the Kahakuloa area in Po'elua Gulch and the adjacent ridge by R. Hobdy (pers. comm.). It is now present in 124 a (50 ha). L.T. Gill (pers. comm.) uprooted one plant about 0.25 mi up the Pu'u Kukui trail in the same year, but no other plants have been found in the area.

Moloka'i

Clidemia was first seen on Moloka'i in January 1973 along the *pali* trail in Wailau Valley between 3,000 and 3,500 ft (1,000-1,150 m) (Wester and Wood 1977). Participants of Sierra Club service trips began removing the plants along the trail in the spring of 1974. In 1978, the plant was reported on Pu'u Ali'i but was not seen the following year. The first report from Pelekunu Valley was the collection of a mature plant along the Manuahi ridge trail below 'Ōhi'alele, at 1,600 ft (530 m) elevation. The population is expanding but not very rapidly. It has also been seen along the Wailau trail on the leeward side of the Island as low as 1,970 ft (600 m) elevation.

A fruiting plant was discovered in The Nature Conservancy of Hawaii Preserve at Kamakou Flats in 1981 and was uprooted immediately. Since then, seedlings have been destroyed, suggesting that the seed bank from the initial invasion is still present. No other populations have been found in the Kamakou Flats area.

I estimate that in 1974, clidemia occupied about 12 a (5 ha) (>1% of the land area of Moloka'i). By 1979, the infestation had spread to 6,175 a (2,500 ha) (<4%), and it now covers 15,560 a (6,300 ha) (9%) of Moloka'i (Fig. 4). In the next few years clidemia will invade most of the available habitat on Moloka'i.

Hawai'i

Hawai'i was the first island other than O'ahu on which clidemia was recorded, yet, until recently, it has not spread as rapidly as on Maui or Moloka'i (Fig. 5). It was first seen in the Waiākea Forest Reserve in 1972 (Davis 1974) and is now firmly established. From there, it spread to other areas in Puna, where it was first recorded below Pu'u Kauka in 1976. Since then the plant has been seen throughout lower Puna, generally in small

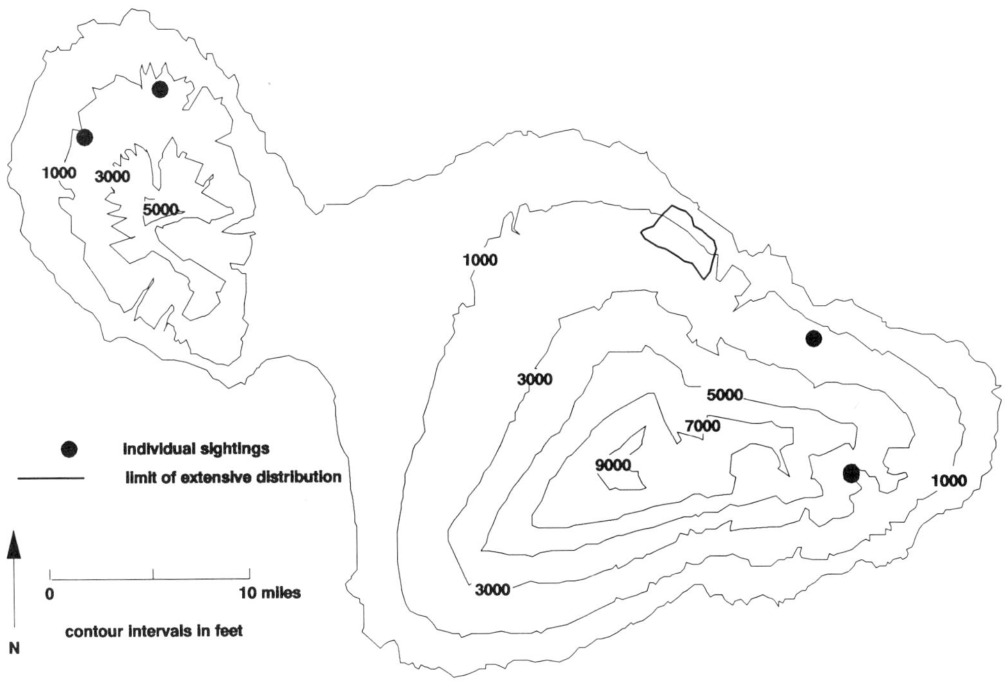

Figure 3. Distribution of *Clidemia hirta* on Maui, Hawaiian Islands.

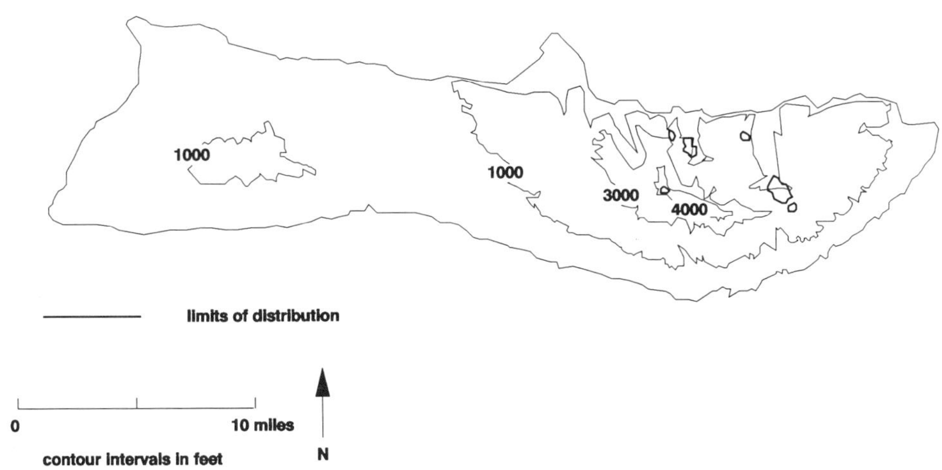

Figure 4. Distribution of *Clidemia hirta* on Moloka'i, Hawaiian Islands.

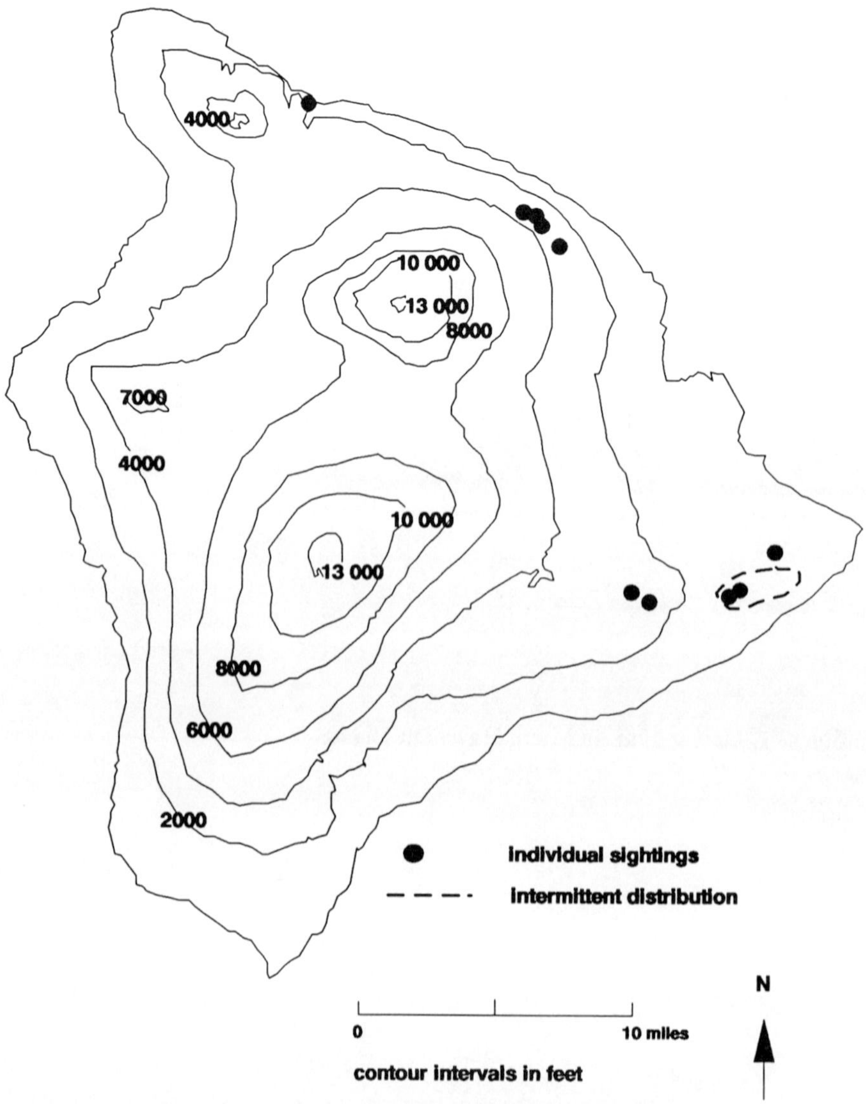

Figure 5. Distribution of *Clidemia hirta* on Hawai'i, Hawaiian Islands.

isolated populations. The latest report, from the area of Pu'u Kia'i, places it three miles from the boundary of Hawaii Volcanoes National Park. In 1983, a separate population was found along the Waimanu trail in the Kohala mountains. It was not there in 1981-1982. Sierra Club trips to control this weed have been unsuccessful to date. Three separate populations above Laupāhoehoe (Kīlau Stream, Humu'ula Trail, and the South Fork of Pāhale Stream) indicate that clidemia is established in that area also.

Lāna'i

Clidemia was reported from Lāna'i in October 1988 by Holt and Connally (Holt 1988). Two plants were found at the beginning of the Hauola Gulch north trailhead, within 20 feet of the Munro jeep trail. The plants were removed.

Kaho'olawe and Ni'ihau

Clidemia has not been recorded from these two islands. Public access to Kaho'olawe and Ni'ihau is severely restricted, which significantly reduces the chance of introduction. Both islands are quite arid, which would also probably prevent establishment.

DISSEMINATION

Long-distance dispersal of clidemia is effected by people who work in or use the forests. No one group is entirely responsible for spread. For example, reasonable circumstantial evidence suggests that the West Maui infestation was introduced by marijuana (*Cannabis sativa*) growers, as the center of the infestation is in some abandoned clearings previously used for that purpose. The first wild population on O'ahu, the Moloka'i infestation, the Nounou and Nā Pali Coast invasions on Kaua'i, and the Waimanu population on Hawai'i were first observed along trails, suggesting that hikers were responsible. Elsewhere, *e.g.*, Kīlauea East Rift zone on Hawai'i Island and the initial infestation in the Wai'anae mountains on O'ahu, it is fairly obvious that the seeds were transported by vehicles. The abundance of clidemia in the Kawailoa and Kahuku military training areas and the Schofield Barracks Military Reservation may have been the result of transportation by soldiers and military vehicles (Anonymous 1976). The remaining infestations, *e.g.*, the Laupāhoehoe populations, are probably the result of pig hunters. It is extremely unlikely that the introduction of clidemia is deliberate, as is the case with banana poka (*Passiflora mollissima*), because this weed is despised by all for its dense, rank growth.

Local dispersal is caused by a variety of organisms. Wester and Wood (1977) stated that frugivorous birds, humans, and perhaps mongooses (*Herpestes auropunctatus*), the latter since confirmed by Burkhart (pers. comm.), are important local disseminators. Seeds are viable after passage through mongooses (Burkhart, pers. comm.). Feral pigs must be added to this list, as they are attracted to the fruit during the peak fruiting season. Seedlings are often seen in dense clusters of several square inches, strongly suggesting that they are growing in the remnants of pig droppings. Pigs also carry seeds on their pelage.

DISCUSSION

Disturbance is a key element in the establishment and invasion of clidemia. In its native environment, plants are confined to open areas and only become dominant about twelve months after disturbance, *e.g.*, in slash-and-burn agricultural areas (Burkhart, pers. comm.). All new range extensions in Hawai'i begin along the open edges of trails or other disturbed areas. Pigs, fires, landslides, and major windstorms all result in the opening of the subcanopy, and if clidemia seeds are present they germinate rapidly. Within two years, a disturbed area can become smothered (Wester and Wood 1977). In order to keep this weed out of an area, the primary management objective should be to minimize if not prevent any disturbance.

There have been many well-meaning attempts by volunteer groups to control the spread of clidemia through pulling it by hand. Unfortunately, unless the invasion is caught in the earliest stage -- that is, before it has set seed -- the effort is likely to fail. The seed bank produced by these plants is colossal over a very short period of time. Each fruit contains well over 100 seeds, and a mature plant produces in excess of 500 fruits each season. Field observations demonstrate that the seeds can remain dormant for up to four years in the soil. These observations suggest that in order to effect control of an infested area by mechanical means, the eradication efforts must be conducted at least once a year and continued rigorously for up to 10 years. Merely uprooting the plants is insufficient, because even the leaves of clidemia form roots in suitable situations, *e.g.*, on the ground in rain forest (A.C. Medeiros, pers. comm.). Only two instances of successful control, Kamakou (Moloka'i) and Pu'u Kukui trail (Maui), are known.

Clidemia is susceptible to a number of herbicides, *e.g.*, 2,4,5-T, 2,4-D, Trichlopyr (Teoh *et al.* 1982). The first two chemicals are banned in most areas of the U.S., and the population soon regenerates from seedlings unless further applications are made. Chemical treatment of clidemia does not appear practical in Hawai'i's native ecosystems, especially those that are difficult of access.

No control effort has been effective in the Islands to date, other than the action of the thrips on ranchlands (Reimer 1985) and the mycoherbicidal action of the fungus *Colletotrichum gloeosporoides* (Trujillo *et al.* 1986). Although thrips are effective only in open areas (Reimer 1985), they are also present in open forested areas heavily infested with clidemia. Perhaps thrips are not able to flourish in these areas due to other environmental constraints, or the plants under shady conditions are sufficiently healthy to resist infection.

The introduction of an effective biological control agent, should one be found, will require very careful evaluation prior to release, particularly on O'ahu. The infestations of clidemia are so complete in some areas that the sudden loss of this plant could result in more serious consequences than currently exist. Two scenarios are possible. The loss of ground cover could be so severe that erosion would become a very serious problem. In contrast, if erosion is not a major problem, the exposure of so much bare ground would provide a good seed bed for a number of other weeds. Of particular concern are the grasses (*e.g.*, Hilo grass, *Paspalum conjugatum*, and broomsedge, *Andropogon virginicus*) and strawberry guava (*Psidium cattleianum*). Grasses would significantly increase fuel loadings in many areas, which could result in further degradation of the native vegetation by fire. Strawberry guava would form impenetrable thickets, which would prevent the reestablishment of native species.

One of the greatest difficulties in documenting the spread of weeds in the Islands is the failure of many field botanists to record their observations for later analysis. Part of the problem is that most people focus on the rarer native and endemic species and just grumble about the weeds. They might lament that a particular alien species is now in a particular locality, but never document it. Herbaria are, understandably, reluctant to accumulate large numbers of verification specimens of new localities of weeds. There is, however, a perfectly good place to deposit information on alien species distribution: newsletters of local botanical, agricultural, or horticultural groups. Misidentifications may be reported, but, if actively pursued, the process can be self correcting.

The rate at which clidemia has spread through the Islands is cause for alarm. Many new infestations occurred even after the considerable publicity in the press between 1975 and 1977. It was almost inevitable, given the mobility of our human population, that this dispersal would happen. What is disquieting is that public discussion on this weed did not result in greater care by the users of our forests. Ignorance is probably not the cause, and laziness about taking the necessary precautions may be only a small component. The major problem is the "couldn't care less" attitude. Our hedonistic society places little value on the quality of natural resources: it is not that society does not understand the general ecological messages of recent years; the public simply does not understand the value of some plants and the harm that can be caused by others. The attitude can best be paraphrased by "Look, the Islands have lots of vegetation -- what more do you want?!" The goal of education should be a reawakening of pride in things Hawaiian, both cultural and biological.

RECOMMENDATIONS

1. The State of Hawaii should intensify its public awareness campaign to prevent the dissemination of clidemia and other weeds to areas where they are absent.

2. Conservationists need to coordinate their programs with one another and reach out to other organizations trying to reinstill a pride in things Hawaiian, improve the quality of life in the Islands, and counteract the dominance of economic considerations as the only criterion in development and politics.

3. Conservationists, scientists, and hikers should record all alien species observed during excursions and report them in the newsletters and other publications of local environmental interest groups.

ACKNOWLEDGMENTS

The maps presented in this paper could not have been constructed without the cooperation of C. Corn, L. Cuddihy, T. Flynn, L. Gill, P. Higashino, R. Hobdy, A. Holt, G. Linney, A. Medeiros, J. Obata, S. Perlman, L. Stemmermann, and V. Tanimoto. I am grateful to R. Burkhart, G. Linney, A. Medeiros, and J. Obata for discussions on the biology of this weed. I also appreciate the helpful comments of R. Hobdy on an earlier draft of this manuscript.

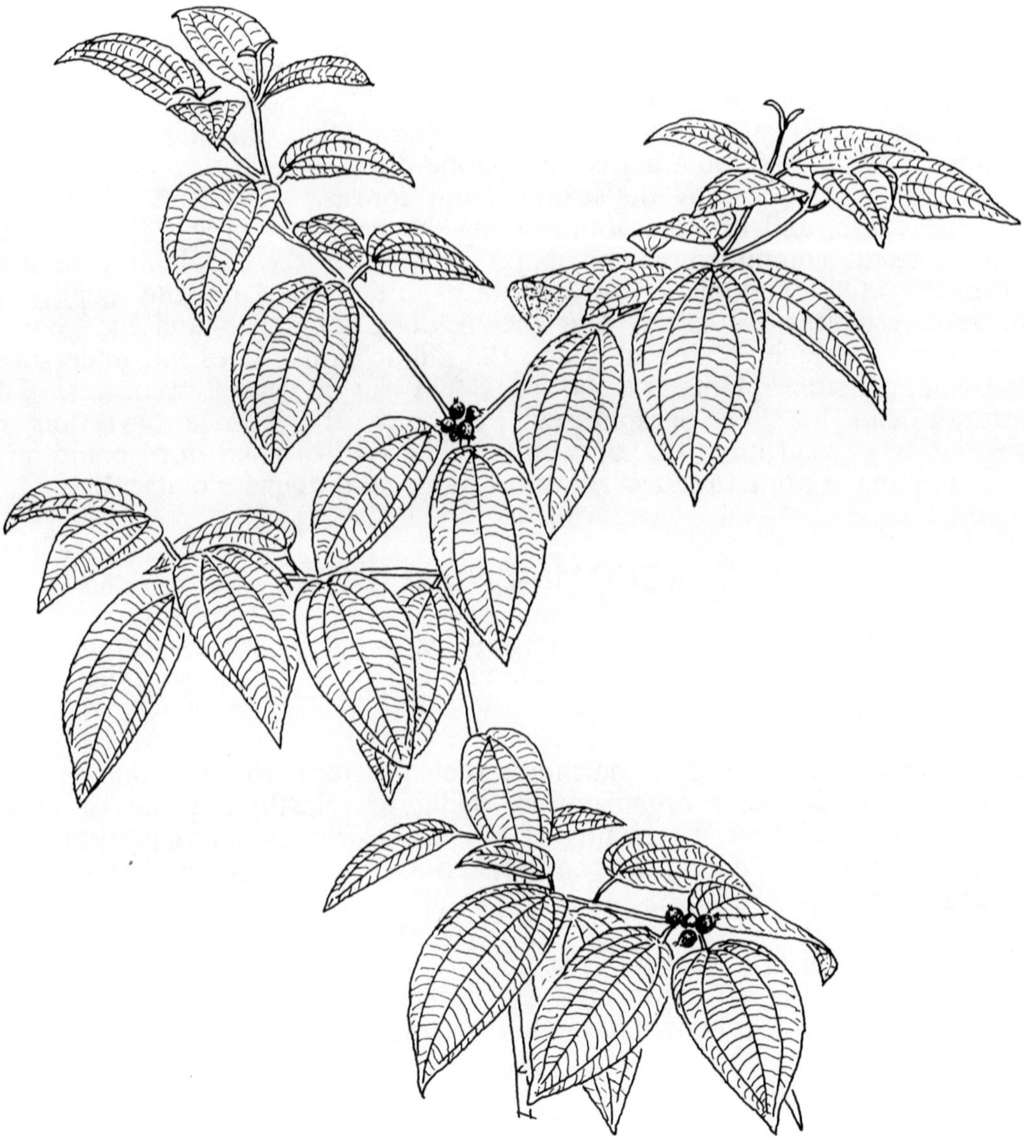

Literature Cited

Anonymous. 1954. Notes and exhibitions. *Proc. Hawn. Entomol. Soc.* 15:263-265, 279-283.

Cook, B.A. 1929. Some notes on the plant associates and habitat of *Clidemia hirta* (L.) D. Don in Trinidad. *Fiji Agric. J.* 2:92-93.

Davis, C.J. 1974. Recent introductions for biological control--XVIII. *Proc. Hawn. Entomol. Soc.* 21(3):355-357.

Hawaii Department of Agriculture. 1975. Report to the Eighth Legislature, State of Hawaii, 1976, in response to Senate Resolution No. 240, SD1, Eighth Legislature, 1975, relating to destruction of native ecosystems. Files, Hawaii State Dep. Agric.

Holt, A. 1988. Letter to C.W. Smith, Oct. 12. Files, Nature Conserv. Hawaii.

Hosaka, E.Y., and A. Thistle. 1954. *Noxious plants of the Hawaiian ranges*. Univ. Hawaii Coop. Ext. Serv. Bull. 62. Honolulu: Univ. Hawaii.

Nakahara, L.M., R.M. Burkhart, and G.Y. Funasaki. [this volume] Review and status of biological control of clidemia in Hawai'i.

Pelly, J. 1975. Clidemia information. Memo to State Forester, May 20. Files, Hawaii State Dep. Land Nat. Resour., Honolulu.

Plucknett, D.L., and B.C. Stone. 1961. The principal weedy Melastomataceae in Hawaii. *Pac. Sci.* 15:301-303.

Reimer, N.J. 1985. An evaluation of the status and effectiveness of *Liothrips urichi* Karny (Thysanoptera: Phlaeothripidae) and *Blepharomastix ebulealis* (Guenee) (Lepidoptera: Pyralidae) on *Clidemia hirta* (L.) D. Don in O'ahu forests. Ph.D. diss., Univ. Hawaii, Honolulu.

Teoh, C.H., P.Y. Toh, and H. Kairudin. 1982. Chemical control of *Asystasia intrusa* (Bl.), *Clidemia hirta* (Don.) and *Elettariopsis curtisii* (Bak.) in rubber and oil plantations. *Proc. Internatl. Conf. Plant Protection in the Tropics*, ed. K.L. Heong, B.S. Lee, T.M. Lim, C.H. Teoh, and Y. Ibrahim, 497-510. Kuala Lumpur, Malaysia: Malaysian Plant Protection Soc.

Thistle, A. 1962. *Clidemia hirta* along Kawailoa Ridge Trail -- North Oahu. Memo to C.J. Davis, Sept. 13. Files, Hawaii State Dep. Land Nat. Resour., Honolulu.

Trujillo, E.E., F.M. Latterell and A.E. Rossi. 1986. *Colletotrichum gloeosporoides*, a possible biological control agent for *Clidemia hirta* in Hawaiian forests. *Plant Disease* 70:974-976.

Wester, L.L., and H.B. Wood. 1977. Koster's curse (*Clidemia hirta*), a weed pest in Hawaiian forests. *Environ. Conserv.* 4(1):35-41.

Whitten, H. 1977. Book on isle plants. *Honolulu Star Bulletin*, March 14.

MULLEIN (VERBASCUM THAPSUS): THE SPREAD AND ADAPTATION OF A TEMPERATE WEED IN THE MONTANE TROPICS

James O. Juvik and Sonia P. Juvik

ABSTRACT

Common mullein (*Verbascum thapsus*) is a rosette, biennial weed native to temperate Eurasia and naturalized worldwide in suitable mid-latitude environments, where it frequently is considered a pest. The species tolerates a wide range of environmental conditions and is particularly adapted as an early successional pioneer in disturbed dry and rocky sites. *Verbascum thapsus* was introduced to the island of Hawai'i sometime around 1900 and has since spread into dry montane environments on the slopes of Mauna Kea, Mauna Loa, and Hualālai volcanoes. On the Indian Ocean island of La Réunion (environmentally similar to Hawai'i), mullein has also recently become established in comparable montane zones. A quantitative survey of roadside mullein populations along the Saddle Road ascending the slopes of Mauna Kea and Mauna Loa resulted in determination of the morphological variation along steep topo-climatic gradients. Over the elevational range of occurrence (5,330 to 10,800 ft or 1,625-3,300 m), roadside mullein densities reached a maximum of 9-17 plants/100 ft^2 (100-180/100 m^2) in the zone between 5,900 and 6,560 ft (1,800-2,000 m). The altitude-density relationship for mullein on the island of La Réunion is nearly identical to that on Hawai'i and demonstrates the vulnerability of other tropical mountain areas to invasion. Over the elevational range of mullein on Mauna Kea the species exhibits striking variation in gross morphology and life history. Both mean rosette diameter and plant height increase with elevation. Above 8,200 ft (2,500 m) "gigantism" is common, with some plants reaching nearly 13 ft (4 m) in height, double the maximum height recorded for this species in its native European habitat. Additionally, with increasing elevation there is a strong tendency toward polycarpy and extreme stem fasciation, resulting in increased woodiness. These morphological trends appear consistent with an adaptive model proposed to explain the development of insular arborescence in rosette weeds. Control and eradication of mullein is made difficult by extensive seed viability (up to 100 years). In order to develop eradication strategies, the spatial configuration and population structure of selected *V. thapsus* colonies were documented prior to removal in order to facilitate long-term study of recolonization and site persistence.

INTRODUCTION

Common mullein (*Verbascum thapsus*: Scrophulariaceae), a monocarpic and biennial Eurasian weed, is widely naturalized in North America and other temperate areas of the world including Australia and New Zealand. Various medicinal properties and other folk culture uses (*e.g*, piscicide) are attributed to the plant and may, in part, explain its widespread introduction (Wilhelm 1974). In its native European habitat and other temperate areas, mullein is an early succession colonizer of disturbed dry and rocky sites and is able to survive under a broad range of environmental conditions (Williams *et al.* 1975; Semenza and Evans 1978; Reinartz 1984a).

The species is generally characterized as an obligate biennial, with flowering dependent on the number of growing seasons required to reach a critical size. However, naturalized populations in North America range from annuals (Georgia) to triennials (Canada) as a function of both environmental and competitive factors (Gross 1981; Reinartz 1984b). The flowering spike (up to 6.5 ft or 2 m in height) may produce 150,000 seeds of comparatively small size (0.03 in. or 0.65-0.75 mm) with no specific adaptations for dispersal (Baskin and Baskin 1981). Indeed, studies have shown that under normal conditions more than 80% of all seedfall occurs within 23 ft (7 m) of the parent plant (Gross and Werner 1978). It appears that the early successional strategy of mullein rests more on extreme seed viability (50-100 years or more) than on long-distance dispersal (Kivilaan and Bandurski 1933; Manning 1965). Mullein cannot compete under situations of dense herbaceous groundcover, but seeds can remain dormant in the soil until vegetation disturbance provides suitable conditions for germination.

Mullein has become an aggressive adventive in most temperate areas of introduction, but little has been written on the impact of this species in the montane tropics. It may be argued that climatic barriers protect tropical mountains from invasion by temperate weed species to some extent. Inadvertent introductions in the tropics would be more likely to occur in the lowlands (rather than isolated mountain peaks), and plants would necessarily face the difficulty of dispersing upward into environmentally favorable montane zones. Wester and Juvik (1983) have shown that in Hawai'i the alien roadside weed flora along a highway ascending Mauna Loa Volcano changes gradually with elevation from a species assemblage dominated by pantropical lowland weeds to one composed primarily of extratropical temperate species at higher elevations. At least in the case of Hawai'i, introduced temperate weed species appear able to successfully colonize the climatically favorable montane zone. Some pest species such as gorse (*Ulex europaeus*) have been directly introduced into montane pastures.

On the island of Hawai'i, mullein provides a classic example of an introduced temperate weed that has successfully spread throughout mid- and high-elevation areas of the Island, achieving serious pest status (Smith 1985). First established in the Kona district of Hawai'i sometime between 1900 and 1910, it is unclear whether the species was an intentional or inadvertent introduction (L.W. Bryan, pers. comm. 1983). The first documented herbarium accession (B.P. Bishop Museum, Honolulu) of mullein in Hawai'i was for specimens collected near the summit of Hualālai Volcano (elev. 8,270 ft or 2,520 m) by L.W. Bryan in 1932. Other Bishop Museum accession sheet comments indicate that the species was "dominant on the South Slope of Hualalai between 7,700-7,900 ft" by 1946 and had spread into the central plateau or saddle area between Mauna Kea and Mauna Loa by 1948. Over the past several decades mullein has expanded its range in both roadside habitats and remote upland areas on Mauna Kea, extending to above 10,800 ft (3,300 m) along the summit access road (Fig. 1).

This study presents results of a quantitative analysis of roadside *V. thapsus* populations along steep altitudinal gradients on the slopes of Mauna Loa and Mauna Kea. The study was undertaken to define both the elevational range and the ecologically optimal zone of the species. In addition, the incidence of morphological abnormalities was recorded for correlation with altitude/climate gradients on the mountain slopes. The Hawai'i data were compared with roadside *V. thapsus* populations sampled along an elevational gradient on the Indian Ocean island of La Réunion (Fig. 2), which is environmentally similar to Hawai'i (Cadet 1974, 1980). The observations are used to illustrate the vulnerability of other tropical montane areas of the world to *V. thapsus* invasion.

This paper also reports on problems associated with *V. thapsus* control and eradication efforts undertaken from 1984 through 1986 on Mauna Loa and Mauna Kea.

Figure 1. *Verbascum thapsus* colonization of roadside areas on Mauna Kea (elevation 2,500 m) near sampling site 6, 1983.

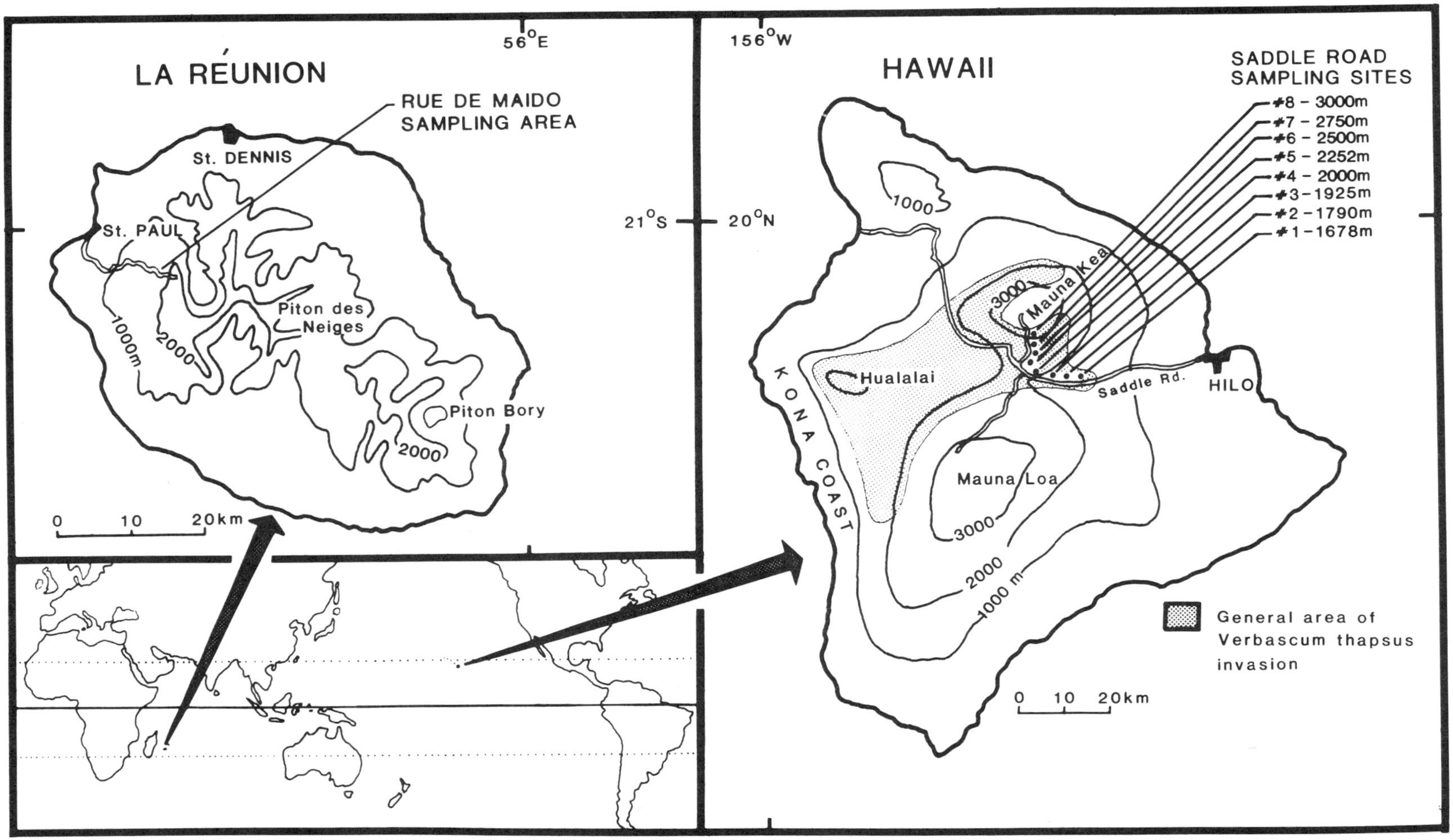

Figure 2. Location of study areas in La Réunion and Hawai'i.

MATERIALS AND METHODS

A review of published and unpublished references and local herbarium accessions of mullein was undertaken to document the early introduction and spread of this species on the island of Hawai'i. The general limits of distribution of the species in the uplands of Hawai'i were extracted from mountain vegetation surveys (114 transects with 7,438 sampling stations) established in conjunction with the U.S. Fish and Wildlife Service Hawai'i Forest Bird Survey (Jacobi and Scott 1985).

Intensive quantitative sampling of roadside mullein populations was undertaken during 1985 at eight sites along the Saddle Road and Mauna Kea summit road between the elevations of 5,500 and 9,850 ft (1,678 m-3,000 m) (Fig. 2). Along the Saddle Road, sample sites were established at odd-numbered mile markers between the lower elevational limit of the species (5,500 ft or 1,678 m at the 21-mile marker) and 6,560 ft (2,000 m) (27-mile marker). Above 6,560 ft (2,000 m) on the Mauna Kea summit road, sampling sites 5-8 were located at altitudinal intervals of 820 ft (250 m). At each sampling site, belt transects 6.6 ft (2 m) wide and of variable lengths were laid out on the roadside verge, parallel to and 1.65 ft (0.5 m) from the pavement. Identical sampling areas were established on both sides of the highway at each site. Total sampling area varied from site to site depending on mullein density and was controlled to insure that each sample included a minimum of 100 live plants and at least 30 mature plants (flowering or fruiting). Site sampling areas ranged from 34,660 ft^2 (3,220 m^2) in areas of sparse mullein occurrence to 2,150 ft^2 (200 m^2) in high-density situations. Within each sampling area, all plants with a rosette diameter of 1.97 in. (5 cm) or larger were measured for rosette diameter, height of flowering spike (if present), reproductive state (budding, flowering, seeding, senescence), presence of any anatomical abnormalities including degree of fasciation, and the presence of polycarpy (basal, medial, or terminal secondary rosettes or flowering spikes).

Sampling in the Rue de Maido area of La Réunion Island (where the species is relatively uncommon as compared with Hawai'i) was restricted to recording mullein abundance in 1.24 mi x 3.24 ft (2 km x 1 m) roadside transects between road kilometer markers over the elevational range 4,920 to 6,890 ft (1,500-2,100 m). No unusual anatomical abnormalities were observed in the La Réunion mullein populations, so morphological data comparable to that collected in Hawai'i were not obtained. Relevant climatological data for the transect areas in Hawai'i and La Réunion were extracted from recent compilations (Hawaii Department of Land and Natural Resources 1970; Juvik *et al.* 1978; Cadet 1980; Hawaii Department of Planning and Economic Development 1985).

RESULTS AND DISCUSSION

General Distribution

The major areas of heavy mullein invasion on the island of Hawai'i generally include the leeward uplands (3,940 to 9,840 ft or 1,200-3,000 m) of Mauna Loa, Mauna Kea, and Hualālai (Fig. 2). The species also occurs sporadically in areas outside this zone, including windward Mauna Loa and even a few sea level locations on the arid west coast of the Island. The area of major mullein invasion and spread corresponds generally to that part of the Island with a temperate, summer-dry climate analogous to that found in the species' native range (Mediterranean Europe). As an example of this striking climatic similarity, Humu'ula, a meterological station within the study area (elevation 6,700 ft or 2,043 m, near sampling site 4), has a mean annual temperature of 52 F (11 C) and annual rainfall (winter concentrated) of 40 in. (1,013 mm). These values are very similar to Mediterranean locations like Milan, Italy (57 F or 13.7 C and 38 in. or 960 mm). The hot, dry summers of the upland interior of Hawai'i result from a strengthening of the subsidence inversion over the Islands, which generally restricts significant summer rainfall to areas below 5,900 ft (1,800 m). However, this summer-dry condition is somewhat mitigated by frequent mist and fog (undetected in the rainfall records) associated with orographic clouds that hug the middle slopes of the mountain (Juvik *et al.* 1978). Mullein appears to thrive in this climatic regime and has successfully invaded an area of over 770 mi^2 (2,000 km^2) in less than 70 years. On the island of La Réunion, *V. thapsus* has not yet achieved widespread distribution and pest status comparable to that found in Hawai'i. No information is available on the date of introduction in La Réunion; however, its limited distribution would suggest a comparatively recent arrival. Currently restricted to roadside areas of the Rue de Maido between 4,920 and 6,890 ft (1,500-2,100 m) elevation, the species has yet to colonize the more extensive volcanic ash and cinder areas surrounding the active volcano at Piton Bory (Fig. 2).

Altitudinal Range and Abundance

In windward areas of the Saddle Road above Hilo, the species reaches its low-altitude limit at 5,330 ft (1,625 m) (Table 1). Below this elevation, mullein appears to be usually excluded from the roadside verge by other aggressive weeds better adapted to warm and wet conditions (Wester and Juvik 1983). The competitive exclusion hypothesis appears confirmed by the fact that mullein has recently been able to sporadically colonize lower windward locations (down to 2,620 ft or 800 m) along the Saddle Road verge where road-widening projects cleared all existing verge vegetation in 1986. Mullein densities in the roadside verge reach 15-18 plants/100 ft^2 (160-190/100 m^2) in the zone between 6,230 and 6,560 ft (1,900- 2,000 m) (Fig. 3). The substantially lower densities (of about 8.3 plants/100 ft^2 or 90 plants/100 m^2) recorded between 7,380 and 8,200 ft (2,250-2,500 m) (sites 5 and 6) may be a result of the open cattle range in this area, where managed pasture grasses may act to reduce mullein densities along the verge. The plant does not appear to be eaten by cattle even under poor forage conditions, as *V. thapsus* colonies were well distributed throughout pastures even in overgrazed areas. Roadside mullein densities diminished rapidly above 9,020 ft (2,750 m) elevation, where the species reaches its present limit at 10,830 ft (3,300 m) on Mauna Kea.

Along the Rue de Maido sampling area on La Réunion, plant densities were much lower than on Hawai'i (Table 2), reaching a maximum of only 0.86 plants/100 ft^2 (9.3/100 m^2) at the road summit (6,890 ft

Table 1. Characteristics of *Verbascum thapsus* populations in roadside habitats on Mauna Loa and Mauna Kea, island of Hawai'i.

Sample Sites[1]	Elevation (m)	Sample Area (m^2)	Live Plants[2]	Density/ 100 m^2	Relative Density[3] (%)	Rosette Diameter:[4] mean and (range) (cm)	Height:[5] mean and (range) (cm)	Extreme Fasciation % of mature plants[6]	Multiple Shoots % of mature plants[7]
1	1,678	3,220	179	5.6	2.9	17.4 (5-46)	79.6 (48-183)	0	11.8
2	1,790	3,220	382	11.9	16.6	17.3 (5-48)	70.3 (37-143)	0	27.5
3	1,925	688	1,107	160.9	83.8	12.6 (5-36)	71.5 (34-194)	0	6.0
4	2,000	200	384	192.0	100.0	12.9 (5-62)	84.5 (61-198)	4.8	14.2
5	2,250	500	474	94.8	49.4	18.7 (5-65)	96.8 (30-220)	7.7	46.1
6	2,500	250	217	88.6	46.1	25.9 (5-110)	140.7 (51-323)	35.1	40.8
7	2,750	250	276	110.4	57.5	20.8 (5-70)	108.7 (25-245)	31.5	57.4
8	3,000	1,000	135	13.5	7.0	27.3 (5-148)	111.1 (34-178)	9.1	59.1

1 Location of sampling sites shown in Figure 2.

2 Only *V. thapsus* plants with a basal rosette diameter >5 cm were recorded. Standing dead plants were also recorded but not included in this tabulation.

3 Relative density = density expressed as a percentage of highest density recorded (Site 4).

4 For the two altitudinal sampling sites (3 and 8) showing the greatest difference in mean rosette diameter, differences are statistically significant ($t = 8.4$; $p < 0.01$). For altitudinally adjacent sampling sites 2-3, 4-5, 5-6, 6-7, and 7-8, differences in mean rosette diameter are also statistically significant ($p < 0.01$).

5 Height data were recorded only for mature plants with an inflorescence spike (including standing dead plants). For the two altitudinal sampling sites (2 and 6) showing the greatest difference in mean plant height, differences were statistically significant ($t = 8.8$; $p < 0.01$). For the altitudinally adjacent sampling sites 5-6, and 6-7, differences in mean plant height were also statistically significant ($p < 0.01$).

6 Extreme fasciation was arbitrarily defined as stem flattening sufficient to produce a maximum cross-sectional diameter more than three times the minimum cross-sectional diameter (non-fasciated stems are generally cylindrical; see Figure 9).

7 Included are plants with either (or both) multiple basal rosettes or multiple flowering spikes.

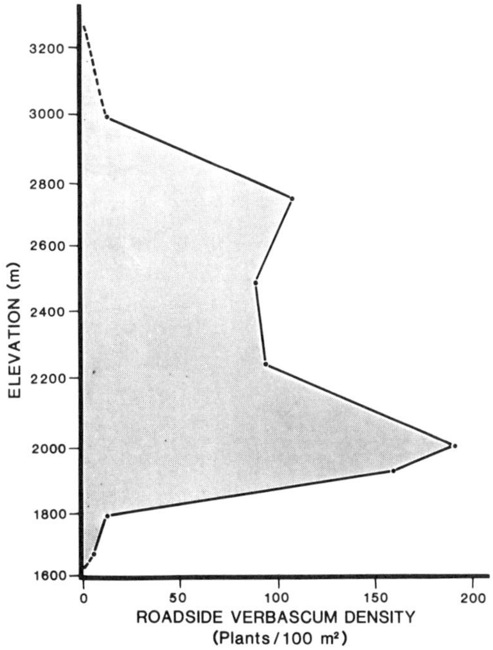

Figure 3. Variation in roadside *Verbascum thapsus* density at different elevations on Mauna Loa and Mauna Kea, 1985.

Table 2. Common mullein density in roadside habitats along the Rue de Maido, La Réunion.

Sample Site*	Elevation** (m)	Live plants	Density/ 100 m^2	Relative Density (%)***
1	1,500	3	0.15	1.6
2	1,664	6	0.30	3.2
3	1,840	32	1.60	17.3
4	1,973	177	8.85	95.6
5	2,085	185	9.25	100.0

* Sampling areas were 2,000 m^2, consisting of roadside verge area 1 m wide x 2 km long.
** Mid-point elevation within each 2-km long sample area.
*** Site density expressed as % of site 5 density.

or 2,100 m). For general comparative purposes, the Hawai'i and La Réunion data were converted to "relative density," where densities at each elevation were expressed as a percentage of the value at the site of the highest density (*i.e.* site 4 at 6,560 ft (2,000 m) in Hawai'i and the summit site at 6,890 ft (2,100 m) in La Réunion). The striking similarities in altitude-relative density relationships between the two sites (Fig. 4) mirror the environmental similarity of these geographically disparate islands and the apparent preference for mid-elevation, tropical montane sites climatically analogous to southern Europe. The parallel invasion patterns in Hawai'i and La Réunion suggest the general susceptibility of other high-elevation tropical montane environments.

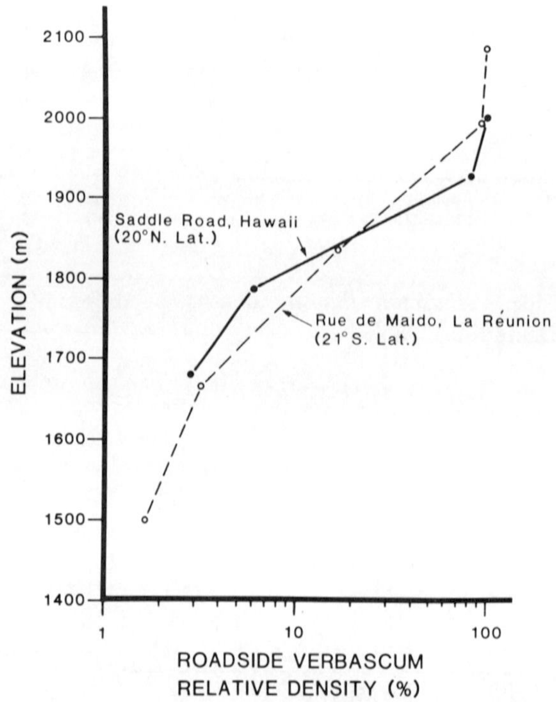

Figure 4. Variation in relative density (see text) of *Verbascum thapsus* along altitudinal transects in Hawai'i and La Réunion.

Plant Size in Relation to Altitude

Mean rosette diameter of mullein increased with elevation to a maximum of 10.75 in. (27.3 cm) at the highest sampling site (8, at 9,840 ft or 3,000 m) in Hawai'i (Fig. 5). This value is more than double the mean rosette diameter at lower elevation sites 3 and 4, and differences are statistically significant (Table 1). While a large number of young plants at some lower elevations might account for the observed trend in sample means, examination of the maximum rosette diameter recorded at each site confirms the same relationship (Table 1). Below 6,560 ft (2,000 m), the largest rosette diameters recorded at different sites were 14.2 to 18.9 in. (36-48 cm), while at 8,200 ft (2,500 m) and above, maximum rosette diameters ranged from 27.6 to 58.3 in. (70-148 cm).

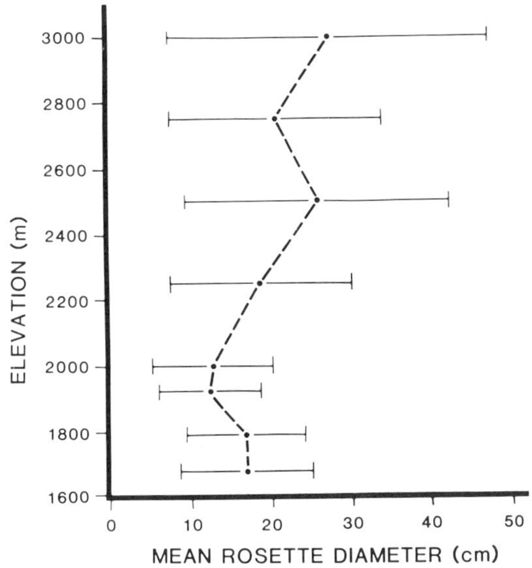

Figure 5. Altitudinal variation (mean ± 1 std. dev.) in mean rosette diameter of *Verbascum thapsus* on Mauna Loa and Mauna Kea, 1985. (Only plants with rosette diameters ≥ 5 cm were measured.)

For mature plants bearing an inflorescence spike, mean plant height data show an altitudinal pattern similar to rosette diameter (Fig. 6). A maximum mean height of 55.4 in. (140.7 cm) was recorded at 8,200 ft (2,500 m), approximately double that recorded for mature plants at elevations below 6,560 ft (2,000 m) (Table 1). An elevational trend toward gigantism is clearly evident in both the height and diameter data. It should be noted that the maximum height of mullein within its native Eurasian range is generally given as 7.6 ft (2.3 m) (Manning 1965; Gross and Werner 1978). On the upper slopes of Mauna Kea specimens exceeding 10 ft (3 m) and occasionally 13 ft (4 m) in height are not unusual. The increase in plant stature with elevation may be related to optimal conditions for increased net photosynthesis in the Hawaiian subalpine zone, since solar radiation increases with elevation (Hawaii Department of Planning and Economic Development 1985), while air temperature (and hence plant respiration losses) decrease.

Morphological Abnormalities

Stem fasciation involves abnormal flattening of the main shoot axis resulting from the gradual expansion of the growing point in one plane. Sinnott (1960) reported fasciation in a wide range of plants resulting from a variety of factors, including mutilation and wounds, application of growth substances, and such environmental factors as short day length. Il'yenko (1982) observed fasciation in *V. thapsus* "adapting to new conditions" under cultivation. On Mauna Kea extreme fasciation of the flowering spike is common at high elevation and frequently produces grotesquely deformed and twisted flowering heads (Fig. 7). No fasciation was noted on mature plants growing below 6,560 ft (2,000 m), while at 8,200 to 9,020 ft (2,500-2,750 m) elevation, extreme fasciation was exhibited by 31-35% of all mature plants (Fig. 8). One of the consequences of extreme

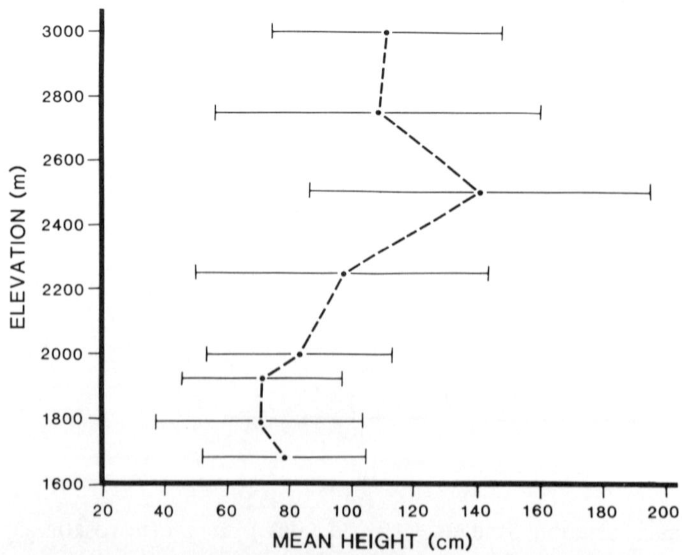

Figure 6. Altitudinal variation in mean height (± 1 std. deviation) of *Verbascum thapsus* flowering stocks on Mauna Loa and Mauna Kea, 1985.

Figure 7. *Verbascum thapsus* plants illustrating extreme fasciation and multiple axillary shoots at high elevation on Mauna Kea.

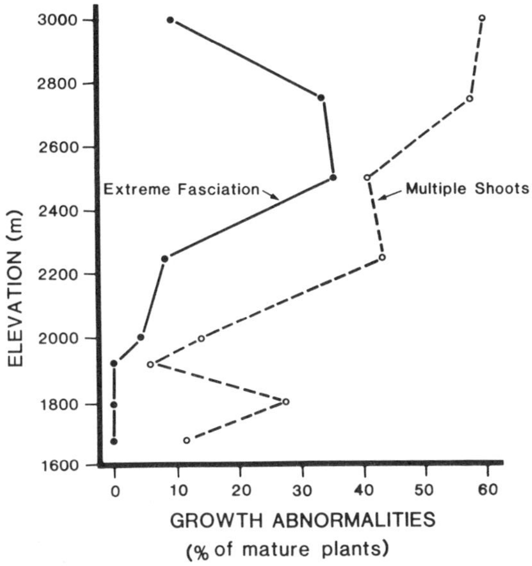

Figure 8. Altitudinal variation in percentage of mature *Verbascum thapsus* plants exhibiting growth abnormalities including extreme fasciation and multiple axillary shoots, 1985.

fasciation is a dramatic increase in stem bulk and woodiness (Fig. 9). Carlquist (1974 and pers. comm. 1986) has reported the effect of severe frost on the growth of several monocarpic Canary Island *Echium* species under cultivation at Claremont, California. Although *Echium* species, native to high elevations, evidenced little frost damage, one mid-elevation species, *Echium pininana* (native to frost-free areas), exhibited leaf damage and subsequent fasciated stem growth by all individuals. *Verbascum thapsus* in its native European range does not normally experience diurnal frosts during the active summer growing season. On Mauna Kea, diurnal frosts may occur throughout the year at high elevation but are rare at elevations below 6,560 ft (2,000 m). The elevational pattern in observed mullein fasciation (Fig. 8) is generally consistent with a frost damage hypothesis.

Although usually described as a monocarpic plant, *V. thapsus* may exhibit a variety of polycarpic growth forms, including multiple basal rosettes and multiple flowering shoots situated basally, medially, or terminally from the primary flowering spike or from separate basal rosette clusters (Fig. 7). The tendency toward a multiple shoot growth habit also shows a strong elevational trend, generally similar to that of fasciation. However, the percentage of mature plants with multiple shoots continues to increase above 9,020 ft (2,750 m), reaching a maximum of 59% near the upper limit of the species range at 9,840 ft (3,000 m) (Fig. 8).

Over the comparatively short period of time since mullein first began colonizing the upper slopes of Mauna Kea in the 1940s, significant

Figure 9. Plane and cross sectional views of *Verbascum thapsus* flowering spikes illustrating normal (lower) and fasciated (upper) forms. Both plants were collected at 2,500 m elevation on Mauna Kea.

morphological responses to the altitudinal/climatic gradient have begun to appear. (An alternative explanation is that founder populations with these adaptations survived.) Characteristics include gigantism, fasciation (with resultant increases in stem woodiness), and polycarpic growth forms, as well as a shift from biennial to perennial habit. All these adaptations conform to the conditions given by Carlquist (1974) as potentially leading to development of insular arborescence in rosette herbs. Carlquist (1974) listed examples of insular arborescence from the family Scrophulariaceae, including *Hebe* in New Zealand and *Lytanthus* in the Canary Islands. At present there are no data to determine whether the elevation-morphology trends of mullein on Mauna Kea are mere environmental variants (as might be expected given the short history of the species on the mountain), or genetically differentiating ecotypes. Studies of naturalized populations of *V. thapsus* in North America indicate that morphological and life history variations along latitudinal gradients are largely genetically based (Reinartz 1984b). Future research on *V. thapsus* in Hawai'i will concentrate on "common garden" experiments to distinguish genetic and environmental components of the observed morphological variations with altitude.

Control and Eradication

With the exception of efforts documented here, few governmental or private efforts to control or eradicate *V. thapsus* in the mountainous areas of Hawai'i have occurred to date. On both Mauna Kea and Mauna Loa, the plant occurs only along the edges of access roads at the upper elevational limit, suggesting that the automobile (tire tread mud) may be a significant mode of disperal at high elevation, since the seeds of mullein are not specifically adapted for long-distance dispersal.

Recent efforts to restrict the spread of *V. thapsus* at high elevation have centered on removal of all roadside plants near the upper altitudinal limits of distribution. From 1984 to 1986, student volunteers from the University of Hawaii at Hilo Geography Club undertook eradication work along 3.1 mi (5 km) of the Mauna Kea summit road above Hale Pōhaku (9,840 to 10,830 ft or 3,000-3,300 m), and 6.2 mi (10 km) of the Mauna Loa access road (6,230 to 6,890 ft or 1,900-2,100 m). More than 5,000 plants were uprooted and destroyed during these eradication efforts (Anonymous 1984). Before removal, the spatial configuration and population structure of selected mullein colonies were fully documented to facilitate long-term studies of site persistence and reinvasion rates, in light of the extremely long seed viability in this species (Juvik and Juvik, unpub. data). Follow-up of seedling growth rates over the next few years should indicate the interval between initial control and first flowering of the subsequent generation, and thus the best retreatment interval for further control efforts. The possibility of biological control for heavily infested areas of Hawai‘i also bears serious investigation. The weevil *Gymnetron tetrum* is a major seed predator of *V. thapsus* throughout its natural range and may prove a viable control agent in Hawai‘i.

Recent comments (Smith 1985) that population size and range of *V. thapsus* have been significantly reduced by a gall-forming insect are not supported by our data. There are presently no significant insect predators, and the species continues to expand its range and population size in favorable upland environments.

CONCLUSIONS

Over the past 70 years, mullein has rapidly invaded montane areas on the island of Hawai‘i. As an early successional colonizer of disturbed sites, the successful spread of *V. thapsus* has been greatly facilitated by native forest and ground cover disturbance associated with the grazing and browsing of both feral and domestic ungulates (Juvik and Juvik 1984). Mullein is continuing to expand its range on Mauna Kea, where the ash and cinder substrates are more favorable than the fresh lava flows of upper Mauna Loa. The potential invasion of summit cinder cones of Mauna Kea would be an aesthetic blight (notwithstanding recent tourist literature that has touted the "particularly photogenic" qualities of the "furry leafed" mullein, *e.g.*, Uprichard 1986) and would have serious ecological consequences. It is particularly ironic that mullein appears to be filling a niche similar to that vacated by Hawai‘i's endemic "rosette species," the silversword (*Argyroxiphium sandwicense*), which unlike the unpalatable *Verbascum* has been nearly eradicated from Mauna Kea by browsing of feral ungulates.

The similar mullein invasion on the Indian Ocean island of La Réunion emphasizes the general vulnerability of other tropical montane areas around the world. In Hawai'i every effort should be made to prevent establishment of this species on the island of Maui, since Haleakalā Crater possesses the altitudinal range (7,220 to 9,890 ft or 2,200-3,000 m) and cinder substrate preferred by *V. thapsus*. The recent discovery and removal (Hawaiian Botanical Society 1986) of an immature mullein plant from the roadside verge at 9,510 ft (2,900 m) in Haleakala National Park confirms the need for continued vigilance to prevent successful establishment of this species in the subalpine zone on Maui.

The altitudinal trends in morphological variation described in this paper may illustrate incipient adaptive stages in a general evolutionary model for the development of woody, arborescent, tropical alpine rosette plants. Mauna Kea mullein clearly evidences gigantism, increased woodiness, and shifts to perennial habit. The genetic significance of these morphological variations must await further study.

ACKNOWLEDGMENTS

We would like to thank J. Jacobi for extracting *Verbascum thapsus* distributional data from unpublished botanical results of the U.S. Fish and Wildlife Service Hawai'i Forest Bird Survey. We further express our appreciation for the field assistance and *V. thapsus* eradication work undertaken by students of the University of Hawaii at Hilo Geography Club. We also thank Sherwin Carlquist for his critical review of the manuscript.

Literature Cited

Anonymous. 1984. Exotic plant eradication. *Hawaii Tribune-Herald* Oct. 21:26.

Baskin, J.M., and C.C. Baskin. 1981. Seasonal changes in germination responses of buried seeds of *Verbascum thapsus* and *V. blattaria* and ecological implications. *Can. J. Bot.* 59:1769-1775.

Cadet, T. 1974. Ētude sur la vegetation des hautes altitudes de l'île de La Réunion (Ocean Indien). *Vegetatio* 29(2):121-130.

Cadet, T. 1980. *La vegetation de l'île de La Réunion.* St. Denis, Réunion: Imprimerie Cazal.

Carlquist, S. 1974. *Island biology.* New York: Columbia Univ. Pr.

Gross, K.L. 1981. Predictions of fate from rosette size in four "biennial" plant species: *Verbascum thapsus, Oenothera biennis, Daucus carota,* and *Tragopogon dubius. Oecologia* 48:209-213.

Gross, K.L., and P.A. Werner. 1978. The biology of Canadian weeds. 28. *Verbascum thapsus* L. and *Verbascum blattaria* L. *Can. J. Plant Sci.* 58:401-413.

Hawaii Department of Land and Natural Resources. 1970. *An inventory of basic water resource data: island of Hawaii.* Rep. 34. Honolulu.

Hawaii Department of Planning and Economic Development. 1985. *Hawaii sunshine map.* Honolulu.

Hawaiian Botanical Society. 1986. Mullein discovered in Haleakala National Park. *Newsletter, Hawn. Bot. Soc.* 25(3):89.

Il'yenko, O.O. 1982. Botanical properties of some species of the Scrophulariaceae family during their cultivated growth. *Ukr. Bot. Zh.* 39(5):24-28 (in Ukranian with Russian and English summaries).

Jacobi, J.D., and J.M. Scott. 1985. An assessment of the current status of native upland habitats and associated endangered species on the island of Hawai'i. In *Hawai'i's terrestrial ecosystems: preservation and management,* ed. C.P. Stone and J.M. Scott, 3-22. Univ. Hawaii Coop. Natl. Park Resour. Stud. Unit. Honolulu: Univ. Hawaii Pr.

Juvik, J.O., and S.P. Juvik. 1984. Mauna Kea and the myth of multiple use: endangered species and mountain management in Hawaii. *Mountain Res. and Devel.* 4(3):191-202.

Juvik, J.O., D.C. Singleton, and G.G. Clarke. 1978. Climate and water balance on the island of Hawaii. In *Mauna Loa Observatory: a 20th anniversary report,* ed. J. Miller, 129-139. Washington, D.C.: U.S. Dep. Commerce.

Kivilaan, A., and R.S. Bandurski. 1933. The ninety-year period of Dr. Beal's seed viability experiment. *Amer. J. Bot.* 60:140-145.

Manning, S.A. 1965. *Systematic guide to flowering plants of the world.* New York: Taplinger Pub. Co.

Reinartz, J.A. 1984a. Life history variation of common mullein (*Verbascum thapsus*). I. Latitudinal differences in population dynamics and timing of reproduction. *J. Ecology* 72:897-912.

Reinartz, J.A. 1984b. Life history variation of common mullein (*Verbascum thapsus*). II. Plant size, biomass partitioning and morphology. *J. Ecology* 72:913-936.

Semenza, R.J., and R.A. Evans. 1978. Influence of light and temperature on the germination and seedbed ecology of common mullein (*Verbascum thapsus*). *Weed Sci.* 26(6):577-581.

Sinnott, E.W. 1960. *Plant morphogenesis*. New York: McGraw-Hill.

Smith, C.W. 1985. Impact of alien plants on Hawai'i's native biota. In *Hawai'i's terrestrial ecosystems: preservation and management*, ed. C.P. Stone and J.M. Scott, 180-250. Univ. Hawaii Coop. Natl. Park Resour. Stud. Unit. Honolulu: Univ. Hawaii Pr.

Uprichard, B. 1986. High atop Mauna Kea. *Hawaii* 5(7):45-48.

Wester, L., and J.O. Juvik. 1983. Roadside plant communities on Mauna Loa, Hawaii. *J. Biogeography* 10:307-316.

Wilhelm, G., Jr. 1974. The mullein: plant piscicide of the mountain folk culture. *Geogr. Rev.* 64(2):235-252.

Williams, G.J. III, R. Lazor, and P. Yourgrau. 1975. Temperature adaptations in the hill reaction of altitudinally and latitudinally diverse populations of *Verbascum thapsus* L. *Photosynthetica* 9(1):35-39.

THE STATUS OF BANANA POKA IN HAWAI'I

Anne Marie LaRosa

ABSTRACT

Banana poka (*Passiflora mollissima)*, a weedy, perennial liana from the Andean highlands, has invaded many of the major upland wet and mesic koa-'ōhi'a (*Acacia koa-Metrosideros polymorpha*) forests on the islands of Hawai'i and Kaua'i. Small, isolated populations have also been found on Maui. Infestations range from scattered individuals with low cover to areas of 100% cover, the latter often inhibiting growth and reproduction of native forest species. According to recent assessments of distribution and abundance in Hawai'i, approximately 200 mi^2 (520 km^2) are infested with this species. As with many tropical lianas, banana poka exhibits an efficient pattern of resource allocation, including staggered germination, rapid growth rates, early reproductive maturity, continuous growth and reproduction, and high seed output. These adaptations and effective dissemination through a number of alien mammals and birds that eat the fruit and seeds have led to the rapid spread of the species in Hawai'i. Annual rainfall in excess of 177-200 in. (4,500-5,100 mm), extremes of elevation, and intensive agriculture appear to limit the distribution of banana poka; very low light levels may check the rate of spread and extent of cover. Establishment and proliferation of the species in low light areas is accomplished through the mechanism of "gap-phase" replacement. Control efforts must address biological and sociopolitical factors. Although banana poka has no known economic value and is a noxious plant in Hawai'i, its close relationship to the edible passion fruit, *P. edulis*, has led to resistance by certain local groups to its control. Control of introduced mammals (especially the feral pig, *Sus scrofa*) and birds that disseminate seeds and often increase the rates of spread and establishment is a necessary adjunct to banana poka control. Deliberate practices that open up closed-canopy forests should also be discouraged, as spread of this climber is tied to disturbance.

INTRODUCTION

Banana poka (*Passiflora mollissima*) is a perennial, fleshy-fruited climber native to the Andean highlands. It occurs over an extensive altitudinal and latitudinal range, from 6,560 to 11,800 ft (2,000-3,600 m) elevation and from Venezuela to Bolivia (Killip 1938; Escobar 1980). The climate of this area may be characterized as a cool, moist, tropical montane environment. Although banana poka is a wide-ranging species that is extensively cultivated throughout the Andes, natural populations are infrequent and limited in size (Pemberton 1982). Natural habitats may be described generally as evergreen, woody formations of relatively low stature containing an abundance of ferns, mosses, and epiphytes (Weberbauer 1936; Cuatrecasas 1953; Escobar 1980). Banana poka also occurs along river courses in the drier inter-Andean valleys, perhaps only where it has escaped from cultivation.

Dispersal by humans has been largely responsible for the widespread distribution of banana poka. Because of its edible fruit and attractive flowers, specimens were introduced into numerous tropical and subtropical areas, including Mexico, California, New Zealand, Australia, Hawai'i, New Guinea, the Kermadec Islands, India, Ceylon, and East Africa (Killip 1938; Martin and Nakasone 1970; Green 1972; deWilde 1975; Sykes 1977; Escobar 1980). Most recently (1987), it was discovered in South Africa (MacDonald 1987). Most information from these areas deals only with banana poka cultivation. Escape from cultivation has been reported in several places, but only in Hawai'i is poka considered a serious pest (Hawaii Volcanoes National Park 1986). Varying land use philosophies within countries may contribute more to the perceived status of the plant than does the biology of the species.

Banana poka was introduced to the Pu'uwa'awa'a area of Hawai'i in the early 20th century (Pung 1971) and has subsequently spread via humans and various birds and mammals. As of 1986 it occupied 200 mi^2 (520 km^2) on the islands of Hawai'i, Kaua'i, and Maui (Fig. 1). Populations occur from 1,640 to 8,200 ft (500-2,500 m) elevation, in areas of 49 to 200 in. (1,250-5,000 mm) annual rainfall and average annual temperatures of 50 to 75 F (11-24 C) (Warshauer *et al.* 1983; Jacobi and Warshauer, this volume).

Only nine of 32 taxa of *Passiflora* introduced into the Hawaiian Islands have naturalized. Only two species, banana poka and sweet granadilla (*P. ligularis*), are known to cause severe impacts to native vegetation (Warshauer *et al.* 1983). Densities of banana poka reported for Hawai'i are often far in excess of those reported for South America (Warshauer *et al.* 1983). The widespread distribution and destructive impact of banana poka in Hawai'i is largely due to several environmental and biological factors, including: favorable environmental conditions similar to those found in the native range, abundance of disturbed habitats suitable for colonization, numerous native and alien dispersal agents, wide degree of environmental tolerance, lack of damaging predators and pathogens, and general plant vigor (LaRosa 1984). The successful pattern of resource allocation, including rapid growth rates, early reproductive

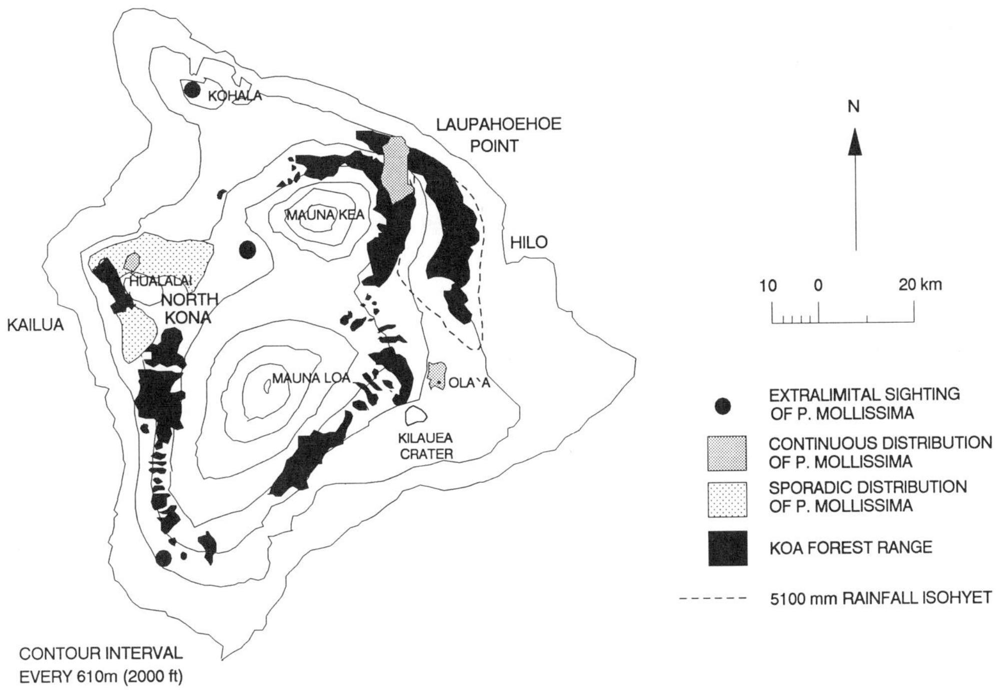

Figure 1a. Distribution of banana poka on the island of Hawai'i.

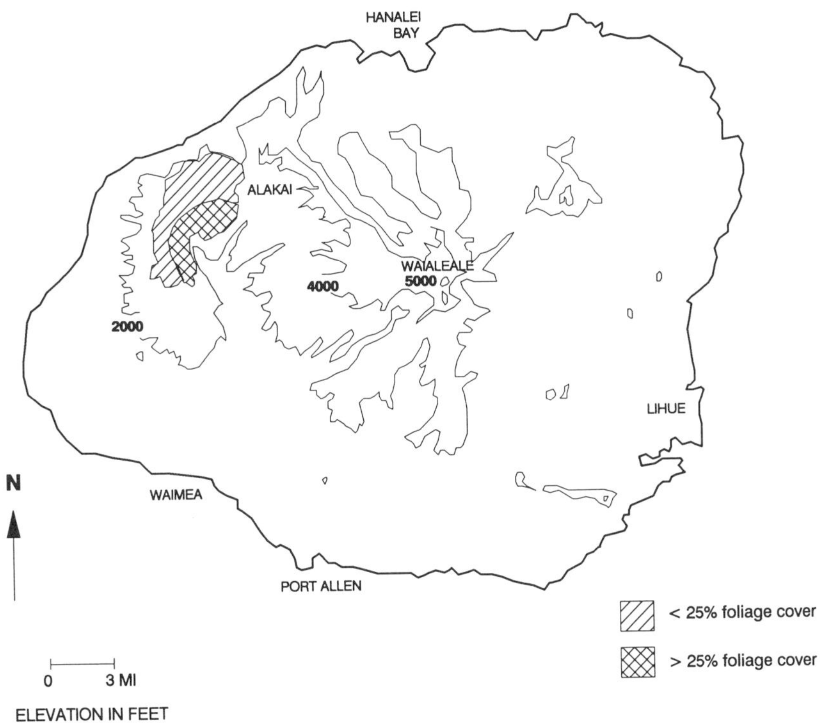

Figure 1b. Distribution of banana poka on the island of Kaua'i.

maturity, continuous reproduction, and high seed output, accompany the favorable environmental conditions. The success of banana poka generally precludes widespread control via mechanical and/or chemical means.

Jacobi and Warshauer (this volume) have demonstrated the potential invasiveness of banana poka in Hawai'i and the immediate need for an effective control program. The impact of the species populations on the koa *(Acacia koa)* logging industry was documented in the late 1970s (Skolmen 1979). An interagency effort to develop an effective biological control program is proceeding. Information on the taxonomy, biology, ecology, and distribution of banana poka, its impacts on Hawaiian ecosystems, and current and past control strategies is available from various sources (Warshauer *et al.* 1983; LaRosa 1984; Jacobi and Warshauer, this volume). This paper summarizes all available information on the subject and formulates management recommendations based on the available information.

TAXONOMY

Some uncertainty and disagreement has existed regarding the identity of the taxon commonly known in Hawai'i as banana poka. It has been variously referred to as *Passiflora mollissima, P. mixta*, and as an introgressive hybrid between banana poka and some unknown species (Fosberg 1975; Mueller-Dombois 1975; Mueller-Dombois *et al.* 1980; LaRosa 1985). Its morphological similarity to *P. tripartita* has also been noted (LaRosa 1985). In its native range, populations of banana poka are morphologically variable and frequently hybridize with other members of the subgenus *Tacsonia*. Introgressive hybrids are common (Escobar 1980).

Hawaiian populations of banana poka exhibit some variability, but little compared with that in native South American populations. The atypical morphology characteristic of Hawaiian populations falls within the range of variation reported for *P. mollissima* (LaRosa 1985). The possibility that this taxon represents an introgressive hybrid has not been ruled out, however. The atypical morphotype of banana poka found in Hawai'i is also found in several other Pacific areas where it has been introduced, suggesting a common ancestry with the Hawaiian introduction (Green 1972; deWilde 1975; Sykes 1977).

BIOLOGY

Information on the biology of banana poka comes principally from measurements and observations of three populations ('Ōla'a Tract, Kaloko-Hualālai, and Laupāhoehoe-Keanakolu) on the island of Hawai'i (LaRosa 1984). For the purposes of the following discussion, the life stages are defined as follows: seedlings -- individuals <3 ft (1 m) in height and having at least one pair of true leaves; juveniles -- non-reproductive individuals >3 ft (1 m) tall; adults -- reproductive individuals of any size.

Life History

Populations of banana poka are density dependent (LaRosa 1984). Although the density of germinants varied by several orders of magnitude, that of established individuals (>4 in. (0.1 m) in height) remained relatively constant (Table 1). High densities of germinants at most sites indicate that site conditions do not generally limit germination. Highest densities occurred near parent vines and in areas of sparse canopy cover and high feral pig (*Sus scrofa*) activity. In areas with little disturbance and greater canopy cover, densities were significantly lower (Table 1).

Seedling (germinant) survival was highly variable and negatively correlated with initial germinant density (Table 1). Most surviving seedlings were vigorous and free of pests, except in wetter areas, where soil-borne organisms (*e.g.*, damping-off fungi) and predators (*e.g.*, slugs) were more common. Growth rates of seedlings were variable but averaged 10 in./yr (25 cm/yr) (Table 2). Many seedlings remain quiescent following germination, forming a seedling bank that can be released under favorable conditions, *e.g.*, when a canopy gap occurs (LaRosa 1984).

The juvenile stage in the life cycle was characterized by fairly high survival (60% survive to reproductive maturity) (Fig. 2) and rapid growth rates (Table 2). Density of juveniles was relatively constant (Table 1). Growth rates in this phase averaged 3 ft (1 m) per year but occasionally exceeded 10 ft (3 m) per year (Table 2), allowing individuals to quickly reach the full sun of the canopy. These growth rates are within the range of rates of other tropical climbers (Bailey 1944; Janzen 1971; Choudhury 1972).

Adults of banana poka are typically climbing but may remain shrubby or become trailing when supports are lacking. Short-term survival records (18 mo) suggest that mortality of adults is low, averaging less than 5% (LaRosa 1984) (Fig. 2). The estimated life span of vines is 15 to 20 years (Schoniger 1969; LaRosa 1984). Growth rates of shoots on adult plants were variable (Table 2) but were composed of two discrete phases: rapidly-growing flushing shoots and slow-growing, non-flushing shoots (LaRosa 1984). Enhanced reproduction usually accompanied the rapid growth of flushing shoots, which were more common in areas of high light intensity, such as the canopy. Plants may reproduce within a single year in areas of full sun or large canopy gaps within closed-canopy forests (Burton 1980; LaRosa 1984).

Reproductive Biology

The reproductive pattern of banana poka combines outcrossing (pollination) early in anthesis (flower opening) with selfing at a later stage (Escobar 1980; LaRosa 1984), ensuring that isolated individuals can produce viable populations from a single colonizer. The natural incidence of selfing observed was low (4.3% fruit set), but a high incidence of fruit set in hand-selfed (pollinated) flowers indicates that flowers of banana poka are potentially highly self-compatible (LaRosa 1984).

Table 1. Estimate of the density (number/hectare) and percent survival of banana poka at seven sites on the island of Hawai'i (LaRosa 1984). Based on population structure analysis, estimates of survival rates can exceed 100%.

	Density (no./ha)				Survival Rate (%)	
		Height classes				
Site	Germinant (< 0.1 m)	Establishment (0.1-1.0 m)	Established (> 1.0 m)	Total	Germinant to Establishment	Establishment to Established
Kaloko 1	3,640	4,610	1,116	9,410	127	24
Kaloko 2	121,600	3,700	1,590	126,890	3	43
Laupāhoehoe 1	4,707	393	387	5,487	8	98
Laupāhoehoe 2	554,000	1,700	775	556,475	0.3	46
Laupāhoehoe 3	267,879	750	458	269,078	0.3	61
'Ōla'a Tract 1	638	625	500	1,763	97	80
'Ōla'a Tract 2	2,957	1,200	700	4,857	40	58

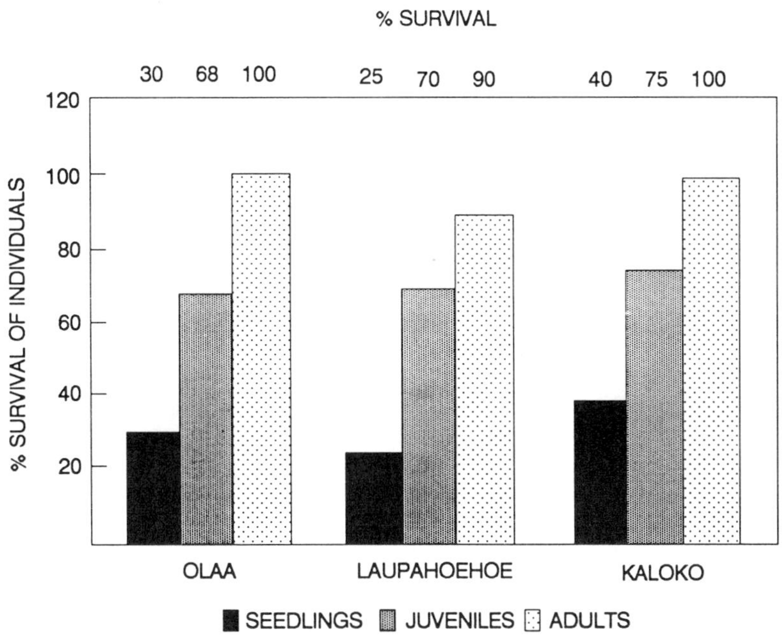

Figure 2. Survival of seedlings, juveniles, and adults of *Passiflora mollissima* over an 18-month period at three sites on the island of Hawai'i.

Table 2. Mean increases in shoot length (cm/yr) at three sites for three ages of banana poka.*

Site	Seedling		Juvenile		Adult	
Kaloko (Hualālai)	14.60	b-2	44.53	c-1	58.77	a-1
Laupāhoehoe (Mauna Kea)	28.83	a-3	101.11	b-1	61.68	a-2
'Ōla'a (Mauna Loa)	33.58	a-2	131.76	a-1	85.04	a-2
Mean	25.55 ± 1.46		91.25 ± 6.20		69.35 ± 5.48	
Overall Range	0 - 80.30		0 - 306.60		0 - 259.15	

*Means in the same column with the same letter are not significantly different by Duncan's Multiple Range Test ($p \leq 0.05$). Means in the same row with the same number are not significantly different ($p \leq 0.05$). Overall means and ranges for all plants measured are also given.

Biotic pollinators played a major role, increasing fruit set in open pollinated flowers to 15.2%. (*Passiflora* pollen is too large and sticky to be wind carried.) Generalist pollinators, including honeybees (*Apis mellifera*) and flower flies (Syrphidae), were the most frequent visitors to flowers in August 1981 (LaRosa 1984). Blowflies (*Calliphora* sp.) and Kamehameha butterflies (*Vanessa tameamea*) were occasional visitors. Pomace flies (*Drosophila* spp.) and thrips (Thysanoptera) were commonly found in flowers but are generally considered too small to be effective pollinators (Nishida 1963). Several species of birds, including the native 'i'iwi *(Vestiaria coccinea)*, 'amakihi *(Hemignathus virens)*, and 'ōma'o *(Myadestes obscurus obscurus)* and the alien Japanese white-eye (*Zosterops japonicus*), regularly visit flowers of banana poka in Hawaiian forests but appear to function principally as nectar robbers (LaRosa 1984). Asexual reproduction contributes little to the overall reproductive rate: apomictic (asexually produced embryos within ovules) fruit set is unknown, and vegetative reproduction is present only rarely, principally in wetter sites.

Banana poka fruits, which ripen in 2.5-3 months, contain an average of 180 seeds and may remain on the vine from several weeks to more than a month. Most ripe fruits are found during the wet months of December to March (LaRosa 1984). Fruit set in Hawai'i is greater than that noted for South America (Escobar 1980; LaRosa 1984). Fruit are reportedly much more abundant in Hawai'i than in most wild populations of banana poka in South America (Pemberton 1982).

Several fungal pathogens, including *Alternaria passiflorae* and *Colletotrichum* sp., attack fruit but generally do not limit fruit production in Hawai'i (LaRosa 1984). Pathogens may also be partially responsible for a condition known as "fruit split" or "fruit crack," which resulted in poor fruit development and abortion of 40% of all fruit in the 'Ōla'a Tract during an 18-month period (LaRosa 1984). Fruit split may also be a result of physiological factors associated with drought and results in significant crop losses of banana poka in South America (Anonymous 1962; Schoninger 1969).

Phenology

Populations of banana poka exhibit continuous growth and reproduction, a common feature of plants growing in the humid tropics (Daubenmire 1972; Wycherly 1973; Stiles 1977; Opler *et al.* 1980) (Fig. 3). Seasonal differences in phenological activity of banana poka in Hawai'i were principally quantititive (changes in magnitude of activity). Some annual variability occurred in phenological activity patterns (LaRosa 1984), but this type of variability is common in tropical species (Richards 1952; Daubenmire 1972; Croat 1975; Walter *et al.* 1975; Opler *et al.* 1980).

All phases of flowering normally exhibited peak activity in the drier months of April to September. Highest levels of fruiting were concentrated in the wetter months of December to March. Variability in phenological patterns occurred among populations, but peak levels of anthesis (flowering opening) and fruiting were fairly synchronous (within 30 days) (LaRosa

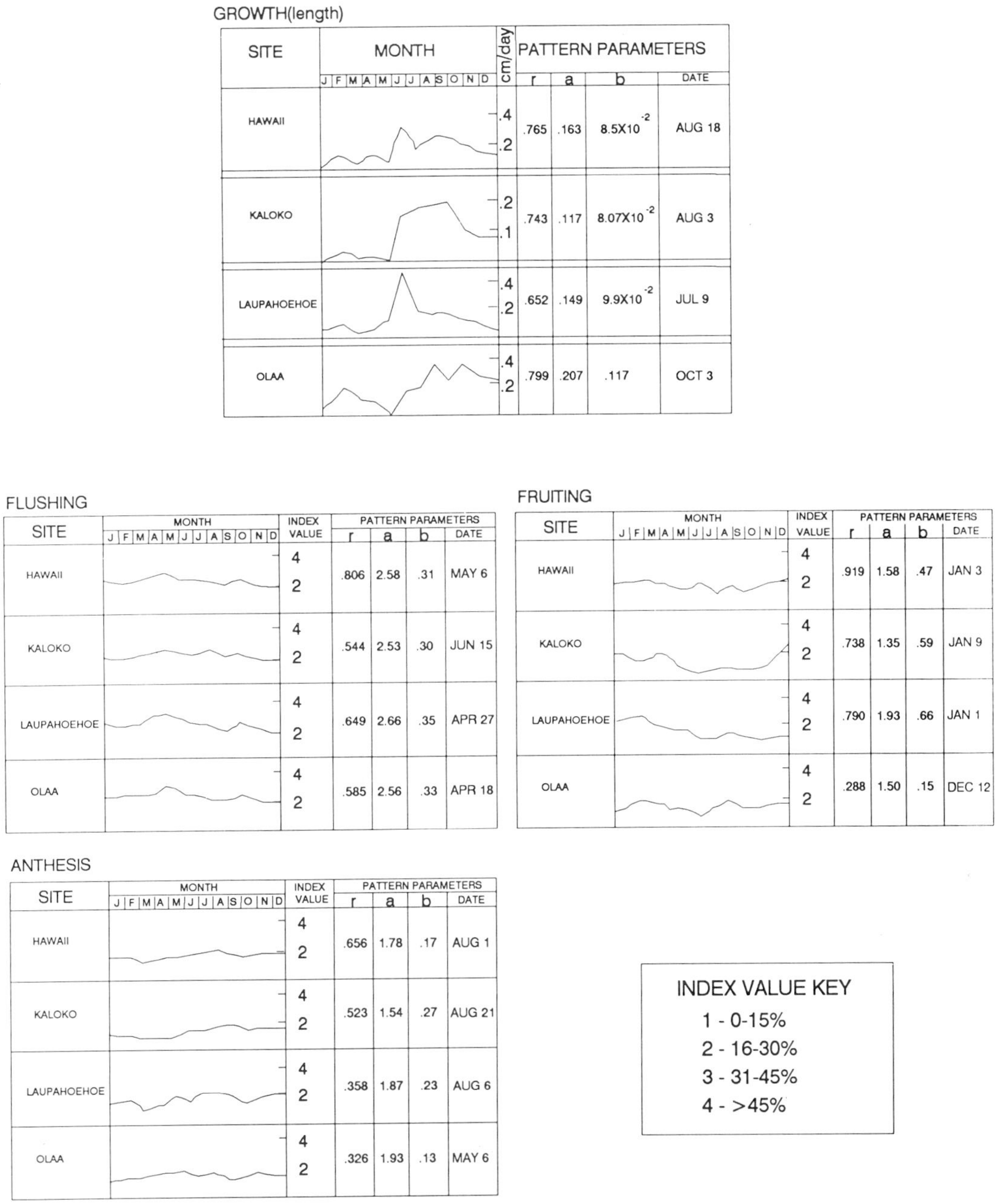

Figure 3. Phenology of banana poka at three sites on the island of Hawai'i: Kaloko (Hualālai), Laupāhoehoe-Keanakolu (Mauna Kea), and 'Ōla'a Tract (Mauna Loa), and all sites combined. Observed values are fitted to a least-squared regression analysis utilizing a sine-based function (Bridges *et al.* 1981). This function examines the annual periodicity of data based on a single peak activity period. r = monthly mean value; b = annual amplitude; c = predicted maximum value. Index values represent percent individuals in a particular phenological state.

1984). Bud initiation (flushing) and growth showed less synchronization among sites (LaRosa 1984).

Growth rates showed greater seasonal fluctuations. Highest levels of shoot growth occurred during the warmer summer months (the highest value in May coinciding with the peak in flushing) at all sites and may be a function of higher summer temperatures. A significant depression in growth occurred from December to April.

Data on climate and seasonality of reproductive activity for Laupāhoehoe, a mesic Hawaiian site, agree well with those for Quito, Ecuador, a site in the range of native populations (Fig. 4) (Herklots 1976; Escobar 1980). At both sites, heightened reproduction occurred during the dry season. A close association exists between seasonality of flowering and pollinator activity for several species, including *Passiflora* spp., in seasonal tropical environments (Janzen 1967, 1968; Gentry 1974; Frankie 1975; Stiles 1977). If this is also true for banana poka, which is largely dependent upon pollinators and occurs in some seasonal habitats in Hawai'i, then the timing of heightened reproduction may be a factor in the relative success of populations. In contrast, reproductive success may be negatively affected by heavy rains, which reduce pollen viability and/or physically damage flowers.

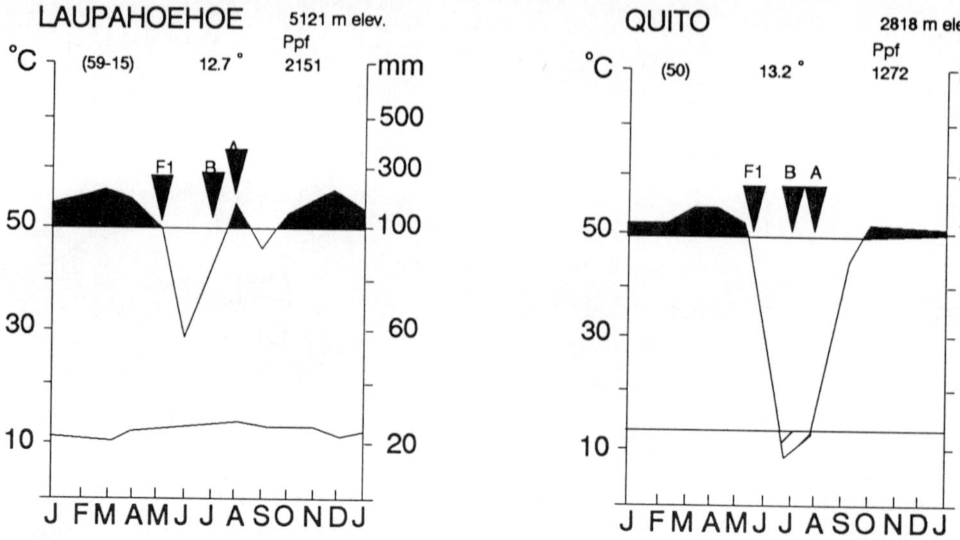

Figure 4. Relationships of climate to phenological activity of banana poka in Laupāhoehoe, Hawai'i, and Quito, Ecuador, South America. Climate diagrams illustrate climatic relationships between mean monthly temperature and rainfall (after Walter *et al.* 1975). Rainfall and temperature data for Hawaii are longtime averages from nearby stations (State of Hawaii 1970) and for South America from the U.S. Department of Commerce (1959, 1966). Arrows indicate peak periods of flushing (Fl), budding (B), and anthesis (A) at each site. Phenological data for Hawaiian site from LaRosa (1984); for South America, from Herklots (1976) and Escobar (1980).

DISPERSAL AND SPREAD

The fleshy fruits of banana poka are well adapted for dispersal by birds and mammals. Hard, impervious seed coats allow adequate seed survival following animal ingestion. While only one-half dozen or so species of birds and mammals are documented dispersal agents in Hawai'i, as many as 20 may disperse seeds of banana poka (Warshauer *et al.* 1983). These dispersal agents may be found in most habitats containing banana poka; therefore, much of the potential habitat is threatened with invasion if these animals colonize new areas (Warshauer *et al.* 1983). Humans may assist in this colonization directly via transplanting or indirectly via the movement of animals (*e.g.*, cattle, *Bos taurus*) from one range or island to another (Warshauer *et al.* 1983).

Two types of dispersal events (short distance and long distance), each having a very different outcome, are important in the spread of banana poka in Hawai'i (Warshauer *et al.* 1983). Most dispersal is over relatively short distances and augments existing populations (*e.g.*, the 'Ōla'a Tract population). Long-distance dispersal is an infrequent event and, when germination and establishment are successful, generally results in the establishment of a new population. Such populations are not subject to the intense intraspecific competition found in dense populations and therefore may expand rapidly if interspecific competition is not too severe. This greatly increases the difficulty of control of outlying populations (Woodall 1981; Mack, this volume); this situation is exemplified in the Kula area of Maui Island, where 15 distinct populations occur within an area of 200 acres (81 ha).

Continuous, prolific fruiting of banana poka provides abundant food for many mammals, particularly feral pigs. Indeed, pig densities are often highest in areas of dense banana poka (Wong 1971; Giffin 1972; Baker 1975). Feral pigs are fairly short-distance dispersal agents. Seed dispersal is limited by their short residence time in the pig's digestive tract (LaRosa 1984) and relatively small home ranges (several miles or less) (Giffin 1972; Baker 1975; Anderson and Stone, in press). Cattle also occasionally disperse seeds over relatively short distances (Warshauer *et al.* 1983). Seed digestion by feral pigs does not enhance germination but provides a fertile medium for germination and seedling growth (LaRosa 1984). Plant establishment is further aided by rooting activities of pigs, which create an environment of low competition.

Several native and alien birds are also important dispersal agents. Seeds of banana poka are used as grit by gallinaceous birds, including wild turkeys (*Meleagris gallopavo*) and Kalij pheasants (*Lophura leucomelana*) (Wong 1971; Warshauer *et al.* 1983). Seeds that pass unharmed through pheasants can germinate (V. Lewin, pers. comm. 1984). Kalij pheasants may be increasingly important dispersers of poka, as they are undergoing rapid range expansion and intensification and are currently found in all mid- and high-elevation forests and forest clearings on Hawai'i except in the Kohala district on Hawai'i Island (V. Lewin, pers. comm. 1982; J. Jacobi, pers. comm. 1987). Songbirds also disperse seeds of banana poka. The common myna (*Acridotheres tristis*) and the Japanese white-eye, wide-ranging alien species, have not been observed eating fruit in the wild but readily ate banana poka fruit in captivity. Seeds passed through the mynas unharmed and successfully germinated, but no germination of the seeds from the white-eyes occurred (LaRosa *et al.* 1985). Fruits of banana poka are also a regular part of the diet of the endangered Hawaiian crow or 'alalā *(Corvus hawaiiensis)* (Giffin 1978, 1980).

Seeds are also passively dispersed along watercourses. This appears to be significant in range expansion in the Kula area on Maui.

Association with Forest Disturbance

Banana poka tolerates both high and low light levels, allowing individuals to colonize small forest gaps as well as open, disturbed areas. Seeds, shed in a dormant state, require a short period of after-ripening. Seed germination is staggered: once dormancy is broken, most seeds within a cohort germinate fairly rapidly (within eight weeks). Others are delayed for eight months or possibly longer, extending the potential for colonization when suitable conditions arise (LaRosa 1984).

Growth and establishment of banana poka increase proportionately to increasing light intensity and associated high temperatures, conditions found in most disturbed areas. Growth rates increase rapidly with increasing light intensity, and optimal growth occurs in full sunlight (LaRosa 1984). This can readily be observed in openings and at forest margins, which generally support the densest infestations of banana poka.

Germinants do not tolerate dense shade, but shade tolerance increases in older seedlings. In low light levels, individuals sacrifice biomass accumulation and leaf expansion for height growth, giving them a competitive advantage in a multi-tiered forest (LaRosa 1984). The ability to remain in the understory as shade-tolerant seedlings yet respond quickly to canopy gaps with rapid growth has allowed this species to persist and proliferate under varying disturbance regimes.

Most small-scale disturbances are naturally caused (including windstorms and the collapse of dead and dying trees) and result in spatial changes in the existing vegetation mosaic. These small gaps are readily colonized by banana poka, as evidenced by the patchwork pattern of its distribution in 'Ōla'a Tract and its response to experimental canopy opening (Burton 1980; LaRosa 1983, 1984). The spread of banana poka via gap invasion is tied to widespread dispersal of propagules by animals. Effective dispersal by animals is evident in the presence of widely scattered seedlings in remote sections of 'Ōla'a Tract where the canopy has not been disturbed.

Within very small gaps, low light levels may limit the success of banana poka, as growth rates may be insufficient for individuals to reach the canopy before gap closure occurs (Burton 1980; LaRosa 1983, 1984). Following canopy closure, growth and flowering of banana poka are largely restricted to the canopy. Few leaves are supported in the dense shade of the understory, and the death of apical shoots is common (LaRosa 1984). Reduced vigor of plants suggests that shade tolerance may be lost with increasing plant age. In larger openings, rapid growth allows individuals to reach the canopy and reproduce before gap closure (Burton 1980; LaRosa 1983). In these situations, the competitive position of banana poka is enhanced, as growth rates of banana poka seedlings exceed those of the dominant forest tree, 'ōhi'a *(Metrosideros polymorpha)* (Burton and Mueller-Dombois 1984). Once the canopy is reached, individuals are no longer inhibited by shade and can proliferate, frequently spreading laterally.

Widespread canopy disturbances often follow large windstorms, which can occur as often as once every three to four years (Blumenstock 1967). The proliferation of banana poka in recent years in the 'Ōla'a Tract on Hawai'i and in the Kula area of Maui is largely a result of a single severe windstorm in January of 1980. 'Ōhi'a dieback (Mueller-Dombois *et al.* 1980) is another naturally occurring, large-scale disturbance that allows for proliferation of banana poka. For example, an area affected by canopy dieback in the Ōla'a Tract was initially colonized by a single (surviving) individual of banana poka. A slow establishment phase and a lag in population expansion closely followed the course of dieback (LaRosa 1984), and the area is now heavily infested with banana poka. The Hawai'i Forest Bird Survey undertaken by the U.S. Fish and Wildlife Service in 1977 recorded the highest density and cover of banana poka in areas of 'ōhi'a dieback in 'Ōla'a Tract (Warshauer *et al.* 1983). Banana poka, however, is currently absent from most areas with extensive dieback, probably indicative of lack of dispersal to many of these areas. Large-scale anthropogenic disturbances, including logging, tree fern (*Cibotium* spp.) harvesting, forest clearing, and cattle ranching, also increase the likelihood of banana poka infestations.

Disturbance of native or alien ground vegetation, particularly by feral pigs, increases colonization and establishment of banana poka (LaRosa 1984). Disturbed areas exhibit much higher seedling densities than surrounding areas where seedlings often occur only as scattered individuals (LaRosa 1984). This is particularly evident where introduced grasses form a dense ground cover. In most areas, ground cover of native species seldom reaches 100%, and these areas are more susceptible to invasion by banana poka than those with high alien species ground cover. Mid-level tree fern canopies can also be heavily disturbed by feral pigs, allowing more light to reach the ground and encouraging banana poka. Increased establishment is not directly related to later survival, however, as populations are density dependent.

Limits to Spread

Banana poka is excluded from few areas in Hawai'i, but several conditions may limit plant vigor and population size and thereby reduce its competitive ability and rate of spread. Dense shade and extremes of rainfall, elevation, and temperature all appear to reduce success (Warshauer *et al.* 1983).

Upper and lower elevational limits have been noted in both Hawai'i and South America (Escobar 1980; LaRosa 1984), but the limiting factors are unknown. Extremes of temperature may be at least partly responsible. High temperatures affect the growth, vigor, and reproduction of banana poka;

cultivation has been unsuccessful at lower elevations in the tropics (Hooker 1845; Paxton 1846; Martin and Nakasone 1970; Howell 1976; Escobar 1980). Markin (pers. comm. 1986) reported that banana poka at elevations under 182 ft (600 m) in Hawai'i shows greater susceptibility to attack by pathogens and nematodes. Increased susceptibility to disease may indicate increased stress. Low temperature (frost) limits the upper elevational spread in South America (Jaramillo 1957; Anonymous 1962; Martin and Nakasone 1970) and may be a factor in some areas of Hawai'i as well.

Extremely wet areas appear to be free of banana poka. A sharp demarcation in the distribution of banana poka in Laupāhoehoe coincides with the 200-in. (5,100-mm) rainfall isohyet and cannot be explained by elevation, temperature, or vegetational changes (Warshauer *et al.* 1983). In Kōke'e on Kaua'i Island, the infestation is concentrated in the more mesic koa forest: no individuals have been seen in the adjacent Alaka'i Swamp. Poor soil drainage may partly explain the distribution patterns noted above.

The occurrence of pests, pathogens and associated damage, shoot abortion, and high seedling mortality was generally higher in wet than in dry sites in Hawai'i (LaRosa 1984). Factors contributing to lowered reproductive success also occur more frequently in wetter areas, including lower pollinator activity and success, loss of pollen viability, physical damage to flowers, and increase in abortion. Conversely, lower growth rates, plant vigor, and population densities occurred at very dry sites, such as the recent, relatively unweathered lava flows of Hualālai (LaRosa 1984). In South America, cultivated crops of banana poka are regularly irrigated in dry areas (Jaramillo 1957).

Light is seldom a limiting factor in the spread of banana poka in Hawai'i. Seedlings reach their light compensation point near 2% of full sun (LaRosa 1984), but light is generally greater than this under relatively sparse canopies in most Hawaiian forests (Burton and Mueller-Dombois 1984). Low light levels found in tree fern forests limit population sizes but, because of frequent disturbances to both the canopy and understory layers, rarely exclude banana poka from an area (Burton and Mueller-Dombois 1984; LaRosa 1984). Dense shade occurs under introduced tree plantations of tsugi pine (*Cryptomeria japonica*) and tropical ash (*Fraxinus uhdei*). On Mauna Kea on Hawai'i Island, these plantations contained few or no individuals of banana poka, while surrounding areas of more open forest supported the most dense populations recorded (LaRosa 1984).

DISTRIBUTION AND STATUS OF BANANA POKA IN HAWAII

Banana poka has successfully invaded areas of diverse climate and vegetation. Colonization has occurred in 28 vegetation types (as defined by Jacobi and Warshauer, this volume) ranging from dry lava flows with sparse, open naio (*Myoporum*) scrub communities to dense, montane 'ōhi'a-tree fern rain forests. Within these communities there is considerable diversity in forest stature, structure, and species composition that reflects variations in moisture conditions and successional states (Warshauer *et al.* 1983). The most heavily impacted vegetation type is the disturbed mesic koa-'ōhi'a forest. Other vegetation types with a high incidence of banana poka are the relatively intact wet 'ōhi'a forest and the wet 'ōhi'a-koa forest with a tree fern subcanopy and native species understory (Jacobi and Warshauer, this volume).

The following discussion of the distribution and status of populations of banana poka in Hawai'i is largely a summary of information presented in Warshauer *et al.* (1983). To determine changes in distribution and abundance, the authors compared information from a 1971 ground survey and map prepared by Pung and Wong (Wong 1971; on file at the Hawaii Division of Forestry Office, Honolulu) with data from ground surveys made between 1976 and 1981 by the U.S. Fish and Wildlife Service Hawai'i Forest Bird Survey (Scott *et al.* 1981), from 1981 aerial surveys of the 'Ōla'a Tract and Mauna Kea populations, and from examination of aerial photographs. Information on survey methods and detailed distribution maps of each major population of banana poka may be found in Warshauer *et al.* (1983). Information gathered during a demographic study from 1979 to 1983 (LaRosa 1984) and incidental observations (R. Hobdy, pers. comm. 1986, 1987) are used to augment the discussion.

In 1986, banana poka inhabited approximately 134,425 a (54,400 ha) on three major islands in Hawai'i (Table 3). Populations are characterized by a core of continuously distributed individuals surrounded by an area of sporadic distribution. The five major populations -- 'Ōla'a Tract, Mauna Kea (Laupāhoehoe), and North Kona (Kaloko), on Hawai'i; Kōke'e on Kaua'i; and Kula on Maui -- all originated from deliberate plantings by humans. In addition, three small, extralimital populations, consisting of from one to several individuals, occur on Hawai'i Island: Kohala, Manukā State Park (Warshauer *et al.* 1983) and Pōhakuloa (Higashino *et al.* 1977) (Fig. 1). More detailed surveys may yield additional isolated populations.

Following is an assessment of the extent and status of each of the major populations from published accounts. Changes in distribution and abundance from 1971 to 1986 are summarized in Table 3.

'Ōla'a Tract, Hawai'i

The most recent and rapidly expanding population of banana poka on Hawai'i is in the 'Ōla'a Tract of Hawaii Volcanoes National Park; this infestation originated from a transplant from Keanakolu (Mauna Kea) to Volcano in 1958 (Pung 1971). Density was generally low (on the order of 2,025 individuals/a or 5,000 individuals/ha) (LaRosa 1984), as are foliage cover values (generally less than 5%) (Warshauer *et al.* 1983). However, cover was much greater locally along forest margins and in forest openings. Habitats occupied by banana poka include closed-canopy, wet 'ōhi'a-tree fern forests, and more open-canopy, wet ohi'a forests (lacking a dense tree fern subcanopy). A small, relict stand of wet koa-'ōhi'a forest is heavily infested with banana poka (Warshauer *et al.* 1983).

Table 3. Summary of the area of continuous distribution and relative abundance of banana poka in five areas in Hawai'i. Continuous distribution in 1986 = 19,357 ha; sporadic distribution in 1978 = 35,000 ha. (Adapted from Warshauer *et al.* 1983.)

Population/ Island	Year of Survey*	Abundance Class** Low	Medium	High	Very High	Total	No. Years between Surveys	% Change in Area of Distribution (%)
'Ōla'a Tract,	1971	142***	--	--	--	142		
Hawai'i	1981	2330***	242	--	--	2572	10	+18.11 (1711)
Mauna Kea,	1971	615	2939	2402	1202	7158		
Hawai'i	1981	1946	1947	3156	325	7374	10	+1.03 (3)
North Kona,	1971	141	3746	262	--	4149		
Hawai'i	1978	n/a	n/a	n/a	n/a	6263	7	+1.51 (50)
Koke'e,	1975	n/a	n/a	n/a	n/a	4900		
Kaua'i	1981	---------2350--------		1050	--	3400	6	-1.44 (30)
Kula,	1971	2	--	--	--	2		
Maui	1986	200	--	1.5	70	200+	15	+100.00
	1988	7		16	77			

* 1971: Kula, Maui: Report by E. Tamura (1978a,b). All others from distribution maps of W. Wong and E. Pung (Wong 1971).
1975: Koke'e, Kaua'i: Report by R. Daehler (1975).
1978: Hawai'i Forest Bird Survey, 1978-81, U.S. Fish and Wildlife Service (Warshauer *et al.* 1983).
1986: Pers. comm., R. Hobdy, Hawaii Div. Forestry.
1988: Memo by R. Hobdy (Dec. 23, 1988) to C.W.S. Lum.

** Percent foliage cover in each abundance class as determined by Warshauer *et al.* (1983):
1971: Low = 10%; Med = 10-30%; High = 30-70%; Very High = 70-100%.
1978: Low = 5%; Med = 5-25%; High = 25-75%; Very High = 75-100%.
n/a: Data for abundance classes not available.

*** Forest margins support greater cover locally.

\+ Main survey detailed acreage actually covered by vines to be 9.05 a (3.6 ha) within a 200-a (81-ha) radius (Hobdy 1988).

The 'Ōla'a Tract population has grown explosively in the last 10 years. The infested area increased 17-fold (1,700%; from 350 to 5,775 a or 142 to 2,338 ha) from 1971 to 1981 (Table 3; Warshauer *et al.* 1983). Most of this increase can be attributed to the severe windstorm of January 1980 (Warshauer *et al.* 1983; LaRosa 1984). Prior to the storm, small seedlings were numerous in the understory, but mature canopy vines were rare. Following the storm, banana poka increased visibly in the many canopy gaps created by the loss of tree fern fronds and canopy trees. Feral pigs played an important role by dispersing seeds throughout the area before and after the disturbance (LaRosa 1984).

Mauna Kea, Hawai'i

Banana poka was introduced to the Keanakolu area of the Laupāhoehoe district in 1928 and spread rapidly. The highest density and cover values (≥202,000 individuals/a or 500,000/ha; 75 to 100% cover) of banana poka occur in the Laupāhoehoe-Keanakolu area of Mauna Kea (Warshauer *et al.* 1983; LaRosa 1984). The heavy infestation may be largely a consequence of historical grazing and subsequent logging. Cattle used the area heavily prior to the introduction of banana poka in 1928 but were removed in the early 1940s (Pung 1971). Following removal, banana poka spread rapidly through the heavily disturbed forest. Areas with less than 30% cover are often grazed by cattle, which reduces most species, including banana poka. Optimal habitats on Mauna Kea include mesic and wet koa-'ōhi'a forests. Rainfall in excess of 175 to 220 in. (4,500-5,100 mm) appears to limit population expansion into wet 'ōhi'a forests on Mauna Kea (Warshauer *et al.* 1983).

Feral pigs, which reach their highest density on Hawai'i on Mauna Kea (Giffin 1972), are now largely responsible for the intensive dispersal of seeds. Other dispersal agents include cattle (feral and escaped livestock), turkeys, gallinaceous birds, and horses (*Equus caballus*).

The Mauna Kea population of banana poka is stable (Warshauer *et al.* 1983): the total area infested by banana poka on Mauna Kea declined 3% between 1971 and 1981 (Table 3). Localized changes in distribution and abundance, however, indicate that population expansion is still occurring within limited areas, but that some population losses have also occurred. For example, the area in the highest cover class declined from 2,970 to 803 a (1,202 to 325 ha). Cattle grazing and the disturbance accompanying koa logging may account for some of the banana poka cover loss.

North Kona, Hawai'i

This infestation is the result of the merging of two populations from separate introductions -- the original introduction to the Hawaiian Islands at Pu'uwa'awa'a in 1921, and a later introduction to Honua'ula in 1928 (Pung 1971). Banana poka is still spreading into favorable habitats in North Kona. The area of continuous distribution increased 51% from 1971 to 1978 (Table 3); 20% of this increase was in the high cover class (30% to 70% cover). In 1981, a 15,565-a (6,300-ha) core area of continuously distributed individuals surrounded a sparsely populated (3-16 individuals/a or 8-40/1,000 ha), 15,565-a (6,300-ha) area with low banana poka cover (Warshauer *et al.* 1983).

While banana poka populations in Kona are as old or older than the Mauna Kea population, the degree of infestation is less. The rate of spread has been slower in Kona, and cover values are less (Warshauer *et al.* 1983). Several factors may explain lower success in Kona, particularly on Hualālai Volcano. Growth of banana poka shoots from Hualālai was slower than for those from 'Ōla'a Tract and Mauna Kea (LaRosa 1984), likely a function of the lower annual rainfall on Hualālai. Other site factors may also be limiting. Hualālai is a composite of geological and plant successional stages, and much of the area is geologically young, with little soil development and early successional vegetation (Warshauer *et al.* 1983). Areas of low-stature 'ōhi'a forest support lower densities and cover of most plants, including banana poka (LaRosa 1984). Areas with banana poka cover in excess of 30% are restricted to more mature forests, such as mesic koa-'ōhi'a or koa-mamane *(Sophora chrysophylla)* (Warshauer *et al.* 1983).

In addition, disturbance factors, including cattle grazing, logging, and forest clearing, have generally been less in Kona than at many areas on Mauna Kea, although disturbance is rapidly increasing. Feral pigs and cattle, Kalij pheasants, and the 'alalā are known to disperse seeds in the Kona area, but dispersal may be less intensive than elsewhere. Coarse lava flows isolate some animal populations, reducing home ranges and resulting in fewer potential sites colonized (Warshauer *et al.* 1983). Also, the density of feral pigs is much lower on Hualālai than on Mauna Kea (Giffin 1972).

Kōke'e, Kaua'i

Banana poka was introduced to the Kōke'e area in 1923 (Wenkham 1967). The infestation currrently extends from 2,790 to 4,265 ft (850-1,300 m) elevation (Daehler 1975; Warshauer *et al.* 1983). Principal habitats invaded include mesic koa-'ōhi'a forests and, to a lesser extent, wet 'ōhi'a forests (Warshauer *et al.* 1983). Alien feral pigs and red jungle fowl (*Gallus gallus*) range throughout the area and, along with passive dispersal of the plant via streams, are largely responsible for its widespread occurrence.

Little quantitative information is available on the distribution and population dynamics of banana poka on Kaua'i. In 1981, approximately one-third of the infested area had high cover values ($\geq$ 30% banana poka foliage cover). Highest cover values occurred in mesic areas such as gulches, while surrounding drier ridges had lower cover values (Warshauer *et al.* 1983). Banana poka is excluded from the wettest habitats (*e.g.*, Alaka'i Swamp).

Comparison of a 1975 population estimate (Daehler 1975) with a 1981 aerial survey (Warshauer *et al.* 1983) suggests that the population declined 30% (Table 1) in six years, although the estimates may not be comparable due to differences in estimation techniques.

In 1982, Hurricane Iwa severely disturbed the vegetation in the Koke'e area and damaged many canopy trees. While its effects on the banana poka infestation have not been quantitatively assessed, a cursory examination of the area suggests a proliferation of banana poka in the ground layer. This likely resulted from extensive canopy disturbance, which allowed more light to reach the forest floor. Vines that were in the canopy prior to the storm were displaced to the ground along with supporting trees (and branches). As the forest recovers, banana poka will be carried into the canopy, increasing its cover in the upper strata. A recent study (LaRosa and Floyd 1988) suggested that banana poka may be interfering with the post-hurricane reproduction of the endemic understory species mokihana *(Pelea anisata)* in a heavily disturbed area of Kōke'e.

Kula, Maui

Prior to the early 1970s, the only known specimens of banana poka on Maui were part of a cultural planting at the Hawaii Agricultural Experiment Station (Wong 1973). However, the most recently documented infestation began presumably as an ornamental planting on private property and when "discovered," in 1971, by Hawaii Division of Forestry personnel, consisted of one adult and approximately 250 seedlings (less than 2 a (0.8 ha) in extent) in the Waiakoa (Kula) area. The potential invasiveness and dominance of banana poka was known to Division of Forestry officials at that time, and they immediately alerted the Hawaii Department of Agriculture, which promptly eradicated the population (Wong 1973). They presumed this was so until 1978, when it was again noted in the same area (Tamura 1978a, 1978b). Little attention was paid to the periodic appearance of seedlings until the early 1980s, when banana poka was noticed spreading through a forest of black wattle (*Acacia mearnsii*).

In 1988, the Waiakoa population was mapped, using aerial and ground surveys, by the Hawaii Division of Forestry (Hobdy 1988). Their data represent the most comprehensive information available on this infestation. In 1989, banana poka continuously covered 9 a (3.6 ha) within a 200-a (80.9-ha) radius (Table 3), largely on privately owned agricultural parcels and homesteads in the Waiakoa area. Fifteen different invasion centers have been mapped. Cover ranges from light (<10%) to heavy (>50%), with 77% of the infestation in the heavy class. Outside of developed areas (agriculture, landscaping), vegetation in the area is primarily a mixture of black wattle and *Eucalyptus* sp. Heaviest densities and cover occur in the gulches.

Comparison of the 1971 and 1988 population estimates indicates an explosive population increase (Table 3). Most of this increase appears to be local population enhancement rather than expansion beyond the 200-acre radius and may be largely a result of banana poka invasion following large-scale vegetation disturbance in the area during the January 1980 windstorm. Additionally, some population expansion is occurring downslope, presumably via passive dispersal down watercourses (gulches). Much of the potential expansion of this population will be held in check by rapid development and utilization of the area, but State-owned lands in the Kula Forest Reserve, located at higher elevation than the infestation, are threatened.

The most encouraging note to date has been the continued support and efforts of the local community, particularly the local chapter of the

Sierra Club, to eradicate this pest from the area. Consequently, in 1988, Senate Bill 599 was passed, appropriating State monies for control of this population.

IMPACTS ON HAWAIIAN ECOSYSTEMS

While little specific information is available on the direct effects of banana poka on host trees, lianas are often detrimental to their hosts and may be considered structural parasites. Dense liana foliage covers tree crowns, thus occluding sunlight and reducing photosynthesis, growth, and reproduction of the host tree (Stevens 1987). Substantial reduction in photosynthetic capacity results in eventual tree death. The physical weight of lianas also causes branch breakage, which often facilitates the entrance of pathogens. Lianas also compete directly with their host tree for soil nutrients (Whigham 1984). They are also recognized as major economic forest pests; increasingly, tropical forests are being replaced by liana thickets following large-scale disturbances, particularly logging (Telford and Childers 1947; Dawkins 1961; Kochummen 1966; Fox 1968; Meijer 1970; Thomas 1980; Putz 1984; Nicholson 1985; Stevens 1987). In open forests, the reduction in light intensity at the forest floor under dense liana cover may inhibit the growth and regeneration of understory species and lead to premature succession (Ogawa *et al.* 1965a, 1965b; Mueller-Dombois *et al.* 1980; Friedland and Smith 1982; Putz 1984).

In Hawai'i, infestations of banana poka have resulted in large-scale disturbances and changes in native vegetation, particularly in the Laupāhoehoe-Keanakolu area of Mauna Kea. Banana poka was identified as the most serious problem in koa silviculture (Skolmen 1979). In logged areas, banana poka dominates koa, completely smothering seedlings and saplings within the first four years of koa forest regeneration (Scowcroft and Nelson 1976). In extreme cases, degradation of watersheds can result from such large-scale changes in vegetation.

Infestations of banana poka have additional impacts on native Hawaiian ecosystems. Shallow-rooted trees, including 'ōlapa *(Cheirodendron trigynum)*, koa, and 'ōhi'a, topple more easily with the added weight of vines, forming gaps in closed-canopy forests, which are further invaded by banana poka (Burton 1980; LaRosa 1983). Banana poka may persist after canopy closure and become a permanent component of the vegetation (Burton 1980; LaRosa 1984). Eventually, the loss of structural integrity and a lowered species richness results (Jacobi and Warshauer 1975; Mueller-Dombois *et al.* 1980; LaRosa 1984; Jacobi and Scott 1985).

Degradation and loss of habitat for native fauna occur concurrently, as in the upper elevation koa-'ōhi'a forest on Hawai'i (Warshauer *et al.* 1983; Jacobi and Scott 1985). Only 13.4% of the original acreage of this forest type remains in Hawai'i, and it is habitat for six endangered forest birds and 11 species of endangered plants (Jacobi and Scott 1985). The wet and mesic 'ōhi'a forest and the wet koa-'ōhi'a forest types also support populations of banana poka and are important habitats for endangered birds and plants. The rich native invertebrate fauna is also seriously affected by alien plants that affect invertebrate host plants (Gagné and Christensen 1985).

HISTORY OF CONTROL EFFORTS

Mechanical, biological, and chemical means of control have all been attempted for banana poka, but to date no comprehensive management strategy exists. In the early 1970s, several small-scale chemical control projects, using picloram (Tordon), 2,4-D (Tordon 22K), a mixture of picloram and 2,4-D (Tordon 212), and Paraquat, were conducted on the island of Hawai'i by the Hawaii Division of Forestry. All herbicides proved effective against banana poka but were nonselective and resulted in undesirable effects on nontarget species (Landgraf 1971; Cusset 1975). The Division of Forestry then looked for biological control agents present in Hawai'i. Several species of the fungus *Alternaria* and a scale insect (*Ceroplastes cerripidiformis*) were identified as potential control agents, but work on these projects was inconclusive (Laemmlen 1971).

In the early 1970s, an alien passion vine butterfly, *Agraulis vanillae*, was introduced to Kōke'e (Kaua'i Island) to control banana poka. Individuals did not readily establish on banana poka and were therefore unsuccessful in its control (Murai 1977; Nakahara 1977; Bianchi 1979). Serious attempts at biological control began in 1981, when the Hawaii State Legislature appropriated funds to begin control work on forest pest plants. This ongoing joint effort of several government agencies is the subject of another paper (Markin *et al.*, this volume).

Hawaii Volcanoes National Park personnel have conducted a herbicide sensitivity test on a small, isolated population of banana poka in the 'Ōla'a Tract (L.W. Cuddihy, pers. comm. 1990). Three herbicides (triclopyr, glyphosate, and a commercial mix of picloram + 2,4-D (Tordon-RTU)) were applied to cut stumps on a very limited basis (three plants per treatment). Early results on the efficacy of herbicide control concur with previous findings: herbicides, and in this case, mechanical means, appear effective in controlling banana poka. More data are needed on nontarget effects, resprouting of vines, and long-term control efficacy.

MANAGEMENT RECOMMENDATIONS

Banana poka has invaded, and is expanding in, a wide range of habitats in Hawai'i. Because of its widespread distribution and dense populations in many areas, overall control of banana poka in Hawai'i can only be achieved by using biological control. But biological control is not a panacea (Howarth 1983), and the potential impacts of introduced control organisms on other native vegetation and the native arthropod fauna must be carefully considered. Follow-up studies will be necessary to monitor the long-term efficacy and effects on the ecosystem.

Because biological control is a long-range goal, if in fact a successful agent (or agents) can be found, it is necessary to weigh the short-term consequences and costs of no action. Following the principles of weed population expansion set forth by Mack (this volume) and Woodall (1981), we may predict several things. For a dense, mature population such as that at Mauna Kea, there would be little additional consequence in waiting for a successful biological control program to become established. Some population expansion might occur on the perimeter of this population, but this would take place at a relatively slow rate.

For populations of limited extent, efforts should be directed toward timely mechanical and/or chemical control. Delaying control measures could be costly. With the potential for rapid rates of banana poka population growth under favorable conditions, the possibility of later control by these means may be quickly lost forever. Rapid increases have probably occurred very recently for the population at Kula, Maui (R. Hobdy, pers. comm. 1986).

When deciding whether to use mechanical or chemical control methods, managers must weigh the success of control efforts against the possible disturbance to the surrounding environment. Herbicide effects on nontarget species and possible drift or leaching of herbicides are especially important. Chemical control can be relatively safe when herbicides are applied carefully, correctly (according to label specifications), and directly to individual target plants. Because stems of banana poka do not readily sprout, mechanical means of control may suffice in selected areas. Some sprouting has occurred on the abaxial cut surface, however (J.T. Tunison, pers. comm. 1986). Careful use of these techniques may be effective for control of small populations in the immediate future.

Following the recommendations of Woodall (1980) and Mack (this volume), current control efforts should be concentrated on outlying populations, as they are the easiest to control and have the greatest potential for spread. Several such populations on the island of Hawai'i were identified by Warshauer *et al.* (1983), including one each in Kohala, Manukā, and Pōhakuloa. Any additional sightings omitted from this most recent distribution survey should be immediately controlled, if possible.

Spread and proliferation of banana poka can also be limited by the control of anthropogenic forest disturbance and introduced animals. The association of banana poka with forest disturbance has been clearly demonstrated. Heavy disturbance has historically occurred in areas where banana poka infestations are presently dense. It is evident from the study of the ecology of this species that control of feral pigs and certain alien birds is necessary to prevent further spread. Studies in pig-free sections of 'Ōla'a Tract support the idea that feral pigs enhance distribution. Studies of the effects of feral pig removal without concommitant treatment (chemical or mechanical) of banana poka on populations in the 'Ōla'a Tract of Hawaii Volcanoes National Park are ongoing (J.T. Tunison, pers. comm. 1989).

Successful biological control programs seldom completely eliminate a pest, but elimination of banana poka from selected areas, such as current efforts in the Special Ecological Areas within Hawaii Volcanoes National

Park (Tunison and Stone, this volume), remains a viable management alternative. Control of banana poka is a long-term problem requiring continued involvement and commitment by government agencies. Coordination is necessary to develop an effective statewide control program, but a single agency needs to take the lead in compiling and updating distribution information and formulating statewide priorities for control of populations. A recent demonstration project to determine the feasibility of chemical/mechanical control solely with the use of volunteers has shown that such an effort, even with the limited population (200 a or 81 ha), requires an intensive, concerted effort preferably coordinated by a governmental agency.

Efforts to predict potential habitat (Jacobi and Warshauer, this volume) are quite helpful in determining future management strategies, and more work should be put into refining the preliminary model. Finally, banana poka can be considered an "attractive nuisance;" increase in public awareness of the problem and proposed solutions are necessary and would decrease chances of intentional introductions to new areas.

ACKNOWLEDGMENTS

This review paper would not have been possible without the assistance and cooperation of the staff of the U.S. Fish and Wildlife Service Hawaii Research Station at Hawaii Volcanoes National Park. In particular, work on the distribution and abundance of banana poka by James D. Jacobi, Fredrick Warshauer, and J. Michael Scott contributed significantly. Data on the distribution and abundance of banana poka were obtained during the U.S. Fish and Wildlife Service Hawai'i Forest Bird Survey.

Research on the biology and ecology of *Passiflora mollissima* fulfilled the requirements for a Master of Science degree by the author and was made possible by grants from the Cooperative National Park Resources Studies Unit at the University of Hawaii (contract CA 8001-2-0001) and the U.S. Forest Service Institute of Pacific Islands Forestry (CPO 40-9231-1-126). Supplemental funding was provided by the Hawaii Division of Forestry, Pacific (now National) Tropical Botanical Garden, and Dr. Dieter Mueller-Dombois at the University of Hawaii Department of Botany. The critical review of an early draft by Dr. Lisa Floyd-Hanna greatly improved the manuscript, and the clerical assistance of Gail Boone and Margie Bouchard was invaluable. The continued support and assistance offered by Dr. Clifford W. Smith is gratefully acknowledged.

Literature Cited

Anderson, S.J., and C.P. Stone. In press. Snaring to control feral pigs (*Sus scrofa*) in a remote Hawaiian rain forest. *Biol. Conserv.*

Anonymous. 1962. Cultivo de la curuba. *Revista Nac. Agric.* 55.

Bailey, R.Y. 1944. Kudzu for erosion control in the Southwest. *USDA Farmers Bull.* 1840.

Baker, J.K. 1975. The feral pig in Hawaii Volcanoes National Park. Pres. Ann. meeting West. Sec. Wildl. Soc. and Calif.-Nevada Chap. Amer. Fisheries Soc., Sacramento. Files, Research Div., Hawaii Volcanoes Natl. Park.

Bianchi, F. 1979. Report to the Bishop Museum, Honolulu. Files, Univ. Hawaii Coop. Natl. Park Resour. Stud. Unit, Honolulu.

Blumenstock, D.I. 1967. *Climates of the states: Hawaii.* Climatography of the United States. Washington D.C.: U.S. Dep. Commerce.

Burton, P.J. 1980. Plant invasion into an 'ohi'a-tree fern rain forest following experimental canopy opening. *Proc. 3rd Conf. Nat. Sci., Hawaii Volcanoes Natl. Park*, 21-39. Honolulu: Univ. Hawaii Coop. Natl. Park Resour. Stud. Unit.

Burton, P.J., and D. Mueller-Dombois. 1984. Response of *Metrosideros polymorpha* seedlings to experimental canopy opening. *Ecology* 65:779-791.

Choudhury, A.K. 1972. Controversial *Mikania* (climber) a threat to the forests and agriculture. *Indian Forester* 98:178-183.

Croat, T.B. 1975. Phenological behavior of habit and habitat classes on Barro Colorado Island (Panama Canal Zone). *Biotropica* 12:42-55.

Cuatrecasas, J. 1953. A sketch of the vegetation of the North-Andean province. *Proc. 8th Pac. Sci. Congr. Pac. Sci. Assn.* Vol. 4, *Botany*, 167-173. Natl. Research Coun. Philippines. Quezon City, Philippines.

Cusset, G. 1975. Correspondence to J. Fujii, Jan. 10. Files, Hawaii Dep. Land Nat. Resour., Div. of For. Honolulu.

Daehler, R.E. 1975. Memorandum to L. Landgraf, June 19. Files, Hawaii Dep. Land Nat. Resour., Div. For. Honolulu.

Daubenmire, R. 1972. Phenology and other characteristics of tropical semi-deciduous forest in northwestern Costa Rica. *J. Ecol.* 60:147-169.

Dawkins, H.C. 1961. New methods of improving stand composition in tropical forests. *Caribbean Forester* 22:12-20.

de Wilde, W.J.J.O. 1975. *Passifloraceae (Flora of tropical East Africa).* London: Whitefriars Pr.

Escobar, L.K. 1980. Interrelationships of the edible species of *Passiflora* centering around *Passiflora mollissima* (HBK.) Bailey subgenus *Tacsonia*. Ph.D. diss., Univ. Texas at Austin.

Fosberg, F.R. 1975. Letter to J. Fujii, April 16. Files, Hawaii Dep. Land Nat. Resour., Div. Forestry. Honolulu.

Fox, J.E. 1968. Logging damage and the influence of climber cutting prior to logging in the lowland dipterocarp forest of Sabah. *Malayan Forester* 31:326-347.

Frankie, G.W. 1975. Tropical forest phenology and pollinator-coevolution. In *Coevolution of animals and plants*, ed. L.E. Gilbert and P.H. Raven, 192-209. Austin: Univ. Texas Pr.

Friedland, A.J., and A.P. Smith. 1982. Effects of vines on successional herbs. *Amer. Midl. Nat.* 108:402-403.

Gagné, W.E., and C.C. Christensen. 1985. Conservation status of native terrestrial invertebrates in Hawai'i. In *Hawai'i's terrestrial ecosystems: preservation and management*, ed. C.P. Stone and J.M. Scott, 105-126. Univ. Hawaii Coop. Natl. Park Resour. Stud. Unit. Honolulu: Univ. Hawaii Pr.

Gentry, A.H. 1974. Flowering phenology and diversity in tropical Bignoniaceae. *Biotropica* 6:64-68.

Giffin, J. 1972. *Ecology of the feral pig.* Final Rep. N-15-3. Hawaii Dep. Land Nat. Resour., Div. Fish and Game. Honolulu.

Giffin, J. 1978. *Alala investigation.* Hawaii Dep. Land Nat. Resour., Div. Fish and Game Project #W-18-R-3. Job #R-11-B. Dep. Land Nat. Resour., Div. Fish and Game. Honolulu.

Giffin, J. 1980. *Alala investigation.* Hawaii Dep. Land Nat. Resour., Div. Fish and Game Project #W-18-R-5. Job #R-11-B. Dept. Land Nat. Resour., Div. Fish and Game. Honolulu.

Green, P.S. 1972. *Passiflora* in Australia and the Pacific. *Kew Bull.* 26:539-558.

Hawaii Department of Land and Natural Resources. 1970. *An inventory of basic water resources, island of Hawaii, state of Hawaii.* Div. Water and Land Dev., Dep. Land Nat. Resour. Honolulu.

Hawaii Volcanoes National Park. 1986. Natural resources management plan and environmental assessment, revised January 1986. Hawaii Volcanoes National Park. Mimeo.

Herklots, G. 1976. *Flowering tropical climbers.* Kent, England: Dawson and Sons, Ltd.

Higashino, P.K., C.H. Lamoureux, L.R. Stemmermann, and F.R. Warshauer. 1977. The botanical survey for the installation. Pohakuloa Training Area. Environmental Impact Statement. U.S. Army Support Command, Hawaii. Dep. Army Corps Engineers, Pac. Div., Honolulu. Honolulu: Environmental Impact Study Corp.

Hobdy, R. 1988. Memorandum to C.W.S. Lum, December 23. Files, Hawaii Dep. land. Nat. Resour., Div. For., Maui.

Hooker, W.J. 1845. *Tacsonia mollissima* Curtis. *Bot. Mag. 1* 3rd ser. tab. 4178.

Howarth, F.G. 1983. Classical biocontrol: panacea or Pandora's box. *Proc. Hawn. Entomol. Soc.* 24(2,3):239-244.

Howell, C.W. 1976. Edible fruited *Passiflora* adapted to South Florida growing conditions. *Proc. Florida State Hortic. Soc.* 89:236-238.

Jacobi, J.D., and J.M. Scott. 1985. An assessment of the current status of native upland habitats and associated endangered species on the island of Hawaii. In *Hawai'i's terrestrial ecosystems: preservation and management*, ed. C.P. Stone and J.M. Scott, 3-22. Univ. Hawaii Coop. Natl. Park Resour. Stud. Unit. Honolulu: Univ. Hawaii Pr.

Jacobi, J.D., and F.R. Warshauer. 1975. A preliminary bioecological survey of the Ola'a Tract, Hawaii Volcanoes National Park. Prep. for Hawaii Natural Hist. Assn. Hawaii Volcanoes National Park. Hawaii Volcanoes National Park Library.

Jacobi, J.D., and F.R. Warshauer. [this volume] Distribution of six alien plants in upland habitats on the island of Hawai'i.

Janzen, D.H. 1967. Synchronization of sexual reproduction of trees within the dry season in Central America. *Evolution* 21:620-637.

Janzen, D.H. 1968. Reproductive behavior in the Passifloraceae and some of its pollinators in Central America. *Behavior* 32:33-48.

Janzen, D.H. 1971. Escape of juvenile *Dioclea megacarpa* (Leguminosae) vines from predators in a deciduous tropical forest. *Amer. Naturalist* 105:97-112.

Jaramillo, A. 1957. Primeros resultados de un ensayo sobre el cultivo de la curuba. *Agric. Trop.* 13:301-308.

Killip, E.P. 1938. *The American species of Passifloraceae*. Bot. Ser. 19. Chicago, Illinois: Field Mus. Nat. Hist.

Kochummen, K.M. 1966. Natural plant succession after farming in S.G. Kroh. *Malayan Forester* 29:170-181.

Laemmlen, F. 1971. Correspondence to W. Wong, May 20. Files, Hawaii Dep. Land Nat. Resour., Div. For. Honolulu.

Landgraf, L. 1971. Memorandum to A. Kawasaki, May 6. Files, Hawaii Dep. Land Nat. Resour., Div. of For. Honolulu.

LaRosa, A.M. 1983. Response of *Passiflora mollissima* (H.B.K.) Bailey to experimental canopy removal: a simulation of natural gap formation in a closed-canopy *Cibotium* spp. forest in Ola'a Tract, Hawaii. *Proc. 4th Conf. Nat. Sci., Hawaii Volcanoes Natl. Park*, 118-134. Honolulu: Univ. Hawaii Coop. Natl. Park Resour. Stud. Unit.

LaRosa, A.M. 1984. The biology and ecology of *Passiflora mollissima* in Hawaii. M.S. thesis, Univ. Hawaii, Honolulu.

LaRosa, A.M. 1985. Note on the identity of the introduced passionflower vine "banana poka" in Hawaii. *Pac. Sci.* 39:369-371.

LaRosa, A.M., and M.E. Floyd. 1988. The effects of Hurricane Iwa on understory regeneration of Koke'e Kauai forests. (Abstract) Pres. Ecol. Soc. Amer. Mtgs., Aug. 1988, Davis, Calif..

LaRosa, A.M., C.W. Smith, and D.E. Gardner. 1985. Role of alien and native birds in the dissemination of firefree (*Myrica faya* Ait. - Myricaceae) and associated plants in Hawaii. *Pac. Sci.* 39:372-378.

MacDonald, I.A.W. 1987. Banana poka in the Knysna forest. *Veld & Flora* 73(4): 133-134.

Mack, R.N. [this volume] Characteristics of invading plant species.

Markin, G.P., P.-Y. Lai, and G.Y. Funasaki. [this volume] Status of biological control of weeds in Hawai'i and implications for managing native ecosystems.

Martin, F.W., and H.Y. Nakasone. 1970. The edible species of *Passiflora*. *Econ. Bot.* 24:333-343.

Meijer, W. 1970. Regeneration of tropical lowland forest in Sabah, Malaysia forty years after logging. *Malayan Forester* 33:204-229.

Mueller-Dombois, D. 1975. Some aspects of island ecosystem analysis. In *Tropical ecological systems -- trends in aquatic and terrestrial research*, ed. F.B. Golley and E. Medina, 353-366. Vol. 11, Ecol. Stud. Ser., Analysis and synthesis. New York: Springer-Verlag.

Mueller-Dombois, D., J.D. Jacobi, R.G. Cooray, and N. Balakrishnan. 1980. *'Ohi'a rain forest study: investigations of the 'ohi'a dieback problem in Hawaii.* Hawaii Agric. Exp. Sta. Misc. Pub. 183. Univ. Hawaii Coll. Trop. Agric. and Human Resour., Honolulu.

Murai, K. 1977. Notes on the life cycle and host range of the passion vine butterfly, *Agraulis vanillae* L. Internal rep., June 24. Files, Hawaii Dep. Agric. Honolulu.

Nakahara, L. 1977. Memorandum to Chief, Entomology Branch, Jan. 17. Files, Hawaii Dep. Land Nat. Resour., Div. For. Honolulu.

Nicholson, D.I. 1985. A review of the natural regeneration in the dipterocarp forests of Sabah. *Malayan Forester* 28:4-25.

Nishida, T. 1963. *Ecology of the pollinators of passion fruit.* Tech. Bull. 55, Hawaii Agric. Exp. Sta., Univ. Hawaii. Honolulu.

Ogawa, H., K. Yoda, K. Ogino, and T. Kira. 1965a. Ecological studies of three main types of forest vegetation in Thailand. II. Plant biomass. *Nat. Life Southeast Asia* 4:49-80.

Ogawa, A., K. Yoda, T. Kira, K. Ogina, T. Shidei, D. Ratanawongse, and C. Apasutaya. 1965b. Comparative ecological studies of three main types of forest vegetation in Thailand. I. Structure and floristic composition. *Nat. Life Southeast Asia* 4:13-48.

Opler, P.A., G.W. Frankie, and H.G. Baker. 1980. Comparative phenological studies of treelet and shrub species in tropical wet and dry forests in the lowlands of Costa Rica. *J. Ecol.* 68:167-188.

Paxton, J. 1846. *Tacsonia mollissima. Paxton's Mag. Bot.* 13:24-26.

Pemberton, R.W. 1982. Exploration for natural enemies of *Passiflora mollissima* in the Andes. U.S. Dep. Agric. Biol. Contr. Weeds Lab., Albany, Calif. Unpub. rep.

Pung, E. 1971. The diet of the 'i'iwi. *Hawai'i Aloha Aina* 2:3-5.

Putz, F.E. 1984. The natural history of lianas on Barro Colorado Island, Panama. *Ecology* 68:77-81.

Richards, P.W. 1952. *The tropical rain forest.* Cambridge: Univ. Pr.

Schoniger, G. 1969. El cultivo de la curuba. *Rev. Esso Agric.* 6:4-12.

Scott, J.M., J.D. Jacobi, and F.L. Ramsey. 1981. Avian surveys of large geographical areas: a systematic approach. *Wildl. Soc. Bull.* 9(3):190-200.

Scowcroft, P.G., and R.E. Nelson. 1976. Disturbance during logging stimulates regeneration of koa. Res. Note PSW-306. Pac. Southwest For. Range Exp. Sta., USDA For. Serv. Berkeley, Calif.

Skolmen, R. 1979. Koa timber management. Pres. Hawaii Forestry and Wildlife Conf., Kahului, Maui, Oct. 1, 1979. Files, U.S. Forest Serv. Inst. Pac. Is. For. Honolulu.

Stevens, G.C. 1987. Lianas as structural parasites: the *Bursera simaruba* example. *Ecology* 68:77-81.

Stiles, F.G. 1977. Co-adapted competitors: the flowering seasons of hummingbird-pollinated plants in a tropical forest. *Science* 1177-1178.

Sykes, W.R. 1977. *Kermadec Islands flora: an annotated checklist.* Bull. 219. Wellington, New Zealand: New Zealand Dep. Sci. Industr. Res.

Tamura, E. 1978a. Corres. to Maui County Noxious Weed Advisory Commision, Dec. 7. Files, Hawaii Dep. Land Nat. Resour., Div. For. Honolulu.

Tamura, E. 1978b. Corres. to J. Raynor, Dec. 2. Files, Hawaii Dep. Land Nat. Resour., Div. For. Honolulu.

Telford, E.A., and N.F. Childers. 1947. Tropical kudzu in Puerto Rico. Circ. 27, USDA Fed. Exp. Sta. Puerto Rico.

Thomas, L.K. 1980. Winter growth of Japanese honeysuckle (*Lonicera japonica* (Thunb.) on Theodore Roosevelt Island. *Proc. 2nd Conf. Sci. Res. Natl. Parks,* Nov. 26-30, 1979, 408-418. Washington D.C.: Natl. Park Serv.

Walter, H., E. Harnickell, and D. Mueller-Dombois. 1975. *Climate diagram maps of the individual continents and the ecological climate regions of the earth.* Berlin: Springer-Verlag.

Warshauer, F.R., J.D. Jacobi, A.M. LaRosa, J.M. Scott, and C.W. Smith. 1983. *The distribution, impact and potential management of the introduced vine* Passiflora mollissima *(Passifloraceae) in Hawai'i.* Tech. Rep. 48, Univ. Hawaii Coop. Natl. Park Resour. Stud. Unit. Honolulu.

Weberbauer, A. 1936. Phytogeography of the Peruvian Andes. In *Flora of Peru,* ed. B.E. Dahlgren, Part I, 1-81. Field Mus. Nat. Hist. Bot. Ser. Vol. 13, Pub. 351. Chicago.

Wenkham, R. 1967. *Kauai and the park country of Hawaii.* San Francisco, Calif.: Sierra Club.

Whigham, D. 1984. The influence of vines on the growth of *Liquidambar styraciflua* L. (sweetgum). *Canad. J. For. Res.* 14:37-39.

Wong, W. 1971. The banana poka problem. Pres. 13th Forestry Conf. Hilo, Hawaii May 13. Files, Hawaii Dep. Land Nat. Resour., Div. For. Honolulu.

Wong, W. 1973. Correspondence to E. Tamura, April 17. Files, Hawaii Dep. Land Nat. Resour., Div. For. Honolulu.

Woodall, S.L. 1981. Integrated methods for *Melaleuca* control. *Proc. Meleleuca Symp.*, ed. R.K. Geiger, 135-140. Tallahassee, Florida: Fla. Div. For.

Wycherly, P.R. 1973. The phenology of plants in the humid tropics. *Micronesica* 9:75-96.

DISTRIBUTION AND SPREAD OF ALIEN PLANTS IN KĪPAHULU VALLEY, HALEAKALA NATIONAL PARK, ABOVE 2,300 FT ELEVATION

Stephen J. Anderson, Charles P. Stone, and Paul K. Higashino

ABSTRACT

The National Park Service interdisciplinary and feral pig research projects, together with the fencing of Kīpahulu Valley in Haleakala National Park for feral pig control, have provided an opportunity for determining seasonal status of alien plants in the Valley over a span of five years (1983-1988). A network of 20 transects and 97, 400-m^2 plots has been established in the 9,000-a (3,640-ha) restricted-entry Scientific Reserve. Nineteen plant communities from the Kalapawili grasslands at 7,800 ft (2,380 m) elevation to the "Dogleg" rain forest area at 2,350 ft (720 m) elevation have been sampled. Comparison of these data with those of past surveys and research efforts have made it possible to assess the threat of alien plants in the Valley, the spread of weeds through time, the impact of research efforts, and the feasibility and prioritization of control efforts. This study, in conjunction with previously constructed exclosures, provided important baseline information on alien plant distribution during feral pig removal and prior to alien plant control efforts in the Valley.

INTRODUCTION

Kīpahulu Valley is located on the eastern slope of Haleakalā Volcano on the island of Maui and is managed as a closed-entry Scientific Reserve within Haleakala National Park. The Valley floor has two levels, referred to as the Upper Plateau and the Lower Plateau; these are separated by a ridge termed the Central Pali (Fig. 1).

The study area extends from 7,800 ft (2,380 m) elevation in the Kalapawili grasslands to a bend in the Valley at 2,350 ft (720 m), referred to as the "Dogleg." Kīpahulu supports some of the best remaining relatively intact forest bird habitat in the State and a diversity of forest bird species, including the endangered Maui parrotbill (*Pseudonestor xanthophrys*), crested honeycreeper (*Palmeria dolei*), Maui nukupu'u (*Hemignathus lucidus affinus*), and Maui 'ākepa (*Loxops coccineus ochraceus*) (Conant and Stemmermann 1980). Native plant species

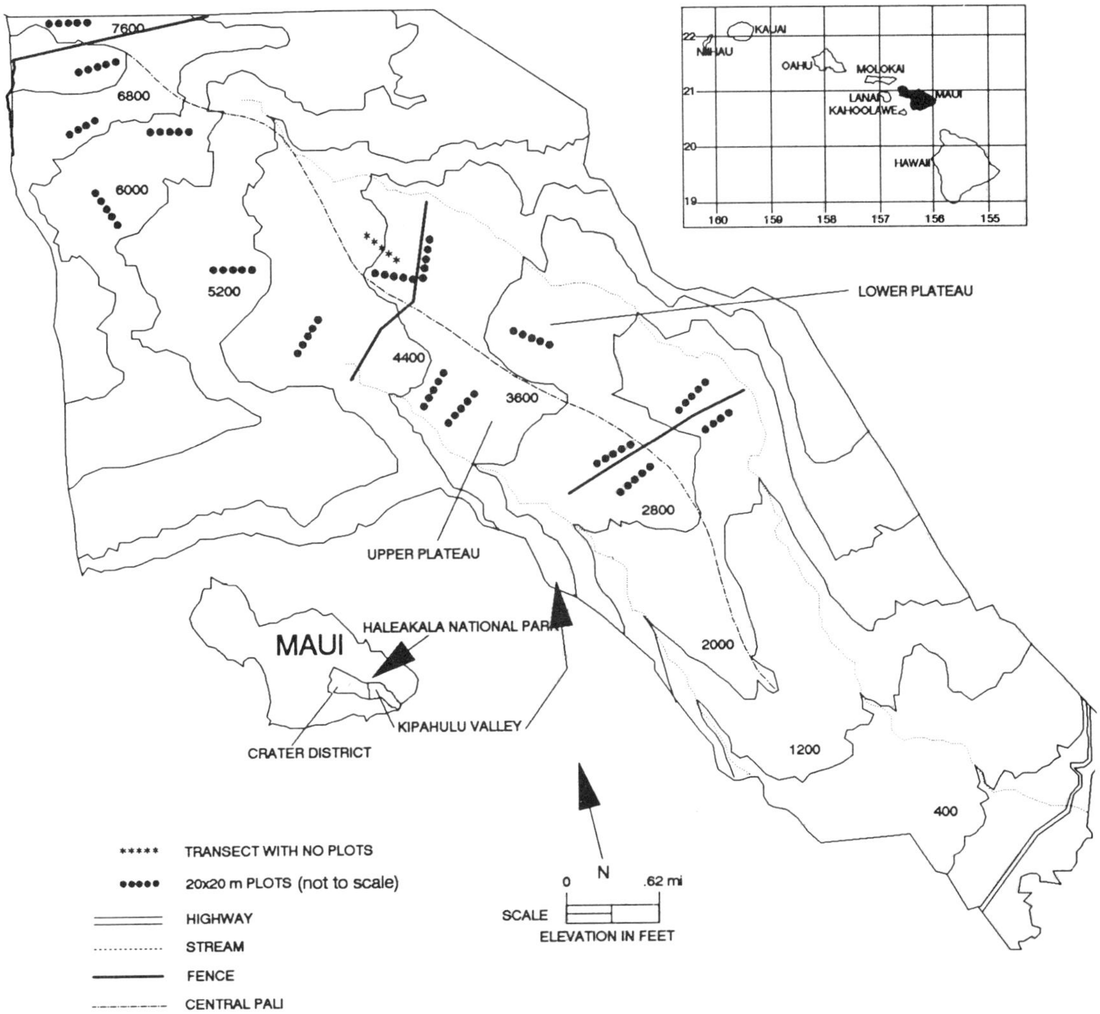

Figure 1. Kīpahulu Valley, Haleakala National Park, island of Maui, Hawai'i.

diversity is also high, with over 89 species of ferns and 209 species of flowering plants, including over 20 species of endemic woody lobelioids (Higashino *et al.* 1988).

The rich ecological resources of Kīpahulu are threatened by feral animal and alien plant invasions. Rooting and trampling of vegetation by feral pigs (*Sus scrofa*) and trampling by goats (*Capra hircus*) create seed beds for the establishment of weeds, which outcompete native species on disturbed sites. Some weedy species do not need disturbance to spread or become established in the Valley. Research and management efforts began in Kīpahulu with the long-term objective of controlling alien plants and animals. The Valley has been partitioned into feral pig and goat management units, with fences constructed across the Valley at the Dogleg, at 4,700 ft (1,430 m) elevation, and above the Valley headwall across the Kalapawili grasslands. Steep cliffs *(pali)* on the sides of the Valley provide topographic barriers that enclose three discrete areas with the cross fencing. Permanent vegetation plots have been established to study rain forest dynamics and the distribution and spread of alien plant populations, particularly between the 4,700-ft elevation fence (Fig. 1) and the Dogleg area, where displacement of the native flora is evident.

Past work on the status of alien plants in Kīpahulu has focused on distribution and occurrence at a given point in time (Warner 1967; Lamoureux and Stemmermann 1976; Yoshinaga 1980). Little was known about the rate of spread or population dynamics of weedy species in the Valley. Such information is critical to managers addressing the questions of prioritization and feasibility of alien plant control. It is equally important to determine the responses of weedy plants to the removal of feral pigs within the variety of plant communities in the Valley.

METHODS

Permanent transects and plots were laid out in stages as field work began in different areas of the Valley. Six 500-m transects were established in 1983 in six vegetation zones between 4,100 and 6,800 ft (1,250-2,070 m) elevation. During the Kīpahulu Interdisciplinary Study (Stone *et al.* 1984; Stone *et al.*, unpub. data), frequency of alien plant species was recorded in 50, 1-m^2 plots spaced at 10-m intervals on each 500-m transect. Presence of pig activity (tracks, feces, rooting, or plant feeding) within 2.5 m of each 10-m interval was recorded. Percent cover of both native and alien plant species was estimated in five 20- x 20-m plots spaced at 100-m intervals equidistant along each transect. Cover was estimated in three vertical layers (0-0.5 m, 0.5-2 m, and >2 m) using the modified Braun-Blanquet Cover-Abundance scale (Mueller-Dombois and Ellenburg 1974). Qualitative observations of weedy species were made for all camp sites, designated helicopter use sites ("helipads"), and trail systems. Frequency of occurrence data and casual observations were recorded on each of the three trips during the interdisciplinary study (Stone *et al.* 1984).

With the onset of the Kīpahulu feral pig project in April 1985, seven additional 500-m transects (Fig. 1) were added to represent the area considered for feral pig control research. The upper elevational limits of the study area were extended to 7,800 ft (2,380 m) in the Kalapawili grasslands, and the lower elevational range was lowered to 2,350 ft (720 m) in the Dogleg area; transects were also added to fill in gaps in the existing sampling scheme. The methods used for the six original transects were continued, with modifications, to increase the alien plant frequency sampling scheme. The 1-m^2 plots were enlarged to a 10-m long x 5-m wide continuous belt transect, the same area used in assessing pig activity. Five intensive 20- x 20-m plots were read on each of the new transects except on the transect at 6,800 ft (2,070 m) elevation; this transect traverses the Valley headwall shrublands and ends abruptly at a steep gulch, allowing space for only four plots. Plots were not established on the 4,400-ft (1,340-m) elevation transect on the Lower Plateau, as the vegetation there is dominated by matted fern or uluhe *(Dicranopteris linearis)*, and plot establishment would negatively impact the area. The vegetation on the Kalapawili transects facilitated the use of a point-frequency frame rather than visual estimates, allowing more accurate estimations of percent cover.

To monitor the effects of the feral pig control fence completed across the Valley in 1985 at the Dogleg, four 500-m transects were established in 1986, one on each side of the fence, for both the Upper and the Lower plateaus. The layout and sampling scheme were similar to those for the existing plots and transects. Percent cover was estimated for only the weed species and native dominants in each vertical vegetation layer. Individuals of the alien tree strawberry guava (*Psidium cattleianum*) were counted in basal diameter size classes to determine population structure and status of the invasion. If the ground cover contained an alien grass component greater than 25%, counts were made of native dominant woody species in basal diameter size classes to determine if seedlings and saplings can grow through the thick grass. Five 20- x 20-m plots were established on each of the transects at 100-m intervals except the fourth transect on the Lower Plateau below the fence. This transect was truncated at a steep gulch and contained only four plots. The 20- x 20-m plots on the 500-m transects were sampled only once, and plant communities were distinguished using Sorenson's quantitative index of similarity.

Three transects were established perpendicular to elevational contours to define the density gradient and location of the invasion front for the weed species invading from lower elevations. These transects followed old trail systems along the Ko'ūko'ūai Stream and the top of the Central Pali from 4,700 ft (1,430 m) to 3,100 ft (950 m) elevation on the Upper Plateau and along the Palikea Stream from 4,000 ft (1,220 m) to 2,350 ft (720 m) elevation on the Lower Plateau. Along the transects, abundance of weed species was recorded for each 100-ft (30-m) decrease in elevation. Abundance ratings used were <u>common</u>: occurring in high numbers or comprising a high percentage cover in an area; <u>occasional</u>: seen in low numbers throughout an area; and <u>sparse</u>: encountered infrequently and in low numbers. Forays off the trail system were used to record weed abundance in previously untraveled areas. A 20- x 20-m plot was established with each 200-ft (60-m) drop in elevation along the transects. Layout and sampling followed the methods used on the Dogleg transect plots (*i.e.*, weed and native dominant cover estimate only).

RESULTS

Frequency of Occurrence

A total of 38 weed species (Appendix) was recorded within the transect frequency plots. Five of these were ferns, 11 were grasses or sedges, 20 were herbs, and two were trees. Only five of the species found in the frequency plots are considered to have major disruptive potential (Appendix). Nine species (five ferns, two grasses, and two herbs) were new records for the study area.

Alien plants were found on each of the 500-m long transects. Highest <u>total</u> frequencies (100%) were recorded on the Kalapawili *(Deschampsia)* grasslands and Dogleg (open and closed *Acacia*) transects, and lowest total frequencies (8%) on the transect at 5,960 ft (1,820 m) elevation in the open-closed ʻōhiʻa *(Metrosideros polymorpha)* forest near the head of the Valley (Figs. 2, 3). The most rapid spread (increase in frequency) of weeds recorded was during the five-month period between November 1985 and March 1986, on the 4,300-ft (1,310-m) elevation transect on the Lower Plateau in wet ʻōhiʻa forest; a 17% increase in alien plant frequency was noted there. The ground cover in these areas is dominated by uluhe, which typically grows under open canopy tree layers and is slow to recover from disturbance. After transect establishment, an increase in pig activity was recorded (Fig. 3). Wind-dispersed Maui pāmakani (*Ageratina adenophora)* and bird-dispersed thimbleberry (*Rubus rosifolius*) (Appendix) were the invading species.

Some fluctuations in weed frequency among trips into the Valley were evident on all but the Kalapawili grasslands and Dogleg transects. Occurrence of velvet grass (*Holcus lanatus*) and gosmore (*Hypochoeris radicata*) remained at frequencies of 100% along the pig-disturbed Kalapawili transects. The constant occurrence of Hilo grass (*Paspalum conjugatum*) and a host of other weeds kept frequencies of weeds in the Dogleg area at 100%. Frequencies on the transects at 5,960, 5,240, and 4,200-ft (1,820, 1,600, and 1,280-m) elevations varied less than 12% during the study. Seasonal patterns in weed distribution and occurrence were not evident on any transect.

An elevational pattern of weed occurrence emerged from the plotting of the frequency transect data (Figs. 2, 3). The central core of the Valley, which contains the thickest forest canopy and has been the least disturbed by pigs, showed the lowest pig and weed frequencies. The open grasslands and shrublands at the head of the Valley, and the thin koa *(Acacia koa)* canopy at lower elevations allow the light needed for the proliferation of many alien plant species. Disturbance by pigs is also greater in these areas. The removal of pigs, resulting in decreased frequency of pig activity over time, did not result in decrease in weed occurrence in these areas. The number of weed species found on the frequency transects (Fig. 4) showed an increase with decreasing elevation.

Percent Cover

The 73 cover plots on the 500-m transects were categorized into 19 plant communities using Sorenson's quantitative index of similarity (Mueller-Dombois and Ellenberg 1974). Tallies of alien plant cover in these communities are presented in Table 1. Although weedy species occur throughout the study area, 10 plant communities sampled between 4,100 and 6,800 ft (1,250-2,073 m) remain relatively pristine. Seven of these communities contain 1% or less alien plant cover, while the remaining three communities contain 3% or less. The weed species that occur in this elevational range (Appendix) are not significant threats to the communities but are indications of disturbance. These 10 communities range from open to closed canopy ʻōhiʻa and/or koa forests, each with a variety of native tree and shrub subcanopies. Ground cover varies from dense mats of uluhe fern to bogs dominated by the native sedge *Carex alligata*, and from

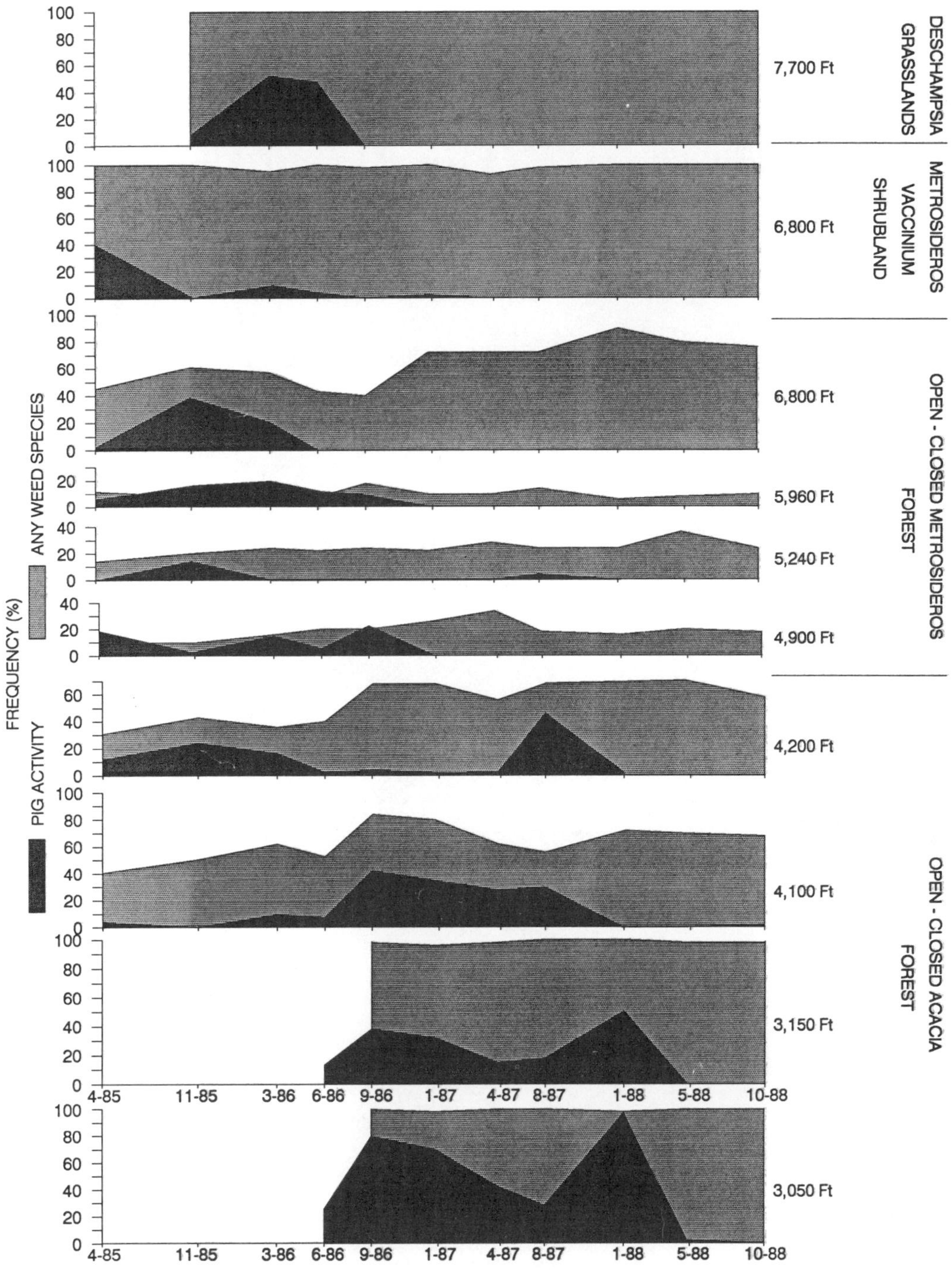

Figure 2. Frequency of pig activity and alien plant species on 10 transects in four vegetation types on the Upper Plateau in Kīpahulu Valley, Haleakala National Park, 1985-1988. (Pig activity data not available for 7,800-ft transect.)

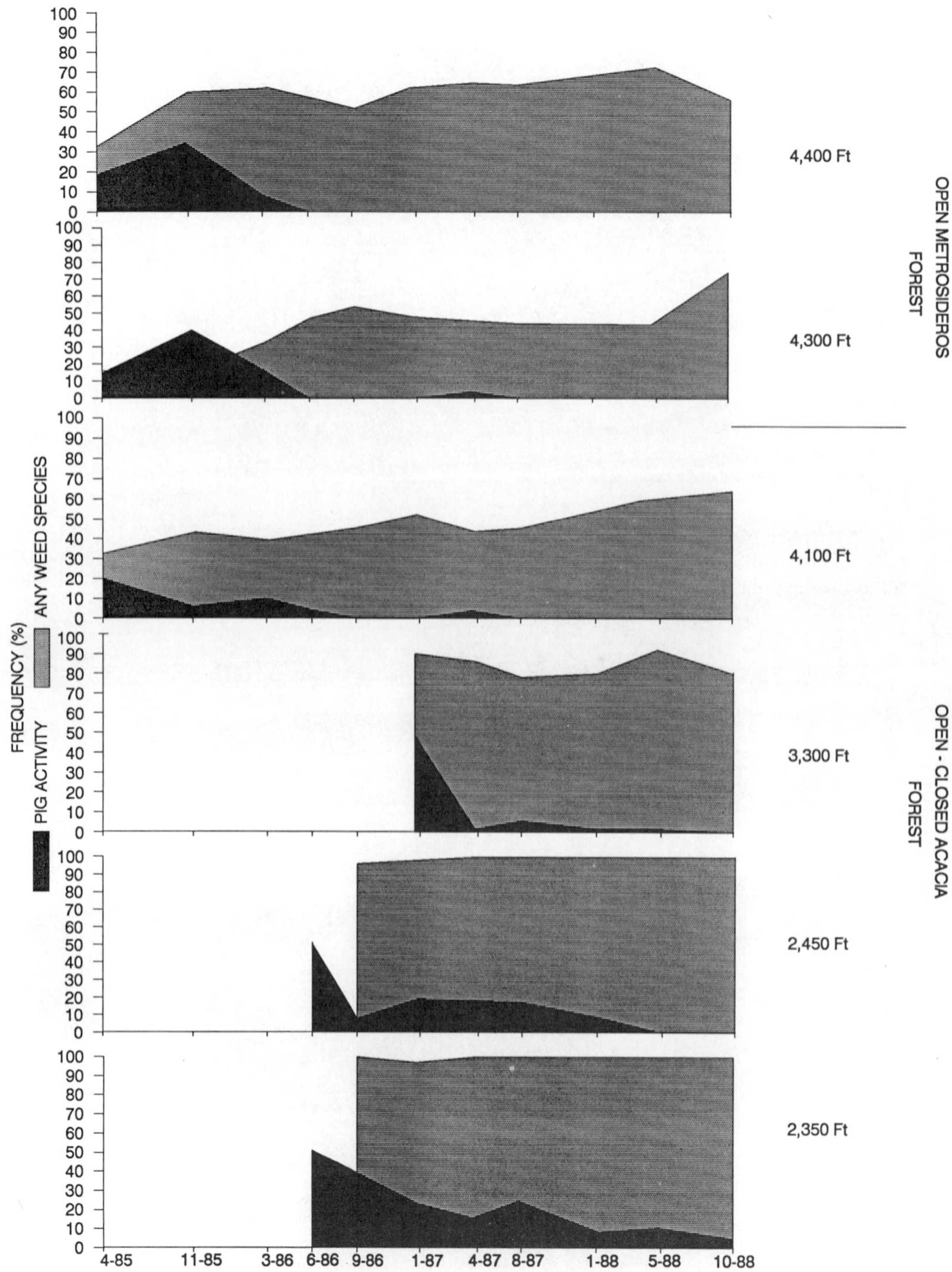

Figure 3. Frequency of pig activity and alien plant species on six transects in two vegetation types in Kīpahulu Valley, Haleakala National Park, 1985-1988.

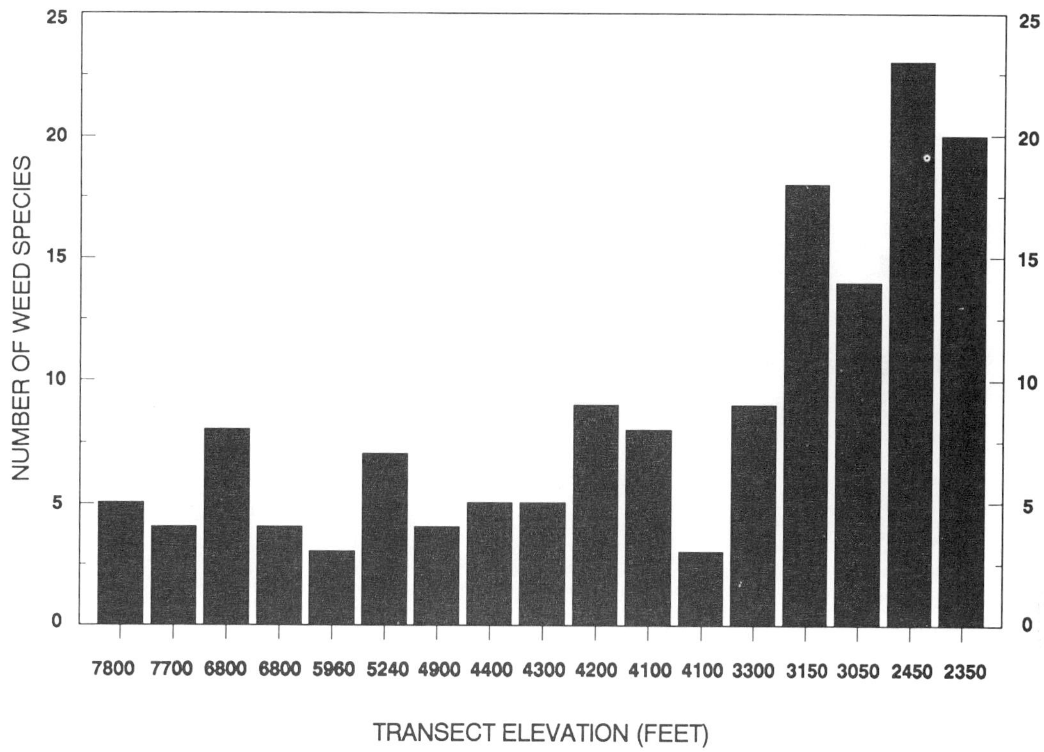

Figure 4. Number of weed species found on transects in Kīpahulu Valley.

thick litter accumulations with sparse vegetation cover to diverse assemblages of native herbs, ferns, and bryophytes.

At upper elevations in the study area at 6,800 ft (2,070 m) elevation and above are several distinct plant communities that vary in alien plant components. The shrubland of open ʻōhiʻa, pūkiawe (*Styphelia tameiameiae*), ʻōhelo (*Vaccinium* spp.), and amaʻu fern (*Sadleria cyatheoides*) on the back wall of the Valley is impacted severely in places by erosion following rooting and trampling of feral pigs and trampling of goats. Exposed areas are invaded by velvet grass, gosmore, and sheep sorrel (*Rumex acetosella*), introduced species that crowd out natives on disturbed sites. These aliens comprise 1 to 5% of the vegetation cover in unexposed areas but may have 100% cover in patches where rooting and erosion have been severe.

The *Deschampsia nubigena*-dominated grassland areas were more severely impacted by pigs and goats. The total percent cover of weed species in the sampled grassland transects was 41% (24% velvet grass, 16% gosmore, 1% sheep sorrel). Grassland plots closest to the top of the Valley headwall, where erosion is most severe, contained the greatest total weed cover (50%).

Weed invasions below 4,200 ft (1,280 m) elevation are most severe due to the amount of light that penetrates the koa canopy, the higher densities of pigs that occurred in these areas, and elevational limits of some of the

Table 1. Percent cover of alien plants in plant communities of Kīpahulu Valley, Haleakala National Park.

Plant Community	Transect Elevation (ft)	No. Plots	Percent Cover of Alien Plants									
			Ferns		Grasses/Sedges		Herbs		Trees		Total	
			$\bar{x}$	sd	$\bar{x}$	sd	$\bar{x}$	sd	$\bar{x}$	sd	$\bar{x}$	sd
Deschampsia Grassland	7,700, 7,800	9	0	--	24	12	17	4	0	--	41	11
Vaccinium-Deschampsia Shrubland	7,700	1	0	--	16	--	21	--	0	--	37	--
Open *Metrosideros-Styphelia-Rubus* Shrubland	6,800	6	0	--	3	5	1	2	0	--	4	7
Open *Metrosideros-Vaccinium* Forest	6,800	1	0	--	0	--	0	--	0	--	0	--
Closed *Metrosideros, Cheirodendron* Forest with *Rubus, Vaccinium,* and *Dryopteris*	6,800	2	0	--	<1	--	<1	--	0	--	<1	--
Open *Metrosideros, Cheirodendron* Forest with *Rubus, Coprosma, Broussaisia, Dryopteris,* and *Athyrium*	5,960	5	0	--	<1	--	<1	--	0	--	<1	--
Open *Metrosideros, Cheirodendron* Forest with *Athyrium* and *Carex*	4,900, 5,240	9	0	--	<1	--	<1	--	0	--	<1	--
Open *Metrosideros, Cheirodendron* Forest with native shrubs and *Dicranopteris*	4,200, 5,240	2	0	--	<1	--	2	1	0	--	2	1
Closed *Metrosideros, Cheirodendron* Forest with native shrubs, dense *Athyrium*, and *Dicranopteris*	4,200, 4,300	4	0	--	<1	--	<1	--	0	--	<1	--

Plant Community	Transect Elevation (ft)	No. Plots	Percent Cover of Alien Plants									
			Ferns		Grasses/Sedges		Herbs		Trees		Total	
			$\bar{x}$	sd	$\bar{x}$	sd	$\bar{x}$	sd	$\bar{x}$	sd	$\bar{x}$	sd
Open *Metrosideros, Cheirodendron* Forest with dense *Dicranopteris*	4,200, 4,300	4	0	--	<1	--	1	1	0	--	1	1
Open *Metrosideros, Acacia* Forest with native shrubs, *Athyrium*, and *Dicranopteris*	4,100	1	0	--	0	--	3	--	0	--	3	--
Open *Acacia, Metrosideros* Forest with native shrubs and dense *Dicranopteris*	4,100	1	0	--	0	--	<1	--	0	--	<1	--
Closed *Acacia, Metrosideros* Forest with native shrubs and native ferns	4,100	8	0	--	<1	--	3	6	0	--	3	6
Open *Acacia, Metrosideros* Forest with dense *Dicranopteris*	3,000, 3,100	3	<1	--	8	4	<1	--	<1	--	9	4
Open *Acacia, Metrosideros* Forest with *Paspalum* and *Dicranopteris*	3,000, 3,100	4	<1	--	43	8	<1	--	<1	--	43	8
Open *Acacia, Metrosideros* Forest with dense *Paspalum*	2,200, 3,000	3	<1	--	75	9	2	1	<1	--	77	8
Closed *Acacia*, diverse understory Forest with dense *Paspalum*	2,400	5	<1	--	57	8	9	3	2	2	67	9

Table 1, continued.

Table 1, continued.

Plant Community	Transect Elevation (ft)	No. Plots	Percent Cover of Alien Plants									
			Ferns		Grasses/Sedges		Herbs		Trees		Total	
			$\bar{x}$	sd	$\bar{x}$	sd	$\bar{x}$	sd	$\bar{x}$	sd	$\bar{x}$	sd
Open *Acacia*, diverse understory Forest with dense *Paspalum*	2,200	3	0	--	27	10	18	14	2	2	46	8
Open *Acacia,Psidium* Forest with dense *Paspalum*	2,200	1	0	--	51	--	33	--	10	--	94	--

weed species. The koa-dominated plant communities (Table 1) (with total weed cover up to 94%) are most vulnerable to invasion by the suite of weed species now present in Kīpahulu Valley. Invasion profiles along transects established perpendicular to elevational contours (Fig. 5) illustrate overall increases in cover percentages in total weed species with decrease in elevation. A total of 24, 20- x 20-m plots was sampled on the three transects between the 4,700 ft (1,430 m) and 2,350 ft (720 m) fence lines. Below 4,100 ft (1,250 m) elevation, the 20- x 20-m plots showed increasing percentage cover and numbers of nonnative species with decreasing elevation. Alien grasses and sedges, primarily Hilo grass and *Cyperus* spp., comprised 84% of the weedy vegetation cover below 3,600 ft (1,100 m) on the Upper Plateau (Ko'ūko'ūai and Central Pali) and 89% below 3,200 ft (975 m) elevation on the Lower Plateau. Above these elevations, alien herbs constituted most of the weedy component of vegetative cover. Alien herbs were less invasive and occupied less cover than weedy grass species. The only vegetation component evidently inhibiting alien plant invasion is matted fern or uluhe, which forms thick, impenetrable thickets.

A particularly weedy plot at 4,200 ft (1,280 m) elevation along the Ko'ūko'ūai Stream reflected the greater extent of the alien grass and herb populations up this major stream gulch. However, the weedy tree species strawberry guava and common guava (*P. guajava*) have spread farther upslope and more continuously on the Lower Plateau and Central Pali than along Ko'ūko'ūai Stream. Total weedy plant cover is greater on the Lower Plateau (75%) than on the Upper Plateau (48%).

Habitat modification by feral pigs is a primary factor in replacement of native by alien plants. Pahole or hō'i'o *(Diplazium sandwichianum)*, a native fern, is particularly vulnerable. This succulent fern probably once dominated large areas of the ground cover at lower elevations in the Valley. The tender shoots and starchy, easily accessible rhizomes are a preferred food of feral pigs. As areas once dominated by this fern are rooted up, a seedbed is prepared for alien plant species such as Hilo grass and strawberry guava.

Major Alien Plant Threats

Three weeds comprise the greatest and most immediate threats to displacement of the native flora in Kīpahulu.

Hedychium gardnerianum **(kāhili ginger)** is a prolific fruit producer that is dispersed by birds as well as spread vegetatively by rapidly growing rhizomes. These plants, which grow over a meter tall in the deepest shade, displace all other plants in the forest understory, making natural forest regeneration an impossibility. The infestation of kāhili ginger in the Valley is at the incipient stage. Isolated individuals have been recorded up to 3,600 ft (1,100 m) on top of the Central Pali and at 2,900 ft (880 m) along Palikea Stream. A population in full fruit was seen at 3,600 ft (1,100 m) in Ko'ūko'ūai Gulch, but the steep terrain prohibited a reliable estimate of size.

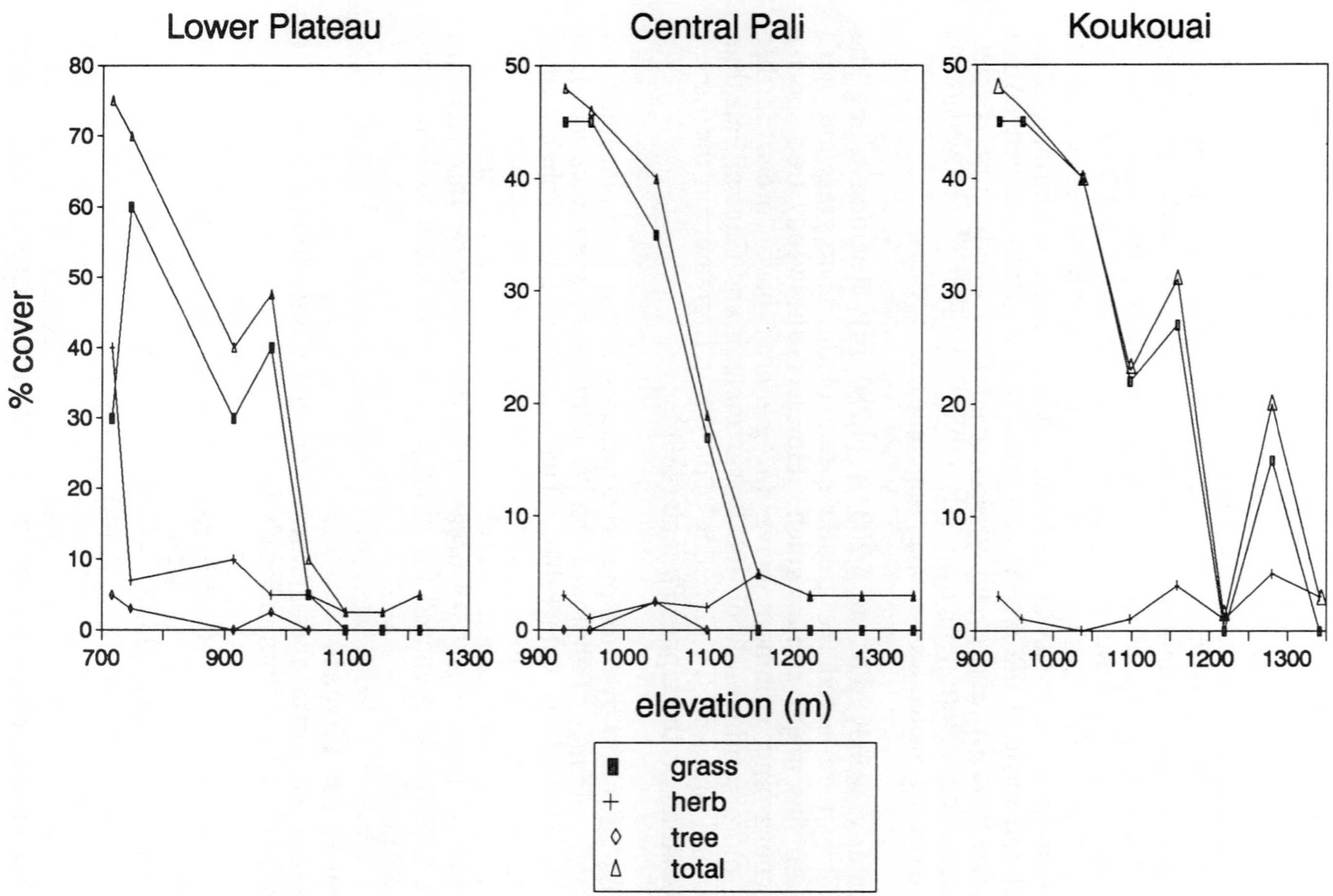

Figure 5. Invasion profiles for alien grass, herb, and tree species along three elevational transects in Kīpahulu Valley.

Paspalum conjugatum **(Hilo grass)** spreads vegetatively by stolons or by seeds that are eaten by birds or stick to humans or other animals. This perennial grass rapidly invades wet habitats from sea level to 6,500 ft (2,000 m) and is suspected of being allelopathic (Smith 1985). It typically invades disturbed areas, forming a dense carpet even under closed-canopy forests. Disturbance is not a prerequisite for establishment, as small patches of Hilo grass can be found in intact areas. The distribution of this species is continuous from 4,400 ft (1,340 m) elevation on the Upper Plateau and 4,000 ft (1,220 m) on the Lower Plateau, downslope to where it dominates the ground cover (>50% cover) at 3,800 ft (1,160 m) on the Upper Plateau and 3,740 ft (1,140 m) on the Lower Plateau. The upper range of Hilo grass is now 4,700 ft (1,430 m) on the Upper Plateau and 4,100 ft (1,250 m) on the Lower Plateau. The thick carpets formed by Hilo grass present a barrier to establishment of seedlings of native woody plants. An exclosure constructed at Upper Dogleg by Yoshinaga in 1979 has been pig free for over 10 years but shows little hope for ground cover and shrub layer regeneration where Hilo grass carpet is established. In contrast to outside the exclosure, alien grass and sedge components other than Hilo grass are absent inside; a pure carpet of Hilo grass now exists there.

Counts of individual native trees were made by size classes for plots on the 3,050 ft (930 m) and 3,150 ft (960 m) elevation transects where ground cover of Hilo grass and other alien species was greater than 25%. Typically, terrestrial shrubs and trees such as pū'aha-nui *(Broussaisia arguta)*, pilo (*Coprosma* spp.), and olomea *(Perrottetia sandwicensis)* were conspicuously absent. Intact native vegetation with abundant reproduction of these species remains on steep hillsides that have not been disturbed by feral pigs. With the exception of koa, only tree species that become established epiphytically demonstrated some reproduction in areas dominated by Hilo grass.

Psidium cattleianum **(strawberry guava)** is a small tree capable of forming dense, monotypic stands even under closed canopy forests. One of the few weed species that can become established in thick uluhe, it requires no disturbance for invasion of an area and can be spread over considerable distances by feral pigs and alien frugivorous birds. The uppermost elevation at which this species was sighted was 4,000 ft (1,220 m) (one individual plant) on the Central Pali. Scattered individuals are present downslope, and this species forms locally thick stands below 3,200 ft (975 m) elevation. Counts of individual strawberry guava trees in plots on the Dogleg transect at the 3,150-ft (960-m) transect elevation were used to extrapolate an estimated density of 2,000 individuals per hectare for this area.

Other Alien Plant Threats

Several other weedy species present in the study area pose additional threats to native communities through direct competition with native plants and their ability to regenerate in disturbed areas.

Andropogon virginicus **(broomsedge)**, a perennial bunchgrass capable of forming a dense, monotypic ground cover excluding natives, spreads readily by wind-borne seeds. Distribution in the study area consisted of one plant in a *Carex* bog at 4,400 ft (1,340 m) on the Lower Plateau, several individuals at 4,100 ft (1,250 m) elevation on the transect in closed-canopy koa-'ōhi'a forest on the Upper Plateau, and a few plants at 3,280 ft (1,000 m) on Palikea Stream and 3,500 ft (1,070 m) in Ko'ūko'ūai gulches and also in closed-canopy 'ōhi'a-koa forest. All of these plants were uprooted. Continuing work in the Valley has led to the discovery of a major population of broomsedge extending along the sheer eastern wall of the Valley from 3,600 ft (1,100 m) to east of Kaumakani Peak.

Anthoxanthum odoratum **(sweet vernalgrass)** invades disturbed sites at higher elevations. It occurs on the 6,800-ft (2,070-m) transect in shrubland areas that have been disturbed by foraging of feral goats and pigs.

Axonopus fissifolius **(narrow-leaved carpetgrass)**, a stoloniferous, mat-forming grass, was found in koa-'ōhi'a forest between 3,800 ft (1,160 m) (in an undisturbed riparian site along Ko'ūko'ūai Gulch) and the lower limits of the study area, where it occurs in patches up to 25 m^2 in size.

Clidemia hirta **(Koster's curse)** was discovered in 1988 in Kīpahulu at 2,800 ft (850 m) elevation along the trail at the base of the Central Pali. One plant 1.5 ft (0.5 m) tall was found; apparently it had never fruited. The plant was removed, and a thorough search of the area revealed no other individuals. This noxious weed was probably dispersed into the Valley by a careless hiker or researcher. Clidemia is recognized as one of the most serious weed pests in Hawai'i, because it spreads rapidly, competes with native understory plants, and invades undisturbed sites.

Cyathea cooperi **(Australian tree fern)** was discovered in the Dogleg area at the base of the Central Pali in 1987 (L.W. Cuddihy, pers. comm.). Once other botanists were informed, isolated individuals and several substantial populations were located. One population covering approximately 20 m^2 was found at 3,300 ft (1,006 m) elevation on the top of the Central Pali. Countless individuals forming a nearly monotypic stand were discovered between 2,400 and 2,600 ft (732-793 m) elevation along Palikea Gulch. Spores carried on winds up the Valley from gardens near sea level began the initial invasion. These alien tree ferns occur in all size classes from several inches to nearly 20 ft (6 m) in height, indicating success of ongoing reproduction in the Valley. We saw no evidence that the alien tree ferns were eaten by feral pigs, whereas native tree ferns (*Cibotium* spp.) are frequently devoured. This competitive advantage, in addition to a faster growth rate than native ferns, could lead to replacement of native tree ferns in the Valley.

Cyperus halpan, a sedge, one of the first colonizers of pig-rooted areas below 3,280 ft (1,000 m), forms thick cover in forest openings. With removal of pigs, other alien species such as Hilo grass are able to compete with and reduce this species.

Ageratina adenophora **(Maui pāmakani)**, a weedy composite from Mexico, is distributed throughout the study area by wind-borne seeds. Primarily a threat at mid elevations in forest openings or riparian stiuations, it forms dense mats with other weeds and hinders native plant regeneration. Biological control agents introduced to reduce this pest in Hawai'i are present in Kīpahulu but fail to significantly affect populations in wet areas (Bess and Haramoto 1972).

Holcus lanatus **(velvet grass)** occurs throughout the study area. In the *Deschampsia* grasslands it invades pig-rooted areas and competes with native grass regeneration (Jacobi 1976). When it grows in forested areas it is shaded out and never gains a competitive advantage.

Hypochoeris radicata **(gosmore)**, a weedy composite of Mediterranean origin, is distributed throughout the study area by wind-borne seeds. Abundantly and widely distributed in the *Deschampsia* grassland areas at upper elevations, it invades pig-dug and eroded areas. Pigs seek out these plants and uproot them for the starchy tap roots. Gosmore cover intensifies and increases in distribution as pigs dig up the roots and create more niches for this plant and other weedy species.

Melinis minutiflora **(molasses grass)** spreads by stolons as well as seeds. Although primarily adapted to drier conditions, it has been found in koa-'ōhi'a forest at 4,200 ft (1,280 m) on the Upper Plateau and forms patches at elevations as high as 3,800 ft (1,160 m) along the Ko'ūko'ūai and 2,950 ft (900 m) along the Palikea gulches.

Ehrharta stipoides **(meadow ricegrass)** is a perennial with the capacity to form a dense monotypic ground cover even under shaded conditions. In ungulate-disturbed areas of Hawaiian rain forests it is spread by barbed fruits that penetrate clothing and animal fur and skin. Several plants discovered at 4,200 ft (1,280 m) elevation along the trail on top of the Central Pali at the onset of this study were removed. Several additional plants found in April 1986 at the 4,700 ft (1,430 m) campsite were also uprooted.

Paspalum dilatatum **and** *P. urvillei* **(dallis grass and vasey grass),** commonly over 3.3 ft (1 m) in height, are perennial grasses that invade disturbed areas with open tree canopies. Copious seed production and the potential height of these species form a barrier to forest gap regeneration. Distribution is limited to the Lower Plateau, with highest elevational sightings at 2,950 ft (900 m).

Pennisetum clandestinum **(kikuyu grass)** is a fast-growing, mat-forming plant that spreads primarily by rhizomes. It is known to release allelopathic substances (Smith 1985) and displace native plants. It was found at the 4,800 ft (1,460 m) feral pig trap site on the upper plateau and removed. It was probably introduced in cattle offal that was used in the trap for bait.

Psidium guajava **(common guava)**, although not as invasive as strawberry guava, forms dense stands in riparian habitat and cleared areas, where it is spread by feral pigs and alien birds. This small tree is growing between sea level and 2,360 ft (720 m) elevation in Kīpahulu. Elsewhere in the Islands it can be found up to 4,920 ft (1,500 m) elevation. On Kapāpala Ranch on the island of Hawai'i it also occurs up to 4,920 ft (1,500 m) elevation.

Rubus argutus **(Florida blackberry)** is a bramble-forming vine or shrub with juicy fruit spread by birds. It is of primary concern in the *Deschampsia* grassland areas, where it has been kept in check by the browsing of feral goats. Goats have now been eradicated from these grassland areas, leaving the blackberry unchecked. Blackberry is capable of changing the unique *Deschampsia* grassland areas into alien shrublands.

Sacciolepis indica **(Glenwood grass)** invades disturbed sites at mid elevations (3,450-4,600 ft or 1,050-1,400 m) in the Valley. This perennial grass appears to be a successional precursor to other species, as it is typically found on recently disturbed sites.

Spathodea campanulata **(African tulip tree)** spreads through wind-borne seeds and floating fruits; it is shade tolerant and capable of attaining heights of 100 ft (30 m) in riparian situations. It is distributed primarily along Ko'ūko'ūai Gulch but has been found up to 2,800 ft (850 m) elevation on the Lower Plateau.

Distribution of relatively innocuous alien species that are present on transects in the study area can be found in the Appendix.

Weeds On and Off Trails

Forays off the Central Pali and Ko'ūko'ūai trail systems were used to assess the impact of researchers on the spread of weeds along trail systems. No apparent differences in weedy vegetation were noted on versus off the trails at the four elevations sampled (Table 2).

Table 2. Number of alien species found on versus off two elevational trail systems in Kīpahulu Valley, Haleakala National Park.

Elevation	Number of Alien Species			
	Central Pali		Ko'ūko'ūai Stream	
(feet)	On Trail	Off Trail	On Trail	Off Trail
4,300	4	2	2	2
3,700	2	4	2	4
3,500	3	3	6	3
3,150	4	4	4	4

DISCUSSION

Two alien plant invasion fronts threaten Kīpahulu Valley: the Kalapawili grasslands-Valley headwall area, and that section of the Valley below the 3,940-ft (1,200-m) elevational contour. The primary catalyst of vegetation change along both these fronts is the feral pig; feral goats were also important in the Valley headwall area. The central to upper portions of the Valley with relatively low feral pig densities remain nearly pristine, with less than one percent cover of alien vegetation. Weedy species now present in this area are innocuous for the most part, as they do not appear capable of dominating vegetation cover in the absence of feral animal disturbance.

Seed sources for the plant invasion on the Valley headwall are Haleakalā Crater and the Kalapawili grassland area. The primary weed species in this area, velvet grass, gosmore, sweet vernalgrass, and sheep sorrel, become established in ungulate-disturbed sites. In *Deschampsia* grasslands, velvet grass colonizes preferentially in pig-rooted areas and increases in cover and abundance as feral pig disturbance continues (Jacobi 1976, 1981; Stone *et al.*, this volume). Grassland and headwall areas of the Valley were fenced in 1986 to facilitate feral animal control. With the elimination of feral pigs and goats in these units, further replacement of native vegetation by weedy species will be minimized. However, removal of feral animals will not solve the alien plant invasion problem in the Valley. Goats and pigs have been removed from the fenced portions of the Kalapawili grasslands and Valley headwalls, and nearly eliminated above the 4,600-ft (1,400-m) elevation fence line. Frequencies of weeds have fluctuated, remained constant, or increased in these areas, but weeds have not decreased even after the near-elimination of pigs and goats.

The primary alien plant invasion front extends from the pasturelands and gardens below the study area upslope to the 3,940-ft (1,200-m) contour across the Valley. The replacement of the native vegetation in this area has been extensive and is proceeding at a rapid rate. Lamoureux and Stemmermann (1976) described Hilo grass as common in 1967 in disturbed wet areas to above 4,000 ft (1,220 m) but stated that only one area (1967 Expedition Base Camp I) had more than a few square meters dominated by Hilo grass above 2,500 ft (760 m) elevation. Personnel on the 1976 Expedition noted dozens of such openings, a marked increase in pig activity, and increased distribution of strawberry guava since 1967 (Lamoureux and Stemmermann 1976). They described the invasion as "probably an emergency." In 1980, Yoshinaga reported Hilo grass to be abundant, especially below 3,280 ft (1,000 m) where it formed solid stands in openings and under thin canopy (Yoshinaga 1980). By 1986, Hilo grass growing in solid stands had replaced the native ground cover in areas up to 3,800 ft (1,160 m) elevation.

Strawberry guava has extended its elevational range upslope 1,738 ft (530 m) from that reported by Lamoureux (1967) in less than 20 years. The potential distribution for this species is unknown, as it invades undisturbed sites under closed canopy tree layers and has been reported to be frost resistant (Popenoe 1935). Feral pigs seek out the fruit and migrate seasonally into areas of heavy fruit production, spreading the seeds over considerable distances (Diong 1983). The potentials of both strawberry guava and feral pigs to alter native communities are independently extreme and synergistically disastrous.

Kāhili ginger is among the most serious of weed threats to the Kīpahulu rain forest. Although it is now only in incipient stages, it has the potential of dominating the forest understory from the lower bounds of the study area to well above 5,000 ft (1,524 m) even in the absence of pig disturbance. In Hawaii Volcanoes National Park, large areas of native rain forest have been invaded, and hope for successful chemical or mechanical control may be limited to areas of light infestation.

MANAGEMENT RECOMMENDATIONS

An alien plant control program must be established for Kīpahulu Valley. Transects, 20- x 20-m plots, and exclosures established during this and previous studies provide precontrol baseline information on plant species distribution and density. Priority species for control are kāhili ginger, strawberry guava, Hilo grass, broomsedge, molasses grass, meadow ricegrass, African tulip tree, Australian tree fern, and Florida blackberry.

Control efforts for target species should start at the upper elevational extent of distribution (4,000 ft or 1,220 m level on both the Upper and Lower Plateaus) and proceed downslope toward the Dogleg fence until alien plant densities are too great to justify control. This will create a buffer zone between the near-pristine upslope areas and the weedy lower elevations. Recommendations were made by Yoshinaga (1980) and Smith (1985) for the establishment of a feral pig and strawberry guava buffer zone to separate the weedy lower elevations below 2,000 ft (610 m) from the more pristine area above. The Valley has been fenced into management units with this objective. A cross-Valley fence is in place at the Dogleg, another has been built at 4,700 ft (1,430 m) elevation, and feral pig control research is complete. Alien plant control needs to be initiated to create a buffer zone against strawberry guava, kāhili ginger, and Hilo grass ingress to the near-pristine upper elevations. Biological control research is not likely for any of the weedy grass species, ginger species, or either guava species, due to the conflict posed by the sugar cane industry and other agricultural interests.

All target species should be searched out and pulled or treated with herbicide. Kāhili ginger, meadow ricegrass, African tulip tree, dallis grass, vasey grass, and molasses grass can be controlled throughout the study area. Kāhili ginger eradication should be given priority, because complete control of this incipient population is still feasible and the potential population explosion and negative effects are horrendous. Hilo grass control will only be possible in the upper half of the buffer zone where it occurs in patches. Methodological tests of strawberry guava control in Kīpahulu Valley have been performed (Gardner 1980), and recent herbicide control methods have been developed (Santos *et al.*, this volume). Sweeps by weed control personnel along parallel belt transects have been used in Hawaii Volcanoes National Park with success and involve an effort of 3.0 person-days per acre (7.5 person-days per hectare) in easily accessible areas with level ground in areas where strawberry guava densities reach 200 plants per acre (500/ha) (J.T. Tunison, unpub. data). Common guava occurs in much lower numbers and can be treated along with strawberry guava when located on control sweeps above the Dogleg.

Florida blackberry should be eradicated from the Kalapawili grasslands, as removal of feral goats and pigs will surely release it from foraging pressure and probably result in a blackberry population explosion. Effective herbicidal methods of control are available for this species (Santos *et al.*, this volume).

Australian tree fern control needs to be researched. The top of one plant was cut off and the remaining stump died. The apical portions will reroot if left in the Valley, however. Removal of the tops of all the Australian tree ferns from the Valley by helicopters would be logistically difficult and may spread spores into new areas.

The remaining invasive species -- sweet vernalgrass, narrow-leaved carpetgrass, *Cyperus halpan*, Maui pāmakani, velvet grass, gosmore, and Glenwood grass -- are not as serious a threat, although frequently they are more widespread than the target species; they are not recommended for control at this time.

It is essential that records of numbers of individuals, locations, and age classes of each alien plant species controlled be recorded. Control programs should be analyzed as they progress and modified as needed. Where densities of target weed species become too great, acceptable limits of control short of eradication will need to be defined. Managers must recognize at the onset of any weed control program that continual, ongoing maintenance programs will be needed to sustain control. Dispersal of weedy propagules into the control area by water, wind, birds, and possibly people will be unceasing, and exhaustion of soil seed banks will take repeated control efforts and continual monitoring. Control of strawberry guava at the Dogleg area, where densities approach 810 plants per acre (2,000/ha), could require 12 person-days per acre (30/ha), at least until densities are considerably reduced.

All workers in the Valley should be aware of target species and remove them on discovery, or at least record distributional information. Monitoring of existing transects, plots, campsites, and trail systems, to detect new weed species and expansion of known weed distributions, should continue. Priority should be given to control of satellite populations of any of the invasive weedy species discussed (Mack, this volume) to limit spread of aliens.

The impact of people in the Valley has been a controversial issue since the 1967 Expedition sponsored by The Nature Conservancy of Hawaii. The Expedition organized in 1976 to assess the frequently implied negative impacts of the 1967 Expedition indicated that the natural resources in the Valley were deteriorating as a result of lack of active management

(Lamoureux and Stemmermann 1976; Smith 1978). Observations on abundance of weedy species made for all campsites, helipads, trail systems, and transects used for research activities showed that spread of alien plants was not restricted to trail systems. The uppermost populations of Hilo grass and all known populations of meadow ricegrass have been treated with herbicide or pulled to prevent spread to higher elevations. The survey of trail systems in the critical area between 4,700 ft (1,430 m) and 2,350 ft (720 m) included forays off the trail to allow comparison of weed invasion on frequently used paths with previously untraversed areas. Until recently, the only new weed species recorded have been in areas not previously visited during this study. The discovery and removal of an individual clidemia plant on a trail in the study area has reinforced the policy of thorough equipment and personal gear inspection and cleaning prior to use in the Valley. Areas where particularly noxious plants (*i.e.*, Koster's curse, meadow ricegrass, kikuyu grass) were removed should be monitored for recurrence and further treatment.

Any work carried out on a Valley-wide basis must proceed from upper elevations to lower, to avoid dispersal of weedy propagules upslope, particularly on helicopter skids. All field workers should check clothing and equipment for weed seeds prior to entry into the Valley. Transects and trails should not be established in areas dominated by uluhe or matted fern, as these areas recover slowly and are subject to invasion by alien plants.

The commitment to control pigs in the Valley has been made. Removal of feral pigs will result in changes in weed species composition and abundance, as evidenced by the monotypic ground cover of Hilo grass now present (after six years' exclusion of feral pigs) in the Dogleg exclosure established by Yoshinaga. In rain forest areas of Hawaii Volcanoes National Park, a dramatic increase in alien grasses was noted after reduction in feral pig densities (Stone *et al.*, this volume). Clearly, the decision to remove feral pigs from the Valley must be matched by a strong commitment toward continued vegetation monitoring, research, and development of a sound alien plant control program.

Kīpahulu Valley represents one of the best opportunities left in Hawai'i for terrestrial ecosystem preservation. But preservation will not be realized without a properly monitored, active, and continual management effort.

Appendix. Frequency of occurrence (%) of alien plants in the Kīpahulu Valley study area, Haleakala National Park.

CATEGORY, SPECIES, AND DATE	Kalapawili Grasslands		Upper Plateau									Lower Plateau					
	7800	7700	6800	6800	5960	5240	4900	4200	4100	3150	3050	4400	4300	4100	3300	2450	2350
	A	AA	WAI	6	5	4	P2	2	1	DL4	DL3	P1A	P1B	3	DLT	DL1	DL2
FERNS																	
Athyriopsis japonica+																	
04/85	*	*								**	**				x	**	**
11/85										**	**				x	**	**
03/86										**	**				x	**	**
09/86										32	36				x	96	86
01/87								8	6						10		8
04/87										38	26		--			83	84
08/87	--	--						2		36	32		6			89	92
01/88								--		68	30	--	--	--		100	59
05/88	--	--								38	20				10	83	81
10/88										36	18		2			87	81
Blechnum occidentale+																	
09/86															x	21	
04/87										6			--			11	3
08/87	--	--								2						9	3
01/88								--				--	--	--		13	
10/88										4						17	3
Christella dentata+																	
04/87													--				5
10/88																2	

LOCATION OF TRANSECT AND ELEVATION IN FEET

Appendix, continued.

Appendix, continued.

CATEGORY, SPECIES, AND DATE	LOCATION OF TRANSECT AND ELEVATION IN FEET: Kalapawili Grasslands		Upper Plateau									Lower Plateau					
	7800	7700	6800	6800	5960	5240	4900	4200	4100	3150	3050	4400	4300	4100	3300	2450	2350
	A	AA	WAI	6	5	4	P2	2	1	DL4	DL3	P1A	P1B	3	DLT	DL1	DL2
Christella parasitica[+]																	
04/87																4	
08/87	--	--														2	3
10/88										2	2						
# *Cyathea cooperi*[+]																	
08/87	--	--														6	
05/88	--	--														4	
10/88																6	
GRASSES AND SEDGES																	
# *Andropogon virginicus*[+]																	
04/85	*	*								**	**	2			x	**	**
11/85										**	**				x	**	**
03/86										**	**				x	**	**
08/87	--	--						2							2		
Anthoxanthum odoratum																	
04/85	*	*								**	**				x	**	**
11/85			5							**	**				x	**	**

CATEGORY, SPECIES, AND DATE	LOCATION OF TRANSECT AND ELEVATION IN FEET: Kalapawili Grasslands 7800 A	7700 AA	Upper Plateau 6800 WAI	6800 6	5960 5	5240 4	4900 P2	4200 2	4100 1	3150 DL4	3050 DL3	Lower Plateau 4400 P1A	4300 P1B	4100 3	3300 DLT	2450 DL1	2350 DL2
Axonopus fissifolius [+]																	
09/86										28	36				x		
04/87										22	12		--				
08/87	--	--								24	10						
01/88								--		30	22	--	--	--			
05/88	--	--								20	12						
10/88										16	8						
Cyperus halpan																	
09/86										54	46				x	6	
01/87										50	52					13	8
04/87										54	76		--			11	5
08/87	--	--								66	78					11	14
01/88								--		78	92	--	--	--		6	8
05/88	--	--								64	84					11	11
10/88										58	74					11	16
Holcus lanatus																	
04/85	*	*	100	8	6	10	2	2		**	**	4	2	4	x	**	**
11/85	100	100	98	8	6	10	2	2		**	**	6			x	**	**
03/86	100	100	80	8	4	12	2			**	**	10	2		x	**	**
06/86	100	100	92	6	6	16	2		2	**	**	--	2	--	x	**	**
09/86	100	100	95		14	10	2	2				2			x		
01/87	100	100	93	6	10	18	4	4				4	2				

Appendix, continued.

Appendix, continued.

CATEGORY, SPECIES, AND DATE	Kalapawili Grasslands 7800 A	7700 AA	Upper Plateau 6800 WAI	6800 6	5960 5	5240 4	4900 P2	4200 2	4100 1	3150 DL4	3050 DL3	Lower Plateau 4400 P1A	4300 P1B	4100 3	3300 DLT	2450 DL1	2350 DL2
Holcus lanatus																	
04/87	100	100	88	16	8	20	2	2	2			10	--		2		
08/87	--	--	93	10	8	10	4	2	2			8			2		
01/88	100	100	98	14	6	14	2	--	2			--	--	--	2		
05/88	--	--	93	14	6	28		2	2			6	2		2		
10/88	100	100	95	10	6	20	2		2			4	2		2		
Kyllinga brevifolia																	
09/86															x	6	
04/87													--			4	
08/87	--	--								2	2					2	
01/88								--				--	--	--		6	
05/88	--	--														6	3
10/88										2	2					6	
# *Melinis minutiflora*																	
09/86															x	4	
01/87																2	5
04/87													--			4	
08/87	--	--								2						2	
01/88								--				--	--	--		2	
05/88	--	--														2	
10/88																2	

CATEGORY, SPECIES, AND DATE	Kalapawili Grasslands		Upper Plateau									Lower Plateau					
	7800	7700	6800	6800	5960	5240	4900	4200	4100	3150	3050	4400	4300	4100	3300	2450	2350
	A	AA	WAI	6	5	4	P2	2	1	DL4	DL3	P1A	P1B	3	DLT	DL1	DL2
Oplismenus hirtellus																	
09/86															x		17
04/87													--				8
08/87	--	--															11
01/88							--					--	--	--			11
05/88	--	--															14
10/88																	16
# *Paspalum conjugatum*																	
04/85	*	*							2	**	**				x	**	**
11/85										**	**				x	**	**
03/86										**	**				x	**	**
06/86									2	**	**	--		--	x	**	**
09/86									4	98	100				x	89	100
01/87									4	96	98				10	98	89
04/87									2	96	100		--		8	100	100
08/87	--	--							6	96	100				8	100	100
01/88								--		98	98	--	--	--	6	98	68
05/88	--	--							2	98	100				10	100	100
10/88								2	2	98	100				6	98	100
Rhynchospora caduca																	
04/87										62	66		--			2	
08/87										54	74					2	

Appendix, continued.

Appendix, continued.

CATEGORY, SPECIES, AND DATE	Kalapawili Grasslands		LOCATION OF TRANSECT AND ELEVATION IN FEET: Upper Plateau									Lower Plateau					
	7800 A	7700 AA	6800 WAI	6800 6	5960 5	5240 4	4900 P2	4200 2	4100 1	3150 DL4	3050 DL3	4400 P1A	4300 P1B	4100 3	3300 DLT	2450 DL1	2350 DL2
Rhynchospora caduca																	
01/88								--		42	62	--	--	--			
10/88										50	58					2	
Sacciolepis indica																	
04/85	*	*							2	**	**				x	**	**
11/85								2	2	**	**				x	**	**
03/86									4	**	**				x	**	**
09/86									8	36	8				x	11	
01/87									4	24					2	19	5
04/87									2	54	34		--			23	5
08/87	--	--							8	72	48				2	23	5
01/88								--	14	56		--	--	--	4		
05/88	--	--							4	72	68				2	40	14
10/88									6	62	44				8	9	11
HERBS																	
Ageratina adenophora																	
04/85	*	*					8	12		**	**	34	4	8	x	**	**
11/85							8	20	4	**	**	44		8	x	**	**
03/86							14	24	2	**	**	58	2	6	x	**	**
06/86						2	20	16	2	**	**	--	2	--	x	**	**
09/86							18	38	6	20		40	6	6	x	11	42

CATEGORY, SPECIES, AND DATE	Kalapawili Grasslands 7800 A	7700 AA	Upper Plateau 6800 WAI	6800 6	5960 5	5240 4	4900 P2	4200 2	4100 1	3150 DL4	3050 DL3	Lower Plateau 4400 P1A	4300 P1B	4100 3	3300 DLT	2450 DL1	2350 DL2
Ageratina adenophora																	
01/87							26	38	4	6	6	50	4	6	30	19	43
04/87							28	22	2	6	4	48	--	2	34	19	32
08/87	--	--					10	26	4	6	4	42	4		22	19	38
01/88							12	--			2	--	--	--	16	17	32
05/88	--	--					16	30		8	6	46	4	6	20	21	38
10/88						2	12	22	2	8	4	42	6	8	26	19	41
Ageratina riparia																	
09/86								2							x		
01/87									2	4							
01/88								--		4		--	--	--			
Ageratum conyzoides																	
09/86										8					x		42
01/87										10							3
04/87										4			--			2	
08/87	--	--								6						2	5
01/88								--		16		--	--	--			22
05/88	--	--								8							3
10/88										8	2						3
Commelina diffusa																	
04/87													--				3

Appendix, continued.

Appendix, continued.

CATEGORY, SPECIES, AND DATE	Kalapawili Grasslands 7800 A	7700 AA	Upper Plateau 6800 WAI	6800 6	5960 5	5240 4	4900 P2	4200 2	4100 1	3150 DL4	3050 DL3	Lower Plateau 4400 P1A	4300 P1B	4100 3	3300 DLT	2450 DL1	2350 DL2
Cuphea carthagenensis																	
09/86										34	34				x	6	36
01/87										30	8						14
04/87										24	12		--			13	3
08/87	--	--								50	26					28	11
01/88								--		32	22	--	--	--		15	19
05/88	--	--								36	30				4	13	11
10/88										30	26					13	19
Drymaria cordata																	
09/86										14	2				x	45	67
01/87										10	2					36	68
04/87										12	2		--			32	70
08/87	--	--								10	2					32	59
01/88								--				--	--	--		36	51
05/88	--	--								10	2					34	51
10/88										10						26	70
Epilobium billardierianum subsp. *cinerium*[+]																	
04/85	*	*	60							**	**				x	**	**
11/85			50	4		2				**	**				x	**	**
03/86		2	50			6				**	**				x	**	**
06/86		2	42			2				**	**	--		--	x	**	**
09/86			63	2		4									x		

CATEGORY, SPECIES, AND DATE	LOCATION OF TRANSECT AND ELEVATION IN FEET																
	Kalapawili Grasslands		Upper Plateau									Lower Plateau					
	7800	7700	6800	6800	5960	5240	4900	4200	4100	3150	3050	4400	4300	4100	3300	2450	2350
	A	AA	WAI	6	5	4	P2	2	1	DL4	DL3	P1A	P1B	3	DLT	DL1	DL2
Epilobium billardierianum subsp. *cinerium* +																	
01/87		2	63			2											
04/87		8	60	4		6							--				
08/87	--	--	55	2													
01/88		6	70	4				--				--	--	--		4	
05/88	--	--	55	2													
10/88	8		63	2		8											
Erechtites valerianifolia																	
04/85	*	*						2	8	**	**				x	**	**
11/85								2	8	**	**	6		4	x	**	**
03/86								6	4	**	**	4			x	**	**
06/86								8	2	**	**	--		--	x	**	**
09/86								16	24	30	56	3	2	4	x	26	50
01/87						2		12	20	56	52	2			12	17	8
04/87								10	10	56	54	2	--		4	19	24
08/87	--	--						12	10	58	66	4			14	30	30
01/88							2	--	2	46	34	--	--	--	20	13	
05/88	--	--					2	18	8	36	68	4	2		20	15	24
10/88								14	2	66	72	2	2		20	19	16
Hydrocotyle verticillata																	
09/86															x	9	17
04/87													--			15	8

Appendix, continued.

Appendix, continued.

CATEGORY, SPECIES, AND DATE	Kalapawili Grasslands 7800 A	7700 AA	Upper Plateau 6800 WAI	6800 6	5960 5	5240 4	4900 P2	4200 2	4100 1	3150 DL4	3050 DL3	Lower Plateau 4400 P1A	4300 P1B	4100 3	3300 DLT	2450 DL1	2350 DL2
Hydrocotyle verticillata																	
08/87	--	--														15	8
01/88								--				--	--	--	2	17	5
05/88	--	--					2								4	11	8
10/88															4	11	11
Hypochoeris radicata																	
04/85	*	*	98	43	6	8				**	**				x	**	**
11/85	100	100	95	62	2	2			2	**	**				x	**	**
03/86	100	100	85	56	6	20		4		**	**				x	**	**
06/86	100	100	98	40	4	14				**	**	--		--	x	**	**
09/86	100	100	95	40	4	16	2	4							x		
01/87	100	100	100	72		14	2	2									
04/87	100	100	90	72	4	22	4	2	2				--				
08/87	--	--	98	72	8	14	4	4									
01/88	100	100	98	90	2	10		--	2			--	--	--		4	
05/88	--	--	100	80	4	16			2								
10/88	100	100	100	76	6	16	6										
Lapsana communis[+]																	
04/85	*	*								**	**				x	**	**
11/85			5			6				**	**				x	**	**
03/86			3							**	**				x	**	**
06/86										**	**	--		--	x	**	**

CATEGORY, SPECIES, AND DATE	Kalapawili Grasslands		Upper Plateau									Lower Plateau					
	7800	7700	6800	6800	5960	5240	4900	4200	4100	3150	3050	4400	4300	4100	3300	2450	2350
	A	AA	WAI	6	5	4	P2	2	1	DL4	DL3	P1A	P1B	3	DLT	DL1	DL2
Lapsana communis[+]																	
09/86			3		2	16									x		
01/87						12				2							
08/87	--	--				10											
05/88	--	--	3			18	4										
10/88			5			20			2								3
Ludwigia octovalvis																	
09/86										2					x	13	31
01/87																6	14
04/87										2	4		--			4	8
08/87	--	--														4	16
01/88								--				--	--	--		6	14
05/88	--	--														2	22
10/88										2						2	16
Lythrum maritimum																	
04/85	*	*								**	**				x	**	**
11/85			8							**	**				x	**	**
03/86										**	**				x	**	**
06/86			8							**	**	--		--	x	**	**
09/86			15												x		
01/87			10														
04/87			20										--				

The table is headed "LOCATION OF TRANSECT AND ELEVATION IN FEET".

Appendix, continued.

Appendix, continued.

CATEGORY, SPECIES, AND DATE	LOCATION OF TRANSECT AND ELEVATION IN FEET																
	Kalapawili Grasslands		Upper Plateau									Lower Plateau					
	7800	7700	6800	6800	5960	5240	4900	4200	4100	3150	3050	4400	4300	4100	3300	2450	2350
	A	AA	WAI	6	5	4	P2	2	1	DL4	DL3	P1A	P1B	3	DLT	DL1	DL2
Lythrum maritimum																	
08/87	--	--	8					2									
01/88			15					--				--	--	--			
05/88	--	--	23														
10/88			25					2									
Musa sp.																	
05/88																	3
Polygonum punctatum																	
09/86															x		31
01/87																9	8
04/87													--			6	11
08/87	--	--														6	16
01/88								--				--	--	--			14
05/88	--	--														13	22
10/88																2	11
Prunella vulgaris																	
04/85	*	*				8				**	**				x	**	**
11/85	2	4				18				**	**				x	**	**
03/86			3			14				**	**				x	**	**
06/86			3			18				**	**	--		--	x	**	**
09/86	2	6	3			20									x		

CATEGORY, SPECIES, AND DATE	LOCATION OF TRANSECT AND ELEVATION IN FEET																
	Kalapawili Grasslands		Upper Plateau									Lower Plateau					
	7800	7700	6800	6800	5960	5240	4900	4200	4100	3150	3050	4400	4300	4100	3300	2450	2350
	A	AA	WAI	6	5	4	P2	2	1	DL4	DL3	P1A	P1B	3	DLT	DL1	DL2
Prunella vulgaris																	
01/87						16											
04/87						22							--				
08/87	--	--	3			18								2			
01/88			3			24		--				--	--	--	4		
05/88	--	--	3			26									2		
10/88	12		3			24											
Rubus rosifolius																	
04/85	*	*						14	36	**	**	4	10	28	x	**	**
11/85								24	48	**	**	10	16	34	x	**	**
03/86								18	60	**	**	20	32	34	x	**	**
06/86								18	50	**	**	--	46	--	x	**	**
09/86								40	80	64	68	16	46	44	x	87	100
01/87								42	76	70	46	8	44	52	72	83	84
04/87							2	32	58	66	54	14	--	44	64	81	95
08/87	--	--					2	46	50	62	50	16	42	46	54	81	95
01/88								--	46	74	44	--	--	--	64	85	100
05/88							2	36	64	68	46	16	38	56	68	81	95
10/88							2	36	66	68	54	12	34	62	56	83	92
Rumex acetosella																	
04/85	*	*	25	8						**	**				x	**	**
11/85	90	90	23	14		2				**	**				x	**	**

Appendix, continued.

Appendix, continued.

CATEGORY, SPECIES, AND DATE	Kalapawili Grasslands 7800 A	7700 AA	Upper Plateau 6800 WAI	6800 6	5960 5	5240 4	4900 P2	4200 2	4100 1	3150 DL4	3050 DL3	Lower Plateau 4400 P1A	4300 P1B	4100 3	3300 DLT	2450 DL1	2350 DL2
Rumex acetosella																	
03/86	100	100	13	10		2				**	**				x	**	**
06/86	88	92	22	8		2				**	**	--		--	x	**	**
09/86	88	84	20	8											x		
01/87	82	88	18	10													
04/87	86	100	18	12		2							--				
08/87	--	--	15	8													
01/88	100	86	8	10				--				--	--	--			
05/88	--	--	15	4													
10/88	84	100	10	10													
Stachytarpheta urticifolia																	
09/86															x		6
04/87													--				3
08/87	--	--															5
01/88								--				--	--	--			5
05/88																	3
10/88																	3
Youngia japonica																	
04/85	*	*						2	2	**	**				x	**	**
11/85										**	**				x	**	**
03/86						20				**	**				x	**	**
06/86						12			2	**	**	--		--	x	**	**

CATEGORY, SPECIES, AND DATE	Kalapawili Grasslands 7800 A	7700 AA	Upper Plateau 6800 WAI	6800 6	5960 5	5240 4	4900 P2	4200 2	4100 1	3150 DL4	3050 DL3	Lower Plateau 4400 P1A	4300 P1B	4100 3	3300 DLT	2450 DL1	2350 DL2
Youngia japonica																	
09/86								2	8	2					x	4	
01/87								6	6	6	2				4	26	
04/87						20		2	4	8			--			45	19
08/87	--	--						4	2	4						19	14
01/88			3			12		--	14	2		--	--	--		21	14
05/88	--	--								8						32	14
10/88										4	4					17	
TREES																	
# *Psidium cattleianum*																	
09/86										26	6				x	26	58
01/87										16	20					28	22
04/87										16	34		--			23	27
08/87	--	--								6	28					28	35
01/88								--		16	52	--	--	--		47	46
05/88										8	18					55	51
10/88										10	2					38	27
# *Psidium guajava*																	
09/86															x	4	17
04/87													--			2	11

Appendix, continued.

Appendix, continued.

CATEGORY, SPECIES, AND DATE	Kalapawili Grasslands 7800 A	7700 AA	Upper Plateau 6800 WAI	6800 6	5960 5	5240 4	4900 P2	4200 2	4100 1	3150 DL4	3050 DL3	Lower Plateau 4400 P1A	4300 P1B	4100 3	3300 DLT	2450 DL1	2350 DL2
# *Psidium guajava*																	
08/87	--	--									2					2	5
01/88								--		12		--	--	--		4	11
05/88	--	--														6	8
10/88										2	26					4	5

Column headings: LOCATION OF TRANSECT AND ELEVATION IN FEET.

+ New record for study area.
* Kalapawili Grasslands transects were not established until 11/85.
** Transects at 960 m and below were not established until 06/86.
-- No data.
x Delta transect was not established until 1/87.
Major disruptive potential.

Literature Cited

Bess, H.A., and F.H. Haramoto. 1972. Biological control of pamakani, *Eupatorium adenophorum*, in Hawaii by a tephritid gall fly, *Procedidochares utilis* -- III. Status of the weed, fly, and parasites of the fly in 1966-1971 versus 1950-1957. *Proc. Hawn. Entomol. Soc.* 21:165-178.

Conant, S., and M.A. Stemmermann. 1980. Birds in the Kipahulu District of Haleakala National Park. *Proc. 3rd Conf. Nat. Sci., Hawaii Volcanoes Natl. Park*, 67-75. Honolulu: Univ. Hawaii Coop. Natl. Park Resour. Stud. Unit.

Diong, C.H. 1983. Population ecology and management of the feral pig (*Sus scrofa* L.) in Kipahulu Valley, Maui. Ph.D. Diss., Univ. Hawaii, Honolulu.

Gardner, D.E. 1980. An evaluation of herbicidal methods of strawberry guava control in Kipahulu Valley. In *Resources base inventory of Kipahulu Valley below 2000 feet*, ed. C.W. Smith, 63-69. Honolulu: Univ. Hawaii Coop. Natl. Park Resour. Stud. Unit.

Higashino, P.K., L.W. Cuddihy, S.J. Anderson, and C.P. Stone. 1988. *Bryophytes and vascular plants of Kipahulu Valley, Haleakala National Park.* Tech. Rep. 65, Univ. Hawaii Coop. Natl. Park Resour. Stud. Unit. Honolulu.

Jacobi, J.D. 1976. The influence of feral pigs on a native alpine grassland in Haleakala National Park. *Proc. 1st Conf. Nat. Sci., Hawaii Volcanoes Natl. Park*, 107-112. Honolulu: Univ. Hawaii Coop. Natl. Park Resour. Stud. Unit.

Jacobi, J.D. 1981. *Vegetation changes in a subalpine grassland in Hawaii following disturbance by feral pigs*. Tech. Rep. 41, Univ. Hawaii Coop. Natl. Park Resour. Stud. Unit. Honolulu.

Lamoureux, C.H. 1967. The vascular plants of Kipahulu Valley, Maui. In *Scientific report of the Kipahulu Valley expedition*, ed. R.E. Warner, 23-54. Honolulu: The Nature Conservancy.

Lamoureux, C.H., and L. Stemmermann. 1976. *Kipahulu expedition 1976.* Tech. Rep. 11, Univ. Hawaii Coop. Natl. Park Resour. Stud. Unit. Honolulu.

Mack, R.N. [this volume] Characteristics of invading plant species.

Mueller-Dombois, D., and H. Ellenberg. 1974. *Aims and methods of vegetation ecology.* New York: John Wiley & Sons.

Popenoe, F.W. 1935. *Psidium*. In *The standard cyclopedia of horticulture*, vol. III, ed. L.H. Bailey, 2847-2849. New York: MacMillan.

Santos, G.L., D. Kageler, D.E. Gardner, L.W. Cuddihy, and C.P. Stone. [this volume] Herbicidal control of selected alien plant species in Hawaii Volcanoes National Park.

Smith, C.W. 1978. *Kipahulu Valley research plan.* Tech. Rep. 22, Univ. Hawaii Coop. Natl. Park Resour. Stud. Unit. Honolulu.

Smith, C.W. 1985. Impact of alien plants on Hawai'i's native biota. In *Hawai'i's terrestrial ecosystems: preservation and management*, ed. C.P. Stone and J.M. Scott, 180-250. Univ. Hawaii Coop. Natl. Park Resour. Stud. Unit. Honolulu: Univ. Hawaii Pr.

Stone, C.P., P.C. Banko, P.K. Higashino, and F.G. Howarth. 1984. Interrelationships of alien and native plants and animals in Kipahulu Valley, Haleakala National Park: a preliminary report. *Proc. 5th Conf. Nat. Sci., Hawaii Volcanoes Natl. Park*, 91-105. Honolulu: Univ. Hawaii Coop. Natl. Park Resour. Stud. Unit.

Stone, C.P., L.W. Cuddihy, and J.T. Tunison. [this volume] Responses of Hawaiian ecosystems to removal of feral pigs and goats.

Warner, R.E., ed. 1967. *Scientific report of the Kipahulu Valley expedition*. Sponsored by The Nature Conservancy. Honolulu.

Yoshinaga, A.Y. 1980. *Upper Kipahulu Valley weed survey*. Tech. Rep. 33, Univ. Hawaii Coop. Natl. Park Resour. Stud. Unit. Honolulu.

PHYSICAL AND CHEMICAL CONTROL

HERBICIDAL CONTROL OF SELECTED ALIEN PLANT SPECIES IN HAWAII VOLCANOES NATIONAL PARK

**Gregory L. Santos, Dina Kageler,
Donald E. Gardner, Linda W. Cuddihy, and Charles P. Stone**

ABSTRACT

Herbicide tests were conducted between April 1984 and June 1986 on seven species of alien plants classified as threats to native ecosystems within Hawaii Volcanoes National Park. The study was designed to be an initial series of tests to develop effective alien plant treatment techniques and to obtain baseline information on the effects of herbicides on native flora. Highly effective treatments were found for olive (*Olea europaea* subsp. *africana*) (TORDON RTU on cut stumps) and for both species of silk oak (*Grevillea banksii* and *G. robusta*) (2.5% GARLON 4 in diesel oil applied in continuous frill cuts). No hazards to native plants were detected. Treatments for glorybush (*Tibouchina urvilleana*) (20% GARLON 4 in diesel oil on cut stumps) and yellow Himalayan raspberry (*Rubus ellipticus*) (40% GARLON 4 in a foliar drizzle spray and 20% TORDON 22K in water on cut stumps) appeared effective; however, further testing is necessary to refine treatments, verify results, and further assess potential harm to native plant species. Kāhili ginger (*Hedychium gardnerianum*) was effectively controlled with TORDON 10K pellets; however, further testing is warranted for several reasons. A 2% foliar spray of ROUNDUP in water was not completely effective on blackberry (*Rubus argutus*), but it did provide a good measure of control and would be useful in selected situations. Further testing on blackberry is necessary to increase treatment effectiveness and to evaluate other treatments. We recommend continued modest emphasis on herbicide research in Hawai'i's native ecosystems to: increase control effectiveness for invasive alien plant species; evaluate the effects of retreatment regimes; increase knowledge of hazards to native plants and animals; broaden the range of safe chemical tools needed to integrate herbicides with other methods of ecosystem restoration; monitor operational herbicide programs for efficacy and cost effectiveness; increase long-term monitoring of treatment effects in selected areas; and determine and enhance responsible management programs in near-native ecosystems.

INTRODUCTION

Of the approximately 400 alien plant species naturalized in Hawaii Volcanoes National Park, 50 present the greatest threats to native ecosystem processes and have been targeted for control. These 50 species were prioritized by the Park Resources Management Division according to the magnitude of the threat they currently pose, the estimated cost and effort of control, and the potential for successful control methods (Hawaii Volcanoes National Park 1984). The purpose of this paper is to summarize information on herbicidal tests conducted from 1984-1986 on seven species classified as disruptive species.

SPECIES OF CONCERN

Kāhili ginger, *Hedychium gardnerianum*, native to the Himalayas, is considered by horticulturists to be a desirable garden plant in Hawai'i. It is a cold-tolerant species found in its native habitat at altitudes to 8,200 ft (2,500 m) (Graf 1978). Although somewhat localized within the Park, kāhili ginger is heavily infesting more than 1,235 a (500 ha) and is a serious threat to native rain forests. A very aggressive, shade-tolerant plant, it can invade and establish in intact native rain forest habitat, where it can form dense monotypic stands 6.5 to 10 ft (2-3 m) in height and displace native understory vegetation.

Olive (*Olea europaea* subsp. *africana*), a relatively recent pest in Hawaii Volcanoes National Park, is found on over 14,825 a (6,000 ha) on the former 'Āinahou Ranch. Its current elevational range in the Park is between 1,640 and 3,280 ft (500-1,000 m) (J.T. Tunison, pers. comm.), and the infestation has expanded rapidly since the removal of cattle (*Bos taurus*) in the early 1970s. Trees 16 ft (5 m) tall have been observed, and it is believed that olive will shade out native species. This olive produces copious fruit, and seedlings beneath a parent tree may exceed 110 individuals/ft^2 (1,200/m^2) (G.L. Santos, unpub. data).

Two species of silk oak (*Grevillea robusta* and *G. banksii*) are currently invading the Park's southwestern boundary. *Grevillea robusta*, a robust, partly deciduous tree native to Australia, was introduced to the State of Hawai'i for reforestation purposes in 1938 (Skolmen 1979). It is found in a wide variety of habitats and from sea level to over 4,265 ft (1,300 m) elevation. *Grevillea banksii*, also native to Australia, is a small tree to 20 ft (6 m) tall and was officially designated a noxious weed by the Hawaii Department of Agriculture in 1978 (Regulation NW 10, updated by Title 4, Chapter 68 Administrative Rules, 1981).

Both species of *Grevillea* are aggressive, drought-tolerant, have the ability to establish in little or no soil, and may form dense, monotypic stands. If unmanaged, these trees may eventually displace dry forest and shrubland in Hawaii Volcanoes National Park.

Florida blackberry (*Rubus argutus*) was introduced to Hawai'i in 1894 (Haselwood and Motter 1984). By 1962, it had infested 43,405 a (17,565 ha) on O'ahu, Kaua'i, Maui, and Hawai'i at altitudes of 1,970 ft (600 m) or higher. It is well adapted to full sun and grows in submontane forests and woodlands, as well as in montane forests and shrub communities. Florida blackberry is found throughout the Park between 2,270 and 6,560 ft (700-2,000 m) elevation (Smith 1985). Areas that are heavily infested are impenetrable due to the tangling growth habit and the sharp thorns on the branches and leaves. The State of Hawai'i officially classified *R. argutus* a noxious weed.

Yellow Himalayan raspberry (*Rubus ellipticus*) is currently widespread throughout the wet forests near Kīlauea Crater and Volcano Village, especially in areas disturbed by either humans or feral pigs (*Sus scrofa*). Yellow Himalayan raspberry is spread by underground sprouting of roots and by seeds that are probably dispersed by alien and native frugivorous birds (Smith 1985; C.P. Stone, unpub. data). The current elevational range of *R. ellipticus* in Hawai'i is between 2,270 and 5,580 ft (700-1,700 m) (Smith 1985). This species is well adapted to the full sun of open canopy forests or pastures and the deep shade of rain forests. Yellow Himalayan raspberry is a major threat to the 'Ōla'a Forest Tract of the Park (Jacobi and Warshauer 1975).

Glorybush or lasiandra (*Tibouchina urvilleana*), a native of Brazil, is a member of the melastome family. It was introduced to Hawai'i as an ornamental, probably for its large, brillant purple flowers, which are borne either singly or in clusters on branch tips (Neal 1965). Glorybush is currently found in wet habitats on O'ahu, Kaua'i, and Hawai'i between 655 and 5,580 ft (200-1,700 m) elevation. This robust shrub is found in disjunct populations in the Park at approximately 3,940 ft (1,200 m) elevation. However, just outside the Park's eastern boundary, glorybush has invaded native rain forests and formed large, dense monotypic stands to 16 ft (5 m) in height by vigorous vegetative reproduction.

HERBICIDES TESTED

Herbicides used in this series of tests and approved for limited use and testing within Park boundaries included:

AMITROL T: (amitrole + ammonium thiocyanate at 2 lbs active ingredient (ai)/gal). Amitrole is a non-selective systemic herbicide used to control both grasses and broad-leafed species.

GARLON 4: (the ester formulation of triclopyr at 4 lbs acid equivalent (ae)/gal). Triclopyr is a selective, systemic herbicide used for control of woody species.

ROUNDUP: (glyphosate as the amine salt formulation at 3 lbs ae/gal). Roundup is a non-selective systemic herbicide.

TORDON 10K Pellets: (picloram as the potassium salt at 10% ae/lb). Picloram is a persistent, non-selective systemic herbicide used primarily to control broadleafed and woody species.

TORDON 22K: (picloram as the potassium salt at 2 lbs ae/gal).

TORDON RTU: (picloram + 2,4-D as amine salts at 3% and 11.2% ae/gal).

Considerations in the selection of treatment techniques used in tests were practicality of field application, life form of target species, applicator safety, cost and energy efficiency, expected residual activity of herbicide, and presumed toxicity to nontarget native plant species. Techniques will be described for tests of particular chemicals on each target species.

KĀHILI GINGER

Materials and Methods

Sixty clumps of kāhili ginger with rhizome masses ranging in diameter from 2.7 to 10.8 ft^2 (0.25-1 m^2) were chosen from populations in a mixed understory wet ʻōhiʻa *(Metrosideros polymorpha)* forest community. Elevation at test sites was 3,940 ft (1,200 m), with mean annual rainfall and temperature of 130 in. (3,300 mm) and 63 F (17 C).

Treatments tested included undiluted GARLON 4, TORDON 22K, AMITROL T, ROUNDUP, and TORDON RTU injected into the rhizomes via the Maujet Micro-Injection system. Maujet injectors are an effective way of introducing insecticide into the vascular systems of pine trees (Johnson *et al.* 1984). Topical rhizome sprays of GARLON 4, AMITROL T, TORDON 22K, and ROUNDUP in a water carrier at 5% concentration volume/volume (v/v), and TORDON 10K pellets at 1.8 lb (4 kg) ae/ha were also tested. A non-herbicidal treatment consisting of removing all top growth followed by removal of all regrowth at six-week intervals was also evaluated. The purpose of the regrowth removal treatment was to deplete the rhizome colony of its energy reserve by removing the emerging shoots before their photosynthetic capacity exceeded energy needed for growth (P. Motooka, pers. comm.).

All treatments were directed at rhizome clumps, as tests by Gardner (1984) indicated that foliar and basal sprays were either ineffective or effective only at rates injurious to surrounding native vegetation. All top growth was removed from each clump with a cane knife immediately prior to chemical treatment. For the Maujet treatments, four injector capsules, each filled with 0.17 oz (5 ml) of the appropriate herbicide, were inserted into each clump at evenly spaced intervals around the outer edge. Injector tubes were inserted into rhizomes by hand, using a pair of vice-grip pliers to hold the tube at the proper depth while the capsule was inserted onto the tube using a rubber mallet.

In the topical spray treatments, the herbicides were sprayed to wet from 16-oz (500-ml) plastic, trigger spray bottles onto the portion of the rhizomes visible above ground (after removal of the vegetation).

TORDON 10K pellets were used at only one site. Two techniques were tested, one in which the pellets were evenly distributed over the surface of the clump, and one in which pellets were dropped into a 1.2-in. (3-cm) diameter hole that had been augered into the center of the clump to soil depth.

All treatments were applied under partially cloudy but dry conditions on September 26, 1984. Monitoring of the 60 ginger plots was conducted for one year at monthly intervals, when rhizome vigor as well as presence or absence of resprouts was noted. Nontarget native species that occurred within a 3.3-ft (1-m) radius of each treated rhizome colony were observed for possible reactions to the treatments.

Results and Discussion

Among ginger plots treated with undiluted herbicides injected into the rhizomes, only the AMITROL T and TORDON 22K treatments showed significant control of ginger (Table 1). With the TORDON 22K treatments, no resprouts were observed on any of the five clumps during the study period. At one year, rhizomes were decayed and control was 100%. With the AMITROL T treatments, clumps resprouted with severely chlorotic shoots. The shoots gradually declined in vigor, and at one year both shoots and rhizomes were in advanced decay. All visible portions of the rhizomes appeared dead at one year, and control effectiveness was rated at 100%. Treatments with the other injected herbicides showed significantly less control of ginger. While GARLON 4, ROUNDUP, and TORDON RTU suppressed regrowth for up to eight months, clumps were producing apparently healthy regrowth at one year. Results of the far less-effective spray and non-chemical treatments are discussed in Santos *et al.* (1986).

The two methods of applying TORDON pellets (surface and auger) differed in effectiveness, with greater control achieved from pellets broadcast evenly over exposed rhizome surfaces. Decay of the rhizome began within two months and control was 100% at one year. When TORDON pellets were dropped into augered holes in rhizomes, only partial death of clumps was recorded in six months, and resprouting was noted after seven months.

Effects on Nontarget Native Species

Effects on nontarget species varied widely (Table 2), apparently as a result of herbicide used, amount of active ingredient applied per plot, and treatment technique. Proximity of the native plant to actual treatment location and different sensitivities to chemicals may also have affected results.

The injected TORDON 22K treatment was most damaging to native vegetation within a 3.3ft (1-m) radius of treated clumps. All pilo *(Coprosma ochracea)*, wāwae'iole *(Lycopodium cernuum)*, and 'ōhi'a were killed or showed signs of necrosis, stunting, or aberrant growth. The only casualty in the injected AMITROL T treatment was one pa'iniu *(Astelia*

Table 1. Response of kāhili ginger (*Hedychium gardnerianum*) to herbicide treatment and removal of regrowth at 6-week intervals thereafter in Hawaii Volcanoes National Park, 1984-1985. (Numbers in table represent plant response category means of individuals treated in each test.)

Treatment*	OCT** (1 mo)	JAN (4 mo)	MAY (8 mo)	SEP (12 mo)
CONTROL	0	0	0	0
GARLON 4				
100% Maujet	2.8	1.8	2.0	0.8
5% Rhizome	2.8	1.4	0.4	0
TORDON 22K				
100% Maujet	2.8	3.0	4.0	4.0
5% Rhizome	3.0	2.6	3.6	2.4
ROUNDUP				
100% Maujet	3.0	0	0	0
5% Rhizome	3.0	1.0	0.2	0
AMITROL T				
100% Maujet	3.0	2.8	3.8	4.0
5% Rhizome	3.0	2.2	2.6	2.8
TORDON RTU 100% Maujet	3.0	2.2	2.2	0.6
TORDON 10K PELLETS				
4 kg ae/ha Augered	2.2	3.0	2.6	2.0
4 kg ae/ha Surface	2.4	3.0	3.8	4.0
TOP GROWTH REMOVAL	0	0	0	0

*n = 5 clumps/treatment.

**Data for intervening months are available from authors; (months) are post treatment.

Plant response categories:

- 0 = no effect, healthy resprouting, growth normal.
- 1 = < 25% rhizome death, resprouts with light abnormalities (chlorosis).
- 2 = 50% rhizome death, moderate resprouting, abnormalities more severe (deformities).
- 3 = 75% rhizome death, light resprouting, low vigor, leaf distortions, stunting.
- 4 = 100% rhizome death, no resprouting.

Table 2. Effects of chemical treatments of kāhili ginger (*Hedychium gardnerianum*) on native vegetation within 1 m of treated clumps (see key).

Species	Control	Treatment										
		Garlon 4		Roundup		Amitrol T		Tordon 22K		Tordon 10K		Tordon RTU
		100%	5%	100%	5%	100%	5%	100%	5%	B'cast	Augured	Maujet
Astelia menziesiana (pa'iniu)		0/1			0/2	-/2						0/2
Cheirodendron trigynum ('ōlapa)									0/1	+/1	0/1	
Cibotium glaucum (hāpu'u pulu)				0/1			0/1	0/1	0/1	0/1	0/1	
Clermontia parviflora ('ōhā)										-/1		
Coprosma ochracea (pilo)	0/2	0/5	0/1	+/4	0/2	+/2*	0/1	-/1	0/1	0/7*	0/4*	+/1
Hedyotis centranthoides			0/1									
Ilex anomala (kāwa'u)									0/1	+/2*	0/4	
Isachne distichophylla ('ohe)						0/1				0/1	0/1	
Lycopodium cernuum (wāwae'iole)		-/1						-/1				0/1
Machaerina angustifolia ('uki)									+/1	0/1		
Metrosideros polymorpha ('ōhi'a)	0/2*	0/5	+/2	0/4	0/2	0/1		-/5	0/4	+/4	+/6	+/4
Myrsine lessertiana (kōlea)				0/1			+/2			0/1	-/1	

Table 2, continued.

Table 2, continued.

Species	Control	Garlon 4 100%	Garlon 4 5%	Roundup 100%	Roundup 5%	Amitrol T 100%	Amitrol T 5%	Tordon 22K 100%	Tordon 22K 5%	Tordon 10K B'cast	Tordon 10K Augured	Tordon RTU Maujet
Sadleria pallida (‘ama‘uma‘u)	+/1				0/1		0/1	0/1		+/1	-/4	+2
Vaccinium spp. (‘ōhelo)	0/2	0/2				0/1					0/1*	

Key:

vigor rating/no. individuals (* = seedlings):
+ = increase in vigor from pretreatment condition.
0 = no change from pretreatment condition.
- = decrease in vigor from pretreatment condition.

menziesiana) individual growing on top of a treated clump. This plant exhibited the typical AMITROL T-induced symptoms of severe chlorosis prior to death.

TORDON 10K pellets caused vigor decline and death within four months in one *Clermontia parviflora* shrub growing 24 in. (60 cm) downslope from a broadcast-treated ginger clump. Similarly, an individual kōlea *(Myrsine lessertiana)* sapling (28 in. (70 cm) tall), located 24 in. (60 cm) from an augered-pellet plot, declined from excellent to poor vigor at 12 months. One kōlea seedling 35 in. (90 cm) from a broadcast-pellet plot remained in good vigor during the study period. Except for the decline in vigor of one wāwae'iole in the 100% GARLON 4 treatment, no adverse effects were observed on nontarget native species associated with any of the remaining treatments for ginger.

Management and Research Recommendations

While injected TORDON 22K treatment effectively controlled kāhili ginger, it cannot be recommended because of the heavy impact it had on native plants. The injected AMITROL T treatment was also very effective, with only minor effects on nontarget plant species; however, the high cost of this treatment (equipment and labor) may preclude its use on a widescale basis. Broadcasted TORDON 10K pellets were an effective and relatively economical treatment and, in general, had only minor effects on the native flora in this test. However, picloram, the active ingredient in TORDON, is a persistent chemical that was shown (in this test) to be lethal to native species when present in sufficient concentrations. Tests in temperate climates have shown that picloram retains soil residual activity (herbicidal concentrations) for at least one year (4 kg ae/ha rate), with sub-herbicidal residues present for at least 2.5 years (1 and 4 lb kg/ha rate) (Anderson 1983). High rainfall, high organic matter in the soil, and microbial activity in the areas where kāhili ginger is found are likely to decrease the residual activity of picloram (P. Motooka, pers. comm.); however, the degree of reduced activity is unknown. High rainfall may also increase the potential threat to native species, as picloram is water soluble and readily transported to other areas in surface runoff and/or in soil solution. A small-scale test in 'Ōla'a Tract indicated that several common native tree and shrub species are very sensitive to TORDON 10K (Hawaii Volcanoes National Park Resources Management Division, unpub. data). Further research is required to address concerns about persistence and mobility.

We recommend that TORDON 10K pellets not be used where water runoff is a problem and where rare plant species are present, until the degree of threat of this chemical to native species has been better established. Tests with increased plot and sample sizes and lower ginger densities are necessary to determine treatment effectiveness and impacts on native species under these conditions. Monitoring should be long term, for target as well as nontarget species, as picloram may persist in the soil for years, and more slowly-growing woody components of native forest systems ('ōhi'a, pilo, etc.) may not show symptoms for a year or more. The effects of picloram on the native microfauna are at present unknown, and that situation should be addressed; however, the cost may be prohibitive.

AMITROL T also appeared promising as a control for kāhili ginger, although it was not completely effective when applied to rhizomes in a 5% solution. Since additional testing of this herbicide and others that are less threatening to native ecosystems than TORDON was warranted, a subsequent test was conducted using a 20% concentration of AMITROL applied as a topical rhizome spray. This treatment resulted in the death of 90% of a sample of 20 kāhili ginger. Two of three additional herbicides tested were also effective against ginger (Santos *et al.* 1986).

OLIVE

Cut-Stump Treatment

Materials and Methods. Three hundred fifty olive shrubs of different size classes were chosen for cut-stump treatment from a healthy population in a dry, very scattered ʻōhiʻa community at 2,740 ft (835 m) elevation with mean annual rainfall and temperature of 77 in. (1,955 mm) and 66 F (19 C).

The 14 treatments tested in this study included an untreated stump control; diesel oil only; TORDON RTU undiluted; TORDON 22K undiluted and as 20% and 5% dilutions in water; ROUNDUP undiluted and as 20% and 5% dilutions in water; and GARLON 4 undiluted and as 20%, 10%, 5% and 2% concentrations in diesel oil. Each treatment was applied to 25 stumps in May 1984, during dry weather punctuated by one brief light rain.

Each shrub was cut with a chain saw as close to the ground as possible (usually less than 6 in. or 15 cm). Treatments were immediately applied to all cut surfaces so that the negative pressure in the xylem could facilitate the deep penetration of herbicides into the stump (Hay 1956).

Stumps were monitored for resprouting, cambium death, and vigor of resprouts. Although herbicides were not directly applied to any native species, nontarget vegetation within a 3.3-ft (1-m) radius was monitored for possible reactions to the treatments.

Results and Discussion. TORDON RTU and undiluted GARLON 4 were the most effective cut-stump control treatments for olive (Table 3). Of the 25 stumps treated with GARLON 4, one had resprouted by the fourth month. This stump produced seven unhealthy shoots, which grew to an average height of 4 in. (10 cm), declined in vigor, and died within a year. A cambium vigor check at one year confirmed that no stumps were alive. Of the stumps treated with TORDON RTU, only one produced a single unhealthy shoot by the fifth month, and this soon died. No viable cambium remained on any of the TORDON RTU-treated stumps at six months. Control effectiveness did not vary among the five size classes of olive for these two treatments or other treatments.

The undiluted TORDON 22K, ROUNDUP, and the 20% GARLON 4 treatments showed good control effectiveness, with only 12% to 20% of the stumps resprouting after one year. In addition, there were few new shoots from those stumps that did resprout. When compared with the stumps in the

Table 3. Response of olive (*Olea europaea* subsp. *africana*) to cut stump control treatments (% plants resprouted) in Hawaii Volcanoes National Park, 1984-1985.

Treatment*	JUN** (1 mo)	SEP (4 mo)	JAN (8 mo)	MAY (12 mo)
CONTROL	36	92	100	100
DIESEL	24	100	100	100
TORDON RTU	0	0	0	0
TORDON 22K				
100%	0	8	16	12
20%	4	40	48	52
5%	0	76	96	96
ROUNDUP				
100%	0	12	8	12
20%	0	28	32	48
5%	0	80	100	96
GARLON 4				
100%	0	4	4	0
20%	0	4	20	20
10%	8	12	28	36
5%	8	20	48	52
2%	8	32	56	60

*n = 25 plants in each treatment.
**Data for intervening months are available from authors; (months) are post treatment.

control group, all of the other nine treatments were relatively ineffective; discussion of these results can be found in Santos *et al.* (1986).

Effects on Nontarget Native Species. Monthly observations of nontarget native vegetation within 3.3 ft (1 m) of treated stumps showed no adverse effects on ‘ōhi‘a, ‘a‘ali‘i *(Dodonaea viscosa)*, and māmane *(Sophora chrysophylla)*. This does not imply that the herbicides tested have no effect on these species; few mamane were exposed, for example. The herbicides applied on the olive stumps apparently remained very localized on the target species and did not volatilize into the atmosphere or leach into the surrounding soil in sufficient concentrations to harm native species in the area.

Foliar Application

Materials and Methods. Sixty healthy olive shrubs ranging in height from 34 to 65 in. (85-165 cm) were selected for foliar treatment. The foliar treatments tested were: 5% aqueous solutions of ROUNDUP, GARLON 4, and TORDON 22K. Each treatment was applied to 15 shrubs.

Plants were sprayed to wet, using hand-pump pressurized sprayers with even, flat-fan nozzles. Pressure was regulated at 20 psi. All treatments were applied on June 18, 1984, under sunny skies with winds estimated at 13 mph (8 km/hr). No rain fell for at least six hours after treatment. Individuals were monitored monthly for a period of one year, although plant responses had peaked by eight months.

No native vegetation was directly sprayed except that growing among the test plants. Monitoring included observations of these native plants as well as natives within a 6.6-ft (2-m) radius that could have been exposed to incidental spraying or drift.

Results and Discussion. The most effective herbicide tested as a foliar spray on olive was the 5% concentration of GARLON 4 in water (Table 4). One month after treatment, 87% of the treated individuals exhibited severely desiccated foliage, which remained attached to branches for several months. Cambium vigor evaluations at eight months and one year indicated complete death of all shrubs.

Effects on Nontarget Native Species. Incidental herbicide drift affected some nontarget plants growing within 3.3 ft (1 m) of target olive shrubs. Most ‘a‘ali‘i receiving herbicide drift of 5% GARLON 4 reacted with rapid leaf desiccation or total defoliation but subsequently recovered. Two ‘a‘ali‘i died; these had received a spray dose equal to that for the target species. ‘A‘ali‘i branches heavily hit with GARLON 4 defoliated and declined to apparent death, while untreated branches on the same shrub remained healthy. This suggests minimal translocation. Pūkiawe *(Styphelia tameiaemeiae)* exhibited localized leaf desiccation with subsequent recovery when exposed to all herbicides tested. ‘Ūlei *(Osteomeles anthyllidifolia)* received only very light, incidental drift from all three herbicides and seemed unaffected.

Treatment of Seedlings

Materials and Methods. Fifty-four 0.25 m^2 plots were chosen under mature olive trees in Hawaii Volcanoes National Park. Prior to treatment, olive seedlings within each plot were counted and height range was recorded for each plot. Six plots were selected for each treatment. Nontarget native plant species growing within the plots were identified and monitored for possible reactions to the treatments. Pretreatment vigor was recorded for both target and nontarget species.

Treatments tested included 1% and 5% concentrations of: ROUNDUP, TORDON 22K, GARLON 4, and AMITROL T. Herbicides were applied with 16-oz (500-ml) trigger spray bottles set for a coarse spray. Treatments were applied at a rate of 63 gal/a of spray solution (600 l/ha) on July 20, 1984, under partially cloudy skies with a 13 mph (8 km/hr) breeze. No rain

Table 4. Plant response to foliar spray tests on olive (*Olea europaea* subsp. *africana*) in Hawaii Volcanoes National Park, 1984-1985. (Numbers in table represent response category means for each test.)

Treatment*	JUL** (1 mo)	OCT (4 mo)	FEB (8 mo)	JUN (12 mo)
CONTROL	1.0	1.0	1.0	1.0
ROUNDUP 5%	4.3	4.4	4.9	4.7
GARLON 4, 5%	4.7	5.7	6.0	6.0
TORDON 22K, 5%	2.1	1.7	1.7	1.6

*n = 15 plants/treatment.

**Data for intervening months are available from authors; (months) are post treatment.

Olive plant response categories:
1 = unaffected, healthy, normal.
2 = < 50% leaves chlorotic.
3 = > 50% leaves chlorotic.
4 = < 50% leaves desiccated/defoliated, cambium viable.
5 = > 50% leaves desiccated/defoliated, cambium viable.
6 = 100% foliar desiccation, cambium dead.

fell for at least six hours after spraying. Observations of all vegetation for changes in vigor and any other reactions were conducted at monthly intervals for six months.

Results and Discussion. Responses of olive seedlings to the treatments peaked between the fourth and sixth month (Table 5). Resprouting had occurred by the sixth month.

Five percent GARLON 4 produced 100% control of olive seedlings by the fourth month. Treated seedlings were defoliated and desiccated, with no viable tissue. No resprouting was observed during the test.

A range of 50 to 90% seedling mortality was achieved with the 5% ROUNDUP, 5% AMITROL T, and 1% GARLON 4 treatments. Within this group the 1% GARLON 4 treatment produced the most consistent results, with 75-90% control in all six plots by the fourth month.

Heavy rains that fell in November 1984 apparently initiated the germination of olive seeds in the test area. Concentration of seeds in the soil was assumed to be the same both inside and immediately outside test

Table 5. Results of herbicide spray trials on olive (*Olea europaea* subsp. *africana*) seedlings, Hawaii Volcanoes National Park, 1984-85. (Numbers in table represent response category means for each test.)

Treatment	AUG* (1 mo)	SEP (2 mo)	NOV (4 mo)	JAN (6 mo)
CONTROL (n = 689)	0.5	0.0	0.0	0.2
GARLON 4				
5% (n = 671)	2.2	3.5	4.0	4.0
1% (n = 671)	1.7	2.3	2.5	3.0
TORDON 22K				
5% (n = 691)	1.2	1.2	1.3	1.0
1% (n = 685)	0.7	0.2	0.0	0.3
ROUNDUP				
5% (n = 683)	1.3	1.5	1.2	2.0
1% (n = 691)	0.7	0.7	0.8	0.8
AMITROL T				
5% (n = 671)	1.0	1.2	1.8	2.3
1% (n = 668)	0.8	0.8	0.8	1.0

*Data for intervening months are available from authors; (months) are post treatment.

Olive seedling response categories:
- 0 = normal, healthy.
- 1 = chlorosis, < 25% death.
- 2 = foliar desiccation, 25-50% death.
- 3 = 50-75% death, defoliation.
- 4 = 100% cambium death.

plots. General observations indicated no residual activity with any of the treatments by this time, as seedlings appeared within all plots with the same apparent frequency as outside.

Effects on Nontarget Native Species. All results are based on chance occurrences of native species within plots and do not represent a designed test for effects on natives (Table 6). Species evaluated were directly sprayed during treatments, as they were growing within test plots. The only native plants treated with 5% GARLON 4, one ʻūlei and one ʻaʻaliʻi seedling, died. Another ʻaʻaliʻi seedling treated with 5% TORDON 22K also died. No other adverse effects were observed in this test, although very low numbers of these plants were exposed to treatments.

Table 6. Effects of herbicide treatments for control of olive (*Olea europaea* subsp. *africana*) seedlings on nontarget native species within 0.25 m^2 treated plots.*

Species	Life Stage**	Treatment: Roundup 5%	Treatment: Roundup 1%	Treatment: Tordon 22K 5%	Treatment: Tordon 22K 1%	Treatment: Garlon 4 5%	Treatment: Garlon 4 1%	Treatment: Amitrol T 5%	Treatment: Amitrol T 1%	Total No. Plants Treated
Carex wahuensis	M		N/1***	N/1					N/2	4
Coprosma ochracea (pilo)	S								N/1	1
Dodonaea viscosa (‘a‘ali‘i)	S		N/1	H/1		H/1		N/1		4
Osteomeles anthyllidifolia (‘ūlei)	S					H/1		N/1		2
Sophora chrysophylla (māmane)	S									1
Styphelia tameiameiae (pūkiawe)	S			N/1	N/1					2

*Plant response categories:
H = Heavy
M = Moderate
L = Light
N = No effect
Blank = Not exposed to treatment.

**Life stage: S = Seedling; M = Mature, capable of sexual reproduction.

***Number of plants exposed in each test.

Management and Research Recommendations

TORDON RTU or undiluted GARLON 4 are effective treatments when applied to cut olive stumps. TORDON RTU is less expensive on a per use basis and presents a lower hazard to the applicator. The oral LD_{50} (the dose that is lethal to 50% of rats tested) of triclopyr (the active ingredient in GARLON 4) is 630 mg of chemical/kg of body weight, and that of picloram is 8,200 mg/kg (the lower the number the higher the toxicity) (Berg 1985). Both herbicides are easy to apply, and residual activity and nontarget hazards do not appear to be a problem. Therefore, TORDON RTU is the preferred treatment on cut stumps of olive. No further research is necessary to improve control techniques for mature olive shrubs. However, it would be desirable to monitor nontarget plants in greater numbers if large-scale management begins.

Although 5% GARLON 4 in water as a foliar spray was found to be an effective control for olive, the potential hazards of this treatment may preclude its use for widescale control efforts. Spraying shrubs greater than 3.3 ft (1 m) in height presents a high probability of exposure by drift to nontarget species as well as to the applicator. Therefore, this treatment is recommended only in situations where olive shrubs are less than 1 m tall and where native plants are not in close proximity. Further research is needed to find the lowest effective concentration of GARLON 4 to minimize cost and possible hazards to the native vegetation and the applicator. In the process of determining efficacy, more data on nontarget responses of native species can be obtained.

GARLON 4 appears to be the most effective herbicide for olive seedlings, with the lowest effective rate somewhere between the two tested rates (1% and 5%). Further testing should be conducted to find the lowest effective rate. More information on hazards to nontarget native plants could be obtained in the process.

The three control methods developed for olive could be combined as part of an overall olive removal strategy. One suggestion would involve a crew of three people using the cut-stump method to eliminate all olive plants taller than 3.3 ft (1 m), followed by a crew of two concentrating on seedlings and saplings, with either the foliar treatments or manual removal, depending on the proximity of native species. Follow-up treatments on seedlings could be scheduled one to two months after a rainy period, to allow time for seed germination, and would continue periodically until the soil seed bank is exhausted. Studies on seed viability in the soil would be needed to determine the length of time necessary to exhaust the seed bank. Studies of the fate of untreated seedlings could shed some light on the importance of seedling treatment as a part of control strategy.

SILK OAK

The purpose of this test was to determine the sensitivity of two species of silk oak (*Grevillea robusta* and *G. banksii*) to a variety of herbicides when applied via basal, continuous frill cuts.

Materials and Methods

One hundred twenty trees of each species were chosen from healthy populations in a dry, scattered ʻōhiʻa community with native shrubs and alien grasses that had been partially converted to pasture. For each of eight treatments, 15 trees of each species were selected; each treatment included members of four diameter classes to parallel size distribution determined by a belt transect in the study area.

Identical treatments were used on both species: GARLON 4 diluted in diesel oil at 2.5%, 5%, and 10% concentrations; TORDON 22K diluted in water at 5%, 10%, and 20% concentrations; and control treatments of diesel oil only and water only. Overlapping rather than spaced frill cuts were used because tests by Leonard (1957) showed them to be more effective. The frill cuts were made with a hatchet around the entire circumference of each trunk less than 6 in. (15 cm) from the ground and deep enough to sever the cambium. A single, solid stream of the appropriate herbicide was squirted from a plastic squeeze bottle into the trough created by the frill cuts. The application dates were August 20, 1984, for *G. banksii* and August 29, 1984, for *G. robusta*. Weather on both days was overcast but dry. Treatments were monitored at monthly intervals for one year.

Results and Discussion

G. banksii. All of the GARLON 4 treatments (2.5%, 5%, and 10% in diesel oil) and two of the TORDON 22K treatments (5% and 20% in water) resulted in 100% mortality of treated individuals within 11 months (Table 7). Slightly less effective was 10% TORDON 22K in water, which killed 87% (13) of the treated trees by the seventh month, with the remaining two trees exhibiting severe leaf desiccation, moderate cambium vigor, and basal resprouting by the ninth month. A characteristic progression of symptoms was apparent in all TORDON 22K and GARLON 4 treatments. Trees demonstrated leaf chlorosis over 50-75% of the canopy within one month, followed by increased leaf desiccation and simultaneous decline of cambium vigor both above and below the frill cut. The highest concentration of both GARLON 4 and TORDON 22K resulted in an accelerated plant reaction and more rapid death of treated individuals. The interval between treatment and death was 5-6 months for the highest concentrations, compared with 11 months for the lower concentrations. It is notable that lower concentrations of GARLON 4 were sufficient to eventually achieve a success rate comparable to higher concentrations of TORDON 22K (*e.g.*, GARLON 4 at 2.5% was as effective as TORDON 22K at 5%). Younger trees (1-3 in. or 3-8 cm basal diameters) were generally more susceptible to all of the herbicides and declined more rapidly than the larger trees.

G. robusta. The three concentrations of GARLON 4 (2.5%, 5% and 10% in diesel oil) produced 100% control of *G. robusta* within eight months. In all three treatments, foliage rapidly desiccated but did not abscise for several months. Decline of cambium vigor both above and below the frill cut was gradual. TORDON 22K, at 20% in water, was slightly less effective, with 93% control of treated individuals at one year.

Table 7. Mean response ratings for *Grevillea banksii* and *G. robusta* subjected to herbicide treatments. (Numbers in table represent response category means for each test.)*

Species and Treatment	OCT** (1 mo)	JAN (4 mo)	MAY (8 mo)	SEP (12 mo)
Grevillea banksii				
DIESEL OIL ONLY	1.0	0.7	1.9	1.6
GARLON 4				
10%	3.3	4.8	5.0	5.0
5%	2.5	4.5	5.0	5.0
2.5%	2.5	4.2	4.7	5.0
TORDON 22K				
20%	3.8	4.7	5.0	5.0
10%	3.4	4.3	4.7	4.7
5%	3.5	4.3	4.8	5.0
WATER	0.6	0.3	0.1	0.2
Grevillea robusta				
DIESEL OIL ONLY	0	0	0.3	0.5
GARLON 4				
10%	2.9	4.9	5.0	5.0
5%	2.4	4.9	5.0	5.0
2.5%	2.4	4.0	5.0	5.0
TORDON 22K				
20%	2.9	4.5	4.9	4.9
10%	2.5	4.1	4.4	4.4
5%	2.0	2.9	3.5	3.6
WATER	0.5	0	0.1	0

*(n = 15 trees/test).
**Data for intervening months are available from authors; (months) are post treatment.

Plant response categories:
0 = no effect, normal healthy growth.
1 = <25% defoliation, cambium alive.
2 = 50% defoliation, cambium alive.
3 = 75% defoliation, cambium alive.
4 = 100% defoliation, cambium alive.
5 = 100% defoliation, cambium dead.

Effects on Nontarget Native Species

Native species within a 3.3-ft (1-m) radius of treated trees included ʻōhiʻa, ʻākia (*Wikstroemia* sp.), and ʻaʻaliʻi. No adverse effects were noted for any of these species.

Management and Research Recommendations

Results indicated that 2.5% GARLON 4 in diesel oil was the lowest concentration of herbicide effective in controlling both species of *Grevillea*. The basal frill application of 2.5% GARLON 4 in diesel oil is therefore recommended. No future research on the control of silk oak is anticipated, and it is recommended that implementation of silk oak control begin as soon as possible.

FLORIDA BLACKBERRY

Materials and Methods

Seventy 32-ft^2 (3-m^2) plots were selected for treatment from a *Rubus argutus* population in a closed canopy, mesic ʻōhiʻa/koa/mānele *(Metrosideros/Acacia/Sapindus)* forest, with mixed native shrubs and alien grasses in the understory. The study site was located at 4,200 ft (1,280 m) elevation, and mean annual rainfall and temperature were 75 in. (1,890 mm) and 59 F (15 C). Untreated buffer zones at least 10 ft (3 m) wide were included around all plots to prevent any overlap of treatments.

Two foliar spray techniques were tested: conventional sprays and drizzle sprays. The drizzle or "magic wand" method developed by Uyeda and described by Motooka *et al.* (1983) is a high-concentration, low-volume, non-atomizing technique using pressure-regulated sprayers fitted with an orifice disk instead of a conventional nozzle tip. The orifice disk produces a single, thin stream of liquid rather than a spray. The conventional sprays included 0.5% and 2% concentrations v/v of GARLON 4, TORDON 22K, ROUNDUP, and AMITROL T. These were applied with water as the carrier at a volume of 20 gal/a (190 l/ha). Pressure was regulated at 30 psi with a spray volume of 1 gal/a (9.5 l/ha) for the drizzle sprays. A No. 20 orifice disk was used, with water as the carrier. A spreader-sticker, TRITON B1956 (Ortho), was added to the conventional sprays at 0.5% v/v and to the drizzle sprays at 1% v/v.

Nontarget species located within the test plots were identified on a presence/absence basis, and vigor was evaluated prior to and one year after treatment. All herbicide treatments were applied sequentially as suitable plots were encountered in the field on October 29, 1984, under partially cloudy skies. A very light rain fell briefly during application, but no rain fell for at least four hours after application.

Results and Discussion

Although no treatment was shown to be totally effective in the control of *Rubus argutus*, the most successful was a conventional spray of 2% ROUNDUP in water (Tables 8 and 9). Plants on all five of the blackberry plots subjected to this treatment reacted with a gradual decline in vigor until the sixth month, when response stabilized. Plants on two of the

Table 8. Response of blackberry (*Rubus argutus*) to foliar herbicide treatments, Hawaii Volcanoes National Park, 1984-85. (Numbers in table represent response category means of individuals treated in each test.)

Treatment*	NOV** (1 mo)	APR (6 mo)	OCT (12 mo)	PRETREATMENT
CONTROL	0.7	0.8	1.0	1.4
GARLON 4				
40% drizzle	4.0	4.4	1.2	1.8
2% conventional	4.0	3.8	1.0	1.4
0.5% conventional	4.0	3.2	1.0	1.2
TORDON 22K				
20% drizzle	3.2	2.6	1.0	1.2
2% conventional	4.0	3.2	1.0	1.0
0.5% conventional	3.4	2.0	1.0	1.2
ROUNDUP				
20% drizzle	2.6	3.8	2.2	1.4
2% conventional	2.4	4.6	4.4	1.6
0.5% conventional	2.6	3.4	1.4	1.6
AMITROL T				
40% drizzle	2.4	1.2	1.0	1.4
2% conventional	2.0	0.8	1.0	1.2
0.5% conventional	2.4	1.2	1.2	1.4

*n = 5 plants/treatment.

**Data for intervening months are available from authors; (months) are post treatment.

Plant response categories:

0 = >90% stems and foliage alive, excellent vigor.
1 = 75-90% stems and foliage alive, excellent vigor.
2 = 50-75% stems and foliage alive, moderate vigor.
3 = 10-50% stems and foliage alive, moderate vigor.
4 = < 10% stems and foliage alive, poor vigor.
5 = 100% death, no resprouts.

Table 9. Cover change of blackberry (*Rubus argutus*) within 3 m^2 plots one year after treatment.*

Plot	AMITROL T			GARLON 4			ROUNDUP			TORDON 22K			UNTREATED CONTROL	
	40%	2.0%	0.5%	40%	2.0%	0.5%	20%	2.0%	0.5%	20%	2.0%	0.5%		
1														
Pretreatment	80	90	80	100	80	90	80	70	80	90	90	90	70	80
Post treatment*	30	70	20	90	30	90	5	10	5	90	50	90	90	90
2														
Pretreatment	90	90	60	90	100	60	90	90	90	100	100	70	80	80
Post treatment	90	90	90	25	70	100	80	0	90	80	80	90	80	80
3														
Pretreatment	80	90	70	80	50	80	90	50	80	90	80	90	80	100
Post treatment	75	80	90	90	10	70	10	0	20	90	70	100	80	100
4														
Pretreatment	90	70	80	80	80	80	100	70	80	80	90	95	90	80
Post treatment	100	70	80	10	80	90	60	<1	80	50	80	100	80	60
5														
Pretreatment	90	100	60	90	90	70	80	100	60	80	60	90	100	80
Post treatment	90	100	30	75	70	50	75	40	5	50	10	90	100	100

*Visual estimates one year post treatment. n = 5 plots/treatment, except 10 plots for control.

five plots exhibited 100% mortality with no resprouting after one year. In a third plot, all sprayed *Rubus* plants were dead, but healthy root sprouts were observed. Plants on the remaining two plots exhibited greater than 90% foliar and stem mortality and a great reduction in blackberry cover. Several stems that survived the herbicide treatment remained dormant for up to 10 months before producing stunted and severely deformed resprouts in tight clusters at the nodes. While these initial resprouts remained stunted, resprouts that appeared during the eleventh month appeared healthy.

Effects on Nontarget Native Species

Plant species monitored for possible reactions to the herbicide treatments included the species listed in Table 10. All nontarget species did not occur within each herbicide treatment plot; therefore, the information presented reflects plant responses to specific herbicides in incidental spray situations. Data are presented to assist in gauging incidental impacts and do not reflect broad-scale, systematic testing of nontarget species. Evaluations were made one year after treatment.

The results in Table 10 indicate that ROUNDUP is more detrimental to native species when applied as low-concentration, high-volume sprays, than with high-concentration, drizzle-spray application. One likely explanation is that high-concentration ROUNDUP physically burns plant tissue to the extent that systemic translocation is not possible. Another possibility is that high-concentration coverage was poor, a drawback to low-volume spray applications in general. It is possible that the poor coverage inherent in drizzle sprays resulted in untreated areas within treated plots. High-volume sprays resulted in uniform coverage, increasing the probability that all vegetation within a plot was treated.

Management and Research Recommendations

Although no treatment was found that provided total control of blackberry, the conventional foliar spray of 2% ROUNDUP in water would be useful in certain applications. This treatment, because of its effectiveness in greatly reducing the shade cover of blackberry for an extended period of time, could be used to assist in the regeneration of native tree species in areas where blackberry may be preventing this. ROUNDUP should not deter the germination of native species, as it is rapidly inactivated in soils (Sprankel *et al.* 1975). The treatment is within labeled application rates when used as prescribed. Post-treatment monitoring would be necessary, as resprouting of blackberry is likely. Large stands may need to be treated in stages, as it may not be possible to reach the central portions of some thickets with the spray. Applicators must be aware of changing wind patterns when spraying blackberry stands taller than 3.3 ft (1 m), to avoid spray drift, which can be hazardous to both the applicator and to surrounding nontarget species. Another critical factor involved in the treatment of blackberry deals with informing the general public where and when this activity is being conducted, to protect berry pickers from any contamination. Blackberries are commonly used as food in Hawai‘i.

Table 10. Responses of nontarget native and introduced plants to herbicide treatments to control blackberry (*Rubus argutus*).

Species	Treatment*											
	GARLON 4			ROUNDUP			AMITROL T			TORDON 22K		
	40%	2%	0.5%	20%	2%	0.5%	40%	2%	0.5%	20%	2%	0.5%
<u>Endemic and Indigenous Species</u>												
Acacia koa (Koa)		+	+	+	+		+	+		+	+	+
Asplenium sp.								+				
Coprosma ochracea (Pilo)					+		+			+		+
Ipomoea congesta (Morning glory)	L-H**	+	+	L**	L-H**	L-H**	H**	+	+	+	+	+
Lythrum maritimum (Pūkāmole)	+	+	+	+	+	H	+	+	+	+	+	+
Microlepia strigosa (Palapalai)					H							
Myoporum sandwicense (Naio)											+	
Pipturus albidus (Māmaki)	+			+								+
Pteridium aquilinum var. *decompositum* (Kīlau)	+	+	+	+	+	+	+	+	+	+	+	+
Sadleria cyatheoides (ʻAmaʻumaʻu)				+								
Sapindus saponaria (Mānele)				H			H**			H		
Sophora chrysophylla (Māmane)								M				
<u>Introduced Species</u>												
Commelina diffusa (Honohono)					+			+			H	
Crocosmia crocosmiiflora (Montbretia)	+											

Table 10, continued.

Table 10, continued.

Species	GARLON 4 40%	GARLON 4 2%	GARLON 4 0.5%	ROUNDUP 20%	ROUNDUP 2%	ROUNDUP 0.5%	AMITROL T 40%	AMITROL T 2%	AMITROL T 0.5%	TORDON 22K 20%	TORDON 22K 2%	TORDON 22K 0.5%
Cynodon dactylon (Bermuda grass)		+			H		+	+	+		+	+
Ehrharta stipoides (Meadow ricegrass)		+	+	+		H					+	+
Fragaria vesca (Wild strawberry)	+			+	+		+			+		
Holcus lanatus (Velvet grass)		+		+	H	H		+				
Paspalum dilatatum (Dallis grass)	+	+	+	+	H	+	+	+	+	+	+	H**
Pennisetum clandestinum (Kikuyu grass)	+		+			+	+		+	+		+
Setaria gracilis (Foxtail grass)			+						+			
Solanum pseudocapsicum (Jerusalem cherry)	L-H		L-H	+	+	L-H	+			+		L-H
Stenotaphrum secundatum (St. Augustine grass)								L				
Verbena litoralis (Weed verbena)		H										
Veronica plebeia (Common speedwell)		+			+			+	+		+	

(Column group heading: Treatment*)

Key:

+ = exposed to treatment, no adverse effects
blank = not present in plot, therefore not exposed to treatment
H = heavily impacted (death)
M = moderately impacted (defoliation/desiccation, severely chlorotic plants partially recovered)
L = light effects (light chlorosis, some defoliation, some plants unaffected, full recovery).

*Drizzle for highest concentrations, spray for middle and lowest concentrations

**Comments:

40% Garlon 4	Four *Ipomoea* vines were treated. Two of these of moderate to poor pretreatment vigor died, while the remaining two, in excellent pretreatment vigor, were unaffected.
20% Roundup	Three *Ipomoea* vines were treated. One declined from good to very poor vigor, while the remaining two, in excellent vigor, were unaffected.
2% Roundup	Three *Ipomoea* vines were exposed. One in good vigor died, one declined from excellent to moderate vigor, and one increased from good to excellent vigor.
0.5% Roundup	Three *Ipomoea* vines in good vigor were treated. One died, one declined to poor vigor, and one increased to excellent vigor.
40% Amitrol T	Three *Ipomoea* vines were treated. One, in poor vigor, died; one, in good vigor died, while the last vine, in excellent vigor, was unaffected. *Sapindus* seedlings in one plot were not found after treatment and are presumed to have died.
2% Amitrol T	One *Sophora chrysophylla* individual reacted to treatment with partial defoliation; however, subsequent resprouts were healthy and vigorous.
20% Tordon 22K	A cluster of *Sapindus* seedlings was not found and is presumed to have died.
0.5% Tordon 22K	One *Paspalum* clump in excellent vigor died, while the others were unaffected.

Further research is definitely needed to improve the methods already tested and to explore new treatments (perhaps combinations of herbicides and/or repeated treatment annually). Larger plot and sample sizes (including treating entire stands) would yield more accurate information about treatment effectiveness. Repeat treatment testing is necessary to determine the number of treatments required to complete control. In a subsequent test of five herbicides applied as a foliar spray, conducted to increase sample size and test additional herbicides, ROUNDUP (2% v/v in water) was more effective at killing and defoliating small individual blackberry plants than GARLON 4, ARSENAL, ESCORT, OR CROSSBOW. As in the test described here, ROUNDUP completely killed only 40% of treated blackberry plants (G.L. Santos, unpub. data). Monitoring of areas where blackberry has been eliminated could yield valuable information about what is likely to colonize these areas. More information is needed about the effects of 2% ROUNDUP on native species, as this treatment is not target specific.

YELLOW HIMALAYAN RASPBERRY

Materials and Methods

Three foliar sprays, four soil applications, and 10 cut-stump treatments were tested on yellow Himalayan raspberry. Each treatment was applied to five plots. For the three foliar and four soil treatments, 35, 32-ft^2 (3-m^2) plots were selected from a *Rubus ellipticus* population in open canopy tall ʻōhiʻa rain forest with hāpuʻu (*Cibotium* spp.) and native shrubs in the understory. Elevation was 3,900 ft (1,190 m), with mean annual rainfall of 98 in. (2,494 mm). Mean monthly maximum temperatures ranged between 66 and 72 F (19-22 C).

Foliar sprays were applied via the drizzle or "magic wand" method in a water carrier at a spray volume of 2.25 gal/a (21 l/ha). The three treatments tested were: 50% concentrations of GARLON 4 and of ROUNDUP, and a 25% concentration of TORDON 22K. Hand-pressurized sprayers fitted with No. 20 orifice disks were used. Pressure was regulated at 30 psi. Triton 1956B (Ortho) was added as a spreader-sticker at 1% v/v.

Four soil applications of TORDON 10K pellets were tested: topical broadcast treatments at 0.08 and 0.15 oz/ft^2 (0.2-0.4 gm ae/m^2), and spot treatments at 0.08 and 0.15 oz/ft^2 (0.2-0.4 gm ae/m^2). The spot treatments involved depositing all pellets in one spot in the center of the 32-ft^2 (3-m^2) plot.

The nine treatments tested on cut stumps included: TORDON RTU; undiluted ROUNDUP, GARLON 4, and TORDON 22K; 5% and 20% dilutions of GARLON 4 in diesel oil; TORDON 22K in water; diesel oil only; and an untreated control. Trunks were severed at a height of <4 in. (<10 cm) either with a hand pruning saw or with pruning shears, depending upon stem size. Treatments were applied immediately to the freshly cut surface with 16-oz (500-ml) plastic squeeze bottles.

All treatments were applied under partially cloudy skies but dry weather with winds estimated at 5 mph (3 km/hr). Treatment application dates were May 5, 1985 for the foliar and soil treatments and July 22, 1985 for the cut-stump tests. No rain fell for at least four hours after treatment. Monitoring was conducted at monthly intervals for one year.

Results and Discussion

Foliar and Soil Treatments. Of the three foliar and four soil treatments tested, only the 50% GARLON 4 drizzle application proved to be effective in controlling yellow Himalayan raspberry (Table 11). After one month, four of the five plots exhibited leaf and stem desiccation on greater than 50% of the treated plants. The remaining plot showed only light chlorosis on a few scattered leaves and recovered by the fifth month. This plot, through an error in application technique, probably received less than the prescribed dose of GARLON 4. Overall reactions to the GARLON 4 treatment peaked by the fourth month after treatment, with greater than 90% defoliation and stem desiccation and no resprouting observed in three of four plots. Plant death in the fourth plot was 75-90%, with one healthy root resprout. No additional resprouting was observed in any of the GARLON 4-treated plots at one year.

Cut Stump Treatments. The most effective of the cut-stump treatments on yellow Himalayan raspberry was the 20% concentration of TORDON 22K in water (Table 12). This resulted in an 80% mortality rate (four of five stumps died with no shoot or root resprouting) at one year. Cambium vigor evaluations at seven months indicated complete mortality of all five stumps. In addition, no resprouting was observed on any of the decaying stumps during the next five months. By the twelfth month, however, resprouting was observed 6 in. (15 cm) from one stump from roots, which had apparently survived the treatment. Leaves that appeared on the two resprouts were deformed, with leaf margins curling upward towards the midrib. In addition, the upper leaf surface was bronze in color, and the shoots appeared stunted. It is not known at this time if these deformities are permanent or if the stump will subsequently produce healthy growth. Data obtained from herbicide tests on another *Rubus* species (*R. argutus* in this report) suggest that the longer the resprouts survive, the greater the likelihood of a resumption of normal growth.

Effects on Nontarget Native Species

Cut Stump Treatments. No adverse effects on any native plant species were observed within a 3.3-ft (1-m) radius of any of the test stumps. Non-target species included ʻōhiʻa, hāpuʻu pulu (*Cibotium glaucum*), ʻamaʻu (*Sadleria pallida* and *S. cyatheoides*), and ʻōlapa (*Cheirodendron trigynum*). Since it is unlikely that any of the species were exposed to the treatments, no conclusions can be drawn as to sensitivity to the herbicides.

Foliar Treatments. Two ʻōhiʻa trees and one hāpuʻu, each 7 to 10 ft (2-3 m) in height, were exposed to the 50% GARLON 4 spray. The hāpuʻu was growing within 3.3 ft (1 m) of one of the test plots, with a single frond hanging within the plot. This frond died within three months of treatment; however, the untreated fronds remained healthy, and new fronds that

Table 11. Response of yellow Himalayan raspberry (*Rubus ellipticus*) to foliar and soil herbicide treatments, Hawaii Volcanoes National Park, 1985-86. (Numbers in table represent mean values for response categories.)

Treatment*	JUN** (1 mo)	SEP (4 mo)	JAN (8 mo)	MAY (12 mo)
ROUNDUP 50%	1.2	1.6	0.6	1.8
GARLON 4, 50%	3.4	4.8	4.4	5.2
TORDON 22K, 25%	1.8	1.4	0.6	0
TORDON 10K PELLETS BROADCAST				
0.4 gm	1.0	1.4	1.0	0.2
0.2 gm	1.0	1.0	0	0
TORDON 10K PELLETS SPOTTED				
0.4 gm	0.2	1.2	1.2	0.4
0.2 gm	1.0	1.0	0.2	0

*n = 5 plants/test.

**Data for intervening months are available from authors; (months) are post treatment.

Plant response category:

0 = healthy, no effect.
1 = light effects, chlorosis < 50% of plant.
2 = chlorosis > 50% of plant.
3 = moderate effects, > 50% leaf desiccation/defoliation.
4 = 50-75% leaf desiccation/defoliation.
5 = 75-90% plant death (cambium, stem desiccation).
6 = 90-100% plant death.

Table 12. Response of yellow Himalayan raspberry (*Rubus ellipticus*) to cut-stump treatments, Hawaii Volcanoes National Park, 1985-86. (Numbers in table represent mean response categories.)

Treatment*	AUG** (1 mo)	NOV (4 mo)	MAR (8 mo)	JUL (10 mo)	% DEAD STUMPS***
UNTREATED	4.4	0	0	0	0
TORDON RTU	5.0	4.6	3.6	3.0	60
ROUNDUP 100%	5.0	3.4	2.8	3.0	40
GARLON 4					
100%	5.0	5.0	5.0	3.6	60
20%	5.0	4.0	3.6	3.4	60
5%	5.0	5.0	4.8	2.6	20
TORDON 22K					
100%	5.0	3.0	3.2	3.0	60
20%	5.0	4.6	5.0	4.4	80
5%	5.0	4.2	0.8	1.4	20
DIESEL OIL	5.0	3.2	1.8	1.4	20

*n = 5 plants/test.

**Data for intervening months are available from authors; (months) are post treatment.

***at one year post treatment.

Plant response category:
- 0 = healthy, normal, resprouts.
- 1 = healthy resprouts w/ light abnormalities (chlorosis).
- 2 = moderate resprouts w/ abnormalities.
- 3 = light resprouts, w/ abnormalities.
- 4 = no resprouts, cambium alive.
- 5 = no resprouts, cambium dead.

appeared after treatment were healthy and normal. The two 'ōhi'a trees were growing within two of the GARLON 4 plots and were directly treated. Although the surrounding raspberry was heavily affected, both 'ōhi'a trees remained healthy and normal and were apparently unaffected by the treatment.

Management and Research Recommendations

Both the cut stump treatment of 20% TORDON 22K in water and the 50% GARLON 4 drizzle spray treatment were moderately effective in controlling *R. ellipticus*. The cut stump treatment is more target-specific than the drizzle spray, but it may involve clearing away considerable amounts of bramble in order to expose the stump for treatment. The difficulty of accessibility to the stem and the presence of native species in close association with yellow Himalayan raspberry are the two major factors in determining treatment feasibility.

Because of the less than satisfactory effectiveness of herbicides used in the cut-stump test, further research was warranted. In a second test, TORDON 22K and five other herbicides were evaluated as cut-stump treatments on a larger sample of yellow Himalayan raspberry plants, and effects on native species were studied systematically. After two years of monitoring, several herbicides were found to be more effective than TORDON 22K for use against yellow Himalayan raspberry (Santos *et al.* 1991).

TIBOUCHINA

Materials and Methods

Two hundred twenty shoots of glorybush with basal diameters ranging from <0.5 to 5 in. (1-13.5 cm) were selected from a population in a wet ʻōhiʻa forest at 4,035 ft (1,230 m) elevation. Mean annual rainfall is 98 in. (2,494 mm), and mean monthly maximum temperatures ranged between 66 and 72 F (19-22 C), with the highest and lowest temperatures in August and February.

Lateral runners, which grow above and below ground on glorybush, made it necessary to trace each stem back to its primary root. Stems were then tagged, diameters measured, and plants were treated. Treatments consisted of severing the stems with a chain saw at a maximum height of 6 in. (15 cm) and applying the herbicide immediately to ensure maximum uptake and translocation throughout the root system. On plants with basal branching, all stems were cut, treated, and monitored, but only the largest stem was measured.

The 11 treatments tested included: 5% and 20% solutions of GARLON 4 in diesel oil; TORDON RTU undiluted; 5% and 20% solutions of ROUNDUP, TORDON 22K, and AMITROL T in water; diesel oil only; and an untreated control. Treatments were applied on the fresh cut so as to cover the entire surface, especially the cambium. The treatment surface was cut as level as possible to reduce runoff prior to absorption of chemical by the stump. The distance between treated areas was as great as practical (usually 10 ft or >3 m) to reduce the chances of cross contamination among connected glorybush roots. Treatments were applied to plants on January 10 and 11, 1985, under sunny and dry weather conditions. No rain fell for at least four hours after treatment on either date. Monitoring, conducted at monthly intervals for one year, consisted of visual observations of resprouting buds.

Results and Discussion

GARLON 4 in a 20% dilution with diesel oil was the most effective treatment tested, with 85% control of glorybush after one year (Table 13). A cambium vigor evaluation of the non-resprouting stumps at one year indicated complete death of all stumps. Resprouting on the surviving stumps was suppressed until the seventh month, when resprouts were etiolated with chlorotic leaves. The deformities were short-lived, and by one year post treatment the resprouts appeared normal.

Table 13. Percentages of glorybush (*Tibouchina urvilleana*) that resprouted after cut-stump treatment at Hawaii Volcanoes National Park, 1985-86.

Treatment*	FEB** (1 mo)	MAY (4 mo)	SEP (8 Mo)	JAN (12 mo)
DIESEL OIL	10	30	40	60
GARLON 4				
20%	0	0	5	15
5%	0	0	25	45
ROUNDUP				
20%	0	45	95	95
5%	5	90	100	95
TORDON 22K				
20%	0	5	10	25
5%	0	0	35	55
TORDON RTU	0	0	5	25
AMITROL T				
20%	10	80	95	95
5%	15	85	85	95
UNTREATED	15	95	95	95

*n = 20 stumps/treatment.

**Data for intervening months are available from authors; (months) are post treatment.

Effects on Nontarget Native Species

None of the native species present within a 3.3-ft (1-m) radius of treated stumps exhibited any adverse reactions to the herbicides. However, during treatment application no nontarget species were directly treated either accidentally or incidentally. Species monitored were: pūkiawe, ʻōhelo or kau-laʻai *(Vaccinium reticulatum)*, ʻōhiʻa, wāwaeʻiole, uluhe *(Dicranopteris emarginata)*, and ʻuki *(Machaerina angustifolia)*.

Management and Research Recommendations

Cut stump treatment is a practical, cost-effective, highly specific treatment method that is well suited for both the growth habit of glorybush and for the forest type in which it is currently found. GARLON 4 at 20% in diesel oil produced a high mortality rate with no adverse effects on native species. Retreatment is strongly recommended to achieve eventual eradication, and follow-up monitoring is necessary. Because glorybush readily reproduces vegetatively, any treatment program will need to include complete removal or spraying of cut branches and slash. Further testing of higher concentrations of both GARLON 4 and TORDON 22K might result in a more effective treatment for glorybush, and therefore require less follow-up monitoring.

CONCLUSIONS

The current restrictions on operational use of TORDON in Hawaiʻi, and the possibility of withdrawal from the market of all TORDON products by Dow Chemical Company in the near future, make it necessary to reevaluate the status of knowledge about control of alien plants considered in this study. For example, without TORDON, effective treatments for kāhili ginger and yellow Himalayan raspberry are compromised wholly or in part. Continual testing of a wide variety of chemicals seems a necessity to ensure availability of safe and effective chemicals. Chemicals that are less persistent than TORDON are known to be available. Determination of retreatment intervals and effective concentrations (and even sequences) of newly available, less persistent chemicals to be used in retreatment also seems desirable. More information on hazards to native species is needed for some application techniques in certain areas, for many chemicals used for many species of alien plants. Especially important is the thorough testing of current and newly available herbicides in near-native systems such as the Special Ecological Areas in Hawaii Volcanoes National Park. Most tests to date have been conducted in areas with heavy concentrations of alien plants. Obviously, less herbicide per unit area is necessary in more pristine areas, but risks to more native individuals and species also increase in these areas.

The use of herbicides on additional alien plant species that threaten the intactness of native ecosystems should be carefully studied. A few examples of such species are banana poka (*Passiflora mollissima*), *Paspalum* spp., pearl flower (*Heterocentron subtriplinervium*), molasses grass (*Melinis minutiflora*), Jerusalem cherry (*Solanum pseudocapsicum*), and meadow ricegrass (*Ehrharta stipoides*). Testing of the effects of herbicide treatment for some of these species and on native plants is needed, and some reports are now (1991) in preparation.

Better data on the persistence of herbicides in soils, the colonization of treated areas by alien and native plants, and the costs of herbicide applications by different techniques in different situations are also needed. Information on the effects of herbicide treatments on invertebrates and birds, in particular, is necessary in more intact ecosystems. Close monitoring of operational usage of chemicals in native systems should be a key element in a management-oriented herbicide research program in a public agency. Long-term monitoring, especially for alien tree or shrub species, is very important in determining effectiveness of chemical treatments.

We believe that herbicides are an effective and essential tool in the restoration of Hawaiian ecosystems. However, much more remains to be learned, and continual research is necessary. A modest investment in research on and monitoring of herbicide use should ensure responsible integration of safe and effective herbicidal control with other approaches used in ecosystem restoration programs.

ACKNOWLEDGMENTS

We would like to thank many individuals who assisted in this project, including Dan Taylor, Tim Tunison, and Chris Zimmer for their encouragement and advice. Thanks also to Casey Baldwin, Kevin McGough, Nelson Ho, and Lou Whiteaker for field assistance. We are very grateful to Paul Higashino and Lani Stemmermann for their help with plant identification. Sincere thanks to Philip Motooka for advice and technical assistance, and to Steve Anderson for help with plant identification and computer technology. Cliff Smith, Dan Taylor, and Tim Tunison provided critical review and comments, and Danielle Stone typed and helped edit the manuscript.

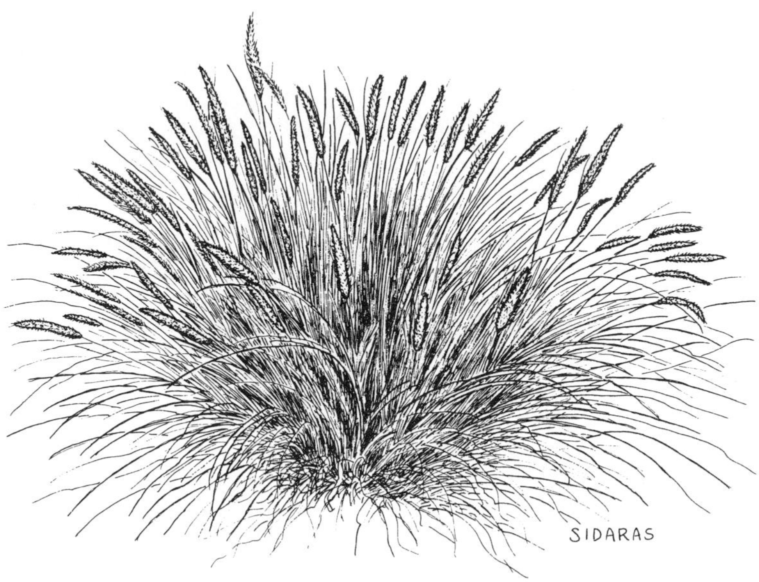

Literature Cited

Anderson, W.P. 1983. *Weed science: principles*, 2nd ed. New York: West Pub. Co.

Berg, G.L., ed. 1985. *Farm chemicals handbook '85.* Willoughby, Ohio: Meister Pub. Co.

Gardner, D.E. 1984. Tests for the control of kahili ginger (*Hedychium gardnerianum*) with GARLON 4 foliar spray and basal applications. Unpub. rep. Files, Hawaii Volcanoes Natl. Park and Coop. Natl. Park Resour. Stud. Unit, Honolulu.

Graf, A.B. 1978. *Tropica, color cyclopedia of exotic plants and trees from the tropics and subtropics.* East Rutherford, N.J.: Roehrs Co.

Haselwood, E.L., and G.G. Motter, eds. 1984. *Handbook of Hawaiian Weeds,* 2nd ed., revised and expanded by R.T. Hirano. Harold L. Lyon Arboretum. Honolulu: Univ. Hawaii Pr.

Hawaii Volcanoes National Park. 1984. Natural resources management management program, November, 1984 update: An addendum to the Natural Resources Management Plan. Div. Resour. Manage. Files, Resources Management Div., Hawaii Volcanoes Natl. Park.

Jacobi, J.D., and F.R. Warshauer. 1975. A preliminary bioecological study of the Ola'a Tract, Hawaii Volcanoes National Park. Prep. for Hawaii Natural Hist. Assn. Hawaii Volcanoes National Park. Univ. Hawaii Dep. Botany, Honolulu. Library, Hawaii Volcanoes Natl. Park. Typewritten.

Johnson, D.R., G.P. Markin, R.C. Reardan, and W.K. Randall. 1984. Injecting Metasystox-R at three spacing intervals to improve seed yield in Douglas fir. *J. Econ. Entomol.* 77:1320-1322.

Leonard, O.A. 1957. Effect of phenoxy herbicide concentrates applied to cuts of sprouting tree species. *Weeds* 5(4):291-303.

Motooka, P., G. Nagai, and L. Ching. 1983. The Uyeda "magic wand" method of herbicide application. *Proc. Asian Pac. Weed Sci. Soc. Conf.*, Suppl. Vol. II, Manila, Philippines, Nov. 28-Dec. 2, 1983, 550-553.

Neal, M.C. 1965. *In gardens of Hawaii.* B.P. Bishop Mus. Spec. Pub. 50. Honolulu: Bishop Mus. Pr.

Santos, G.L., L.W. Cuddihy, and C.P. Stone. 1991. *Control of yellow Himalayan raspberry (*Rubus ellipticus *Sm.) with cut-stump herbicide treatments in Hawai'i.* Tech. Rep. 80, Univ. Hawaii Coop. Natl. Resour. Stud. Unit. Honolulu.

Santos, G.L., D. Kageler, D.E. Gardner, and C.P. Stone. 1986. *Herbicidal control of selected alien plant species in Hawaii Volcanoes National Park: a preliminary report.* Tech. Rep. 60, Univ. Hawaii Coop. Natl. Park Resour. Stud. Unit. Honolulu.

Skolmen, R.G. 1979. *Plantings on the forest reserves of Hawaii 1910-1960.* Inst. Pac. Islands For., Pac. SW For. Range Exp. Sta., U.S. For. Serv., Honolulu.

Smith, C.W. 1985. Impact of alien plants on Hawai'i's native biota. In *Hawai'i's terrestrial ecosystems: preservation and management*, ed. C.P. Stone and J.M. Scott, 180-250. Univ. Hawaii Coop. Natl. Park Resour. Stud. Unit. Honolulu: Univ. Hawaii Pr.

Sprankle, P., W.F. Meggitt, and E. Penner. 1975. Rapid inactivation of glyphosate in the soil. *Weed Sci.* 23(3):224-228.

FOUNTAIN GRASS CONTROL IN HAWAII VOLCANOES NATIONAL PARK: MANAGEMENT CONSIDERATIONS AND STRATEGIES

J. Timothy Tunison

ABSTRACT

Fountain grass (*Pennisetum setaceum*), perceived as one of the most disruptive alien species in Hawaii, has threatened native ecosystems below 3,940 ft (1,200 m) elevation in Hawaii Volcanoes National Park for about 30 years. Because this species has spread rapidly to high densities on the leeward side of Hawai'i Island, invaded bare lava flows (which results in disruption of primary succession), has a broad elevational range (sea level to 6,500 ft (2,000 m) elevation), and has a tendency to raise fuel loadings, efforts to control this hardy bunchgrass have been part of the program of resource management at Hawaii Volcanoes since the early 1960s. In an attempt to increase efficacy of fountain grass control, data on distribution of the grass, treatment effectiveness, and work load requirements were collected beginning in 1979. On the basis of the information obtained from observation and control attempts, five management strategies were proposed in 1986; three of these have since been adopted to control this threat to Park ecosystems.

INTRODUCTION

Fountain grass (*Pennisetum setaceum*), native to Africa, was introduced to Hawai'i Island in the Kona district in the early part of the 20th century. It is now well established on the leeward side of the Island, with highest densities in North Kona, North Kohala, and the Pōhakuloa area. Scattered populations occur in windward areas, mostly in roadside habitats. Fountain grass is readily dispersed by vehicles, humans, wind, water, and possibly birds.

The species is perceived as one of the most disruptive alien plants in Hawai'i (Smith 1985). This large bunchgrass can form monospecific stands, is stimulated by fire, and enhances fuel loadings, thus endangering native woody plant communities it invades. Fountain grass differs from most other nonnative grasses in that it colonizes bare or sparsely vegetated lava flows, thereby disrupting primary succession. It grows in xeric and mesic habitats from sea level to above 8,990 ft (2,740 m) elevation (Jacobi, pers. comm.). Potential distribution in Hawaii Volcanoes National Park may include all areas outside closed-canopy rain forest. Therefore, a control program in the Park has also had to address fountain grass distribution in the vicinity of the Park.

Control programs have gradually expanded since Park managers first perceived fountain grass as an alien plant threat in the 1960s. At that time, a control program along Park roadsides was started. In 1979, efforts were expanded to a 900-a (400-ha) infestation adjacent to the Kamoʻoaliʻi lava flows, an area in the southwestern part of the Park with the highest density of fountain grass. The primary treatment method consisted of uprooting plants by hand and destroying inflorescences to exhaust the soil seed bank. From 1979 to 1983 this control program was expanded to include treatment of roadside plants outside the Park, isolated populations near Kūʻēʻē Ruins and Pit Craters, and scattered populations found incidental to goat (*Capra hircus*) hunting by helicopter. Approximately $125,000 was spent from 1976 to 1983 on fountain grass control. Helicopter reconnaissance indicated that the range of fountain grass in the Park was much wider than previously thought.

The fountain grass control and monitoring program was again intensified in 1983 to achieve the following objectives:

1. Prevent the spread of fountain grass into other areas.
2. Determine the distribution of fountain grass in the Park and immediate vicinity.
3. Quantify the work load needed to control this weed.
4. Assess the effectiveness of treatments on population levels.
5. Monitor growth rate and onset of flowering.

Nearly $75,000 was spent during 1984 and 1985 to collect data incidental to control work. Managers were then able to evaluate the costs, impacts, and feasibility of different management strategies for fountain grass.

A management decision was made in 1985 to abandon control efforts within the interior portions of the major infestation in the Park lowlands, although control efforts were continued on known outlying populations, roadside populations, portions of a 0.6 to 1.25-mi (1-2-km) wide band at the periphery of the major infestation, and in Special Ecological Areas (Tunison and Stone, this volume). Funds were requested to expand the control program. Additionally, fountain grass was identified as a possibility for biological control research.

This paper outlines the decision-making process used in 1986 to formulate a new strategy for managing fountain grass, based on treatment effectiveness, distribution, and work load data collected in 1984-1985, and management priorities at that time.

MANAGEMENT CONSIDERATIONS

Distribution in the Park and Vicinity

Fountain grass distribution was mapped from helicopter, horseback, and on foot. The southwestern lowlands of Hawaii Volcanoes National Park were systematically searched on foot and horseback below 2,300 ft (700 m) elevation and west of the Mauna Ulu lava flows. The remaining dry habitats in the Park were scouted by helicopter at 100 to 330 ft (30-100 m) above the ground, from which larger plants could usually be detected, especially those on mostly bare lava flows. All developed roads surrounding the Park from Whittington Beach Park (outside the southwestern boundary of the Park) to Kapoho (outside the eastern boundary) were searched, as well as Powerline Road and some jeep trails outside the western boundary in the Ka'ū district near the Park. Local residents were queried about the presence of fountain grass on their lands. Some areas outside the Park, away from roads and adjacent to the infested areas, were surveyed by helicopter, while others that may support fountain grass were not. The numbers of plants in fountain grass populations (clusters of one to several thousand plants disjunct from other clusters of plants) were estimated or counted. All populations were mapped on 1:24,000 U.S. Geological Survey orthophotoquads.

In 1986, fountain grass away from roadsides occurred in two major infestations totalling approximately 19,800 a (8,000 ha) (Fig. 1). By far the largest group of populations occurred over 18,530 a (7,500 ha) in the southwestern corner of the Park and adjacent State land in Ka'ū, although densities were low throughout most of the Park. The smaller infestation of 1,235 a (500 ha) occurred north of Keauhou Landing. The area of highest density in the two infestations contained 5.9 plants/a (14.7 plants/ha) (after treatment), areas of moderate density 0.24 plants/a (0.6 plants/ha) (before treatment), and areas of low density, 0.12 plants/a (0.4 plants/ha) (before treatment). Two hundred ninety-eight small, isolated populations were located in areas of moderate and low density in the two infestations (Fig. 2). Fifty-six percent had one individual (when first discovered); 20% had two individuals; only 32 populations (11%) had six or more individuals.

Fourteen isolated populations with one or two plants each were located disjunct from the two large infestations (Fig. 1). Fountain grass was also scattered in 80 populations along roadsides inside and outside the Park in the districts of Ka'ū and Puna (Park population illustrated); 40 occurred in the Park, almost all along Chain of Craters Road and Highway 11 in Ka'ū, and 40 occurred outside the Park (not illustrated). A 1.24-a (0.5 ha) population occurred in Kalapana Village. Two populations in Ka'ū outside the Park and two populations along Chain of Craters Road inside the Park appeared to have spread 250 to 330 ft (75-100 m) from the roadside.

Although some fountain grass control work occurred from the mid-1960s to 1983, formal and systematic mapping was not conducted. Therefore, distribution data presented here serve as a baseline for the future, and densities shown in Figure 1 should be considered a conservative estimate. The range of fountain grass is therefore probably greater than indicated.

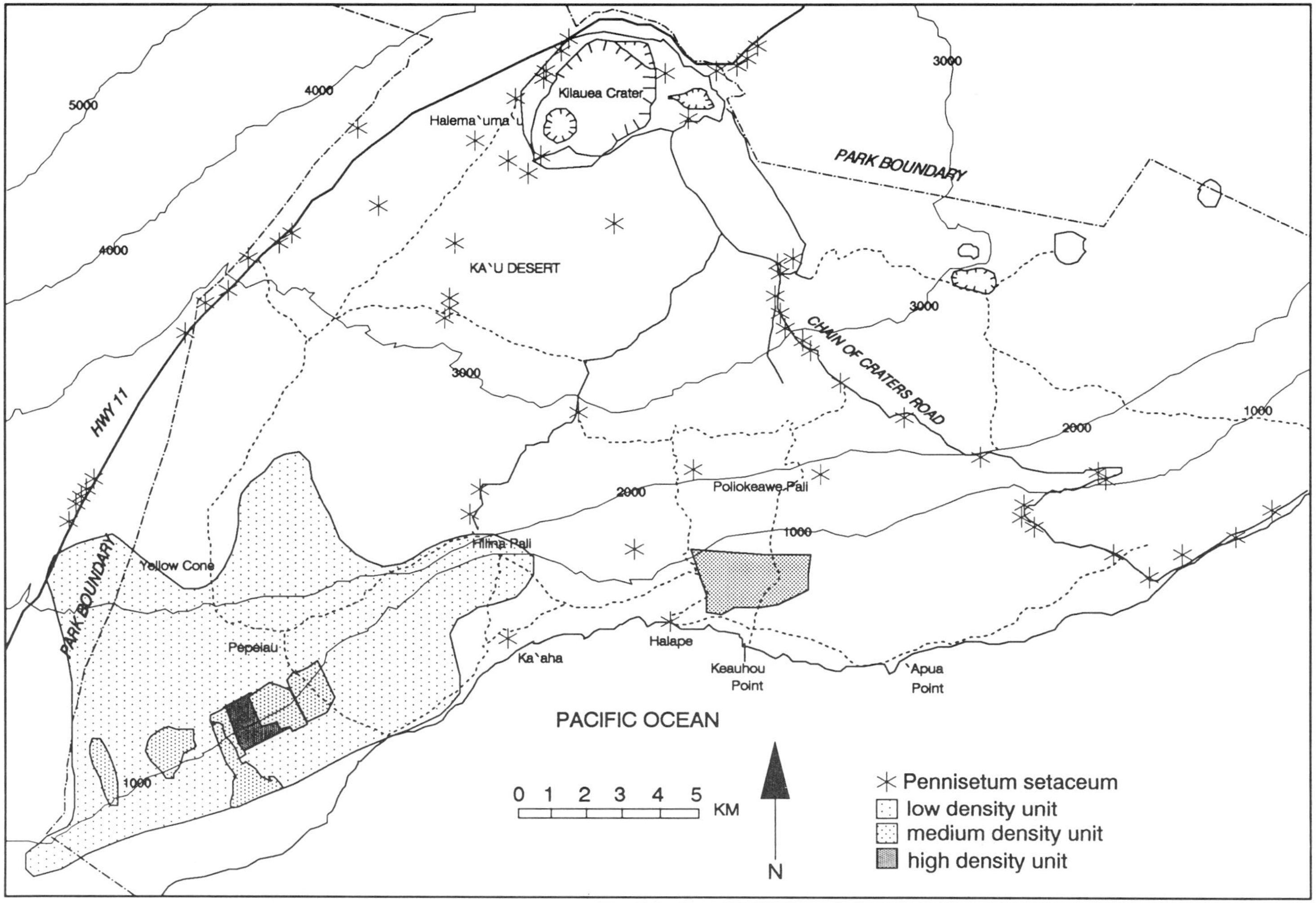

Figure 1. Distribution of fountain grass in Hawaii Volcanoes National Park and immediate vicinity, 1986.

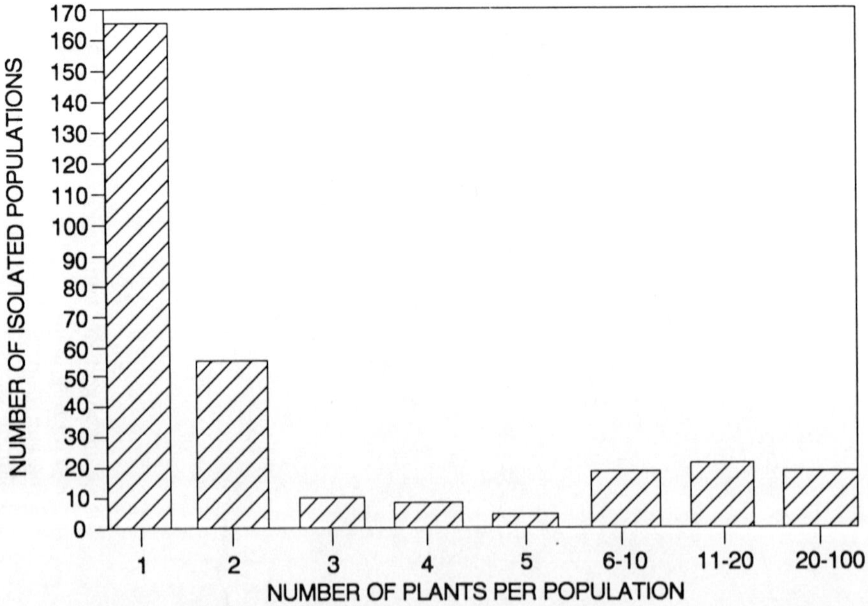

Figure 2. Number of fountain grass plants per population in 298 isolated populations in Hawaii Volcanoes National Park, 1986.

In the Park, large plants 5 to 20 years in age were found over the 1986 range of the plant, suggesting long establishment over a wide area. Search technique efficiency data (see below) indicate that many plants are overlooked during reconnaissance. Helicopter search was found to be only partially effective as a survey technique (see below).

The distribution pattern in the southwestern lowlands of the Park (Fig. 1) indicates widely scattered, small populations surrounding two central areas of higher population density (one on the Kamoʻoaliʻi flows and one above Keauhou Landing). This distribution pattern suggests that fountain grass has been spreading from two centers of distribution, and that densities in the outlying portions of the range will increase over time.

Growth Rate and Phenology

Growth rates and onset of flowering of fountain grass were monitored to determine the time interval between treatments that precludes flowering, and to allow interpretation of search technique efficiency and distribution of plants encountered. The growth rate and phenology of 50 naturally occurring fountain grass plants were monitored for 49 months. The plants grew in a site with a light ash over pāhoehoe substrate at 820 ft (250 m) elevation adjacent to the Kamoʻoaliʻi flows. They were located several meters from adjacent fountain grass plants in an area of moderate densities of Natal redtop (*Rhynchelytrum repens*) and pili grass (*Heteropogon contortus*). This area supported very large fountain grass plants at high density before control measures were taken. The mean length of monitored plants at the outset of the study was 20.7 in. (52.8 cm) (standard error = 3.60 (9.14)). Plants were monitored each year from 1985 to 1989, usually in the summer. The length of the longest leaves from ground level was measured. In addition, basal diameters were measured with a diameter tape at the soil line. The average basal diameter at the beginning of the study was 1.0 in. (2.7 cm) (standard error = 0.09 (0.23)). The date of onset of flowering and size of plant at flowering were also noted.

Thirteen of the 50 monitored plants were accidentally removed by field workers or died during the study. Their measurements were not used in the following calculations. The monitored plants grew in length an average of 4.6 in. (11.6 cm)/year (standard error = 1.40 (3.55)) (n = 37). Basal diameters grew an average of 3.0 in. (7.74 cm)/year (standard error = 0.31 (0.77)) (n = 37). Fifteen of the 37 plants flowered during the course of the study. The mean length when first monitored after flowering was 41.5 in. (105.5 cm) (standard error = 2.48 (6.30)). The mean basal diameter at the time of flowering was 2.5 in. (6.5 cm) (standard error = 0.12 (0.33)).

Qualitative observations by field workers indicated that fountain grass seems to grow much more rapidly and flower much earlier than shown by the growth rate and phenology data. This disparity is explainable by ineffectiveness of search techniques: field personnel miss plants. Growth rates may be greater in areas other than the study site, and growth rates may increase with age; however, growth rate data are not available from areas other than Kamo'oali'i. The growth rate data indicated that the age of the large plants found in the study area was greater than five years. These plants were typically 36 to 48 in. (91-122 cm) in length and 8 to 12 in. (20-30 cm) in basal diameter.

Search Technique Efficiency

The effectiveness of search techniques was tested in five areas by repeated scouting of the searched areas more intensively within a short time, or scouting again with the same or different search techniques. Helicopter searching was evaluated by surveying the target area from a helicopter and subsequently on foot. Search techniques on foot and horse were checked by searching surveyed areas again along more closely spaced transects. The spacing of transects between searchers on initial surveys from horse and on foot varied from 80 to 165 ft (25-50 m); the spacing on subsequent searches ranged from 35 to 80 ft (10-25 m). Growth rate data (see above) indicate that large fountain grass plants may be 5 to 20 years old; therefore, finding large plants in an area searched and controlled a few weeks previously was considered a reliable indicator of search technique efficiency.

In an area of low fountain grass density, where 24 populations were found by helicopter in late February 1984, 14 additional populations were found six weeks later. One week after that, a ground crew found five more populations in the same area. These populations consisted of 1-10 large plants. A four-person crew on foot found only one population in another area of low fountain grass density in which a two-person crew had found five populations two weeks previously. This four-person crew found six additional populations in an area of moderate density west of Pepeiau Shelter, where the two-person crew had found six populations three weeks before. Two searchers on horseback found no additional populations in another area of moderate fountain grass density at Kūkalau'ula that they had surveyed two weeks before. A five-person crew on horseback found 46 plants in a third area of moderate fountain grass density above Keauhou Landing. Two months later, a four-person crew on the ground found 56 additional large plants. Three months after that, a three-person crew on horseback located 43 large plants in this same area.

Efficiency data indicated that helicopter searching is the least effective search method. Only large plants can be seen from helicopter, as evidenced by the failure to find smaller plants in areas of high fountain grass density at Kamo'oali'i. A difference in lighting conditions is thought to account for finding a large number of populations on a follow-up helicopter survey. The data did not indicate that searching from horseback is more effective than searching on foot. Additional plants were found on follow-up surveys using either method.

Fountain grass plants are readily overlooked, even by the most effective searchers surveying along closely spaced transects from helicopter, on foot, or on horseback. Pāhoehoe tumuli, large aa boulders, and vegetation, particularly grasses, readily conceal small fountain grass plants. As a result, search transects need to be closely spaced and all areas within the infested areas should be searched systematically.

Treatment Effectiveness

Fountain grass was treated by uprooting plants and destroying inflorescences. Treatment effectiveness was determined by numbering and marking small populations on site, delineating larger ones with landmarks, and tracking the numbers and sizes of plants found at successive treatments.

Small plants (<2 ft or 0.6 m tall) found during follow-up treatments were presumed to be germinants of seed produced by plants of the isolated population monitored, if they occurred within 330 ft (100 m) of large plants found on first treatment. Larger plants or smaller plants found at greater distances were considered to be overlooked plants or new populations. Three hundred thirty feet (100 m) was used as a criterion because this distance may approximate the maximum for natural dispersal of fountain grass. This distance is substantiated by the pattern seen in three instances within the Park, in which distinctly disjunct large individuals were found during the survey with numerous small plants downwind for approximately 330 ft (100 m). Along stream drainages, apparent germinants occurred more than 330 ft downstream. Treatment effectiveness could not be monitored on a population by population basis in areas of continuous distribution. In these areas, the number of plants found and uprooted within the area of continuous distribution was noted.

The approximately 990-a (400-ha) infestation at Kamo'oali'i has been treated more or less regularly at three- to six-month intervals since 1979.

The number of plants found during each treatment has been recorded since 1982. A general decrease in number of plants has been found with successive treatments, with minor fluctuations associated with time of year (Fig. 3). Worker effort varied from treatment to treatment but was sufficient to find and uproot plants encountered in the block. In isolated populations, approximately 41% of the populations originally with one individual and 27% of the populations with two individuals did not have germinants when observed a second time after 2 to 20 months (Fig. 4). Isolated populations originally with three or more individuals almost always had germinants when observed during the second treatment (Fig. 4). The number of germinants dropped slightly with a third treatment in populations with one or two individuals (Fig. 5).

Short-term treatment effectiveness appears to be encouraging for small, isolated populations. However, one cannot reliably predict from favorable short-term responses that the seed bank has been exhausted in small populations. The lack of germinants may reflect recent conditions unfavorable for germination. More germinants may eventually be recruited from the seed bank. For example, 12 of the 56 small populations treated three times had no germinants the second time but germinants at the third treatment.

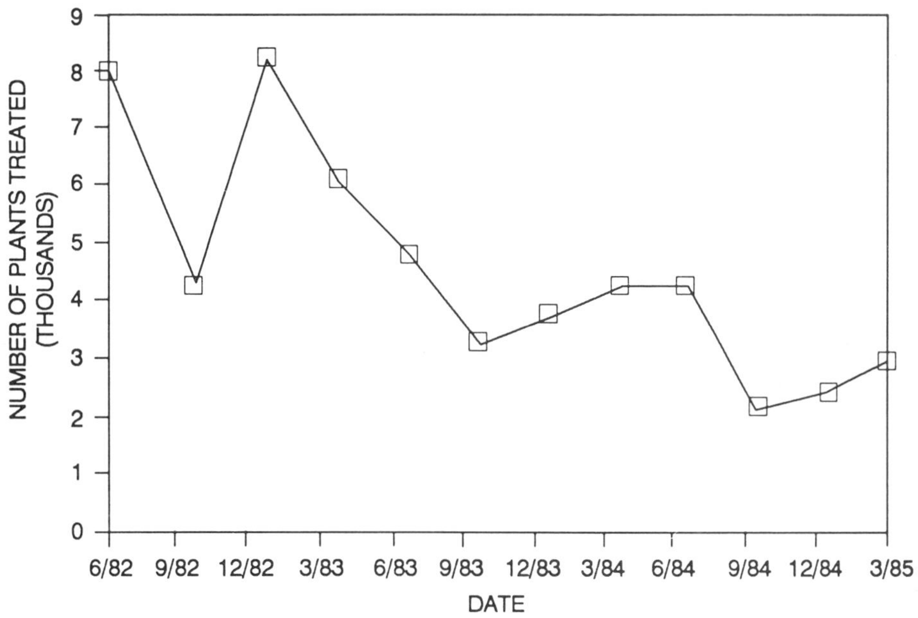

Figure 3. Number of plants through 1985 found at successive treatments in the 400-ha infestation adjacent to the Kamo'oali'i lava flows. The populations declined with control efforts at 3-4 mo intervals. Fewer plants were found in September of each year of the observation period, a trend perhaps reflecting germination during the dry summer months prior to the September treatment.

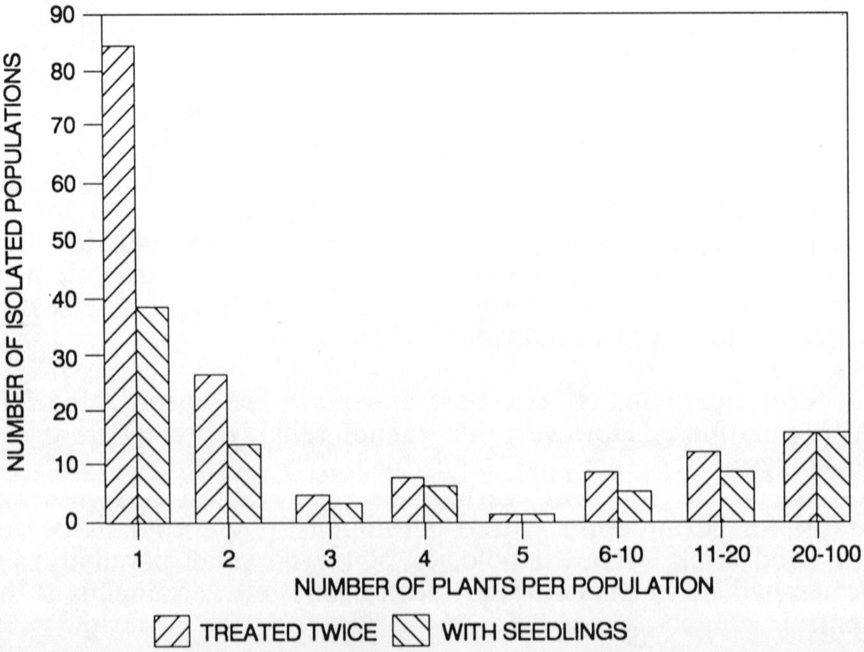

Figure 4. Number of plants found during a second treatment of isolated fountain grass populations, 1986. The left column of each pair indicates the number of populations of each size class treated two times. The right column indicates the number of populations of that size class found to have germinants during a second treatment.

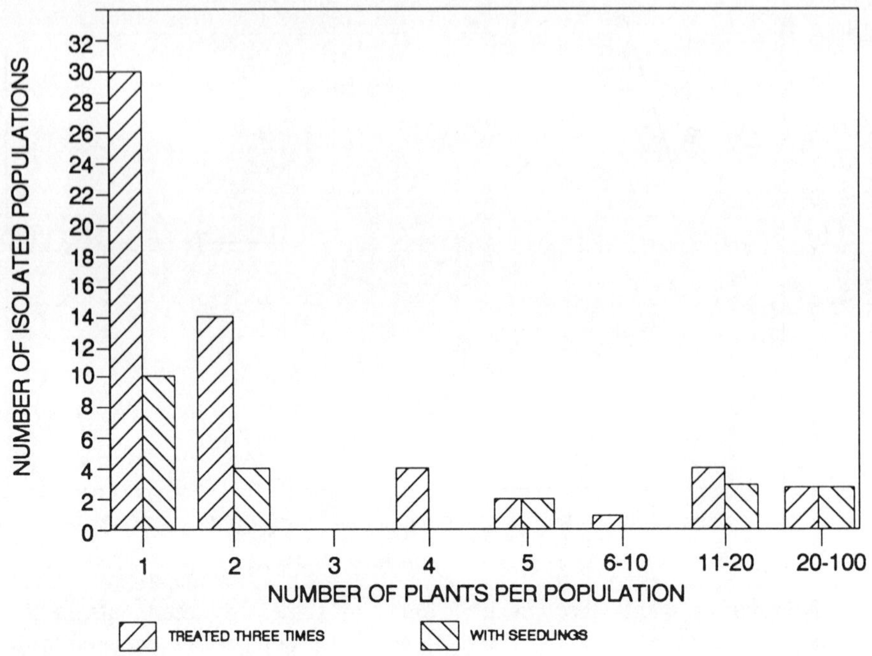

Figure 5. Number of plants found in 1986 during a third treatment of isolated fountain grass populations. The left column of each pair indicates the number of populations treated three times. The right column indicates the number of populations with germinants found at a third treatment.

The seed viability of fountain grass suggests that seed banks may persist for six or more years. Data are not available on viability of fountain grass seeds in soil. A short-term experiment at Hawaii Volcanoes National Park, terminated by an insect infestation at 18 months, showed a decline in viability from 80% to 44% of seeds stored in the laboratory under dry conditions. Treatment data at Kamo'oali'i suggested that fountain grass seed may be viable in the soil for several years. The great majority of the seedlings found at this site was probably recruited from the local seed bank because surrounding areas are of low fountain grass density. Approximately 1% of the plants at each treatment were flowering. The number of seeds contributed by these flowering plants was probably very low because of regular treatments, which precluded maturation and proliferation of inflorescences. The number of inflorescences per plant, typically immature spikes, rarely exceeded one or two.

Work load Requirements

The time required to systematically search and uproot fountain grass once in all infested areas was determined by recording the time spent systematically treating large sections of the infested area and extrapolating to the entire area. The spacing between parallel transects covered by adjacent searchers was 30 to 80 ft (10-25 m) for surveys on foot and 65 to 100 ft (20-30 m) for surveys on horseback. Distances between searchers varied with terrain and vegetation. Approximately 50% of the study area was surveyed with closely spaced transects.

Each searcher, working in teams of two to six individuals, was able to cover an average of 85 a (35 ha) per day. Surveys on horseback were slightly more efficient (95 a or 38 ha per day) than surveys on foot.

Work load requirements per year are based on three components: 1) The number of worker-days required to survey and treat fountain grass in infested areas one time; 2) the number of treatments needed per year; 3) the number of years that treatments are required.

The number of worker days required to survey and treat fountain grass was quantitatively determined. Determining the number of treatments needed per year and the number of years that treatments would be required would take several years of data collection from control programs. Growth rate data and onset of flowering do not address the question because workers tend to overlook plants, thus requiring more frequent searches to prevent flowering. The best estimate is that three or four searches would be needed per year, based on the frequency at Kamo'o'ali'i, which reduced population levels and percentage of plants with flowers to acceptable levels. However, less frequent searches may work in areas with lower densities of fountain grass.

Establishing the work load requirement is an empirical determination for which only a partial data base is currently available. The appropriate thoroughness of surveys and number of treatments needed per year can be adjusted on the basis of apparent treatment effectiveness as more data accumulate.

MANAGEMENT STRATEGIES: FEASIBILITY, IMPACTS, AND COSTS

Five strategies for managing fountain grass were considered by the Division of Resources Management at Hawaii Volcanoes National Park after three years of gathering the information and experience described above. The predicted biological effects and costs of each alternative are described.

Abandon conventional control efforts

The eventual range and ecological impact of fountain grass, if left unchecked, are uncertain. On the one hand, the soils, vegetation, and climatic regime in the Park differ from those in areas in which fountain grass has reached high densities over broad elevational and vegetational gradients. For example, high densities of other grasses in much of the submontane and montane seasonal habitat in the Park may preclude establishment of fountain grass at high densities. Furthermore, the Park has a policy of mandatory fire suppression, which may help limit the intensification of fountain grass. On the other hand, locally dense populations at Kamo'oali'i, Kalapana Village, and Punalu'u suggest that fountain grass in roadside sites in coastal areas can become very dense, reaching up to 50% cover. The spread of fountain grass populations in the Park and nearby areas to date may not provide an appropriate indication of spread potential; many populations may have established relatively recently and have not yet reached equilibrium levels.

A range of probable outcomes following abandonment of fountain grass control can be predicted. One scenario is that fountain grass will probably intensify along roadsides and spread from roadsides into surrounding areas, with continued expansion of infestation in Park lowlands. As a result, fountain grass would expand into most coastal lowland, submontane seasonal, montane seasonal, and subalpine environments at low densities. The plant will become a minor to important component of plant communities other than the coastal lowlands in Ka'ū District. Locally dense populations may occur on favorable substrates with adequate ash and low densities of other grasses. The occurrence of fire may be the most important factor in converting sparse populations to dense populations.

Another scenario is that fountain grass would expand into all plant communities except closed-canopy rain forest; areas of high fountain grass density would be common, and widespread fires will rapidly make fountain grass a dominant plant in many areas.

Rely Solely on Biological Control

Fountain grass is considered an undesirable, widely distributed weed by all interests in Hawai'i and is therefore an appropriate target for biological control efforts. Some limitations to successful biological control are:

1. Biological control agents may not be found. Markin *et al.* (this volume) indicated that success rates of biological control programs are

approximately 50%. Gardner and Davis (1982) indicated that no pathogens for fountain grass are reported in the literature.

2. Permits to import potential biocontrol agents may not be granted by State authorities because fountain grass is congeneric with a valued forage grass (kikuyu grass, *Pennisetum clandestinum*).

3. Biological control research and development may require 10 years or more (Markin *et al.*, this volume) and be only partially effective. Abandoning conventional control efforts in the meantime may permit considerable fountain grass range expansion and ecological disruption. Costs of development and application may average $1 million (Markin *et al.*, this volume).

4. Facilities and research personnel will be occupied with other target species unless priorities are rearranged, although some work on fountain grass may be started opportunistically.

M. Isherwood (pers. comm.) of the Hawaii Department of Agriculture has evaluated prospects for biological control of fountain grass and indicates that this approach is not feasible because of a poor scientific understanding of the species and the location of its natural range in politically unstable developing nations.

Control Fountain Grass Only in Special Ecological Areas

Under this strategy, fountain grass will not be controlled outside areas chosen to represent the most intact, diverse, representative, and manageable areas in Hawaii Volcanoes National Park. This is the same management strategy used with firetree (*Myrica faya*), kāhili ginger (*Hedychium gardnerianum*), strawberry guava (*Psidium cattleianum*), and other widespread species not controllable on a parkwide basis (Tunison and Stone, this volume). If effective, this management strategy would maintain fountain grass populations at very low levels in some of the richest, most intact plant communities representative of the range of vegetation types present in the Park. Fountain grass would spread and intensify in most areas outside of the Special Ecological Areas.

Predicting the feasibility and work load of controlling fountain grass in Special Ecological Areas is difficult for these reasons:

1. The eventual densities of fountain grass in areas surrounding Special Ecological Areas are uncertain; therefore, the magnitude of seed dispersal into these areas, and seedling establishment, cannot be predicted.

2. Colonization potential and rate of spread of fountain grass in the range of vegetation types found in Special Ecological Areas is unknown.

3. Designation of Special Ecological Areas has not been completed; consequently, the total area subject to fountain grass invasion is not known.

Fountain grass control in Special Ecological Areas surrounded by high densities of fountain grass may become highly labor intensive. However, large boundary-to-area ratios may result in lower seed dispersal rates into these areas; alternatively, small Special Ecological Areas may require small work loads commensurate with their size. The feasibility of controlling fountain grass in Special Ecological Areas is best determined by attempting it. Assuming that populations in Special Ecological Areas reach levels similar to those currently found in the infested area, controlling fountain grass by systematic searching at three- to four-month intervals in the entire 22,240 a (9,000 ha) of candidate or designated Special Ecological Areas (outside of closed rain forest) would require approximately as many worker days as controlling fountain grass in the existing infestation (Table 1): 9,000 ha at 35 ha per worker-day x 3.5 treatments/yr = 900 worker days. This figure may be reduced by treating fountain grass concurrently with firetree, because the most time-consuming part of fountain grass control is surveying on foot or horseback.

Confine Fountain Grass to the Lowlands

This management strategy consists of these elements:

1. Develop a fountain grass-free buffer zone around the upper elevations (3,940 ft or 1,200 m) of the Park through cooperation with neighboring ranchers.

2. Systematically and regularly control fountain grass in a buffer zone adjacent to the large infestation in the lowlands of the Park.

3. Treat all known outlying populations in the Park, especially those in Kalapana Village.

4. Treat fountain grass along all roadsides inside and immediately outside the Park.

5. Regularly scout within the Park above the buffer zone.

6. Control Kalapana area populations.

The feasibility of a quarantine strategy is based on the present distribution of fountain grass in the Park lowlands. Apparently, fountain grass is currently spreading by short-distance dispersal. The greatest disjunction of an isolated population within the main infestation in this Park lowlands infestation is approximately 0.6 mi (1 km).

Weaknesses of a quarantine strategy are:

1. As densities increase in the infested area, the probabilities of dispersal into uninfested areas increase, as does the work load of maintaining a buffer zone. Prevailing trade winds tend to blow seeds downhill and toward the coast. Occasional strong kona (southwesterly) winds may blow seeds uphill and across a buffer zone. The feasibility of a buffer zone adjacent to the major infestation can only be determined by attempting to maintain a buffer zone for a number of years.

2. Fountain grass may already occur outside the proposed buffer zone adjacent to the current infestation. These areas have not been thoroughly searched. More intensive surveys in 1986 indicate additional hitherto- unknown populations of fountain grass in the upper Ka'ū Desert.

3. Neighboring landowners have not agreed to control fountain grass on their lands. Control efforts are not economically feasible, and they believe that current grazing practices will limit the densities of fountain grass to acceptable levels. Even if they do control fountain grass, acceptable population levels may permit flowering and therefore a range expansion into the Park. Jacobi and Warshauer (this volume) have predicted the expansion of fountain grass into the upper-elevation plant communities on Mauna Loa between infested areas in Kona and Ka'ū and the Park.

Work load estimates (Table 1) can be calculated if it is assumed that populations in buffer zones will reach levels similar to those in the currently infested area; and that a 0.6-mi (1-km) wide buffer zone is regularly and systematically surveyed.

Table 1. Estimated work load for confining fountain grass to lowlands of Hawaii Volcanoes National Park.

Control Project	Worker-Days /Treatment	Treatments/Year	Worker-Days/Year
500 ha Buffer zone adjacent to infestation	43	3-4	130-170
Roadsides	10	3-4	60-80
Scouting outside known range	100	2	200
Outlying populations	50	3-4	150-200
Kalapana Village population	25	3-4	75-100
Total Field Time	228	----	615-750*

*150 worker-days of data management and administrative support should be added to this figure to indicate total work load involved.

Control Fountain Grass Parkwide

Treatment effectiveness data suggest that the sizes of large populations may be reduced. Control efforts on isolated populations are promising, but they have not been conducted over sufficient time to demonstrate that small, isolated populations can be eradicated. Seed viability and treatment data at Kamo'oali'i indicate that control programs must be carried out for at least five or six years, and possibly 10 years or longer. A major drawback to controlling fountain grass in the Park lowlands is that fountain grass also occurs adjacent to the Park and is continually introduced from these areas. These unmanaged infestations can be expected to intensify, expand, and spread into the Park.

The work load for controlling fountain grass parkwide, discussed in the section on Work load Requirements, is beyond current budgetary support (Table 2).

Table 2. Work load estimates for controlling fountain grass in the existing infestations in Hawaii Volcanoes National Park and vicinity.

Control Project	Worker-Days /Treatment	Treatments/Year	Worker-Days/Year
7,500 ha infestation (outside Kamo'oali'i)	215	3-4	645-860
Kamo'oali'i infestation	50	3-4	150-200
Roadsides	10	3-4	60-80
Scouting outside known range	100	2	200
Kalapana Village population	25	3-4	75-100
Total Field Time	400	----	1,130-1,440*

*150 worker-days of data management and administrative support should be added to this figure to indicate total work load involved.

MANAGEMENT OF FOUNTAIN GRASS

The fountain grass control program developed in 1986 included three of the five strategies considered. Control was undertaken in the following areas:

1. Along roadsides inside and immediately outside the Park, at four- to six-month intervals.

2. In Keamoku Special Ecological Area in the upper Kaʻū Desert, Kīpuka Puaulu Special Ecological Area and buffer zone, and ʻĀinahou buffer zone, at 12- to 18-month intervals incidental to firetree control work. In all areas of Mauna Loa above Kīpuka Kī along pig activity transects at irregular intervals.

3. Along jeep roads in the upper Kaʻū Desert, every six months.

4. Populations disjunct from the main infestation, every three to four months. This includes populations above Keauhou Landing and the upper Kaʻū Desert.

5. All populations on the periphery of the main infested area in the Park and adjacent State lands. Much of the periphery of the infestation can be surveyed systematically on an annual basis. The intent is that, as control of these populations occurs, control efforts will be extended toward the center of the infestations.

6. Areas in the Park outside the main infestation. These were poorly surveyed in previous efforts and may indicate a distribution pattern with a bearing on the feasibility of a quarantine approach to controlling fountain grass.

The above strategy was implemented for two years, and by 1988 it was apparent that control of outlying populations and in the buffer zone was a very effective approach. Several systematic searches were needed to locate all five small populations in the buffer zone, but once these were located and uprooted, there was little seedling recruitment. The work load thus dropped significantly, permitting managers to expand control efforts parkwide. By 1991, approximately 75% of the range of fountain grass was being regularly searched. The current (1991) management goal is control of fountain grass parkwide by continuing to expand efforts toward the center of the infestations.

RECOMMENDATIONS FOR FUTURE MANAGEMENT

Future management of fountain grass will depend on funding levels and spread of the species both inside and outside the Park. Fountain grass control should be conducted aggressively at the highest level possible without jeopardizing progress in feral animal control, localized alien plant control, and management of Special Ecological Areas. Parkwide fountain grass control should continue to be a high-priority funding request. The species should be controlled in Special Ecological Areas, along roadsides, and, to the extent that funding is available, in populations on the periphery of the main infestation. Control on the periphery should be made systematic, thereby creating a buffer zone to confine the lowland infestation to its current range. As funding becomes available, the buffer zone should be extended toward the central portions of the infested area. The infestation in Kalapana Village should be controlled.

Fountain grass has been targeted in the Resources Management Plan update for biological control, but it is a lower priority for biological control than banana poka (*Passiflora mollissima*), *Rubus* spp., and fire-tree. A request for additional funding to fully support conventional fountain grass control parkwide (Strategy 1) has been made, and an extensive monitoring program to detect changes in the density of fountain grass in infested areas has been established.

Monitoring the effectiveness and impacts of fountain grass control should continue. Densities within the untreated infested area should be followed, and colonization of new areas above the buffer zone should also be studied to determine the effectiveness of this approach. Upland areas on Mauna Loa outside the boundaries of Hawaii Volcanoes National Park should also be checked for encroachment of fountain grass. Changes in isolated populations within the treated zone should be evaluated to determine the feasibility of eradicating small, isolated populations. Progress in controlling the species parkwide will be described in a subsequent paper.

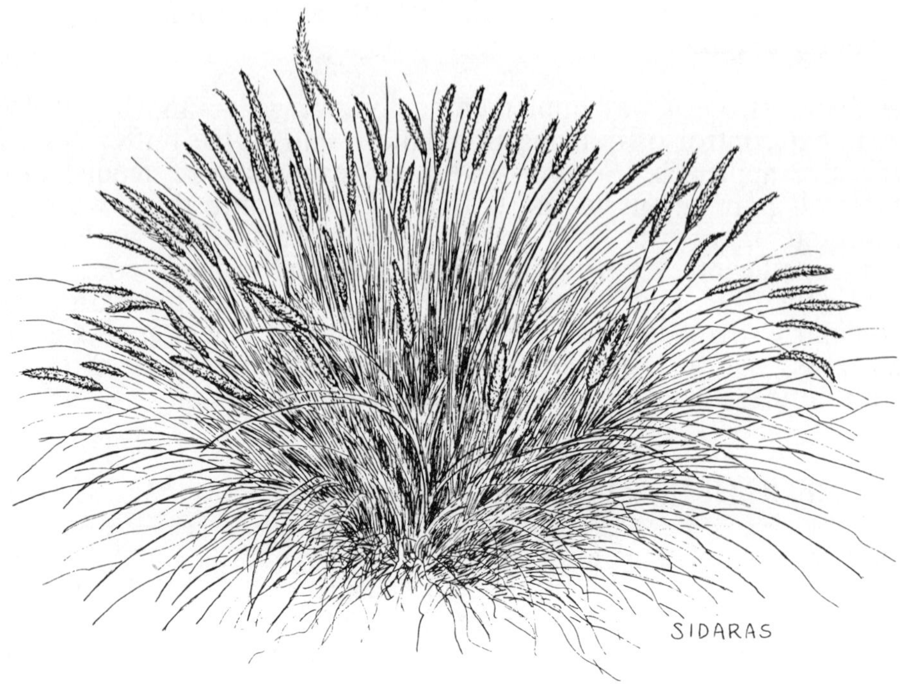

Literature Cited

Gardner, D.E., and C.J. Davis. 1982. *The prospects for biological control of nonnative plants in Hawaiian national parks.* Tech. Rep. 45, Univ. Hawaii Coop. Natl. Park Resour. Stud. Unit. Honolulu.

Jacobi, J.D., and F.R. Warshauer. [this volume] Distribution of six alien plant species in upland habitats on the island of Hawai'i.

Markin, G., P.-Y. Lai, and G.Y. Funasaki. [this volume] Status of biological control of weeds in Hawai'i and implications for managing native ecosystems.

Smith, C.W. 1985. Impact of alien plants in Hawai'i's native biota. In *Hawai'i's terrestrial ecosystems: preservation and management* , ed. C.P. Stone and J.M. Scott, 180-250. Univ. Hawaii Coop. Natl. Park Resour. Stud. Unit. Honolulu: Univ. Hawaii Pr.

Tunison, J.T., and C.P. Stone. [this volume] Special Ecological Areas: an approach to alien plant control in Hawaii Volcanoes National Park.

FIRE AND ALIEN PLANTS IN HAWAI'I: RESEARCH AND MANAGEMENT IMPLICATIONS FOR NATIVE ECOSYSTEMS

Clifford W. Smith and J. Timothy Tunison

ABSTRACT

Fire, an important evolutionary determinant in many continental ecosystems, probably has not played a significant role in the State. The most common natural ignition source is lava flows, but the majority of fires are now started by humans. Some native plant species, *e.g.*, koa (*Acacia koa*), 'a'ali'i (*Dodonaea viscosa*), 'ōhi'a (*Metrosideros polymorpha*), and 'ūlei (*Osteomeles anthyllidifolia*), show some tolerance of fire, which may be a legacy of their pre-insular biology. Most species, *e.g.*, pūkiawe (*Styphelia tameiameiae*), are severely set back, killed, or compete poorly with alien vegetation stimulated by fire. The natural fire regime in Hawai'i has been markedly altered by the rapid spread of alien grasses, particularly broomsedge (*Andropogon virginicus*), molasses grass (*Melinis minutiflora*), bush beardgrass (*Schizachyrium condensatum*), and fountain grass (*Pennisetum setaceum*). These alien grasses are stimulated by fire. As a result, fire frequencies and sizes have increased, resulting in the intensification of alien plants to the detriment of native species. Fire is now a very powerful disturbance factor in Hawaiian ecosystems, not as extensive as the depredations of feral animals but certainly as destructive. Fire frequency will probably continue to increase and affect natural areas more seriously because of the spread of alien grasses. Fire, therefore, cannot be used as a management tool in the restoration or preservation of native Hawaiian ecosystems. The protection of natural areas can be enhanced by aggressive fire suppression and presuppression planning, including construction of fuel breaks, preparation of fuels distribution maps, strengthening of interagency fire agreements, and fire prevention.

INTRODUCTION

Fire, in Europe and the U.S., was traditionally thought of as a negative factor in natural ecosystems (Pyne 1982). This philosophy was epitomized by the highly effective U.S. Forest Service "Smokey the Bear" advertising campaign. However, in the last 20 years, fire has been acknowledged to play a vital, natural role in many ecosystems. In fact, periodic low-intensity fires maintain certain forest types that would otherwise be highly altered by the occasional major conflagration (Kilgore 1981). Other ecosystems are subject to infrequent, high-intensity fires without long-term harm (Heinselman 1981).

The natural ecological role of fire is well understood in temperate areas, particularly North America, and in the subtropics, principally Australia. In wet tropical areas, there is little evidence that fire plays a role in the maintenance of natural ecosystems (Mueller-Dombois 1981). Fire, usually started by humans, degrades tropical forests. Pyrophytic grasses typically invade burned sites, and grass-dominated areas expand due to increasingly frequent and destructive subsequent fires. The vast majority of savannah ecosystems are fire-dependent and almost certainly man-induced, although herbivores also play a significant role in maintaining them (Buechner and Dawkins 1961). A few grasslands, however, are not fire-dependent (Sarmiento and Monasterio 1974). Fire has not been reported as an important natural factor in most island ecosystems. However, once humans colonize an island, fire is used aggressively as a vegetation management technique (Kirch 1982), and accidental fire frequency increases.

In this paper, we characterize changes in the natural fire regimes in Hawai‘i and discuss the current ecological effects of fire in natural ecosystems. These observations serve as the basis for recommendations about fire management and research in natural areas throughout the State. The information and recommendations in this paper are based mostly on studies and experiences of researchers and fire managers working in Hawaii Volcanoes National Park, where fire history records are relatively complete, fire effects studies have been conducted, and active programs of fire suppression and fire prevention have been in place since the mid-1970s.

THE NATURAL FIRE REGIME IN HAWAI‘I

The natural fire regime in Hawai‘i is difficult to reconstruct in detail. Accounts of natural fires are rare and contain little ecological information. Lack of annual tree growth rings precludes developing a precise fire history based on fire scar chronologies. Radiocarbon dating of charcoal found in soil and bogs has the potential to indicate broad patterns of fire occurrence, but most carbon-dating studies have been conducted by geologists interested in dating lava flows or by anthropologists interested in cultural sites. Moreover, carbon in volcanic sites may indicate fires along the immediate edge of flows or phreatic eruptions, rather than free-burning fires. The published literature on carbon-dating of charcoal deposits in Hawaiian bogs is meager and not useful for detailed fire history studies. A serious study of buried carbon as an indicator of natural fire frequency is needed.

J. Juvik and P. Scowcroft (pers. comm.) have confirmed the probable occurrence of fire prior to human colonization. Juvik found carbon at depth in a core from an O'ahu bog revealing a 26,000-year record of fire occurrence. Scowcroft reported undated carbon burned at a depth of 0.5 m in Laupāhoehoe Forest Reserve on Hawai'i Island, a location that suggests the occurrence of wildfire rather than volcanic activity. J.P. Lockwood (pers. comm.) found evidence of natural fire in a widely distributed pattern of carbon dates on the northern, eastern, and southeastern slopes of Mauna Loa, all approximately 400 years old and not consistent with local flows. This pattern suggests the possibility of a large fire, possibly ignited by Keamoku lava flows of that age.

The paucity of carbon in subsurface soils formed before human colonization and away from volcanically active areas implies a low fire frequency. Carbon dating in Hawaii Volcanoes National Park, which should have the highest occurrence of soil carbon because of volcanic activity, confirms this pattern. Soil profiles from three Park sites, Kīpuka Ki, Kīpuka Puaulu (Mueller-Dombois and Lamoureux 1967), and 'Ōla'a (Mueller-Dombois 1981), indicate the absence of fires over the last 2,000 years, although charcoal is associated in these sites with phreatic eruptions (J.P. Lockwood, pers. comm.). However, these are mesic and wet sites, and soil pits in dry sites might indicate a different pattern.

A low natural frequency of fire is also implied by the infrequency of natural ignition sources, low flammability of native vegetation, and discontinuous distribution of natural fuels (Mueller-Dombois 1981). Vogl (1969) suggested that lightning was a more frequent source of fire than commonly believed. However, lightning-initiated fire is uncommon (Sorensen 1979; Mueller-Dombois 1981), although strikes without fire spread are responsible for tree mortality in rain forest (Sorenson 1979; J.T. Tunison and J. Leialoha, unpub. data), and lightning strikes are known to have started a few recent grassland fires (Kartinawata and Mueller-Dombois 1972; Tunison and Leialoha 1988). Fires started by lava flows are common during volcanic eruptions, at least in alien grass fuels. However, volcanism is highly localized in Hawai'i, being confined mostly to one island at any one point in time (Mueller-Dombois 1981).

Flammable, fire-dependent, or fire-maintained vegetation did not characterize most original, undisturbed Hawaiian vegetation types. The low flammability of native fuels implies poor adaptation to fire (Mutch 1970). Few native species have oils or resins frequently associated with fire-adapted plants, although *Dubautia* and *Dodonaea* may have. Moreover, most vegetation types were dominated by open stands of trees and shrubs (Mueller-Dombois 1981). Alpine and subalpine plant communities were characterized by a sparse to open cover of low shrubs with localized patches of grass. Before disturbance, the submontane seasonal vegetation zone was dominated by woodlands with an understory of open shrubs, sedges, lichens, and bryophytes. The rain forest vegetation zone is almost invariably wet, and litter usually recyles rapidly, although uluhe *(Dicranopteris linearis)*-dominated rain forest can carry fire after extended dry periods. Pre-Polynesian coastal vegetation is not well understood. Pili grass *(Heteropogon contortus)*, possibly indigenous to

Hawai'i, may have been abundant in some areas. Dryland forest, which may have been much more widespread, does not support fine fuels to carry fire.

The apparent fire tolerance of some native plants is less clearly indicative of past natural fire regimes than fire history, ignition sources, flammability, and fuel distribution. Some native species do not recover well from fire, but these species may eventually recover from seed and be important late in plant succession. The fact that other native plants can resprout after fire does not imply that they are specifically adapted to fire. Mueller-Dombois (1981) suggested that adaptations to various natural stress factors provides a good explanation for fire tolerance of some native species. In addition, ancestral populations of certain species may have been adapted to fire prior to their establishment in Hawai'i. The ability to resprout after fire is found in species characteristic of plant communities unlikely to carry fire. For example, many rain forest woody plants (*e.g.*, kāwa'u, *Ilex anomala*) and tree ferns (*e.g.*, *Cibotium glaucum*) resprout vigorously after fire (J.T. Tunison, unpub. data). Yet, many lowland trees in Hawai'i, *e.g.*, lama (*Diospyros sandwicensis*) and wiliwili (*Erythrina sandwicensis*), which grow in relatively dry environments, do not resprout. Recovery of native plants from fire is also difficult to evaluate in Hawaiian ecosystems because it is strongly influenced by competition from alien vegetation. Native vegetation may recover adequately from fire, but it is often masked by the prolific spread of alien plants.

In conclusion, the natural fire regime in many areas of Hawai'i is best characterized as fire-independent (Vogl 1975). Pre-settlement fires were probably irregular, infrequent, short-term ecological perturbations from which vegetation eventually recovered; fires were thus insignificant as long-term evolutionary factors (Mueller-Dombois 1981).

FIRE REGIME SINCE THE ARRIVAL OF MAN

Polynesians colonized Hawai'i approximately 1,600 years ago. They used fire extensively to manipulate vegetation in the lowlands in a "slash-and-burn" form of agriculture, which created an anthropogenic landscape (Kirch 1982; Cuddihy and Stone 1990). The Polynesian landscape was characterized by open grasslands, and pioneer species predominated below 1,475 ft (450 m) in the Hilo area and below 1,970 ft (600 m) in the Ka'ū area (McEldowney 1979). Continuous burning through the years resulted in the formation of the open savannah observed by several early visitors (Kirch 1982). A. Menzies, for example, observed a fire burning in the coastal grasslands of Kaua'i in 1792, which the local inhabitants said was set to stimulate the growth of pili grass, used for thatching. Apparently since European contact, fire frequency and size have generally been low, although large fires have occurred occasionally (Hall 1904; Vogl 1969). The majority of fires are now started by people, either intentionally or inadvertently.

The fire regime in Hawai'i has changed dramatically in recent decades, with a considerable increase in fire frequency and probably fire size. An altered fire regime is reflected in the fire history of Hawaii Volcanoes

National Park, the only natural area with continuously recorded fire history over the last 60 years. Most of the fires (73%) in the Park have been caused by humans. Twenty-five percent were ignited by lava flows in an area with the highest incidence of volcanism in the State, and 2% were caused by lightning (National Park Service 1990).

Fire frequency and size increased greatly in Hawaii Volcanoes after 1968. Twenty-seven fires, averaging 10 a (4 ha), were recorded between 1920 and 1967. Since then, 58 fires have burned an average of 507 a (205 ha) each. The increase in fire frequency and size can be attributed to an accumulation of fine fuels, primarily alien grasses, which spread and intensified in the 1960s and 1970s. Photopoints and exclosure studies indicate that increase of grasses in the coastal lowlands coincided to the removal of feral goats (*Capra hircus*) in the early 1970s (Mueller-Dombois and Spatz 1975; Mueller-Dombois 1980). During this period, the proliferation of perennial grasses such as thatching grass (*Hyparrhenia rufa*), Natal redtop (*Rhynchelytrum repens*), pili grass, and molasses grass (*Melinis minutiflora*) produced a many-fold increase in biomass and a fuel bed capable of supporting fire. Broomsedge (*Andropogon virginicus*) and bush beardgrass (*Schizachyrium condensatum*) invaded and spread throughout the submontane seasonal zone during the 1960s and 1970s, in spite of the presence of feral goats. The first fire carried by these grasses was reported in 1968 (Smathers, pers. comm.). Increased volcanic activity and greater numbers of visitors may have also contributed to the increased number and average size of fires in Hawaii Volcanoes National Park (National Park Service 1990).

Wildland fires in Hawai‘i have been mostly confined to mesic and dry habitats, although the largest historic fire in the State occurred in rain forest (Vogl 1969). As alien bunchgrasses invade wetter habitats, fires will burn more frequently in these areas (Mueller-Dombois 1973).

EFFECTS OF FIRE IN HAWAIIAN ECOSYSTEMS

As a modifier of natural vegetation, fire is equal in importance to, but not as widespread in natural communities as, feral ungulates and alien plants. Twenty-three of the 86 alien plants that Smith (1985) characterized as disruptive weeds in native Hawaiian ecosystems resprout after fire. Fourteen are stimulated by fire, and only eight are eliminated by fire. Native plants, adapted to a very low frequency of fire, recover to some degree after burns, but typically to populations far less than those found in pre-burn conditions because they typically must compete with fire-stimulated alien grasses. Fires establish a destructive cycle by increasing fuel loadings of flammable alien grasses, which in turn increase fire frequency and fire size. The current ecological role of fire in Hawai‘i is consistent with other tropical areas, where fire, usually of anthropogenic origin, type-converts tropical forest to savannah or grassland. Pyrophytic grasses typically invade burned sites, thus rendering them more fire-prone. Subsequent burns further inhibit the reestablishment of tree species (Mueller-Dombois 1981).

Not much published fire effects information exists in Hawai'i, particularly for long-term successional patterns. However, a comprehensive study of fire effects in Hawaii Volcanoes National Park is under way, and some tentative conclusions about fire effects can be made from the limited literature available and unpublished data of the authors and other researchers:

1. Alien grasses usually recover to pre-burn levels and often intensify after fire. Fire facilitates the spread of broomsedge, bush beardgrass, molasses grass, and fountain grass. These are fire-adapted alien grasses from the tropics, subtropics, and warm-temperate areas and are the most important widespread alien fuels in Hawai'i. Smith *et al.* (1980) described increases in broomsedge (*Andropogon* spp.) and molasses grass after fire. Hughes (1989) found an increase in cover of alien grasses and a shift in dominance from *Andropogon* spp. to molasses grass. Fire thus promotes conditions that increase fire frequency and size by increasing the continuity and/or fuel loadings of alien grasses. Hughes (1989) also found that a second fire at the same site can further degrade native Hawaiian vegetation, intensifying alien grass and fuel loadings.

2. Woody alien plants usually invade burned areas only to a limited degree and are typically early successional species. Alien shrubs such as partridge pea (*Chamaecrista nictitans*), indigo (*Indigofera suffruticosa*), sourbush (*Pluchea symphytifolia*), and yellow Himalayan raspberry (*Rubus ellipticus*) invade burned sites immediately after fire but appear to be early successional species. Other common trees and shrubs, such as common guava (*Psidium guajava*), lantana (*Lantana camara*), koa haole (*Leucaena leucocephala*), faya tree (*Myrica faya*), and Java plum (*Syzygium cumini*), resprout vigorously. Koa haole also establishes by seed.

3. A number of native plant species are fire-tolerant, although no native plant is fire-resistant. In other words, aerial portions of all native plant species are readily killed by fire, but some resprout or recover from seed. Māmane (*Sophora chrysophylla*) resprouts from the base (Mueller-Dombois, 1981), but the cover of the tree is reduced significantly for at least five years. Seed viability is diminished by fire (C.W. Smith, unpub. data). Koa (*Acacia koa*), although killed by high-intensity fire, regenerates profusely from seed (Scowcroft and Wood 1976) and/or vegetatively by means of root suckers (Warshauer 1975). Wester and Wood (1977) found that weeds, particularly Koster's curse (*Clidemia hirta*), outgrew koa seedlings within two years after fire. T. Parman and C.W. Smith (unpub. data) noted an inverse relationship between the severity of canopy scorching and sucker growth. Mueller-Dombois (1981) concluded that koa will coexist with fire as long as the forest is reestablished before the next conflagration. Naio (*Myoporum sandwicense*) also resprouts after fire as long as the fire is not too intense. Regrowth is slow and the original canopy cover takes over five years to regenerate.

The seeds of 'a'ali'i (*Dodonaea viscosa*) germinate readily after fire, but it is not known if they are fire-stimulated (Hughes 1989). 'A'ali'i may reach higher densities after fire than before (Hughes 1989).

Sandalwood (*Santalum* spp.) appears to reach its highest densities in recently burned areas. It recovers well vegetatively (Warshauer 1974) and appears to become established from seed. Several species of ʻōhelo (*Vaccinium* spp.) also regenerate rapidly from epicormic and root sprouts (L.L. Loope and C.W. Smith, unpub. data). However, they take at least five years to reestablish their cover and stature.

Most native bunchgrasses recover rapidly after burning. Pili grass resprouts readily with a significant increase in cover following low-intensity fire (J.T. Tunison, unpub. data). The aboriginal Hawaiians used fire to stimulate fresh growth in pili (Kirch 1982). ʻEmoloa *(Eragrostis grandis)* (Smith *et al.* 1980) and the native tussock grass *Deschampsia nubigena* (L.L. Loope and C.W. Smith, unpub. data) also resprout after fire.

Three native ferns regenerate after burning. Bracken (*Pteridium aquilinum*) resprouts rapidly after fire, often appearing within 10 days. Its cover increases dramatically during the first three months but then declines rapidly to a level comparable with the pre-burn level. Hāpuʻu *(Cibotium glaucum)* and ʻamaʻu *(Sadleria cyatheoides)* are known to lose their fronds after fire, but most plants soon produce new leaves.

4. A number of widespread native plants are sensitive to fire and do not recover well from seeds or sprouts. Pūkiawe *(Styphelia tameiameiae)*, an important shrub in many communities in the submontane and montane seasonal zones, is readily killed by fire (Smith *et al.* 1980) and does not become established from seed (Hughes 1989). Uluhe fern is very fire-sensitive and recovers slowly, if at all. ʻĀkia *(Wikstroemia sandwicensis)* and lama in the coastal lowlands are also easily killed by fire.

5. The fire tolerance of ʻōhiʻa *(Metrosideros polymorpha)* varies with fire intensity. ʻŌhiʻa recovers by resprouting epicormically or more usually from the root crown; seedlings are rare. Few ʻōhiʻa recovered from high-intensity fire (Hughes 1989). Smith *et. al.* (1980) noted approximately 50% recovery, with higher recovery rates correlated with the number of boles in a coppice. J.T. Tunison (unpub. data) observed 70% recovery in a low-intensity burn and less than 10% recovery in a high-intensity burn.

In Hawaiʻi, fire effects and fire potential vary among ecosystems. Fire appears to be more deleterious to native ecosystems below 4,600 ft (1,400 m) elevation than above. Fire potential is greatest in environments subject to seasonal drought, especially those invaded by alien grasses and previously affected by fire.

Subalpine Ecosystems

In subalpine ecosystems, fires are rare natural events, but a few fires of human origin have occurred. The māmane/naio forest tolerates fire to a certain extent (P. Scowcroft, pers. comm.), but recovery of the native species is slow, and alien bunchgrasses rapidly invade the opened environment.

'Ōhelo-pūkiawe scrub does not tolerate fire (L.L. Loope and C.W. Smith, unpub. data). Pūkiawe is killed and is slow to become reestablished. When fire intensity is low, 'ōhelo shoots resprout, but recovery after high-intensity fire is only from root suckers. The native tussock grass *Deschampsia nubigena* recovers slowly but may achieve higher cover after fire (D. Mueller-Dombois and J.T. Tunison, unpub. data). 'Ama'u fronds are killed, but new ones are produced rapidly. A few other species, *e.g.*, kūkaenēnē *(Coprosma ernodioides)* and strawberry (*Fragaria chiloensis*), recover within a year or so after fire. However, all native species subsequently have lower cover values following fire, as compared with alien species, particularly velvet grass (*Holcus lanatus*) and sweet vernalgrass (*Anthoxanthum odoratum*). The fire potential of subalpine ecosystems is increasing because alien grasses, particularly sweet vernalgrass and velvet grass, are invading the naturally discontinuous fuel bed.

Montane Seasonal Ecosystems

The montane parkland ecosystem, a mosaic of koa stands, shrublands, and grasslands, demonstrates considerable adaptation to fire (Mueller-Dombois 1981), with rapid recovery of native plant species. Koa is sensitive to fire but regenerates very rapidly (Scowcroft and Wood 1976) and, with 'emoloa and bracken, produces abundant regrowth within a month. The rapid regeneration of koa from suckers often reaches beyond the original aerial boundary of parent trees. The native *Deschampsia nubigena* and mountain pili *(Panicum tenuifolium)* recover vegetatively. 'A'ali'i is one of the first native shrubs to reinvade burned areas, whereas pūkiawe recovers much more slowly. In montane seasonal communities elsewhere, fire has contributed to the spread of alien grasses, as demonstrated by the invasion of fountain grass in the Pōhakuloa area of Hawai'i Island.

Fire potential is relatively high in the montane seasonal zone because of the continuous arrangement of grasses and abundance of shrubs. The invasion of alien grasses such as fountain grass, *Andropogon, Schizachyrium,* and *Paspalum* spp. is increasing the fire potential in this ecological zone.

Montane Rain Forest Ecosystems

Documented fires in montane rain forest ecosystems are rare (Sorensen 1979; Mueller-Dombois 1981), and fire-effects studies in this zone have been limited to uluhe-dominated rain forest (J.T. Tunison, unpub. data). After fire, 'ōhi'a, kāwa'u, and other native trees typically resprout from the root crown. Hāpu'u and 'ama'u also recover vegetatively. However, the other common rain forest fern, uluhe, is eliminated from early successional stages, and alien species such as broomsedge and yellow Himalayan raspberry invade burned sites and become dominant. Fire potential is relatively low in rain forest because fuels rarely dry out and organic matter is rapidly recyled.

Submontane Seasonal Ecosystems

Many of the ecosystems in this zone have been disturbed significantly by fire. The lower elevations of this zone were occupied by the indigenous Hawaiians, who probably burned the area inadvertently or deliberately to clear land for agriculture, thereby selecting for fire-tolerant species. In the 1800s, many lowland areas were abandoned as the native Hawaiian populations were decimated by disease and the survivors moved. Fuel loadings were low in these principally native sclerophyllous woodlands because of the paucity of grasses. However, invasions of alien bunchgrasses have greatly increased fire frequency and size. Fire suppresses most native shrubs, particulary pūkiawe, ‘ulei (*Osteomeles anthyllidifolia*), and ‘ākia, and increases alien grass cover, particularly bush beardgrass and molasses grass (Smith *et al.* 1980; Hughes 1989). ‘A‘ali‘i is an early colonizer and may reach higher densities than in pre-fire communities. The fire potential of the submontane seasonal zone is now very high because of alien grasses.

Dry Lowland Ecosystems

This zone was the most heavily impacted by aboriginal Hawaiians, and alien plant species now dominate many coastal lowland areas. Many native species, for example, wiliwili and lama, remaining in this environment are not adapted to fire. For this reason, relictual native woodlands are now threatened by fine fuel loading from alien grasses.

Kartawinata and Mueller-Dombois (1972) observed that fire in the coastal lowlands of O‘ahu maintained and extended alien grass communities. Broomsedge, bush beardgrass, fountain grass, and thatching grass, as well as the native pili grass, have replaced shrub communities in Hawaii Volcanoes National Park after fire (J.T. Tunison, unpub. data).

There are extensive remnants of a forest dominted by hala *(Pandanus odoratissimus)*, ‘ōhi‘a, and uluhe inland of Kolo Point, Hawai‘i. These forest remnants have been burned on several occasions recently. Hala and the alien shrub Malabar melastome *(Melastoma candidum)* disappeared from the community very rapidly, and the aerial portions of ‘ōhi‘a have been killed. Alien broomsedge, bamboo orchid *(Arundina graminifolia)*, melochia (*Melochia umbellata*), and native hi‘aloa *(Waltheria americana)* commonly invade after each fire, whereas the native uluhe only reinvades as long as the forbs and shrubs are not too dense.

Koa haole and kiawe *(Prosopis pallida)* communities are frequently burned. The recovery rate of kiawe is approximately 20%. Fountain grass and buffel grass (*Cenchrus ciliaris*) rapidly invade or take over the area. Fire intensifies the cover of koa haole, which resprouts from the base and becomes established from seed. The fire potential of the coastal lowlands is very high because of low rainfall and continuously arranged fine fuels.

Coastal Strand Ecosystems

Coastal strand environments are rarely subjected to fire because most of the communities have insufficient fuel to carry it. The heavy salt loading from ocean spray may also retard fires. However, when these communities are burned the majority of the species are killed. The succulent species, such as naupaka kahakai (*Scaevola sericea*), are particularly sensitive. Other species, such as *Chamaesyce degeneri, Myoporum,* and

Wikstroemia, resprout. Coastal strand ecosystems are not readily invaded by alien grasses.

CONCLUSIONS AND RECOMMENDATIONS

The use of natural fire is not appropriate to the management of native ecosystems in Hawai'i. Studies and observations to date indicate that fires of all intensities, timing, and causes are harmful to most native ecosystems in Hawai'i. Alien species, particularly grasses, recover rapidly and fill in the spaces left by native plants, which recover more slowly, if at all. The intensification of fire-prone alien grasses following fire assures the increased incidence of fire, thereby establishing a self-perpetuating alien grass/fire cycle. The effects of fire argue strongly that all agencies responsible for the management of fire in natural areas in Hawai'i should aggressively suppress all fires, whether caused by lava flows, lightning, or humans. Fire size should be minimized, even though limited suppression resources may necessitate confinement or containment strategies, with control lines established at a distance from the fire, rather than direct attack.

Prescribed burning is not, on the whole, an effective management tool in Hawai'i, although it may have potential limited management and research value. On the U.S. Mainland, prescribed burning is useful in restoring natural ecological processes, rectifying anthropogenically altered vegetation patterns, reducing hazardous fuel loadings, enhancing wildlife habitat, and in a few cases controlling alien plants. However, fire is not useful in controlling alien plant species in native Hawaiian ecosystems, although it may be useful in controlling selected alien plants in managed systems dominated by alien plants. Kiawe and melastomes, for example, are relatively sensitive to fire and may be controlled by burning.

Prescribed fire in Hawai'i may be useful in accelerating the recovery of 'a'ali'i shrublands and koa forest or woodlands, because these two species are stimulated by fire. However, before fire is used for this purpose, the effects of fire on stimulating alien species, damaging sensitive native plants in these communities, and natural recovery should be considered. Fire may have been a natural ecological factor in montane seasonal shrublands, grasslands, and koa forest.

Prescribed fire may also be useful in managing habitat for the Hawaiian goose or nēnē *(Nesochen sandwicensis)*. Wildfire at Hawaii Volcanoes National Park provided an attractive, albeit short-term, food source in the form of resprouting bush beardgrass, broomsedge, and other alien grasses. Wildfire also created an open habitat favorable for nesting. There were twelve nesting attempts in 1987-1988 in the Uila Burn, an area with less than 17 attempts in the six years previous to the burn (H. Hoshide, pers. comm.). Fledgling success was apparently limited by mongoose (*Herpestes auropunctatus*) predation, which is controllable by diphacinone bait stations (H. Hoshide, pers. comm.).

Prescribed fire is essential in developing customized fuel models, which enable fire suppression personnel to predict fire behavior (rate of spread, fire intensity, fire size). This information is invaluable in developing fire suppression strategies. Prescribed fire is also important in understanding fire effects because pre-fire conditions can be characterized.

Suppression capabilities for wildland fire in natural areas are limited in Hawai‘i by poorly developed detection, fire suppression, and logistical limitations in the State. Historically low fire frequency has resulted in complacency on the part of natural area managers; however, nowadays fires may become quite large and destructive. Some of these limitations can be overcome by following these measures:

1. Develop an aggressive program of fire prevention. Most fires are human-caused in Hawai‘i. Alien grass fuels are spreading in the State at the same time that residential development is expanding into rural areas, and both trends increase the threat of fire to natural areas. A fire prevention program in Hawai‘i should be developed around local symbols and themes to appeal to school-age children. A fire prevention consciousness among younger people can be raised by educational contacts from fire personnel in the schools, media messages, and fire prevention signing in fire-prone areas.

2. Formulate strong interagency agreements that provide rapid movement of fire fighters and resources to wildfires on one island or jurisdiction to another. This is especially important in Hawai‘i, where limitation of personnel and equipment necessitate sharing resources, and islands with the most extensive natural areas have the smallest populations. Interagency agreements will develop a mobile, well-equipped, trained cadre of fire fighters from the Hawaii Department of Land and Natural Resources, National Park Service, Department of Defense, The Nature Conservancy, and counties. A unified fire organization can be developed by use of the Incident Command System, which is practiced widely on the U.S. Mainland and provides for common organizational roles, terminology, standards, and training. Interagency agreements can be reinforced by common training sessions sponsored by different agencies and a statewide coordinating committee that meets annually to discuss mutual problems and share information. The national system of fire danger rating should be used throughout Hawai‘i.

3. Expand the scope of county fire departments to wildland fire. This measure will increase the cadre of fire suppression personnel and protect natural areas under county fire jurisdiction.

4. Allocate more funding to wildland fire to maintain high standards of preparedness, training, and effectiveness. More fire management specialist positions, unencumbered by collateral duties, should be established.

5. Establish fuel breaks in the most fire-prone areas to break up extensive fuel beds or protect particularly valuable biological resources. Hazardous fuels should be routinely reduced in areas of probable human-caused ignitions.

6. Develop a program of presuppression planning, rather than simply responding to fires once they occur. Presuppression planning includes all those actions taken to minimize fire size and improve suppression capabilities prior to a fire. In addition to establishing fuel breaks, it includes mapping the distribution of fuel types, fuel breaks, special biological resources, water sources, and helicopter landing sites. These maps serve as the basis for developing suppression strategies and tactics prior to fires.

The most important fire research needs in Hawai'i are an understanding of fire effects and the development of local fuel models. Fire effects studies have been conducted principally in Hawaii Volcanoes National Park and Pu'uwa'awa'a Ranch and should be undertaken in other natural areas in Hawai'i to indicate the response of fire in the variety of vegetation types present in the State. Studies at Hawaii Volcanoes have begun to determine why alien grasses are more successful than native plants after fire. The effects of fire exclusion on Hawaiian vegetation that evolved with fire, such as the montane seasonal type in Hawaii Volcanoes, should be investigated. In Mainland ecosystems, the exclusion of fire may result in destructive accumulations of fuel and the loss of species and community diversity.

Research should be conducted to establish fuel models for the State. Fuel models will allow rapid prediction of rates of spread, fire intensity, and size, the most important information in planning suppression or presuppression strategies. Preliminary work on characterizing Hawaiian fuels was undertaken by the U.S. Forest Service in the early 1970s (Fujioka 1976; Fujioka and Fujii 1980). These investigations and preliminary studies at Hawaii Volcanoes indicate that the standard Fire Behavior Models do not describe fuels beds dominated by broomsedge, bush beardgrass, and molasses grass. Fire behavior data from local fuels are insufficient to determine if the standard models accurately describe local fire behavior. If not, customized models can be constructed and revised based on fuel bed characteristics and, more importantly, actual fire behavior observed. A series of small prescribed fires is needed to provide a range of wind speed and fuel moisture conditions and measurement of fire and fuels under controlled conditions. The utility of customized fuel models needs to be closely considered before prescribed fire is used. The most important considerations include whether training of local fire managers is sufficient to use the modern fire behavior prediction system, how valuable fuel models are in grass fuels that have generally very high rates of spread from head fires and similar rates of spread on backing fire, impacts on native vegetation, low frequency of fire in Hawai'i, and costs to managers to conduct prescribed fire.

Literature Cited

Buechner, H.K., and H.C. Dawkins. 1961. Vegetation change induced by elephants and fire in Murchinson Falls National Park, Uganda. *Ecology* 42:752-766.

Cuddihy, L.W., and C.P. Stone. 1990. *Alteration of native Hawaiian vegetation: effects of humans, their activities and introductions.* Univ. Hawaii Coop. Natl. Park Resour. Stud. Unit. Honolulu: Univ. Hawaii Pr.

Fujioka, F.M. 1976. Fine fuel moisture measured and estimated in dead *Andropogon virginicus* in Hawai'i. USDA For. Serv. Res. Note PSW-317.

Fujioka, F.M. and D.M. Fujii. 1980. Physical characteristics of selected fine fuels in Hawai'i -- some refinements on surface area-to-volume calculations. USDA For. Serv. Res. Note PSW-348.

Hall, W.L. 1904. *The forests of the Hawaiian Islands.* Bull. 48. U.S. Dep. Agric., Bureau Forestry. Washington, D.C.: U.S. Govt. Pr. Ofc.

Heinselman, M.L. 1981. Fire intensity and frequency as a factor in the distribution and structure of northern ecosystems. *Proc. Conf. Fire Regimes and Ecosystem Properties*. U.S. Forest Service General Tech. Rep. WO-26. pp. 7-57.

Hughes, F. 1989. The coupled effects of grass invasion and fire on the submontane seasonal ecosystem of Hawaii Volcanoes National Park. Senior Honor's Thesis, Stanford Univ., Stanford, California.

Kartinawata, K., and D. Mueller-Dombois. 1972. Phytosociology and ecology of the natural dry-grass communities on Oahu, Hawaii. *Reinwardtia* 8(3):369-472.

Kilgore, B.M. 1981. Fire in ecosystem distribution and structure: western forests and scrublands. *Proc. Conf. Fire Regimes and Ecosystem Properties.* U.S. For. Serv. Gen. Tech. Rep. WO-26, 7-57.

Kirch, P.V. 1982. The impact of the prehistoric Polynesians on the Hawaiian ecosystem. *Pac. Sci.* 36(1):1-14.

McEldowney, H. 1979. Archeological and historical literature search and research design. Lava flow control study, Hilo, Hawaii. Manuscript No. 050879, Dept. of Anthropology, B. P. Bishop Museum, Honolulu, Hawai'i. Prepared for U.S. Army Corps of Engineers, Contract No. DACWB4-77-C-0019.

Mueller-Dombois, D. 1973. A non-adapted vegetation interferes with water removal in a tropical rain forest area in Hawaii. *Trop. Ecol.* 14:1-18.

Mueller-Dombois, D. 1980. Spatial variation and vegetation dynamics in the coastal lowland ecosystem, Hawaii Volcanoes National Park. *Proc. Third Conf. Nat. Sci., Hawaii Volcanoes National Park*, 235-248. Honolulu: Univ. Hawaii Coop. Natl. Park Resour. Stud. Unit.

Mueller-Dombois, D. 1981. Fire in tropical ecosystems. In *Fire regimes and ecosystem properties. Proc. of the Conf.*, 137-176. U.S. Forest Serv. General Tech. Rep. WO-26.

Mueller-Dombois, D., and C.H. Lamoureux. 1967. Soil-vegetation relationships in Hawaiian kipukas. *Pac. Sci.* 21(2):286-299.

Mueller-Dombois, D., and G. Spatz. 1975. The influence of feral goats on the lowland vegetation in Hawaii Volcanoes National Park. *Phytocoenologia* 3(1):1-29.

Mutch, R.W. 1970. Wildland fires and ecosystems: a hypothesis. *Ecology* 51:1046-1051.

National Park Service. 1990. Wildland fire management plan and environmental assessment (Hawaii Volcanoes National Park). U.S. Dep. Int. Natl. Park Serv.

Pyne, S.J. 1982. *Fire in America: a cultural history of wildland and rural fire.* Princeton, New Jersey: Princeton Univ. Pr.

Sarmiento, G., and M. Monasterio. 1975. A critical consideration of the environmental conditions associated with the occurrence of savanna ecosystems in Tropical America. In *Tropical ecological systems*, ed. F.B. Golley and E. Medina. Ecol. Stud. Ser. 11:223-250. New York: Springer-Verlag.

Scowcroft, P.G., and H.B. Wood. 1976. Reproduction of *Acacia koa* following wildfire. *Pac. Sci.* 30(2):177-186.

Smith C.W. 1985. Impact of alien plants on Hawai'i's native biota. In *Hawai'i's terrestrial ecosystems: preservation and management*, ed. C.P. Stone and J.M. Scott, 180-250. Univ. Hawaii Coop. Natl. Park Resour. Stud. Unit. Honolulu: Univ. Hawaii Pr.

Smith, C.W., T. Parman, and K. Wampler. 1980. Impact of fire in a tropical submontane seasonal forest. *Proc. 2nd Conf. Sci. Res. Natl. Parks*, Vol. 10, *Fire ecology*, 313-324. San Francisco, Nov. 26-30, 1979. Natl. Park Serv.

Sorensen, J. 1979. Fire, lightning and 'ohi'a decline. *Newsletter, Hawn. Bot. Soc.* 18:9-23.

Tunison, J.T., and J.A.K. Leialoha. 1988. The spread of fire in alien grasses after lightning strikes in Hawaii Volcanoes National Park. *Newsletter, Hawn. Bot. Soc.* 27(3):102-109.

Vogl, R.J. 1969. The role of fire in the evolution of the Hawaiian flora and vegetation. In *Proc. Tall Timbers Fire Ecology Conf.*, April 10-11, 1969, Tallahassee, Florida.

Vogl, R.J. 1975. Fire: A destructive menace or a natural process? In *Recovery and restoration of damaged ecosystems*, ed. J. Cairns, Jr., K.L. Dickson, and E.E. Herricks. Charlottesville, Virginia: Univ. Pr. Virginia.

Warshauer, F.R. 1974. Biological survey of Kealakomo and vicinity affected by 1969-1973 lava generated wildfires, Hawaii Volcanoes National Park. Files, Resources Management Div., Hawaii Volcanoes Natl. Park. Unpub. rep.

Warshauer, F.R. 1975. Vegetation report on the Six Tanks Fire. Files, Resources Management Div., Hawaii Volcanoes Natl. Park. Unpub. rep.

Wester, L.L., and H.B. Wood. 1977. Koster's curse (*Clidemia hirta*), a weed pest in Hawaiian forests. *Environ. Conserv.* 4(1):35-41.

BIOLOGICAL CONTROL

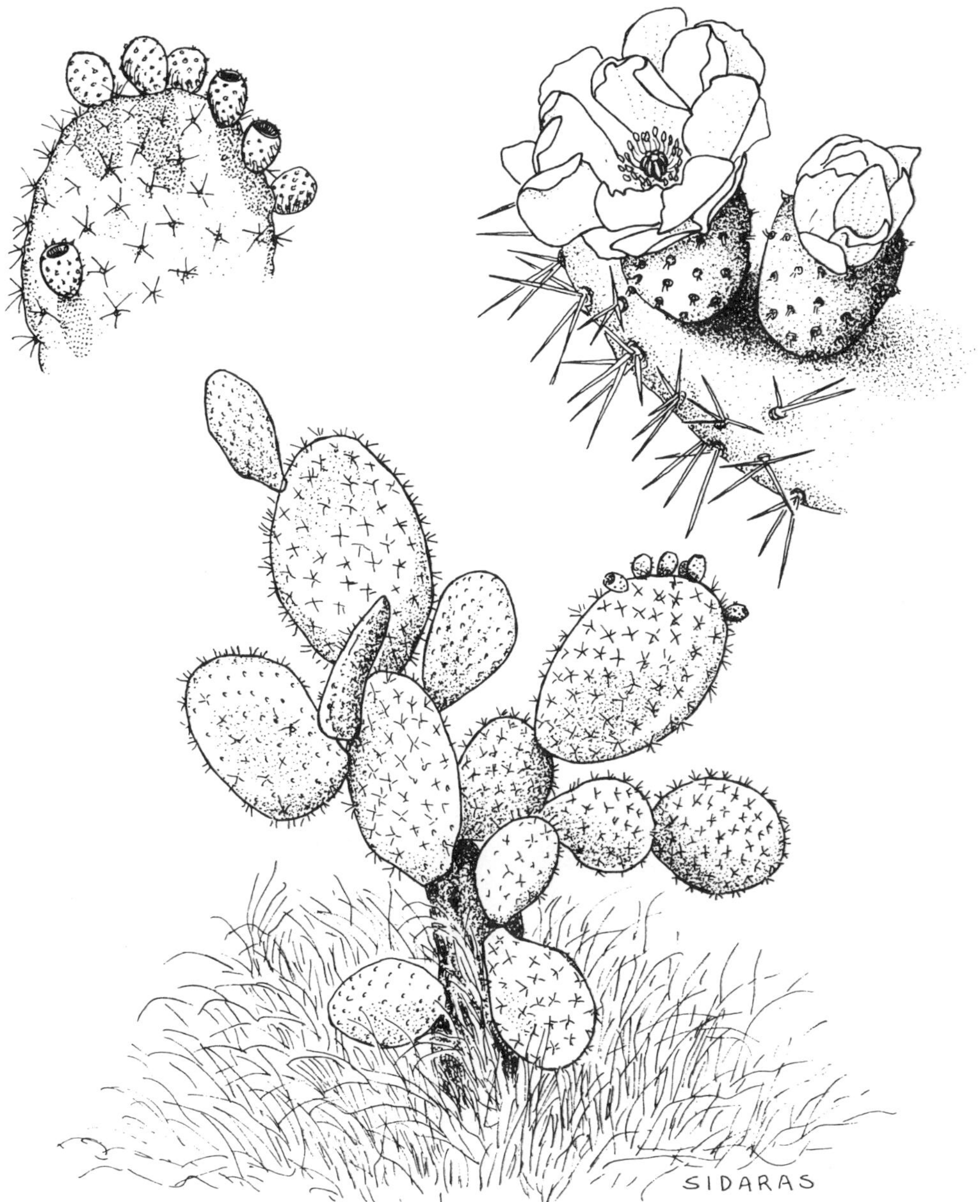

BIOLOGICAL CONTROL OF LANTANA, PRICKLY PEAR, AND HĀMĀKUA PĀMAKANI IN HAWAI'I: A REVIEW AND UPDATE

Clifton J. Davis, Ernest Yoshioka, and Dina Kageler

ABSTRACT

The biological control of noxious weeds in Hawai'i has been carried on intermittently since 1902, when insects and diseases of lantana (*Lantana camara*) were sought in Mexico by the Territorial Board of Agriculture and Forestry (now Hawaii Department of Agriculture). This approach was subsequently employed for the control of 20 other noxious weed pests between the 1940s and 1970s. Lantana was the first weed to be controlled by this method in the U.S. Results were very dramatic in some areas of the State, especially after later introductions by Hawaiian and Australian entomologists resulted in heavy stress on lantana. In addition to lantana, excellent results have been obtained in the biological control of cacti (*Opuntia* spp.), and Hāmākua pāmakani (*Ageratina riparia*). Prior to the introduction of cactus insects in 1949, 66,000 a (26,400 ha) of Parker Ranch range lands on Hawai'i Island were infested with cacti. By 1965, 7,610 a (<3,080 ha) remained infested, the result of three introduced insects and an accidentally introduced fungus disease; the red-fruited variety of cactus is particularly susceptible to the fungus. A spineless variety of the cactus occurs in the 'Āinahou-Poliokeawe Pali sector of Hawaii Volcanoes National Park, and biocontrol efforts are in progress. With the introduction of insects from Mexico and a foliar fungus disease from Jamaica, Hāmākua pāmakani is under excellent control on many ranch as well as privately owned and government lands on Hawai'i Island. The pathogenic fungus introduced into Hawai'i specifically for Hāmākua pāmakani represents the first successful attempt to establish a disease of this noxious weed.

INTRODUCTION

Biological control has been attempted on 20 species of alien plants in Hawai'i. Of these, 10, including the three examples discussed herein, have been judged successful in terms of control of the target weed. Biocontrol insects have established on eight additional weeds, but not enough stress on the target plant has been achieved for good control. Two attempts (firetree (*Myrica faya*) and Russian thistle (*Salsola kali*)) are considered failures so far, since no recoveries of the biocontrol insects on target plants have been made. Biological controls achieved for the three weeds discussed below represent classical examples of success.

BIOLOGICAL CONTROL OF LANTANA

Introduction

Lantana (*Lantana camara*) is native to subtropical and tropical Central America and has become established as a weed in many countries of the world. It was introduced to Hawai'i as an ornamental in 1858 (Hillebrand 1888), and by 1902 very large areas were occupied, especially lower, drier areas (Perkins 1924). Perkins stated that "many of these dry areas are said to have afforded good pasture during the wet season before the coming of lantana." He attributed the spread of lantana on all islands largely to two introduced birds, the spotted or lace-necked dove, *Streptopelia chinensis*, and the Indian myna, *Acridotheres tristis*, both of which feed on the aromatic berries. Encroachment of this noxious weed into pasture lands was of much concern to ranchers, as it displaced valuable forage grasses, was costly to control, and reduced the carrying capacity of the range (Perkins 1924). A purposefully introduced plant without its native enemies and under very favorable climatic conditions was spreading throughout the Hawaiian Islands.

Insects on Lantana in Hawai'i Prior to Biocontrol Introductions

An alien scale insect known as the greenhouse orthezia, *Orthezia insignis*, was observed damaging lantana by Maui ranchers about the turn of the century; they named it Maui blight (Fullaway and Krauss 1945). Cattlemen spread the insect on O'ahu, but it was not host specific and became quite a nuisance on cultivated ornamentals. Other insects such as noctuids, probably *Chrysodeixis chalcites*; tortricids (leaf rollers); a native geometrid, *Scotorythra* sp. (looper) (Perkins 1924); and a phycitid, *Cryptoblabes aliena* (Fullaway and Krauss 1945), were recorded, but these were polyphagous (generalist herbivores) and had little if any effect on lantana.

Biological Control Efforts

In 1902, Albert Koebele, then an entomologist under the Commissioner of Agriculture of the Provisional Hawaiian Government, was sent to Mexico to investigate insect and disease enemies of lantana, with the object of introducing into Hawai'i insect species that he considered safe for trial. This was the first time the control of a noxious plant using insect enemies was attempted in what is now the United States. Large numbers of insects were found in Mexico attacking lantana flowers, seeds, flower stalks, leaves, stems, terminal branches, and roots. By the close of 1902, Koebele had shipped 23 different insect species to Hawai'i, many of which were found heavily parasitized upon arrival. Host specificity studies were not conducted, as Koebele shipped only those insects he considered safe for release (Swezey 1924). Of those insects liberated, eight species became established in the vicinity of Honolulu and from there spread to the other islands (Table 1). To quote Swezey (1924), "Just how they were distributed to the other Islands is not recorded but it is presumed that this was done shortly after they became established in Honolulu." Natural spread of insects to neighbor islands has been recorded (Davis 1978). By 1905, introduced lantana insects were generally distributed throughout the Islands, and by 1924 the area occupied by lantana appeared to be much less than previously (Swezey 1924).

Swezey (1924) noted that "each of the eight introduced insects attacks the lantana plant in a somewhat different way and their combined attack results in greatly checking the production of seed and thus minimizes its chances for further spreading or reoccupying land from which it had been cleared." Although Spencer (in Hardy and Delfinado 1980) questioned the effectiveness of the seed fly *Ophiomyia lantanae* in controlling lantana, Swezey's reasoning for the combined effects of insect attack appears valid. Swezey regarded the larvae of the tortricid moth *Epinotia lantana* as probably the most beneficial of all insects in checking production of berries, because it bores into the flower stems, feeds in the receptacles of the flower clusters, and eats flowers and fruit.

Table 1. Insects introduced in 1902 for lantana control that became established in the Hawaiian Islands.

Order	Species	Activity
Diptera (flies)	*Ophiomyia lantanae* (Froggatt) (agromyzid)	Fruit infesting
	Eutreta xanthochaeta Aldrich (tephritid)	Stem galling
Lepidoptera (moths)	*Epinotia lantana* (Busck) (tortricid)	Flower, shoot, seed feeding
	Cremastobombycia lantanella Busck (gracillariid)	Leaf mining
Lepidoptera (plume moth)	*Lantanophaga* (= *Platyptilia*) *pussillidactyla* (Walker) (pterophorid)	Flower feeding
Lepidoptera (butterflies)	*Strymon echion* (L.) (lycaenid)	Flower feeding
	S. bazochii gundlachianus (Bates) (lycaenid)	Flower feeding
Hemiptera (lace bug)	*Teleonemia scrupulosa* Stal (tingid)	Leaf feeding

Lantana was still a problem on many ranches in 1952. Surveys by Hawaiian and Australian entomologists indicated that earlier biological control agents were not providing sufficient stress for effective control. This may have been due to the effects of parasitism on some of the introduced insect species such as the braconid *Opius lantanae* on *Ophiomyia lantanae* (Bridwell 1919) and *Bracon mellitor* (Braconidae), *Perisierola emigrata* (Bethylidae), and *Pristomerus hawaiiensis* (Ichneumonidae) on *Epinotia lantana* (Zimmerman 1978). *Pachodynerus nasidens*, an eumenid wasp, may be an important predator of *E. lantana* (Fullaway and Krauss 1945). The eggs of the lantana butterflies *Strymon echion* and *S. bazochii gundlachianus* are often heavily parasitized by a tiny wasp, *Trichogramma* sp. (Fullaway and Krauss 1945).

Further exploration for biocontrol agents was undertaken in 1952 by the entomologist N.L.H. Krauss, who carried on investigations intermittently from 1952 to 1965 throughout tropical America and other regions. This work was followed by a cooperative arrangement in 1953 among Fiji, Trust Territory of the Pacific Islands; Queensland, Commonwealth of Australia; and Hawai'i. N.L.H. Krauss and J. Mann of Australia (during 1953-54) were the principal explorers (Davis and Krauss 1962a).

Thirteen candidate insects were received in the Hawaii Department of Agriculture quarantine facility, Honolulu, as a result of these explorations. Ten species were recommended for release after extensive host range tests by the Hawaii Department of Agriculture Advisory Committee and approved for release by the Board of Agriculture (Table 2). Three of the candidates, a cossid, *Langsdorfia franckii*, and the foliar chrysomelids *Octotoma plicatula* and *O. gundlachi*, failed to propagate sufficiently and died in quarantine. Of the seven remaining approved insects, six became established statewide. In addition, a foliar-feeding tingid from Peru, *Leptobyrsa decora*, was received in 1969 from K.L.S. Harley, an Australian entomologist, and is well established in certain areas on Kaua'i, Maui, and the island of Hawai'i. Strains of the tingid *Teleonemia scrupulosa* were introduced from Brazil, British Honduras, Trinidad, and Florida (Weber 1955) but are not included in the table because they are conspecific with the previously established strain from Mexico.

Caterpillars of the pyraustid *Syngamia haemorrhoidalis* denuded many acres of lantana at Mokule'ia, O'ahu, beginning in 1957. Up to 1958, this species was the only outstanding lepidopterous skeletonizer-defoliator of lantana (Davis and Krauss 1962a). The depredations caused by this insect were followed by the purposeful introduction of another lepidopterous leaf feeder, *Hypena strigata*. Establishment of *Hypena* was somewhat slower, but population irruptions occurred in Lāwa'i Valley, Kaua'i, in late 1958. Foliar devastation in thousands of acres throughout the State occurred between 1959 and 1961. In most lantana infestations, *Hypena* replaced *Syngamia* as the dominant control agent, building up very large populations during the winter months and leveling off during the summer, when *T. scrupulosa* replaced *Hypena* and *Syngamia* in importance from sea level to approximately 2,520 ft (763 m) elevation. In 1960, five

Table 2. Insects introduced into Hawai'i 1952-1969 to control lantana.

Name	Family	Date Introduced & Origin	Established	Site of Activity
Teleonemia vanduzii Drake	Tingidae	1952 (Cuba)	-	foliar
Blepharomastix acutangulalis (Snellen)	Pyraustidae	1953 (Mexico)	-	foliar
Octotoma gundlachi Suffrain	Chrysomelidae	1953 (Cuba)	-	foliar
Plagiohammus spinipennis (Thomson)*	Cerambycidae	1954-60 (Mexico)	+	stem
Octotoma plicatula (Fabricius)	Chrysomelidae	1954 (Honduras)	-	foliar
Octotoma scabripennis (Guer)*	Chrysomelidae	1955-59 (Mexico)	+	foliar
Aerenicopsis championi Bates*	Cerambycidae	1955 (Mexico)	-	branches
Catabena esula (Druce)	Noctuidae	1955 (California)	+	foliar
Langsdorfia franckii Hubner	Cossidae	1955 (Mexico)	-	roots
Syngamia haemorrhoidalis Guenee	Pyraustidae	1956 (Cuba, Florida)	+	foliar
Hypena strigata F.	Noctuidae	1957 (Kenya, E. Africa)	+	foliar
Uroplata girardi Pic	Chrysomelidae	1961 (Brazil)	+	foliar
Diastema tigris Guenee	Noctuidae	1962 (Panama Canal Zone)	-	foliar
Leptobyrsa decora Drake	Tingidae	1969 (Peru)	+	foliar

*Previously introduced in 1902 but did not become established.

years after its introduction from California, *Catabena esula*, another lepidopterous defoliator, began to increase on Hawai'i, Maui, and O'ahu and later caused extensive defoliation at South Point, Kohala, and other localities on Hawai'i Island (Davis and Krauss 1962b).

The impact of these lepidopteran introductions in the late 1950s and 1960s began to have an effect on lantana in some localities. Dieback to ground level was observed in the drier parts of the island of Hawai'i. At Kēōkea, Maui, lantana was defoliated continuously over many hundreds of acres despite heavy rains. Leaves that managed to appear temporarily on some plants were aborted; this appeared to be a prelude to dieback. On West Moloka'i, drought and heavy insect pressure by *Hypena* killed many acres of lantana (Davis and Krauss 1962a).

In the meantime, the foliar chrysomelids *Octotoma scabripennis* and *Uroplata girardi*, as well as the stem-boring cerambycid *Plagiohammus spinipennis*, became well established on the island of Hawai'i. These caused considerable damage to lantana in the wetter areas of Kona and Ka'ū (Davis and Krauss 1966). The tingid *Leptobyrsa decora* also attacks leaves, adding to the foliar problems of its host, and noticeable damage was reported recently at Pu'uanahulu, North Kona (Yoshioka, 1986 unpub. data).

Biocontrol of lantana by most established insects appeared to have attained an equilibrium by 1969. This may have been due to unfavorable weather conditions or lepidopterous parasitism, especially to the pupae of the California noctuid, *C. esula*, by the ichneumonid *Ecthromorpha fuscator*, and egg parasitism of the East African noctuid *Hypena strigata* by *Trichogramma* sp. (Yoshioka 1970).

The discovery of the stem- and trunk-boring cerambycid *Plagiohammus spinipennis* and the leaf-mining chrysomelid *Octotoma scabripennis* at 2,260 ft (914 m) elevation at Kahauloa and Pu'ulehua parcels of land, Kona, in February 1969 was a significant development. This represented a considerable natural spread from the original release point at lower Kahauloa, 680 ft (274 m) in elevation, a linear distance of approximately 6 to 10 miles (9.6-16 km) (Davis 1970). Thus far, there are no records of parasitism on *O. scabripennis* and *Uroplata girardi*. Larval parasitism on *P. spinipennis* by the braconid *Doryctes palliatus* (Cameron) has been recorded (Harley 1967).

Lantana in Hawaii Volcanoes National Park

Lantana occurs from sea level to approximately 2,350 ft (950 m) elevation in Hawaii Volcanoes National Park (much of it in dry habitats). It is generally scrubby in growth form. Most of the lantana insects are present, and at Hōlei Pali the butterfly *Strymon bazochii* has been frequently observed on lantana, where it feeds on the flowers and seeds (Williams 1980). In February 1986, severe foliar damage was observed at Poliokeawe Pali (1,520 ft or 615 m elevation) and at the approach to the 'Āinahou Ranch house (2,350 ft or 950 m elevation) (Davis, unpub. data). Defoliation was caused by *Hypena strigata* and *O. scabripennis*. At Poliokeawe Pali, sporadic dieback of lantana was also observed and was perhaps due to incessant foliar damage by *T. scrupulosa*, *H. strigata*, and possibly *C. esula*. Stem galls initiated by *Eutreta xanthochaeta* were also observed, but the incidence of gall infestation was low and not considered important.

Present Status

Lantana is generally under partial to substantial control in the drier areas of Hawai'i (Table 3), but there are some areas such as Hawaii Volcanoes National Park and possibly others where biocontrol could be improved (Fig. 1). The foliar feeding bug *L. decora* has not been observed in the Park, and the chrysomelids *O. scabripennis* and *U. girardi* have not occurred at damaging population levels to date. In the wetter areas of Hawai'i, control has been improved by the work of *P. spinipenis*, *O. scabripennis*, *U. girardi*, *H. strigata*, and *L. decora*; but more continuity of depredation is needed.

Recommendations

For the State at large, no further introductions for biological control in the drier areas are recommended. For the wetter areas, the reintroduction of the chrysomelids *O. plicatula* and *O. gundlachi* and the introduction of a host-specific fungus are recommended. *O. gundlachi* was previously collected in Cuba, and unless it occurs in other

Table 3. Optimum climatic conditions for lantana insects.

Control Agent	Dry (20-40 in.)	Wet (50-100+ in.)	Wet and Dry Areas
Teleonemia scrupulosa Stal.	X		
Ophiomyia lantanae Froggatt	X		
Epinotia lantana Busck			X
Lantanophaga (=*Platyptilia*) *pussillidactyla* (Walker)	X		
Strymon echion (L.)	X		
Strymon bazochii gundlachianus (Bates)	X		
Octotoma scabripennis (Guerin)			X
Uroplata girardi Pic			X
Cremastobombycia lantanella Busck		X	
Eutreta xanthochaeta Aldrich			X
Plagiohammus spinipennis (Thomson)		X	
Hypena strigata Fab.		X	
Syngamia haemorrhoidalis Guenee			X
Leptobyrsa decora Drake			X
Catabena esula Druce	X		

countries, it may not be available. Since the established leafmining beetles are apparently free of parasites and predators, the beetles proposed for reintroduction would most likely not be parasitized and would put more stress on lantana when damaging population levels are reached.

The foliar bug *L. decora* should be collected and distributed in Hawaii Volcanoes to augment and provide additional stress to lantana in dry areas where other biocontrol agents are established.

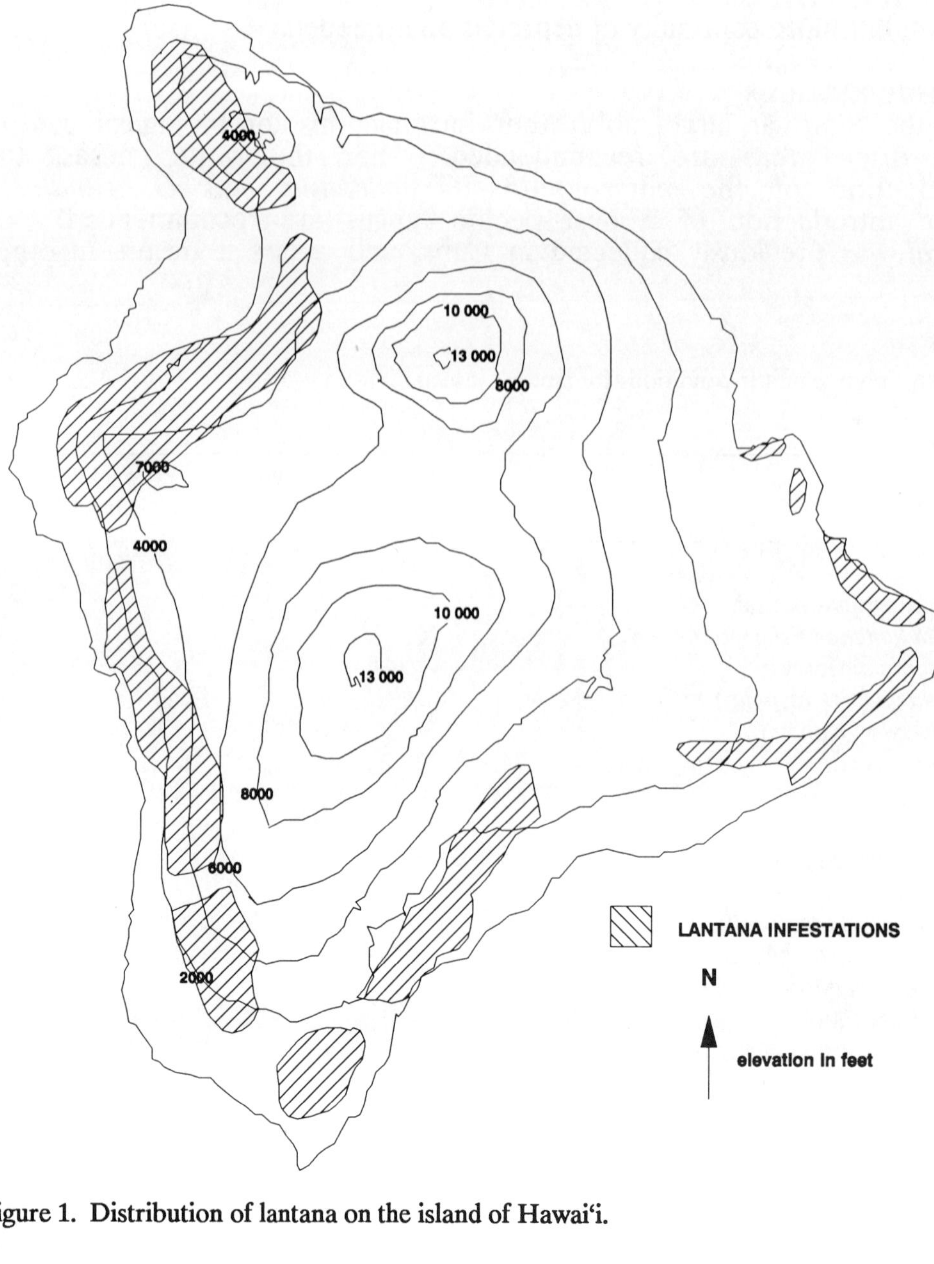

Figure 1. Distribution of lantana on the island of Hawai'i.

BIOLOGICAL CONTROL OF PRICKLY PEAR CACTUS (*OPUNTIA* SPP.)

Introduction

The prickly pear, or pānini as it is locally called, is represented by two species in Hawai'i, *Opuntia ficus-indica* (= *megacantha*) and *O. cordobensis*. These species are native to the drier parts of tropical America. Don Marin probably introduced *O. ficus-indica* to Hawai'i from Mexico about 1809 (Hosaka and Thistle 1954).

Cacti generally occur in the dry lowland areas on all islands, occasionally up to 2,640 ft (1,068 m) elevation such as on Parker Ranch, Hawai'i Island. Thick stands once occurred in the coastal areas between Puakō and Māhukona, inland from Kawaihae to lower Waimea, and from Waimea to Ke'āmuku, the last up to 2,640 ft (1,068 m) elevation. Prickly pear was also common in Ka'ū District, between Honu'apo and Punalu'u, on Hawai'i Island. A spineless cactus was introduced from the King Ranch, Texas, in 1937 and planted in the lower Keauhou-'Āinahou tract of Hawaii Volcanoes National Park at approximate elevations of 1,640 to 2,070 ft (500-630 m) (T. Lindsey, pers. comm. 1986).

Botanists disagree as to the identity and correct nomenclature of the *Opuntia* species common in Hawai'i. St. John (1973) listed *O. megacantha* as the common, established prickly pear cactus in Hawai'i. This plant has also been known by the name *O. tuna* (Hillebrand 1888). However, Benson (1982) considered *O. megacantha* to be a synonym of *O. ficus-indica* (India fig), which he recorded as escaped and naturalized on the dry leeward side of the Hawaiian Islands. He reported that spineless horticultural forms of *O. ficus-indica* are common and widespread. Much of the cactus in Hawaii Volcanoes is a spineless form, presumably a cultivar of *O. ficus-indica*. There is, however, some question as to the identify of some cacti in the Park (L.W. Cuddihy, pers. comm.), as hybrid forms may have been introduced into the 'Āinahou area of the Park during the time it was used as a cattle ranch.

The cochineal cactus, *Nopalea cochenillifera*, is also present in Hawai'i (Degener 1946). The cladodes or "pads" of this species are usually spineless. Luther Burbank introduced this plant to Santa Rosa, California, from the Hawaiian Islands and Mexico for his experiments with spineless cactus (Burbank 1914).

Uses

There are many uses for cacti in tropical America, such as food, drink, and medicine; in Mexico a commercially valuable red dye is derived from the cochineal insect *Dactylopius coccus*, which breeds on *Nopalea cacti*, *N. nopalnocheztli*, and *N. nocheznopalli* (Ross 1986). In the low, dry areas of Hawai'i, cattle used the common wild cactus *O. ficus-indica* for moisture, and the Hawaiians made a fermented drink from the fruits and also ate them raw (Neal 1965).

Recorded Enemies of Cacti Prior to Hawai'i State Introduction of Biocontrol Agents

A form of the fungus *Fusarium oxysporum* destroys the red-fruited cactus and was recorded from Kaua'i, Ni'ihau, O'ahu, Moloka'i, and Maui (Carpenter 1944). A cactus scale, *Diaspis echinocacti*, infests *Opuntia* spp. (Zimmerman 1948) but is relatively unimportant. A dipterous scavenger, *Copestylum* (= *Volucella*) *tamaulipana*, was frequently found in rotting cactus pads (Swezey 1932). An undetermined scab (fungus) is present on spineless cactus, sometimes occupying much of the pad surface, but is also relatively unimportant (C.J. Davis, pers. observ. 1985).

The Problem

Prior to 1949, tree cactus, *O. ficus-indica*, spread at an alarming rate on the island of Hawai'i, covering the lowlands almost solidly in many areas and spreading into upper elevations and better pastures. Attention was drawn to this menace as early as 1930 by the former manager of Parker Ranch, W.W. Carter. The mechanical destruction of cactus appeared too costly in those days. Hearing of the great success of the Australian government in tackling the cactus problem there by the biological control method, Carter turned to the local government agency, the Hawaii Board of Agriculture and Forestry (now Hawaii Department of Agriculture) to conduct a similar program, if feasible (Fullaway 1952).

Introductions for Biological Control

At first, ranchers responsible for lands with poor water supplies opposed the introduction of cactus insects. They believed that cactus was an asset to them in times of drought. It took years to overcome this opposition. In the interim, a joint project to effect destruction of the red-fruited variety of cactus by the fungus disease spores of *Fusarium oxysporum* was undertaken by Parker Ranch and the Board of Agriculture and Forestry. This project continued until 1949 (Fullaway 1954).

In 1949, the Board of Agriculture and Forestry approved the introduction, testing, and release of a number of biocontrol agents for Hawai'i Island only (Table 4). In addition, two species of coreid bugs, *Chelinidea tabulata* and *C. vittiger*, were received from Riverside, California in June 1949. However, in host range testing, positive results on pineapple (*Ananas comosus*) were obtained, and the remaining bugs were ordered destroyed.

The 66,000 a (26,400 ha) of cactus infestation on Parker Ranch in 1949 was reduced to 7,610 a (3,080 ha) by 1965 (H. Kimura, pers. comm. 1986). Sequential photo documentation of cactus destruction by *Cactoblastis cactorum* and *Dactylopius opuntiae* between 1954 and 1965 showed partial to complete destruction of cacti at Ke'āmuku between 2,000 and 3,500 ft (609-1,067 m) elevation (Fig. 2A and 2B). *Opuntia* spp. were totally eliminated from Kawaihae to the lower limit of Waimea (*ca.* 1,700 ft or 518 m elevation) in three to four years by *Dactylopius* and *Cactoblastis*. *Archlagocheirus funestus* replaced *Cactoblastis* and *Dactylopius* in effectiveness off the Kohala Road at 3,000 ft (914 m) elevation and below the Kawaihae-Kohala Junction. In Ka'ū between

Figure 2. Destruction of cactus at Ke'āmuku between 1954 (upper) and 1979 (lower) by *Cactoblastis cactorum* and *Dactylopius opuntiae*.

Table 4. Biocontrol agents introduced to control cactus in Hawai'i.

Date Intro.	Name	Source	Results
1949	*Dactylopius* sp.	California	Failed to establish
1949	*Dactylopius opuntiae* CKLL	Australia, "Mexican strain"	Established
1949	*Melitara doddalis* Dyar	Texas	Failed to establish
1949	*M. prodenialis* Walker	Texas	Failed to establish
1950	*Cactoblastis cactorum* (Berg)	Australia (Uruguay)	Established
1950	*Moneilema armatum* Le Conte	Texas	Failed to establish
1950	*M. crassa* Le Conte	Texas	Failed to establish
1950	*Archlagocheirus funestus* Thomson	Australia	Established

Honu'apo and Punalu'u, pānini has nearly disappeared, due to biocontrol agents. The 1987 distribution of Cactaceae on the island of Hawai'i is much reduced (Fig. 3).

Current Efforts toward Biocontrol of *Opuntia* in Hawaii Volcanoes National Park

Opuntia sp. occurs in the Park in the lower 'Āinahou Ranch area from nearly sea level to 3,100 ft (945 m) elevation. The single most-dense stand occurs on the slope of Poliokeawe Pali and covers an area of approximately 161 ft^2 (15 m^2). Outlying individuals occur singly or in small clumps and are sparsely distributed in the open grassland areas of lower 'Āinahou. Considering the difficult terrain and widely scattered plants, control by means other than biocontrol would seem economically unfeasible.

Methods. *Opuntia* pads heavily infested with the larvae of *C. cactorum* and *D. opuntiae* were collected from the Parker Ranch area of Hawai'i Island and were subsequently set out by National Park Service personnel at two release sites. The primary release site was at Poliokeawe Pali, on the Ka'ū side of the Keauhou Trail in the lower 'Āinahou area of Hawaii Volcanoes National Park at 2,000 ft (610 m) elevation. The second release site was on a single *Opuntia* individual occurring approximately 130 ft (40 m) south of the 'Āinahou Ranch house at an elevation of 3,100 ft (945 m). Four releases totalling approximately 40 infested pads occurred through January 1987 at the primary site. Releases were made in May and September 1985, October 1986, and January 1987. A single release of approximately seven infested pads was made at the 'Āinahou Ranch house site in September 1985. At both sites, pads of collected infested cactus were lashed with twine or flagging tape directly to apparently healthy pads so that infested and healthy surfaces were in maximum contact. Additionally, at the 'Āinahou Ranch house site, holes were drilled into host pads with a brace and bit. Infested pads were lashed onto the drilled pads so that

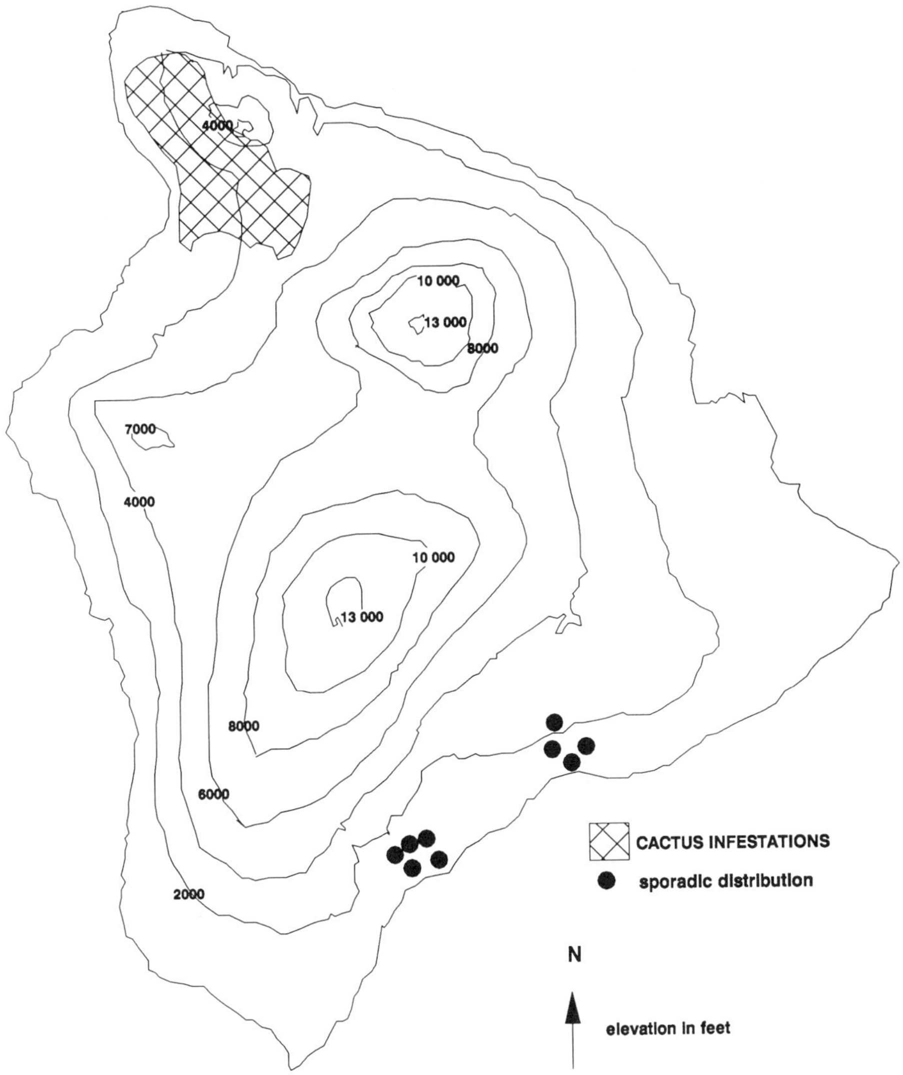

Figure 3. Distribution of cactus on the island of Hawai'i.

newly introduced *Cactoblastis* larvae could transfer to the new host through the holes provided.

Cactoblastis moths lay their eggs on the spines of cactus (Fullaway 1954). Since the cactus in the Park is primarily a spineless variety, prevention of the establishment of *Cactoblastis* populations was a concern. To enhance establishment, wooden toothpicks were placed in two host cactus pads in close proximity to larval release sites. Round and flat toothpicks were tested as surrogate spines. Toothpicks were cut in half with clippers, and the blunt end of the wooden pick was stuck through the epidermal layer to a depth of 0.4 to 0.8 in. (1-2 cm). Approximately 10 surrogate spines were inserted in the test pads.

Six permanent photopoints were established at the Poliokeawe site to monitor the long-range vigor or decline of the cactus plants and to follow the spread of the insect populations.

Results. Poliokeawe Site -- Results at the primary release site on Poliokeawe Pali at nine months showed fresh *Cactoblastis* infestations apparent in at least six pads, and *Dactylopius* colonies active on approximately 12 pads. Furthermore, fresh *Cactoblastis* activity was observed on an outlying cactus individual 165 ft (50 m) from the primary release site. However, 13 months after first release, neither species had become well established. At that time, three old *Cactoblastis*-infested pads were observed, but there was no evidence of fresh activity. Similarly, *Dactylopius* infestation was minimal, with only one colony observed. At 20 months after the first release and at the time of the fourth release, fresh *Cactoblastis* activity was observed in three cactus individuals located 65 to 100 ft (20-30 m) from the primary release site. No fresh *Cactoblastis* activity was observed within the release site. In the approximately 10 colonies observed, low levels of *Dactylopius* activity were noted in the release site.

'Āinahou Ranch House Site -- Three *Cactoblastis* egg sticks were observed on cactus pads at the 'Ainahou Ranch house site in August 1986, 15 months after release. Two of the egg sticks were approximately 1.6 in. (4 cm) in length, while the third egg stick was approximately 0.8 in. (1.5 cm) in length. The egg sticks occurred on the tips of small spines and were not observed on the surrogate spines. At 20 months after release, fresh or recent *Cactoblastis* activity was apparent in four pads, with approximately 12 pads showing old activity. Fewer than five pads remained uninfested. A single *Cactoblastis* larva was observed on a detached pad on the ground beneath the host cactus. No *Dactylopius* colonies were observed.

Discussion. Reasons for the fluctuation in populations of *Cactoblastis* and *Dactylopius* at the Poliokeawe Pali site are not known. Possible factors may include seasonally or climatically controlled reproductive cycles of the biocontrol agents, low population levels, or other conditions. Additionally, a *Dactylopius* predator, *Cryptolaemus montrouzieri* (family Coccinellidae, order Coleoptera), was observed at the release site and may be a contributing factor in the fluctuation of *Dactylopius* populations.

Additional introductions of *Cactoblastis* and *Dactylopius* will probably increase probabilities of distribution and mating among biocontrol agents. Observations at release sites will continue, and further documentation of biocontrol efforts will be forthcoming. Since most of the cacti in Hawaii Volcanoes National Park are spineless, the cochineal insect on spineless cactus in Mexico, *Dactylopius coccus*, may be more suitable for control than *D. opuntiae* and should be considered for introduction.

BIOCONTROL OF HĀMĀKUA PĀMAKANI

Introduction

Hāmākua pāmakani, or spreading mistflower (*Ageratina riparia* (= *Eupatorium riparium*), is an aggressive, fast-spreading, noxious weed of no forage value on ranchlands. Pāmakani means "windblown" in the Hawaiian language and originally referred to Maui pāmakani, *A. adenophora*, a former pest of Maui rangelands. Hāmākua pāmakani is a native of Mexico, and as with Maui pāmakani, the seeds are windblown.

Hāmākua pāmakani was first observed naturalized in the vicinity of Hilo, Hawai'i in 1926. By 1930 it had spread to Mountain View and along the Hāmākua Coast (Degener 1946). It was recorded as infrequent in the Hawaii Volcanoes National Park residential area and at Makaopuhi Crater (Fagerlund and Mitchell 1944). By 1960, this noxious weed was prevalent in many moist areas of the Park as well as the Volcano Village district and in Ka'ū and Kona, especially in upper ranch areas. In addition to Hawai'i Island, *A. riparia* is present on O'ahu and Maui but is not considered a problem. A large population exists on Tantalus on O'ahu. In Hawaii Volcanoes it is now an undesirable weed, competing with native plants and occupying disturbed areas.

Data assembled by the Hawaii Cattlemen's Association supplemented by Hawaii Department of Agriculture personnel indicated that in 1972, 128,500 a (52,000 ha) were infested with Hāmākua pāmakani, of which 98,840 a (40,000 ha) were ranchlands and 29,650 a (12,000 ha) were governmental or other privately held lands (Matayoshi 1981). As the carrying capacity of ranges deteriorated due to invasion by *A. riparia*, chemical control became impractical for most if not all ranchers, and other solutions were sought by the Cattlemen's Association. A spittle bug, *Philaenus spumarius*, was recorded (E. Yoshioka, slide taken at Hualālai Ranch, Kona, May 1977), but damage by this bug and other general feeders such as *Brevipalpus phoenicis*, a red and black flat mite, was insignificant. *B. phoenicis* was found on *A. riparia* in Hawaii Volcanoes National Park at 2,165 ft (660 m) elevation in 1986 and apparently had not been recorded from this host previously.

Biological Control

Ranchers impressed by the successful control of lantana, cacti, and Maui pāmakani urged the State to follow a similar approach for Hāmākua pāmakani. At a meeting held in Honolulu attended by representatives of the Cattlemen's Association, the Hawaii Department of Agriculture, and University of Hawaii personnel, it was unanimously agreed that further exploration for natural enemies be undertaken, with funds provided by ranchers and with a general plan of action, including: 1) insect exploration, by Hawaii Department of Agriculture; 2) plant disease search, University of Hawaii; 3) propagation, testing, and distribution of approved insects and pathogens, Hawaii Department of Agriculture and University of Hawaii.

Exploration in Mexico and neighboring countries was undertaken by a Hawaii Department of Agriculture entomologist and a University of Hawaii plant pathologist beginning in 1972. A plume moth, *Oidaematophorus beneficus*, was found in Mexico, tested in Department quarantine facilities, approved for release by the Board, and liberated on a Kona ranch in October 1973 (Nakao and Funasaki 1976). Subsequent releases were made in the Volcano area and other localities on Hawai'i Island.

In 1973, a stem-galling fly, *Procecidochares alani*, from Vera Cruz, Mexico (*ca.* 2,000 ft or 610 m elevation), was brought to Hawai'i by a Hawaii Department of Agriculture entomologist. Propagation of the fly was very successful in the Honolulu quarantine facility, and upon the conclusion of the host range tests, approval for release was obtained. The first release was made in Kona in April 1974 at 3,400 ft (1,036 m) elevation, and subsequent island-wide releases were made (Nakao and Funasaki 1976).

A phytopathogenic fungus, *Cercosporella* (now *Entyloma* sp., according to Trujillo *et al.* 1988), was collected on *A. riparia* in Jamaica, in 1974 by the University exploratory pathologist. After host range tests it was approved for release by the Board of Agriculture, and cooperative liberations were made by University and Hawaii Department of Agriculture personnel beginning in November 1975 (Nakao and Funasaki 1979).

Results

The establishment of the plume moth defoliator and the stem gall fly in Volcano in 1974 led to a steady build-up of both insects, and by 1975 Hāmākua pāmakani was under heavy stress. At one particular site (3,800 ft or 1,165 m elevation), the weed began to disappear (Davis, pers. observ. 1975). The foliar pathogenic fungus *Cercosporella* (now *Entyloma*) sp., which was released in November 1975 (Trujillo 1985), is established in Volcano, but disease formation at this elevation appears to be limited. However, the disease has done exceptionally well in some Kona ranching areas and is especially devastating between 2,000 and 3,500 ft (610-1,067 m elevation) (Matayoshi 1981) (Fig. 4A and 4B). These agents have significantly reduced infestation of Hāmākua pāmakani on Hawai'i Island (Fig. 5).

CONCLUSIONS

Based on year-round observations of lantana and its imported enemies on ranch, government, and private lands statewide, and on sequential photo documentation of cactus and Hāmākua pāmakani on the Island of Hawai'i at fixed photographic stations, biological control of all three weeds has been dramatic. For lantana, this control ranges from partial to substantial; for cacti, substantial to complete; and for Hāmākua pāmakani, substantial to complete. In addition to purposely introduced insects, an introduced pathogen has contributed to the successful biocontrol of Hāmākua pāmakani.

All three weeds are present in Hawaii Volcanoes National Park. Most of the introduced insects are present on lantana, but the degree of control has not been determined. Introduced insects for spineless cactus at Poliokeawe are barely established, and the status of control cannot be assessed at this time. Hāmākua pāmakani is under partial control at 'Āinahou at 3,100 ft (945 m) elevation, and control is substantial to complete at 3,800 ft (1,159 m) elevation.

Figure 4. Biocontrol of an infestation of Hāmākua pāmakani in North Kona (1974, upper photo) by *Entyloma* sp. (1978, lower photo).

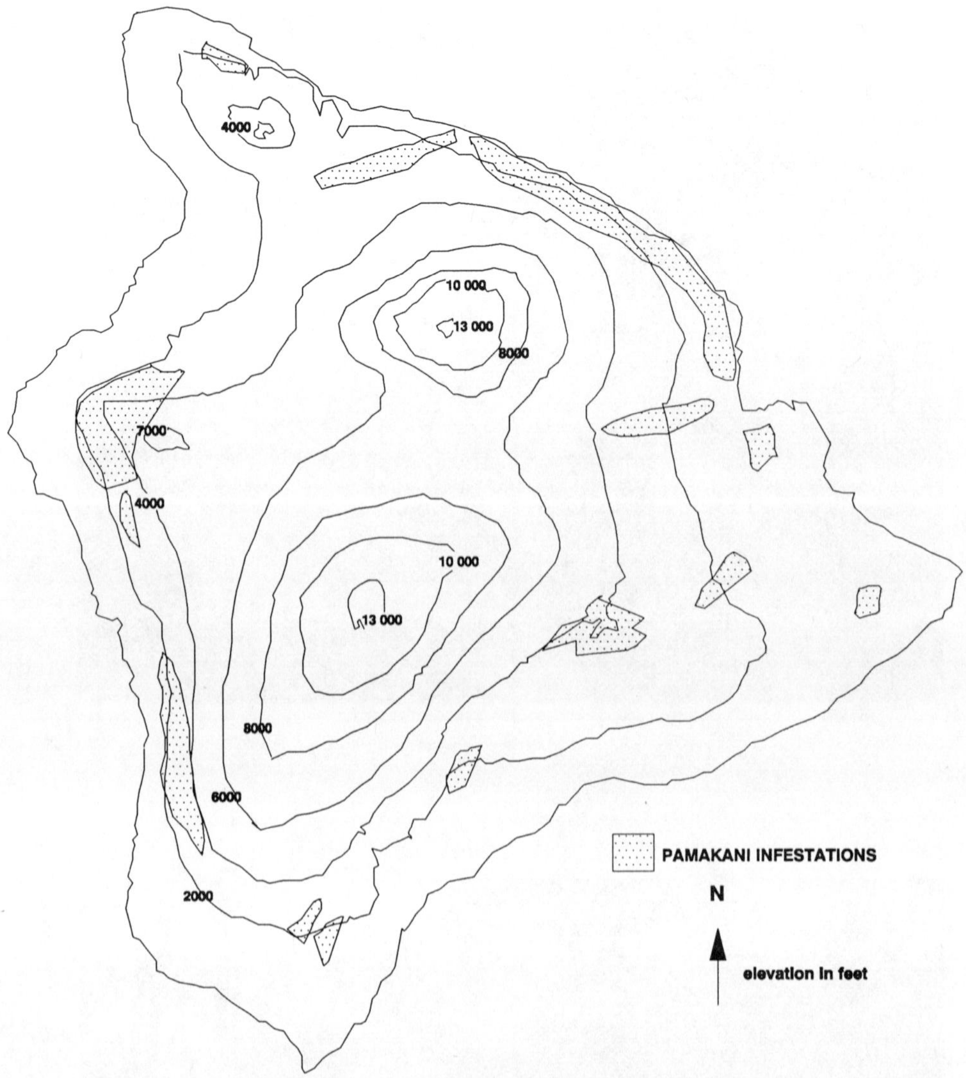

Figure 5. Infestation of Hāmākua pāmakani on the island of Hawai'i, 1987.

ACKNOWLEDGMENTS

The authors express appreciation to the following people for their support and assistance in preparation of the manuscript and presentation: Colleen Iwata, State of Hawaii Department of Agriculture, Hilo, Hawai'i, for typing portions of the manuscript; Danielle Stone, Administrative Assistant for Research at Hawaii Volcanoes National Park, University of Hawaii Cooperative National Park Resources Studies Unit, for computer transcriptions; and the Hawaii Department of Agriculture for loan of slides.

Literature Cited

Barreto, R.W., and H.C. Evans. 1988. Taxonomy of a fungus introduced into Hawaii for biological control of *Ageratina riparia* (Eupatorieae; Compositae), with observations on related weed pathogens. *Trans. Brit. Mycol. Soc.* 91(1):81-97.

Benson, L. 1982. *The cacti of the United States and Canada.* Stanford, Calif.: Stanford Univ. Pr.

Bridwell, J.C. 1919. Descriptions of new species of hymenopterous parasites of muscoid Diptera with notes on their habits. *Proc. Hawn. Entomol. Soc.* 4:166-179.

Burbank, L. 1914. *His methods and discoveries.* Vol. 8: 193-198. New York: Luther Burbank Pr.

Carpenter, C.W. 1944. *Fusarium* disease of the prickly pear. *Hawn. Plant. Rec.* 48:59-63.

Davis, C.J. 1970. Recent introductions for biological control in Hawaii--XV. *Proc. Hawn. Entomol. Soc.* 20:521-525.

Davis, C.J. 1978. Establishment of some recent immigrant insects in Hawaii Volcanoes National Park. *Proc. 2nd Conf. Nat. Sci., Hawaii Volcanoes Natl. Park,* 98-104. Honolulu: Univ. Hawaii Coop. Natl. Park Resour. Stud. Unit.

Davis, C.J., and N.L.H. Krauss. 1962a. Recent developments in the biological control of weed pests in Hawaii. *Proc. Hawn. Entomol. Soc.* 18(1):65-67.

Davis, C.J., and N.L.H. Krauss. 1962b. Recent introductions for biological control in Hawaii--VII. *Proc. Hawn. Entomol. Soc.* 18(1):125-126.

Davis, C.J., and N.L.H. Krauss. 1966. Recent introductions for biological control in Hawaii--XI. *Proc. Hawn. Entomol. Soc.* 19(2):201-204.

Degener, O. 1946. *Flora Hawaiiensis.* Books 1-4. Families: 257, 344. Private pub.

Fagerlund, G.O., and A.L. Mitchell. 1944. *A checklist of the plants -- Hawaii National Park, Kilauea Mauna Loa section with a discussion of the vegetation.* Nat. Hist. Bull. 9, Hawaii National Park.

Fullaway, D.T. 1952. Cactus eradication on the island of Hawaii 1943-1952. Honolulu: Hawaii Bd. Agric. and For. Mimeo.

Fullaway, D.T. 1954. Biological control of cactus in Hawaii. *J. Econ. Entomol.* 47(4):696-700.

Fullaway, D.T., and N.L.H. Krauss. 1945. *Common insects of Hawaii.* Honolulu: Tongg Pub. Co.

Hardy, D.E., and M.D. Delfinado. 1980. *Ophiomyia lantanae* (Froggatt). *Insects of Hawaii* 13:199-200.

Harley, K.L.S. 1967. Notes on parasitism of *Plagiohammus spinipennis* Thomson. *Proc. Hawn. Entomol. Soc.* 19(3):344.

Hillebrand, W. 1888. *Flora of the Hawaiian Islands, a description of their phanerogams and vascular cryptogams*. London: Williams & Norgate.

Hosaka, E.Y., and A. Thistle. 1954. *Noxious plants of the Hawaiian ranges*. Univ. Hawaii Coop. Ext. Serv. Bull. 62. Honolulu: Univ. Hawaii.

Matayoshi, S. 1981. Report on Hamakua pamakani. Hawaii Dep. Agric. Mimeo.

Nakao, H.K., and G.Y. Funasaki. 1976. Introductions for biological control in Hawaii, 1974. *Proc. Hawn. Entomol. Soc.* 22(2):329-331.

Nakao, H.K., and G.Y. Funasaki. 1979. Introductions for biological control in Hawaii. *Proc. Hawn. Entomol. Soc.* 23(1):125-128.

Neal, M.C. 1965. *In gardens of Hawaii*. B.P. Bishop Mus. Spec. Pub. 50. Honolulu: Bishop Mus. Pr.

Perkins, R.C.L. 1924. The introduction into Hawaii of insects that attack lantana. *Bull. 16, Hawaii Agric. Exp. Sta. (Entomol. ser.)*, 1-4.

Ross, G.N. 1986. The bug in the rug. *Nat. Hist.* 95(3):66-73.

St. John, H. 1973. *List and summary of the flowering plants in the Hawaiian Islands*. Mem. 1. Lawai, Kauai, Hawaii: Pac. Trop. Botan. Gdn.

Swezey, O.H. 1924. Present status of lantana and its introduced insect enemies. *Bull. 16, Hawaii Agric. Exp. Sta. (Entomol. ser.)*, Append. C, 72-83.

Swezey, O.H. 1932. Note. *Proc. Hawn. Entomol. Soc.* 8(1):28.

Trujillo, E.E. 1985. Biological control of Hamakua pamakani with *Cercosporella* sp. in Hawaii. *Proc. VI Internatl. Biol. Conf., Vancouver, Canada*, 661-671.

Trujillo, E.E., M. Aragaki, and R.A. Shoemaker. 1988. Infection, disease development, and axenic culture of *Entyloma compositarum*, the cause of Hamakua pamakani blight in Hawaii. *Plant Disease* 72(4):355-357.

Weber, P.W. 1955. Recent liberations of beneficial insects in Hawaii--IV. *Proc. Hawn. Entomol. Soc.* 15(3):635-636.

Williams, J. 1980. Native vs. exotic woody vegetation recovery following goat removal in the eastern lowlands of Hawaii Volcanoes National Park. *Proc. 3rd Conf. Nat. Sci., Hawaii Volcanoes Natl. Park*, 373-382. Honolulu: Univ. Hawaii Coop. Natl. Park Resour. Stud. Unit.

Yoshioka, E. 1970. Notes and exhibitions, *Trichogramma semifumatum* (Perkins) ex eggs of *Hypena strigata* F. *Proc. Hawn. Entomol. Soc.* 20(3):480.

Zimmerman, E.C. 1948. *Insects of Hawaii*. Vol. 5, *Homoptera*. Honolulu: Univ. Hawaii Pr.

Zimmerman, E.C. 1978. *Insects of Hawaii*. Vol. 9, pt. I, *Microlepidoptera*. Honolulu: Univ. Hawaii Pr.

PLANT PATHOGENS AS BIOCONTROL AGENTS IN NATIVE HAWAIIAN ECOSYSTEMS

Donald E. Gardner

Abstract

Several examples, some spectacular, are available of successful applications of weed biocontrol approaches with plant pathogenic agents in agricultural and rangeland situations. The overall percentage rate of success of all such programs attempted cannot yet be documented, however. Both the "mycoherbicide" method, in which large quantities of inoculum are directly applied to target plants, and the "conventional" method, in which the pathogen is allowed to distribute itself from an introduction point, are used. Two successful biocontrol programs in Hawai'i using plant pathogens are the fungus *Fusarium oxysporum* control of prickly pear cactus (*Opuntia ficus-indica*), and Hāmākua pāmakani (*Ageratina riparia*) control by the fungus *Entyloma ageratinae* or *E. compositarum*. A third project, in which the fungus *Colletotrichum gloeosporioides* has been released on Koster's curse (*Clidemia hirta*), is currently under evaluation. Little precedent exists for management use of biocontrol with plant pathogens in native habitats with the goal of preservation of intact ecosystems, as distinct from agricultural applications, however.

Certain limitations, both of a biological and an administrative or political nature, must be considered in National Park Service biocontrol programs. Perhaps the most important potential biological limitation may be the unavailability of suitable control agents. Suitable agents must be both host specific and sufficiently virulent to effectively control the target species.

Conflicts between National Park Service and agricultural (including horticultural and rangeland) management objectives may also occur. Possibilities for success must be evaluated for each target species individually, taking into account all biological factors as well as other factors involved. Biocontrol offers certain potential advantages over herbicidal or mechanical methods, including longevity and self-dispersal of the agent in the environment; disadvantages include slower, often less perceptible, and less complete control as compared with herbicidal or mechanical removal. Biocontrol, with insects or diseases, may be the only feasible approach for certain widely distributed alien species, however.

The following pathogenic fungi are currently being evaluated under controlled conditions as potential biocontrol agents for target alien plants in Hawaiian National Park Service areas: *Fusarium oxysporum* f. sp. *passiflorae* for banana poka (*Passiflora mollissima*); *Gymnoconia nitens* and *Kuehneola uredinis* for brambles (*Rubus* spp.); and *Ramularia destructiva*, *Nectria galligena*, and *Cryphonectria* sp. for firetree (*Myrica faya*). None of these agents has yet been released in the field.

INTRODUCTION

Biocontrol research on weeds has most frequently been conducted by entomologists and has focused on the use of phytophagous (plant-feeding) insects as control agents. However, the potential of plant pathogens for this purpose, although as yet relatively unexplored, is becoming increasingly apparent (Shrum 1982). Biocontrol research with plant pathogens (fungi, viruses, and nematodes) is currently under way in at least 13 countries in temperate to tropical zones and in 18 states of the U.S. (Templeton 1982).

Two approaches to weed biocontrol with plant pathogens have been pursued: the use of mycological herbicides (termed "myco-" or "bioherbicides"), and the use of organisms, usually from a foreign country, in conventional, or "classical," programs (Templeton 1982). The mycological herbicide technique was developed largely by plant pathologists in Arkansas and Florida and is applicable mainly in annual monoculture crop production (Templeton 1982). It involves the selection of diseases that already occur on target weeds, and artificially culturing the pathogen in large quantities. Masses of fungal spores are then collected, sometimes combined with an inert carrier, and sprayed or dusted on the field in much the same manner that a chemical herbicide is applied. Mass-produced inoculum of the endemic fungus *Colletotrichum gloeosporioides* f. sp. *aeschynomene* was sprayed in Arkansas rice fields infested with northern jointvetch, *Aeschynomene virginica* (Daniel *et al.* 1973; Smith *et al.* 1973). Success rates in the control of this weed averaged as high as 99% by this method (Daniel *et al.* 1973). For large acreages, the mycoherbicide may be applied with crop duster aircraft. Weed control by this approach, as with conventional herbicide application, usually requires the reapplication of the mycoherbicide on a regular schedule each growing season. Success depends on the amplification of otherwise insignificant effects of pathogens through concentration to abnormally high levels. Some mycoherbicides are sufficiently effective to be patented, commercially produced, and marketed under trade names (Bowers 1982).

The conventional approach to biocontrol with plant pathogens usually involves the introduction of a new pathogenic agent into the environment, followed by natural dispersion from the introduction point(s) (Templeton 1982). Little or no attempt is made to artificially increase inoculum amounts or to apply infective material widely to target plant populations. The success of this approach is enhanced by the ability of large quantities of inoculum to survive unfavorable environmental conditions and to efficiently disperse (Shrum 1982).

As might be expected, conventional biological control would be usually most applicable in natural areas or other non-agricultural situations in which weeds are widely scattered and difficult to locate or to approach for treatment on an individual basis. Even in natural areas, however, alien grasses and woody species may become established to the exclusion of other vegetation types in monoculture-like stands. The alien grasses *Andropogon virginicus* and *Schizachyrium condensatum* form such stands in Hawaii Volcanoes National Park.

SUCCESSFUL APPLICATIONS OF BIOCONTROL WITH PLANT DISEASES IN HAWAI'I

Two significant weed biocontrol programs have been undertaken in Hawai'i in the past. Carpenter (1944) reported a notable example of the successful application of a fungal disease agent already present in the environment for biocontrol of an unwanted plant. A variety of the pathogenic fungus *Fusarium oxysporum*, which causes a soft rot, was used to control prickly pear cactus (*Opuntia ficus-indica*). Researchers manually inoculated the plants by creating wounds in the cladodes (pads) into which small portions of fungal cultures were introduced. This approach, which was particularly well-suited to the growth habit and distribution of prickly pear in Hawai'i, may perhaps be considered a hybrid biocontrol technique since it involved elements of both the mycoherbicidal and the conventional concepts. A naturally-occurring pathogen was cultured and its successes in distribution and infection were greatly enhanced through artificial manipulation. However, the pathogen was not produced in large quantities for widespread, general application. This is among the most successful of any biocontrol work thus far attempted in Hawai'i (Gardner and Davis 1982; Davis *et al.*, this volume).

A second outstanding success in Hawai'i has been achieved with *Entyloma ageratinae* or *E. compositarum* (Barreto and Evans 1988; Trujillo *et al.* 1988), a pathogenic leaf-spotting fungus in the control of Hāmākua pāmakani, *Ageratina riparia* (= *Eupatorium riparium*) (Trujillo 1976; Templeton 1982; Davis *et al.*, this volume). The weed is from Mexico, and the fungus was brought to Hawai'i from Jamaica specifically for the purpose of biocontrol. Upon establishment in the field, this pathogen has caused dramatic reductions to the point of complete elimination of host populations in many areas between 1,500 and 6,500 ft (455-1,980 m) elevation. Resprouting has been observed in wetter areas, but other plants shade out young seedlings. The pathogen was capable of distributing itself by natural means, particularly at lower elevations, and thus required little subsequent manipulation (Trujillo 1976; C.J. Davis, pers. comm.). The self-distribution of the Hāmākua pāmakani fungus to infestations of this weed in natural areas is a good example of the effectiveness of biocontrol with plant pathogens in native communities.

A third project, directed at the control of the forest weed commonly known as Koster's curse or clidemia, *Clidemia hirta*, which was introduced to Hawai'i from tropical America, is under way (Trujillo *et al.* 1986). An isolate of the fungus *Colletotrichum gloeosporioides*, which causes anthracnose of *Clidemia hirta*, was collected on this weed in Panama by E.E. Trujillo of the University of Hawaii. A special form of *Colletotrichum gloeosporioides* was mentioned above as a mycoherbicide for control of northern jointvetch in Arkansas. Pathogenicity tests with the Panama isolate were conducted on *Clidemia hirta* at the U.S. Department of Agriculture Foreign Disease-Weed Science Research Unit at Fort Detrick, Maryland. The fungus was subsequently released at a number of clidemia-infested sites in Hawai'i, where it shows initial promise, but its effectiveness is still under evaluation.

SUCCESSES ELSEWHERE WITH PLANT PATHOGENS AS BIOCONTROL AGENTS

Plant disease agents have proven successful or are showing promise in biocontrol programs elsewhere. Approximately 83 pathogens under current study for the control of 54 target weed species throughout the U.S. and in other countries were listed by Templeton (1982). The attractiveness of fungi as biocontrol agents, in particular, focuses on their general ubiquitousness, high degree of host specificity, destructiveness to the host, persistence, dispersal efficiency, and ease of culture and maintenance in the laboratory (U.S. Department of Agriculture 1978).

The plant rust fungi are typically wind dispersed, may infect directly through the host epidermis or through stomata and do not require wounds, are virulent, and are highly host specific. Members of this group of fungi are therefore particularly well suited as biocontrol agents (Quimby 1982; Gardner, pers. observ.). On the other hand, rust fungi are obligate parasites (*i.e.*, they cannot be cultured on artificial media), a possible disadvantage (Quimby 1982).

Successful and well-documented examples of fungi, particularly rusts, in biocontrol programs include the use of *Puccinia chondrillina* to control rush skeleton weed (*Chondrilla juncea*) in the U.S. and Australia (Hasan 1972; Cullen *et al.* 1973; Hasan and Wapshere 1973; Wapshere 1975; U.S. Department of Agriculture 1978; Emge *et al.* 1981). Rush skeleton weed, a perennial weed of Eurasian origin, has become a serious pest in farm and range land in the western U.S., as well as in the wheatlands of Australia, where it is a strong competitor for moisture and nitrogen. Infection with *P. chondrillina* has been shown to drastically reduce seed production, seed germination, and plant vegetative growth. A gradual reduction in the weed population results from the disease (Emge *et al.* 1981).

The rust fungus *Phragmidium violaceum* in Germany was recognized as a potential biocontrol agent for two bramble (*Rubus*) species, *R.*

constrictus and *R. ulmifolius*, in Chile (Oehrens 1977; Oehrens and Gonzalez 1977) and was introduced to Chile for evaluation. The subsequent success of this approach has led to the further investigation of *P. violaceum* for control of brambles in other countries as well.

Successful control of yellow nutsedge, *Cyperus esculentus*, under experimental conditions by another rust fungus, *Puccinia canaliculata*, has been reported in the southeastern U.S. (Phatak *et al.* 1983). Yellow nutsedge occurs on a worldwide basis and is considered one of the most troublesome weeds. In recent years it has spread rapidly throughout the U.S. and in 1981 was considered the most important perennial weed in most of the Midwest (Stoller 1981). The fungus inhibited nutsedge flowering and new tuber formation, and dehydrated and killed plants.

Uromyces rumicis is credited with the suppression of curly dock (*Rumex crispus*) in Europe (Inman 1971; U.S. Department of Agriculture 1978). Curly dock is a common pasture weed throughout the U.S. and is a significant problem in certain southern states. Based upon the severe attack of *U. rumicis* on curly dock in Europe, and the apparent host specificity of this pathogen, research efforts to evaluate its introduction into the U.S. showed promise (Inman 1971). Current reports of the study could not be located, however.

Wilson (1969), Zettler and Freeman (1972), Goeden *et al.* (1974), Hasan (1974, 1980), and Charudattan and Walker (1982) have provided comprehensive discussions and literature reviews of the subject of weed biocontrol with plant pathogens.

CONSIDERATIONS IN USING BIOCONTROL IN NATIONAL PARK SERVICE AREAS

From the above examples of successful applications, it is evident that little precedent exists for the concept of biocontrol in natural environments. The weed control objective of the National Park Service in preserving native ecosystems differs significantly from that of agricultural, including horticultural or range management, objectives. Whereas the goal of other biocontrol research programs discussed has been the preservation of particular crop or range species in a severely manipulated agricultural environment, a purpose of the National Park Service is to maintain, as nearly as possible, the overall integrity of native environments. Thus, cultural practices recommended for agricultural applications, such as frequent cultivation or favorable timing of crop plantings, are not applicable to weed control in natural systems.

Alien plants in Hawai'i's national parks for which biocontrol is considered are those with little or no value as crops, ornamentals, or forage species elsewhere in the Islands, and those which are not so closely related to economically valuable plants that introduced pathogens or insects would pose a threat (Gardner and Davis 1982; Markin and Yoshioka, this volume). Once released into the environment, biocontrol agents may become naturally distributed throughout their biological ranges and are not limited by political boundaries. The movement of soil-borne pathogens, or of those transmitted by insects or rain, is presumably more easily controlled than is that of wind-dispersed fungi (Shrum 1982). Potential biocontrol agents would most likely, but not necessarily, be found in the country of origin of the problem alien plant (Charudattan 1982; Shrum 1982). Locally occurring pathogens also may be evaluated as biocontrol agents in situations where artificial manipulation and application appear feasible.

Experience indicates that insect agents may exhibit a significant difference in activity between the old and new habitats due to predation and parasitism in the former. Thus, subdued activity observed in the native habitat does not necessarily indicate a potential ineffectiveness of the insect in the new environment where it may be free of its enemies. However, while hyperparasitism is known among some plant pathogens, this is not usually an important factor in limiting the activity of these agents. Therefore, given the consistency of other environmental factors, the observed effectiveness of a disease in the native habitat is more nearly an indication of its projected effectiveness in the new habitat.

Procedures and Limitations

Biocontrol programs must focus on knowledge of suitable agents that attack the target plant, and perhaps equally important, closely related species. Although preliminary review of the scientific literature is necessary, little information is available on alien plants in Hawai'i (Gardner and Davis 1982). Problem alien plants in Hawai'i are often of little significance in their native habitats, and few entomologists or phytopathologists are available in developing countries who can devote attention to research projects of less than high priority in those regions.

Firsthand exploration in the native regions of the target plant to discover its insects and pathogens may be essential to the success of a biocontrol program (Charudattan 1982). Strawberry guava (*Psidium cattleianum*) is an aggressive, widely distributed alien species invading National Park Service areas in Hawai'i and is targeted for biocontrol (Hawaii Volcanoes National Park 1985). Literature searches have revealed no reports of insects or diseases on this species in its native South American habitats. Neither diseases nor insects of apparent significance have been observed on this plant in Hawai'i. However, during recent exploration in Brazil, sponsored by the National Park Service, C.S. Hodges found several apparently unreported insects and diseases on the yellow-fruited variety of strawberry guava. Some of the insects, in particular, showed promise as possible biocontrol agents and should be further investigated (Hodges 1988).

Scientists undertaking exploratory trips should be accompanied by local scientists, if possible, who are familiar with the region and can facilitate research efforts. Characteristics and activities of the potential agent(s) under natural conditions can thereby be observed and information on host ranges, life cycles, fruiting habits, distribution, and abundance gathered. Although these data are useful, caution should be exercised since the agent, like the target plant itself, may behave differently in the new environment than in the native ecosystem (Wilson 1969; R. Burkhart, pers. comm.; D. Gardner, unpub. data).

Ideally, exploration should be conducted during different seasons to provide the greatest opportunity for observation of potential biocontrol agents and their effects. Initial evaluation and screening of a potential agent is best conducted in the native country before importation into the U.S. (Charudattan 1982). When this is not possible, agents not already present in the U.S. have to be tested under quarantine. They should be hand carried by the researcher into the U.S., to the quarantine facility, to insure proper care and the greatest chance of survival (G. Markin, pers. comm.).

The researcher may find it necessary to extend his or her stay in a foreign country; six-month visits are normal. Use of laboratory facilities, supplies, equipment, and other needs at a local research establishment may be required, or contract funding for local scientists, if available, to conduct the work may be necessary.

Foreign exploration presupposes the cooperation of the foreign government in permitting access by U.S. researchers. Likewise, conditions of political unrest, dangers associated with drug trafficking activities, and similar situations that cause travel in particular countries to be unsafe for Americans may lead the U.S. Government to declare such localities off limits to foreign travel. The U.S. Forest Service imposed such a prohibition on travel by its personnel to Colombia in 1985, where exploration is currently under way for biocontrol agents for banana poka (*Passiflora mollissima*) in Hawai'i (G. Markin, pers. comm.).

U.S. Department of Agriculture Animal and Plant Health Inspection Service and Hawaii State Department of Agriculture regulations require that a permit be obtained prior to the importation of foreign insects or pathogens not already present into the U.S. and into Hawai'i (Charudattan 1982; State of Hawaii Department of Agriculture Chapter 71 Administrative Rules). Requirements specify that the potential biocontrol agent must be maintained in a certified quarantine facility to prevent premature escape (Klingman and Coulson 1983). Since pathogenic fungi and other plant disease agents are microscopic or produce microscopic spores that are often air-borne, specifications for quarantine facilities in which pathogens are to be contained are more stringent than are those for facilities in which insects are maintained (Melching *et al.* 1983; Gardner and Smith 1985).

Relatively few certified facilities for entomological research are in operation throughout the U.S., presumably due to the high cost of construction. The new quarantine facility at Hawaii Volcanoes National Park is among these. Even fewer containment facilities certified for work with plant pathogens are in operation (Charudattan 1982; Melching *et al.* 1983). The Hawaii Department of Agriculture has recently constructed a small biocontrol facility for plant pathogens and added a plant pathologist to its staff (P.-Y. Lai, pers. comm.). A full description of the U.S. Department of Agriculture quarantine facility for plant pathogens, located at Frederick, Maryland, has been published (Melching *et al.*

1983). Kahn (1983) published a comprehensive discussion of the proper design and maintenance of quarantine facilities for all types of biocontrol programs.

For Hawai'i, as is presumably true for other states that impose restrictions on importation of foreign organisms, the enforcement authority is with the local agriculture department. Until recent years, all biocontrol work in Hawai'i was focused on the protection of agricultural or horticultural products. Current involvement of other State and federal agencies, including the National Park Service, in biocontrol research has complicated the objectives. The Interagency Steering Committee for Biological Control of Forest Pests, chaired by the Hawaii Department of Land and Natural Resources, was created to alleviate this problem. The University of Hawaii, Hawaii Department of Agriculture, U.S. Forest Service, and National Park Service representatives also serve on the committee.

The purpose of the Steering Committee is to coordinate biocontrol interests among agencies so that conflicts and duplications of effort are minimized, and to provide a forum where priorities are set and cooperative efforts facilitated. The Steering Committee serves only to advise the Hawaii Department of Agriculture, however, since that agency retains final authority in biocontrol research matters. Following review by the Hawaii Department of Agriculture, permit requests for importation or release of foreign organisms are further reviewed by representatives of major agricultural industries, such as those of the Hawaiian Sugar Planters' Association. Upon issuance of a State permit, application for a similar U.S. Department of Agriculture Animal and Plant Health Inspection Service permit must be initiated. All restrictions and conditions imposed by these agencies must be complied with. Importation requirements can often be more readily addressed if information is available on the characteristics of the organism to be imported prior to introduction.

Whereas a fully certified insect containment facility for biocontrol research is currently in operation at Hawaii Volcanoes, a comparable National Park Service facility for plant pathogens is not yet available. Biocontrol research with plant pathogens has been conducted at the University of Hawaii under confined laboratory conditions by both University and National Park Service researchers working independently of one another. The Park Service work is presently conducted in Plexiglas plant growth chambers designed and constructed for this purpose. Future research accommodations remain to be determined, although it is possible that development of specific cooperative agreements will enable use of the new Hawaii Department of Agriculture facility within constraints imposed by that agency on a space-available basis (P.-Y. Lai and G. Funasaki, pers. comm.).

Expectations from Successful Biocontrol Programs

Unlike other methods of weed control, such as the use of herbicides, the effects of a successful biocontrol program may be subtle and not immediately or readily perceptible, particularly to persons not closely involved with the program. Gradual reduction in stands of lantana

(*Lantana camara*) has become apparent over a period of several years, as has been the decline in prickly pear cactus populations on ranchland on the island of Hawai'i (Davis *et al.*, this volume). A program may be regarded successful if reduction in vigor, growth rate, or ability of the alien plant to reproduce occurs and may, over time, allow native species to successfully compete where they could not previously (C.W. Smith, pers. comm.). Attack by disease agents may only reduce to manageable levels, rather than eliminate, target weed populations. Alien plant populations, while usually maintained at low and controllable levels by biocontrol agents operating successfully in the environment, may be expected, on occasion, to irrupt. These periodic rapid increases may correspond to environmental conditions which unusually favor the host and/or disfavor the pathogen (Holcomb 1982; Shrum 1982).

Although, as indicated above, some successful biocontrol programs have depended on a single control agent, the possibility of integrating biocontrol approaches with other methods of plant control should not be overlooked (Smith 1982). Thus, any complementary effects of insects, pathogens, mechanical, cultural, and chemical methods that can be applied to particular alien species are utilized.

Potential Advantages and Disadvantages of Biocontrol

Although the potential for success of biocontrol in natural ecosystems is not known, National Park Service resource managers consider this approach an important long-term solution to widespread, aggressive aliens (Gardner 1982; Hawaii Volcanoes National Park 1985). The National Park Service discourages use of chemical pesticides where other methods will suffice, but considerable use occurs for many reasons. A 1981 memorandum from Park Service Director Russell Dickenson, updating the 1978 National Park Service Management Policies Handbook, stated:

> Chemical pesticides of any type will be used only where feasible alternatives are not available or acceptable. The Service's use of all pesticides shall be approved by the Director. Application shall be in accordance with applicable laws, Departmental and Service guidelines, and Environmental Protection Agency and Occupational Health and Safety Administration regulations.
>
> Service policy does not prohibit, as such, the use of chemical pesticides. However, chemical controls are to be allowed only if (a) there is a clear and present danger to the health and safety of man; and/or (b) there is danger of destruction of significant property or resources and a determination has been made that the control methods of no action, mechanical, cultural and/or biological control are non-existent, unavailable, or unacceptable.

Since, as stated, mechanical, cultural, and biocontrol approaches are considered preferable to use of chemical pesticides, these methods should of necessity be thoroughly evaluated in any Park Service alien plant control program. As with other insular systems, the problem of invasion by alien species in Hawai'i is particularly severe as compared with that of continental systems (Gardner 1982). For this reason, although official pesticide usage regulations still apply, these guidelines may be relaxed in actual practice in favor of expeditious alien plant control in Hawaiian National Park Service areas. Manual removal of alien plants, sometimes in combination with herbicidal application, has proven effective in local, relatively restricted populations of alien plants (Tunison and Zimmer, this volume).

Notwithstanding this, labor-intensive approaches have not proven successful for widespread, well-established alien plant species over large areas. Cultural techniques, which involve manipulation of the environment to favor the desired and/or disfavor the undesired entity, appear more applicable under conditions of crop or ornamental cultivation than to natural habitats.

On the other hand, the field of biocontrol, especially as it is applied to native ecosystems, is controversial. Introduction of foreign organisms, including biocontrol agents, into native habitats is considered undesirable environmental tampering by some individuals concerned with environmental preservation (Howarth 1983). The long-term effects of introduced organisms in fragile ecosystems are unpredictable, possibly causing more harm than they were introduced to prevent. Howarth (1983) stated "experience has shown that there are serious risks to the natural environment involved when foreign organisms are purposefully established in a new land." Introduction of foreign organisms, especially indiscriminately, causes "biological pollution" (Howarth 1983) and results in a native environment becoming nonnative by definition. Some of the objections to biocontrol approaches are presumably less applicable to agricultural situations, in which there is no intention of maintaining environmental integrity, than to natural systems, but agricultural agents may spread to native systems.

A major potential problem in relying on biocontrol approaches for particular alien species prior to an investigation is the lack of assurance that suitable control agents exist at all. Since such agents must be both sufficiently host-specific and virulent to be effective, it is quite possible that pathogens or insects found on a given alien species meet neither, or only one, of the requirements but fail to meet both. Problems associated with locating potential agents may also pose a significant drawback.

A related difficulty is the loss of research time if, after considerable time has been invested in host testing, the candidate agent is shown to attack desirable plants and must be discarded. As an example, in 1955 F.A. Bianchi, a Hawaiian Sugar Planters' Association entomologist, traveled under the sponsorship of the Hawaii Territorial Board of Agriculture to the Azores, Madeira, and the Canary Islands to search for biocontrol agents of firetree (*Myrica faya*) (Hodges and Gardner 1985; F. Bianchi, unpub. corres.). The most promising discovery was that of a fungus, *Dothiorella* sp., which caused significant shoot dieback of firetree. However, host range testing by a plant pathologist in Portugal eventually indicated that the fungus would also attack several trees of value in Hawai'i, including mango (*Mangifera indica*) and avocado (*Persea americana*). It was subsequently necessary to discontinue the research (Azevedo 1960; Hodges and Gardner 1985; F. Bianchi, unpub. corres.).

Costs associated with a biocontrol program are initially high, and results are usually not quickly evident. Although the testing process in quarantine is designed to be comprehensive, it is not possible to evaluate the behavior of a potential biocontrol agent under every combination of environmental condition and host plant that it may encounter in the field. Due to space limitations, host range testing in quarantine is often restricted to seedlings or small plants, whereas certain plants are known to be more susceptible to particular pathogens at certain developmental stages than at others (Shrum 1982). Misleading results arising from these variables may presumably lead to unwarranted release of the agent. Mutability among microorganisms following release and innate genetic heterogeneity among plant populations are also unpredictable dangers (Leonard 1982). Thus, although the problems are minimized, even extensive testing under quarantine cannot fully assure continued reliability of the released organism.

On the other hand, although initial research expenses may appear prohibitively high, the long-term cost of a successful biocontrol program may be much lower and less labor intensive, requiring a small effort relative to conventional programs. An efficiently self-dispersed agent does not cause the detrimental environmental effects incurred by the activities of laborers with their equipment in conventional control efforts in sensitive ecosystems. Tactical problems, such as transportation of large numbers of laborers to remote areas, with their supplies and equipment, are also lessened in biocontrol programs.

The long-term environmental impacts of neither biocontrol, for the above stated reasons, nor herbicidal use can be fully assessed with current information. Although manufacturers are required to specify permissible uses on the labels of each commercially available pesticide, these are usually agriculture-oriented applications; little information is currently available on persistence and activities of these materials, or their degradation products, in native Hawaiian ecosystems. Proven parkwide control methods for the particular, widespread, well-established alien species in Hawaiian National Park Service areas, whether chemical, mechanical, or biological, therefore do not appear to exist at present.

CURRENT STATUS OF NATIONAL PARK SERVICE RESEARCH IN BIOCONTROL WITH PLANT PATHOGENS

Banana Poka

Banana poka, generally known as *Passiflora mollissima* in Hawai'i (LaRosa 1985), although some researchers consider *P. tripartita* the correct name (E. Trujillo, pers. comm.), is among the highest priority forest weeds designated by Hawaii Volcanoes National Park natural resource managers for biocontrol research (Hawaii Volcanoes National Park 1985). The ability of this aggressive vine to invade and destroy native forests, where it covers the canopy with dense mats of foliage, is also recognized by the Hawaii Department of Land and Natural Resources, which is participating in biocontrol efforts against this plant (Tanimoto and Char, this volume).

Literature reports were found of a destructive vascular wilt disease affecting *Passiflora edulis* forma *edulis*, the purple-fruited passion fruit variety favored for human consumption, in Queensland, Australia (Purss 1954, 1958). The causal fungus was *Fusarium oxysporum* f. sp. *passiflorae*, a form considered specific to this host. A later report indicated that effective control of this soil-borne disease could be obtained by grafting *P. edulis* f. *edulis* scions to *P. edulis* f. *flavicarpa* rootstocks, which were disease resistant (Purss 1958; Inch 1978). *Passiflora edulis* f. *flavicarpa* is the yellow-fruited variety more common in Hawai'i, known locally as liliko'i. It is of some commercial value in Hawai'i. Varieties of passion fruit resistant to the *Fusarium* wilt disease (*e.g.*, Redlands Triangular) have been developed and are currently available to the passion fruit industry (Inch 1978). No genera of Passifloraceae occur in Hawai'i other than *Passiflora*, and no members of this genus are native. With the exception of liliko'i, none of the approximately 30 species of *Passiflora* in Hawai'i, and their varieties, are of any commercial value. Banana poka is presumed to have been introduced as an ornamental that has escaped cultivation. Whereas banana poka fruit may be consumed as human food in the native countries of this species (E. Trujillo, pers. comm.), it is not used for this purpose in Hawai'i.

The special forms of *F. oxysporum* are characterized by their host specificity, often attacking only one or a few closely allied host species (Gordon 1965). The likelihood was not considered great that *F. oxysporum* f. sp. *passiflorae* would attack banana poka. Even though both are members of the genus *Passiflora*, *P. edulis*, classified in the subgenus *Passiflora*, and *P. mollissima*, subgenus *Tacsonia*, are less closely allied. It was nevertheless desirable to confirm this assumption by inoculation tests.

Upon obtaining Hawaii Department of Agriculture and U.S. Department of Agriculture Animal and Plant Health Inspection Service permits, cultures of *F. oxysporum* f. sp. *passiflorae* were obtained from Queensland, Australia. Seedlings of banana poka, purple liliko'i, yellow liliko'i, *P. suberosa*, *P. ligularis*, and *P. foetida* were inoculated by a root-dipping method, whereby the seedling was removed from the growth medium, the roots were immersed in a spore/water suspension, and the seedling was replanted. Although the root-dip method is an artificial method not applicable in the field, it is useful as a direct, uniform screening approach to determine basic resistance or susceptibility. Similarly, seedlings of many crop and ornamental species, unrelated to Passifloraceae and representing a number of plant families, were tested by the root-dip method to establish host specificity.

Mortality rates among representatives of the Passifloraceae tested (Table 1) have shown banana poka to be significantly susceptible, with more

Table 1. Mortality of *Passiflora* spp. seedlings inoculated by dipping roots in a spore suspension of *Fusarium oxysporum* f. sp. *passiflorae*.

Species	% Mortality rate	Sample size
P. mollissima	69	459
P. mollissima (older seedlings)	46	307
P. edulis f. *edulis*	37	250
P. edulis f. *edulis* (older seedlings)	30	63
P. edulis f. *flavicarpa*	0.29	345
P. edulis f. *flavicarpa* (older seedlings)	0	101
P. foetida	32	233
P. ligularis	26	70
P. suberosa	0	224

than 90% of seedlings killed in some tests. None of the inoculated seedlings of species not in the family Passifloraceae became diseased, nor did ratoons of sugar cane (*Saccharum officinarum*) and pineapple (*Ananas comosus*), Hawai'i's most important agriculture crops.

Banana poka was further tested by a more natural method, in which seeds were sown in soil amended with inoculum (Gardner 1989). Resulting mortality reached 66% (of 79 emerged seedlings). The fungus was re-isolated from stem tissue of diseased plants and recultured on artificial medium. No disease-related mortality was obtained in the non-inoculated control treatment (Gardner 1989). As is common among vascular wilt diseases associated with *F. oxysporum* (Gardner and Burge 1980), the disease in infected *Passiflora* spp. plants was severe, and no infected plants recovered.

Although *Fusarium* wilt disease occurred in larger, fruit-bearing plants in Australia (Purss 1954, 1958), space limitations in the growth chambers did not allow testing of large *P. mollissima* plants in the National Park Service study. Current results with seedlings suggest, however, that *F. oxysporum* f. sp. *passiflorae* has a promising degree of virulence on banana poka. Furthermore, these tests indicate that this fungus may not pose a serious threat to *P. edulis* f. *flavicarpa* in Hawai'i, as demonstrated by the use of *Passiflora edulis* f. *flavicarpa* for resistant rootstocks in Queensland. Field testing under the actual environmental conditions at which banana poka occurs in Hawaiian forests, with particular reference to soil and air temperatures, is necessary before further conclusions concerning its effectiveness can be drawn.

A high-pressure stem injection method of inoculation, developed for biocontrol experimentation with a related vascular wilt disease (Gardner and Burge 1980), should also be tested on large banana poka plants in the field.

Firetree

Among the most intensive of any mechanical and chemical control efforts have been directed against firetree during the past two decades by Hawaii Volcanoes National Park natural resource management (Gardner and Davis 1982). Through constant surveillance, the plant is now controlled in limited areas of the Park, designated as Special Ecological Areas, but is spreading out of control elsewhere. There are no native or otherwise valued members of the Myricaceae family in Hawai'i, making this plant a good candidate for biocontrol. In previous investigations of biocontrol of firetree, Krauss (1964), a Hawaii Department of Agriculture entomologist, explored the Azores, Madeira, and the Canary Islands in 1960 and 1962 for potential control agents. He also searched for insects on other species of *Myrica* in the southeastern and northeastern U.S., Mexico, Costa Rica, and various parts of Africa. A list of insects and fungi he found was published (Krauss 1964). Attempts to establish several of the insect species on firetree in Hawai'i were not successful, however (Krauss 1964; Hodges and Gardner 1985).

In 1984, C.S. Hodges of the U.S. Forest Service and the author, of the National Park Service, traveled to the Azores, Madeira, and Canary Islands in a further attempt to discover potential biocontrol agents for this plant in Hawai'i (Gardner 1984; Hodges and Gardner 1985). A second exploratory trip to the Azores and Madeira was undertaken in 1987 (Gardner *et al.* 1988). Three fungus diseases were found, which appeared to be sufficiently important to warrant further study. However, the diseases did not appear to significantly limit populations of firetree in the native environment (Gardner 1984; Hodges and Gardner 1985). A stem canker disease associated with an unidentified species of *Cryphonectria*, however, was not widespread, and its potential should be further investigated (Gardner *et al.* 1988). Insects found, including the fruit-boring species *Carposina atlanticella*, are reported by Krauss (1964) and discussed elsewhere.

Florida Blackberry

Florida blackberry (*Rubus argutus* = *R. penetrans*) is a high-priority designated species for biocontrol (Hawaii Volcanoes National Park 1985; L.L. Loope, pers. comm.). This thorny, aggressive bramble forms dense, impenetrable thickets in native forests. Other alien species of *Rubus* pose similar threats but are more limited in distribution at present. Although blackberry fruit is sometimes consumed by individuals, none of the alien species of *Rubus* in Hawai'i are of commercial value. Two endemic species, *R. hawaiiensis* and *R. macraei*, occur in native forests on the islands of Kaua'i, Maui, and Hawai'i, however. The protection of these species is of primary concern in any biocontrol program against the alien species of *Rubus*.

Successful biocontrol with the rust fungi *Kuehneola uredinis* and *Phragmidium violaceum* of other species of *Rubus* in South Africa and

South America was described in the literature (Oehrens 1977; Oehrens and Gonzalez 1977; Hasan 1980). The probability of *P. violaceum* attacking any of the alien species of *Rubus* in Hawai'i was not considered great, since *Rubus* is an extremely genetically diverse genus throughout the world. A small amount of *P. violaceum* inoculum was provided the National Park Service, under permit, by E. Oehrens in Chile, but upon its arrival the fungus was apparently nonviable. No further work with *P. violaceum* has been attempted.

Kuehneola uredinis was discovered to be already present in Hawai'i, where it attacks leaves of *R. argutus* (Gardner and Hodges 1985). Subsequent field observation and inoculation experiments showed the pathogen to be capable of producing small, isolated sori on leaves of both *R. hawaiiensis* and *R. macraei*. The extent of infection appeared insignificant on both native species, however, and caused no damage.

In further work, root cuttings of the two endemic species and four prominent alien species (*R. argutus, R. ellipticus, R. rosifolius, R. glaucus*) occurring in Hawaiian National Park Service areas were established at North Carolina State University. A small number (usually two to four) of plants of each species was inoculated with *Gymnoconia nitens*, a rust fungus capable of causing severe systemic infection (as distinct from leaf infection only) on species of *Rubus* in the Southeast (Gardner 1984). Inoculation was accomplished by spraying a spore/water suspension on the leaves and maintaining plants under conditions of high humidity for two days. The native plants did not thrive in the new environment and eventually died, but without evidence that they had become infected. Among the alien species of *Rubus*, infection occurred only on *R. argutus*, some plants of which died from the disease (J. Ballington, pers. comm.). Further work to establish conclusively the pathogenicity of *G. nitens* on *R. argutus* from Hawai'i and to evaluate its effects on the native species is under way in North Carolina. Also, *K. uredinis*, which is reported to cause systemic infection that significantly impacts the development of species of *Rubus* in the Southeast (C. Hodges, pers. comm.), should be tested on the native and alien species from Hawai'i. Even though this fungus in Hawai'i is considered to be the same species as that in the Southeast, the possibility exists that a more virulent strain occurs in the Southeast that may be useful in Hawai'i.

CONCLUSIONS

The overall percentage rate of success among all biocontrol efforts undertaken with plant pathogenic agents cannot yet be documented (Templeton 1982). Several examples of successful, and even spectacular, results of biocontrol approaches with plant pathogenic agents in agricultural situations are known, and success with Hāmākua pāmakani in natural areas of Hawai'i has been acknowledged (Davis *et al.*, this volume). These successes indicate more possibilities for application of biocontrol in natural environments. A prerequisite of most biocontrol programs is that target species be of little or no value outside designated control areas, since biocontrol agents usually can neither be contained within political boundaries nor recovered once released. Furthermore, suitable agents may not be available for all species for which biocontrol would be desirable. For these reasons, opportunities for application of biocontrol techniques may be limited and controversial, and other alien plant control approaches should also be used. The future possible use of genetic engineering techniques to manipulate the virulence or host specificity of pathogens and thus create suitable agents for biocontrol purposes may have merit.

ACKNOWLEDGMENTS

The technical assistance of Donna Ikuta is appreciated.

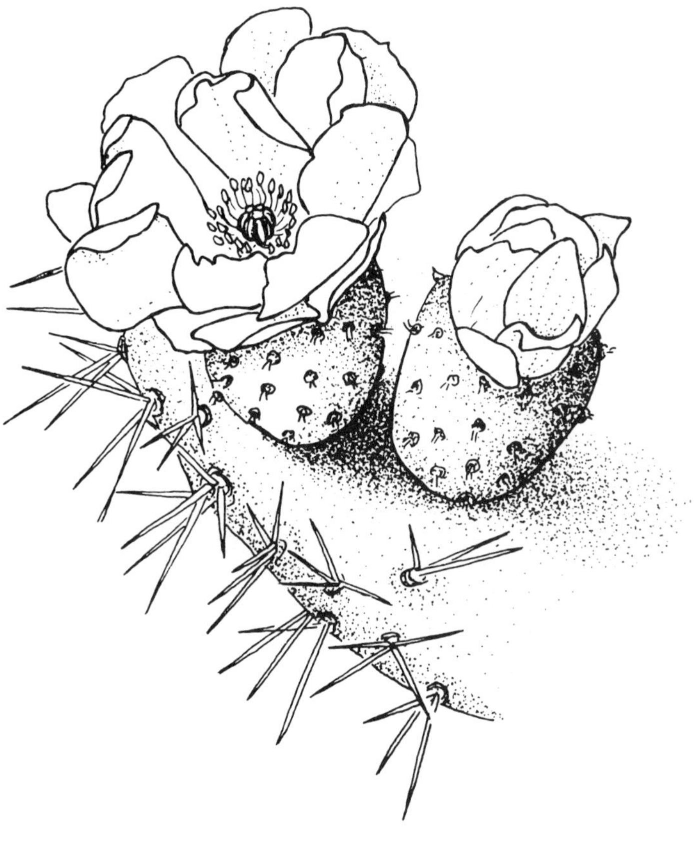

Literature Cited

Azevedo, N. 1960. *Dothiorella* sp. agente de murchidao em *Myrica faya* Ait. *Pub. Dir. Serv. Flor. Agric.* 27:101-115.

Barreto, R.W., and H.C. Evans. 1988. Taxonomy of a fungus introduced into Hawaii for biological control of *Ageratina riparia* (Eupatorieae; Compositae), with observations on related weed pathogens. *Trans. Brit. Mycol. Soc.* 91(1):81-97.

Bowers, R.C. 1982. Commercialization of microbial biological control agents. In *Biological control of weeds with plant pathogens*, ed. R. Charudattan and H.L. Walker, 157-173. New York: John Wiley & Sons.

Carpenter, C.W. 1944. *Fusarium* disease of the prickly pear. *Hawn. Plant. Rec.* 48:59-63.

Charudattan, R. 1982. Regulation of microbial weed control agents. In *Biological control of weeds with plant pathogens*, ed. R. Charudattan and H.L. Walker, 175-188. New York: John Wiley & Sons.

Charudattan, R., and H.L. Walker, eds. 1982. *Biological control of weeds with plant pathogens*. New York: John Wiley & Sons.

Cullen, J.M., P.F. Kable, and M. Catt. 1973. Epidemic spread of a rust imported for biological control. *Nature* 244:462-464.

Daniel, J.T., G.E. Templeton, R.J. Smith, Jr., and W.T. Fox. 1973. Biological control of northern jointvetch in rice with an endemic fungal disease. *Weed Sci.* 21:303-307.

Emge, R.G., J.S. Melching, and C.H. Kingsolver. 1981. Epidemiology of *Puccinia chondrillina*, a rust pathogen for the biological control of rush skeleton weed in the United States. *Phytopathology* 71:839-843.

Gardner, D.E. 1982. Exotic plants in Hawaii's national parks: a major challenge. *Park Sci.* 2(2):18-19.

Gardner, D.E. 1984. Current biocontrol investigations: exploration for control agents in the native habitats of firetree, and evaluation of rust diseases of *Rubus* spp. in the southeastern U.S. *Proc. Fifth Conf. Nat. Sci., Hawaii Volcanoes Natl. Park*, 53-60. Honolulu: Univ. Hawaii Coop. Natl. Park Resour. Stud. Unit.

Gardner, D.E. 1989. Pathogenicity of *Fusarium oxysporum* f. sp. *passiflorae* to banana poka and other *Passiflora* spp. in Hawaii. *Plant Disease* 73:476-478.

Gardner, D.E., and R.E. Burge. 1980. Approaches to the biocontrol of certain exotic trees using vascular wilt fungi. *Proc. 2nd Sci. Conf. Res. Natl. Parks* 8:447-451. Washington, D. C.: Natl. Park Service.

Gardner, D.E., and C.J. Davis. 1982. *The prospects for biological control of nonnative plants in Hawaiian national parks.* Tech. Rep. 45, Univ. Hawaii Coop. Natl. Park Resour. Stud. Unit. Honolulu.

Gardner, D.E., and C.S. Hodges, Jr. 1985. Leaf rust caused by *Kuehneola uredinis* on native and nonnative *Rubus* species in Hawaii. *Plant Disease* 67:962-963.

Gardner, D.E., G.P. Markin, and C.S. Hodges, Jr. 1988. *Survey for potential biological control agents for* Myrica faya *in the Azores and Madeira.* Tech. Rep. 63, Univ. Hawaii Coop. Natl. Park Resour. Stud. Unit. Honolulu.

Gardner, D.E., and C.W. Smith. 1985. Plant biocontrol quarantine facility at Hawaii Volcanoes. *Park Sci.* 6(1):3-4.

Goeden, R.D., L.A. Andres, T.E. Freeman, P. Harris, R.L. Pienkowski, and C.R. Walker. 1974. Present status of projects on the biological control of weeds with insects and plant pathogens in the United States and Canada. *Weed Sci.* 22:490-495.

Gordon, W.L. 1965. Pathogenic strains of *Fusarium oxysporum*. *Can. J. Bot.* 43:1309-1318.

Hasan, S. 1972. Specificity and host specialization of *Puccinia chondrillina*. *Ann. Appl. Biol.* 72:257-263.

Hasan, S. 1974. Recent advances in the use of plant pathogens as biocontrol agents of weeds. *Pest Artic. News Summ.* 20:437-443.

Hasan, S. 1980. Plant pathogens and biological control of weeds. *Rev. Plant Pathol.* 59(8):349-356.

Hasan, S., and A.J. Wapshere. 1973. The biology of *Puccinia chondrillina*, a potential biological control agent of skeleton weed. *Ann. Appl. Biol.* 74:325-332.

Hawaii Volcanoes National Park. 1985. Natural resources management plan and environmental assessment, revised 1985. Mimeo.

Hodges, C.S. 1988. *Preliminary exploration for potential biological control agents for* Psidium cattleianum. Tech. Rep. 66, Univ. Hawaii Coop. Natl. Park Resour. Stud. Unit. Honolulu.

Hodges, C.S., Jr., and D.E. Gardner. 1985. *Myrica faya: potential biological control agents*. Tech. Rep. 54, Univ. Hawaii Coop. Natl. Park Resour. Stud. Unit. Honolulu.

Holcomb, G.E. 1982. Constraints on disease development. In *Biological control of weeds with plant pathogens*, ed. R. Charudattan and H.L. Walker, 61-71. New York: John Wiley & Sons.

Howarth, F.S. 1983. Classical biocontrol: panacea or Pandora's box. *Proc. Hawn. Entomol. Soc.* 24(2,3):239-244.

Inch, A.J. 1978. Passionfruit diseases. *Queensl. Agric. J.* 104:479-484.

Inman, R.E. 1971. A preliminary evaluation of *Rumex* rust as a biological control agent for curly dock. *Phytopathology* 61:102-107.

Kahn, R.P. 1983. A model plant quarantine station: principles, concepts, and requirements. In *Exotic plant quarantine pests and procedures for introduction of plant materials*, ed. K. G. Singh, 303-333. Serdang, Selangor, Malaysia: ASEAN Plant Quarantine Centre and Training Institute.

Klingman, D.L., and J.R. Coulson. 1983. Guidelines for introducing foreign organisms into the United States for the biological control of weeds. *Bull. Entomol. Soc. Amer.* 29:55-61.

Krauss, N.L.H. 1964. Insects associated with firebush (*Myrica faya* Aiton). *Proc. Hawn. Entomol. Soc.* 18(3):405-411.

LaRosa, A.M. 1985. Note on the identity of the introduced passionflower vine "banana poka" in Hawaii. *Pac. Sci.* 39:369-371.

Leonard, K. J. 1982. The benefits and potential hazards of genetic heterogeneity in plant pathogens. In *Biological control of weeds with plant pathogens*, ed. R. Charudattan and H.L. Walker, 99-112. New York: John Wiley & Sons.

Melching, J.S., K.R. Bromfield, and C.H. Kingsolver. 1983. The plant pathogen containment facility at Frederick, Maryland. *Plant Disease* 67:717-722.

Oehrens, E. 1977. Biological control of the blackberry through the introduction of rust, *Phragmidium violaceum*, in Chile. *FAO Plant Prot. Bull.* 25:26-28.

Oehrens, E., and S. Gonzalez. 1977. Dispersion, ciclo biologico y danos causados por *Phragmidium violaceum* (Schulz) Winter en zarzamora (*Rubus constrictus* Lef. et M. y *R. ulmifolius* Schott.) en las zonas centro-sur y sur de Chile. *Agro Sur* 5(2):73-85.

Phatak, S.C., D.R. Sumner, H.D. Wells, D.K. Bell, and N.C. Glaze. 1983. Biological control of yellow nutsedge with the indigenous rust fungus *Puccinia canaliculata*. *Science* 219:1446-1447.

Purss, G.S. 1954. Identification of a species of *Fusarium* causing wilt in passion vines in Queensland. *Queensl. J. Agric. Sci.* 11:79-81.

Purss, G.S. 1958. Studies of the resistance of species of *Passiflora* to *Fusarium* wilt (*F. oxysporum* f. *passiflorae*). *Queensl. J. Agric. Sci.* 15:95-99.

Quimby, P.C., Jr. 1982. Impact of diseases on plant populations. In *Biological control of weeds with plant pathogens*, ed. R. Charudattan and H.L. Walker, 47-60. New York: John Wiley & Sons.

Shrum, R.D. 1982. Creating epiphytotics. In *Biological control of weeds with plant pathogens*, ed. R. Charudattan and H. L. Walker, 113-136. New York: John Wiley & Sons.

Smith, R.J., Jr. 1982. Integration of microbial herbicides with existing pest management programs. In *Biological control of weeds with plant pathogens*, ed. R. Charudattan and H.L. Walker, 189-203. New York: John Wiley & Sons.

Smith, R.J., Jr., J.T. Daniel, W.T. Fox, and G.E. Templeton. 1973. Distribution in Arkansas of a fungus disease used for bio-control of northern jointvetch in rice. *Plant Dis. Rep.* 57:695-697.

Stoller, E.W. 1981. Yellow nutsedge: a menace in the corn belt. *U.S. Dep. Agric. Tech. Bull.* 1642.

Templeton, G.E. 1982. Status of weed control with plant pathogens. In *Biological control of weeds with plant pathogens*, ed. R. Charudattan and H.L. Walker, 29-44. New York: John Wiley & Sons.

Trujillo, E.E. 1976. Biological control of Hamakua pamakani with plant pathogens. *Proc. Amer. Phytopath. Soc.* 3:298.

Trujillo, E.E., M. Aragaki, and R.A. Shoemaker. 1988. Infection, disease development, and axenic culture of *Entyloma compositarum*, the cause of Hamakua pamakani blight in Hawaii. *Plant Disease* 72(4):355-357.

Trujillo, E.E., F.M. Laterrell, and A.E. Rossi. 1986. *Colletotrichum gloeosporioides*, a possible biological control agent for *Clidemia hirta* in Hawaiian forests. *Plant Disease* 70:974-976.

U.S. Department of Agriculture. 1978. *Biological agents for pest control. Status and prospects*. Washington, D.C.: U.S. Gov. Print. Off.

Wapshere, A.J. 1975. A protocol for programmes for biological control of weeds. *Pest Artic. News Summ.* 21:295-303.

Wilson, C.L. 1969. Use of plant pathogens in weed control. *Ann. Rev. Phytopathol.* 7:411-434.

Zettler, F.W., and T.E. Freeman. 1972. Plant pathogens as biocontrols of aquatic weeds. *Ann. Rev. Phytopathol.* 10:455-470.

REVIEW AND STATUS OF BIOLOGICAL CONTROL OF CLIDEMIA IN HAWAI'I

Larry M. Nakahara, Robert M. Burkhart, and George Y. Funasaki

ABSTRACT

Efforts to control clidemia (*Clidemia hirta*) in Hawai'i with phytophagous insects began in 1952. Several attempts have been made since then to introduce potential biological control agents to reduce the spread of this weed into forest areas, and other control measures have been tried by various groups and agencies. It was not until recently, however, through the enactment of significant legislation and subsequent funding, that explorations and studies were conducted specifically on clidemia insects in Trinidad, West Indies. Fourteen species of insects were evaluated, including *Carposina bullata, Mompha trithalama*, a midge or cecidomyiid, two *Eurytoma* species, *Piesmopoda* sp., and *Compsolechia seductella*, which feed on the flowers and fruits; *Lius poseidon, Antiblemma acclinalis, Druentia* sp., prob. *inscita, Ategumia matutinalis* (formerly *Blepharomastix ebulealis* but originally thought to be *Sylepte matutinalis), Penestes* n. sp., and a leaf beetle or chrysomelid, which feed on the leaves; and a long-horned cerambycid beetle, which bores into the stem. Studies indicated that several species were sufficiently host specific to warrant introduction into Hawai'i for the biological control of clidemia.

INTRODUCTION

In Hawai'i, clidemia or Koster's curse (*Clidemia hirta*) (Melastomataceae), a fast-growing, heavy-seeding, tropical American shrub, has colonized forest clearings, trailsides, and burn sites and intruded into the understories of forests that were formerly free of introduced, or alien, weeds (Wester and Wood 1977). The species has spread rapidly throughout the State and represents a serious threat to our native forests. First observed on O'ahu in 1941, it has since spread to Hawai'i (1972), Moloka'i (1973), Maui (1977), and Kaua'i (1982).

Clidemia hirta is generally not considered a pest in its native range. Interspecific competition may be a major factor for its low density in Trinidad, where at least 100 species of melastomes and other aggressive weeds compete for space. Clidemia is commonly found along trails, roadsides, and at the edges of forests there, and unlike Hawai'i, it is rarely found in non-disturbed forest areas with dense canopies. Plants in Trinidad are shorter and have thinner stems than those in Hawai'i, and few seedlings are found under mature plants (Burkhart 1983).

The manner in which clidemia developed into a weed pest in Hawai'i, the problems encountered while trying to control it, the strategies employed, and the reliance on biological control measures as the only possible cost-effective, long-lasting, management solution are similar to many other weed problems we face in Hawai'i. The cooperation and support provided by concerned individuals and agencies in controlling this weed are also indicative of what can be accomplished when people work together to solve a common problem. Early attempts made to reduce clidemia and the current status of additional insects being studied for possible introduction into Hawai'i to aid in its biological control are reviewed in this paper.

BIOLOGICAL CONTROL ATTEMPTS IN HAWAI'I

Clidemia was recognized as a serious potential weed pest following a survey in the forest reserve near Wahiawā, O'ahu in 1952 (Krauss 1954). Because of the problems experienced in Fiji (Simmonds 1933), the major concern was that the weed would spread over wide areas into pastures, cultivated lands, and forest areas in Hawai'i (Krauss 1952). Officials of the Board of Commissioners of Agriculture and Forestry initiated surveys on the neighbor islands (*i.e.*, the main Islands other than O'ahu) (Thistle 1952) and distributed information sheets with sketches of the plant, background information, and a chemical control recommendation (Hosaka and Thistle 1953). A biological control program was begun in 1952 by soliciting comments from the Committee on Beneficial Insect Introductions; this Committee reviewed the merits of introducing the thrips *Liothrips urichi* into Hawai'i. The Hawaii Department of Agriculture (hereafter called the Department) has continued to evaluate biological control of clidemia off and on for the past 34 years.

In April 1953, after testing, review, and approval for introduction by other entomologists employed by the pineapple and sugar industries and the University of Hawaii, *L. urichi* was imported into Hawai'i from Fiji to control clidemia (Fullaway 1954). Although extensive host range tests had been conducted on *L. urichi* before release in Fiji, there was concern that the thrips could be a vector of the spotted wilt virus that affected pineapple (*Ananas comosus*) in Hawai'i (Sakimura 1953). Tests showed that *L. urichi* did not transmit the virus, and the first release of the thrips was made on June 1, 1953, in Wahiawā Forest Reserve. Eight days later, on June 9th, the thrips were found breeding on clidemia plants in the area and were considered to be established.

Liothrips urichi was effective in preventing the spread of clidemia into open pastures and cultivated lands. It did not, however, prove to be effective in shaded areas (Davis 1969; Reimer 1985) and was thus ineffective under forest conditions (Davis and Chong 1969). Recent studies by Reimer (1985) showed that while *L. urichi* was effective in killing or reducing the growth of juvenile clidemia plants, female thrips laid significantly fewer eggs in shady than in sunny areas. Predation also affected thrips populations. The beneficial anthocorid *Montandoniola moraguesi*, purposely introduced into Hawai'i in 1964 to control the Cuban laurel thrips, *Gynaikothrips ficorum*, was found to feed on *L. urichi*. There was no significant correlation between *M. moraguesi* and *L. urichi* populations, however. Further cage exclusion experiments showed that predation on *L. urichi* by the alien big-headed ant, *Pheidole megacephala*, was significantly greater than that by *M. moraguesi*.

Several species of insects were found attacking the melastome *Dissotis hensii* during explorations in Africa in 1956. These insects showed promise in attacking clidemia and another weed, *Melastoma candidum*, and were shipped to Hawai'i and studied in quarantine. Larvae of one unidentified moth among these species were found to feed on clidemia. The caterpillars also fed on princess flower or lasiandra (*Tibouchina urvilleana*), however, which was considered an attractive ornamental by most people during the 1950s, so the stock was destroyed (Nakao 1975).

The search for additional natural enemies of clidemia was reemphasized when the ineffectiveness of *L. urichi* under forest conditions became more apparent during the late 1960s. Chemical control and mechanical clearing were not considered feasible due to the rugged, mountainous terrain in which the weed occurred (Uehara 1968). During 1968-69, an arctiid moth, *Selca brunella*, and a pyralid moth, *Bocchoris fatualis*, were tested for possible clidemia control on O'ahu. Both species had been released earlier on Hawai'i and Kaua'i for control of *Melastoma candidum*. Neither species was able to survive on clidemia (Au 1968; Nakao 1969).

Another pyralid moth, *Ategumia matutinalis*, was introduced to Hawai'i from Puerto Rico and Trinidad and studied in the Hawaii Department of Agriculture's quarantine facility in 1969. This moth had been collected from *Heterotrichum cymosum* (family Melastomataceae) in Puerto Rico and from a species of *Clidemia* in Trinidad. Host specificity studies were conducted on 31 species of plants from 29 families (Nakao and Suzukawa 1970). Trace feeding occurred on rose, cucumber, hibiscus, and passion fruit (*Rosa* spp., *Cucumis sativus, Hibiscus* spp., *Passiflora edulis*), but the larvae were unable to survive on these plants. *Ategumia matutinalis*, however, completed its life cycle on all melastomes that were tested, including princess flower, which was achieving weed status by the late 1960s. After review of the host range tests by the Advisory Committee, a request for the release from quarantine of *B. ebulealis* was made to the Board of Agriculture in September 1970 (Uehara 1970). Approval was given the following month (Davis 1971), and the first release of the moth was made in December 1970 after mass propagation in the Department's insectary facility. With the help of Sierra Club (Hawai'i Chapter) and Hawaiian Trail and Mountain Club members, thousands of adult moths were liberated on O'ahu to control clidemia and on Hawai'i for control of other melastomes (Davis 1974). No recoveries of the moth, however, were made until October 1974, when light infestations and feeding damage were observed on clidemia along a two-mile stretch on the Kawai Iki Trail on O'ahu (Nakao and Funasaki 1976). Infestations were subsequently observed on the Hau'ula and Pōamoho trails and in Waiāhole Valley on O'ahu in the mid 1970s. All infestations were light, and parasites were thought to be reducing the effectiveness of this leaf roller (Fujii 1977).

Recent studies by Reimer (1985) have shown that parasites are an important mortality factor for *Ategumia matutinalis*; the moth averaged 43% parasitization in sampled sites. The following four species of parasites have been reared from *A. matutinalis* larvae. The chalcid *Brachymeria obscurata* was purposely introduced into Hawai'i in 1895 to control two native insects considered to be serious pests at that time, the coconut leaf roller, *Hedylepta blackburni*, and the sugar cane leaf roller, *Hedylepta accepta*. The braconid *Meteorus laphygmae* was purposely introduced in 1942 to control armyworms, then plaguing the Islands. Finally, two species of ichneumonids, *Trathala flavoorbitalis* and *Casinaria infesta*, were reared. Both species are aliens that have established in the Islands since arrival in 1910 and 1921.

Caterpillars of a fruit-feeding lepidopteran were collected from a species of *Clidemia* and shipped to Hawai'i during a 1969 exploration in Trinidad. The collection was heavily parasitized, and no adult moths were recovered in quarantine.

A pseudococcid, *Crisicoccus* sp., was found infesting clidemia in 1977, while explorations were being conducted in El Salvador. Specimens were shipped to Hawai'i and a colony was established in the Hawaii Department of Agriculture's quarantine facility. In a joint project with the University of Hawaii, host specificity studies of this mealybug were conducted on various economically important crops. The mealybug bred successfully on avocado, coffee, mountain apple, and gardenia (*Persea americana, Coffea arabica, Syzygium malaccense, Gardenia jasminoides*), and the colony was destroyed (Gon 1978; Lai 1978).

As biological control efforts continued on O'ahu, other control methods were being applied by concerned individuals, government agencies, and other groups on the neighbor islands. Infestations at Wailau on Moloka'i were periodically rogued (controlled) by Sierra Club members and Hawaii Division of Forestry and Wildlife personnel (Shinbara 1977). In the Puna and Waiākea Forest Reserves on the island of Hawai'i, a combination of roguing and herbicide treatments was applied by the Hawaii Division of Forestry and Department of Agriculture staff members (Kami 1977). Efforts to eradicate or control localized infestations through these methods proved unsuccessful, however, due to difficulties with the terrain and the need to remove all clidemia seedlings and the seed bank.

SIGNIFICANT LEGISLATION AND FUNDING

In the past, biological control efforts by the Department were spread over a number of serious agricultural, forest, and urban pest problems. These pests were not only economically important but also affected the health and well-being of the people in the State. During the 25-year period from 1953 to 1978, many biological control organisms were introduced into the Department's quarantine facility for screening, rearing, and host specificity testing. Introductions were made to combat 85 pests, including 14 species of weeds (Lai and Funasaki 1986). Many of these candidate control agents were destroyed in quarantine after studies were conducted. Eventually, 283 species were approved for release, of which 98 actually became established. Funds and effort to control clidemia during this period had to be shared with projects on 84 other pests.

The enactment of significant legislation during the late 1970s enabled State agencies to focus specific funding on clidemia control programs; this resulted in an in-depth exploration program directed at clidemia insects. In 1975, the Legislature adopted Senate Resolution No. 240, which requested the Department to compile data on the weed and to propose control measures and funding requirements. This resulted in a joint Hawaii Department of Agriculture, Department of Land and Natural Resources, and University of Hawaii proposal to 1) conduct a biological and ecological study of clidemia in the tropics; and 2) initiate exploration, screening, and evaluation of biological organisms by an entomologist and a plant pathologist.

Through the efforts of legislators, the Hawaii Chapter of the Sierra Club, and other concerned parties, funds were appropriated in 1977 for the biological control of clidemia. Because of spending restrictions in 1978, the funds allocated could not be used. That same year, however, the Governor designated the Department as lead agency in the biological control effort and directed that current resources be used, with additional funding available if needed (Ariyoshi 1978). An exploration specifically for clidemia insects was planned for the following year.

Exploration plans had to be revised in late 1978 with the discovery of the spiraling whitefly, *Aleurodicus dispersus*, in September of that year. The whitefly spread rapidly on O'ahu, posing a serious threat to many facets of Hawai'i's economy because of its host diversity (over 100 species of plants) (Hawaii Department of Agriculture 1981; Kumashiro *et al.* 1983). Top priority was given to the control of this new pest, and clidemia control was relegated to secondary status. Nevertheless, a three-month exploration, partially funded by the Department of Land and Natural Resources, was conducted in tropical America in 1979, and several species associated with clidemia were collected and shipped to Hawai'i for screening. Continuous propagation of the insects in quarantine was unsuccessful due to problems in growing host plants for the rearing of these insects under laboratory conditions to the flowering and fruiting stages.

Since 1980, the Hawaii Department of Agriculture and the Department of Land and Natural Resources have jointly funded explorations to evaluate natural enemies of clidemia in Trinidad. In 1980, a six-month expedition was conducted there to search for control agents of agricultural pests as well as to survey insects attacking *C. hirta*. A joint agreement was signed in 1982, whereby the Department of Agriculture would search for additional biological control agents and do basic studies, while the Department of Land and Natural Resources would provide the necessary funding. All studies would be done in areas where *C. hirta* is native, due to difficulties of propagating the weed, other test plants, and the candidate insects in the Honolulu quarantine facility. Twelve species of insects were identified from June 1982 to June 1983 as showing a potential to control *C. hirta.* Of particular interest were the insects that caused an estimated 95% reduction in seed production. This group of insects could be very useful in Hawai'i, where clidemia occupied only a small percentage of its possible range. Preliminary host range studies indicated that feeding by most of the 12 species was confined to melastomes (Burkhart 1983). The June 1984 to June 1986 exploration emphasized studying the biology and ecology, developing rearing techniques, and completing host range testing of the selected candidate biological control agents.

STRATEGIES AND OBJECTIVES

The objectives of the biological control program are to search for and evaluate insects that may be effective in destroying or placing stress on *C. hirta* plants, especially in shady areas, as well as those insects and pathogens that are effective in destroying its seeds. Potential biological control agents are screened with particular attention to 1) the ability of *C. hirta* to aggressively invade new areas because of its abundant seed production, and 2) its ability to establish itself primarily in shaded areas.

The Department of Land and Natural Resources has actively coordinated a strategy of using both insects and pathogens to control *C. hirta.* Work with pathogens at the University of Hawaii has been fruitful, with the discovery in Panama of a leaf spot fungus that attacks clidemia foliage. The fungus, *Colletotrichum gloeosporioides* f.s. *clidemiae*, was introduced to Fort Detrick, Frederick, Maryland, in September 1985. After preliminary host range testing, permission was obtained for its introduction into Hawai'i, where further host range studies were conducted (Trujillo 1986a; Trujillo *et al.* 1986).

SCREENING PROCEDURES

Studies have been conducted in Trinidad since 1982, where the availability of plants, insects, and the facilities of the Commonwealth Institute of Biological Control provide the necessary combination for the required studies and host range tests. Field data were collected from 40 sites, most of them in the Northern Range, which represents the widest range of ecological conditions for *C. hirta* in that country. The elevations of these sites range from 2 to 3,090 ft (0.5-936 m), mean minimum-maximum temperatures ranged from 95.4 to 110.7 F (21.0-29.5 C) (1984-1985), and mean annual precipitation varied from 45 to 110 in. (1,120-2,920 mm) among sites (Burkhart 1986).

Species determinations are sought for all insects attacking *C. hirta* as well as for natural enemies of the insects, yet the identities of many are unknown. Laboratory studies are conducted to determine basic biological information, such as life history, adult longevity, mating and oviposition requirements, and feeding behavior. These studies enable the development of rearing techniques for further testing, as well as for the development of mass-rearing procedures should the insect be approved for release in Hawai'i. Other biological and ecological field observations relating to habitat preference, seasonal abundance, natural enemies, and natural host plants provide clues to the potential importance of each species as a candidate biocontrol agent. Finally, further testing in the lab determines the host specificity of each insect under laboratory conditions. For some species, a starvation test may be all that is necessary to determine if the plant-feeding stage(s) of the insect can survive on test plants. For others, the tests may be more involved to determine if the candidate insect can complete its life cycle on test plants. Sometimes the degree of preference when candidate insects are given a choice of several host plants is determined.

Host specificity studies for flower- and fruit-feeding insects are conducted by bagging the insects with the susceptible reproductive stage of test plants. Leaf-feeding insects are exposed to mixed bouquets of freshly cut test plant shoots. Plant selection for studies is based on their economic importance in Hawai'i as well as their taxonomic relationship to *C. hirta* (Burkhart 1983, 1986).

POTENTIAL BIOCONTROL CANDIDATES

Fourteen species of insects have been evaluated in Trinidad (Table 1). These insects can be grouped according to the type of damage they inflict on the plant (Burkhart 1983, 1986).

Flower and Fruit Feeders

Carposina bullata. Larvae feed internally in the newly formed flower buds. The life cycle is approximately one month. The host range is limited to plants in the family Melastomataceae, particularly in the genera *Miconia* and *Clidemia*. Studies are completed.

Mompha trithalama. The larvae feed internally on both flowers and berries. When flowers are attacked, no seeds are formed. Host range is limited to the Melastomataceae family. Studies are near completion.

Unidentified midge. The larvae feed on the anthers of flowers, causing some to abort. More studies are needed.

Eurytoma **sp. (black)**. The larvae feed individually on young flowers, resulting in no seed production. More studies are needed.

Eurytoma **sp. (red)**. Larvae feed in developing berries, causing the formation of galls, which reduces seed production. More studies are needed.

Table 1. Species of insects being evaluated for biocontrol of *Clidemia hirta* by Hawaii Department of Agriculture.

Species (Order: Family)	Site of Damage	Type of Damage	Comments
Carposina bullata Meyr. (Lepidoptera: Carposinidae)	Flower / Fruit	Within fruit/Flower	Non gall-forming
Mompha trithalama Meyr. (Lepidoptera: Momphidae)	Flower / Fruit	Within fruit/Flower	Non gall-forming
Unidentified midge (Diptera: Cecidomyiidae)	Flower / Fruit	Within fruit/Flower	Non gall-forming
Eurytoma sp. (black) (Hymenoptera: Eurytomidae)	Flower / Fruit	Within fruit/Flower	Non gall-forming
Eurytoma sp. (red) (Hymenoptera: Eurytomidae)	Flower / Fruit	Within fruit	Gall-forming
Piesmopoda sp. (Lepidoptera: Pyralidae)	Flower heads	Flower feeder	
Compsolechia seductella Walk. (Lepidoptera: Gelechiidae)	Flower heads	Flower/Leaf feeder	
Lius poseidon Napp (Coleoptera: Buprestidae)	Leaves	Leaf feeder	
Antiblemma acclinalis Hubner (Lepidoptera: Noctuidae)	Leaves	Leaf feeder	
Druentia sp. prob. *inscita* (Schaus) (Lepidoptera: Lascosomidae)	Leaves	Leaf feeder	
Ategumia matutinalis (Guenee) (Lepidoptera: Pyralidae)	Leaves	Leaf feeder	
Penestes sp. (Coleoptera: Curculionidae)	Leaves	Leaf feeder	
Unidentified leaf beetle (Coleoptera: Chrysomelidae)	Leaves	Leaf feeder	
Unidentified long-horned beetle (Coleoptera: Cerambycidae)	Stem	Stem borer	

Piesmopoda **sp.** The larvae are voracious feeders, consuming one or more flower clusters, resulting in the destruction of many seeds. In starvation tests, larvae fed on flowers of other plant families. Under field conditions, feeding was confined to plants in the genera *Clidemia* and *Miconia*. More studies are needed.

Compsolechia seductella. The larvae cut the growing leaves and flower clusters and feed on the dead plant mass. In the lab, some feeding occurred on plants outside of the family Melastomataceae. Under field conditions, only plants in the genera *Miconia* and *Clidemia* are recorded hosts. More studies are needed.

Leaf Feeders

Lius poseidon. Both the adult and larval stages of this leaf-mining buprestid feed on *C. hirta* leaves. The shiny black adult feeds externally on the leaves, producing a ragged feeding strip. The egg is laid singly on the upper surface of a leaf. After the egg hatches, the young larva tunnels into the leaf, producing a blotch mine. Pupation occurs in the mine, which forms a "blister" on the leaf, and the adult emerges 40 days after the egg is laid. *Lius poseidon* favors plants in light shade but never in heavy shade and rarely in open fields. In field studies, *L. poseidon* adults fed on 13 species of plants in four genera, while the larvae were recorded from seven species in three genera, all in the family Melastomataceae. Most of the damage was observed on three species of *Clidemia*, including *C. hirta*. In starvation tests, *L. poseidon* did not feed on another 35 species of plants in 27 families. Studies are completed.

Antiblemma acclinalis. Larvae feed on leaves of young *C. hirta* plants, primarily in cool, moist locations with moderate to heavy shade. The life cycle is about 36 days. Problems with mass rearing this species limited the number of tests. In host range tests, *A. acclinalis* did not feed on 18 non-melastomataceous species from 15 families. In the field, host range was limited to four species of melastomes, with 95% occurring on *C. hirta* and *C. debilis*. Studies are completed.

Druentia **sp., prob.** *inscita*. Larvae cut and fold the leaves of *C. hirta* to form a feeding shelter. The life cycle is approximately two to six months long, due to the ability of the species to aestivate in the pupal stage during the dry season. *Druentia* is found over a wide range of elevations but was common in the upper-elevation sites. It occurred in various habitats, except under conditions of full sun. Field studies show that feeding by this species is restricted to only 13 species of melastomes in four genera, despite the presence of 39 species of melastomes in four genera in the surveyed area. Clidemia, the second-most preferred host plant, was used half as often as *Miconia acinodendrum*, the most preferred. No larval feeding damage was observed on non-melastomes in the field or on 15 species of non-melastomes from 15 families tested in the lab. Studies are near completion.

Ategumia matutinalis. This pyralid leaf roller feeds on only a few species of Melastomataceae. It is adapted to a wide range of ecological conditions but prefers moist, shaded conditions. Studies are continuing.

Penestes **sp.** Adults were most common at higher elevations and in cool, damp locations in deep shade, where they caused as much as 50% leaf damage. In the field, adults feed only on melastomes, species in the genera *Clidemia* and *Miconia* being preferred. After an intensive search, larvae were found breeding in the succulent petioles of a few species of *Miconia*, particularly *M. mirabilis*. No larvae were found breeding in clidemia petioles. Adults fed on eight species of non-melastomes, with heavy feeding on *Begonia*. Although this species was rejected from further studies for *C. hirta* control, its possible role in controlling *Miconia* species would require further work.

Unidentified leaf beetle. Adults caused considerable foliar damage to *C. hirta* leaves. Larvae were not found on *C. hirta*, and adults fed on all non-melastomes tested with extensive feeding on eggplant, cucumber, and bell or sweet pepper (*Solanum melongena, Cucumis sativus, Capsicum annuum*). The candidate was rejected from further studies.

Stem Borers

Unidentified long-horned beetle. Larvae of this unidentified cerambycid beetle were occasionally found in stems of *C. hirta*, with plants showing signs of stunting and dieback. No adults have been recovered. More studies are needed.

Many of the above lepidopterous species and the buprestid beetle are heavily parasitized in Trinidad by hymenopterous species not found in Hawai'i. While some of Hawai'i's locally established parasites may adapt to these insects were they to be released in the Islands, without their natural enemies, *C. hirta* insects would become established in Hawai'i, probably resulting in higher infestation levels than currently seen in Trinidad.

FUTURE CONSIDERATIONS

A request for approval to release *Colletotrichum gloeosporioides* f.s. *clidemiae* for *C. hirta* control is currently being reviewed by the Board of Agriculture's Advisory Subcommittee on Microorganisms (Trujillo 1986a). The pathogen was found to infect only *C. hirta* in inoculation tests on more than 40 other plant species representing 28 families (Trujillo 1986b). In addition, requests for approval to introduce and release *Lius poseidon*, *Antiblemma acclinalis*, and *Carposina bullata* will be made in the near future, following a review of requirements for further testing on native plants found in Hawai'i. Other requests on *Mompha trithalama* and *Druentia* sp. prob. *inscita* would also follow. Requests for the release from quarantine of these candidate biological control insects must first be reviewed by the State Advisory Subcommittee on Entomology and the State Advisory Committee on Plants and Animals; their recommendations are forwarded to the Board of

Agriculture. Field releases may be made if approval is granted by the Board of Agriculture.

Based on the studies conducted so far with the plant pathogen and phytophagous insects, the outlook for the eventual management of clidemia in Hawai'i's forests appears promising. However, it cannot be predicted to what extent these organisms may suppress the continuing spread of *C. hirta* in the forests or reduce its current area of establishment. If *C. hirta* is successfully controlled, it will be because of the cooperation, concern, and efforts of many individuals, groups, and agencies.

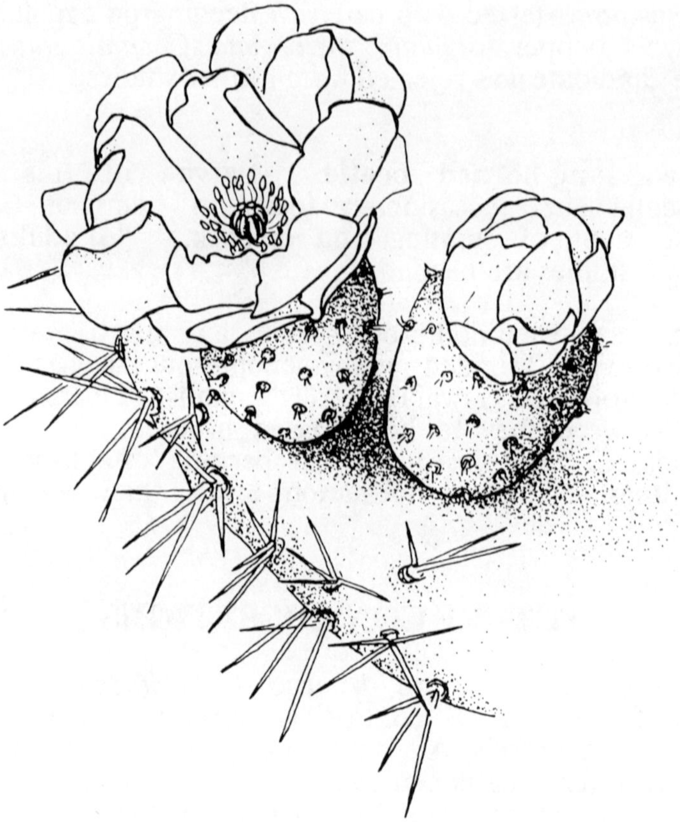

Literature Cited

Ariyoshi, G.R. 1978. Memorandum - *Clidemia hirta* - biocontrol program. Executive Chamber. Files, Hawaii Dep. Agric., Honolulu. Typescript.

Au, S.H. 1968. Memorandum - *Selca brunella* Hampson test on *Clidemia hirta*. Files, Hawaii Dep. Agric., Honolulu. Typescript.

Burkhart, R.M. 1983. Progress report on exploratory studies on *Clidemia hirta* in Trinidad, West Indies, June 1982 - June 1983. Files, Hawaii Dep. Agric., Honolulu. Unpub. rep.

Burkhart, R.M. 1986. Progress report on exploratory studies on *Clidemia hirta* in Trinidad, West Indies, June 1984 - June 1985. Files, Hawaii Dep. Agric., Honolulu. Unpub. rep.

Davis, C.J. 1969. Notes and exhibitions. *Proc. Hawn. Entomol. Soc.* 20(3):277.

Davis, C.J. 1971. Recent introductions for biological control in Hawaii -- XVI. *Proc. Hawn. Entomol. Soc.* 21(1):59-62.

Davis, C.J. 1974. Recent introductions for biological control -- XVIII. *Proc. Hawn. Entomol. Soc.* 21(3):355-358.

Davis, C.J., and M. Chong. 1969. Recent introductions for biological control in Hawaii XIV. *Proc. Hawn. Entomol. Soc.* 22(2):317-322.

Fujii, J.K. 1977. Notes and exhibitions. *Proc. Hawn. Entomol. Soc.* 22(3):394-395.

Fullaway, D.T. 1954. Notes and exhibitions. *Proc. Hawn. Entomol. Soc.* 15(2):280.

Gon, S. 1978. Feeding tests on plants of economic importance with *Crisicoccus* sp. (Homoptera: Pseudococcidae) to determine its host specificity. Hawaii Dep. Agric., Univ. Hawaii Joint Project, Honolulu. Unpub. rep.

Hawaii Department of Agriculture. 1981. *Hawaii Pest Report* I(5). Plant Pest Control Branch, Hawaii Dep. Agric. Honolulu.

Hosaka, E.Y., and A. Thistle. 1953. *Clidemia hirta* (L.) D. Don (Melastomataceae), a new and potentially serious weed pest of agricultural lands in Hawaii. Bd. Commissioners Agric. and For. Honolulu. Mimeo.

Kami, R. 1977. Memorandum - *Clidemia hirta*. Hawaii Dep. Agric., Honolulu. Typescript.

Krauss, N. 1952. Memorandum on the biological control of the weed *Clidemia hirta* in Fiji. Bd. Commissioners Agric. and For. Honolulu. Typescript.

Krauss, N. 1954. Notes and exhibitions. *Proc. Hawn. Entomol. Soc.* 15:264.

Kumashiro, B.R., P.Y. Lai, G.Y. Funasaki, and K.K. Teramoto. 1983. Efficacy of *Nephaspis amnicola* and *Encarsia ? haitiensis* in controlling *Aleurodicus dispersus* in Hawaii. *Proc. Hawn. Entomol. Soc.* 24(2 & 3):261-269.

Lai, P.-Y. 1978. Memorandum - Destroy and disposition of *Crisicoccus* sp. from quarantine insectary. Files, Hawaii Dep. Agric., Honolulu. Typescript.

Lai, P.-Y., and G.Y. Funasaki. 1986. List of beneficial organisms purposely introduced for biological control in Hawaii: 1890 to 1985. Revised. Hawaii Dep. Agric., Honolulu. Unpub. rep.

Nakao, H.K. 1969. Memorandum - Feeding test with *Bocchoris* sp. on *Clidemia hirta*. Files, Hawaii Dep. Agric., Honolulu. Typescript.

Nakao, H.K. 1975. Memorandum - Bio-control efforts against *Clidemia hirta* in Hawaii. Files, Hawaii Dep. Agric., Honolulu. Typescript.

Nakao, H.K. and G.Y. Funasaki. 1976. Introductions for biological control in Hawaii, 1974. *Proc. Hawn. Entomol. Soc.* 22(2):329-331.

Nakao, H.K., and R. Suzukawa. 1970. Host specificity tests with *Blepharomastix ebulealis* (Guenee) (family Pyralidae) from *Clidemia hirta*. Files, Hawaii Dep. Agric., Honolulu. Unpub. rep.

Pelly, J. 1975. Clidemia information. Memo to State Forester, May 20. Files, Hawaii State Dep. Land Nat. Resour., Honolulu.

Reimer, N.J. 1985. An evaluation of the status and effectiveness of *Liothrips urichi* Karny (Thysanoptera: Phlaeothripidae) and *Blepharomastix ebulealis* (Guenee) (Lepidoptera: Pyralidae) on *Clidemia hirta* (L.) D. Don in O'ahu forests. Ph.D. diss., Univ. Hawaii, Honolulu.

Sakimura, K. 1953. Memorandum - Innoculated test plants with spotted wilt virus for feeding test with *Liothrips urichi*. Pineapple Research Inst. Hawaii, Honolulu. Typescript.

Shinbara, B.H. 1977. Memorandum - Status of *Clidemia* infestations on Molokai and Hawaii. Files, Hawaii Dep. Agric., Honolulu. Typescript.

Simmonds, H.W. 1933. The biological control of the weed *Clidemia hirta* (L.) D. Don in Fiji. *Bull. Entomol. Res.* 24:345-348.

Thistle, A. 1952. Memorandum - *Clidemia hirta*. Bd. Commissioners Agric. and For., Honolulu. Typescript.

Trujillo, E.E. 1986a. Memo - Request approval to release *Colletotrichum gloeosporioides* f.s. *clidemiae* from quarantine for use as a biological control agent of *Clidemia hirta* (Koster's curse) in Hawaii forest. Univ. Hawaii, Honolulu. Typescript.

Trujillo, E.E. 1986b. Host specificity test with *Colletotrichum gloeosporioides* f.s. *clidemia* Cda. (family Melanconiaceae, Fungi Imperfecti). A biological control candidate for *Clidemia hirta* (L.) D. Don. Univ. Hawaii, Honolulu. Unpub. rep.

Trujillo, E.E., F.M. Latterell, and A.E. Rossi. 1986. *Colletotrichum gloeosporioides*, a possible biological control agent for *Clidemia hirta* in Hawaiian forests. *Plant Disease* 70:974-976.

Uehara, S. 1968. Memorandum - Meeting on *Clidemia hirta*. Hawaii Dep. Agric. Honolulu. Typescript.

Uehara, S. 1970. Memorandum - Release of the *Heterotrichum* leaf roller, *Blepharomastix ebulealis* (Guenee) from the quarantine laboratory. Hawaii Dep. Agric., Honolulu. Mimeo.

Wester, L.L. and H.B. Wood. 1977. Koster's curse (*Clidemia hirta*), a weed pest in Hawaiian forests. *Environ. Conserv.* 4(1):35-41.

STATUS OF BIOLOGICAL CONTROL OF WEEDS IN HAWAI'I AND IMPLICATIONS FOR MANAGING NATIVE ECOSYSTEMS

George P. Markin,
Po-Yung Lai, and George Y. Funasaki

ABSTRACT

Biological control of weeds in Hawai'i dates back almost 90 years. Seventy-three species of insects and one disease organism have been introduced in an effort to control 21 species of weeds. Forty-three of these introduced species of insects and one species of fungus have become established on 19 weeds. In 11 cases, populations of the phytophagous organisms have been sufficiently high and persistent enough that they succeeded in partially or completely eliminating the host weeds as significant ecosystem components. The majority of biocontrol programs have been aimed at controlling agricultural weeds. The problem of weeds in native ecosystems was serious enough to require initiation of a biological control program on weeds in Hawaiian forests in 1983. Biological control of weeds has several drawbacks: 1) it can be expensive; 2) implementation may take 10 years or more; 3) chances of success are about 50%; 4) a potential ecological hazard exists when introducing a foreign organism into an ecosystem; 5) it may be difficult to justify, economically and ecologically; and 6) conflicts of interest can arise between individuals or groups with different views about whether a plant is a problem or not. When weed control is restricted to agricultural concerns, most of these problems can be overcome. However, if biological control of weeds is extended to natural ecosystems, several of these problems become critical and can prevent progress or greatly reduce their efficacy. Biological control of weeds in native ecosystems should be considered a management tool to be used in conjunction with all other management tools available to the resource manager, rather than an overall solution to weed problems.

INTRODUCTION

Hawai'i has one of the oldest and most extensive biological control programs in the world. Others have reviewed the more successful projects against several species of weeds in Hawai'i (e.g., Davis *et al.*, this volume). These projects, however, represent only a small number of the programs thus far initiated. In this paper, biological control of weeds in

Hawai'i is reviewed, the potential of biological control as a management tool for protecting native ecosystems is examined, and the problems that can be expected when initiating a new program are summarized. Emphasis will be on biological control of weeds with insects, although some work with pathogens will be mentioned.

STATUS OF BIOLOGICAL CONTROL OF WEEDS IN HAWAI'I

By 1982, on a worldwide basis, biological control had been attempted against 82 species of weeds through release of 189 potential biocontrol agents, including insects, plant pathogens, nematodes, and mites (Julien 1982). Of these, 73 insects and one plant pathogen had been introduced for control of 21 weeds in Hawai'i (Table 1). Approximately half of Hawai'i's target species were cosmopolitan weeds. In these cases, the insects were first tested, released, and proven effective against the host weeds in other countries.

In Hawai'i, biological control of weeds in agricultural situations is not a new, experimental, or untested management tool. It is a proven technique with a long history of development and a respectable rate of success, and some unforeseen ecological problems (Howarth 1983). Biological control is now a long-term goal of most management programs for agricultural weeds and natural areas in Hawai'i.

Biological control of weeds in Hawai'i until now has focused on those weeds on agricultural lands, primarily rangelands. However, several weeds such as firetree (*Myrica faya*), blackberry (*Rubus argutus*), clidemia (*Clidemia hirta*), and glorybush (*Tibouchina urvilleana*) have invaded Hawaiian forests, and programs against lantana (*Lantana camara*) and prickly pear cactus (*Opuntia ficus-indica*) have possibly reduced these weeds as major problems in some natural areas.

BIOLOGICAL CONTROL OF WEEDS IN NATIVE HAWAIIAN ECOSYSTEMS

Interest in the biological control of weeds that affect natural areas in Hawai'i has been renewed with increasing recognition of the value of the areas and the importance of impacts of weeds in them. A biological control research program begun in 1983 is jointly funded by the Hawaii Department of Land and Natural Resources, the U.S. Department of the Interior National Park Service, and the U.S. Department of Agriculture Forest Service. A new quarantine facility dedicated to studying biological control of introduced weeds with insects in Hawaiian forests has been constructed in Hawaii Volcanoes National Park.

The quarantine building was completed in the spring of 1984, certified by the Animal and Plant Health Inspection Service and the State of Hawai'i in October, and received its first shipment of foreign insects in December of that year. The facility is an escape-proof insect quarantine building,

Table 1. Introduction of insects and diseases for the biological control of weeds in Hawai'i.*

Target weed	Common name	Origin	Release dates	Total species released	Total species established	Success of program**
Ageratina adenophora (Sprengel) King and Robinson	Maui pāmakani	Central America	1944	1	1	Partially effective; excellent control in drier areas, ineffective in most moist areas
Ageratina riparia (Regal) King and Robinson	Hāmākua pāmakani	Mexico	1955 1973 1974	4***	3***	Very successful; plant has been eliminated as a weed over most of its range
Clidemia hirta (L.) D. Don	Clidemia, Koster's curse	Tropical America	1953 1969	2	2	Effective in open pasture, but not effective in shade areas where greatest problem exists
Cyperus rotundus L.	Purple nutsedge	Eurasia	1922 1925	2	2	No detectable effect
Elephantopus mollis Humboldt, Bonplaud Kunth	Elephant's foot	Central America	1961	1	1	No detectable effect (Davis & Krauss 1962)
E. scaber		Cosmopolitan		1	?****	No effect; not believed established

Target weed	Common name	Origin	Release dates	Total species released	Total species established	Success of program**
Emex spinosa (L.) Campdera and *E. australis* Steinhell	Emex	N. Africa Western Europe South Africa	1957 1962 1962	3	1	Very successful control of both plants by the same insect
Hypericum perforatum L.	Klamath weed	Eurasia N. Africa	1965- 1966	3	3	Very successful; plant almost impossible to find
Lantana camara L.	Lantana	Tropical America	1802- 1974	27	15	Very successful; few surrounding patches in very dry areas
Myrica faya Aiton	Firetree	Island of Atlantic Ocean	1955	1	1	No detectable effect
Opuntia cordobensis Spegazzini		Argentina	1949	1	0	Not established (Weber 1951)
Opuntia ficus-indica (L.) Miller (= *O. megacantha* (L.))	Prickly pear, pānini	Central America	1949 1950 1951	8	4	Very successful; few patches left at higher elevations
Pluchea symphytifolia (Mill.) Gillis	Sourbush	Tropical America	1955 1959	2	2	No detectable effect

Table 1, continued.

Table 1, continued.

Target weed	Common name	Origin	Release dates	Total species released	Total species established	Success of program**
Rubus argutus Link (=*R. penetrans* Bailey or *R. lucidus* Rydberg)	Blackberry	North America (?)	1962-1968	5	3	Not considered effective
Schinus terebinthifolius Raddi	Christmas berry	South America	1932 1954 1961	3	2	No detectable effect
Salsola kali L. (=*S. australis* R. Brown)	Russian thistle	Eurasia	1980	2	0	No effect since insects not established
Tibouchina urvilleana L.	Glorybush	SE Asia 1964	1938	3	2	No detectable effect
Tribulus terrestris L. and *T. cistoides* L.	Puncture vine	Cosmopolitan Tropical	1962 1963	2	2	Very successful; complete control both formed by same two insects
Ulex europaeus L.	Gorse	W. Europe	1927 1958 1961	3	1	Destroys 50% of seeds, but no detectable decrease in spread of plant (Markin 1984)

Target weed	Common name	Origin	Release dates	Total species released	Total species established	Success of program**	
Total 21				74	45	Very successful	8
						Partially successful	3
						No effect	10

*Release data from Lai and Funasaki (1983) unless noted. Some of the data differ slightly from summaries presented by Julien (1982) and Goeden (1978).

**Ranking of success based on interview of Hawaii Department of Agriculture entomologist and weed control personnel (1984-85).

***One agent was a disease.

****Fate unknown, no further reference in literature.

in which insects from foreign countries can be received, screened to remove diseased or parasitized individuals, studied, and host-tested against native and agricultural plants to determine if their feeding is specific enough to qualify as suitable biological control agents. Both the quarantine facility and the research program are operated under the supervision of the Hawaii Department of Agriculture, with additional support from the National Park Service, the Department of Entomology at the University of Hawaii, and the U.S. Department of Agriculture Animal and Plant Health Inspection Service. A Steering Committee, comprised of representatives of each of these agencies, sets priorities for the program.

For the first time, a program for biological control of weeds that threaten native ecosystems has been specifically set up and funded. Presently (1987), the facility is screening and host-testing potential biological control agents of banana poka (*Passiflora mollissima*), considered Hawai'i's major forest weed (Warshauer *et al.* 1983). As time, money, and availability of candidate insect biological control agents allow, expansion of the program should include other forest weeds such as firetree, yellow Himalayan raspberry (*Rubus ellipticus*), and glorybush, as well as participation with other agencies in programs dealing with *Clidemia*, gorse (*Ulex europaeus*), and other species.

PROBLEMS WITH BIOLOGICAL CONTROL OF WEEDS

An examination of past biological control programs in Hawai'i and elsewhere in the world indicates that biocontrol is not the simple solution to weed problems that many resource managers presume it to be. Considerations in setting up a program, particularly for weeds in native ecosystems, include: a predicted success rate of about 50%; high program costs; time required to test, establish, and evaluate agents; ecological hazards; economic justifications; and conflicts of interest. These will be discussed below.

50% Success Rate

In Hawai'i, 73 insects and one plant pathogen have been introduced for biocontrol of 21 plants (Table 1) (Lai and Funasaki 1983). The target weed was completely eliminated as a problem except in a few isolated patches in eight cases; weeds were eliminated in limited areas in three cases; and in the remaining 10 cases, the insects have failed to become established, or if established, have failed to exert any noticeable effect. The success rate for biocontrol in Hawai'i is thus approximately 50%.

DeBach (1974) estimated that complete or substantial biocontrol success was achieved for weeds about 55% of the time, worldwide. The most recent summary of the status of biological control of weeds (Julien 1982) reviewed and evaluated the effectiveness of programs against 82 species of terrestrial plants. Since each of his sources reported the results differently, it is difficult to summarize and compare accurately the different programs. However, the following interpretive conclusions can be drawn.

Worldwide, biological control programs on lantana have been implemented in at least 18 countries or island groups, using most of the same insects that were effective in a successful control program in Hawai'i (see Davis *et al.*, this volume). However, outside of Hawai'i, no country reported satisfactory control, and only five countries indicated partial control -- a success ratio of only 28%. This ratio seems unusually low, given that biocontrol of lantana is one of the oldest and most extensive programs in the world.

In contrast, 10 species of cactus have become established throughout arid regions of 11 countries or island groups. Efforts to control these plants have been very successful, with at least nine, or 83%, of the control programs reporting partial to complete success. The most successful control programs worldwide for any weed (Julien 1982) are against Klamath weed (*Hypericum perforatum*). Biological control has been initiated against this weed in seven countries or island groups, with partial to complete success reported by all.

Of the remaining 70 species of weeds discussed by Julien (1982), 97 control programs have been initiated worldwide. Of these, partial to complete success was reported for 33 (47%). (The success ratio may now be higher, since at least 25% of the programs were listed as ongoing or unevaluated at the time of Julien's report.)

Worldwide, the success rate for weed biocontrol programs resulting in at least partial to substantial or complete control of the target weed is thus about 50%, comparable to the success rate in Hawai'i. However, as noted by Harris (1980, 1981), we may be underestimating the success achieved. Normally, a program is considered successful if ranchers or land managers can report disappearance of the weed from an area. However, insects, once established, may not kill the plant outright but may induce stress sufficient to prevent further spread, reduce regeneration, or reduce competitiveness of the weed in a particular ecosystem. In this sense, biocontrol of weeds in Hawai'i may have been more successful than recognized in the past. Unfortunately, quantitative monitoring of regeneration and reduction or spread of weeds is usually not a strong component of biological control programs.

High Program Costs

A prerequisite to conducting a biological control program with insects is access to an escape-proof facility in which foreign insects can be quarantined while their biology and feeding preferences are studied. Besides the two quarantine facilities in Hawai'i, nine state or Federal facilities are located on the U.S. Mainland (Klingman and Coulson 1982). These are the only facilities certified to receive direct shipments of foreign insects. Any potential agent entering the U.S. must be cleared through one of these quarantine facilities before being sent to other university, Federal, or state laboratories. The cost of constructing and equipping a new quarantine facility is fairly high. The smallest, simplest design that meets both U.S. Department of Agriculture and state certification (such as the one recently constructed at Hawaii Volcanoes National Park) today costs over $150,000. A larger facility has just been completed by the U.S. Department of Agriculture at Albany, California, at a cost of about $1 million. Of course, a biological control program aimed at a specific weed does not require its own quarantine facility if it can use an existing one. But most facilities are already dedicated to specific programs, so the waiting list for new programs may be quite long. In addition, a new program must assume its share of overhead and operating costs.

Quarantine facility costs are only a minor part of a biological control program. The insects attacking the target weed in its native range (probably a foreign country) must be inventoried by scientists in order to select promising agents. If possible, preliminary biological studies of the insects should be conducted in their natural ranges to determine their life histories and association with the weed; feeding tests in the field, to determine the variety of hosts of the promising agent, should also be done. If foreign scientists with suitable facilities can be located, much of this work can be contracted. Otherwise, U.S. scientists must be assigned to the foreign country on a part- or full-time basis. For the banana poka control project, approximately $40,000 a year is budgeted for foreign exploration, including expenses of hiring local help and periodic visits by U.S. scientists. On the *Clidemia* project, a scientist has been stationed in Trinidad for two years, at a cost of about $48,000 per year (Nakahara *et al.*, this volume).

The operation of the quarantine facility where most the work will be conducted, including the salaries of at least one scientist and support personnel, costs approximately $100,000 a year. For the program in Hawai'i, presently aimed at banana poka, foreign exploration, quarantine work, and equipping and setting up the quarantine facility is estimated at about $1 million over a five-year period. Additional effort beyond five years will likely be required for successful control.

Andres (1977) estimated that the total cost of a biocontrol program for a single weed species ranges between $1 and $2 million. Harris (1979) stated that in Canada, a complete biological control program for a single weed would require 18.8 to 23.7 scientist years, would cost between $1.2 and $1.5 million, and would require the establishment of 2.3 biocontrol agents. These costs are relatively small compared with the annual cost of applying herbicides to control many agricultural weeds. However, when biological control is applied to non-agricultural weeds, particularly in natural ecosystems, $1 to $2 million could be considered excessive by some in relation to the intangible values we are trying to protect.

Time Requirements

The time necessary to set up and conduct a successful biological control program for a weed is considerable. Construction and approval of a quarantine facility or obtaining permission to use an existing structure can be expected to take a minimum of two years. Once a facility is in operation, given that a guaranteed supply of foreign insects for testing has been arranged, it can take three to five years to evaluate 10 or more insects; five of these may be selected for release. The establishment of insects in the field usually requires a rearing facility to propagate the insects, which are then released at different locations and times of year, the goal being to achieve the right combination of variables for establishment. This can take an additional two to three years. Thus, 7-10 years may be required for successful establishment of five insects.

Once insects have been released and established in the field, it may take quite some time before their populations have built up and spread sufficiently to cause appreciable impact on the target weed. In Hawai'i, impact on target weeds has been quite rapid; this is attributable to the moderate climate and use of insects that produce many generations per year. In 1973-74, for example, two insects and one plant pathogen were released to control Hāmākua pāmakani (*Ageratina (= Eupatorium) riparia*); by 1980, almost complete control had resulted (Nakao *et al.* 1981; Lai *et al.* 1982). In 1965-66, agents were introduced to control Klamath weed, and by 1972, the target weed was nearly eliminated (Davis 1971, 1972). In 1962 and 1964, agents were introduced to control two species of puncture vine (*Tribulus terrestris* and *T. cistoides*). Within four years, the weeds were completely eliminated in many areas (Davis and Krauss 1966). Even more spectacular were the results obtained in controlling emex (*Emex australis*); the weed was nearly eliminated in most areas within three years (Davis 1961; Davis and Krauss 1962).

The impacts of cactus insects in Hawai'i have been more gradual. Species were introduced from 1949 through 1951 and within five years had destroyed much of the cactus at low elevations. However, nine years later they were reported still spreading and only causing appreciable damage at altitudes above 3,000 ft (900 m) (C.J. Davis, pers. comm., 1986).

The small weevil *Apion ulicis* was successfully introduced to Hawai'i in 1958 to destroy the seeds of gorse. In 1966, the weevil was reported established (Goeden 1978). Yet by 1972, its populations still were very low, with only 1.5% of the gorse pods attacked. By 1984, the population had increased and 84% of the pods were infested (Markin 1984).

Hawai'i's most famous biocontrol program, for lantana, began in 1902. A fair degree of control was obtained by the early 1950s but release of new insects continued through that decade (Krauss 1962). It is generally accepted that lantana was not controlled until the early 1960s, a period of almost 60 years after introduction.

In general, insects introduced for weed control have become established in Hawai'i in 10 years or less, often with significant destruction of the plant in as short a time as three to four years. Even under optimum conditions, however, a program can take a minimum of 10-15 years from inception until significant control is obtained.

Ecological Hazards

In the biological control of weeds, starving adult insects have been known to move to nearby species, usually closely related to the weed, once the host weeds have been decimated. The best-known example is *Teleonemia scrupulosa* (Hemiptera: Tingidae), introduced into Africa in the early 1950s to control lantana (Davies and Greathead 1967; Greathead 1971). The insect readily became established and rapidly built up high populations which, combined with drought, brought about complete defoliation of the lantana. Without a food source, the starving adults moved to adjacent fields of sesame (*Sesamum indicum*), on which they tried to feed. The adults were able to live on the alternate plant but were unable to reproduce, and in a short time the population of *T. scrupulosa* also died.

Additional cases have been reported when insects introduced for control of puncture vine and Klamath weed attacked related plants on which they were able not only to feed but to successfully reproduce (Andres 1980, 1985). Adults of insects introduced to Hawai'i to control Klamath weed have also been found feeding on the related plant *Hypericum degeneri* (Davis 1971). The small leaf rolling moth *Croesia zimmermanii*, introduced to control blackberry, was discovered feeding on 'ākala (*Rubus hawaiiensis*), one of two endemic *Rubus* species in Hawai'i (Gagné 1972). In none of these cases has the introduced insect caused appreciable damage to the new host, but these cases emphasize that there is an element of risk when introducing any new biological control agent.

Economic Justifications

To determine whether the expense and effort of a biological control program is justified, the economic and ecological impacts of any weed should be fully evaluated before any biological program is initiated. Biological weed control programs can be expensive and may not be justifiable if the weed is actually causing few or no problems. In some cases, alien plants have probably been designated as weeds and control programs initiated against them without fully understanding their economic impact to agriculture. The same difficulty can occur for weeds in native ecosystems, where the impact is not economic but ecological; however, determining ecological values may sometimes be difficult.

Another consideration is that nothing is gained by biological control if another disruptive weed invades the habitat from which the target species has been removed. In the United States, Klamath weed was successfully controlled by the introduction of its natural enemies but was replaced by the weed *Elymus caput-medusae* in many areas (Huffaker 1959). A similar situation may have occurred on the island of Hawai'i, where a successful biological control program eliminated lantana over large tracts of rangeland on the western side of the island; some of this rangeland is now unusable since it has been invaded by Brazilian peppertree or Christmas berry (*Schinus terebinthifolius*). If the biocontrol program for banana poka in Hawaiian forests is successful and this weed is eliminated, will native trees and shrubs regenerate, or will weeds such as alien *Rubus* spp., firetree, strawberry guava (*Psidium cattleianum*), or introduced grasses dominate?

One argument against biological control presumes that the weed will disappear if left alone long enough. This is possible since many weeds are primary invaders that take advantage of disturbed ecosystems resulting from grazing, fire, logging, or soil disturbance. If areas are managed properly so that no further disturbances occur, natural succession may take its course, and eventually climax species will replace and eliminate the weed. This solution has been proposed for the control of gorse in New Zealand, where it has been suggested that this weed acts as a nurse crop for native forest trees and shrubs. If left alone, in 15 to 50 years gorse will mature, die, and be replaced by native climax forest (Hackwell 1980). On most oceanic islands, however, this scenario is far less likely to occur, especially where invading plants change successional patterns, soil nutrients, light and moisture regimes, and other processes in favor of alien species and to the disadvantage of natives (Vitousek, this volume).

Conflicts of Interest

Conflict of interest is always a potential threat to any biological control program. Many plants that need to be controlled on rangeland or in native ecosystems are important to other interest groups that claim the same areas for their own uses or objectives (for example, watersheds vs. hunting in Hawai'i). Conflicts of interest also occur where a weed is a serious problem in one area but is beneficial in adjacent areas. In Hawai'i, for example, the biological control program against the prickly pear cactus has been one of the most successful (Davis *et al.*, this volume). The program was first opposed by ranchers who occupied lands poorly supplied with water and feed and considered the cactus an asset during droughts (Fullaway 1954). In New Zealand, gorse is considered a prime weed problem, but some Angora and Cashmere goat farmers on marginal land can profitably use this shrub to establish and support their herds (Krause *et al.* 1984).

Conflict of interest seems almost continual between landowners who attempt to control noxious weeds and beekeepers who view the same weeds as a source of nectar and pollen. In Australia, beekeepers not only opposed efforts to control *Echium plantagineum* (Delfosse 1985), but initiated legal actions to stop the biocontrol program (Cullen and Delfosse 1985). A similar conflict presently exists in New Zealand, where beekeepers view gorse as a critical pollen source for overwintering hives. Their opposition has delayed the release of several very promising biological control agents (R. Hill, pers. comm. 1985).

As a final example, in Hawai'i, attempts to initiate a program to control banana poka may encounter opposition because hunters know that feral pigs (*Sus scrofa*) and the Kalij pheasant (*Lophura leucomelana*) use the fruit as a food source. Environmentalists might argue that because a native honeycreeper (Drepanidinae), the 'i'iwi (*Vestiaria coccinea*), is now using its floral nectary as a food source (Pung 1971), successful elimination of banana poka may impact 'i'iwi populations.

FUTURE OF BIOLOGICAL CONTROL IN HAWAI'I

None of the problems reviewed here are new. All have existed since the inception of biological control but have been unimportant as long as control was limited to a few major agricultural weeds. The benefits of a successful program were considered great enough to outweigh potential harm, and landowners and agricultural interests were politically strong enough to support programs they believed were needed. This is no longer the case worldwide. The release of any new biological control agent must be approved by one or more review committees, which usually contain representatives of most interested groups. Also, if an opposing group felt ignored, it could initiate legal actions against the agency or scientist conducting the biocontrol program. This happened recently in Australia (Delfosse 1985), where the conflict over the control of weeds became sufficient to force the legislature to pass a series of laws to attempt to resolve it (Australian Commonwealth Government 1984; Field and Bruzzege 1984; Cullen and Delfosse 1985).

Biological control is still one of Hawai'i's more viable management tools to combat introduced plants. Questions, however, persist about the ecological hazards, effectiveness, and need for biocontrol programs, and more vigorous opposition can be expected when conflicts of interest occur. All individuals involved in biological control must strive to respond adequately to outside criticism. To this end, researchers must concentrate on more vigorous examination and documentation of results of past programs and the progress of ongoing ones. More rigorous selection and testing procedures are required before new agents are released, to ensure that they are the best available, have the highest chance of containing the host weed, but have minimal or at least acceptable adverse impacts on other plants. Host testing of native species (as well as agricultural species) will probably have to increase. Monitoring the success and failure of releases must receive more attention and support. As conflicts of interest will continue and probably increase, public education and involvement of user groups during planning should become an important part of any new program.

RECOMMENDATIONS FOR BIOLOGICAL CONTROL OF WEEDS AS A MANAGEMENT TOOL FOR NATIVE ECOSYSTEMS

In native Hawaiian forests, biological control can play an integral role and provide acceptable control of alien plants with minimal impact. For natural areas in Hawai'i as well as elsewhere in the world, biological control of weeds definitely should be considered as a management tool in planning any control or containment program. However, before embarking on such a program, all considerations of need, cost, impacts, etc. should be addressed. A successful program can completely and permanently eliminate the target weed as a significant ecological component, but many variables are likely to prevent this. Before undertaking a biological control program in a natural area, the following questions should be considered:

1. Are **funds** sufficient to set up the program and operate it to completion? Biological control is not cheap. A conservative estimate of a complete biological control program for one weed is at least $1 million.

2. Is sufficient **time** available to set up and oversee a biological control program to its completion? An estimate based on experience in Hawai'i is that a new program takes 10 to 15 years to succeed.

3. Do researchers, managers, and administrators acknowledge that the **chance of success** for a biological control program is about 50%? Do they also acknowledge that even in the most successful programs, pockets of the target weed will persist where the ecological range of a weed is often greater than that of its natural enemies?

4. Is the **ecological risk** of introducing a foreign insect into a natural area acceptable? The chance that a properly studied introduced insect would move to a nontarget native or agricultural plant (after screening tests have been conducted) is very low, but nevertheless exists.

5. Are those advocating biocontrol willing to accept challenges, **conflicts of interest**, and opposition to the new program? As biological control expands to natural areas, public lands become increasingly involved. Other interest groups have valid uses for those lands or adjacent areas where the target weed may be considered an asset. Conflicts of interest can be expected.

If, after reviewing the above questions, the answers are all "yes", then biological control may provide one solution to a weed problem. However, it is not the only possible solution. In view of the chance that a biological control program may not succeed, the 50% success rate of past programs, and the possibility of a program being halted before it can be completed, biological control should be considered as <u>one</u> alternative in a weed management program. If a biocontrol program is successful, the weed problem may be permanently solved; but more likely, other methods of alien plant management (chemical, manual, mechanical, cultural, and ecological control) will be needed to supplement biocontrol, at least in some areas.

ACKNOWLEDGMENTS

We thank the staff of the Hawaii Department of Agriculture (Plant Pest Control Branch), in particular Larry Nakahara, Ernest Yoshioka, and Clifton C.J. Davis (Branch Chief, retired), for critically reviewing this manuscript.

Literature Cited

Andres, L.A. 1977. The economics of biological control of weeds. *Aquatic Botany* 3:111-123.

Andres, L.A. 1980. Conflicting interests and the biological control of weeds. *Proc. V, Internatl. Symp. Biol. Contr. Weeds,* 11-20. Brisbane, Australia.

Andres, L.A. 1985. Interaction of St. Johnswort insects and North American *Hypericum* species. *Proc. VI, Internatl. Symp. Biol. Contr. Weeds,* 235-239. Vancouver, Canada.

Australian Commonwealth Government. 1984. Biological control act 1984, No. 139 or 1984 an act to make provision for the biological control of pests in Australian Capitol territory and for related purposes. Australian Commonwealth Govt. Printer.

Cullen, J.M. and E.S. Delfosse. 1985. *Echium plantagineum*: catalyst for conflict and change in Australia. *Proc. VI, Internatl. Symp. Biol. Contr. Weeds,* 249-292. Vancouver, Canada.

Davies, J.C. and D.J. Greathead. 1967. Occurrence of *Teleonemia scrupulosa* on *Sesamum indicum* Linn. in Uganda. *Nature* 213:102-103.

Davis, C.J. 1961. Recent introductions for biological control in Hawaii -- VI. *Proc. Hawn. Entomol. Soc.* 17:389-93.

Davis, C.J. 1971. Recent introductions for biological control in Hawaii -- XVI. *Proc. Hawn. Entomol. Soc.* 21(1):59-62.

Davis, C.J. 1972. Recent introductions for biological control in Hawaii -- XVIII. *Proc. Hawn. Entomol. Soc.* 21:187-90.

Davis, C.J. and N.L.H. Krauss. 1962. Recent developments in the biological control of weed pests in Hawaii. *Proc. Hawn. Entomol. Soc.* 18(1):65-67.

Davis, C.J. and N.L.H. Krauss. 1966. Recent introductions for biological control in Hawaii -- XI. *Proc. Hawn. Entomol. Soc.* 19(2):201-204.

Davis, C.J., E. Yoshioka, and D. Kageler. [this volume] Biological control of lantana, prickly pear, and Hāmākua pāmakani in Hawai'i: a review and update.

DeBach, P. 1974. *Biological control by natural enemies*. London: Cambridge Univ. Pr.

Delfosse, E.S. 1985. *Echium plantagineum*: a program halted by legal injunction in Australia. *Proc. VI, Internatl. Symp. Biol. Contr. Weeds,* 293-299. Vancouver, Canada.

Field, R.P., and E. Bruzzese. 1984. Biological control of blackberry -- resolving a conflict of Australia. *Proc. VI, Internatl. Symp. Biol. Contr. Weeds,* 341-349. Vancouver, Canada.

Fullaway, D.T. 1954. Biological control of cactus in Hawaii. *J. Econ. Entomol.* 47(4):696-700.

Gagné, W.C. 1972. Notes and exhibitions: *Apotoforma* sp. *Proc. Hawn. Entomol. Soc.* 21:158-59.

Goeden, R.D. 1978. Part II. Biological control of weeds. In *Introduced parasites and predators of arthropod pests and weeds: a world review*, ed. C.P. Clausen, 357-545. U.S. Dep. Agric. Handb. No. 480. Washington, D.C.: U.S. Govt. Print. Office.

Greathead, D.J. 1971. Progress in the biological control of *Lantana camara* in East Africa and discussion of problems raised by the unexpected reaction of some of the more promising insects to *Sesame indicum*. *Proc. II, Internatl. Symp. Biol. Contr. Weeds*, 89-92. Rome, Italy.

Hackwell, K. 1980. Gorse: a helpful nurse plant for regenerating native forests. *Forest and Birds* 13:25-28.

Harris, P. 1979. Cost of biological control of weeds by insects in Canada. *Weed Sci.* 27:242-50.

Harris, P. 1980. Evaluating biocontrol of weeds projects. *Proc. V, Internatl. Symp. Biol. Contr. Weeds,* 345-353. Brisbane, Australia.

Harris, P. 1981. Stress as a strategy in the biological control of weeds. In *Biological control in crop production*, ed. G.C. Popowizas, 330-340. Symp. Agric. Res., May 18-21, 1980, Beltsville, Maryland. London: Allanheld, Osmun.

Howarth, F.G. 1983. Classical biocontrol: panacea or Pandora's box. *Proc. Hawn. Entomol. Soc.* 24:239-244.

Huffaker, C.B. 1959. Biological control of weeds with insects. *Ann. Review Entomol.* 4:251-276.

Julien, M.H. 1982. *Biological control of weeds, a world catalogue of agents and their target weeds*. Commonwealth Inst. Biol. Contr. Farnham Royal, Slough, U.K.: Commonwealth Agric. Bur.

Klingman, D.L. and J.K. Coulson. 1982. Guidelines for introducing foreign organisms into the United States for biological control of weeds. *Weed Sci.* 30:661-667.

Krause, M.A., A.C. Beck, and J.B. Dent. 1984. *The economics of controlling gorse in hill country: goats versus chemicals*. Res. Rep. 149. New Zealand Agricultural Economics Research Unit.

Krauss, N.L.H. 1962. Biological control investigation on Lantana. *Proc. Hawn. Entomol. Soc.* 18:134-136.

Lai, P.-Y., and G.Y. Funasaki, 1983. List of beneficial organisms purposely introduced for biological control in Hawaii, 1890 to 1982. Plant Pest Control Branch, Hawaii Dep. Agric. Unpub. rep.

Lai, P.-Y., G.Y. Funasaki, and S.J. Higa. 1982. Introductions for biological control in Hawaii: 1979 and 1980. *Proc. Hawn. Entomol. Soc.* 24(1):109-113.

Markin, G.P. 1984. Biological control of the noxious weed gorse *Ulex europaeus* L.: a status report. [abstract] *Proc. Fifth Conf. Nat. Sci., Hawaii Volcanoes National Park*, 77. Honolulu: Univ. Hawaii Coop. Natl. Park Resour. Stud. Unit.

Nakahara, L.M., R.M. Burkhart, and G.Y. Funasaki. [this volume] Review and status of biological control of clidemia in Hawai'i.

Nakao, H.K., G.Y. Funasaki, S.Y. Higa, and P.Y. Lai. 1981. Introduction for biological control in Hawaii 1977 and 1978. *Proc. Hawn. Entomol. Soc.* 23:425-30.

Pung, E. 1971. The diet of the i'iwi. *Hawaii Aloha Aina* 2:3-5.

Rosenthal, S.S., D.M. Maddox, and K. Brunetti. 1985. *Principles of weed control in California*, 65-94. Fresno, Calif.: Thomson Pub.

Vitousek, P.M. [this volume] Effects of alien plants on native ecosystems.

Warshauer, F.R., J.D. Jacobi, A.M. LaRosa, J.M. Scott, and C.W. Smith. 1983. *The distribution, impact, and potential management of the introduced vine* Passiflora mollissima *(Passifloraceae) in Hawai'i.* Tech. Rep. 48, Univ. Hawaii Coop. Natl. Park Resour. Stud. Unit. Honolulu.

Weber, P.W. 1951. Recent liberations of beneficial insects in Hawaii. *Proc. Entomol. Soc.* 14:327-330.

AGENCY, ORGANIZATION, AND LANDOWNER APPROACHES

ALIEN PLANT CONTROL STRATEGIES IN HAWAII VOLCANOES NATIONAL PARK*

J. Timothy Tunison

ABSTRACT

Alien plants have greatly altered some native plant communities and threaten others in Hawaii Volcanoes National Park. The Park contains approximately 405 alien and 375 native vascular plant species. Desultory efforts were made to control alien plants until the 1980s, at which time a systematic program was initiated. The following strategies or actions have been taken to protect native species assemblages: 1) controlling feral pigs and goats; 2) excluding fire; 3) controlling localized alien plants throughout the Park; 4) controlling all disruptive alien plants in Special Ecological Areas (the most diverse and intact areas in the Park); 5) confining one widespread species, fountain grass (*Pennisetum setaceum*), to the area it currently infests; 6) developing herbicidal control methods for target species; 7) developing biological controls for some widespread species; 8) mapping the distribution of important alien plants; 9) researching the ecology, seed biology, and phenology of important alien plant pest species; 10) educating the public to the importance of alien plant control; and 11) working with other agencies and groups in alien plant management. Control of new introductions and incipient infestations and protection of some of the most diverse and intact areas in the Park have been very successful strategies. Biological control programs, now in the research phase, provide potential long-term solutions to the control of some widespread plant species.

INTRODUCTION

Alien plant problems have been recognized by park managers at Hawaii Volcanoes National Park since the 1920s. Desultory efforts were made to control alien plants until the 1980s, at which time alien plant management

---

*Adapted from Tunison 1991.

intensified in response to the rapid spread of several obviously disruptive alien plant species in the Park. During this period, the National Park Service also began to deal more forcefully with natural resource problems nationwide. By 1982, the most important alien plant species in Hawaii Volcanoes had been listed in order of priority, and an interagency cooperative agreement with State and other Federal agencies was drafted for initiating biological control research on some of the most disruptive alien plants in forests. Twenty-six species in the Park were targeted for mechanical and chemical control programs, and control efforts were applied to 14 of these. However, no species was considered controlled, and management was not applied to the entire geographical range of any target species in the Park.

The alien plant control program expanded and became more systematic in the mid-1980s. By 1988, work was in progress on 48 target species using herbicide or mechanical control measures, and biological control research on three target species was under way. Alien plant control accounted for approximately one-third of the Park's Division of Resources Management budget, which supported the equivalent of four full-time personnel per year for alien plant control.

In this paper I describe the current (1988) alien plant program at Hawaii Volcanoes National Park, emphasizing control strategies. The effectiveness of control programs, briefly summarized here, is discussed in detail in other papers in this volume and elsewhere (Cuddihy *et al.* 1988; Tunison, this volume; Tunison and Zimmer, this volume; Tunison 1991).

ALIEN PLANT PROBLEMS

Alien plants have invaded many modified plant communities in Hawaii Volcanoes National Park and are serious threats to the remaining intact communities. Smith (1985, 1991) has characterized alien plant problems in Hawai'i, delineated impacts, and identified disruptive species. In Hawaii Volcanoes, which has a native vascular plant flora of about 375 native species (Higashino *et al.* 1988), approximately 405 alien plant species are now naturalized. The Park's flora includes 34 of the 86 most disruptive introduced plant species in Hawai'i (Smith 1985) (Table 1). Park plant communities vary from essentially alien-dominated to native-dominated, with the degree of intactness increasing with elevation; in general, the coastal lowlands are dominated by alien plants, whereas the upper montane seasonal, subalpine, and alpine zones (above 4,620 ft or 1,400 m elevation) are dominated by natives. The submontane seasonal and montane rain forest zones between the two elevational extremes are a mosaic of native and partly native plant communities.

Disruptive alien plant species are capable of replacing native vegetation, often forming monospecific stands or significantly altering fire and nutrient cycling regimes (Smith 1985). Some of the most disruptive species such as firetree (*Myrica faya*), banana poka (*Passiflora mollissima*), bush beardgrass (*Schizachyrium condensatum*), and broomsedge (*Andropogon virginicus*) have expanded

Table 1. Important alien plant species in Hawaii Volcanoes National Park and their management.

Species	Common Name	Family	Distribution in Park	Management Status
*Acacia confusa**	Formosan koa	Leguminosae	Localized	Controlled
*Acacia mearnsii**	Black wattle	Leguminosae	Localized	Controlled; reduced to seedling stages
Acacia melanoxylon[+]	Blackwood acacia	Leguminosae	Localized	Possibly eradicated
Agave americana	Agave	Amaryllidaceae	Localized	Possibly eradicated
Agave sisalana	Sisal	Amaryllidaceae	Localized	Mostly controlled
Albizia sp.	Albizia	Leguminosae	Localized	Possibly eradicated
*Andropogon virginicus**	Broomsedge	Gramineae	Widespread	Not controlled; targeted for biological control research; comments for *Schizachyrium condensatum* apply
*Anthoxanthum odoratum**	Sweet vernalgrass	Gramineae	Widespread	Not managed
Ardisia crenata	Hilo holly	Myrsinaceae	Localized	Controlled
Arthrostema ciliatum	Arthrostema	Melastomataceae	Localized	Partially controlled
Benicasa hispida	Chinese melon	Cucurbitaceae	Localized	Probably eradicated

Table 1, continued.

Table 1, continued.

Species	Common Name	Family	Distribution in Park	Management Status
*Casuarina equisetifolia**	Ironwood	Casuarinaceae	Localized	Controlled with some resprouting
*Ehrharta stipoides**	Meadow ricegrass	Gramineae	Widespread	Not managed, but research in progress for chemical control in localized sites in SEAs
Eleagnus umbellata	Oleaster	Eleagnaceae	Localized	Controlled
Eriobotrya japonica	Loquat	Rosaceae	Localized	Controlled; reduced to seedling stages
Eucalyptus globulus[+]	Eucalyptus	Myrtaceae	Localized	Controlled except where planted in a campground
Feijoa sellowiana	Guavasteen	Myrtaceae	Localized	Probably eradicated
Fraxinus americana	White ash	Oleaceae	Localized	Controlled; reduced to seedling stages
*Fraxinus uhdei**	Tropical ash	Oleaceae	Localized	Probably eradicated
*Grevillea robusta**	Silk oak	Proteaceae	Widespread	Controlled in SEAs;[#] funding requested for parkwide control

Species	Common Name	Family	Distribution in Park	Management Status
Hedera helix	English ivy	Araliaceae	Localized	Controlled
*Hedychium gardnerianum**	Kāhili ginger	Zingiberaceae	Widespread	Controlled in SEAs; funding requested for parkwide control
*Holcus lanatus**	Velvet grass	Gramineae	Widespread	Not managed
Hyparrhenia rufa	Thatching grass	Gramineae	Widespread	Not managed
*Lantana camara**	Lantana	Verbenaceae	Widespread	Partially controlled by biological control agents introduced to state
*Leucaena leucocephala**	Koa haole	Leguminosae	Widespread	Not managed; funding requested for parkwide control
Lupinus hybridus	Lupine	Leguminosae	Localized	Controlled; reduced to seedling stages
*Melaleuca quinquenervia**	Paperbark	Myrtaceae	Localized	Probably eradicated
*Melochia umbellata**	Gunpowder tree	Sterculiaceae	Localized	Controlled; reduced to seedling stages

Table 1, continued.

Table 1, continued.

Species	Common Name	Family	Distribution in Park	Management Status
*Melinis minutiflora**	Molasses grass	Gramineae	Widespread	Targeted for biological control research, but permission to import biocontrol agents problematical
*Myrica faya**	Firetree	Myricaceae	Widespread	Controlled in SEAs; biological control research under way
Olea europaea subsp *africana**	Olive	Oleaceae	Widespread	Controlled in SEAs; funding requested for parkwide control
Opuntia ficus-indica	Prickly pear cactus	Cactaceae	Localized	Biocontrol agents introduced to Park from west side of Island without noticeable effects
Paederia scandens	Maile pilau	Rubiaceae	Localized	Possibly eradicated
*Paspalum conjugatum**	Hilo grass	Gramineae	Widespread	Not managed; research in progress for chemical control in localized sites in some SEAs

Species	Common Name	Family	Distribution in Park	Management Status
Paspalum dilatatum	Dallis grass	Gramineae	Widespread	Not managed; comments for *P. conjugatum* apply
Paspalum urvillei	Vasey grass	Gramineae	Widespread	Not managed; comments for *P. conjugatum* apply
*Passiflora mollissima**	Banana poka	Passifloraceae	Widespread	Controlled chemically in SEAs; biological control research in progress, and first release in 1988
*Pennisetum clandestinum**	Kikuyu grass	Gramineae	Widespread	Controlled in SEAs and other localized sites
*Pennisetum setaceum**	Fountain grass	Gramineae	Widespread	Controlled in SEAs, in outlying populations, and on periphery of main infestation
Phormium tenax	New Zealand flax	Liliaceae	Localized	Controlled with some resprouting
Pinus caribaea	Slash pine	Pinaceae	Localized	Controlled to seedling stages

Table 1, continued.

Table 1, continued.

Species	Common Name	Family	Distribution in Park	Management Status
Pittosporum undulatum	Pittosporum	Pittosporaceae	Localized	Possibly eradicated
*Prosopis pallida**	Kiawe	Leguminosae	Localized	Controlled with some resprouting
*Psidium cattleianum**	Strawberry guava	Myrtaceae	Widespread	Controlled in SEAs; targeted for biocontrol research
*Psidium guajava**	Common guava	Myrtaceae	Widespread	Controlled in some SEAs
Pueraria lobata	Kudzu	Leguminosae	Localized	Controlled with some resprouting
Ricinus communis	Castor bean	Euphorbiaceae	Localized	Controlled to seedling stages
Rubus argutus	Blackberry	Rosaceae	Widespread	Biocontrol research in progress
Rubus ellipticus	Himalayan raspberry	Rosaceae	Widespread	Targeted for biocontrol research
Rubus glaucus	Raspberry	Rosaceae	Widespread	Funding requested for parkwide control

Species	Common Name	Family	Distribution in Park	Management Status
*Schinus terebinthifolius**	Christmas berry	Anacardiaceae	Widespread	Not managed, but funding requested for control in part of Park
*Schizachyrium condensatum**	Bush beardgrass	Gramineae	Widespread	Not controlled; targeted for biocontrol but permission to import biocontrol agents is problematical
*Setaria palmifolia**	Palmgrass	Gramineae	Widespread	Control only in SEAs
Solanum pseudocapsicum	Jerusalem cherry	Solanaceae	Widespread	Control only in SEAs
Soliva sessilis		Compositae	Localized	Controlled
*Syzygium cumini**	Java plum	Myrtaceae	Widespread	Not managed
*Syzygium jambos**	Rose apple	Myrtaceae	Widespread	One disjunct population controlled
*Tibouchina urvilleana**	Glorybush	Melastomataceae	Localized	Partially controlled
Trema orientalis	Trema	Ulmaceae	Localized	Controlled to seedling stages

Table 1, continued.

Table 1, continued.

Species	Common Name	Family	Distribution in Park	Management Status
Tropaeolum majus	Nasturtium	Tropaeolaceae	Localized	Controlled
*Verbascum thapsus**	Common mullein	Scrophulariaceae	Localized	Not managed

*Considered by Smith (1985) to be serious plant pests in natural communities in Hawai'i.

[+]Considered by Smith (1985) to be potential plant pests in natural communities in Hawai'i.

[#]Special Ecological Areas (see text).

their ranges significantly in the Park within the last 25 years. These and many other alien weeds continue to expand and to increase in density (Warshauer *et al.* 1982; Whiteaker and Gardner 1985; Tunison *et al.* 1992). Twenty-nine disruptive alien plant species in the Park are now too widespread to control with herbicides or mechanical measures, at least at current funding levels (Table 1).

The solutions to alien plant problems in the Park are complicated by the following factors: 1) Most species are too widespread to be controlled islandwide by conventional means; reinvasion of the Park is inevitable in spite of successful parkwide control programs. 2) Quarantine programs for the State are inadequate; the Islands are visited by millions of tourists yearly (in addition to military and others), and newly introduced alien plant species are anticipated. 3) Various user groups (such as horticulturists, landscapers, local residents) do not understand the magnitude of the alien plant threat. 4) Alien plant seed banks in the Park are substantial. 5) Accurate knowledge of alien species distributions and control of alien plants are beyond current funding levels. 6) Control techniques are not known for many species and/or are not permitted by the National Park Service or other authorities.

ALIEN PLANT CONTROL STRATEGIES

The immediate objective of the vegetation management program at Hawaii Volcanoes is to protect native species assemblages from alien plants (National Park Service 1988). The long-range objective is to protect and promote natural processes, the evolutionary potential of native species, and the diversity of genotypes, species, and communities. Since alien plant species are well established in Hawai'i and the Park, managers must realistically accept some alien species, but only at population levels compatible with the survival of native species assemblages. Alien plant control is directed only at alien species that disrupt native ecosystems; the great majority of alien plant species probably do not significantly alter native plant populations or natural processes.

Alien plant management at Hawaii Volcanoes consists of the following approaches or actions: 1) controlling feral pigs and goats; 2) excluding fire; 3) mapping the distribution of important alien plant species; 4) controlling localized alien plant species throughout the Park; 5) developing biological controls for some widespread species; 6) controlling all disruptive alien plants in Special Ecological Areas; 7) confining one widespread species, fountain grass (*Pennisetum setaceum*), to the area it currently infests; 8) developing herbicidal control methods for target species; 9) researching the ecology, seed biology, and phenology of important alien plant pest species; 10) educating the public to the importance of alien plant control; 11) working with other agencies and groups in alien plant management.

Controlling Feral Pigs and Feral Goats

The success of alien species has been attributed, in part, to the fact that island ecosystems have developed in isolation from many of the selective pressures found in continental systems (Mueller-Dombois *et al.* 1981; Loope and Mueller-Dombois 1988). Feral ungulates are probably the most important introduced selective force or disturbance factor (other than humans) in Hawaiian ecosystems. Hawaiian vegetation evolved in the absence of ungulates and lacks defenses against herbivory of introduced mammals. Alien ungulates browse and graze native vegetation, disperse alien plant seeds, and disturb the soil, favoring the establishment of alien plants.

A management strategy of removing ungulates to prevent or mitigate alien plant problems is predicated on observations of natural exclosures and data from experimental exclosures (Loope and Scowcroft 1985; Scowcroft and Conrad, this volume). Natural exclosures such as steep slopes or pit craters tend to be areas of pristine vegetation with few alien plants, suggesting that damage by ungulates is a prerequisite for many large-scale invasions of alien plants (Mueller-Dombois *et al.* 1981). Patterns of vegetation recovery in experimental exclosures from which ungulates are removed are less definitive. Native vegetation tends to hold its own or even recover dramatically in areas that are predominantly native. In areas that are mostly dominated by aliens, alien plants tend to persist or even increase, inhibiting the invasion of native vegetation (Loope and Scowcroft 1985).

Hawaii Volcanoes has achieved considerable success in eliminating feral goats (*Capra hircus*), and feral pig (*Sus scrofa*) control in management units is progressing slowly but effectively. Feral goats were essentially eradicated in the Park below 6,000 ft (2,000 m) elevation by the early 1980s, and removal of the few remaining goats above this elevation is planned. Feral pigs have been reduced to low levels in approximately 30% of the pig habitat in the Park (nine management units) totalling approximately 17,300 a (7,000 ha). Pigs will eventually be removed from additional areas in the Park, but not all pig habitat in the Park will be cleared, partly because some areas have degenerated to the point where plant restoration after pig removal would be too expensive.

The effect of ungulate control in reducing alien plants over large areas is not as well documented as for exclusion from small experimental exclosures. In most cases, feral goats and pigs have only recently been removed and trends are not yet apparent. Documenting the recovery of vegetation following ungulate removal is one of the most important research and monitoring activities in the Park.

Excluding Fire

The spread and intensification of alien grasses in the coastal lowlands and submontane seasonal zone have dramatically altered the natural fire regime in much of Hawaii Volcanoes in the last 20 years (Smith and Tunison, this volume). To prevent the loss of native vegetation and stimulation of alien plants, the Park has adopted a total fire suppression policy and does not allow natural fire and prescribed burning (National Park Service 1985). An active fire prevention program, emphasizing manual hazardous-fuel reduction along roadsides and construction of fuel breaks, was started in 1988. Distribution maps of fuels are being prepared, and preliminary customized fuel models are being developed. An intensive effort to assess fire effects in both recent and old burns has been undertaken to determine the ecological impacts and the necessity for fire exclusion. The effectiveness of these research and management programs in reducing the impacts of fire in the Park cannot yet be predicted.

Distribution Mapping of Important Alien Plant Species

A systematic program to map the distribution of 38 alien plant species was conducted from 1983 to 1985 (Tunison *et al.* 1992). Prior to this, the paucity and quality of distribution maps hampered the alien plant control program. Systematic mapping was undertaken to achieve the following objectives: 1) locate every population of a target species to assure that all plants are treated once a control program commences; 2) determine the feasibility of controlling alien plant species targeted for management in the Resources Management Plan; 3) develop a strategy for controlling target species; 4) establish a baseline for assessing the spread of individual alien plant populations; 5) assess the effectiveness of control programs.

Distribution maps are very effective in helping understand the scope of the problem, develop control strategies, and establish workload requirements (Tunison *et al.* 1992). Mapping efforts allowed classification of the status of 12 species, changed the priority of currently targeted species, and indicated the locations of untreated populations. Finally, distribution mapping indicated that eight species were too widespread for control with the resources available. Control efforts for these species were shifted from a parkwide emphasis to control in selected areas with high biological values and a greater chance of success (Tunison and Stone, this volume).

Monitoring of alien plants continues through systematic distribution mapping. Research and Resources Management personnel are collecting data on approximately 75 alien plant species incidental to other field work. Research personnel collect distribution data on all alien plant species in rain forests while conducting surveys to help determine new areas to be managed and monitoring vegetation recovery after ungulate removal. Alien plants are also quantitatively monitored in a number of Special Ecological Areas by Research and Resources Management personnel.

Controlling Localized Alien Plant Species

Forty-one potentially disruptive localized alien plant species are controlled throughout Hawaii Volcanoes (Table 1). These species are represented by populations with a few to several thousand individuals that are localized in distribution. They are either new introductions to the Park or more-established populations now viewed as incipient infestations. Many alien species were planted for range or reforestation in a former inholding used as a sheep and cattle ranch, and many ornamental species were planted near Kīlauea Caldera, the main visitor attraction in the Park.

Localized alien plant species are controlled to remove them from Park ecosystems and forestall future costly control programs. The spread of kāhili ginger (*Hedychium gardnerianum*), an aggressive perennial herb now well established in Park rain forests, might have been curtailed in much of the Park if it had been removed when the first few plants were noticed in the Park residential area 40 years ago (Fagerlund 1947).

Target species are selected for localized control primarily on the basis of invasiveness in the Park or other parts of the State. Other considerations include viney habit, taxonomic affinities to invasive species, and potential for dispersal. Many aggressive aliens become established initially in ruderal habitats, from which they spread to natural communities. Therefore, roads, trails, and developed areas are surveyed at least annually.

Control efforts have been highly effective on 26 of the 41 target species (Tunison and Zimmer, this volume). Seed production and plant populations have been markedly reduced. Nine of the 26 species have apparently been eradicated in the Park. Seven others have been reduced to seedling stages, with eradication awaiting depletion of the seed bank. Control efforts have been partially successful on the balance of the 41 target species, with reductions in populations and seed production (Table 1).

Developing Biological Controls for Some Widespread Species

Biological control (biocontrol) is theoretically the best approach for alien plants that are widespread in Hawaii Volcanoes and/or present outside and have the potential to continually reinfest the Park (National Park Service 1988). For example, biological control is perceived as a long-range solution for banana poka and firetree infestations in Hawaii Volcanoes. These species are already too widespread in the Park for parkwide control by herbicides or other means, and reinvasion is inevitable because they are widely distributed on Hawai'i Island. Biological control programs have been effective in Hawai'i in controlling several alien plant species (Davis *et al.*, this volume).

A Memorandum of Understanding has been established among the National Park Service, U.S. Forest Service, Hawaii Departments of Land and Natural Resources and Agriculture, and the University of Hawaii to intensify biological control efforts on forest pests in the State. A quarantine facility for entomological studies has been constructed in Hawaii Volcanoes and is staffed by the U.S. Forest Service and the State of Hawai'i. Studies on biocontrol with pathogens have been started by a National Park Service Research Scientist stationed at the University of Hawaii; however, facilities are inadequate for continued studies. Eight target species have been identified for biological control research, work is in progress on four of these species (Table 1), and one release of a biological control agent in the Park has been made.

Controlling All Disruptive Alien Plants in Special Ecological Areas

Since 1985, all disruptive alien plant species, including those too widespread for parkwide control, are now controlled in intensive management and research units called Special Ecological Areas (Tunison *et al.*

1986; Tunison and Stone, this volume). The Special Ecological Areas approach to alien plant control developed from these perceptions:

1. The impossibility of managing widespread alien plant species on a parkwide basis. Some alien species are too dense and/or widespread for control by herbicidal or mechanical means. The establishment of Special Ecological Areas is partly a concession that the alien plant problem in the Park is too advanced for conventional means of control except in localized areas, with current or expected funding.

2. The need to protect the most intact, most representative, and most biologically rich sites of the Park from disruptive alien species, and to begin control efforts in these areas while alien species are still localized and manageable by conventional methods.

3. The need to integrate alien plant control with feral animal control, with research on small mammals, arthropods, and ecosystem processes, with monitoring of key species, and with interpretation of biological resources to Park visitors.

4. The awareness that the most effective management of alien biota has occurred incrementally at Hawaii Volcanoes. As with feral ungulates, once alien plant populations reach maintenance levels, areas can then be expanded or new Special Ecological Areas can be managed. Special Ecological Areas can serve as models of successful management.

Control of widespread alien plants in Special Ecological Areas was started in 1985. All disruptive alien plant species controllable by chemical or mechanical means were targeted for control. Even species targeted for biological control were managed in Special Ecological Areas because the effectiveness and timeliness of biological control agents is uncertain at this time (Markin *et al.*, this volume). Considerable, and possibly irreversible, ecological damage may occur prior to effective biological control without deployment of other management strategies.

Eight Special Ecological Areas totalling 12,954 a (5,266 ha) (of approximately 172,900 a or 70,000 ha in Hawaii Volcanoes capable of supporting vegetation) are now managed. After four years of Special Ecological Areas management, populations of the eleven target plant species (Table 1) have been reduced to very low levels, seed production has been greatly reduced, and recruitment is low (Tunison and Stone, this volume). Workload requirements have dropped 73%.

Confining One Widespread Species (Fountain Grass) to the Existing Infestation

Widespread species are targeted for biological control and/or for conventional control in Special Ecological Areas. The management of one widespread species, fountain grass differs in that it also includes an attempt to confine this species to the existing infestation in the Park (Tunison, this volume). Most of the 19,760-a (8,000-ha) infestation in Hawaii Volcanoes is localized in the southwestern corner of the Park. Populations also occur along roadsides and in a few scattered outlying populations. The management strategy for fountain grass is to control all outlying populations, scout the areas between these populations from a helicopter, and control the periphery of the main infestation. Fountain grass is also controlled in Special Ecological Areas. The fountain grass control program requires one-half the worker resources of the Special Ecological Areas management program.

The management strategy for fountain grass is an exception to the rule of controlling widespread species in Special Ecological Areas or through biocontrol for these reasons:

1. Fountain grass is capable of invading all Park plant communities, outside of closed rain forest, from sea level to 8,250 ft (2,500 m) elevation. It increases fuel loadings, thus increasing fire potential. It invades and reaches high densities on barren lava flows, which normally are pristine sites with few aliens, thus upsetting primary succession at an early stage.

2. Securing permits to import biocontrol agents is unlikely for any grass in Hawai'i (Markin *et al.*, this volume).

3. Fountain grass is mostly confined to one area of the Park and occurs at low densities. Almost all outlying populations, and those populations on the periphery of the main infestation, consist of one or two plants; average densities are two plants per 100 acres (5/100 ha).

4. Fountain grass appears to be dispersed short distances except along roads; thus, the zone of active population expansion is on the periphery of the main infestation.

5. Control of fountain grass was found to initially require approximately four to five full-time personnel equivalents per year, far beyond the means of Park management without major funding increases or reversals of long-standing priorities.

Results to date, using a strategy of confining fountain grass to existing locations, are somewhat promising. Workloads and populations are declining noticeably in portions of the control area on the periphery of the infestation managed for more than two years. There is only modest recruitment in these small populations. No long-range dispersal has been found. Control efforts will extend toward the central portion of the infestation or over its entire range in the Park as progress in control is made or additional funding becomes available.

Developing Herbicidal Control Methods for Target Species

Herbicides are needed to control most target species, because mechanical control methods are not feasible in native or semi-native communities in a park setting; because most target species resprout when cut; and because effective biocontrol agents are not available. Chemical means of control are not known for many target species in the Park or, in some cases, are known but not permitted in National Parks. Frequently, treatment methods must be developed by National Park Service researchers, who attempt to find ecologically safe, sound, and efficient chemicals and application methods (Santos *et al.*, this volume).

Treatment methods have been developed by Park Service researchers for nine target species, with work in progress on six others. Effects on non-target plant species have also been investigated. In all but two cases, cut-stump techniques have been found to localize and minimize pesticide application; less than 9.25 gal (35 l) of herbicide were used in alien plant control and research combined in 1987.

Research on the Ecology, Seed Biology, and Phenology of Important Alien Plant Species

Understanding the biology of disruptive alien plant species is important to: 1) identify species requiring control; 2) predict the ranges and densities of invasive species; 3) identify vulnerable links in the life cycles of managed species to formulate control strategies; and 4) monitor the impacts of alien plants and control programs on park ecosystems. These objectives can be achieved by understanding features such as dispersal, factors limiting establishment, alien plant population dynamics, and interactions of alien species and native species.

One alien species, firetree, has been studied in depth in the Park (Whiteaker and Gardner 1986; Vitousek *et al.* 1987; Vitousek and Walker 1989). Researchers have studied biological factors responsible for the success of this species such as demography, growth rates, phenology, and seed biology (including dispersal and viability in the soil), and soil nutrient changes resulting from the invasion of this nitrogen-fixing species. Vitousek and other researchers are now assessing the impacts of firetree on the vegetation of invaded sites and determining what may limit its distribution in the Park. Studies of banana poka have also been conducted (see LaRosa, this volume), and research addressing limiting factors are in progress for fountain grass (R.N. Mack, pers. comm.). Population and genetic studies of strawberry guava (*Psidium cattleianum*) are in progress (Huenneke 1991; C.W. Smith, pers. comm.). Population data have been collected for a number of rain forest species while monitoring recovery of vegetation following ungulate removal, or comparing rain forest areas in setting priorities for management of alien biota. Seed viability and germination studies have been conducted on a number of alien plant species.

The biology of other alien plant species has not been as thoroughly studied as that of firetree. As a result, managers have established priorities for management and developed control strategies subjectively from distribution maps, qualitative observations, and the advice of local botanists. However, alien plant population dynamics, impacts, and even limiting factors are evident in some cases, or can be determined through monitoring control programs. Minimal research seems warranted to effectively manage target species that are localized in distribution and amenable to mechanical control. However, thorough weed ecology studies are needed for widespread disruptive species requiring a great investment of resources to manage.

Educating the Public and Organizations

Resource managers and researchers try to educate the public about alien plant problems so that they will develop an understanding of and support for alien plant control programs. Alien plant problems are a minor theme of some park interpretive programs that emphasize geological, cultural, and other biological messages. The Resources Management Division in the Park has an active volunteer program but has been able to generate little sustained interest in alien plant control activities; volunteers prefer assignments with wildlife (*e.g.*, nēnē (*Nesochen sandwicensis*) or turtle programs) or fire (Misaki and Tunison, this volume). On the whole, it has been very difficult to interest the public in alien plant problems and the need to control alien plants. One reason is that Hawai'i Island has a mostly rural and small-town culture with little interest in conservation. A shortcoming of the alien plant control program has been the inability to find a way to make alien plant problems important to the public.

In contrast, education of conservation groups and scientists about alien plant problems has been exemplary. Alien plant problems were the theme of a three-day symposium held in Hawaii Volcanoes in 1986 (results published in this volume), and a number of papers presented at the biennial science conferences and symposia held at the Park since 1976 have addressed alien plants.

Working with Other Agencies and Organizations

Alien plants are a statewide problem in Hawai'i, not only in natural areas, but also on agricultural lands, rangelands, and watersheds. Some alien plant pests are common to all these areas. However, some species are disruptive to native ecosystems, yet beneficial on rangelands or otherwise considered desirable plants in Hawai'i. Cooperation among managers of natural areas and most State agencies is not expected for these species.

The Hawaii Department of Agriculture, the lead State agency in alien plant control, employs a small crew to manage some alien plant species using mechanical or chemical means. This agency confines most of its efforts to species affecting rangelands. The State is also the lead agency in an interagency effort to control gorse (*Ulex europaeus*) Statewide with herbicides, grazing, reforestation, and biocontrol.

Common interests have developed in biocontrol of some alien plants disruptive to rangelands or forests (*e.g.*, banana poka, firetree, and gorse). State cooperation is needed in biocontrol research because the State determines which biocontrol agents may be imported; it does not allow importation of biocontrol agents for target alien plant species that are closely related to economically valuable species or desirable plants. On the whole, cooperation with local agencies has improved in the 1980s, particularly in biocontrol.

CONCLUSIONS

The long-term effectiveness of control programs at Hawaii Volcanoes National Park cannot be predicted at this time because of uncertainties about the effectiveness of biological control, potential invasion of new alien plants, the eventual impacts of uncontrollable plant species, and the uncertainty of support for continued programs. The alien plant control program is still growing but lags behind successful ungulate control programs in funding and success. However, the short-term effectiveness of localized alien plant control and Special Ecological Areas management has given managers considerable hope where little progress had been made previously, where the best biological resources of the Park are threatened, and where alien plant populations are expanding very rapidly. Productive approaches could be expanded to larger areas of the Park with similar effectiveness, given sufficient funding.

ACKNOWLEDGMENTS

I thank all those researchers and managers who have taken an interest in alien plants and contributed ideas towards formulating management strategies at Hawaii Volcanoes. The most important include Linda Cuddihy, Chuck Stone, Dan Taylor, Chris Zimmer, and Cliff Smith.

Literature Cited

Cuddihy, L.W., C.P. Stone, and J.T. Tunison. 1988. Alien plants and their management in Hawaii Volcanoes National Park. *Trans. West. Sect. Wildl. Soc.* 24:42-46.

Davis, C.J., E. Yoshioka, and D. Kageler. [this volume] Biological control of lantana, prickly pear, and Hāmākua pāmakani in Hawai'i: a review and update.

Fagerlund, G.O. 1947. *The exotic plants of Hawaii National Park.* Nat. Hist. Bull. 10, Hawaii National Park.

Higashino, P.K., L.W. Cuddihy, S.J. Anderson, and C.P. Stone. 1988. *Checklist of vascular plants of Hawaii Volcanoes National Park.* Tech. Rep. 64, Univ. Hawaii Coop. Natl. Park Resour. Stud. Unit. Honolulu.

Huenneke, L. 1991. Population biology of an invading tree, *Psidium cattleianum*, in Hawaii Volcanoes National Park. *Proc. Symp. Exotic Pest Plants*, Miami, Fla., Nov. 1988, 177-188. Washington, D.C.: U.S. Dep. Int. Natl. Park Serv.

Loope, L.L., and D. Mueller-Dombois. 1989. Characteristics of invaded islands, with special reference to Hawaii. In *Biological invasions: a global perspective*, ed. J. Drake, F. di Castri, R. Groves, F. Kruger, H. Mooney, M. Rejmanek, and M. Williamson, 257-280. Chichester, U.K.: John Wiley & Sons.

Loope, L.L., and P.G. Scowcroft. 1985. Vegetation response within exclosures in Hawai'i: a review. In *Hawai'i's terrestrial ecosystems: preservation and management*, ed. C.P. Stone and J.M. Scott, 377-402. Univ. Hawaii Coop. Natl. Park Resour. Stud. Unit. Honolulu: Univ. Hawaii Pr.

Markin, G.P., P.-Y. Lai, and G.Y. Funasaki. [this volume] Status of biological control of weeds in Hawai'i and implications for managing native ecosystems.

Mueller-Dombois, D., K.W. Bridges, and H.L. Carson, eds. 1981. *Island ecosystems: biological organization in selected Hawaiian communities.* Stroudsburg, Penn.: Hutchinson Ross Pub. Co.

National Park Service. 1985. *Fire Management Plan: Hawaii Volcanoes National Park.* Washington, D.C.: Natl. Park Serv.

National Park Service. 1988. *Resources Management Plan, Hawaii Volcanoes National Park.*

Santos, G.L., D. Kageler, D.E. Gardner, and C.P. Stone. 1986. *Herbicidal control of selected alien plant species in Hawaii Volcanoes National Park: a preliminary report.* Tech. Rep. 60, Univ. Hawaii Coop. Natl. Park Resour. Stud. Unit. Honolulu.

Scowcroft, P.G., and C.E. Conrad. [this volume] Alien and native plant response to release from feral sheep browsing on Mauna Kea.

Smith, C.W. 1985. Impact of alien plants on Hawai'i's native biota. In *Hawai'i's terrestrial ecosystems: preservation and management*, ed. C.P. Stone and J.M. Scott, 180-250. Univ. Hawaii Coop. Natl. Park Resour. Stud. Unit. Honolulu: Univ. Hawaii Pr.

Smith, C.W. 1991. The alien plant problem in Hawaii. *Proc. Symp. Exotic Pest Plants,* Miami, Fla., Nov. 1988, 327-338. Washington, D.C.: U.S. Dep. Int. Natl. Park Serv.

Smith, C.W., and J.T. Tunison. [this volume] Fire and alien plants in Hawai'i: research and management implications for native ecosystems.

Stone, C.P., L.W. Cuddihy, and J.T. Tunison. [this volume] Responses of Hawaiian ecosystems to removal of feral pigs and goats.

Tunison, J.T. [this volume] Fountain grass control in Hawaii Volcanoes National Park: management considerations and strategies.

Tunison, J.T. 1991. Strategies and successes in controlling alien plants in an Hawaiian National Park. *Proc. Symp. Exotic Pest Plants*, Miami, Fla., Nov. 1988, 353-376. Washington, D.C.: U.S. Dep. Int. Natl. Park Serv.

Tunison, J.T., and E. Misaki. [this volume] The use of volunteers for alien plant control at Hawaii Volcanoes National Park and The Nature Conservancy of Hawaii's Kamakou Preserve.

Tunison, J.T., and C.P. Stone. [this volume] Special Ecological Areas: an approach to alien plant control in Hawaii Volcanoes National Park.

Tunison, J.T., C.P. Stone, and L.W. Cuddihy. 1986. SEAs provide ecosystem focus for management and research. *Park Sci.* 6(3):10-13.

Tunison, J.T., L.D. Whiteaker, L.W. Cuddihy, A.M. LaRosa, D.V. Kageler, M.R. Gates, N.G. Zimmer, and L. Stemmermann. 1992. *The distribution of selected localized alien plant species in Hawaii Volcanoes National Park.* Tech. Rep. 84, Univ. Hawaii Coop. Natl. Park Resour. Stud. Unit. Honolulu.

Tunison, J.T., and N.G. Zimmer. [this volume] Control of localized alien plants in Hawaii Volcanoes National Park.

Vitousek, P.M., L.R. Walker, L.D. Whiteaker, D. Mueller-Dombois, and P.A. Matson. 1987. Biological invasion by *Myrica faya* alters ecosystem development in Hawaii. *Science* 238:802-804.

Vitousek, P.M., and L.R. Walker. 1989. Biological invasion by *Myrica faya* in Hawai'i: plant demography, nitrogen fixation, and ecosystem effects. *Ecol. Monogr.* 59:247-265.

Warshauer, F.R., J.D. Jacobi, A.M. LaRosa, J.M. Scott, and C.W. Smith. 1983. *The distribution, impact, and potential management of the introduced vine* Passiflora mollissima *(Passifloraceae) in Hawai'i.* Tech. Rep. 48, Univ. Hawaii Coop. Natl. Park Resour. Stud. Unit. Honolulu.

Whiteaker, L.D., and D.E. Gardner. 1985. *The distribution of* Myrica faya *Ait. in the state of Hawai'i.* Tech. Rep. 55, Univ. Hawaii Coop. Natl. Park Resour. Stud. Unit. Honolulu.

Whiteaker, L.D., and D.E. Gardner. 1987. *The phenology and stand structure of* Myrica faya *Ait. in Hawai'i.* Tech. Rep. 62, Univ. Hawaii Coop. Natl. Park Resour. Stud. Unit. Honolulu.

SUCCESS IN CONTROLLING LOCALIZED ALIEN PLANTS IN HAWAII VOLCANOES NATIONAL PARK

J. Timothy Tunison and Nicholas G. Zimmer

ABSTRACT

Alien plant management at Hawaii Volcanoes National Park emphasizes control of localized alien plant species on a parkwide basis and control of widespread alien species in Special Ecological Areas. The purpose of the localized alien plant control program is to prevent the spread of potentially disruptive nonnative species while populations are still manageable. The program has intensified in the last five years, with control efforts currently directed at 41 species. Some of the more significant target species are black wattle (*Acacia mearnsii*), castor bean (*Ricinus communis*), nasturtium (*Trapaeolum majus*), New Zealand flax (*Phormium tenax*), and Formosan koa (*Acacia confusa*). Target species were identified by weediness in other areas, expansions of range or densities in the Park, affinities to known invasive taxa in Hawaii, viney habit, and/or readily dispersed dissemules. Target species were mapped by surveys from 1983 to 1986 along roadsides, trails, developed areas, and in approximately 39,500 a (16,000 ha) of backcountry incidental to other management tasks. Effective treatment methods, developed mostly from trial and error practice, are described. Target populations were monitored to determine the effectiveness of control practices by counting individuals surviving treatments or estimating aerial cover following treatment. Control efforts have been effective on 21 of 41 species under management. Reproduction has been curtailed, and populations have been markedly reduced. Eleven of these species were not observed for over one year, but it is premature to conclude that they have been eradicated. Nine other species were reduced to a population of seedlings (or a few sprouts). Population size and/or reproduction were reduced in 17 other species, suggesting partial control. Work loads are declining on all species partially or completely controlled. Reproduction or population size were not affected on three species. The ability of managers to correctly identify potentially disruptive species, together with restriction of reconnaissance for new introductions to a fraction of the total area in the Park, may result in preventing establishment and spread of new alien plants in the Park.

INTRODUCTION

Localized Alien Plant Control Program

Alien plant management at Hawaii Volcanoes National Park emphasizes control of localized plant species on a parkwide basis, control of some widespread species in localized areas called Special Ecological Areas (Tunison and Stone, this volume), and biological control for suitable widespread species. Only localized alien plant species with the potential to disrupt native vegetation or prevent the recovery of native plant communities are targeted for control in the program described in this paper.

Localized alien plant species to be managed are selected on the basis of plant aggressiveness in Hawai'i and elsewhere, affinites with aggressive taxa, possession of readily dispersed dissemules, and/or viney habit. The advice of local botanists is sought. Some localized alien plant species are threats to a number of native plant communities; others affect limited areas or vegetation types (Table 1). A few are targeted for control because they became established recently in the Park and appear to be easily eradicable.

The localized alien plant control program started in 1982 with two target species and expanded rapidly; 12 species were added in 1983, 10 in 1984, 15 in 1985, and only one in 1986. The reason for the program expansion was not the recent establishment of target species, but a growing emphasis on control of potentially disruptive introduced plant species with currently limited ranges. Although Park collection records are incomplete, they suggest that most target species have occurred in the Park for 10 to 60 years (Fosberg 1975; Division of Resources Management, Hawaii Volcanoes National Park, unpub. data).

Surveys for new weeds along roadsides are now made every three months incidental to fountain grass (*Pennisetum setaceum*) control, annually in Special Ecological Areas in conjunction with other weed control efforts, and twice yearly in developed areas and along trails.

Target species

Approximately 405 nonnative plant species grow in the Park (Higashino *et. al.* 1988). Many naturalized plant species are localized and are comprised of a few to a few hundred individuals distributed over a few to several hundred hectares, or as many as 100,000 individuals over a few hectares (*e.g.*, wire vine (*Muehlenbeckia axillaris*) -- three individual vines in two locations; Formosan koa (*Acacia confusa*) -- approximately 370 individuals scattered over 1,235 a (500 ha); and sisal (*Agave sisalana*) -- over 100,000 individuals concentrated in 12 a or 5 ha).

Twenty-seven of the 40 target species are probably deliberate introductions to the Park, most planted as ornamentals in the Kīlauea Crater Rim area or as reforestation, shade, or forage species at 'Āinahou Ranch, a former inholding now included as National Park Service land. Six species are unintentional introductions along roads, apparently from fill (Table 1).

Table 1. Characteristics of localized alien plant species currently managed in Hawaii Volcanoes National Park.

Target Species	Family	Origin	Fruit	Habit	Introduction to Park	Reason for Controlling
Acacia confusa (Formosan koa)	Legum.	Philippines, Taiwan	Dry pod	Tree	Intentional: shade	Spreading in Park
Acacia mearnsii (black wattle)	Legum.	Australia	Dry pod	Tree	Intentional: reforestation	Invasive in Hawai'i; spreading in Park
Acacia melanoxylon (blackwood acacia)	Legum.	Australia	Dry pod	Tree	Intentional: reforestation	Invasive in Hawai'i; spreading in Park
Agave americana (century plant)	Amary.	Trop. Amer.	Capsule	Per. herb	Intentional: ornamental	Spreading in Park
Agave sisalana (sisal)	Amary.	Trop. Amer.	Bulbils	Per. herb	Intentional: ornamental	Invasive in Hawai'i; spreading in Park
Albizia sp. (albizia)	Legum.		Dry pod	Tree	Accidental: road fill	Invasive in Puna
Ardisia crenata (Hilo holly)	Mryta.	S. Asia	Red berry	Low	Intentional: ornamental	Invasive in lowlands; manageable
Arthrostema ciliatum (arthrostema)	Melas.	S. Amer.	Many-seeded capsule	Herb	Accidental	Invasive in Hawai'i; invasive family; manageable

Target Species	Family	Origin	Fruit	Habit	Introduction to Park	Reason for Controlling
Benicasa hispida (Chinese melon)	Cucur.	Trop. Asia	Large berry	Vine	Intentional: food	Spreading in Park
Casuarina equisetifolia (ironwood)	Casua.	Trop. Pacif.	Winged seeds	Tree	Intentional: reforestation	Invasive in Hawai'i
Desmodium cajanifolium	Legum.	Trop. Amer.	Dry pod	Perenn. herb	Accidental: road fill	Invasive in Hawai'i
Eleagnus umbellata	Elea.		Red berry	Low shrub	Intentional: ornamental	Invasive in Volcano Village; manageable
Eriobotrya japonica (loquat)	Rosac.	China/Japan	Pome	Small tree	Intentional: food	Spreading in Park
Eucalyptus spp. (eucalyptus)	Myrta.	Australia	Many-seeded capsule	Tree	Intentional	Spreading in Park
Feijoa sellowiana (guavasteen)	Myrta.	S. Amer.	Berry	Small tree	Intentional: food	Spreading in Park; manageable
Ficus pumila (climbing fig)	Morac.	Japan/China	Drupe	Vine	Intentional: ornamental	Viney habit; manageable
Fraxinus americana (white ash)	Oleac.	N. Amer.	Winged seeds	Tree	Intentional: reforestation	Spreading in Park

Table 1, continued.

Table 1, continued.

Target Species	Family	Origin	Fruit	Habit	Introduction to Park	Reason for Controlling
Fraxinus uhdei (tropical ash)	Oleac.	Mexico	Winged seeds	Tree	Intentional: reforestation	Invasive in Hawai'i
Fuchsia paniculata (fuchsia)	Onagr.	Mexico	Fleshy black berry	Shrub	Intentional: ornamental	Fruit dispersable by birds; manageable
Hedera helix (English ivy)	Arali.	Eurasia	Black berries	Vine	Intentional: ornamental	Fruiting in Park; fruits attractive to birds; spreading in Park
Heterocentron subtriplinervium (pearl flower)	Melas.	Mexico	Many-seeded capsule	Shrub	Intentional? ornamental	Invasive family; population spreading from roadside
Luculia gratissima	Rubia.	S. Asia	Winged	Small Tree	Intentional: ornamental	Seeds readily dispersed
Lupinus sp. (lupine)	Legum.		Dry pod	Herb	Accidental: road fill?	Manageable
Melaleuca quinquenervia (paperbark)	Myrta.	Australia	Many-seeded capsule	Tree	Intentional: reforestation	Invasive in Hawai'i
Melochia umbellata (melochia)	Sterc.	India	Winged seeds in capsules	Tree	Accidental: road fill	Invasive in lowlands; manageable

Target Species	Family	Origin	Fruit	Habit	Introduction to Park	Reason for Controlling
Muehlenbeckia axillaris (wire vine)	Polyg.	New Zealand	Nut	Vine	Intentional: ornamental	Viney habit
Opuntia ficus-indica (prickly pear)	Cacta.	Mexico	Fleshy berry	Stem succulent	Intentional: forage	Invasive in Hawai'i
Paederia scandens (maile pilau)	Rubia.	S. Asia	Berry	Vine	Accidental: road fill	Spreads in lowlands; Viney habit; manageable
Phoenix dactylifera (date palm)	Palma.	N. Africa	Drupe	Tree	Intentional?	Manageable
Phormium tenax (New Zealand flax)	Lilia.	New Zealand	Capsule	Herb	Intentional	Invasive in Hawai'i
Phyllostachys sp. (bamboo) (possibly two spp.)	Grami.			Woody grass	Intentional: ornamental	Invasive in Hawai'i
Pinus sp. (pine)	Pinac.		Winged seeds	Tree	Intentional: wood, shade	Spreading in Park
Pittosporum undulatum	Pitto.	Australia	Orange capsule; red seeds	Small tree	Intentional: ornamental	Spreading in Park

Table 1, continued.

Table 1, continued.

Target Species	Family	Origin	Fruit	Habit	Introduction to Park	Reason for Controlling
Prosopis pallida (kiawe)	Legum.	S. Amer.	Dry pod	Tree	Accidental	Invasive in Hawai'i; spreading in Park
Pueraria lobata (kudzu)	Legum.	Asia	Dry pod	Vine	Intentional: livestock forage	Invasive on Mainland
Ricinus communis (castor bean)	Eupho.		Capsule	Shrub	Accidental	Invasive in Hawai'i; spreading in Park
Soliva sessilis	Compo.	N. Amer.	Achene	Herb	Accidental	Forms dense ground cover in Calif; manageable
Syzygium jambos (rose apple)	Myrta.	Trop. Asia	Berry	Tree	Accidental	Invasive in Hawai'i; manageable
Tibouchina urvilleana (glorybush)	Melas.	Brazil	Capsule(no seeds seen)	Tall shrub	Intentional: ornamental	Invasive in Hawai'i
Trema orientalis (gunpowder Tree)	Ulmac.	Malaysia	Drupe	Tree	Accidental; in roadside fill	Invasive in lowlands
Tropaeolum majus (Nasturtium)	Tropae.	S. Amer.	Capsule	Vine	Intentional: ornamental	Monospecific stands in montane seasonal

METHODS

Distributions of localized alien plant species were delineated by systematic mapping. Surveys were made along roads, developed areas, trails, and neighboring subdivisions (L. Stemmermann, unpub. data), in seven Special Ecological Areas (6,670 a or 2,700 ha), and in ʻĀinahou Ranch (3,705 a or 1,500 ha). Park staff examined pig activity transects in approximately 1,360 a (550 ha) of ʻŌlaʻa Tract (S.J. Anderson, unpub. data) and 8,645 a (3,500 ha) of the Mauna Loa Strip (L.K. Katahira, unpub. data). About 29,840 a (12,000 ha) of coastal lowlands and submontane seasonal environment are also surveyed incidental to fountain grass control work.

Treatment methods developed by Park Resources Management staff were applied to 36 target species (Table 2). Park researchers developed methods for one species, and control methods for another were available from the literature. Introduction of known effective biocontrol agents was attempted for a third species (Davis *et al.*, this volume). Uprooting was used exclusively for three herbaceous species and for seedlings of most species. Herbicides were used to control at least one life stage of 36 species and were applied by foliar spray, cut stump, or frill (notching the bole of a standing tree) techniques (Table 2).

Effectiveness of treatments was monitored at three- to six-month intervals, and retreatments were applied as needed. The retreatments required were determined by counting numbers of target individuals, or estimating the aerial cover of target species prior to, and at intervals after, treatments were applied. Effects on nontarget (native) species were not consistently determined.

RESULTS

Eleven treated plant species were not observed over one year after treatment: albizia (*Albizia* sp.), Chinese melon (*Benicasa hispida*), eleagnus (*Eleagnus umbellata*), guavasteen (*Feijoa sellowiana*), tropical ash (*Fraxinus uhdei*), fuchsia (*Fuchsia paniculata*), *Luculia gratissima*, paperbark (*Melaleuca quinquenervia*), maile pilau *(Paederia scandens)*, New Zealand flax (*Phormium tenax*), and rose apple (*Syzygium jambos*) (Table 2). One other species, *Desmodium cajanifolium*, was not observed six months after treatment.

Populations of nine species were reduced to seedling stages (or sprouts) with significantly reduced population numbers: black wattle (*Acacia mearnsii*), from 240 to 30 seedlings (when last monitored); blackwood acacia (*Acacia melanoxylon*), from 20 plants to 8 seedlings; ironwood (*Casuarina equisetifolia*), from 20 plants to 4 sprouts; loquat (*Eriobotrya japonica*), from 6,750 plants to 42 seedlings; white ash (*Fraxinus americana*), from 76 plants to 2 seedlings; date palm (*Phoenix dactylifera*), from 16 plants to 2 seedlings; *Pittosporum undulatum*, from 25 plants to one seedling; kiawe *(Prosopis pallida)*,

Table 2. Changes in target alien plant populations with treatment; most effective treatments used in Hawaii Volcanoes National Park.

Species	Population Changes		Most Effective Treatment Used		Comments
	Status/Start	Current Status	Appl. Technique	Herbicide	
Acacia confusa (Formosan koa)	370 plants (12/84)	50 plants (11/86)	Basal bark (sapl.) Cut stump (mature) Uproot (seedlings)	5% Garlon 4/ diesel; Tordon RTU	Remainder to be removed by contractor.
Acacia mearnsii (black wattle)	240 plants (1/81)	30 plants (4/87)	Basal bark (mature) Uproot (seedlings)	5% Garlon 4/ diesel	Population reduced to seedling stage.
Acacia melanoxylon (blackwood acacia)	20 plants (2/85)	8 plants (11/86)	Cut stump (mature) Uproot (seedlings)	Tordon RTU	Population reduced to seedling stage.
Agave americana (century plant)	25 plants (1/85)	5 plants (4/87)	Foliar (mature) Uproot (seedlings)	5% Garlon 4	Some plants in hard to reach area on wall of Kīlauea Caldera.
Agave sisalana (sisal)	>100,000 plants	<25,000 plants	Foliar	5% Garlon 4	Effective treatment found 2/87; method only partially effective on small plants.
Albizia sp. (albizia)	1 plant (12/85)	0 plants (4/87)	Cut Stump	2% Garlon 4/ diesel	One plant found along roadside killed.
Ardisia crenata (Hilo holly)	2 plants (1/86)	2 plants (4/87)			No effective treatment found to date.

Species	Population Changes Status/Start	Current Status	Most Effective Treatment Used Appl. Technique	Herbicide	Comments
Arthrostema ciliatum (arthrostema)	>100 plants (6/84)	<25 plants (11/85)	Uproot		
Benicasa hispida (Chinese melon)	1 ha colony (7/83)	No plants (4/87)	Uproot		Apparently eradicated; no plants found since 10/84.
Casuarina equisetifolia (ironwood)	20 plants (3/83)	4 plants (12/86)	Basal bark	2-5% Garlon 4/ diesel	Reproduction curtailed; retreating persistent sprouts.
Desmodium cajanifolium	1 plant (11/86)	0 plants (4/87)	Uproot		One plant found along roadside.
Eleagnus umbellata	1 plant (10/84)	0 plants 11/86)	Cut stump	Tordon RTU	One plant at Volcano House killed.
Eriobotrya japonica (loquat)	6,750 plants (7/84)	42 plants (4/87)	Cut stump Uproot (seedlings)	2% Garlon 4	Population mostly seedlings and saplings; small numbers of seedlings recruited.

Table 2, continued.

Table 2, continued.

Species	Population Changes Status/Start	Current Status	Most Effective Treatment Used Appl. Technique	Herbicide	Comments
Eucalyptus spp. (eucalyptus)	30 plants (5/83)	3 plants (1/87)	Cut stump	2% Garlon 4 Tordon RTU	Retreating persistent sprouts; effectiveness of treatments increasing as methods refined; both Garlon and Tordon RTU only partially effective.
Feijoa sellowiana (guavasteen)	20 plants (11/85)	0 plants (4/87)	Cut stump (mature) Uproot (seedlings)	Tordon RTU	This species apparently eliminated from Park.
Ficus pumila (climbing fig)	3 plants ($900\ m^2$) (1/85)	3 plants ($90\ m^2$) (5/86)	Foliar	5% Garlon	Possibly eradicated.
Fraxinus americana (white ash)	76 plants (1/83)	2 plants (4/87)	Frill (mature) Foliar (saplings) Uproot (seedlings)	5% Garlon 4 5% Garlon 4	Population reduced to seedling stages.
Fraxinus uhdei (tropical ash)	1 plant (9/84)	0 plants (3/87)	Cut stump	Tordon RTU	Probably eradicated.
Fuchsia paniculata (fuchsia)	3 plants (12/85)	0 plants (12/86)	Cut stump (mature) Uproot (small)	Tordon RTU	Population probably eradicated.

Species	Population Changes Status/Start	Current Status	Most Effective Treatment Used Appl. Technique	Herbicide	Comments
Hedera helix (English ivy)	8 plants (6/85)	7 plants (4/87)	Foliar (mature) Uproot (seedlings)	2% Garlon 4	Large population at ʻĀinahou Ranch not yet treated; seedling found in Thurston; treatment results inconsistent.
Heterocentron subtriplinervium (pearl flower)					Control efforts not started.
Luculia gratissima	2 plants	0 plants	Cut stump	Tordon RTU	Probably eradicated.
Lupinus sp. (lupine)	35 plants (3/84)	4 plants (4/87)	Uproot		Population reduced to flowering stages.
Melaleuca quinquenervia (paperbark)	6 plants (6/83)	0 plants 4/87)	Cut stump (mature) Uproot (seedlings)	Tordon RTU	Probably eradicated.
Melochia umbellata (melochia)	9 plants (10/83)	5 plants (4/87)	Cut stump (mature) Uproot (seedlings/ saplings)	Tordon RTU	Continues to be introduced along Park boundary.
Muehlenbeckia axillaris (wire vine)	3 plants (2/83)	1 plant (11/86)	Foliar	2% Garlon 4	Two of three plants probably eradicated.

Table 2, continued.

Table 2, continued.

Species	Population Changes		Most Effective Treatment Used		Comments
	Status/Start	Current Status	Appl. Technique	Herbicide	
Opuntia ficus-indica (pānini cactus)	100 plants/ 30 populations	100 plants/ 30 populations			Two biological control agents, *Cactoblastis cactorum* and *Dactylopius opuntiae*, introduced to local populations; some plants declining.
Paederia scandens (maile pilau)	1 colony (3/85)	0 plants (4/87)	Foliar	5% Tordon 22K	Probably eradicated.
Phoenix dactylifera (date palm)	16 plants (3/85)	2 plants (4/87)	Cut stump	100% Banvel	Reduced to seedling stages.
Phormium tenax (New Zealand flax)	3 colonies (12/82)	0 plants (4/87)	Foliar (mature) Uproot (sprouts)	2% Roundup	Probably eradicated from Park.
Phyllostachys sp. (bamboo)	2 colonies (12/83)	1 colony (4/87)	Foliar	5% Garlon 4	One colony eradicated; the other resistent to all treatments tried; possibly two species.
Pinus sp. (pine)	110 plants (4/83)	5 plants (5/86)	Cut (mature) Uproot (seedlings)		Reproduction away from main population treated; main population not treated.

Species	Population Changes Status/Start	Current Status	Most Effective Treatment Used Appl. Technique	Herbicide	Comments
Pittosporum undulatum	25 plants (12/85)	0 plants (11/86)	Cut stump (uproot) Uproot (seedlings)	Tordon RTU	Premature to consider eradicated.
Prosopis pallida (kiawe)	105 plants (3/85)	3 plants (4/87)	Cut stump (mature) Uproot (seedlings)	100% Roundup 5% Garlon 4/ diesel	5% Garlon 4/diesel partially effective.
Pueraria lobata (kudzu)	210 shoots (9/84)	42 shoots (4/87)	Dig up root		Recommended method (2% Roundup) only partially effective.
Ricinus communis (castor bean)	600 plants (3/85)	200 plants (3/87)	Cut stump (mature) Uproot (seedlings)	100% Banvel	Population reduced to seedling stages; Banvel nearly 100% effective.
Soliva sessilis	125 plants (4/85)	5 plants (4/87)	Uproot		Reduced to seedling stages.
Syzygium jambos (rose apple)	2 plants (12/85)	0 plants (5/86)	Cut stump	Tordon RTU	Two plants in Kīlauea Caldera below Volcano House apparently killed.
Tibouchina urvilleana (glorybush)					Some populations not treated to date.

Table 2, continued.

Table 2, continued.

Species	Population Changes: Status/Start	Population Changes: Current Status	Most Effective Treatment Used: Appl. Technique	Most Effective Treatment Used: Herbicide	Comments
Trema orientalis (gunpowder tree)	200 (3/83)	25 (3/87)	Basal Bark	5% Garlon 4/ diesel	Plants invade roadsides.
Tropaeolum majus (nasturtium)	67% cover in monitoring plots (6/84)	2% cover (6/87)	Foliar	0.4% Garlon 4	Mapped 9/83; new populations found since then; no populations eradicated but cover reduced in populations by 23%.

from 76 plants to 2 sprouts; and castor bean (*Ricinus communis*), from 600 to 200 seedlings (Table 2).

Reproduction and/or population size were reduced in 18 of the 20 other taxa: Formosan koa (*Acacia confusa*), century plant (*Agave americana*), sisal (*A. sisalana*), *Arthrostema ciliatum*, eucalyptus (*Eucalyptus* sp.), climbing fig (*Ficus pumila*), English ivy (*Hedera helix*), lupine (*Lupinus* sp.), melochia (*Melochia umbellata*), wire vine (*Muehlenbeckia axillaris*), pānini cactus (*Opuntia ficus-indica*), pine (*Pinus* sp.), bamboo (*Phyllostachys* sp.), kudzu (*Pueraria lobata*), *Soliva sessilis*, gunpowder tree (*Trema orientalis*), glorybush (*Tibouchina urvilleana*), and nasturtium (*Tropaeolum majus*) (Table 2). Reproduction was curtailed but population size was not affected in Hilo holly (*Ardisia crenata*). Reproduction and population size were not reduced in pearl flower (*Heterocentron subtriplinervium*).

The number of worker days allocated to localized alien plant control projects increased significantly from 1983 through 1985 as the number of target species increased. Forty-seven worker days were expended in 1983, and 299 in 1985. The work load decreased sharply thereafter as the time required to treat each species declined. Seventy-five worker days were expended in 1987. Surveys to locate target plants, labor-intensive initial treatments of mature individuals, and larger populations required relatively greater time in the first stages of control of most weed species than in later stages.

DISCUSSION

Eradication

The 11 target species not observed more than one year after treatment may be eradicated. However, seed banks may be present, and resprouting may occur. Populations of 21 target species were very small. Two of these were trees or shrubs consisting of one individual each. Populations of six other species were comprised of less than seven individuals. One tree population was 20 individuals. Another species was an herbaceous vine, possibly consisting of one individual colonizing a 5-a (2-ha) area (Table 2).

More species of localized alien plants were not eradicated for the following reasons:

1. Control work had begun only recently at the time. Treatments of most species were started between 1983 and 1985 (Table 2). Most treatment methods were discovered by trial and error procedures, and typically 6 to 12 months were required to determine the effectiveness of the methods. Only a few populations of six target species have been treated (Formosan koa, century plant, sisal, eucalyptus, English ivy, pine, and glorybush). Control efforts were not started for one species, pearl flower.

2. Effective treatment methods were not known at the outset of control efforts, and effective methods are still not available for pearl flower and pānini cactus (Table 2).

3. Soil seed banks or extensive underground root systems may require a number of years to deplete.

4. Some species continue to be reintroduced along roadsides, *e.g.*, gunpowder tree and melochia.

Rationale for Control Program

The main rationale for the localized alien plant control program is the cost effectiveness of controlling alien plant species prior to expansion. Expensive management programs on species following range expansions are thus avoided. Moreover, controlling localized species reduces further degradation of native plant communities and Special Ecological Areas by alien species.

Localized alien plant control may not seem worthwhile when disruptive, widespread alien plants continue to spread throughout most of the Park (except in Special Ecological Areas). For example, the value of controlling castor bean in the lowlands may be questioned if fountain grass cannot be controlled. However, we believe that control of localized species is justified because the potential range and ecological impact of many widespread species cannot be predicted with great certainty. Moreover, biological control may eventually become available for some widespread species.

Limitations of Control Program

Effective management of localized alien plant species on a parkwide basis is limited by factors that make it impossible to prevent the successful introduction of all new weed species in the Park:

1. Reconnaissance to detect new introductions is restricted to a small percentage of the Park. Systematic and regular scouting for new introductions has focused on ruderal habitats such as roadsides, trails, and developments, areas known to support many weed species (*e.g.*, ʻĀinahou Ranch), and designated Special Ecological Areas. Some weeds become established in areas not regularly reconnoitered for new weeds. In many cases, botanists are not available for survey work. Consequently, most areas of the Park are not being adequately surveyed for the establishment of new weeds.

2. Managers are not able to correctly predict with certainty the invasiveness of localized alien plants, so potential threats may not be correctly identified. Most localized alien plant species are not currently managed. Additional species can be anticipated to invade the Park, and it is likely that some of these will become disruptive. Compounding the problem of correctly identifying alien plant threats is the fact that disruptive species may occur at low population levels for many years before spreading. For example, Christmas berry (*Schinus terebinthifolius*) is known from collections in Florida in the 1850s,

yet it did not become a significant problem until the 1950s (Ewell *et al.* 1982). Moreover, it is not possible to predict with great certainty the behavior of alien plants from other parts of Hawai'i. For example, olive (*Olea europaea* subsp. *africana*) has spread rapidly over several thousand hectares in the 'Āinahou Ranch area of the Park, but it is not normally invasive in Hawai'i.

3. There is high potential for invasions from adjacent areas. This results from increasing introduction of species due to ineffective control and quarantine (Smith 1985) and encroachment of development on Park boundaries. To date, reinvasion appears to be a problem only with a few species along roadsides.

CONCLUSIONS

Effective treatment methods and parkwide control were implemented for 21 of 40 target species. Control was effected by markedly reducing populations and curtailing reproduction; work loads decreased. It is premature to conclude that any species have been eradicated because of the presence of seed banks or the potential for resprouting. Seventeen additional target species are partially controlled in that populations and reproduction are reduced. Three species have not been controlled at all. Success of the first five years of the program suggests that uncontrolled species will be controlled once effective treatment methods become available and control efforts are extended to all populations in the Park.

One of the most significant findings of the localized plant control program is that successful control requires a regime of consistent and timely followup treatments. Moreover, alien plant species, even when comprised of only one to several plants, may require a number of years for successful control, and probably much longer for eradication.

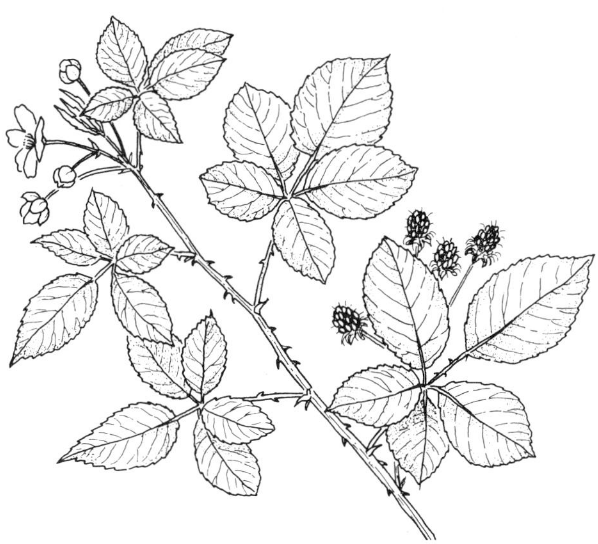

Literature Cited

Davis, C.J., E. Yoshioka, and D. Kageler. [this volume] Biological control of lantana, prickly pear, and Hāmākua pāmakani in Hawai'i: a review and update.

Ewell, J.J., D.S. Ojiwa, D.A. Karl, and W.F. DeBusk. 1982. Schinus *in successional ecosystems of Everglades National Park.* Tech. Rep. T-676, Natl. Park Serv. So. Fla. Res. Cent., Everglades Natl. Park. Homestead, Florida.

Fosberg, F.R. 1975. *Revised checklist of vascular plants of Hawaii Volcanoes National Park.* Tech. Rep. 5, Univ. Hawaii Coop. Natl. Park Resour. Stud. Unit. Honolulu.

Higashino, P.K., L.W. Cuddihy, S.J. Anderson, and C.P. Stone. 1988. *Checklist of vascular plants of Hawaii Volcanoes National Park.* Tech. Rep. 64, Univ. Hawaii Coop. Natl. Park Resour. Stud. Unit. Honolulu.

Smith, C.W. 1985. Impact of alien plants on Hawai'i's native biota. In *Hawai'i's terrestrial ecosystems: preservation and management*, ed. C.P. Stone and J.M. Scott, 180-250. Univ. Hawaii Coop. Natl. Park Resour. Stud. Unit. Honolulu: Univ. Hawaii Pr.

Tunison, J.T., and C.P. Stone. [this volume] Special Ecological Areas: an approach to alien plant control in Hawaii Volcanoes National Park.

CONTROL OF ALIEN PLANTS ON NATURE CONSERVANCY PRESERVES

R. Alan Holt

ABSTRACT

The Nature Conservancy of Hawaii is currently (1989) engaged in active alien plant control on two large montane forest and shrubland preserves on the islands of Moloka'i and Maui. The national Conservancy organization is in the process of developing comprehensive guidelines for the use of pesticides in alien species control and is producing an Element Stewardship Abstract database to assemble pertinent information on management of key alien species. Strategy for alien plant control in Hawai'i follows that of Hawaii Volcanoes National Park in prioritizing control tasks but is highly reliant on volunteer labor and interagency collaborative research effort. Conservancy managers view effective ungulate control as a necessary prerequisite to long-term alien plant control. Current alien plant control priorities are localized infestations of highly invasive weeds (*e.g., Rubus argutus* and *Clidemia hirta* on Moloka'i) and development of control methods for more widespread invasive plants that threaten preserve integrity. Interagency cooperation to eliminate new alien plant populations before they spread to key natural areas and research on revegetation and weed control through use of native plants are recommended.

INTRODUCTION

The Nature Conservancy is a private, nonprofit conservation organization dedicated to establishing managed, protected habitat for the full array of the earth's remaining natural communities and species. To that end, the Conservancy has set aside more than 31 million acres (12 million ha) of land in the United States, Canada, Latin America, and the Caribbean since the early 1950s. Today The Conservancy manages more than 1.3 million acres (526,000 ha) itself in about 1,630 Conservancy nature preserves nationwide. Programs of the Conservancy for the control of alien plant species on nature preserves nationally and in Hawai'i are described in this paper.

ORGANIZATIONAL POLICIES AND PROGRAMS

The control of alien plants is a serious concern of land stewards on many Nature Conservancy preserves. The Conservancy has been active in management applications of alien plant control methods in native ecosystems, with the greatest attention to date given to the maintenance and restoration of native grasslands and savannah in the western, midwestern, and southeastern U.S., through mowing, grazing, and prescribed burning. Organizational policy states that the use of chemical herbicides for alien plant control is to be viewed as a "last resort" after manual and other methods have been exhausted, and that herbicides used must be the most biologically safe available (The Nature Conservancy 1984). Conservancy managers are currently developing a more detailed and comprehensive set of policies and guidelines for pesticide use and are considering those of the National Park Service as a model from which to start.

The Conservancy is also developing a nationwide database of biological and other management-related information on priority alien plant species. Information on each species is compiled in an "Element Stewardship Abstract," which can then be shared by the resource management community. Such abstracts have been completed for several dozen species, including a handful that are important to Hawai'i. Probably the greatest benefit of the Element Stewardship Abstract database to Hawai'i is the abstract format itself, which local managers may use to organize and share important information.

THE CONSERVANCY'S APPROACH TO ALIEN PLANT CONTROL IN HAWAI'I

The Nature Conservancy of Hawaii is currently engaged in active alien plant control work at the 2,744-a (1,110-ha) Kamakou Preserve on Moloka'i and the 5,230-a (2,117-ha) Waikamoi Preserve on Maui. With extremely limited staffing (one staff person per preserve between 1983 and 1986, three or four per preserve since 1987), limited funding, and other urgent resource management demands, alien plant control efforts to date have concentrated on 1) establishing an organizational framework for effective alien plant control, and 2) addressing the most pressing weed problems with the help of volunteers.

The Conservancy's strategy assumes that most alien plant species have difficulty becoming established or spreading rapidly in an intact native plant community (Juvik and Juvik, this volume; Stone *et al.*, this volume; Tunison, this volume). The majority of aliens appear to rely on disturbance of the native vegetation by feral ungulates or humans to create or maintain the opportunity for invasion. Exclosure studies have also demonstrated that in some cases, native vegetation can recover area lost to alien plant species once ungulate and direct human disturbance of the plant cover is halted (Loope and Scowcroft 1985).

For these reasons, the alien plant control program on Conservancy preserves in Hawai'i is closely linked with ungulate control efforts. Ungulate control is seen as a necessary prerequisite to effective alien plant control. Preserves must also be protected from trampling or other direct damage by humans.

These assumptions about the relationship between disturbance and alien plant problems have prompted us to organize alien plant control along three lines. First, like the National Park Service, we have prioritized the list of alien plant species occurring in our preserves according to their invasiveness. This divides alien plants into two broad categories: 1) those that appear to be dependent on some disturbance of the native cover to become established or to maintain a place in the plant community, and 2) those that exhibit an ability to invade native vegetation even in the absence of such disturbance. Our top priority must be to control or eradicate the latter, more disruptive, species in preserve areas. A list of these and other target weeds being worked on at Waikamoi and Kamakou Preserves is given in Table 1.

Second, we have divided the alien plant control program on each preserve according to feasibility, or the chance to make real progress in control. We elect not to work on species that are very widespread. Likewise, we will seize any opportunity to eradicate the only population of a moderately invasive weed before it can become a problem. For disruptive species for which no adequate control method exists, we seek outside help to develop appropriate methods.

Third, we divide the alien plant control problem geographically. As in the National Park Service (Tunison and Stone, this volume), we give priority to controlling weeds in pristine areas and areas especially rich in rare species, although portions of a preserve already dominated by alien plants and containing no rare native species may eventually be marked for restoration.

Using the guidelines of species invasiveness, feasibility of control, and location of weed populations within the preserve, Conservancy preserve managers have developed an Alien Plant Control Job List, which sets the direction for volunteer and staff work schedules and for collaborative research. Volunteers have accounted for the majority of manpower devoted to alien plant control at the Kamakou Preserve in the past, and after six years they can be credited with effective control of several populations of serious invasive species (Tunison and Misaki, this volume). Where access to problem weed populations is difficult or when control methods that are potentially hazardous to the environment or workers are used (*e.g.*, application of restricted chemicals), reliance on volunteers may not be feasible. This has been the case at Waikamoi Preserve in most aspects of gorse (*Ulex europaeus*) and blackberry (*Rubus argutus*) control.

The current staff and volunteer work activity of Conservancy preserves is not adequate to meet the full challenges of alien plant control. Present programs are especially inadequate in dealing with widespread or high-biomass weed problems (*e.g.*, alien tree plantations or widespread

Table 1. Current target alien plants at Kamakou and Waikamoi Preserves.

Species	Common name	Kamakou	Waikamoi	Management Status	Prevalence
Acacia mearnsii	Black wattle	X		Controlled with some resprout	Localized
Acacia melanoxylon	Australian blackwood		X	Not managed	Localized
Casuarina equisetifolia	Ironwood	X		Not managed	Localized
Cirsium vulgare	Bull thistle	X	X	Not managed/ not managed	Localized/ localized
Clidemia hirta	Clidemia	X		Mostly controlled	Localized
Corynocarpus laevigatus	Karakanut	X		Controlled; reduced to seedling stage	Localized
Ehrharta stipoides	Meadow ricegrass		X	Not managed	Localized
Eucalyptus robusta	Swamp mahogany	X		Not managed	Localized
Eucalyptus saligna	Sydney blue gum		X	Not managed	Localized
Fraxinus uhdei	Mexican ash	X	X	Not managed/ controlled w/ some resprout	Localized/ localized
Grevillea banksii	Kāhili flower	X		Not managed	Localized
Grevillea robusta	Silk oak	X		Not managed	Localized

Species	Common name	Kamakou	Waikamoi	Management Status	Prevalence
Hedychium coronarium	White ginger	X		Ongoing control	Widespread
Hedychium flavescens	Yellow ginger	X		Ongoing control	Widespread
Hedychium gardnerianum	Kāhili ginger		X	Ongoing control	Localized
Lophostemon confertus	Vinegar tree	X		Ongoing control	Localized
Melaleuca quinquenervia	Paperbark	X		Controlled in one population	Localized
Opuntia ficus-indica	Prickly pear	X		Not managed	Localized
Passiflora edulis	Liliko'i	X		Mostly controlled	Localized
Pennisetum clandestinum	Kikuyu grass		X	Mostly controlled	Localized
Phormium tenax	New Zealand flax	X		Not managed	Localized
Pinus spp.	Pines	X	X	Not managed/ mostly controlled	Localized/ localized
Psidium cattleianum	Strawberry guava	X		Not managed	Localized
Psidium guajava	Common guava	X		Not managed	Widespread
Rubus argutus	Blackberry	X	X	Controlled/ not managed	Localized/ widespread

Table 1, continued.

Table 1, continued.

Species	Common name	Kamakou	Waikamoi	Management Status	Prevalence
Schinus terebinthifolius	Christmas berry	X		Not managed	Localized
Syncarpia glomulifera		X		Not managed	Localized
Syzygium jambos	Rose apple	X		Controlled	Localized
Ulex europaeus	Gorse		X	Ongoing control	Localized

grasses). The Conservancy has not yet undertaken major alien plant removal from sizable areas of rich native habitat, focusing limited resources instead on localized populations of the most worrisome alien plant species.

The Nature Conservancy of Hawaii has enjoyed the cooperation of State and Federal research agencies in field programs to develop or refine treatment methods for weeds in native plant communities. The Conservancy does not have the capability to conduct these and other research activities on its own and will continue to rely on this sort of interagency support. Field trials have been conducted or are under way for treatment of blackberry, gorse, strawberry guava (*Psidium cattleianum*), tropical ash (*Fraxinus uhdei*), and several other species under the auspices of the University of Hawaii Cooperative Extension Service. The U.S. Soil Conservation Service has provided valuable input on gradual removal of alien tree plantations, and the Hawaii Volcanoes National Park Division of Resources Management has shared information on control methods and organization of workloads (Taylor, this volume; Tunison and Zimmer, this volume). Volunteers from Colorado College and from Chicago's McHenry County Conservation District have conducted extensive field surveys of priority weeds and have studied the population biology of tropical ash. The Conservancy hopes to establish long-term collaborative relationships with these and other institutions to bring funds and research personnel to bear on critical alien plant management problems in Hawai'i by offering appropriate Conservancy preserves as field research stations for interested workers.

CURRENT PRIORITIES FOR ALIEN PLANT CONTROL ON TWO PRESERVES IN HAWAI'I

The Kamakou and Waikamoi preserves differ in the character and extent of their alien plant control problems. At Kamakou, 169 of the 442 known vascular plant species (38%) are aliens. Of these, 16 are regarded as dangerously invasive and requiring eradication or intensive control, while several others (mostly plantation trees) require some control but do not immediately threaten the integrity of the preserve. Most of these 16 weeds are accessible to volunteer or staff crews, and many are still localized enough that complete eradication is a feasible goal.

At Waikamoi Preserve on Maui, 71 (21%) of the 343 known vascular plant taxa are aliens. Eight of these are considered to be habitat-modifying weed species that should be controlled and monitored to prevent their further spread. Unlike Kamakou, much of Waikamoi is too rugged or inaccessible for volunteer work parties, although volunteers do provide the majority of labor for control projects in the accessible management units.

Alien plants on Conservancy preserves that are of special interest are documented below. The listing is not comprehensive, but it is intended to familiarize other workers in alien plant control with the range of preserve control programs. A complete list of priority weeds on these two preserves is given in Table 1.

Blackberry (*Rubus argutus*)

Blackberry is the alien plant of greatest concern at Waikamoi Preserve on Maui, where the species is found scattered throughout subalpine shrublands and portions of koa *(Acacia koa)* and 'ōhi'a *(Metrosideros polymorpha)* forests. Managers are uncertain of the effects that goat (*Capra hircus*) and pig (*Sus scrofa*) removal will have on this pest. Manual control followed by treatment of new growth with glyphosate (ROUNDUP) is used to keep trails clear and to control small populations in otherwise native vegetation. On Moloka'i, this weed is abundant in a fairly restricted area (about 500 a or 200 ha) of forest reserve near the Kamakou Preserve. Small, localized patches have been found and destroyed in the Preserve, indicating that blackberry is being disseminated into the island's native forests.

Tropical Ash (*Fraxinus uhdei*)

Occurring on both Preserves, tropical ash is a serious threat to mesic koa-'ōhi'a forests at Waikamoi and especially the diverse mesic forest remnants in the gulches of Kamakou Preserve. Ash escaped from tree plantations in Kamakou Preserve, is spreading rapidly along water courses, and is forming dense, nearly monotypic stands. Control efforts will be focused on preventing ash establishment in new drainages and on development of herbicide injection treatments for mature trees. A serious concern is disposal of the large numbers of dead trees. Logging followed by revegetation with natives and treatment of ash seedlings may be the only feasible approach to large-scale ash control in established stands.

Koster's Curse or Clidemia (*Clidemia hirta*)

A few individuals of this species have been found on Kamakou Preserve since 1979. All were located in a heavily-used public area of the Preserve, indicating that the original introduction may have been by humans. Otherwise, clidemia is presently widespread in Wailau and Pelekunu valleys on Moloka'i. Without development of an effective control method for large, established populations and a coordinated, island-wide control strategy, this species will probably overrun the understory in much of Moloka'i's native upland forest.

Karakanut (*Corynocarpus laevigatus*)

Kamakou Preserve contains what is believed to be the only population of karakanut on Moloka'i. This invasive, alien, subcanopy tree has become a serious localized weed on Kaua'i. At Kamakou, chemical control (with ROUNDUP applied on cut stumps or frilled trunks) of mature trees and pulling of seedlings and saplings by volunteers over the past three years have nearly eliminated this species.

New Zealand Flax (*Phormium tenax*)

This native of New Zealand has spread over about 10 a (4 ha) of very wet forest terrain at Kamakou. Volunteer crews have cut back and injected with ROUNDUP at least 60% of the original infestation and have eliminated several outlying smaller populations.

Molasses Grass (*Melinis minutiflora*)

While no control efforts are currently under way, this alien grass is perhaps the principal concern in preservation and restoration of dry and mesic shrublands and forests in Kamakou Preserve below about 3,700 ft (1,125 m) elevation. As the Conservancy and other managers of natural areas look toward protection of remaining dry lowland areas, control of molasses grass and the wildfire that is part of its ecology will be a critical stewardship issue.

Pinus spp.

Control of pines and other conifers escaping from established stands in both Kamakou and Waikamoi Preserves is important to maintenance of both low-elevation and subalpine shrublands. While removal of individual trees is technically quite simple, the control effort is made difficult by the steep terrain the trees are invading. In the long run, complete removal of source tree stands from preserves may be the only effective solution. The Conservancy's concern is that these stands be removed in a way that allows native vegetation to become reestablished and avoids accumulation of large pine slash and detritus fuel loads. At Kamakou, The Conservancy is contemplating gradual thinning of stands, with thinned trees sold or donated for biomass energy products or rough lumber. Most trees in these stands are not of millable size.

RECOMMENDATIONS

Clearly, the chief concern of Conservancy managers, like managers of any natural area, is to prevent the introduction of any new, uncontrollable weeds. Unfortunately, the control of species not yet established in preserves is something natural resource managers have very little ability to effect. Managers are generally limited in their jurisdiction by the legal boundaries of the property they manage, while weeds are not held to the same limits. For this reason, it is recommended that resource managers from private and government entities cooperatively control priority alien plant species or populations. These include weeds that threaten forests and other native ecosystems and that are not currently the focus of agricultural weed control efforts. Such an interagency task force would cooperate to eliminate localized infestations before they become well established on an island or in a nature preserve, natural area reserve, or park. The task force, to be effective, must be allowed to address serious weed problems wherever they occur. The idea is to not restrict alien plant control along political boundaries that the target species does not observe. Such an approach could result in a more effective statewide control effort and allow for efficient pooling of limited management resources and knowledge. Such an interagency group is the Interagency Commission on Biocontrol of Forest Pests, now developing strategy and funding for biological control of a few weeds.

As managers face the many years of alien plant control ahead in Hawai'i, it is important that control methods be developed that minimize nontarget negative impacts on the native biota and the overall ecological health of the Islands. Research is needed to develop methods of revegetation with native species in formerly weed-dominated areas, and it should consider the use of invasive native species to outcompete alien populations. The latter idea requires further understanding of the roles of native species in plant succession.

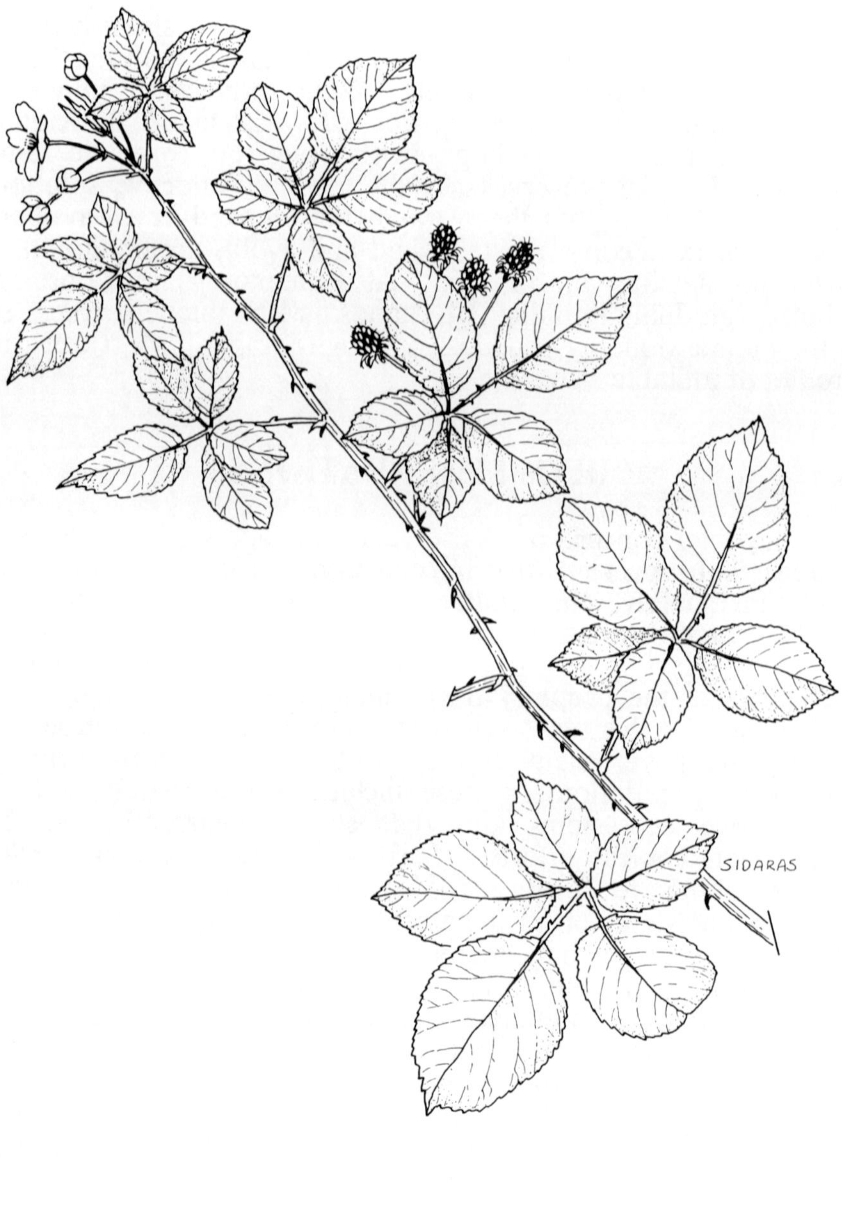

Literature Cited

Juvik, J.O., and S.P. Juvik. [this volume] Mullein (*Verbascum thapsus*): the spread and adaptation of a temperate weed in the montane tropics.

Loope, L.L., and P.G. Scowcroft. 1985. Vegetation response within exclosures in Hawai'i: a review. In *Hawai'i's terrestrial ecosystems: preservation and management*, ed. C.P. Stone and J.M. Scott, 377-402. Univ. Hawaii Coop. Natl. Park Resour. Stud. Unit. Honolulu: Univ. Hawaii Pr.

Stone, C.P., L.W. Cuddihy, and J.T. Tunison. [this volume] Responses of Hawaiian ecosystems to removal of feral pigs and goats.

Taylor, D. [this volume] Controlling weeds in natural areas in Hawai'i: a manager's perspective.

The Nature Conservancy. 1984. *Stewardship manual.*. Arlington, Virginia: Private pub.

Tunison, J.T. [this volume] Alien plant control strategies in Hawaii Volcanoes National Park.

Tunison, J.T., and E.T. Misaki. [this volume] The use of volunteers for alien plant control at Hawaii Volcanoes National Park and The Nature Conservancy's Kamakou Preserve.

Tunison, J.T., and C.P. Stone. [this volume] Special Ecological Areas: an approach to alien plant control in Hawaii Volcanoes National Park.

Tunison, J.T., and N.G. Zimmer. [this volume] Success in controlling localized alien plants in Hawaii Volcanoes National Park.

ALIEN PLANT CONTROL ON STATE LANDS INCLUDING NATURAL AREAS

Victor M. Tanimoto and Winona P. Char

ABSTRACT

The priorities for weed control by the Hawaii Division of Forestry and Wildlife are clidemia (*Clidemia hirta*) and banana poka (*Passiflora mollissima*), at present (1989). A discussion of the selection system for weed control projects and the types of control actions employed is presented, along with suggestions for improving the forest pest management program. In 1978, the Natural Area Reserves System was authorized by Chapter 195 of the Hawaii Revised Statutes, to preserve and protect in perpetuity unique and representative samples of Hawai'i's natural environment. Management policy was adopted in March 1984. Efforts were initially focused on identifying and setting aside natural area sites for preservation. Greater effort is now being directed towards management of these reserves, which number 18 and encompass 108,461 acres (43,894 ha). Alien plant management programs for two Natural Area reserves on O'ahu (Mt. Ka'ala and Pahole) are described herein. Different management approaches were taken due to the differences in size and accessibility of each reserve and the differences in the severity of nonnative plant infestations. Specific management problems and solutions and research needs of the two reserves are emphasized.

INTRODUCTION

Alien plant management on lands managed by the Hawaii Division of Forestry and Wildlife in the Department of Land and Natural Resources can be separated into Forest Reserve programs and Natural Area Reserves System programs. Forest reserves are managed for consumptive uses such as forestry, grazing, and hunting, while Natural Area reserves are managed for preservation of native ecosystems. Hunting of alien ungulates and birds is allowed on Natural Area reserves at present to provide recreational opportunities, as a public relations gesture, and to achieve control (of ungulates) where possible.

FOREST RESERVES

Alien plant control on Forest reserves consists of: 1) small-area actions, such as trail and arboretum maintenance, and clearing of tree-planting areas; and 2) extensive-area treatment of widespread weeds with biological control methods. Small-area actions are accomplished routinely by Division field office personnel. Extensive-area treatments are the responsibility of the Division forest entomologist and presently emphasize control of the forest weeds clidemia (*Clidemia hirta*) and banana poka (*Passiflora mollissima*). Other weeds chosen for control by Division field offices are listed by island in Table 1. Progress for these species will be discussed in this paper only for Natural Area reserves.

Forest Pest Management

Forest pest management became a program within the Division during the early 1970s, when the position of Forest Entomologist was created through a cooperative agreement with the U.S. Forest Service. The Agreement, executed in November 1968, provides for the prevention, detection, evaluation, and suppression of forest insect pests and plant diseases. Originally written for the implementation and management of control programs for these forest pests, the agreement now extends to control programs for other forest pests as well, including weeds. The objective of Forest Pest Management is to reduce (or prevent) damage, loss, and impacts caused by pests (insects, plant diseases, vegetation, drought, or mammals) on all state-owned forests to levels commensurate with resource management objectives (U.S. Forest Service 1981). The Forest Pest Management staff is to provide overall leadership in the Integrated Pest Management process to maximize pest management effectiveness and efficiency while minimizing adverse impacts on humans and the environment (U.S. Forest Service 1981).

The Forest Pest Management program is based in large part by the following laws and documents:

1. Chapter 344, Title 19 Health, Hawaii Revised Statutes 1976, The State Environmental Policy:

> It shall be the policy of the State, through its programs, authorities, and resources to conserve the natural resources, so that . . . natural resources are protected by . . . preserving or augmenting natural resources, and by safeguarding the State's unique natural environmental characteristics.

2. Chapter 183, Title 12 Conservation and Resources, Hawaii Revised Statutes 1976:

> The Department of Land and Natural Resources shall . . . have the care, custody, control, and regulation of all lands which may be set apart as forest reservations . . . [and] devise ways and means of protecting, extending, increasing, and utilizing the forests and forest reserves, more particularly for protecting and developing the springs, streams, and sources of water supply.

Table 1. Weeds listed for control by four Division of Forestry and Wildlife field offices, 1975-present.

Species	Island				
	HI	MA	MO	K	O
Acacia confusa (Formosan koa)		X			
Acacia farnesiana (klu)		X			
Acacia mearnsii (black wattle)		X	X		
Agave sisalana (sisal)				X	
Ageratina adenophora (Maui pāmakani)		X			
Ageratina riparia (Hāmākua pāmakani)	X				
Andropogon virginicus (broomsedge)					X
Ardisia elliptica (ardisia)		X			
Asclepias physocarpa (balloon plant)	X				
Bambusa vulgaris (feathery bamboo)		X			
Caesalpinia decapetala (cat's claw)					X
Clidemia hirta (clidemia, Koster's curse)	X				
Dicranopteris linearis (false staghorn fern)				X	
Hibiscus tiliaceus (hau)		X	X	X	
Ipomoea spp. (morning glory)					X
Lantana camara (lantana)	X	X		X	X
Melastoma candidum (Indian rhododendron)	X			X	
Melastoma sanguineum (fox-tongue melastoma)	X			X	
Melinis minutiflora (molasses grass)					
Melochia umbellata (melochia)	X				
Myrica faya (firetree)	X	X		X	X
Paederia scandens (maile pilau)	X				
Panicum repens (torpedo grass)	X				
Paspalum urvillei (vasey grass)				X	
Passiflora mollissima (banana poka)	X				
Pennisetum setaceum (fountain grass)	X				
Pluchea symphytifolia (pluchea)	X				
Psidium cattleianum var. *cattleianum* (strawberry guava)	X		X	X	X
Psidium cattleianum f. *lucidum* (waiawī)	X	X		X	X
Psidium guajava (common guava)	X				
Pyracantha angustifolia (firethorn)		X			
Rhodomyrtus tomentosa (downy rose myrtle)				X	
Rubus argutus (blackberry)	X	X	X	X	
Schinus terebinthifolius (Christmas berry)	X	X		X	
Senecio mikanioides (German ivy)	X				
Setaria palmifolia (palmgrass)	X				X
Stachytarpheta dichotoma (cayenne vervain)	X				
Syzygium cumini (Java plum)	X				
Tibouchina urvilleana (glorybush)	X				
Ulex europaeus (gorse)	X	X			

3. "A Multiple Use Program for the State Forest Lands of Hawaii," summarizing a change in the forestry program from an almost-exclusively watershed protection emphasis to one emphasizing sustained yields and a variety of resources (Hawaii Department of Agriculture 1962).

4. "A Program for the State Forest Lands of Hawaii," summarizing a new forest management plan that defines the objectives needed to carry out the concepts advocated in the publication "A Multiple Use Program for the State Forest Lands of Hawaii" (Hawaii Department of Land and Natural Resources 1975).

The Hawaii Department of Agriculture administers the state's weed laws, which are regulatory in function. The Division of Forestry and Wildlife is partner in four cooperative agreements that deal directly with weed control:

1. An agreement with the Hawaii Department of Agriculture in 1980, emphasizing working for the "common good [of] the flora and fauna of the State."

2. An agreement with the Hawaii Department of Agriculture in 1984, emphasizing the study of potential biological control insects of clidemia.

3. An agreement with the University of Hawaii in 1983, emphasizing the exploration for potential biological control insects and plant diseases for the control of banana poka.

4. A Memorandum of Understanding in 1983 with the U.S. Forest Service, National Park Service, University of Hawaii, and Hawaii Department of Agriculture emphasizing the biological control of forest pests on public lands in Hawai'i.

The biological control programs for clidemia and banana poka necessitate multi-agency approval. Within the context of pest control cooperation, the Forest Entomologist acts as a broker for the Division of Forestry and Wildlife, seeking funds, encouraging cooperative work, and fostering cooperation.

Integrated Pest Management

Integrated Pest Management is a part of forest management that deals with regulation of pest populations to minimize their effects on resources such as water, soil, and recreation opportunities in an ecologically sound manner. Integrated Pest Management consists of a decision-making process and action alternatives based on the following considerations: 1) the ecology of the host and its pests throughout forest succession; 2) land management objectives; and 3) economic values of the subject resource. Monitoring of pest populations and environmental factors that influence them is required in order to determine when action is needed to reduce excessive losses of the resource. Action is also taken in response to chronic or catastrophic losses.

Improving the Forest Pest Management Program

The Forest Pest Management Program has made good progress in the control of clidemia and banana poka in the 15 years since its inception. Although the staff has learned much about pest control in Hawai'i's forests and about program administration, improvement or refinement in some areas is necessary. Suggestions are as follows:

Total program management:

1. Review and adjust periodically the overall strategy of the Forest Pest Management Program.

2. Collect, collate, and implement all feasible suggestions and ideas that are provided (*e.g.*, Stone and Stone 1984; Hawaii Department of Land and Natural Resources 1985; Smith 1985).

3. Improve the "Information and Education" system for disseminating information about weed problems to the public, legislators, private land managers, and other government agencies. This is important to assure funding of control programs.

4. Promote international cooperation and participation in weed control programs, especially biological control (*e.g.*, New Zealand on the biological control of gorse, *Ulex europaeus*).

5. Implement research that would answer such questions as: Can forests be managed to increase the water supply? What effects do weeds have on each forest value? What are the economics of weed control?

Prevention of alien plant introduction:

1. Work closely with Hawaii Department of Agriculture in strengthening weed laws and rules and plant importation procedures.

2. Prepare fact sheets on plant species that are known to be pests elsewhere and that have potential for becoming pests in Hawai'i.

3. Improve the data-gathering system to detect alien plant problems.

Biological evaluation:

1. Investigate the effects of various weeds on the water supply, both in quality and quantity, and protocol to deal with space created suddenly when a biological control agent does its work swiftly.

2. Develop a system to rate weeds on the basis of their detrimental effects for comparative purposes.

Control of weeds:

1. Develop adequate technical knowledge on the control of all weeds, using all methods of control.

2. Develop ideas and methods to control seed dispersal by birds, mongooses (*Herpestes auropunctatus*), feral pigs (*Sus scrofa*), and other

agents. (See Hawaii Department of Land and Natural Resources 1984 and Hawaii Department of Land and Natural Resources 1988.)

NATURAL AREA RESERVES SYSTEM

Presently (1986), 18 reserves (Table 2), totalling approximately 108,461 acres (43,894 ha), are administered by the Natural Area Reserves System. Many reserves were originally in the State Forest Reserve System discussed in the preceding section. The reserves encompass an array of diverse native ecosystems that include freshwater, marine, and coastal environments, wet, mesic, and dry forests, bogs, grasslands, and alpine scrub. The Natural Area Reserves System preserves the critical habitat of a number of rare and endangered plant and animal species.

Legal and Administrative Framework

The Natural Area Reserves System was established by Hawaii Revised Statutes Chapter 195 in 1970 to "preserve in perpetuity specific land and water areas which support communities, as relatively unmodified as possible, of the natural flora and fauna, as well as geological sites, of Hawaii."

The System is managed by the Hawaii Division of Forestry and Wildlife (the Division). Any state-owned land under the jurisdiction of the Department of Land and Natural Resources (the Department) may be designated a Natural Area Reserve, subject to the approval of the Governor by executive order. New Natural Area reserves may also be established by approval of gift, device, or purchase; by eminent domain; and by setting aside of other state-owned land by the Governor. The Department is responsible for making and enforcing all rules and regulations governing the use, control, and protection of the areas included within the Reserves System.

Chapter 209 of Title 13, Administrative Rules, regulates activities within the reserves. Everything within a Natural Area Reserve is protected except game mammals and birds, which may be hunted according to Department rules. No form of plant or animal life, except hunting dogs, may be introduced to a Reserve. Use of reserves is restricted to non-destructive activities, such as hiking, nature study, photography, and camping without tents or other temporary structures. A Special-Use Permit to conduct activities otherwise prohibited within a Natural Area Reserve for research, education, and management purposes may be issued by the Department.

An 11-member Natural Area Reserves Commission recommends to the Governor and the Department areas suitable for inclusion within the System, as well as policies regarding required controls and permitted uses of areas that are part of the Reserves System. The Commission also advises the Governor and the Department on any matter relating to the preservation of Hawai'i's unique natural resources; and develops ways and means of extending and strengthening presently established preserves, sanctuaries, and refuges within the State. The Natural Area Reserves program provides administrative and technical staff assistance to the Commission. The

Table 2. Reserves in the State of Hawaii Natural Area Reserves System (as of June 1986).

Natural Area Reserve	Acres	Primary Feature
Island of Hawai'i		
Kīpāhoehoe	5,583	Mesic forest, 'ōhi'a and koa wet forests
Laupāhoehoe	7,894	'Ōhi'a wet forest
Manukā	25,550	Dry and mesic forests, 'ōhi'a wet forest
Mauna Kea Ice Age	3,894	Geological: Ice Age and Lake Waiau
Pu'u Maka'ala	12,106	'Ōhi'a and koa wet forests
Pu'u o 'Umi	10,182	'Ōhi'a wet forest, montane bog, headwaters of six perennial streams
Waiākea 1942 Lava flow	640	Developing wet forest
Wao Kele o Puna	16,844	'Ōhi'a wet forest
Island of Maui		
'Āhihi-Kīna'u	1,238 (land) 807 (water)	Anchialine pools, last lava flow on Maui; marine ecosystem
Hanawī	7,500	'Ōhi'a wet forest, cloud forest, alpine grassland
West Maui:		
Kahakuloa Section	3,275	Pili grassland, mesic forest,
Honokōwai Section	750	'ōhi'a wet forests, cloud forests,
Pana'ewa Section	1,717	montane bogs, headwaters and upper
Līhau Section	960	reaches of four perennial streams
Island of Moloka'i		
Oloku'i	1,635 (approx.)	Mesic forest, wet shrub, 'ōhi'a wet forest
Pu'u Ali'i	1,340 (approx.)	Wet shrub and 'ōhi'a wet forest
Island of O'ahu		
Ka'ena Point	12	Coastal dune ecosystem
Mt. Ka'ala	1,100	Mesic and cloud forests, montane bog
Pahole	658	Dry and mesic forests
Island of Kaua'i		
Hono o Nā Pali	3,150	Mesic forest, 'ōhi'a wet forest
Ku'ia	1,636	Dry and mesic forests

Natural Area Reserves System program was staffed by an executive secretary, a half-time clerk-typist, and two biologists on temporary-hire status (1987). Additional positions have been created and filled since that time.

Management Approach

In the initial stages of development, the Natural Area Reserves System program focused primarily on identifying and setting aside natural area sites for preservation. Having set aside a number of reserves, greater effort is now being directed towards their management. Management of the Natural Area Reserves System is guided by a Management Policy Plan, adopted in March 1984. Its purpose is to direct formulation and implementation of site-specific operational plans. Management plans have never been prepared for most reserves under contract with The Nature Conservancy of Hawaii. Two principles determine the course of management activities for the System: 1) Natural processes should control ecosystems; and 2) Native species are resources that should be conserved for esthetic, heritage, educational, and scientific values.

Management objectives for Natural Area reserves fall into three categories: managing biological disturbances, development of a database to facilitate the accomplishment of this, and managing human impacts. The major requirements in managing biological disturbances are: 1) Eliminate ungulate populations on reserves. Hunting may be used as an interim measure in accordance with Department hunting rules and policies in certain areas, but eradication of ungulates in the reserves is the goal. 2) Reduce or eliminate established alien biota in accordance with a site-specific operational plan. 3) Restrict establishment of alien biota in accordance with a site-specific operational plan.

Database and human impact information requirements are: to develop a natural resource inventory, to monitor areas of heavy human use or areas where changes are likely to occur, and to monitor the occurrence and impact of alien plant and animal species. Gathering baseline information on the resources of each reserve is the first step in management. One cannot manage an area for alien species until one knows just what species are present, their abundance and status, and their effects upon the native components.

Alien Plant Management

The general approach to management of alien plant species on the reserves can be summarized in four words -- gather, evaluate, propose, and implement. Vegetation mapping, species inventory, and collecting data on phenology and population structure are necessary activities for obtaining baseline information. Data from baseline surveys are then evaluated, and alien plants that threaten, or have the potential to threaten, the continuing quality of the reserve are targeted for management. The alien plant species selected for management can usually be placed in three groups: 1) Incipient threats, which have a limited distribution and are found in small numbers. 2) Minor threats, which appear to be more or less stable in population size and distribution; they do not appear to be a threat to the native components but should be monitored. 3) Major threats, which aggressively compete with native species, occur in large numbers, and/or are widespread throughout the reserve.

The baseline information that has been gathered and evaluated is then used to propose a site-specific program for monitoring, control, and eradication of alien plant infestations. Implementation of the proposed program is the final step in management.

Initially, the Natural Area Reserves System management program was directed towards gathering baseline data on the distribution and abundance of alien plant species and evaluating their immediate or potential threat on two Oʻahu reserves. A comprehensive list of species and a map and descriptions of the major vegetation types were prepared for these reserves. Baseline surveys of this type have now been conducted on most of the 18 reserves. A summary of management accomplishments on the Kaʻala and Pahole Natural Area reserves on Oʻahu follows.

Mt. Kaʻala Natural Area Reserve -- Summit Plateau. The Mt. Kaʻala Natural Area Reserve summit plateau remains primarily a native ecosystem, with the alien plant species confined largely to areas disturbed by humans or feral pigs. Two vegetation types are recognized there. Wet scrub forest covers most of the summit plateau and consists of a stunted native forest, roughly 6 to 10 ft (2-3 m) in height, that is closed and densely vegetated. ʻŌhiʻa *(Metrosideros polymorpha)* makes up 40 - 50% of the forest cover, with a rich array of shrubs and ferns forming the remaining cover. Liverworts and mosses are also abundant. Roadside vegetation, dominated by alien species, comprises the second type. It is found along the paved access roads and the radar/communication facilities located on the summit plateau. Roadsides are periodically maintained and consist of open, grassy areas with scattered shrubs and ferns.

Of the 131 plant species inventoried during the survey, 66 (50%) are alien. Fifty of these alien species are restricted to the roadside habitat, 13 occur in both the scrub forest and roadside habitat, and three occur only in the scrub forest. Nine of the 66 alien species (13.7%) were identified as immediate or potential threats to the native scrub forest. These species can be categorized in three groups, as discussed above:

1. Incipient Group: These plants occur in small numbers and are found in or near disturbed areas that are easily accessible (e.g., along trailsides and roadsides). Species include: clidemia or Koster's curse, Hāmākua pāmakani (*Ageratina riparia*), tea tree (*Leptospermum scoparium*), koa haole *(Leucaena leucocephala)*, strawberry guava (*Psidium cattleianum*), and montbretia (*Tritonia crocosmiflora*).

2. Monitor Group: Maui pāmakani (*Ageratina adenophora*) and kikuyu grass (*Pennisetum clandestinum*), the two plants in this group, are found in the roadside vegetation or in the scrub forest adjacent to the roadside where the forest had been disturbed sometime in the past. The two species appear to be more or less stable. The tephritid gall flies introduced to control Maui pāmakani seem to be effectively repressing the growth of this species on the summit.

3. Major Threat Group: Blackberry (*Rubus argutus*) is the only alien species on the plateau that is actively competing with and displacing the native plant growth. Dense thickets are found adjacent to roadsides

and other disturbed areas. Infestation becomes less dense away from these areas.

A nonnative plant eradication program for the Reserve's summit plateau has been drafted, and management efforts have begun on incipient infestations. All species in the incipient group, with the exception of montbretia, have either been uprooted and removed or severely cut to ground level. The sites where these plants occurred are being monitored at six-month intervals for new seedlings or resprouting.

Unless an effective biological control agent is found, present management plans for blackberry on the Reserve's summit will require the use of herbicides. Areas with heavy infestations are targeted for an intensive eradication program; a more selective approach will be used in light to very light infestations.

Pahole Natural Area Reserve. The ʻōhiʻa dieback phenomenon, which began in the early 1970s on this Reserve, has had a tremendous impact on the quality and species composition of the mesic forest. This mesic forest was, at one time, considered one of the best examples of this forest type on the island of Oʻahu. Approximately 187 acres (76 ha) of the reserve's 658 acres (266 ha), or 28%, contains large tracts of standing dead ʻōhiʻa trees. The dieback is generally confined to the older trees, and large tracts of standing dead trees are associated with the more or less broad and gentle slopes. Smaller, scattered patches of dieback are generally confined to the upper slopes of the more narrow ridges, and there is poor regeneration of ʻōhiʻa. Christmas berry (*Schinus terebinthifolius*) often quickly invades dieback areas once the canopy is opened, and the species forms dense thickets. At lower elevations, dry forest species are scattered among Java plum (*Syzygium cumini*), common guava (*Psidium guajava*), and Christmas berry. The alien species are dominant at lower elevations.

The vegetation on the Reserve is in a state of flux. Unfortunately, the prognosis is grim. The trend is towards replacement of the native elements by alien species, principally Christmas berry or a mixture of Christmas berry, common guava, strawberry guava, and lantana (*Lantana camara*) in mesic forests, and Java plum, common guava, and Christmas berry in the dry forests. Pockets of more or less undisturbed dry and mesic forest exist on the steep slopes of certain gulches and ridges. Because the terrain is so steep, these areas have sustained very little pig damage, and alien species are few and scattered. While the Mt. Kaʻala Summit Plateau Management Plan focuses on the eradication of alien species, the Pahole Reserve Plan will focus on managing those areas with remnant pockets of native forests and endangered plant species.

Pahole Gulch has been recommended as critical habitat for two candidate endangered plant species. Cyanea (*Cyanea superba*) is currently being reviewed by the U.S. Fish and Wildlife Service and is expected to be officially listed soon (D. Herbst, pers. comm.). A status packet for mehamehame *(Flueggea neowawraea)* has been completed and is currently being processed. Preservation of critical habitat calls for an active management program eliminating pigs and controlling aggressive alien plant species.

Two hundred fifty-two plant taxa were inventoried during the survey of Pahole Natural Area Reserve. Of these, 108 (43%) are alien species. Twenty-six (24%) of the 108 species are categorized as immediate or potential threats on the Reserve, as shown in Table 3. The seven alien species in the major threat group are present in large numbers and are widespread throughout the Reserve. Control of these species on a reserve-wide basis would be cost and time ineffective. Incipient species, however, can be controlled by physical removal or by limited herbicide application for the larger trees and shrubs.

Natural Area Reserves Research and Management Needs

The first steps in active management of alien species were initiated on two of the State's Natural Area reserves in 1986. Baseline surveys and site-specific management plans are now being completed for all 18 reserves within the System. The effectiveness of management will depend on applying the information gathered from baseline surveys, research, and management experience, to produce a site-specific program for each area, which, when implemented, will reduce or eliminate alien species. Four fundamental needs of the Natural Area Reserves System management program are: 1) long-term commitment and support from the State; 2) research on alien species; 3) research on the interaction between native and alien species; 4) coordination of herbicide, research, and management information.

Long-term commitment and support from the State. The future of the Natural Area Reserves management program depends on a long-term commitment from the State. A total commitment to the perpetual management of the reserves is necessary. Lack of support has been mentioned previously (Holt and Fox 1985; Smith 1985; Stone and Scott 1985). Unfortunately, the Natural Area Reserves program is at times a political football. An active alien plant management and research program is discouraged by a cumbersome, time-consuming bureaucracy (Smith 1985). Aggressive alien species continue to increase and spread despite budgetary constraints and lack of support; the longer alien plants are allowed to continue unmanaged, the longer and more expensive the remedy and the less likely damaged ecosystems will recover. An intensive Natural Area Reserves System management program will need the "muscle" and equipment support from the Division of Forestry and other Department divisions. The Natural Area Reserves System itself lacks a logistical support system.

Research on alien species. Weedy alien species are discussed in a number of weed manuals, reports on plant surveys, and flora descriptions (Hosaka and Thistle 1954; Hawaii Department of Agriculture 1962, 1979; Neal 1965; Pope 1968; Haselwood and Motter 1983). However, the literature is largely for use by agriculturists, ranchers, and gardeners. Smith (1985) provided a useful annotated list of 86 alien plant species considered pests in Hawaiian native ecosystems. For the majority of these species, there is a lack of basic background information, such as the behavior of these species in their native habitat, their ecological tolerances, mode of reproduction, and dispersal agents. The methods used for management will

Table 3. Immediate and potential alien plant threats to Pahole Natural Area Reserve.

Incipient Group:
- *Acacia mearnsii* (black wattle)
- *Cenchrus ciliaris* (buffelgrass)
- *Clidemia hirta* (Koster's curse)
- *Flindersia brayleyana* (silkwood)
- *Fraxinus uhdei* (tropical ash)
- *Melaleuca quinquenervia* (paperbark)
- *Melia azedarach* (pride of India)
- *Montanoa hibiscifolia* (montanoa)
- *Samanea saman* (monkeypod)
- *Schefflera actinophylla* (octopus tree)
- *Setaria palmifolia* (palmgrass)
- *Spathodea companulata* (African tulip tree)

Monitor Group:
- *Ageratina adenophora* (Maui pāmakani)
- *Ageratina riparia* (Hāmākua pāmakani)
- *Andropogon virginicus* (broomsedge)
- *Buddleia asiatica* (Asiatic butterfly bush)
- *Ficus microcarpa* (Chinese banyan)
- *Lantana camara* (lantana)
- *Leucaena leucocephala* (koa haole)

Major Threat Group:
- *Grevillea robusta* (silk oak)
- *Melinis minutiflora* (molasses grass)
- *Psidium cattleianum* (strawberry guava)
- *Psidium cattleianum* var. *littorale* (waiawī)
- *Psidium guajava* (common guava)
- *Schinus terebinthifolius* (Christmas berry)
- *Syzygium cumini* (Java plum)

depend largely on knowledge of "the lifestyles of the noxious and aggressive" alien species. Knowledge of the life cycle of a species can be used to pinpoint the weakest link in that cycle.

There is also a need to be able to predict which of an ever-increasing number of new or incipient species have the potential to become a threat to native ecosystems. With the limited resources available, administrators must make hard decisions on choosing which of these species to control (Bridges, this volume; Taylor, this volume).

Research on the interactions between native and alien species. More management-oriented research is needed on native ecosystems and the role of alien plant species. We need to know why and how a supposedly stable ecosystem begins to collapse. We need to know what effects, if any, alien species have on ecosystem collapse, and if and how alien species affect recovery potentials of damaged ecosystems.

Management-related research such as that by Mueller-Dombois and his students (Gerrish and Mueller-Dombois 1980; Mueller-Dombois *et al.* 1980; Jacobi *et al.* 1983) on ‘ōhi‘a dieback should be supported. Information from these studies helps us better understand what is presently happening to the ‘ōhi‘a forests on the Pahole Natural Area Reserve.

Feral pigs are the dispersal agents for the three guava species on the Pahole Reserve, while the Japanese white-eye (*Zosterops japonicus*) is primarily responsible for the spread of Christmas berry. The increasing numbers of red-vented bulbul (*Pycnonotus cafer*) on the Reserve will also hasten the dispersal of Christmas berry. Studies on the role of alien fauna on the modification of native ecosystems and the spread of alien plant species should be supported.

Coordination of Herbicide, Research, and Management Information. A need exists for some sort of clearing house of information on alien plant research and management programs. Information on the types of control strategies that have been tried or are currently being employed is, at times, fragmented. We need to improve communication and share data and findings. With limited resources (a characteristic of many management programs), we do not each individually have to "re-invent the wheel." Ineffective strategies that waste time and money can be avoided. Knowledge gained through considerable effort elsewhere should be used (Stone 1985).

The University of Hawaii Agricultural Extension Service provides a yearly summary of herbicide trials for pasture, range, and non-crop weed control through its Hawaii Institute of Tropical Agricultural and Human Resources publication. Observations on the responses of treated plants to different herbicides, concentrations, and application methods used, as well as data record sheets, are presented. The Extension Service has developed and standardized herbicide application methods and a rating system of weed response to herbicides. Similarly, yearly reports summarizing preliminary data by the various agencies and organizations involved in alien plant research and management would greatly improve communication and coordination of information.

Literature Cited

Bridges, K.W. [this volume] Rule-based decision making: a way to determine which alien species to control.

Gerrish, G., and D. Mueller-Dombois. 1980. Behavior of native and non-native plants in two tropical rain forests on Oahu, Hawaiian Islands. *Phytocoenologia* 8(2):237-295.

Haselwood, E.L., and G.G. Motter, eds. 1983. *Handbook of Hawaiian weeds*, 2nd ed., revised and expanded by R.T. Hirano. Harold L. Lyon Arboretum. Honolulu: Univ. Hawaii Pr.

Hawaii Department of Agriculture. 1962. Noxious weeds of Hawaii. Plant Pest Control Branch, Hawaii Dep. Agric. Honolulu. Mimeo.

Hawaii Department of Agriculture. 1979. Foreign noxious weed survey. Plant Pest Control Branch, Hawaii Dep. Agric. Honolulu. Mimeo.

Hawaii Department of Land and Natural Resources. 1975. *A program for the State forest lands of Hawaii*. Div. Forestry, Hawaii Dep. Land Nat. Resour. Honolulu.

Hawaii Department of Land and Natural Resources. 1984. *Hawai'i's wildlife plan*. Div. Forestry and Wildlife, Hawaii Dep. Land Nat. Resour. Honolulu.

Hawaii Department of Land and Natural Resources. 1985. *Hawaii renewable resources research plan for the eighties*. Div. Forestry and Wildlife, Hawaii Dep. Land Nat. Resour. Honolulu.

Hawaii Department of Land and Natural Resources. 1988. *Threatened and endangered species plan for wildlife, plants and invertebrates*. Div. Forestry and Wildlife, Hawaii Dep. Land Nat. Resour. Honolulu.

Holt, R. A. and B. Fox. 1985. Protection status of the native Hawaiian biota. In *Hawai'i's terrestrial ecosystems: preservation and management*, ed. C.P. Stone and J.M. Scott, 127-141. Univ. Hawaii Coop. Natl. Park Resour. Stud. Unit. Honolulu: Univ. Hawaii Pr.

Hosaka, E.Y., and A. Thistle. 1954. *Noxious plants of the Hawaiian ranges*. Univ. Hawaii Agric. Ext. Serv. Bull. 62. Honolulu: Univ. Hawaii.

Jacobi, J.D., G. Gerrish, and D. Mueller-Dombois. 1983. 'Ohi'a dieback in Hawai'i: vegetation changes in permanent plots. *Pac. Sci.* 37:327-337.

Mueller-Dombois, D., J.D. Jacobi, R.G. Cooray, and N. Balakrishnan. 1980. *'Ohi'a rain forest study: investigations of the 'ohi'a dieback problem in Hawaii*. Hawaii Agric. Exp. Sta. Misc. Pub. 183. Univ. Hawaii Coll. Trop. Agric. and Human Resour., Honolulu.

Neal, M.C. 1965. *In gardens of Hawaii*. B.P. Bishop Mus. Spec. Pub. 50. Honolulu: Bishop Mus. Pr.

Pope, W.T. 1968. *Manual of wayside plants of Hawaii.* Rutland, Vermont: C.E. Tuttle Co.

Smith, C.W. 1985. Impact of alien plants on Hawai'i's native biota. In *Hawai'i's terrestrial ecosystems: preservation and management*, ed. C.P. Stone and J.M. Scott, 180-250. Univ. Hawaii Coop. Natl. Park Resour. Stud. Unit. Honolulu: Univ. Hawaii Pr.

Stone, C.P. 1985. Alien animals in Hawai'i's native ecosystems: toward controlling the adverse effects of introduced vertebrates. In *Hawai'i's terrestrial ecosystems: preservation and management*, ed. C.P. Stone and J.M. Scott, 251-297. Univ. Hawaii Coop. Natl. Park Resour. Stud. Unit. Honolulu: Univ. Hawaii Pr.

Stone, C.P., and J.M. Scott. 1985. Hawai'i's native ecosystems: importance, conflicts, and suggestions for the future. In *Hawai'i's terrestrial ecosystems: preservation and management*, ed. C.P. Stone and J.M. Scott, 495-516. Univ. Hawaii Coop. Natl. Park Resour. Stud. Unit. Honolulu: Univ. Hawaii Pr.

Stone, C.P., and D.B. Stone. 1984. The "10 Most-Wanted" management actions for terrestrial Hawaiian ecosystems: a survey. *'Elepaio* 45(6):41-46.

Taylor, D. [this volume] Controlling weeds in natural areas in Hawai'i: a manager's perspective.

U.S. Forest Service. 1981. *Proc. Forest Pest Management Workshop*, Kokee State Park, Kauai, Hawaii, March 16-18, 1981. San Francisco: Pac. Northwest For. Range Exp. Sta., USDA For. Serv.

ALIEN PLANTS IN HALEAKALA NATIONAL PARK

Lloyd L. Loope,
Ronald J. Nagata, and Arthur C. Medeiros

ABSTRACT

The alien plant problems of Haleakala National Park are substantial but may be fewer than those of other Hawaiian natural areas, partly because most of the Park is at high elevation (75% above 5,000 ft or 1,525 m). However, potential problems with alien plants are still not clear because of the suppression of all vegetation by intense herbivory by feral goats (*Capra hircus*), which has occurred until very recently. Alpine cinder desert and dense rain forest above 4,000 ft (1,220 m) have so far proved relatively resistant to invasion. High-elevation grasslands, bogs, and shrublands have undergone degradation and alien plant invasion as long as feral animals have been present, but these areas will undergo at least partial recovery as long as these ungulates are under control. In the absence of feral goats, native and alien plant species are rapidly increasing in leeward Kaupō Gap. The threat of wildland fire due to fuel buildup by spreading alien grasses has the potential to negate recovery. Control of feral pigs (*Sus scrofa*) may slow the spread of alien plants below 4,000 ft (1,220 m) in Kīpahulu Valley, but dispersal of strawberry guava (*Psidium cattleianum*) and other weeds will continue, and active vegetation management is needed to prevent the decline and disappearance of rare native species. The following alien plant species must be controlled to avoid future irreversible vegetation changes: strawberry guava (above 3,000 ft or 910 m), molasses grass (*Melinis minutiflora*), blackberry (*Rubus argutus*), Australian tree fern (*Cyathea cooperi*), kāhili ginger (*Hedychium gardnerianum*), gorse (*Ulex europaeus*), pines (*Pinus*, three species), and *Eucalyptus globulus*. Clidemia (*Clidemia hirta*), banana poka (*Passiflora mollissima*), and fountain grass (*Pennisetum setaceum*), among the most invasive alien species in Hawai'i overall, are present on Maui and threaten Haleakala National Park. These threats will be reduced but not eliminated as feral animals are controlled in the Park.

INTRODUCTION

Although alien plants create serious problems for preserving native ecosystems in Haleakala National Park, these problems are reduced somewhat because much of the Park is at high elevation (Fig. 1). About 75% is located above 5,000 ft (1,525 m); 80% is above 4,000 ft (1,220 m). Haleakalā has a rich native vascular flora in spite of its relatively small area (12% as large as Hawaii Volcanoes National Park) but has substantially fewer alien plant species (Table 1). Haleakala National Park largely lacks extensive areas of highly flammable alien grasses (*Andropogon virginicus, Schizachyrium condensatum*, and others), although approximately 1,235 a (500 ha) in Kaupō Gap and the western edge of the Manawainui planeze are vulnerable. The Park also lacks large areas susceptible to dominance by low-elevation alien tree and shrub species such as koa haole *(Leucaena leucocephala)*. Having stated that the alien plant problem at Haleakalā is relatively manageable, we believe real problems do exist and urgently require active management to avoid loss of species and ecosystems. In this paper, we attempt to assess the Park's alien plant problems and describe a program to deal with the worst of them.

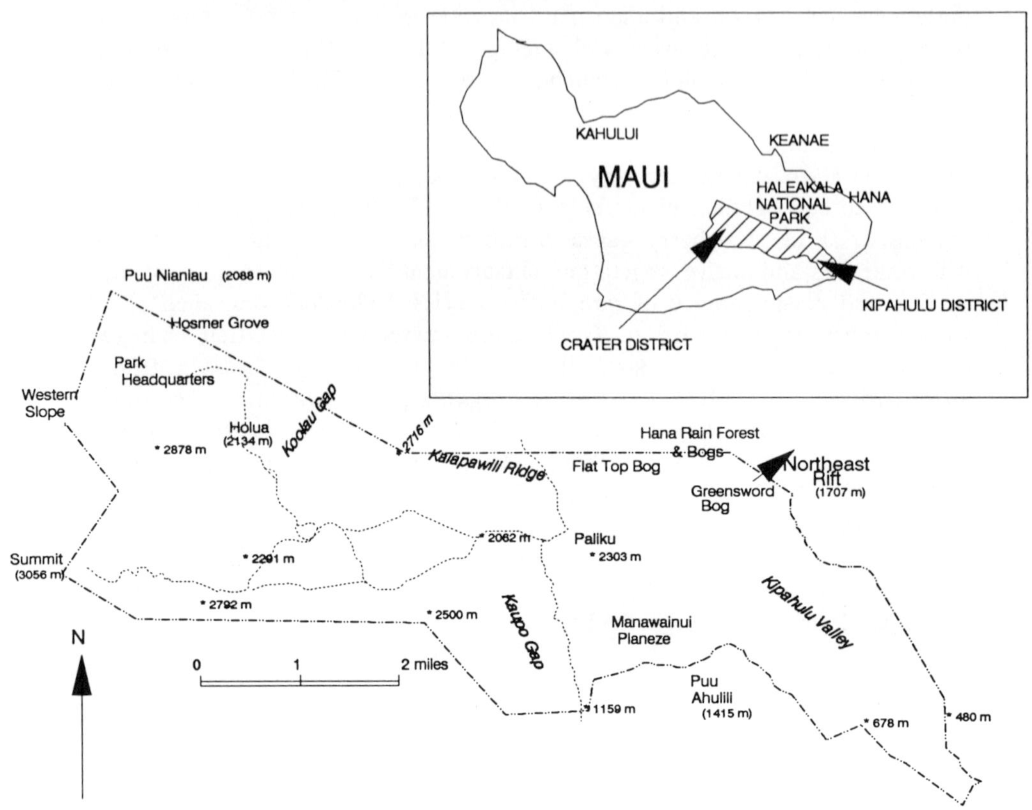

Figure 1. Haleakala National Park (elevations in meters).

Table 1. Comparison of vascular floras of Haleakala and Hawaii Volcanoes National Parks.*

	Haleakala	Hawaii Volcanoes
Park area (ha)	11,400	93,000
% of park area above 6,000 ft (1,500 m)	ca. 75	ca. 20
Number of ferns and fern allies		
Indigenous	41	35
Endemic	63	60
Historic Introductions	12	14
Number of gymnosperm species (all introduced)	19	14
Number of flowering plant species		
Indigenous	57	44
Endemic	189	174
Polynesian Introductions	21	20
Historic Introductions	282	574
Total native vascular plant species	334	313
Total introduced vascular plant species	334	602

*Based on compilation of data from Medeiros and Loope for Haleakala and from Higashino *et al.* 1988 a for Hawaii Volcanoes. Taxonomic status used is that of Wagner *et al.* 1990.

DISTURBANCE HISTORY AND MODES OF INTRODUCTION

The Crater district of Haleakala National Park has a history of human disturbance, probably dating back to Polynesian practices of burning and harvesting birds for food and feathers. Disturbance was greatly accelerated with the introduction of goats (*Capra hircus*) and cattle (*Bos taurus*) to Maui in the 1790s. Wilkes (1845) reported seeing "a few goats" as well as "bullock" tracks and those of wild dogs (*Canis domesticus*) near the summit of Haleakalā Volcano in 1841. Domestic cattle were brought into the Crater beginning in the late 1800s and were at least occasionally taken up to the Kalapawili grasslands to graze (von Tempski 1940). Degener (1930) lamented the damage to native vegetation by cattle on his 1927 trip into Koʻolau Gap. The dominance of alien grasses in the pasture-like areas below Hōlua and at Palikū probably had its origin in this grazing during the early 1900s. Many alien plant species were undoubtedly introduced to the Crater with cattle and horses (*Equus caballus*). Cattle grazing ceased in the 1930s, soon after the Park was designated.

The first feral pigs (*Sus scrofa*) were seen in the Crater in the 1930s, apparently having come up through Ko'olau Gap from piggeries near the coast (Diong 1983); pigs later became widespread in the Crater and surrounding forests. The mid-elevation (2,700 to 5,400 ft or 825-1,650 m) portion of Kīpahulu Valley was reportedly near-pristine, showing little or no pig activity when a 1967 Kīpahulu Expedition explored the area; however, pigs were sighted in the upper and lower parts of the Valley (Warner 1968; Diong 1983). East Maui's montane bogs were nearly pristine when first explored in the early 1970s. When Vogl and Henrickson (1971) visited Flat Top Bog in 1969, gosmore (*Hypochoeris radicata*) was the only alien plant recorded, with 0.1% cover. When the larger bogs were first visited in the early 1970s, there was very low cover of alien plants, although 10 alien species were present (Loope *et al.*, in press).

Much of the Kīpahulu district below 2,000 ft (610 m) was used intensively for agriculture by the pre-European contact Hawaiians. These lower slopes were used for sugar (*Saccharum officinarum*) and pineapple (*Ananas comosus*) production and eventually converted to pasture (Krauss 1980).

The experimental plantings by the forester Ralph Hosmer in 1910 at 6,800 ft (2,075 m) and 8,500 ft (2,590 m) provided a seed source for alien species, especially blue gum (*Eucalyptus globulus*). Eucalyptus has spread extensively in a watercourse downslope from the 8,500 ft elevation grove. Extensive plantings of pines and other conifers for "watershed improvement" (R. Hobdy, pers. comm.) on ranchland below Hosmer Grove in the early 1940s were the apparent source of invasive pines. Three planted pine species (*Pinus radiata*, *P. patula*, and *P. pinaster*) perpetually establish fast-growing seedlings on Park lands, and if unchecked these species would eventually convert large expanses of native shrubland to alien coniferous forest.

Military use of the area below Park headquarters during World War II, accompanied by transport of materials from O'ahu and other parts of the world, was a significant source of alien plant introductions. Also contributing to weed introductions has been the presence of horse stable facilities in the same area. Another major contribution to the introduction of alien species into the Park was the extensive road construction and resurfacing in the late 1970s and early 1980s. Many alien plants have been recorded only in the area below Park headquarters or along the highway embankment and areas of construction (*e.g.*, the water catchment and purification unit). About 20 new alien plant species appeared in 1978-81 during road and water catchment construction (Table 2). Few persisted as of 1991, and none are believed to constitute a significant threat to native ecosystems. Nevertheless, the potential threat of such alien plant introductions is difficult to accurately assess in advance.

Table 2. Plant species introduced to the western slope of Haleakalā in 1976-81 during widening and resurfacing of Haleakalā Highway.* As of April 1991, only *Vicia sativa* persists.

Species	Location	Date First Observed
Amaranthus spinosus	roadsides	06/81
Amaranthus viridis	roadsides near 7,000 ft	summer 1981
Atriplex suberecta	roadsides near 7,000 ft, stables	summer 1981
Capsicum annuum	headquarters	1981
Chenopodium carinatum	roadsides	06/81
Crotalaria incana	roadsides	1981
Datura stramonium	headquarters, stables	08/81
Descurainia sophia	stables area	11/78
Lamium amplexicaule	headquarters, stables	1981
Leucaena leucocephala	headquarters	1981
Lycopersicon pimpinellifolium	roadside	08/81
Malvastrum coromandelianum	headquarters	10/81
Melilotus alba	near summit	11/78
Nicotiana glauca	occasional roadside weed	summer 1981
Polygonum aviculare	stables, roadsides, headquarters	08/81
Polygonum convolvulus	stables, landfill	1981
Portulaca oleracea	roadside	08/81

Table 2, continued.

Table 2, continued.

Species	Location	Date First Observed
Sisymbrium altissimum	stables and headquarters	1980
Verbesina encelioides	roadside	08/81
Vicia sativa	stables, headquarters	09/81
Xanthium strumarium	roadsides, stables	06/81

*The source of most is believed to have been heavy equipment or fill brought from low elevation near Kahului, Maui. Some equipment came from Washington State. Only one species persisted for more than 10 years, and none are believed to present a serious threat to native ecosystems. The potential for introductions in this manner is obvious, however. Two species (*Descurainia sophia* and *Lamium amplexicaule*) were new state records.

DYNAMICS OF MAJOR PLANT COMMUNITIES WITH AND WITHOUT FERAL ANIMALS

The current strategy for preservation of native ecosystems of Haleakala National Park involves fencing, control of feral animals, and strategic removal of alien plants. As of 1991, all fencing was complete. Feral animal control is well advanced in both districts of the Park. With the results of exclosure studies, we are able to better understand and predict trends of vegetation dynamics. In the absence of definitive research, such "best guess" predictions are necessary for the early guidance of crucial management decisions. Continued monitoring is essential to allow evaluation and adjustment of management strategies and priorities.

Alpine Cinder Deserts

Because of its relatively harsh environment (particularly so for seedling establishment), this is one of the least modified of Haleakalā's vegetation types. Extreme conditions of this ecosystem include potential freezing temperatures any night of the year, large diurnal temperature fluctuations with periodically high soil surface temperatures, and a desiccated layer of surface soil. Although no alien plant species thrives throughout this habitat, sorrel (*Rumex acetosella*), gosmore, and a few other species are established sporadically. The monocarpic Haleakalā silversword (*Argyroxiphium sandwicense* subsp. *macrocephalum*) exists virtually without competition from alien plant species on cinder cones in Haleakalā Crater and in relatively barren sites near the summit. However, mullein (*Verbascum thapsus*), abundant on upper Mauna Kea (Juvik and Juvik, this volume), is a potential invader of this zone. Two mullein plants (the first records of this species on Maui) were found and destroyed on Haleakala National Park roadsides at 6,800 ft (2,070 m) and 9,000 ft (2,750 m) in 1986-87.

Subalpine Shrubland

The subalpine shrubland has been modified by years of impact of goats, cattle, and pigs. The often dense mat of alien grasses such as velvet grass (*Holcus lanatus*), sweet vernalgrass (*Anthoxanthum odoratum*), and orchard grass (*Dactylis glomerata*) inhibits reproduction by seed of native shrubs. Pines, eucalypts, gorse (*Ulex europaeus*), and blackberry (*Rubus argutus*) are also potentially serious invaders of this zone.

Subalpine Grassland

Pigs, digging for bracken fern (*Pteridium aquilinum* var. *decompositum*), gosmore rhizomes, and invertebrates, uproot large areas of native vegetation, which are invaded by gosmore and by velvet grass (Jacobi 1981). By mid-1989, pigs and goats were eliminated from a 270-a (110 ha) exclosure in the grasslands, and areas dug up by pigs were revegetated. Initially, the removal of pigs resulted in large increases of gosmore. However, in the absence of continual disturbance, the endemic grass *Deschampsia nubigena* overtops alien grasses and forbs and may eventually eliminate them except in drainages, where velvet grass appears capable of persisting. The most serious problem in the Kalapawili grasslands is the spread of blackberry, dispersed by pigs and birds. Alien blackberry continues to increase, and control appears necessary.

Montane Bogs

Most bogs in the Park were damaged (some severely) by pig digging, beginning in the late 1970s. Introduced plants began to invade these bogs in pig-rooted areas, and some alien species achieved substantial cover. Flat Top Bog at 7,400 ft (2,260 m) was fenced in 1979 but showed continued degradation through spread of already-abundant alien species. In 1969, the cover of alien plant species in Flat Top Bog was 0.1% (Vogl and Henrickson 1971); by 1983, cover of velvet grass and gosmore had increased to 12% in the center (wettest area) and 42% on the grassland margin of the bog (Medeiros and Loope, unpub. data).

On the other hand, Greensword Bog, at 6,100 ft (1,860 m), recovered quickly with native sedges and grasses after near destruction of the bog by pigs in 1981. The bog was fenced in June 1981 in an emergency attempt to protect it. Three months after fencing, the most disturbed central portion of the bog had 96% bare ground. By October 1984, bare ground had been reduced to only 10% by the recovery of native sedges *Oreobolus furcatus* and *Carex echinata* and grasses *Deschampsia nubigena* and *Dichanthelium cynodon*. Alien species had not become established by the time the bog was fenced and have not invaded since (Loope *et al.* 1991b).

Prior to protection by fencing in 1987-88, the two largest bogs (Mid Camp and Big Bog) at 5,400-5,440 ft (1,650-1,660 m) were undergoing chronic pig damage and substantial invasion of alien plant species. Permanent vegetation plots, monitored in 1982, 1984, 1986, and 1988, showed progressive increase in alien species, especially velvet grass, Glenwood grass *(Sacciolepis indica), Juncus planifolius, Cyperus halpan,* Asiatic pennywort (*Centella asiatica*), and sheepgrass (*Eragrostis brownei*). These species have the potential of displacing natives by preventing establishment of native seedlings. The fencing of bog areas is expected to slow the invasion and allow native vegetation to at least partially recover (Medeiros *et al.* 1991).

ʻŌhiʻa-dominated Rain Forests (above 4,000 ft)

The cool, damp, shaded understories of high-elevation East Maui rain forests have up to now been relatively resistant to invasion by alien plants. ʻŌhiʻa (*Metrosideros polymorpha*)-dominated rain forests of upper Kīpahulu Valley, the northeast rift, and the Manawainui planeze have been highly disturbed locally by feral pigs in recent years, with the reduction of ground bryophytes and certain preferred vascular plant species. Alien plants invading these forests include blackberry, velvet grass, gosmore, thimbleberry *(Rubus rosifolius), Erechtites valerianifolia*, selfheal (*Prunella vulgaris*), and Maui pāmakani (*Ageratina adenophora*). Up to now, their overall impacts on the system appear to be minimal, but their presence appears to inhibit the establishment of native species in stream courses. Stream courses are much more susceptible to invasion by alien species than is the surrounding forest, probably due to a regime of more frequent disturbance and higher levels of incident sunlight.

Like montane bogs, high-elevation rain forests, under regimes of limited disturbance by feral pigs, have proved relatively resistant to invasion by alien plants. After removal of feral pigs, these ecosystems undergo rapid recovery of dominant species, *e.g., Diplazium sandwichianum, Athyrium microphyllum,* and other ferns in rain forests, and *Oreobolus* and native *Carex* spp. in bogs. It remains to be seen whether drastically depleted rare species can recover after removal of feral pigs.

Koa-dominated Rain Forests (below 4,000 ft)

Compared to the situation in high-elevation ʻōhiʻa forests, alien plant species pose a much more formidable problem for long-term preservation of native ecosystems below 4,000 ft (1,220 m) in Kīpahulu Valley. Two reasons for this are: 1) there is greater disturbance in lower-elevation forests, and 2) more alien species are adjacent to and thrive in lower-elevation rain forests. Alien strawberry guava (*Psidium cattleianum*) locally grows in such dense stands that native shrubs and herbs are shaded and crowded out and koa *(Acacia koa)* seedlings are unable to establish. Hilo grass (*Paspalum conjugatum*) establishes dense mats and also appears to inhibit establishment of native species. These two aliens are still spreading in koa-dominated forest up to 3,500-4,000 ft (1,070-1,220 m). Rose apple (*Syzygium jambos*), fortunately found only below 2,200 ft (670 m) elevation so far, is even more aggressive than strawberry guava in its tendency to dominate forest stands. Feral pigs are believed the primary agent for the dispersal and disturbance that have allowed alien species to become so well established. Even with control of pigs, alien plant species may persist and spread, but at a slower rate. A recent discovery is an invasion of an Australian tree fern (*Cyathea cooperi*) into this habitat (Higashino *et al.* 1988b).

Dry Forest and Shrubland

The rich flora of leeward East Maui has undergone much damage from feral and domestic animals and alien plants. Of the 239 native angiosperm species in the flora, 46 have already been extirpated (Medeiros *et al.* 1986). Of all the native ecosystems of Haleakala National Park, the dryland forest flora is the most threatened. Due to the low number of surviving individuals of many rare dryland forest species and the degree to which this ecosystem has been altered, some extirpations will almost certainly occur here during the coming decades. Other species, however, may recover to varying degrees. The lower Kaupō Gap portion of the Park, descending to 3,800 ft (1,160 m), provides a relatively secure refuge for most of the 119 leeward species found within Park boundaries. However, their survival is by no means assured through the control of feral goats and pigs.

A small, particularly rich, grove of rare leeward forest trees is concentrated on a single ridge on the eastern wall of Kaupō Gap. Alien kikuyu grass (*Pennisetum clandestinum*), Maui pāmakani, and (potentially) molasses grass (*Melinis minutiflora*) create a formidable barrier to reproduction of many dryland forest tree species in eastern Kaupō Gap. In western Kaupō Gap, alien grasses can carry fire, increasing the potential threat to the reestablishment of native vegetation. Without the effects of wildland fire, however, ‘a‘ali‘i (*Dodonaea viscosa*), pūkiawe (*Styphelia tameiameiae*), and ‘ūlei (*Osteomeles anthyllidifolia*) have recovered rapidly in a 10-year-old exclosure at 4,000 ft (1,220 m) in western Kaupō Gap that was previously almost barren of vegetation (A.C. Medeiros and L.L. Loope, unpub. data).

Highly Disturbed Vegetation below 2,000 ft in Kīpahulu District

In this area, only the coastal strand vegetation is primarily native, although stands of relatively intact mesic lowland forest, unique within the Park, are scattered locally above 1,000 ft (305 m). The pasture is dominated by the alien rattail grass (*Sporobolus africanus*) and kikuyu grass, with dense stands of common guava (*Psidium guajava*), Christmas berry (*Schinus terebinthifolius*), Java plum (*Syzygium cumini*), and mango (*Mangifera indica*) occupying areas where rocks have been piled from the days of field clearing for planting of sugar cane. In recent years, with minimal pasture maintenance, the alien trees have gradually invaded the open pasture. Black bamboo (*Phyllostachys nigra*) dominates the northeastern section of the Valley above the pasture area, interspersed with stands of native forest species being gradually replaced by strawberry guava, common guava, rose apple, Java plum, and African tulip tree (*Spathodea campanulata*). This constantly changing vegetation was well mapped by Smith *et al.* (1985).

THE WORST ALIEN PLANT INVADERS OF HALEAKALĀ

As with the previous section, much of the following discussion of individual species is based on preliminary data and anecdotal information and needs to be reevaluated periodically through monitoring.

Strawberry guava, *Psidium cattleianum*

Varying from shrub to large tree, depending on density of stocking and habitat conditions, strawberry guava establishes dense stands from primarily pig-dispersed seed (Diong 1983) and tends to displace native species. Its elevational range in Kīpahulu Valley is 300-4,000 ft (90-1,220 m), but it is abundant only up to about 3,200 ft (975 m). It is shade tolerant and grows in nearly impenetrable thickets in Kīpahulu Valley at 1,500-2,500 ft (460-760 m). Strawberry guava potentially threatens numerous low- to mid-elevation rain forest plant species with extirpation from Haleakala National Park through displacement of reproduction. Species particularly threatened in this way include *Antidesma platyphyllum, Claoxylon sandwicense, Cyanea grimesiana, Joinvillea gaudichaudiana, Nothocestrum longifolium,* and *Strongylodon ruber*. These species are rare, with some very near extirpation. The rare tree *Antidesma* is one of the primary host trees for the very rare arboreal snail *Partulina porcellana*, an East Maui endemic (M. Hadfield, pers. comm.).

Strawberry guava reaches its current maximum development on East Maui on the southeast flanks of the Volcano below Pu'u 'Ahulili, where it occurs as high as 4,600 ft (1,400 m). In this area, at 3,300-3,900 ft (1,000-1,190 m), it occupies over 25% of the forest understory. Below 3,300 ft (1,000 m), it occupies over 75% of the understory, rising into the canopy and attaining 60 ft (18 m) heights and 12 in. (30 cm) diameters (J. Jacobi, pers. comm.). On Haleakalā's north slope, strawberry guava commonly occurs above 3,000 ft (915 m) and reaches 3,900 ft (1,190 m) below Pu'u o Kaka'e near Waikamoi Stream (J. Jacobi, pers. comm.).

Molasses grass, *Melinis minutiflora*

After exclusion of feral goats from the lower elevations of Hawaii Volcanoes National Park, this mat-forming grass spread explosively throughout formerly open goat-browsed areas, developing local populations with nearly 100% cover. A similar situation occurred on East Maui in the Healani exclosure in goat-damaged koa forest at 4,000 ft (1,220 m) about 1.5 km south of the Haleakala National Park boundary. Within the exclosure, molasses grass developed dense mats that inhibited reproduction of koa and other native species (Scowcroft and Hobdy 1986). By 1988, after removal of feral goats from the Crater District of the Park, molasses grass had increased its biomass and distribution substantially in Kaupō Gap and on the western edge of the Manawainui planeze at 4,000-5,500 ft (1,220-1,680 m).

The most critical aspect of the biology of this species is its ability to fuel, carry, and subsequently quickly recover from wildfires. Wildfires are one of the most serious threats to the long-term perpetuation of dryland forest on leeward Haleakalā. Survival within the Park of such rare species as *Nothocestrum latifolium, Pelea hawaiensis, Pouteria sandwicensis, Pleomele auwahiensis, Sanicula sandwicensis, Viola chamissoniana,* and *Zanthoxylum kauaense* is increasingly threatened as fuel increases. Experimental control of molasses grass with glyphosate was initiated at Haleakalā in late 1990, with the hope of preventing the (unfortunate) scenario that occurred with molasses grass in the lowlands of Hawaii Volcanoes from recurring in the middle elevations of the leeward Crater District of the Park. The outcome of this experimental control is still being evaluated as of April 1991.

Hilo grass, *Paspalum conjugatum*

Hilo grass was first recorded in the Park as a component of the koa forest understory in the 1919 field notes of C.N. Forbes (A.C. Medeiros and L.L. Loope, unpub. data). It now occurs up to 4,450 ft (1360 m) in Haleakalā's Kīpahulu Valley, where it forms a mat, (sometimes) thick and dense, under koa forest at 2,200-3,400 ft (670-1,040 m). Although this sprawling grass may inhibit seedling establishment of koa and other native species, these seedlings are still able to establish on decaying logs and other raised microsites. Hilo grass threatens the survival of many of the same rare native species of koa forests of Kīpahulu Valley previously mentioned as threatened by strawberry guava. Despite its negative effects on native biota, no action is planned for the immediate future due to its pervasiveness.

Blackberry, *Rubus argutus*

Spread of this thorny species native to the southeastern U.S. presents potentially severe problems in high-elevation grasslands and rain forests of Haleakalā, where large fruits are formed at irregular intervals. Pig control may slow its spread in the grasslands, since pigs disperse it and it becomes established primarily in areas of pig digging. However, birds also eat the fruit and disperse the seeds, with new plants sometimes appearing as far away as 0.6 mi (1 km) from the nearest established population (A.C. Medeiros, unpub. data). Once established, its primary method of spread in an area is vegetative, by underground rhizomes. Backcountry personnel in the Park have been combating blackberry for years in the Palikū pasture and have succeeded in keeping it localized. Extensive stands occur along the lower edge of the Kalapawili grasslands between Pu'u 'Alaea and Flat Top Bog. Large infestations occur near the Park in the Ko'olau Forest Reserve, Hanawī Natural Area Reserve, and Waikamoi Preserve below Pu'u 'Alaea. A smaller population occurs in Waikamoi Preserve in eastern Ko'olau Gap.

Australian tree fern, *Cyathea cooperi*

Native to eastern Australia and Tasmania, the Australian tree fern is widely planted in Hawai'i, as it is faster growing than native Hawaiian tree ferns (*Cibotium* spp.) (G. Wescott, pers. comm.). First reported in Kīpahulu Valley in 1988 (Higashino *et al.* 1988b), A.C. Medeiros (unpub. data) counted 279 tree ferns of this species in 1989 with trunk diameters greater than 3 in. (7.5 cm) in three populations covering several thousand square meters at 2,400-3,400 ft (730-1,040 m) elevation. The average height of these ferns was 3-6 ft (1-2 m), with some individuals over 12 ft (4 m) tall. In 'ōhi'a and koa forests of Kīpahulu Valley, Australian tree fern has achieved impressive local dominance and shown considerable invasive potential. *Cyathea* is considered an aggressive alien species that should be controlled if feasible.

Unlike native *Cibotium* tree ferns, Australian tree fern does not support epiphytic native species. The Australian tree fern also acts as a host for an alien herbivorous beetle, the Australian fern weevil (*Syagrius fulvitarsus*), which occurred in epidemic numbers in Hawaii Volcanoes National Park in the 1950s, defoliating endemic ʻamaʻu ferns (*Sadleria cyatheoides*). It is quite possible that as Australian tree fern increases, this alien insect may begin to seriously impact nearby native ferns. An exploratory attempt to control this species in Kīpahulu Valley was initiated in April 1991.

Kikuyu grass, *Pennisetum clandestinum*

Introduced to Maui as a pasture grass in the 1920s-1940s, kikuyu grass (from East Africa) has filled that role successfully on East Maui ranchlands, spreading in disturbed mesic habitats from near sea level to 10,000 ft (3,050 m), although uncommon above 7,000 ft (2,130 m). At high elevations, periodic freezing temperatures damage it and generally prevent it from competing well. It dominates in certain leeward areas at elevations of 3,000-5,000 ft (915-1,525 m) and typically forms dense mats (up to 3 ft or 1 m thick) that can overwhelm all low-growing vegetation and prevent any native seedling establishment. It presents a serious problem in Kaupō Gap, where it first spread from ranchlands about 1945 (Mitchell 1945) and is still expanding its range. Kikuyu grass can be killed by application of glyphosate (a non-persistent organophosphorus herbicide) at a sufficiently low concentration that native species are not harmed (Gardner and Kageler 1983). Thus, the potential exists for eliminating this species from areas even where it is well established and possibly allowing for the reestablishment of native species.

Kāhili ginger, *Hedychium gardnerianum*

This large (to 6 ft or 2 m tall) ginger with bright golden-yellow flowers is an aggressive invader of low- to mid-elevation rain forest in Kīpahulu Valley up to about 4,000 ft (1,220 m) elevation. Kāhili ginger is dispersed, sometimes long distances apparently, by birds that eat its large, fleshy, orange fruits. Once kāhili ginger establishes at a site, it spreads vegetatively, forming large, continuous clumps that displace nearly all other understory vegetation. It is especially aggressive in wet, disturbed, well-lit areas, such as open-canopied forest understory and along streambeds. An especially large population of kāhili ginger occurs along the Koʻūkoʻūai stream in southwestern Kīpahulu Valley at 3,500-3,600 ft (1,070-1,100 m) (Anderson *et al.*, this volume).

Velvet grass, *Holcus lanatus*

This perennial grass from Europe is abundant in most open shrubland habitats of East Maui at 6,000-9,000 ft (1,830-2,745 m). It is particularly dense in the lower elevation subalpine shrubland of northwestern Haleakalā between Hosmer Grove and Park Headquarters. Wherever it grows it displaces seedling establishment by native species, such as the important shrub/tree māmane *(Sophora chrysophylla)*. Velvet grass is also the major alien species invading pig-damaged montane bogs of northeastern Haleakalā. Though it is currently a serious threat, control of the widespread velvet grass is impractical.

Maui pāmakani, *Ageratina adenophora*

This subshrub, native to Mexico, was brought to Hawai'i before 1900 and spread quickly on Maui, coming to dominate large areas by 1913 (see Plates 89, 108, 147, and 173 in Rock 1974). In some areas, it formed pure stands, reaching up to 5 ft (1.5 m) in height. A stem gall fly (*Procecidochares utilis*), introduced to Maui in 1945-1947 for biocontrol purposes from the native habitat of this species, aided in bringing it under control (Bess and Haramoto 1972). In many pastureland areas of Maui, it was replaced by another alien, kikuyu grass, planted to revegetate the barren areas. Maui pāmakani still occurs throughout much of the Park; most populations are infected by the biocontrol gall fly, but not at sufficient levels to eliminate the weed. In eastern Kaupō Gap at elevations of 4,000-6,000 ft (1,220-1,830 m), it grows in locally dense monospecific stands. It also grows in open areas in rain forest and along stream courses in Kīpahulu Valley. In other parts of Haleakalā Crater (*e.g.*, Pu'u Māmane) at elevations up to ca 8,000 ft (2,440 m), it is sparsely present and until recently was typically cropped by goats. This is one alien species that may increase markedly in certain areas of the Crater District now that goats are under control.

Black bamboo, *Phyllostachys nigra*

This large bamboo occurs in dense, monospecific stands below 2,000 ft (610 m) in the Kīpahulu District. We are unaware of the reproductive biology of this species, although it may, like many bamboos that flower at intervals of a century or longer, eventually flower, set seed, and die synchronously (Janzen 1976). Up to now spread has presumably been vegetative. Since no clones on East Maui are known to extend above 2,000 ft (610 m), this species may not threaten forests above that level. Alternatively, the species may have the potential for spreading but is slowed by its purely vegetative reproduction. Its status should be carefully monitored since bamboo thickets result in elimination of virtually all native plant and animal species. Bamboo is one of the most difficult alien plant species to control, even locally.

Broomsedge, *Andropogon virginicus*

Although well established in many low-elevation areas of East Maui, including wet slopes and clearings in Kīpahulu Valley, broomsedge is still very rare in Kaupō Gap. Because of its well-known negative effects in Hawaii Volcanoes National Park, linked to fire, at elevations comparable to Kaupō Gap, it warrants early control in Haleakalā. Interestingly, this bunchgrass, generally associated with leeward, drier sites, has recently been recorded in the wet montane bogs of Haleakalā's Northeast Rift. Opportunistic control has prevented its further spread there and should be continued.

Gorse, *Ulex europaeus*

On East Maui, this spiny shrub from Europe is abundant in the Olinda area and is locally common just within and outside the northwestern boundaries of Haleakala National Park. Seed weevils and a flower-feeding caterpillar introduced in the past for biocontrol reduced the reproductive potential of gorse by 73% on Maui (Markin 1984). A major gorse biocontrol program is under way, and additional insects are being released on Maui, beginning in 1989 with the moth *Agonopterix ulicitella*, which feeds on new shoots. Long-time persistence of dormant seeds makes early control important before populations become well established. Its potential invasiveness in subalpine shrublands of the Park warrants prompt control, which thus far has been attempted within the Park, largely by volunteer groups.

Pines, *Pinus* spp.

Pinus radiata from California, *Pinus patula* from Mexico, and *Pinus pinaster* from southern Europe are the only conifers (of many planted species in the Hosmer Grove area) to aggressively establish seedlings in native shrubland of the Park along the boundary adjacent to an experimental watershed improvement project on ranchland. On favorable sites these species can grow in height at a rate of about 12 in./yr (30 cm/year) and start to produce cones after six to eight years.

Rose apple, *Syzygium jambos*

Rose apple is a very aggressive invader of low-elevation forests in Hawai'i. It is found up to 2,200 ft (670 m) in Kīpahulu Valley, often forming impenetrable thickets. Its spread should be carefully monitored, since within its lower elevation range it forms even more dense monospecific stands than strawberry guava.

Chinese banyan, *Ficus microcarpa*

Chinese banyan is a strangling, aggressive invader on rocky walls of low-elevation stream courses and sea cliffs in lower Kīpahulu. The several dozen known established plants present in the Park should be removed as soon as possible in order to prevent this species from taking over these habitats.

Christmas berry, *Schinus terebinthifolius*

Stemmermann *et al.* (1981) noted the presence of a single large individual of Christmas berry in western Kaupō Gap. By the mid-1980s, individuals of *Schinus* were scattered within the pūkiawe/'a'ali'i shrubland of lower central Kaupō Gap. The species is common on lower-elevation Haleakalā outside the Park but is seldom found above 4,000 ft (1,220 m) and consequently poses little threat to the Park's Crater District except in extreme lower Kaupō Gap. Christmas berry is abundant in the highly disturbed pastures of lower Kīpahulu, but because of its poor shade tolerance, as well as a preference for mesic rather than wet sites, it presents little threat to the native forests in Kīpahulu Valley.

Juncus planifolius

This species was found by Medeiros *et al.* (1991) to be among the more aggressive weeds invading montane bogs in Haleakala National Park. First recorded in 1982, by 1988 *Juncus planifolius* had a mean cover of 8.4% in seven 10 m x 10 m plots within *Carex echinata*-dominated bogs at 5,400-5440 ft (1,650-1,660 m). It is also an aggressive invader of Alaka'i and Wai'ale'ale bogs on Kaua'i.

Glenwood grass, *Sacciolepis indica*

This grass is an aggressive invader of East Maui bogs, first recorded there in 1973 (Loope *et al.* 1991a). Over a period of six years from 1982 through 1988, it tripled its frequency (from 9 to 30%) in *Carex*-dominated bog plots at 5,400-5,440 ft (1,650-1,660 m) elevation (Medeiros *et al.* 1991). Once established, it spreads rapidly both by seed and vegetatively, often colonizing areas dug up by pigs and water courses. Like velvet grass and *Juncus planifolius*, it has the potential for displacing endemic bog species such as *Dichanthelium cynodon*, *Selaginella deflexa*, and *Viola maviensis*. In bogs protected from pigs, native grasses and sedges recover well, and this pioneer species may be reduced. We recommend no control measures, but continued monitoring, at this time.

Rattail grass, *Sporobolus africanus*, and Natal redtop, *Rhyncheletrum repens*

These grasses are abundant on leeward Haleakalā at 3,000-6,000 ft (915-1830 m) as well as in pastures of lower Kīpahulu. Currently, these two grass species make up much of the alien grass cover of western Kaupō Gap. They produce sufficient combustible material to carry wildfires once feral goats are controlled. Fire would enormously retard recovery of native shrubs and herbs. However, the threat of these grasses is considerably less than that presented by molasses grass.

THE REMAINDER OF THE ALIEN FLORA

The Haleakala National Park flora includes more than 300 additional introduced species beyond those already discussed as the worst alien plant species. Many of these are confined to the highly modified area below 2,000 ft (610 m) in the Park's Kīpahulu district and present little threat to upland native ecosystems. Many more of those occurring in the Crater and Kīpahulu districts above 2,000 ft (610 m) may represent little more than aesthetic intrusions. Some remaining species warrant special mention and concern, however.

Common guava (*Psidium guajava*) occurs together with, but at lower densities than, strawberry guava in the rain forest of lower Kīpahulu Valley to above 4,100 ft (1,250 m) elevation. Its effects are similar to those of its more threatening congener, and any efforts at control of strawberry guava should be applied to common guava in the same area.

Gosmore (*Hypochoeris radicata*), a yellow-flowered composite, is almost ubiquitous in non-forest vegetation from dry sites of Haleakalā Crater to bogs of the northeast rift. Although its total ground cover at a given site rarely exceeds 10% in sites not recently disturbed and averages 5% or less, it occupies a significant area that might otherwise be available to seedlings of native plants. In recent pig diggings, cover of this species may sometimes approach 50%. This species is very conspicuous after removal of feral pigs and goats in the Kalapawili grasslands. It appears to be an early successional pioneer forb that decreases with time. However, it may persist especially in areas of frequent disturbance, *e.g.*, trailsides, streambeds, and landslides.

A spiny Eurasian thistle, *Cirsium vulgare*, has been controlled locally by backcountry personnel and volunteers for several years with some success. At present it is unclear what the long-term effect of goat removal on this species will be. Goats eat it but may also encourage its establishment through grazing and trampling of native vegetation. We expect that it may increase initially following goat removal but may decline and perhaps even disappear eventually (except along trails, in other disturbed areas, and on scree slopes) as goat impacts are eliminated at Haleakalā. We interpret the thistle problem at Haleakala National Park primarily as an aesthetic one and for that reason place it lower in priority for control. Volunteer effort may help reduce thistle populations but is unlikely to completely eliminate it from areas with continuing disturbance.

Coffee senna (*Senna occidentalis*) is an alien shrub found in lower eastern and central Kaupō Gap below 4,800 ft (1,400 m). It should be controlled before it spreads farther.

Castor bean (*Ricinus communis*) occurs in a small area of eastern Kaupō Gap at about 4,100 ft (1,250 m). Though perhaps a Polynesian introduction, castor bean is potentially invasive and should eventually be eliminated through repeated mechanical and/or herbicidal treatment.

African tulip tree (*Spathodea campanulata*) is a large tree that is established sporadically in lower and middle Kīpahulu Valley and on adjacent Kaumakani Planeze. It has become aggressively invasive in some rain forest areas of Hawai'i (Smith 1985). It should be monitored and opportunistically controlled in conjunction with systematic control of strawberry guava and other rain forest weeds.

Sweet fennel (*Foeniculum vulgare*) occurs sporadically in the Park at present on the lower western slope (about 7,000 ft or 2,130 m) and in the Hōlua area of Haleakalā Crater. Because of its tall growth form, it is a conspicuous aesthetic intrusion in the shrubland ecosystem. It should be opportunistically or systematically controlled before it becomes more widespread.

The Park has over 50 species of alien grasses and sedges, some of which surely displace native species. Beyond those already mentioned, the most abundant include sweet vernalgrass, orchard grass, hairy oatgrass (*Danthonia pilosa*), brome fescue (*Vulpia bromoides*), annual bluegrass (*Poa annua*), Kentucky bluegrass (*Poa pratensis*), and *Rhynchospora* spp. Although their combined effects may be significant, they are generally overshadowed by the dominant alien grasses already mentioned.

THE THREAT OF FURTHER INVASION

A number of species not yet in Haleakala National Park but found on Maui pose severe problems in other parts of the State. Other species already on Maui are well known for their aggressive behavior elsewhere in the world.

These species have the potential to eventually cause problems in Haleakala National Park.

Clidemia or Koster's curse, *Clidemia hirta*

This densely branching shrub (to 13 ft or 4 m tall) is native to the Neotropics (southern Mexico and West Indes to Argentina). It has become an aggressive alien in many parts of Africa, Asia, and the Pacific islands. A particularly severe invasion in Fiji was controlled by the intentional introduction of a thrips (*Liothrips urichi*) native to Trinidad (Wester and Wood 1977). This same organism has been introduced to Hawai'i but has proved effective in controlling clidemia only in open (non-shaded) habitats. Other biological control agents, more recently brought to Hawai'i, show promise of increased control (Smith, this volume; Nakahara *et al.*, this volume).

Clidemia was first introduced in the Hawaiian Islands on O'ahu in 1941. Since then it has become very widespread on that Island, dominating large areas of rain forest understories. It was first recognized as a pest in the Islands in the 1950s, at which time *Liothrips* was introduced. However, the severity of the problem was not generally recognized by government agencies until the 1970s. Clidemia seeds are believed to be dispersed by the abundant alien Japanese white-eye (*Zosterops japonicus*) (Wester and Wood 1977) and probably by other alien birds, but mongooses (*Herpestes auropunctatus*) disperse it as well (A.C. Medeiros, unpub. data). Clidemia was first noted on northern East Maui in 1977 (Nāhiku District) and 1980 (Makaīwa District) on lands owned by the State of Hawai'i and East Maui Irrigation Company. By 1988, clidemia had spread along watercourses and established along the main Hāna highway in numerous drainages (Medeiros *et al.* 1989). A single plant was found (and destroyed) by L. Cuddihy and G. Santos at 2,800 ft (850 m) in Kīpahulu Valley in October 1988. Clidemia may provide a major threat to rain forests below about 5,000 ft (1,525 m) on Park lands in the future. As of 1991, clidemia was not known to be established in Kīpahulu Valley but had reached as far east on Nāhiku in lower-elevation East Maui.

Banana poka, *Passiflora mollissima*

This passionflower vine, locally known as banana poka, was introduced to the Hawaiian Islands as an ornamental in the early 1900s. It is uncommon in its native habitat in the Andes, where it is controlled by numerous species of co-evolved insects and where humans are destroying its habitat. Lacking natural herbivores in Hawai'i, it has become established in more than 190 mi^2 (500 km^2) of native forest on the islands of Hawai'i and Kaua'i. In some areas, it has become so dense that the vines drape from tree to tree, smothering large tracts of native forest. It occupies elevations of 2,000-5,000 ft (610-1,525 m) and thrives where mean annual precipitation is between 20 and 200 in. (500-5000 mm). Feral pigs are its primary dispersal agent, but alien birds spread it as well (Warshauer *et al.* 1983; LaRosa 1984; LaRosa, this volume).

In 1971 a group of three mature banana poka plants was reported from a farm lot in Kula, Maui. Hawaii Department of Agriculture personnel on Maui periodically attempted eradication in that area but had difficulty destroying all plants. The problem was still considered manageable until November 1984, when State Forestry crews combatting a persistent fire in the Kula Forest Reserve noted the unsuspected spread of this weedy vine throughout several hundred hectares of dense black wattle (*Acacia mearnsii*, alien) forest in upper Kula (Waiakoa, ʻAlae, at about 3,500-4,000 ft or 1,070-1,220 m). Prospects for control of this infestation are complicated by its occurrence on private lands. A well-organized education and eradication campaign was mounted, however, by L. Harrison of Makawao, Maui, in 1987-88. This effort led directly to funding by the State Legislature for mechanical control of banana poka through the Department of Land and Natural Resources in 1989-90. A cooperative effort is under way by Federal and State agencies to screen organisms from *P. mollissima* native habitat in South America for introduction of biocontrol agents to Hawaiʻi (Gardner, this volume; Markin, this volume). In April 1991, two herbivorous moths (*Cyanotricha necyria* and *Pyrausta perelegans*) were released for control of banana poka in the Kula area. Unless control measures succeed, the plant poses a threat to Kīpahulu Valley koa forests as well as to many other natural areas on Maui.

Fountain grass, *Pennisetum setaceum*

This large bunchgrass from northern Africa has spread aggressively throughout leeward Hawaiʻi Island during the past two decades, becoming potentially uncontrollable in Hawaii Volcanoes National Park (Tunison, this volume). Fountain grass creates an exceptionally large standing fuel source and promotes the spread of fires more than any other grass yet introduced to the Hawaiian Islands. Fountain grass is present on Maui only in a small area of the sand hills of southeastern Wailuku. However, it poses a serious threat to rangelands of southern East Maui with lightly vegetated, young volcanic substrates as well as to the largely barren, relatively undisturbed ecosystems of upper Haleakalā. Based on its occurrence as high as 9,000 ft (2,740 m) on Mauna Kea (J. Jacobi, pers. comm.), it must be regarded as a potential invader of Haleakalā Crater. Persistent control efforts by the Maui weed control office of the Hawaii Department of Agriculture for over 20 years have kept fountain grass confined to two small populations, one a residential area where the invasive populations apparently originated from an ornamental planting. A larger secondary population in a nearby dump area occupies more favorable habitat on relatively open, sandy sites. These populations are controlled by hand pulling of seedlings and mature plants at one- to two-month intervals, with bagging of seeding inflorescences for disposal at a nearby landfill. An average of about 2,000 plants/year were removed from 1983 to 1985 (E. Tamura, Hawaii Department of Agriculture, pers. comm.). In May 1986, population numbers were low and apparently decreasing due to a declining seed bank after several years of concerted effort at control of young plants before seed was set. As of early 1991, however, a small population persisted.

Firetree, *Myrica faya*

One of the worst invaders in Hawaii Volcanoes National Park and on Kauaʻi (Whiteaker and Gardner 1985), this tree from the Azores, Madeira, and the Canary Islands fixes nitrogen in root nodules and has great potential for massive alteration of ecosystems (Vitousek, this volume). A large infestation occurs on Haleakalā's western slope, particularly in the Kula Forest reserve at 3,200-6,400 ft (975-1,950 m) (Whiteaker and Gardner 1985). The upper firetree populations approach the Park boundary.

Miconia, *Miconia calvescens*

Miconia calvescens, native to middle-elevation (980 to 5,900 ft or 300-1,800 m) forests of Central and South America, has large, dark green leaves with maroon undersides. Because of its perceived attractiveness as an ornamental, it has been introduced recently through the horticultural industry to at least three Hawaiian Islands -- Hawai'i, O'ahu, and Maui. Miconia has gained special notoriety for its spread in Tahiti and Moorea, where in the past two decades it has extensively invaded forests (Gaubert and Florence 1991). The species is believed to have first reached windward eastern Maui in the early 1980s, but founding individuals have already grown to over 30 ft (10 m) tall and produced abundant seedlings locally. Tiny, bird-dispersed seeds are produced after a few years of vegetative growth; seedlings become established in open sunlight or dense shade. Growth is rapid, so that trees over 30 ft tall are produced after as few as five years. As of May 1991, there are seven known foci of this species on East Maui, but we are still in the information-gathering stage. A cooperative effort to eradicate it is being developed (Gagné *et al.* 1991).

It can be confidently predicted that propagules of additional aggressive plant species from various parts of the world will arrive on Maui in the future. We know of no way to accurately predict which species will become future invaders, but likely candidates include aggressive plants already occurring in the Hawaiian Islands (*e.g., Coccinia grandis, Oxyspora paniculata*). Indeed, alien plant problems now may represent merely the "tip of the iceberg" in relation to what potential invaders exist. Efforts to prevent as many additional introductions of aggressive plant species as possible would appear highly cost effective. Admittedly, documenting the prevention of an introduction may usually be so difficult that cost/benefit analyses may be impossible. Eradication of a species already introduced is easier to document. Strangely, however, there appear to be few, if any, documented cases where an aggressive weed has been eradicated. Maui, as well as other Hawaiian Islands, needs to develop and refine interagency cooperation to avoid the absurd, but probably common, situation in which aggressive species become established because no agency believes that it has the mandate or responsibility to stop them.

CONTROL EFFORTS TO DATE

Sustained efforts at control of gorse, pines, eucalyptus, thistle, and blackberry have been carried out at Haleakalā over the past decade, but only on a very limited basis and primarily through use of volunteers supervised by Park employees. In addition, isolated individuals or small populations of many species have been removed, including Christmas berry, castor bean, black wattle, and many other species not yet established in the Park. Pine and eucalyptus control on the northwestern slope of

Haleakalā has consisted of removing seedlings and saplings three to four times a year. Gorse has been monitored and patches reduced opportunistically with volunteers. Herbicide has been applied to these populations every three to four years. Thistles and blackberry have been controlled in limited areas of the backcountry whenever volunteers are available. These efforts have been highly successful in slowing invasions and will help to achieve eventual complete control. Although the use of volunteers can serve as a meaningful supplement to the Park's alien plant control effort, it cannot substitute for a full-fledged program, especially in remote rain forest sites. Many major problem species have continued to increase during past years. In addition, some problems can be anticipated with effective removal of goats in the Park and with species already present on Maui known to be detrimental elsewhere. Efforts must be accelerated in the near future for the highest priority species before they reach the irreversible stage in which future effort and expenditures cannot compensate for past neglect.

CONTROL NEEDS, PRIORITIES, AND STRATEGIES FOR THE FUTURE

Control of feral goats and pigs is currently the highest resource management priority at Haleakalā. Strategy has involved fencing the perimeter of the Crater District, to prevent ingress of animals from adjacent areas, and constructing three barrier fences across Kīpahulu Valley to establish management units for pig control. Control of feral pigs is a very important step toward controlling the establishment and spread of alien plants. Pigs are believed to be important for the spread of strawberry guava, molasses grass, Hilo grass, blackberry, kāhili ginger, velvet grass, rose apple, common guava, *Juncus planifolius*, and Glenwood grass. As ungulate control programs require less effort, funding should shift to alien plant control programs. Priorities for alien plant control at Haleakala National Park follow.

1. Strawberry guava provides the greatest long-term threat to the Park. It has the potential to affect the rain forests of Kīpahulu Valley below 4,000-4,600 ft (1,220-1,400 m) and could drastically transform this species-rich system, eliminating some native species and sharply reducing the rest. Banana poka and clidemia, not yet established in the Park, have similar potential. The spread of all three of these alien species (and most others as well) is greatly aided by feral pigs. If control of pigs can continue to succeed in Kīpahulu Valley, we expect much less rapid spread of banana poka and clidemia. Strawberry guava should be controlled above the "Dogleg" area (2,800 ft or 850 m elevation) through an aggressive, well-funded program involving cutting and application of herbicide to the stumps. (Santos *et al.* 1989 recommended use of Garlon for control of this species.)

2. Molasses grass control in Kaupō Gap should be continued, with the objective of at least delaying the spread of this fire-promoting grass until succession of native shrubs produces a more stable native-dominated vegetation that is relatively resistant to fire.

3. Australian tree fern and kāhili ginger problems should be addressed as soon as possible. Management strategies should be developed and action taken.

4. Pine, gorse, and eucalyptus control on the western slope near Park Headquarters has a very high ratio of benefit to cost. Without persistent control of these species, they would gradually become established and perhaps eventually gain dominance within the otherwise largely native vegetation of the subalpine shrubland.

5. Blackberry control in the Kalapawili grasslands and adjacent forest openings is another needed and very cost effective (at least in the long run) project. After a major initial effort at control, maintenance may be relatively easy with only limited local seed sources and absence of pigs. Without a major blackberry control effort in the next decade, this species may become uncontrollable. Biocontrol is another possibility (see below).

6. Strategic control of alien grasses and forbs will prove useful in certain small-scale (probably no more than several acres each) "special ecological areas" (similar to those described for Hawaii Volcanoes National Park by Tunison and Stone, this volume) after control of feral animals. Areas considered for such selective control include bogs, prime leeward forest areas, and sites threatened by Hilo grass in Kīpahulu Valley.

7. Kikuyu grass control in eastern Kaupō Gap could greatly stimulate reestablishment of koa forest. Control of Christmas berry and *Senna occidentalis* in eastern Kaupō Gap would assist recovery from goat browsing of native vegetation in that area.

8. Helicopter landing sites, especially in rain forest areas, are major sites for introduction and spread of alien plant species. "Helipads" where alien species are gaining a foothold should be periodically sprayed with glyphosate, with continual followup monitoring.

9. State and private control efforts for banana poka, fountain grass, firetree, and other species outside the Park should continue to be supported.

10. The Park needs to develop a balanced and sustained program to stop introduction and spread of new alien species before they become established on Maui, using direct attack, research and monitoring, and education of the public. We also need to work harder to gain support of local and state governments. The truly threatening alien invasions on the Island should be mapped using a Geographic Information System to develop management tactics and to educate constituencies. This work should be closely coordinated with the County of Maui and with State (Department of Land and Natural Resources, Department of Agriculture), Federal (Animal and Plant Health Inspection Service, U.S. Fish and Wildlife Service), and private (The Nature Conservancy, Maui Land and Pineapple Company) agencies and organizations.

Priorities for research and monitoring to aid alien plant control efforts at Haleakalā follow.

1. Monitoring of vegetation change, including changes in populations of rare plant species, throughout the Park following removal of feral animals is an essential adjunct to all efforts at ecosystem management through control of alien species. Lower Kaupō Gap, middle Kīpahulu Valley, and the montane bogs are particularly critical sites for monitoring, but a comprehensive network of sites throughout the Park is needed. Ehrlich (1983) and Soulé (1983) have pointed out the crucial value of long-term monitoring in any effort to formulate strategies for preventing extinction. Along similar lines, Macnab (1983) pointed out the shortsighted failure of most manipulative wildlife management efforts to adequately evaluate the effects of the management. Such monitoring may at times simply verify the appropriateness of management actions, but monitoring often detects unforeseen trends that may lead to rethinking management strategies and priorities.

2. Explore methodology for a campaign to prevent introduction of new aggressive nonnative species. Develop protocols for cooperative monitoring of alien plants outside the Park.

3. Experimental use of herbicidal treatments to determine optimal methods for control of strawberry guava, molasses grass, blackberry, kāhili ginger, and other invasive alien species is needed. Although efforts from Hawaii Volcanoes are applicable, additional research in Haleakalā will probably be necessary because of elevation, vegetation, soil, and climatic differences.

4. Continued research on and monitoring of biocontrol agents for banana poka and clidemia are also necessary. Biocontrol is the only hope for defense against these aggressive rain forest invaders. These aliens will probably spread throughout forests of East Maui in the coming decades, including those of the Park, unless efforts at biocontrol are successful.

5. Biological control of blackberry may become an important tool. Efforts already under way could prove valuable for control of this alien invader. We should continue to support these efforts.

6. Biological control should be investigated for strawberry guava, Haleakalā's primary alien plant problem.

7. A study of vegetation and faunal dynamics in Kīpahulu Valley below 4,000 ft (1,220 m) should be conducted. The resources and alien species that threaten them are still largely unexplored. Such research is essential to develop realistic and effective resource management strategies.

8. The biology of black bamboo should be studied, and monitoring of bamboo in lower Kīpahulu Valley should be started.

Literature Cited

Anderson, S.J., C.P. Stone, and P.K. Higashino. [this volume] Distribution and spread of alien plants in Kipahulu Valley, Haleakala National Park, above 2,300 ft elevation.

Beardsley, J.W. 1980. *Haleakala National Park Crater District resources basic inventory: insects*. Tech. Rep. 31, Univ. Hawaii Coop. Natl. Park Resour. Stud. Unit. Honolulu.

Bess, H.A., and F.H. Haramoto. 1972. Biological control of pamakani, *Eupatorium adenophorum* in Hawaii by a tephritid gall fly, *Procecidochares utilis* -- III. Status of the weed, fly and parasites of the fly in 1966-1971 versus 1950-1957. *Proc. Hawn. Entomol. Soc.* 21: 165-178.

Degener, O.H. 1930. *Plants of Hawaii National Park*. Ann Arbor, Michigan: Edward Bros.

Diong, C.H. 1983. Population ecology and management of the feral pig (*Sus scrofa* L.) in Kipahulu Valley, Maui. Ph.D. Diss., Univ. Hawaii, Honolulu.

Ehrlich, P.R. 1983. Genetics and the extinction of butterfly populations. In *Genetics and conservation: a reference for managing wild animal and plant populations*, ed. C.M. Schonewald-Cox, S.M. Chambers, B. MacBryde, and W.L. Thomas, 152-163. Menlo Park, Calif.: The Benjamin/Cummins Pub. Co.

Gagné, B.H., S. Anderson, L. Loope, and A. Medeiros. 1991. *Miconia calvescens*: a threat to native forests of the Hawaiian Islands. (Abstract) *Pac. Sci. Congr.*, Honolulu, 30 May.

Gardner, D.E. [this volume] Plant pathogens as biocontrol agents in native Hawaiian ecosystems.

Gardner, D.E., and V.A.D. Kageler. 1983. *Glyphosate in the control of kikuyugrass and its effects on associated native and non-native plants in Hawaiian National Parks*. Tech. Rep. 45, Univ. Hawaii Coop. Natl. Park Resour. Stud. Unit. Honolulu.

Gaubert, H., and J. Florence. 1991. Alien plants in French Polynesia with the peculiar example of *Miconia calvescens* D.C. in Tahiti. (Abstract) *Pac. Sci. Congr.*, Honolulu, 30 May.

Higashino, P.K., L.W. Cuddihy, S.J. Anderson, and C.P. Stone. 1988 a. *Checklist of vascular plants of Hawaii Volcanoes National Park*. Tech. Rep. 64, Univ. Hawaii Coop. Natl. Park Resour. Stud. Unit. Honolulu.

Higashino, P.K., L.W. Cuddihy, S.J. Anderson, and C.P. Stone. 1988 b. *Bryophytes and vascular plants of Kipahulu Valley, Haleakala National Park.* Tech. Rep. 65, Univ. Hawaii Coop. Natl. Park Resour. Stud. Unit. Honolulu.

Jacobi, J.D. 1981. *Vegetation changes in a subalpine grassland in Hawai'i following disturbance by feral pigs*. Tech. Rep. 41, Univ. Hawaii Coop. Natl. Park Resour. Stud. Unit. Honolulu.

Janzen, D.H. 1976. Why bamboos take so long to flower. *Ann. Rev. Ecol. and Syst.* 7:347-391.

Juvik, J.O., and S.P. Juvik. [this volume] Mullein (*Verbascum thapsus*): the spread and adaptation of a temperate weed in the montane tropics.

Krauss, B.H. 1980. Ethnobotanical resources of Kipahulu Valley below 2000 feet. In *Resources base inventory of Kipahulu Valley below 2000 feet*, ed. C.W. Smith, 71-82. Sponsored by The Nature Conservancy. Honolulu: Univ. Hawaii Coop. Natl. Park Resour. Stud. Unit.

LaRosa, A.M. 1984. *The biology and ecology of* Passiflora mollissima *in Hawaii.* Tech. Rep. 50, Univ. Hawaii Coop. Natl. Park Resour. Stud. Unit. Honolulu.

LaRosa, A.M. [this volume] The status of banana poka in Hawai'i.

Loope, L.L., A.C. Medeiros, and B.H. Gagné. 1991a. *Aspects of the history and biology of the montane bogs of Haleakala National Park.* Tech. Rep. 76, Univ. Hawaii Coop. Natl. Park Resour. Stud. Unit. Honolulu.

Loope, L.L., A.C. Medeiros, and B.H. Gagné. 1991b. *Recovery of vegetation of a montane bog in Haleakala National Park following protection from feral pig rooting.* Tech. Rep. 77, Univ. Hawaii Coop. Natl. Park Resour. Stud. Unit. Honolulu.

Macnab, J. 1983. Wildlife management as experimentation. *Bull. Wildl. Soc.* 11(4):397-401.

Markin, G.P. 1984. Biological control of the noxious weed gorse *Ulex europaeus* L.: a status report. [abstract] *Proc. Fifth Conf. Nat. Sci., Hawaii Volcanoes Natl. Park*, 77. Honolulu: Univ. Hawaii Coop. Natl. Park Resour. Stud. Unit.

Markin, G.P., and E. Yoshioka. [this volume] Evaluating proposed biological control programs for introduced plants.

Medeiros, A.C., R.W. Hobdy, and L.L. Loope. 1989. Status of *Clidemia hirta* on Haleakala. *Newsletter, Hawn. Bot. Soc.* 28(1):3-4.

Medeiros, A.C., L.L. Loope, and B.H. Gagné. 1991. *Degradation of vegetation in two montane bogs of Haleakala National Park: 1982-1988.* Tech. Rep. 78, Univ. Hawaii Coop. Natl. Park Resour. Stud. Unit. Honolulu.

Medeiros, A.C., L.L. Loope, and R.A. Holt. 1986. *Status of native plant species on the south slope of Haleakala, East Maui.* Tech. Rep. 59, Univ. Hawaii Coop. Natl. Park Resour. Stud. Unit. Honolulu.

Mitchell, A.L. 1945. Checklist of higher flowering plants, grasses, sedges, rushes and ferns of the Haleakala section, Hawaii National Park. Files, Haleakala Natl. Park library. Unpub. rep.

Nakahara, L.M., R.M. Burkhart, and G.Y. Funasaki. [this volume] Review and status of biological control of clidemia in Hawai'i.

Rock, J.F. [1913] 1974. *The indigenous trees of the Hawaiian Islands.* Rutland, Vermont, and Tokyo, Japan: Charles E. Tuttle Co.

Santos, G.L., L.W. Cuddihy, and C.P. Stone. 1989. Cut-stump, frill, and basal bark treatments of triclopyr on strawberry guava. In *1989 Research progress report*, Western Soceity of Weed Science, Honolulu, Hawaii, March 14-16, 1989, 134.

Scowcroft, P.G., and R. Hobdy. 1986. Recovery of montane koa parkland vegetation protected from feral goats. *Biotropica* 19:208-215.

Smith, C.W. 1985. Impact of alien plants on Hawai'i's native biota. In *Hawai'i's terrestrial ecosystems: preservation and management*, ed. C.P. Stone and J.M. Scott, 180-250. Univ. Hawaii Coop. Natl. Park Resour. Stud. Unit. Honolulu: Univ. Hawaii Pr.

Smith, C.W. [this volume] Distribution, status, phenology, rate of spread, and management of clidemia in Hawai'i.

Smith, C.W., J.E. Williams, and K.E. Ascherman. 1985. *Vegetation map and resource management recommendations for Kipahulu Valley (below 700 metres), Haleakala National Park.* Tech. Rep. 53, Univ. Hawaii Coop. Natl. Park Resour. Stud. Unit. Honolulu.

Soulé, M.E. 1983. What do we really know about extinction? In *Genetics and conservation: a reference for managing wild animal and plant populations*, ed. C.M. Schonewald-Cox, S.M. Chambers, B. MacBryde, and W.L. Thomas, 111-124. Menlo Park, Calif.: The Benjamin/Cummins Pub. Co.

Stemmermann, L., P.K. Higashino, and C.W. Smith. 1981. *Haleakala National Park Crater District resources basic inventory. Conifers and flowering plants.* Tech. Rep. 38, Univ. Hawaii Coop. Natl. Park Resour. Stud. Unit. Honolulu.

Tunison, J.T. [this volume] Fountain grass control in Hawaii Volcanoes National Park: management considerations and strategies.

Tunison, J.T., and C.P. Stone. [this volume] Special Ecological Areas: an approach to alien plant control in Hawaii Volcanoes National Park.

Vitousek, P.M. [this volume] Effects of alien plants on native ecosystems.

Vogl, R.J., and J. Henrickson. 1971. Vegetation of an alpine bog on East Maui, Hawaii. *Pac. Sci.* 25:475-483.

von Tempski, A. 1940. *Born in paradise*. New York: Duell, Sloan and Pierce.

Wagner, W.L., D.R. Herbst, and S.H. Sohmer. 1990. *Manual of the flowering plants of Hawai'i.* Bishop Mus. Spec. Pub. 83. Honolulu: Univ. Hawaii and Bishop Mus. Pr.

Warner, R.E., ed. 1968. *Scientific report of the Kipahulu Valley Expedition.* Sponsored by The Nature Conservancy.

Warshauer, F.R., J.D. Jacobi, A.M. LaRosa, J.M. Scott, and C.W. Smith. 1983. *The distribution, impact and potential management of the introduced vine* Passiflora mollissima *(Passifloraceae) in Hawai'i.* Tech. Rep. 48, Univ. Hawaii Coop. Natl. Park Resour. Stud. Unit. Honolulu.

Wester, L.L., and H.B. Wood. 1977. Koster's curse (*Clidemia hirta*), a weed pest in Hawaiian forests. *Environ. Conserv.* 4(1):35-41.

Whiteaker, L.D., and D.E. Gardner. 1985. *The distribution of* Myrica faya *Ait. in the state of Hawaii.* Tech. Rep. 55, Univ. Hawaii Coop. Natl. Park Resour. Stud. Unit. Honolulu.

Wilkes, C. 1845. *Narrative of the United States Exploring Expedition during the years 1838, 1839, 1840, 1841, 1842.* Vol. I, pp. 252-256.

THE U.S. DEPARTMENT OF AGRICULTURE'S RURAL DEVELOPMENT APPROACH TO ALIEN PLANT CONTROL IN HAWAI'I: A CASE STUDY

Mike Tulang

Alien plant control programs provide no guarantees of success. In Hawai'i today we are faced with many plants that are extremely threatening to our native ecosystems, and gorse (*Ulex europaeus*) is one of them. Gorse also presents a fire hazard to private property (primarily on Maui now); it invades critical watersheds (*e.g.*, 20% of the water we drink in Hilo comes from the Hilo watershed, where gorse grows as low as 1,500 ft (460 m) elevation); and it threatens agricultural and grazing lands.

In this paper, the gorse problem on the island of Hawai'i is described and an approach to its solution detailed. The strategy involves interagency cooperation and a task force and management team. A similar approach is currently being used in dealing with the problem of firetree (*Myrica faya*) invasions on several islands.

THE PROBLEM

Gorse was inadvertently introduced sometime before 1910 with the establishment of the wool industry in Hawai'i. By 1966, upcountry Maui above Makawao was infested; a few thousand acres of gorse were present by 1986. Gorse now threatens to expand into suburban areas in upper Kula on Maui. On Hawai'i Island, the sheep industry produced mutton and wool for U.S. West Coast markets by the early 1900s. The major sheep (*Ovis aries*) raising area was at Humu'ula. With the fall of wool prices during the Depression and competition from foreign markets, ranchers converted the Humu'ula land division to cattle (*Bos taurus*) ranching. Gorse, until that time effectively controlled by sheep grazing, began to increase, and each succeeding drought seemed to contribute to the infestation as foraging animals were temporarily removed. Gorse now is found on some 20,415 a (8,262 ha) at Humu'ula. The infestation on Maui has followed a similar course, with total infested area estimated at 14,800 a (5,985 ha).

The current gorse infestation at Humu'ula (Fig. 1) affects three major ranch operations: Parker Ranch, Pu'u 'Ō'ō Ranch, and Pua 'Ākala Ranch. It also impacts the Wailuku watershed area. Gorse has effectively removed

7,500 a (3,035 ha) of land from forage production in the area. The bulk of the land infested is leased from the Hawaii Department of Hawaiian Home Lands. As of spring 1987, the total known area of distribution, excluding isolated plants located in the Mauna Kea and Hilo forest reserves, was estimated at 20,415 a (8,262 ha), with a heavy concentration of plants on 6,200 a (2,509 ha) located along Keanakolu Road at Humu'ula. The lower boundary of gorse infestation was generally the upper edge of the Hilo and Mauna Kea forest reserves at 2,400 m (7,870 ft) elevation.

On Maui, the northernmost part of the 14,800-a (5,985-ha) infestation was at an elevation of 2,070 ft (630 m) near Pu'u Pi'iholo, southeast of the community of Makawao. The northernmost and highest point of the infestation at 7,280 ft (2,220 m) is located at the boundary of Haleakala National Park (Fig. 2). The heaviest concentration of plants is located within a half-mile radius of the Olinda Endangered Species Captive Propagation Facility at 3,530 ft (1,025 m). An isolated patch of plants was located at 2,070 ft (630 m) elevation along Māliko Gulch.

Temporary control of gorse was accomplished in the 1970s, with manpower provided through the funding of the Comprehensive Employment Training Act and the use of 2,4-5 T, or Agent Orange. Termination of the Training program and a subsequent ban of 2,4-5 T by the U.S. Environmental Protection Agency allowed a major increase.

THE GORSE TASK FORCE

Program and Objectives

The gorse control program in Hawai'i now consists of a coordinated management approach featuring a containment program, a long-term biological control program, and implementation of an integrated native species reforestation and grazing management program. In order to manage and coordinate the program, a management team and a Gorse Task Force, consisting of affected land owners, ranchers, and government agency representatives, were established. The Task Force is sponsored and supported by the Mauna Kea Soil and Water Conservation District. This District draws its legal authority from Chapter 180, Hawaii Revised Statutes, and all land owners in the District are represented on the Task Force. At the onset, landowners and operators realized that they could not control gorse without the assistance of the Hawaii Department of Agriculture and the U.S. Department of Agriculture.

The Task Force was developed to plan a strategy for an immediate containment program and also to combat gorse over the long term. There were four program considerations: 1) A strategy for securing adequate funding had to be developed. (Markin *et al.*, this volume, alluded to the need for at least a million dollars for insect biological control agents alone.) 2) Research and development had to be conducted throughout the project term. (The Gorse Task Force decided on a seven-year project.) 3) Approaches to resolving the gorse problem had to be integrated with long-term land use objectives of the land owners. 4) Cooperation among

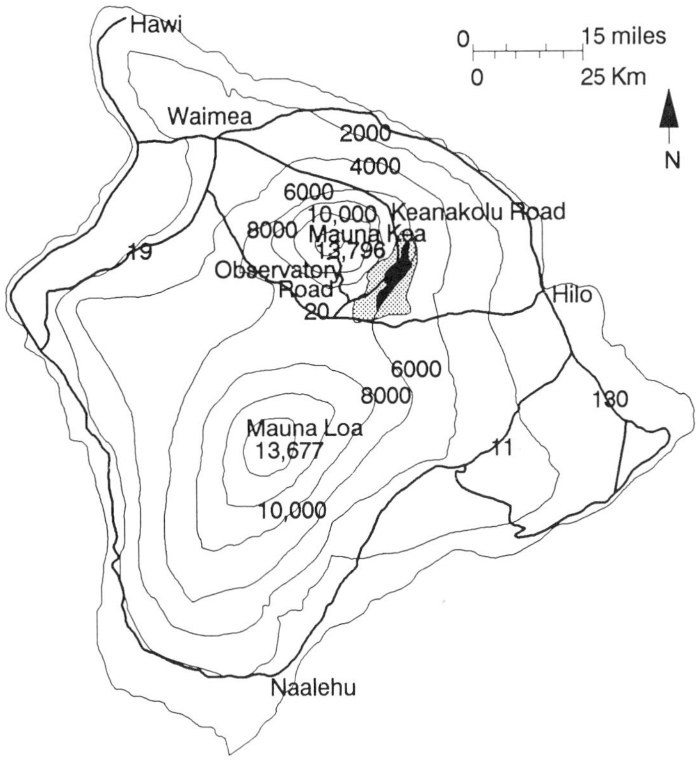

Figure 1. Gorse infestation areas on the island of Hawai'i are indicated by shading (sparse) and solid-fill (abundant). Contour lines are at 2,000 ft (610 m) intervals.

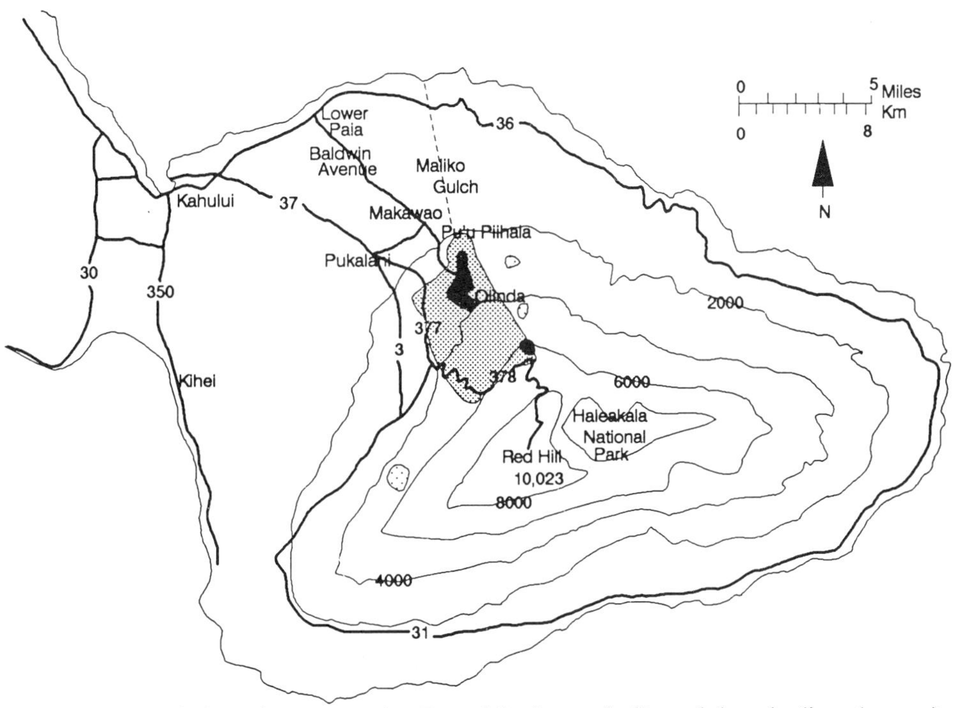

Figure 2. Gorse infestation areas in East Maui are indicated by shading (sparse) and solid-fill (abundant). Contour lines are at 2,000 ft (610 m) intervals.

governmental agencies, special interest groups, and private industry had to be initiated and maintained during program operation.

Present planning objectives of the Task Force are to: 1) contain the gorse infestation on Hawai'i Island to its present extent at Humu'ula and prevent its spread to other areas; 2) reduce gorse within the project area to a level that ranchers can economically justify controlling; 3) restore the productive capacity of affected rangelands; 4) protect the Hilo watershed and the surrounding forest areas by introducing native trees such as koa (*Acacia koa*) and māmane (*Sophora chrysophylla*) in suitable sites within the project areas.

Planning for the first five years of the program is completed, and all phases of the program (below) were implemented in the field by October 1986. The search for and testing of potential biological control insects began simultaneously with the herbicide spraying of gorse patches located along the borders of the project area and 100 ft (33 m) on each side of Keanakolu Road. The Tordon 22K spraying plan was designed to eliminate any possibility of seed transport to other areas of Hawai'i.

Task Force Plan Components

The biological control component of the plan features research on pathogens in the first year. Late in 1986, a researcher sponsored by private industry was sent to Western Europe (England, Spain, France, and Portugal), the native habitat of gorse, to develop a list of gorse pathogens. A list of 20 pathogens was reviewed, and at least one should be released within the next several years. Pathogens must be carefully screened to assure that they will not attack nontarget plants. A $1.5 million pathogen lab has been constructed in Honolulu, and a research pathologist has been hired.

Under a cooperatively funded contract managed by the Hawaii Department of Agriculture and the Ministry of Agriculture and Forestry, Government of New Zealand, host testing of biological control insects of gorse were undertaken in England. Candidates considered were the gorse mite *(Teranycus lintearious)*, the gorse thrips (*Sericothrips staphylinum*), and the spine-feeding moth (*Scythris grandipennis*). In addition, a new colony of moths (*Agonopterix ulicitella*) was secured as new genetic material for mass rearing. During 1984, the gorse seed weevil *Apienulicis* was released and continued to expand its range and increase its population through the gorse population on Mauna Kea. To date, the weevil is well established and spreading rapidly to all parts of the gorse infestation in the Humu'ula area on the island of Hawai'i.

The reforestation component of the program is jointly pursued by the Hawaii Division of Forestry and Wildlife and the U.S. Forest Service. Experiments using various alien and native forest tree species to suppress gorse growth by shading and the release of plant toxins are in progress on the islands of Maui and Hawai'i. Seven experiments are ongoing: one at Olinda, Maui, and six in the Humu'ula area of Hawai'i Island. Interim reports indicate that slash pine (*Pinus elliottii*) and koa may be the most promising candidates. The slash pine, after 10 years' growth, has completely eliminated any gorse growth within the reforested ara. The koa, after five years, appears to be outgrowing and suppressing gorse by "shading out" the gorse plants in the experimental plots.

The chemical component of the program is really a containment operation that is designed to keep the gorse infestation in check and reduce populations until gorse can be controlled by a biological control program. Herbicides, which are expensive management tools, provide quick suppression but not permanent control. The Hawaii Department of Agriculture has undertaken an aggressive program to control gorse growth in a corridor along the Keanakolu road on the island of Hawai'i. The Department has also executed cooperative herbicide spraying agreements with all the ranches in the gorse-affected areas in Hawai'i.

The animal impact component of the gorse control program is managed by the University of Hawaii Cooperative Extension Service through a grant received from the county of Maui. The grant of $10,000 has established an ongoing experiment on Maui using goats (*Capra hircus*) as a potential grazing management for gorse control without the use of herbicides. A grazing management trial was also undertaken by the Agricultural Research Division, Ministry of Agriculture and Fisheries, of New Zealand in January 1981. This trial compared the effects of goats or sheep grazed alone, or in mixtures with two goats equivalent to one sheep (all dry stock), with either rotational grazing or set stocking (Rolston *et al.* 1981; Radcliffe 1985). An extra-fast rotation with sheep was included. Three years of grazing after a gorse burn and oversowing resulted in both goat-only management treatments and the sheep and goat mixtures rotationally grazed reducing gorse to negligible levels. If the experiment on Maui proves to be equally successful, the Gorse Interagency Task Force plans to expand the goat trials to 600 a (240 ha) in the Laumai'a area located within the gorse infestation at Humu'ula, Hawai'i.

Why don't we control gorse by burning it? Gorse, believe it or not, is a very combustible plant. In Europe, it has been burned for hundreds of years to control it and encourage reproduction, but seed banks are, unfortunately, encouraged by fire. One gorse fire that we deliberately set in June 1985 was visible 25 mi (35 km) away in Hilo and burned very rapidly. The response of gorse to this fire is being monitored.

DISCUSSION AND CONCLUSIONS

The alien plant problem is largely one of rural communities and natural areas. To be successful, programs dealing with such problems must have adequate financing. Researchers, administrators, and resource managers concerned with alien plants must develop sound strategies and collective approaches. Equally important, rural areas must do a good job of program management in order to compete with urban programs for much-needed funding.

Resource managers, researchers, and administrators have to recognize the realities of gorse control. Alien plant problems are long term and require long-term and expensive solutions. They require considerable research, commitment of personnel, and real continuity of support. Control is very expensive, especially when requirements of international travel, communication, and coordination of a biological control program are considered. Alien plant control is highly political, and coordination, planning, and an organized approach to obtain support are necessary. Political actions must be well timed, and friendships and knowledge must be cultivated before a well-documented plan goes into effect.

A cooperative approach using a variety of control methods is the most feasible way to address problems of such complexity and expense. A well-designed program demands the cooperation and support of all participants. A program to control aggressive and oppressive plants like gorse is best coordinated by a task force, which can identify and marshall all the resources that relate to the alien plant. Politicians must be lobbied to obtain legislative action. We must enjoin the support of government at all levels -- county, State, and Federal -- and establish positive working relationships. We must develop a well conceived strategy that is supported by effective public information programs. And lastly, we must not forget that we are part of a large community and must try to work within that framework. Our task is too important to accommodate adversary relationships.

ACKNOWLEDGMENTS

Sincere appreciation is expressed to members of the Hawaii Interagency Committee on Gorse Control for their technical assistance and valued cooperation: G.P. Markin, U.S. Forest Service; H. Ooka, M. Isherwood, and E. Yoshioka, Hawaii Department of Agriculture.

Literature Cited

Markin, G.P., P.-Y. Lai, and G.Y. Funasaki. [this volume] Status of biological control of weeds in Hawai'i and implications for managing native ecosystems.

Radcliffe, J.E. 1985. Grazing management of goats and sheep for gorse control. *New Zealand Journal Exp. Agric.* 13:181-190.

Rolston, M.P., M.G. Lambert, D.A. Clark, and B.P. Devantier. 1981. Control of rushes and thistles in pastures by goat and sheep grazing. *Proc. 34th New Zealand Weed and Pest Control Conf.* 117-121.

INTRODUCED PLANT CONTROL ON PRIVATE LANDS IN HAWAI‘I

Carl "Soot" Bredhoff

ABSTRACT

Introduced plant control as practiced by ranchers in Hawai‘i is briefly discussed. Mechanical control is usually used to open new areas or maintain pastures heavily overgrown with brush. Herbicides have been applied in many different ways by ranchers, and more knowledge is needed. The use of foraging animals to control introduced weeds, especially woody plants, has been used for years in Hawai‘i, but fire has been less successful because of long-term regime effects on rangelands. More biological research is needed to address rangeland weeds.

INTRODUCTION

When I was in school, my soils professor used to remind us not to refer to soil as dirt. "Dirt," he said, "is matter out of place." Similarly, when alien or introduced plants get out of place, they become pests that must be eliminated or controlled. Lantana (*Lantana camara*), for instance, is a desirable ornamental at the Keāhole Airport on Hawai‘i Island or in your local shopping center, but in the pasture it is an undesirable pest. The same is true of Christmas berry (*Schinus terebinthifolius*), common guava (*Psidium guajava*), kāhili ginger (*Hedychium gardnerianum*), and many other plants.

Ranchers in Hawai‘i, along with other private landowners, have tried over the years, with varying degrees of success, to control introduced plants. This paper is written from the perspective of a cattle rancher controlling alien plants on private lands. Some of the methods used by ranchers are outlined in this paper.

MANUAL CONTROL

It all started with the axe, pick, saw, mules, and horses. These were used until better methods came along and labor costs went up.

MECHANICAL CONTROL

Mechanical controls usually are a primary step in taming pasturelands that have become overgrown and heavily infested with alien plants, or in opening up new areas for establishing pastures. After the primary step, eventually other methods are used to control alien plants.

Bulldozing, or grubbing, is usually used where large alien plants such as Christmas berry have invaded. Chain dragging can be used on woody plants such as koa haole *(Leucaena leucocephala)*, the native 'a'ali'i *(Dodonaea viscosa)*, and lantana. The chain either breaks off the plants or pulls them up by the roots. Usually, an anchor chain is pulled by a bulldozer. The loop of the chain is held open by a heavy bar, or the chain can be pulled by two bulldozers.

Rolling, or the use of a roller-chopper pulled by a bulldozer, is also effective on woody plants and shrubs. In addition, a smooth roller can prepare the ground, especially rocky ground, for mowing later. The two most common types of rollers used here in Hawai'i are the smooth Krajewski roller and the roller-chopper. The Krajewski weighs about 15 tons and is fabricated from used sugar mill rollers. Many ranchers on the island of Hawai'i use this type. The roller-chopper has blades on it and can be filled with water to make it heavier. It does a good job of chopping woody brush. At Kaupō Ranch on Maui, we had good results controlling lantana and koa haole with a roller-chopper.

Mowing can be used to control introduced plants if the terrain permits. The widths of mowers vary, but mowers usually create a five- to seven-foot swath. Mowers are pulled either by crawler or rubber tire tractors. At Kahuku Ranch in Ka'ū on Hawai'i Island, the lower pastures in the Kamā'oa area were first rolled with a Krajewski roller to smooth out the terrain. Lantana was then controlled by mowing.

BIOLOGICAL CONTROL

Biological control was successful with cactus (*Opuntia* spp.) several years ago and more recently with Hāmākua pāmakani (*Ageratina riparia*) on the island of Hawai'i (Davis *et al.*, this volume). On the other hand, several insects have been introduced to control lantana (Davis *et al.*, this volume), but it continues to thrive on our rangelands. I hope that more work will be done on biological control of alien plants. It is certainly an economical method of control, once developed.

It was through the efforts of local cattlemen, working with the University of Hawaii, that the biological control of Hāmākua pāmakani was made possible just a few years ago. The present efforts of Maui cattlemen towards biological control of the grass webworm (*Herpetogramma licarsisalis*) is another example. Ranchers have a role in getting biological control research started, through lobbying legislators and so forth.

HERBICIDES

Over the years, herbicides have often produced very good results for ranchers. However, herbicides must be used with caution and with the safety of the environment and the user in mind. There is a need for continuing research into what chemicals, or combinations of chemicals, to use and the best ways to apply them. Some methods of application that have been used over the years are paint brush, pellets, knapsack sprayer, spray rig with nozzles or hose attachments, aerial spraying from fixed-wing plane or helicopter, and the rope-wick applicator which is brushed over the leaves of the plant. Some herbicides are sold over the counter in stores, while others are restricted in use and require a license for purchase. Information on herbicides can be obtained from the Cooperative Extension Agents in the localities where use is planned.

ANIMALS

By carefully grazing pastures with cattle, sheep, or goats (*Bos taurus, Ovis aries, Capra hircus*), we can control the introduced plants that are palatable to the grazing animals. Insufficient grazing causes grasses to become rank, stemmy, and unpalatable, and to lose much of their nutritional value. At the same time, we must try not to overgraze the desired species, or soon grass cover will decrease, and the pastures will become infested with weeds.

In addition to controlling introduced plants through general grazing, animals have been used for years in Hawai'i to open up brush land in order to develop grassland. At Kapāpala Ranch on the island of Hawai'i, a herd of goats is being used to control common guava and Christmas berry. Cattle can be used to control certain species of introduced plants by using high-density stocking rates for a short period of time. Intensive grazing may control certain problem weeds and in fact has been compared for effectiveness with herbicidal control (Anonymous 1987). The Voison or short-duration grazing system is useful in brush control, and very briefly it works as follows: grazing, hoof action, and movement of the cattle serve to break and trample the vegetation, allowing sunlight and moisture to reach desirable grasses and legumes; the hoof action breaks up any soil capping and makes places to trap rain; and the cattle deposit dung and urine. All of this has a beneficial effect on the pasture and assists in brush control. Increasingly, ranchers are putting part or all of their pasture lands into intensive grazing units for better pasture utilization and greater production. A book entitled *Intensive grazing management: forage, animals, men, profits* addresses situations in Hawai'i and is available locally (Smith *et al.* 1986).

FIRE

Fire has been used, with varying degrees of success, to control introduced plants on pasture land and to clear off old, rank grasses prior to broadcasting seed. When fire is used, timing is very important. Rain is essential shortly after a fire so that the desirable grasses can make a quick recovery, getting a jump on the species being controlled and covering the ground exposed by the fire. But fire has a number of negative effects. For instance, it disturbs the microenvironment of the topsoil and decreases the aeration. It can also be a hazard. Many a pasture fire has burned out of control. Overall, the long-term detrimental effects of burning overshadow the short-term benefits, so this method is probably the least used for control of introduced plants on private lands in Hawai'i.

CONCLUSION

State and Federal agencies help private landowners by disseminating information, setting up trials and experiments, conducting research, and giving direct assistance such as controlling fountain grass (*Pennisetum setaceum*) along the highway in Ka'ū on the Hawai'i Island. These agencies include the University of Hawaii Extension Service, the U.S. Department of Agriculture Soil Conservation Service, the Hawaii Department of Agriculture Plant Pest Control Branch, and local cattlemen's associations.

This has been a very brief and general summary of how introduced plants are controlled on private ranchlands in Hawai'i. Each ranch has unique weed problems, and the type of control used depends on location, economics, and the landowner's situation.

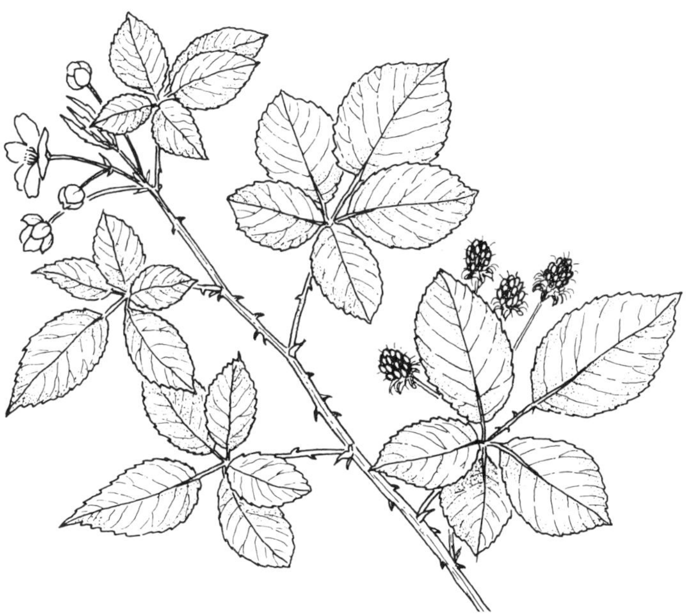

Literature Cited

Anonymous. 1987. Intensive grazing may control problem weed. *Hay & Forage Grower*, January, p. 19.

Davis, C.J., E. Yoshioka, and D. Kageler. [this volume] Biological control of lantana, prickly pear, and Hāmākua pāmakani in Hawai'i: a review and update.

Smith, B., P. Leung, and G. Love. 1986. *Intensive grazing management: forage, animals, men, profits*. Kamuela, Hawai'i: Grazers Hui.

PUBLIC/PRIVATE SECTOR PARTNERSHIPS TO PRESERVE NATIVE ECOSYSTEMS -- THE DEPARTMENT OF HAWAIIAN HOME LANDS EXPERIENCE

Hardy Spoehr

ABSTRACT

The Department of Hawaiian Home Lands manages 190,000 acres for native Hawaiians in Hawai'i. Homestead leases are provided for about 99 years, with renewal provisions, and general revenue-producing leases are also available. Eight biologically sensitive areas have been identified on Department lands, each with present and potential threats. Major land stewardship programs for preservation and management of native ecosystems are not yet possible, but cooperative arrangements with other organizations and individuals are being emphasized.

INTRODUCTION

This paper will focus on public sector involvement in native ecosystem preservation in Hawai'i. Examples for discussion are drawn from activities being conducted by the Department of Hawaiian Home Lands (the Department). However, before embarking on a brief review of Department activities, a few personal observations regarding public sector involvement are in order.

DEALING WITH BUREAUCRACY

One of the major problems conservationists have in dealing with the public sector is structural bureaucracy. Individuals in public bureaucracies sometimes have different perspectives, not about basic needs or concerns, but rather about the methods to achieve desired ends. Bureaucrats, by definition, have rules according to which they must make decisions. They are often confined to specific areas, and this is not always understood by those outside the bureaucracy. Bureaucratic decision-making sometimes manifests itself in negative reasoning, *i.e.*, telling people all the reasons why something "can't be done." To counter this, there has to be an active awareness -- by bureaucrats and environmentalists alike -- of each other's <u>constraints</u> and <u>restraints</u>. Constraints often are brought about by mandates from legislative acts or judicial opinions. To remove constraints, legal action is usually necessary. Restraints, however, are self imposed and may become the basis for non-action. To remove restraints, legal action should not be necessary. Statements like "the boss won't like it" or "it's not popular" are examples of restraints, not constraints. Those who support change must distinguish between these two. If restraints are problems, it is our responsibility as bureaucrats or conservationists to reevaluate our thinking, to educate ourselves, and to be open to new ideas and thoughts. One of the ways this can begin to occur is through dialogue. Whether we are bureaucrats or conservationists, there needs to be a basic agreement in values. Discussions of disagreements caused by constraints should eventually receive our major emphasis.

Within public agencies, decisions are made at two levels -- at the policy level (boards, commissions, trustees, etc.) and at the staff level (department, etc.). How conservationists integrate and involve people from these two levels is important. Individuals from the policy level need to attend symposiums, workshops, and conferences; yet, all too often, they don't. Policy makers may be attorneys, business people, educators, or accountants, and they may have little or no professional training or education in the sciences. It is the task of conservationists to begin educating these people, to develop in them an awareness and appreciation of Hawai'i's environment, and to get them to attend and contribute to meetings concerned with conservation. Similarly, public agency staffs need to be encouraged to attend meetings and contribute. Staff people can make things happen. From secretaries to administrators, all should have some degree of knowledge about Hawai'i's natural heritage.

In brief, conservationists have two challenges with regard to public agency personnel. First, they must raise awareness levels about environmental issues in ways that are meaningful; second, they must include and involve public agency people in meaningful projects and activities.

The public sector has a responsibility to be reactive rather than proactive. This often is frustrating for conservationists, who tend to be action oriented. But the wheels of government bureaucracy turn slowly, and, more often than not, bureaucrats are concerned with what is about to or has already occurred. Conservationists should keep this in mind when approaching the public sector for support. Conservation projects must be developed completely and should fall into already-existing public agency mandates. If they do not, perhaps private sector support should be attempted.

Impacts on the integrity and vitality of Hawai'i's native ecosystems are occurring daily. Yet the job of protecting the environment has become too large and too critical for only the public or private sector alone. Mother Nature does not make the distinction between lands owned by public and private interests. It will undoubtedly take a concerted effort by all parties through public-private sector partnerships to successfully preserve for future generations the native ecosystems that can serve us in so many ways. The Department of Hawaiian Home Lands is attempting to establish such partnerships.

PROGRAMS, POLICIES, AND PARTNERSHIPS

The Department of Hawaiian Home Lands is unique among state agencies in that it was established by U.S. Congressional mandate. The Hawaiian Homes Commission Act of 1920, as amended, set aside some 190,000 a (76,923 ha) for native Hawaiian "rehabilitation" (Fig. 1). Although the Act neglected to define "rehabilitation" directly, the term has come to mean providing for a homesteading process, whereby those individuals of at least 50% Hawaiian ancestry are provided residential, agricultural, and/or pasture allotments for 99 years. In addition, the Department issues general leases that generate revenue to support the Department's operating expenses. The legal status of Department land ownership changed with statehood from that of the Federal government to that of the State, but Department lands are not to be confused with those administered by the Department of Land and Natural Resources. Hawaiian Home Lands are trust lands specifically set aside for the use and benefit of native Hawaiians. While legal questions still exist about whether or not specific Federal, State, and/or county laws and codes apply to Department lands, the Department does have a great deal of latitude in the use of its lands. For example, Department lands are not subject to county zoning or state land use designations if used for homesteading, and the Department has the authority to condemn "surplus water" for homesteader use. These provisions are in the Hawaiian Homes Commission Act, as amended. The Act serves as the Department's constraint. It is the mandate under which it operates, and conservationists who do business with the Department must keep this in mind. If environmental concerns can fit within the parameters of native Hawaiian "rehabilitation," then there is ground for mutual discussion and benefit.

Department of Hawaiian Home Lands policy is established by a nine-member Commission, which meets monthly. None of the current commissioners has a science background, but all have some awareness of native ecosystems as they relate to Hawaiian culture. The expertise of Department staff is focused primarily on land development either for native Hawaiian homestead use or for commercial development. Departmental services include loans and loan servicing for homestead development, and land development activities such as engineering and planning. It may be difficult to see where native ecosystem preservation fits into this picture, but it does. The trick is knowing who does what in the organization. It then becomes a matter of bringing the individuals responsible for native ecosystems up the learning curve, if necessary, and having those individuals interact with appropriate policy makers.

The Department does not operate in a vacuum. As a landowner, it attempts to know its land resources. To date, the Department of Hawaiian Home Lands has identified nine major areas that are biologically sensitive or that have unique ecosystems. Many people have helped in the identification process, especially Jim Jacobi, Sonny Kaniho, Cameron Kepler, Lloyd Loope, Art Medeiros, Mike Scott, and Rick Warshauer. The nine biologically sensitive areas so far identified on Department lands are shown in Table 1.

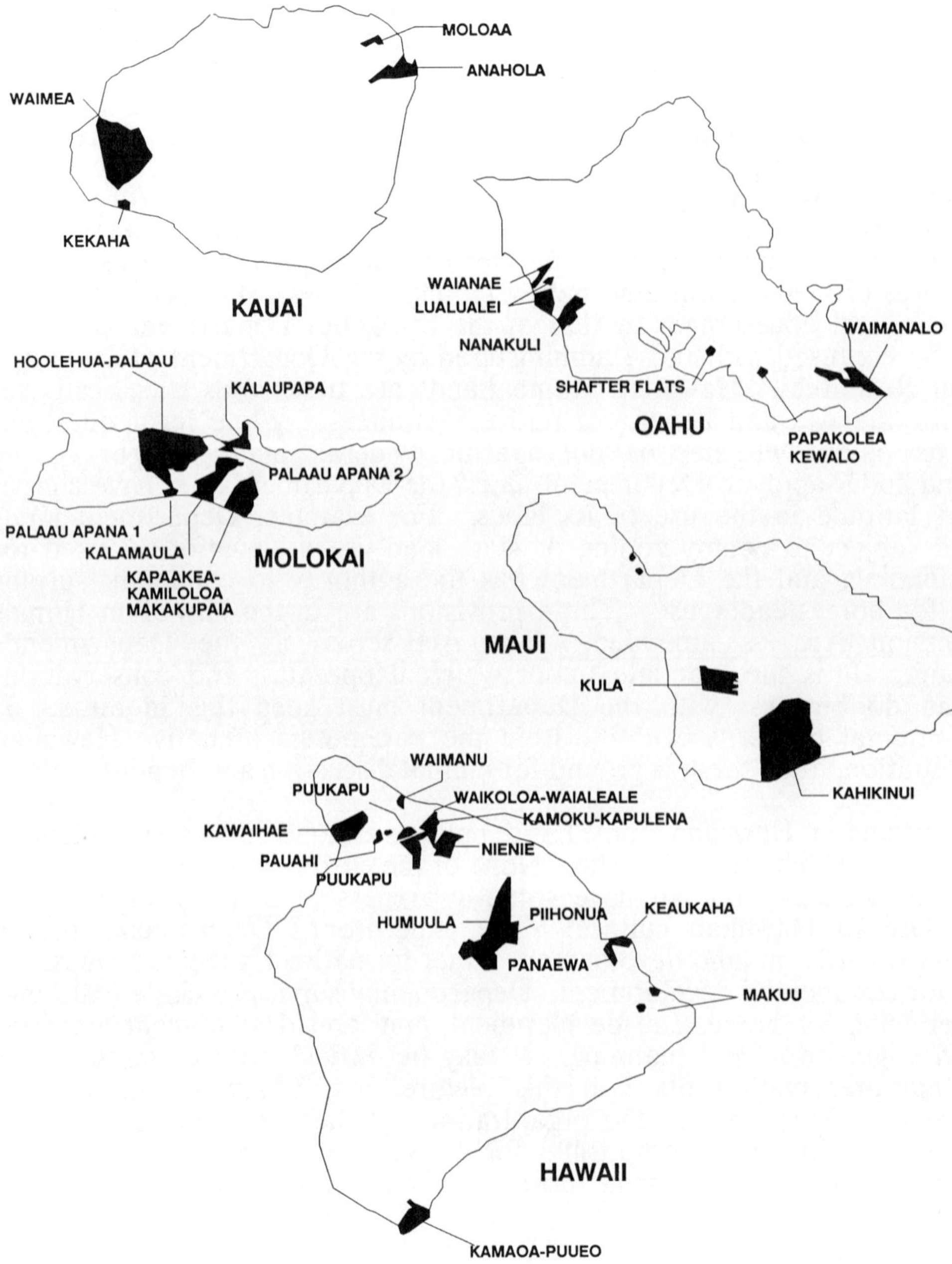

Figure 1. Hawaii Home Lands in the state of Hawai'i.

Table 1. Biologically sensitive areas included in Department of Hawaiian Home Lands holdings.

Area	Location	Biological value	Size (a)
Waimanu Valley	Kohala, Hawai'i	Water estuary	200
Wailuku River headwaters	Humu'ula, Hawai'i	Mauna Kea silverswords	50
Upper Pi'ihonua	Pi'ihonua, Hawaii	Koa forest	1,000
Mauna Kea Forest Reserve	Humu'ula, Hawai'i	Māmane forest	1,500
Pu'u o Kali	Kula, Maui	Native dryland forest	300
Kahikinui Forest Reserve	Kahikinui, Maui	Native forest plants	1,200
Mo'omomi Beach	Ho'olehua, Moloka'i	Beach strand vegetation	25
Waimea gullies	Waimea, Kaua'i	Native dry shrublands	500
Nanakuli Valley	Nanakuli, O'ahu	Native plants	10

There are many present and potential threats to these areas, which the Department cannot begin to control by itself. There will have to be help from concerned and informed agencies and individuals. Gorse (*Ulex europaeus*), banana poka (*Passiflora mollissima*), clidemia (*Clidemia hirta*), lantana (*Lantana camara*), and Christmas berry (*Schinus terebinthifolius*) are already present on Department lands. The Department has begun to build partnerships with public and private sector landowners who have similar problems and concerns. Because of its overriding mandate, the Department is not able to conduct major land stewardship programs that focus on preservation and management of native ecosystems. However, it can assist other groups that are willing to do so. As examples, the Department is participating in the Hawaii Steering Committee on Gorse Control (Tulang, this volume) and has begun discussions with The Nature Conservancy of Hawaii on ways in which the Conservancy's Heritage Program can assist in identifying Federally Threatened and Endangered species on Department lands.

The work of alien plant control on many lands in Hawai'i is just beginning. The preserves of The Nature Conservancy of Hawaii, the U.S. Fish and Wildlife Service Hakalau Forest National Wildlife Refuge, the State Natural Area Reserves, the National Park Service lands, and the many unique ecosystems on private lands cannot be sustained or maintained without cooperative public-private efforts. We must build on current efforts, create new joint ventures where necessary, and together continue the job of native ecosystem identification, preservation, and management.

INTRODUCED PLANTS IN THE ABSENCE OF UNGULATES

RELATIONSHIPS AMONG NATIVE AND ALIEN PLANTS ON PACIFIC ISLANDS WITH AND WITHOUT SIGNIFICANT HUMAN DISTURBANCE AND FERAL UNGULATES

Mark D. Merlin and James O. Juvik

ABSTRACT

The native plants of remote tropical islands have been frequently characterized as poor competitors against seemingly more aggressive alien species. Does this "weak competitor" characterization relate to some real adaptive consequences of island isolation and endemism, or does the generally concurrent presence of introduced ungulates and other forms of recurrent human disturbance also act to encourage alien plant dominance? A comparison of tropical islands with and without introduced ungulates suggests that some insular plant species competitively resist alien displacement in the absence of ungulates.

INTRODUCTION

For millions of years remote tropical islands in the Pacific Ocean have provided a variety of ecological opportunities for plant species that reached them through long-distance dispersal mechanisms. Many species that successfully established themselves on far-flung oceanic islands gave rise to extraordinary endemic forms, examples of adaptive radiation, and unusual adaptive shifts. Evolutionary developments occurred on isolated islands largely because of the limited numbers and kinds of colonizing taxa and varying environmental diversity within islands or groups of islands.

Among the structural and physiological adaptations that frequently occur in remote island environments is the disappearance of typical defensive mechanisms such as poisons, strong odors, thorns, deep tap roots, and tough stems and branches in insular plant species. Some of these adaptations left many species especially vulnerable to a variety of alien ungulates introduced in the historic period (Fosberg 1965; Mangenot 1965; Mueller-Dombois 1975). In the Hawaiian Islands, for example, only a very small fraction of the endemic species of plants produce poisons, thorns, or other defensive strategies against herbivory (Carlquist 1974, 1980).

The depredations of introduced hoofed mammals in isolated, tropical island ecosystems are well documented in the literature (*e.g.*, Skottsberg 1957; Cuddihy and Stone 1990) and are starkly exemplified in many island ecosystems by an array of habitat deterioration and soil erosion problems. What is less clear is the role of competition (without associated disturbance by humans or introduced animals) among alien and native plants in insular ecosystems. In other words, do differential survival rates generally occur for native and alien plants without the influence of free-ranging ungulates and other forms of direct or indirect human environmental disturbance?

If native island plant communities (especially those with high percentages of endemism) contain intrinsically weak competitors, then aggressive alien plant species should ultimately prevail under competition even without sustained human disturbance or alien herbivores (Dasmann *et al.* 1973; Gade 1985). By contrast, if insular ecosystems are reasonably resilient and native plants are well adapted and competitive, these communities should generally persist in the face of continued alien plant introductions as long as alien herbivores are absent or controlled (Mueller-Dombois *et al.* 1981; Merlin 1985, 1991). Research in the Hawaiian Islands (Egler 1942; Hathaway 1952; Mueller-Dombois 1975; Wirawan 1978), and on the raised coral island of Nauru (Manner *et al.* 1985) has shown that native vegetation can sometimes regain dominance after the removal of unnatural disturbance.

The question of alien-native plant competition in the absence of ungulates is of more than theoretical interest; it has important implications in developing management strategies for protection of intact native ecosystems and rehabilitation of disturbed insular environments. If lack of adaptation to grazing results in the replacement of native plant species by aliens, then animal removal may be sufficient to promote native ecosystem regeneration (barring irreversible changes caused by ungulates, *e.g.*, severe soil erosion, extinction of pollinators, or altered nutrient cycles derived from alien plants that fix nitrogen (Allan 1936; Bates 1956; C.W. Smith, pers. comm. 1990). However, if the native flora is composed of comparatively weak competitors, then management of alien plants will pose a continuing problem even after ungulate removal.

In many contemporary oceanic island settings, environmental conditions appear to be stacked heavily on the side of alien biota. Alien plants are often strongly adapted to survival in association with direct human disturbance and browsing, grazing, and rooting ungulates. In this paper, we review the literature and describe our own comparative field work on islands where plant competition among alien and native species may be assessed in the absence of free-ranging ungulates or other forms of continual human disturbance. Our present analysis is based on the following Pacific island cases (Fig. 1): Nihoa and Kaho‘olawe in the Hawaiian Islands; Hatuta‘a and Eiao in the northern Marquesas; the Kermadec Islands; Henderson and Nauru Islands; and the southern Cook Islands. These island pairs were chosen to contrast disturbed and relatively undisturbed (by ungulates) regimes.

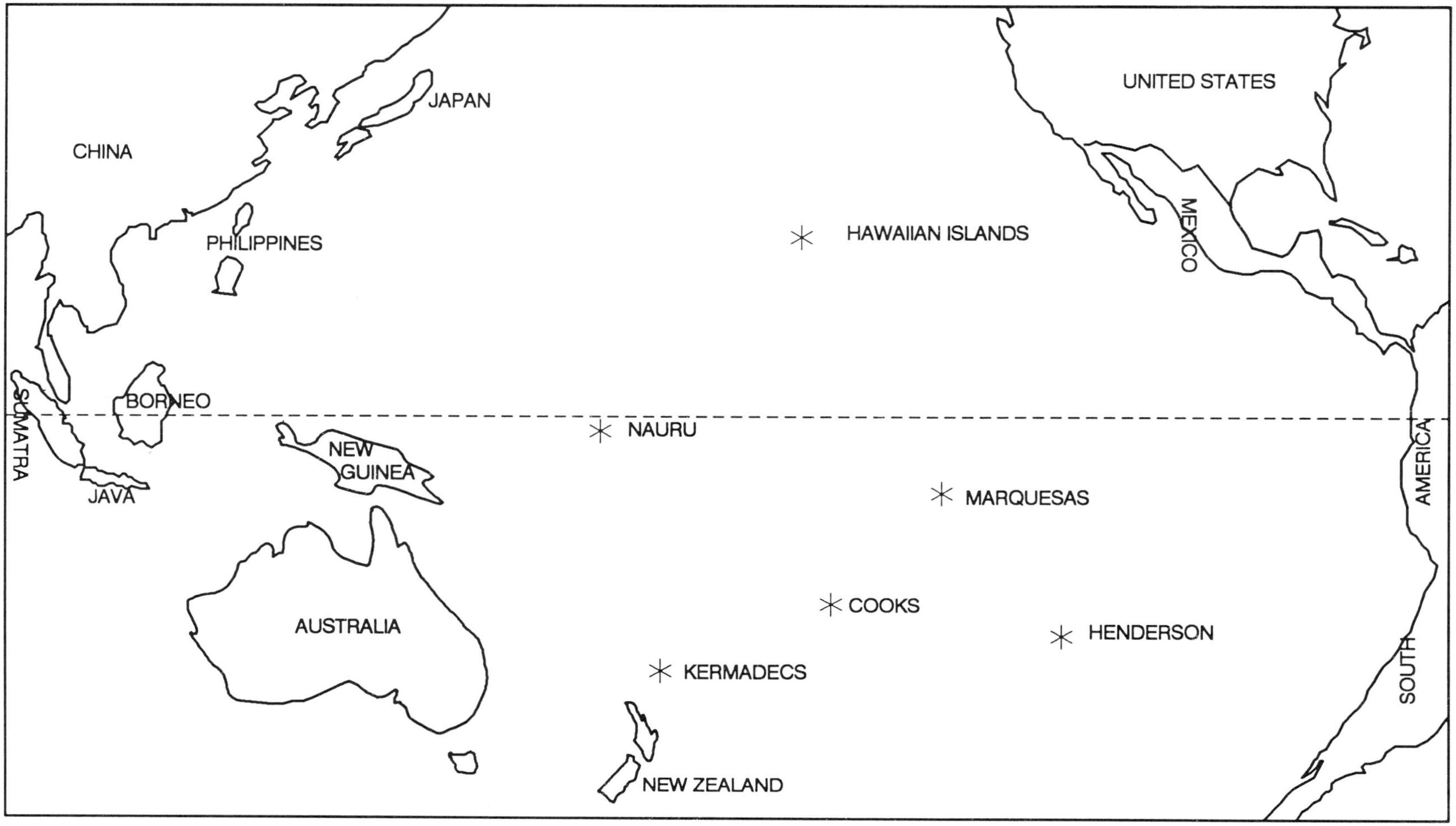

Figure 1. The Pacific Basin, showing relative positions of the island groups and islands discussed in this paper.

Nihoa and Kaho'olawe

Nihoa is a small (190 a or 77 ha), steeply sloped volcanic island of the Hawaiian Archipelago that reaches a maximum elevation of 910 ft (277 m). It lies about 155 mi (250 km) northeast of Kaua'i (Fig. 2), is relatively dry, geographically remote, and has very poor landing access. Although ample evidence exists that prehistoric humans used the Island, the duration of their occupation is unknown. Based on observations of archaeological remains, Emory (1928) speculated that as many as 200 people may have lived there at one time. Although Nihoa had been abandoned prior to European contact, there can be little doubt that prehistoric Polynesians transformed much of the ecosystem, given the small size of the Island and the abundance of archaeological sites. However, in spite of early human impacts (burning, farming, and possible pig (*Sus scrofa*) rooting), the present recovered condition of the native ecosystems on Nihoa is especially noteworthy.

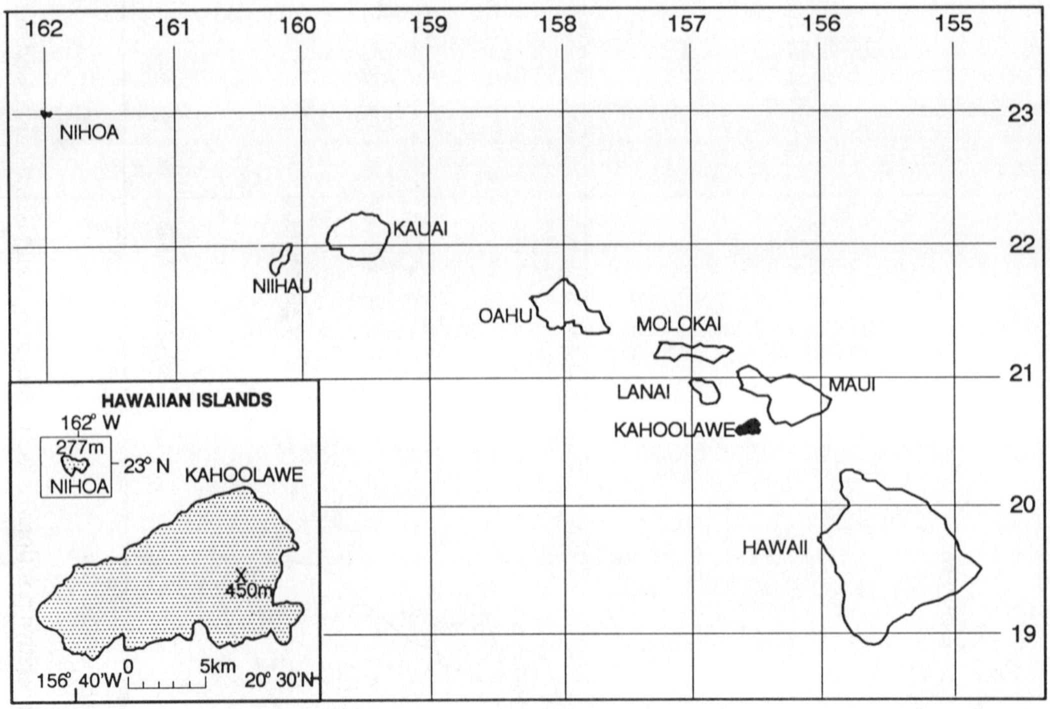

Figure 2. The southeastern Hawaiian Islands (Nihoa to Hawai'i), showing placement of Nihoa and Kaho'olawe in relation to the others.

Nihoa has apparently been little, if at all, disturbed by ungulates, and consequently it has retained a good deal of its native biota. According to Conant (1985), Nihoa not only has the greatest integrity and variation in native species composition of any of the Northwestern Hawaiian Islands, it is also probably "one of the only places remaining in the entire Hawaiian Archipelago where there is a diverse and relatively intact low-elevation dry-land ecosystem, with its complement of native terrestrial plants, arthropods, and birds" (Fig. 3). Conant's recent research on Nihoa indicates that 21 plant species are now found on the Island, of which five are endemic; most of the rest are indigenous. Six alien species have been recorded, and only one of these (a small herbaceous plant, *Portulaca oleracea* or pigweed) is considered common on the Island. The predominantly native plant communities of Nihoa support many endemic animals, including numerous terrestrial arthropods, several land snails, and two land birds, as well as large populations of breeding seabirds (Conant *et al.* 1984).

Kaho'olawe is the eighth-largest of the Hawaiian Islands (Fig. 2), with an area of 45 mi^2 (117 km^2) and a maximum elevation of 1,490 ft (450 m). Located in the leeward rain shadow of the larger island of Maui, Kaho'olawe possesses a semiarid climate similar to that of Nihoa. Recent botanical surveys indicate that the vegetation of the Island is overwhelmingly alien (Environmental Impact Study Corporation 1979). The disappearance of the great majority of native plant life can probably be explained by the lengthy presence of domesticated and feral ungulates on Kaho'olawe; their ecological impact has undoubtedly imperiled all native plants save those in the most steeply-sloped cliff-face refugia.

Figure 3. The island of Nihoa, despite an early period of Polynesian settlement (note stone wall in center left of photo), today remains one of the few places in the Hawaiian Islands with an intact native dry coastal ecosystem (note thickets of the endemic *Pritchardia* palm). Feral ungulate populations were never established on Nihoa. (Photo: S. Conant.)

Recent archaeological research on Kaho'olawe indicates that significant human alteration of the native vegetation and soils probably started during the prehistoric period. A distinctive, widespread layer of burned grasses and ash up to 0.8 in. (2 cm) thick has been identified; this layer was probably produced through a series of different events over a time span of as much as 200 years before the arrival of the first Europeans: "It is suggested that dryland swiddening practices, wherein large areas of virgin forest were burned off to make room for the cultivation of plants, may have been the primary initiating factor in the degradation of the interior island soil cover" (Barrera 1979). Subfossil land snail assemblages found in erosional deposits or associated with the widespread burn layer on Kaho'olawe have also been used to support the assumption that prehistoric Polynesians severely altered dry forest ecosystems on this island (Hommon 1980). Fires that appear to have caused significant alteration of the pristine dry forest ecosystems may have been initiated purposefully to promote slash and burn agriculture. Fires may also have been used to extend (and later maintain) the range of pili grass (*Heteropogon contortus*), a culturally important thatch and mulch source (Yen *et al.* 1972; Kirch 1982).

Even if the prehistoric occupants converted much of the original dry forest ecosystem on Kaho'olawe into anthropogenic grassland, alien hoofed mammals introduced in the Post-Cook period were also responsible for "the massive erosion and landscape degradation" of the Island (Spriggs 1986). Devastation of the soil and vegetation (especially on the higher parts of the Island) occurred primarily during the last part of the 19th and the early 20th centuries, when large herds of livestock and feral animals consumed much of the remaining vegetation and exposed the Island to increased soil erosion.

The first ungulates to reach Kaho'olawe were probably goats (*Capra hircus*) given to a Maui chief by Captain George Vancouver. The chief is said to have sent some animals to Kaho'olawe in 1800 with hopes that they would multiply (Barrera 1979). Pigs were also reported on Kaho'olawe during the first half of the 19th century. These alien mammals had developed sizable populations by the 1850s (Allen 1858).

By the late 1890s, about 15,000 sheep (*Ovis aries*) and a large number of goats were present. Less than 20 years later (1909), the territorial legislature of Hawaii voted not to renew a 50-year lease that had been granted to two ranchers using the Island. The justification was severe deterioration of the vegetation and related soil erosion attributed to ungulate populations that were by then ravaging the remaining vegetation on Kaho'olawe (Lebarron and Walker 1970). Limits were subsequently set on a new ranching lease, which required that goats and sheep be eradicated and that the number of cattle, horses, and mules (*Bos taurus, Equus caballus, Equus caballus x asinus*) that could be brought to Kaho'olawe be restricted. Between 1912 and 1918, about 5,000 goats were eliminated. However, even with these new measures, there was little evidence of vegetation regeneration. As of 1918, large-scale removal of the native vegetation by alien ungulates had opened up the greater part of the central plateau (area above 600 ft or 200 m) to massive soil erosion, exposing the reddish, saprolitic hardpan one still encounters in this part of the Island (Barrera 1979). Although sheep, cattle, and horses were eliminated from Kaho'olawe during this century, a few goats still inhabit the Island, even after a number of concerted, but unsuccessful, efforts to eradicate them.

The impoverished vegetation found on Kaho'olawe today is dominated by alien species with defense mechanisms against herbivory. Defenses were either absent or weakly expressed among the native plants. For example, although a few living specimens of the native wiliwili (*Erythrina sandwicensis*) may now be found growing on steep-sided ravines, this tree was probably much more common in the past. Although it does produce some thorns on its trunk, wiliwili does not have the same degree of thorny protection from ungulates that has helped the alien kiawe (*Prosopis pallida*) become the dominant tree species in all but the almost barren hardpan desert on the summit plateau. "Tree tobacco" (*Nicotiana glauca*), now found in severely eroded areas, is also resistant to ungulates (Fig. 4). Toxic substances produced by this tree generally discourage feral goat browsing.

Figure 4. The central plateau of Kaho'olawe has been stripped of both vegetation and soil under the impact of almost two centuries of large domestic and feral ungulate populations. The alien "tree tobacco" (*Nicotiana glauca*) has survived the intense browse pressure by virtue of its unpalatable (toxic) leaves, a defense mechanism generally lacking in native Hawaiian plants. (Photo: M. Merlin.)

Hatuta'a and Eiao

The old volcanic island of Hatuta'a (referred to as Hatutu on several charts) is located 335 mi (102 km) northwest of Nukuhiva in the northwestern group of the Marquesas Archipelago (Figs. 1, 5). Its surface area is about 6.2 mi^2 (16 km^2), and it has a maximum elevation of 1,395 ft (425 m) (Decker 1970). The Island projects prominently from the sea but does not have sufficient height or size to induce much orographic rainfall; thus, the climate of Hatuta'a is similar to that of the surrounding ocean, resulting in markedly xeric conditions.

Domesticated or wild ungulates have not been introduced on Hatuta'a because of difficult access and insufficient fresh water (Decker 1973). Hatuta'a is an important rookery for many species of seabirds, including boobies, sooty terns, white terns, and frigate birds (*Sula* sp., *Sterna fuscata, Gygis alba,* and *Fregata minor*); it also provides a final refuge for the small Marquesan ground dove (*Gallicolumba rubescens*). A number of endemic invertebrate species can still be found on this remote Marquesan island (Perrault 1978). The largely undisturbed xerotropical vegetation of Hatuta'a is dominated by clumps of native scrub and tussock grass (*Eragrostis xerophila*). The flora includes 30 known species, with six endemic to the Marquesas Islands (Fig. 6).

Prehistoric Polynesians used Hatuta'a, as evidenced by a number of stone platforms along the central crest of the Island beneath groves of *Pisonia grandis*, the only tree on Hatuta'a. The Island is occupied by some alien species, including a rat and a number of pantropical weeds (*e.g.*, coffee senna, hairy spurge, wild water lemon, and pigweed (*Senna occidentalis, Chamaesyce hirta, Passiflora foetida, Portulaca oleracea*); yet Hatuta'a can be described as "a pristine terrestrial ecosystem -- the only sizeable one left undisturbed in the central Pacific dry zone" (Decker 1973).

The brief, but insightful, ecological observations on Hatuta'a made by Decker over 30 years ago (April 1960) were generally confirmed more recently by Montgomery *et al.* (1980). These authors described the "thick, green mantle of grasses with clumps of *Pisonia* trees" that covered the Island. Although they noted the presence of "curious and bold rats (probably *Rattus exulans*)," they also observed many seabirds still nesting on Hatuta'a. Furthermore, they reported that the native Marquesan ground dove was still common and appeared to be unalarmed by the presence of humans until approached at close range.

Eiao is located 1.5 mi (2.5 km) northeast of Hatuta'a (Fig. 5). It is volcanic in origin but is much larger (20 mi^2 or 52 km^2) and higher (1,890 ft or 575 m) than Hatuta'a. Eiao has accessible landing spots and a number of freshwater springs. The history of human land use and impact on Eiao has differed greatly from that on Hatuta'a, especially after the arrival of Europeans. For example, livestock released on Eiao have had a devastating effect on the native vegetation and soil conditions on Eiao in the past hundred years. When Decker visited Eiao in February 1960, he reported that the Island was "a barren gullied desert of rock and orange clay" (Decker 1973). He pointed out that the small amount of native forest that still remained on the Island was limited to topographic refugia that persisted because they were "inaccessible to the feral sheep that run freely over the island, starving and preyed upon by feral swine" (Fig. 7). The sheep were so enfeebled that they "could be run down and captured by hand" (Decker 1973).

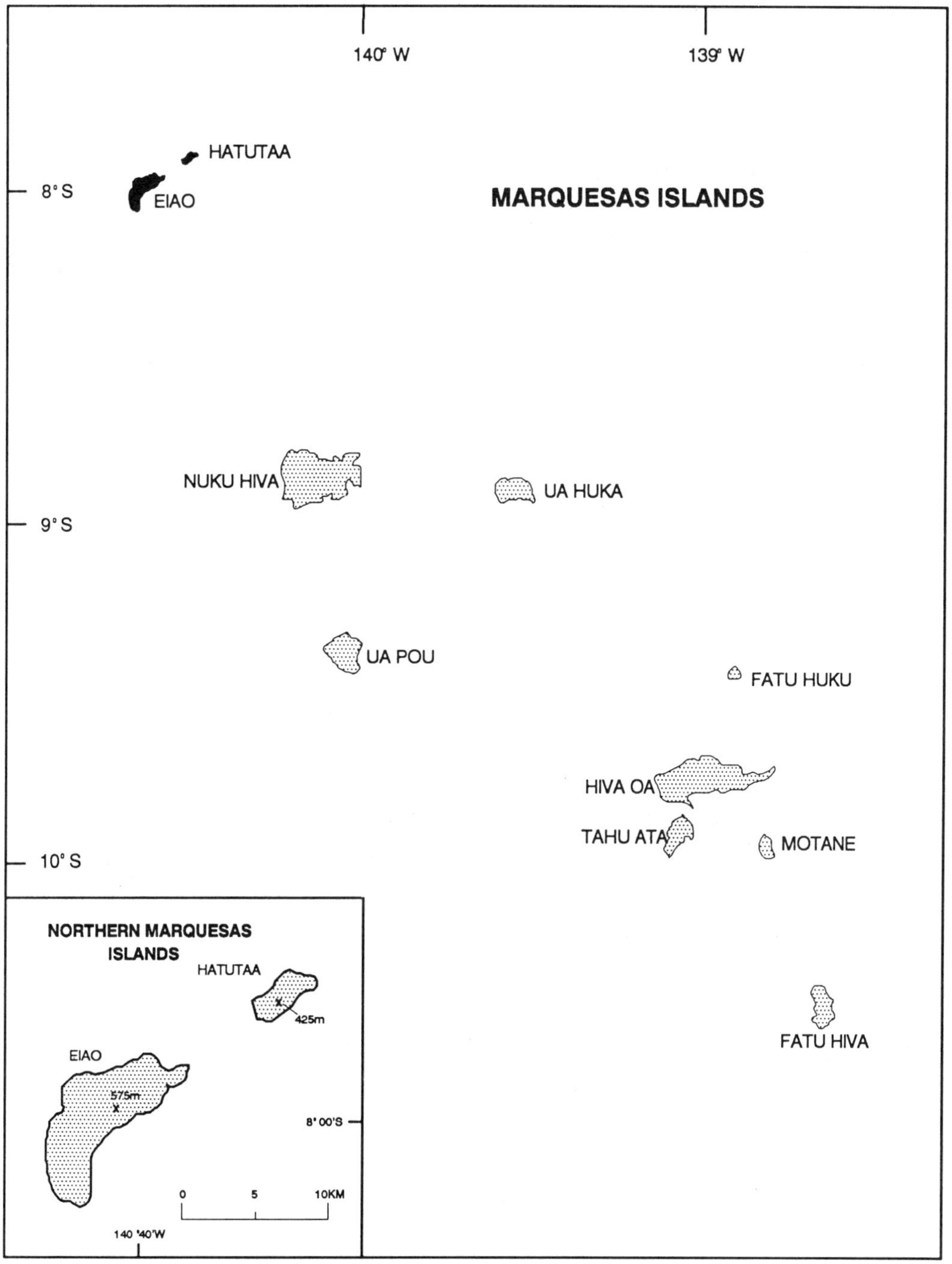

Figure 5. Map of Marquesas Archipelago, showing Hatuta'a and Eiao.

Figure 6. The Marquesan island of Hatuta'a with nearby Eiao in the background. Ungulate-free Hatuta'a supports a native dryland ecosystem dominated by tussocks of *Eragrostis xerophila* and the shrubs *Waltheria lophanthus* and *Cordia lutea*. Windswept trees in background are *Pisonia grandis*. (Photo: B. Decker.)

Observations made during a more recent expedition to Eiao in 1978, (W.C. Gagné, pers. comm. 1986) indicated that many native species may still be found on steep cliffs out of reach of feral sheep and pigs. These refugia would likely be greatly diminished if goats were introduced to the Island. Presently the area covered by "primitive forest" on the extensive plateau above 1,475 ft (450 m) is extremely reduced (Perrault 1978). The extreme habitat deterioration on Eiao is indicated by the condition of feral sheep, which were even observed consuming poisonous plants such as coffee senna, a tropical American shrub that is rare on Hatuta'a but relatively common on Eiao. Although normally toxic to livestock, parts of the plant were found in the stomach of a sheep caught and killed on Eiao in 1960 (Decker, pers. comm. 1986).

The relatively remote, adjacent islands of Hatuta'a and Eiao in the semiarid northern Marquesas provide dramatic illustration of the persistence of native ecosystems with and without feral ungulates. Feral sheep and pigs have devastated much of Eiao, and the native flora has been largely supplanted by alien species. On nearby Hatuta'a the lack of permanent water sources frustrated human settlement and the successful establishment of feral ungulates. The native vegetation on Hatuta'a has

Figure 7. The island of Eiao has been ravaged by feral sheep, leading to native ecosystem destruction and massive soil erosion. (Photo: B. Decker.)

not been significantly modified, although some of the same alien plant species that have become common in the ungulate-disturbed ecosystem of nearby Eiao are present in small numbers as casual adventives on Hatuta'a (*e.g.*, coffee senna and pigweed). We hypothesize that the lack of early successional habitats on undisturbed Hatuta'a has limited the successful spread of adventives.

Kermadec Islands

The Kermadecs are a chain of small, basaltic islands lying northeast of New Zealand (Figs. 1, 8). The northern group includes volcanically active Raoul (by far the largest island in the chain, with an area of 7,270 a (2,943 ha) and eight small outliers comprising the Meyer and Herald islets (combined area 100 a or 40 ha). The southern Kermadecs include Macauley (800 a or 324 ha), volcanically active Curtis (130 a or 53 ha), and the smaller islets of Cheeseman and L'Esperance (20 and 12 a, or 8 and 5 ha).

Although these islands were uninhabited at the time of European discovery (1788), direct archaeological and indirect botanical evidence indicate prehistoric Polynesian presence on Raoul, where adzes of a type common in the southern Cook Islands have been found (Duff 1968); several plant species of probable Polynesian introduction occur (*e.g.*, candlenut, *Aleurites moluccana*, and ti, *Cordyline fruticosa*).

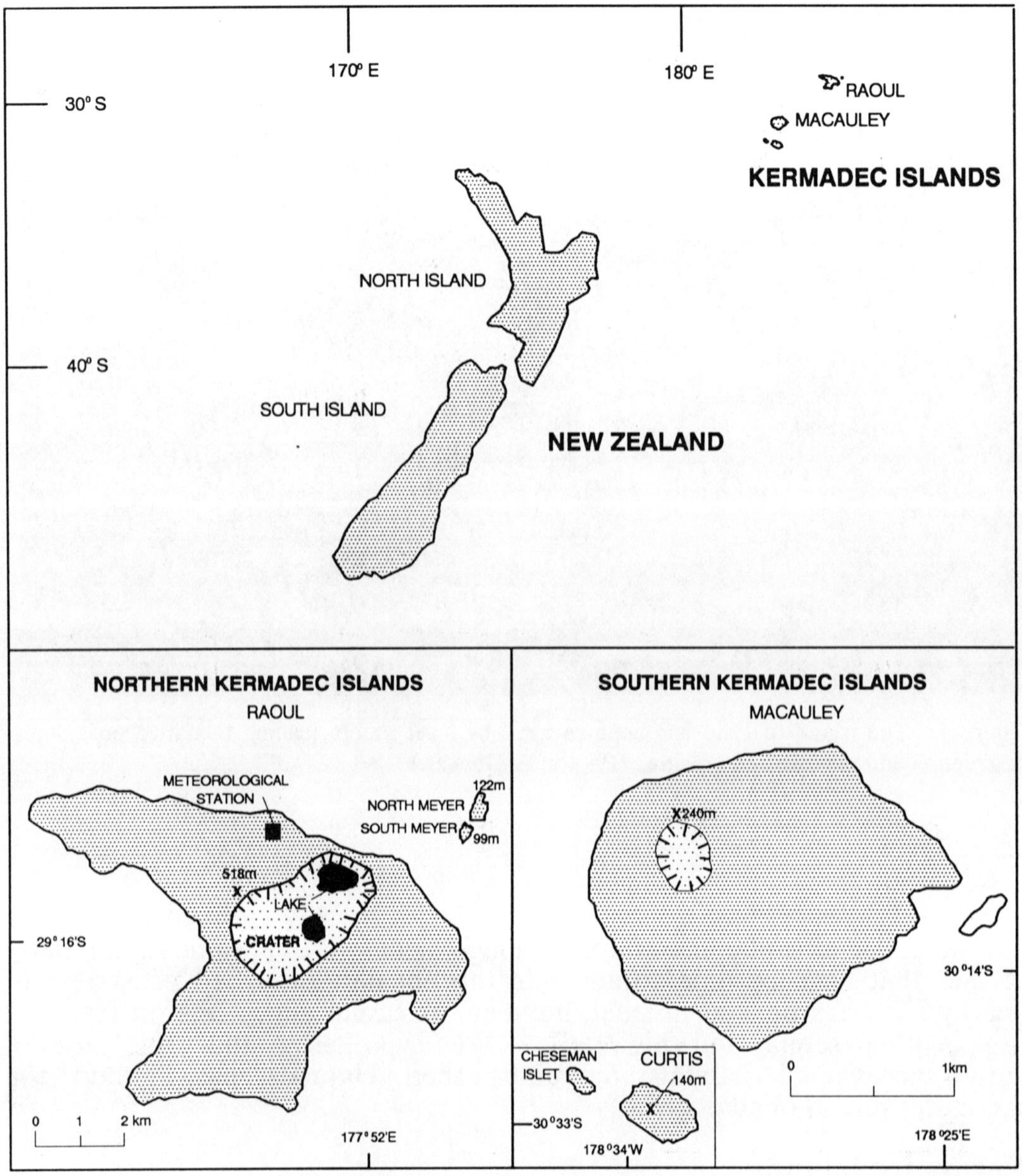

Figure 8. Map of Kermadecs showing Raoul and other islands discussed.

Periodic European settlement and farming attempts occurred on Raoul beginning in 1836 (Morton 1964). A permanently manned meteorological station has been in operation since 1937. All other islands in the Kermadecs have remained uninhabited.

Feral ungulate populations became established in the Kermadecs early in the 19th century but were restricted to Raoul and Macauley, where both goats and pigs were feral by 1836 (Straubel 1954). Pigs disappeared from Macauley by 1840 and only occurred in limited numbers on Raoul until their final eradication in 1966 (Sykes 1969). Feral goats have remained the only significant alien ungulate to impact native ecosystems on Raoul and Macauley over the past 150 years.

The interactions among native and alien plants on goat-ravaged Raoul and Macauley may be compared with the situation on the nearby outliers of North and South Meyer (1.2 mi or 2 km from Raoul) and Curtis (2 mi or 3.5 km from Macauley). The outliers have never supported ungulate populations but do share a number of alien plant species with their larger neighbor islands. Numerical comparisons of alien and indigenous plants in the Kermadecs are drawn from Sykes' (1977) definitive flora of the Islands.

The native vascular flora of the Kermadecs includes 133 species, of which 23 taxa (20.3%) are considered endemic. Although there are a few tropical Pacific elements, the overwhelming affinities of the native flora are with temperate New Zealand. A detailed comparison of native and alien elements in the flora of Raoul and the nearby Meyers is presented in Table 1. The total vascular flora of Raoul is 237 species, with 133 aliens accounting for 56% of this total. On the ungulate-free Meyer islets, there are 59 species, of which 25 (42%) are aliens.

At first glance, the percentage of aliens on the Meyers may seem high given lack of human disturbance and absence of feral ungulates; however, large colonies of nesting seabirds provide substantial natural ground disturbance (and probable dispersal mechanisms) favoring early successional aliens. The impacts of the seabird colonies in the Meyers may be compared with similar conditions of natural disturbance and plant dispersal created by relatively large seabird colonies on Nihoa in the Northwestern Hawaiian Islands and Hatuta'a in the northern Marquesas (see discussion of these islands above).

Comparative analysis of 21 alien species occurring on both Raoul and the Meyers (Table 1) shows that on Raoul 15 (71%) of the shared species can be described as "common or widespread" (qualitative assessment in Sykes 1977), whereas on the Meyers, only 7 (33%) of the shared species were described as common. The remaining 14 species on the Meyers exist only as rare or casual adventives. From these data it would appear that alien plants are twice as likely to establish widespread populations where goats had been present (Raoul), as where absent (Meyers).

On Raoul, the feral goat population increased to several thousand from the initial liberation of a few individuals in the early 19th century. Efforts to rid Raoul of goats began in 1937, and since then, about 15,000 animals have been killed (Parkes 1984a). After 1971, annual organized hunting expeditions steadily accelerated the decline in numbers of feral goats. Parkes (1984a) reported that feral goats were close to eradication, and Sykes (pers. comm. 1986) noted that the last goat was exterminated in 1984. The evidence from Raoul indicates that native species are now more common, especially those that were most palatable and previously quite rare due to severe ungulate foraging pressure. For example, since the significant decline in the number of goats on Raoul, the woody endemic *Coprosma acutifolia* has become a dominant understory species again in some local forests on the Island. Furthermore, rare species such as

Table 1. Composition of Raoul and North and South Meyer islands in the northern Kermadecs, with special reference to shared alien species.

	Raoul Island	North & South Meyer
Island area (ha)	2,943	12
Maximum elevation (m)	518	122
No. indigenous vascular plant species	104	34
No. alien plant species	133	25
Total flora (species)	237	59
Percent aliens in total flora	56	42
No. shared alien species on Raoul and Meyers	21	21
Rare or casual shared aliens	6	14
Common or widespread shared aliens*	15	7
Percent common and widespread species in shared alien flora	71	33

*Common and widespread shared aliens: *Cassia floribunda, Polycarpon tetraphyllum, Conyza bonariensis, Sonchus oleraceus, Sida rhombifolia, Oxalis corniculata, Rumex brownii, Portulaca oleracea, Alocasia macrorrhiza, Cyperus brevifolia, Bromus catharticus, Eleusine indica, Paspalum dilatatum, Sporobolus africanus, Stenotaphrum secundatum*. (Extracted from Sykes 1977.)

Homalanthus polyandrus, Hebe breviracemosa, and *Boehm australis* var. *dealbata* are now seen more often than before the goat removal program (Parkes 1984b). Goat browsing had reduced *H. breviracemosa* from its relatively common status in 1854 to a very rare species by 1908. The species was not observed again until a single young individual was seen in 1983.

A generally similar picture prevails in the comparison of Macauley (with goats) and Curtis (goat-free) in the southern Kermadecs. Macauley has a total vascular flora of 53 species, of which 14 (36%) are aliens. (Seven of the aliens are common or widespread.) Curtis Island, lying 22 mi (35 km) to the south of Macauley (Fig. 8), supports a vascular flora of 19 species, which includes only three aliens (15.8%). These aliens (*Oxalis corniculata* var. *repens, Solanum nigrum,* and *Conyza bonariensis*) are shared with Macauley. Only the composite herb *Conyza bonariensis* is common on both islands, associated with the disturbed ground of seabird colonies and pioneering on cinder substrates in the active volcanic crater on Curtis. *Oxalis corniculata* is common on Macauley but rare on Curtis, while *Solanum nigrum* is abundant on the recent volcanic substrates of Curtis but largely replaced on Macauley by the native *Solanum nodiflorum*. In spite of significant differences in outlier distance from the main islands of Raoul and Macauley, the proportions of widespread adventives in the total flora of the Meyers (7/59 or 11.8%) and Curtis (2/19 or 10.5%) are strikingly similar.

Goats were finally eradicated from Macauley in 1970, although all except two had been killed by 1966 (W.R. Sykes, pers. comm. 1986). Further vegetation change on Macauley and potential re-equilibration among alien and native elements should provide valuable information on the persistence and resilience of insular flora and vegetation. Woody plants were almost completely eliminated on Macauley through browsing by goats, with only a few *Myoporum* left growing on the inaccessible cliffs of the Island. However, regeneration of native herbaceous plants such as *Mariscus*, *Scirpus*, *Solanum*, and *Conyza* was already under way by 1970 (W.R. Sykes, pers. comm. 1986).

Now that ungulates have been removed from Raoul, the chances are good that a number of endemic species, such as those mentioned above, may once again extend their distributions and develop larger populations. What is still lacking from the Kermadecs are quantitative, long-term, micro-distributional studies of plant species interactions that could be used to test the generally subjective conclusions discussed above. The Kermadecs provide an exceptional opportunity for studies of competition between alien and native elements.

Henderson and Nauru

Although not in close proximity, both Henderson and Nauru islands, in the tropical Pacific (Fig. 1), are elevated limestone islands known as makatea. Henderson has been uninhabited and only rarely visited during recent centuries, but Nauru has supported a human population for over a millenium. During the 20th century, Nauru has experienced the radical environmental effect of phosphate mining.

The great majority of the approximately 24 or more of these known raised coral or makatea islands in the world's tropical oceans has been severely modified by long-term human settlement, including the mining of phosphate deposits in the 20th century. Aldabra in the western Indian Ocean and Henderson in the South Pacific are the only islands of this raised limestone type that are still largely unaltered. Henderson has been called one of the most scientifically fascinating of all coral islands (Fosberg *et al.* 1983).

Henderson Island, like Hatuta'a in the northern Marquesas and Nihoa in the Hawaiian Islands, is a small tropical Pacific landfall where relatively inaccessible and dry conditions have precluded establishment of domestic livestock or feral ungulate populations. More remote than either Hatuta'a or Nihoa, Henderson (Figs. 1, 9) is a raised atoll located 125 mi (200 km) east-northeast of Pitcairn, the closest inhabited island. It covers about 14 mi^2 (37 km^2) in land area and reaches a maximum elevation of 110 ft (33 m). Like other elevated atolls, Henderson has surfaces that are extraordinarily eroded. Due largely to solution of its limestone substrate, the surface is commonly pitted, often presenting deep holes and jagged pinnacles, which, with isolation, location, and almost complete lack of freshwater resources, have limited human use of this island.

Preliminary archaeological research on Henderson in 1971 revealed traces of prehistoric Polynesian occupation that may have lasted 300 years (Sinoto

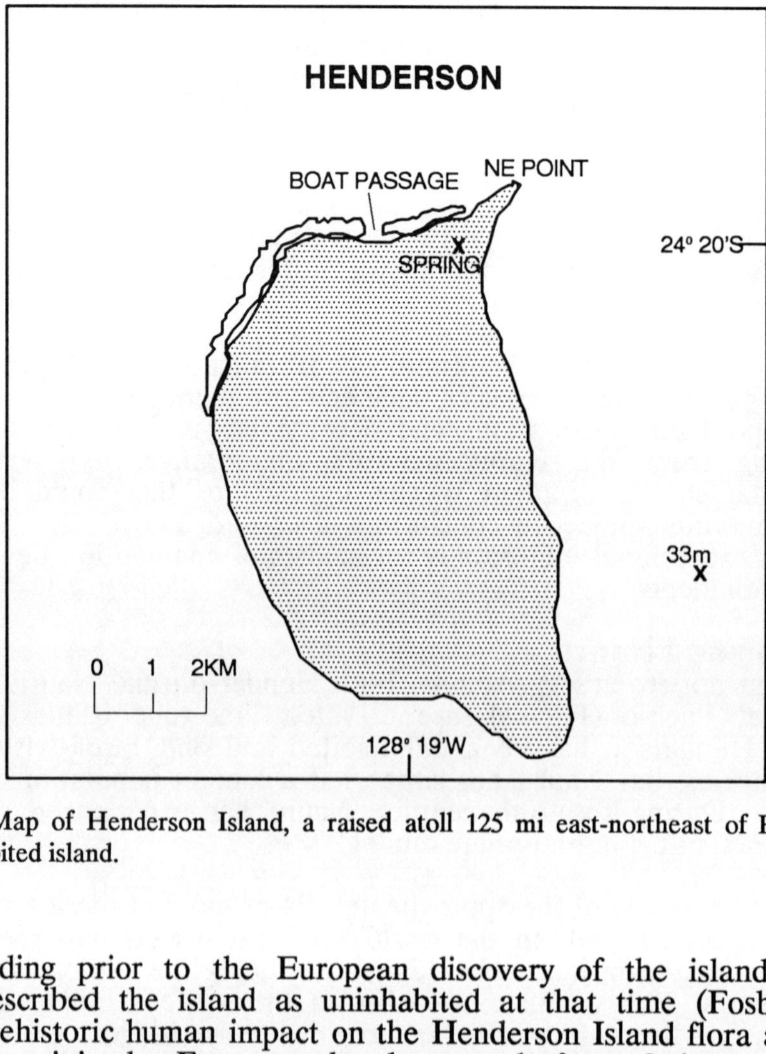

Figure 9. Map of Henderson Island, a raised atoll 125 mi east-northeast of Pitcairn, the closest inhabited island.

1983), ending prior to the European discovery of the island in 1606. Quiros described the island as uninhabited at that time (Fosberg *et al.* 1983). Prehistoric human impact on the Henderson Island flora appears to have been minimal. For example, the records from Quiros's expedition indicate that the only plant producing edible fruit was a native species of hala (*Pandanus tectorius*). If the early Polynesians brought in and cultivated any of their traditionally important cultigens, or inadvertently introduced any alien weeds, only two of these species, ti and candlenut, have ever been recorded from Henderson, and it is possible that both were brought in by Pitcairn islanders in the 19th century. Ti was collected only during the Mangarevan Expedition in 1934, when Fosberg recorded just six plants, and candlenut was collected only during the Whitney Expedition in 1922 (Fosberg *et al.* 1983).

Although Henderson has its share of native birds (including four extant endemic taxa of land birds) (Bourne and David 1983), herpetofauna, insects, and land snails (18 species, according to Cooke and Kondo 1960), no native terrestrial mammals are found on the Island. The Polynesian rat (*Rattus exulans*), first recorded from Henderson in 1819 (King 1820), was probably brought to the Island by prehistoric Polynesians. At least three species of birds, a storm-petrel (*Nesofregetta fuliginosa*) and two pigeons (*Ducula* spp.), no longer are found on Henderson. They have been identified in archaeological sites and are thought to have been extirpated by prehistoric Polynesians (Steadman and Olson 1985).

If pigs were brought to Henderson by prehistoric Polynesians, they almost certainly disappeared long before the first European contact. D.R. Tait released some pigs in 1912, but no record of them exists on Henderson after this date (Fosberg *et al.* 1983). Brodie (1851) reported that Pitcairn islanders brought several goats with them on their first visit to Henderson in 1843 but failed to release them. The only report of an actual release of goats on Henderson indicated that three individuals were put on the Island in 1923 (Beck 1923). However, they reportedly had vanished by the time of the Mangarevan Expedition in 1934 (Fosberg *et al.* 1983). Perhaps the very dense vegetation and treacherous terrain precluded the establishment of feral goat or pig populations on Henderson.

In 1982 a wealthy American citizen proposed to the British Government to build a "holiday home with an airstrip" on Henderson in exchange for development assistance on Pitcairn (Bourne and David 1983). At that time there were no formal conservation measures that applied to Henderson or the other islands in the Pitcairn group. However, strong concern was expressed within the scientific community. Noting the advantages of its small size, isolated position, and "simple enough biota that its relationships, processes and functioning may possibly be understood with adequate investigation," Fosberg *et al.* (1983) argued that the proposed settlement of Henderson should not be approved since it would undoubtedly "permit the destruction of the only remaining intact example of one of the most important classes of islands." A resolution to this effect was adopted at the XV Pacific Science Congress, February 1-11, 1983, in Dunedin, New Zealand (Pacific Science Association Information Bulletin 1983). The Government of the United Kingdom later rejected the lease and development proposal in mid-December of 1983 (Pacific Science Association 1984).

Most of Henderson Island is today covered by a thickly vegetated tangled scrub and scrub-forest, with emergent *Pandanus* trees rising above the general canopy. In the interior of the depression that once was a former lagoon site, as well as parts of the makatea, the vegetation is more sparsely distributed (Fosberg *et al.* 1983). Except for some of the taller parts of the scrub-forest where the canopy is more complete, it is extremely difficult for humans, (or presumably, hoofed mammals) to move about freely because of very dense vegetation and the dissected limestone substrate.

The flora of Henderson (St. John and Philipson 1962) includes 8 ferns and 55 angiosperms. Fosberg *et al.* (1983) recorded two additional species of flowering plants and listed six species and five varieties of angiosperms as endemic to Henderson. The alien plant species that were listed included coconut, *Cocos nucifera* (Pitcairn islanders are reported to have started planting this species on Henderson as early as 1843; see Brodie 1851), ti, candlenut, and *Achyranthes aspera*. Fosberg *et al.* (1983) suggested that *A. aspera*, collected just once (1912), was "the only one of the four [adventives that was] not a deliberate introduction." However, it is possible that *A. aspera* is a relatively rare indigenous species on Henderson, since its fruits become attached to birds that could have dispersed it to this remote island.

In any case, it is safe to say that Henderson Island today probably contains one of the least disturbed ecosystems of any tropical oceanic island in the Pacific. Except for a very small number of species, the alien plants and mammals that have been consciously or inadvertently brought to Henderson have not become naturalized. The rugged terrain has probably played a major part in suppressing the establishment of feral animal populations and consequently has allowed the native vegetation to remain largely intact.

Nauru is a small (8.5 mi^2 or 22 km^2) makatea island in the Central Pacific 26 mi or 42 km south of the equator (Figs. 1, 10). The highest spot on the Island is no more than 100 ft (30 m) above sea level. Although the Island is normally wet (average rainfall is 80 in. or 2,030 mm/yr), annual precipitation exhibits great variation from 41 to 180 in. (1041-4,572 mm) per year (Carter 1981). The discovery of high-quality tricalcic phosphate rock in the central plateau of Nauru in 1906 led to the initiation of surface strip mining, which has continued to the present. Phosphate extraction involves complete removal of surface vegetation and soil, exposing new coralline substrate. Manner *et al.* (1985) recently investigated both the composition of remaining undisturbed plateau forest

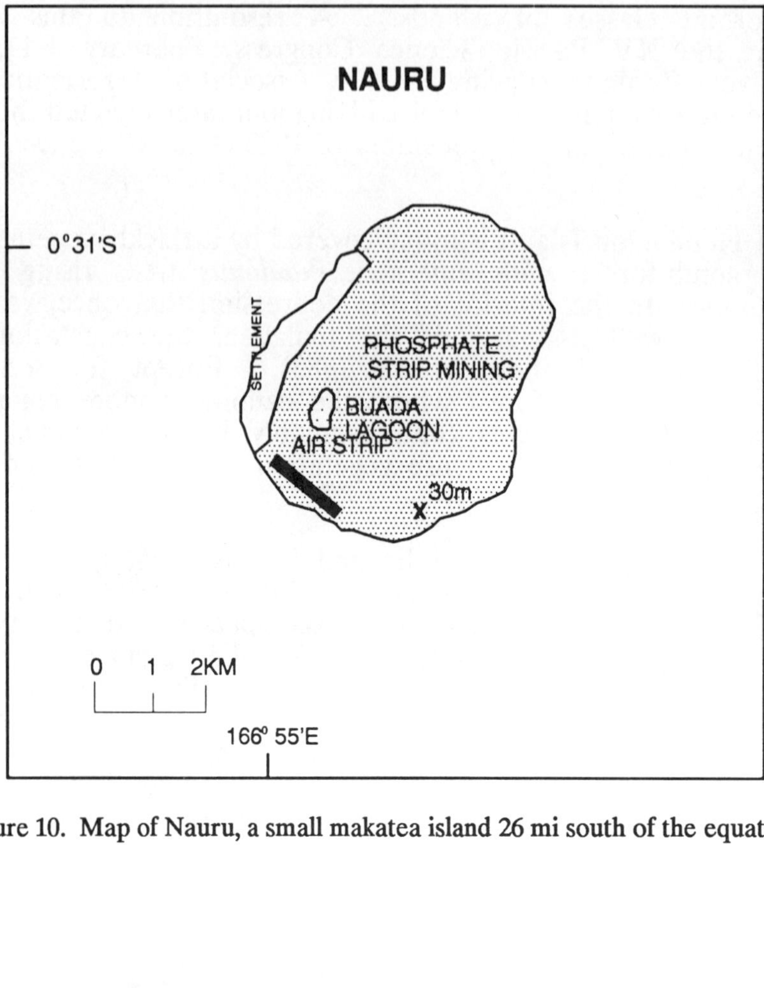

Figure 10. Map of Nauru, a small makatea island 26 mi south of the equator.

on Nauru and post-mining successional sequences that have developed over the past 75 years (Fig. 11).

Figure 11. Open-face mining of phosphate rock on Nauru involves removal of all vegetation, soil, and mineral deposits, leaving newly exposed barren coralline rock with "pit and pinnacle" micro-topography (bottom photo). Vegetation succession on these mined surfaces, although initially involving some aliens, returns after 20-40 years to native species-dominated shrub and tree communities (*e.g., Dodonaea viscosa, Ficus prolixa*) (top photo). (Photos: H. Manner.)

Although the Nauru flora includes 467 species, of which more than 90% are aliens of prehistoric Polynesian or recent introduction, the majority of these species (261) are introduced ornamentals that have not become naturalized and are restricted to home gardens and cultivation (Manner *et al.* 1985). A total of 126 species, including 81 adventive aliens and the native flora of 45 species, constitutes the potential pool of successional colonizers on disturbed areas of the central plateau. The natural forest vegetation of the central plateau is dominated by *Calophyllum inophyllum* (canopy height to 52 ft or 16 m), and in the most severe limestone areas by *Ficus prolixa* (canopy height to 65 ft or 20 m). Manner *et al.* (1985) found that in areas mined within the past 20 years, alien species such as *Tridax procumbens*, *Vernonia cinerea*, *Eragrostis tenella*, and *Phyllanthus amarus* were dominant. The native ferns *Microsorium (Polypodium) scolopendria* and *Nephrolepis biserrata* were also abundant in this early successional stage. At sites mined 20 to 40 years ago, native shrubs and sedges (*Scaevola taccada*, *Dodonaea viscosa*, and *Fimbristylis cymosa*) have substantially replaced pioneer aliens. At sites mined over 40 years ago, native trees and shrubs, including *Ficus prolixa* and noni or Indian mulberry (*Morinda citrifolia*), become dominant. Alien tree species such as *Casuarina littoralis* and common guava (*Psidium guajava*) do not figure significantly in intermediate successional stages. These results from Naura suggest a resilience in the native flora and a clear competitive advantage where human disturbance is not continual and where feral ungulates are absent.

Southern Cook Islands

The islands of Atiu (10 mi^2 or 27 km^2), Mitiaro (8.5 mi^2 or 22 km^2), Mauke (7 mi^2 or 18 km^2), and Mangaia (20 mi^2 or 52 km^2) are all located in the southern region of the Cook Islands (Figs. 1, 12). The first three islands lie relatively close together and are collectively known as the Ngapotoru group. This group and Mangaia (110 mi or 180 km to the south) are not nearly as isolated as Henderson Island. They also differ from Henderson in that they all have central volcanic areas and a continuity of human occupation extending back at least a thousand years. On the other hand, like Henderson, these four southern Cook Islands have relatively wide areas of makatea, which circumscribe the central volcanic regions (Fig. 12).

In spite of the fact that ungulates have long been present on Atiu, Mitiaro, Mauke, and Mangaia, the topographically rugged, raised reef regions of these four islands, like that on Henderson, remain largely dominated by native species (Sykes 1980; Merlin 1991). One has only to look at the central volcanic regions of these islands, now largely (in some areas completely) devoid of native plant species (Fig. 13), to see what centuries of cultivation and animal husbandry have done (Sykes 1980). Although a large number of weedy plants have become naturalized in the central volcanic interiors of Atiu, Mitiaro, Mauke, and Mangaia, aliens have generally not been able to spread into the makatea areas. The native species on raised reefs appear to be more than adequate competitors when faced with aliens in the absence of hoofed mammals or other sustained human disturbances that might create early successional conditions.

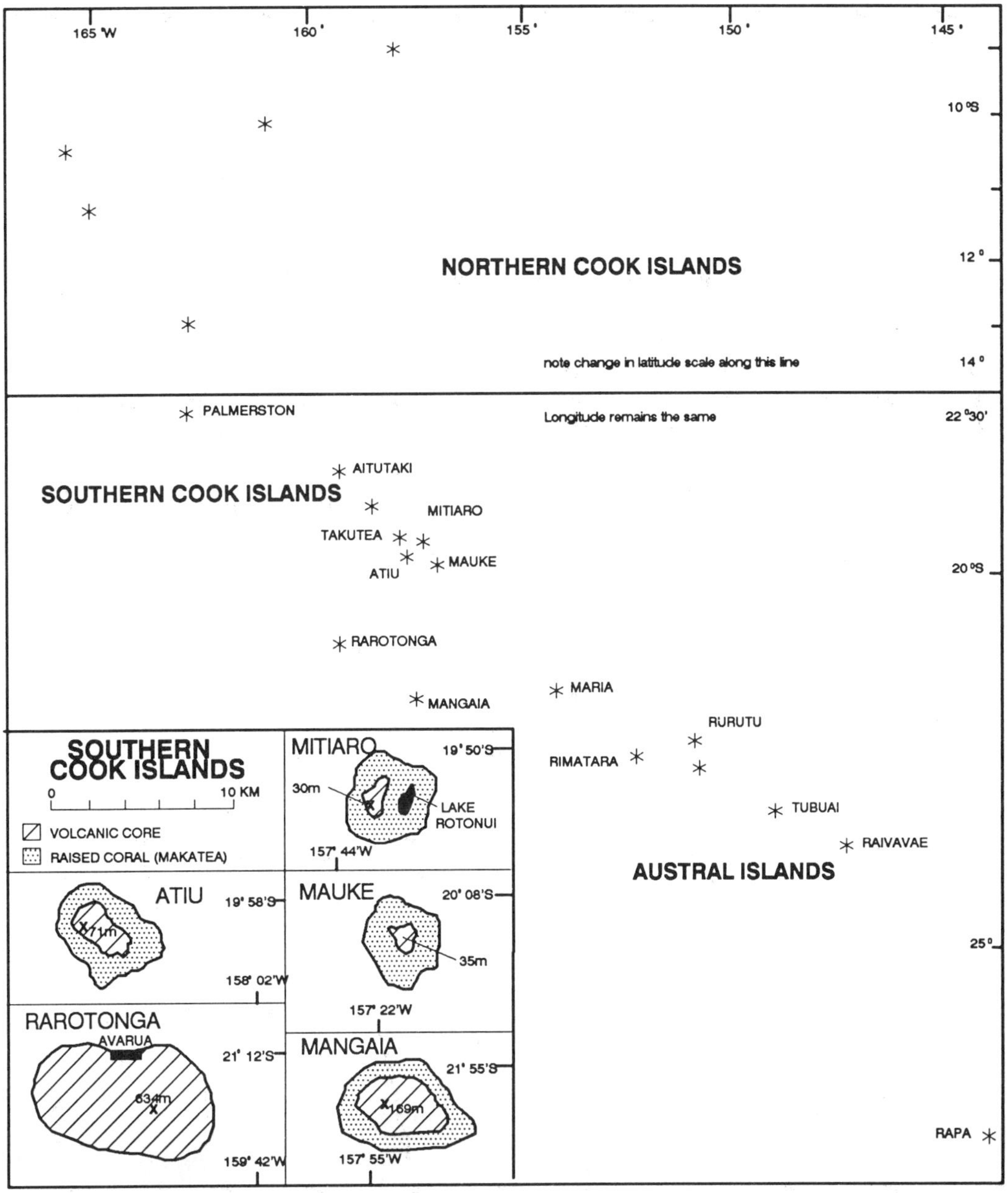

Figure 12. Map of the southern Cook Islands.

Rarotonga (Fig. 12), the largest (25 mi^2 or 64 km^2), highest (2,140 ft or 652 m), and most populated of the Cook Islands (approximately 10,000 people), has only a very small area of coastal makatea. Most of the coastal lowlands are composed of alluvial and colluvial deposits from the eroded basaltic interior of the Island. Although the restricted area of raised coral on Rarotonga is still dominated by native species, the vegetation in the coastal and lowland zones of the Island today is dominated by alien species, including a host of cultivated crops, ornamentals, and a wide range of weedy adventives (Sykes 1980). The

Figure 13. Atiu in the Southern Cook Islands retains an intact native makatea forest on coastal limestone, while the more hospitable (with respect to soils and topography) central volcanic core of the island (at right) has been largely transformed by centuries of agricultural use. (Photo: M. Merlin.)

anthropogenic vegetation that nearly covers the entire extent of the low-lying areas inland of the coastal strand (and often deep into the interior of the Island along the valley bottoms) is very much like the severely altered vegetation one finds on the much-weathered volcanic interiors of Atiu, Mauke, Mitiaro and Mangaia islands.

A number of introduced woody species have become naturalized in parts of the uplands of Rarotonga. These include candlenut, found in some areas up to about 1,310 ft (400 m); ti, usually seen on shady slopes up to about 1,475 ft (450 m); and hau (*Hibiscus tiliaceus*), which is quite common along the many streams in the interior of the Island. In addition to these three plants (probably introduced during the pre-European period), some alien woody species have become naturalized in the montane areas of Rarotonga. These include common guava, *Syzygium* sp., *Solanum mauritianum*, lantana (*Lantana camara*), and *Ardisia humilis* in the lower elevations; and African tulip (*Spathodea campanulata*), strawberry guava (*Psidium cattleianum*), and *Cecropia palmata* in the lower and upper areas of the uplands. A number of introduced herbaceous weeds also in the uplands of Rarotonga include *Elephantopus mollis*, *Mikania micrantha*, and *Paspalum conjugatum* (Hilo grass).

Recent research in the upland region of Rarotonga provides quantitative evidence that native species continue to dominate the vegetation of the rugged interior (Merlin 1985). For example, over 92% of all the woody plants (dbh 1.0 in. or $\geq$ 2.5 cm) sampled along upland forest transects were either indigenous or endemic to Rarotonga. In addition, native plants accounted for 95.6% of the basal area covered by the arborescent vegetation in the montane study area. On the highest parts of the Island (above 1,475 ft or 450 m), native species accounted for more than 99% of the woody vegetation. Casual observations in the uplands as a whole also indicated that herbaceous vegetation is heavily dominated by native species (Merlin 1985).

Alien plants in the upland regions of Rarotonga are almost exclusively found in environments intentionally or accidentally disturbed by humans. Undoubtedly the lack of easy access, steepness of slope, and high rainfall in the montane areas of Rarotonga has greatly reduced the degree of human interference there. Native vegetation in the uplands has also probably not been significantly replaced by alien species because of the general absence of ungulates. In all likelihood, truly feral populations of pigs, goats, cattle, or other potentially disruptive hoofed animals have never been present in the montane region of Rarotonga. "The native vegetation in the rugged interior of the island therefore has not suffered from the deleterious effects of mammalian grazing, trampling, scarification, or seed dispersal of exotic weeds" (Merlin 1985). The evidence from the uplands of Rarotonga thus tends to support the hypothesis that in conditions of competition between alien and native plants without associated disturbance (on volcanic soils), native plant species are also able to hold their own against a diverse assemblage of alien species.

CONCLUSIONS

Generalizations drawn from the above qualitative examples of competitive interactions among native and alien flora must be evaluated cautiously in light of the many unconsidered variables and unique environmental histories associated with each of the islands discussed. It would appear however, that the competitive ability of insular floras (at least against early successional adventives), in the absence of severe or persistent feral ungulates, is probably sufficient to resist displacement by many alien plant species. One of the best examples of this can be found in the uplands of Rarotonga. Moreover, there is evidence from a number of oceanic islands, including Nauru in the Pacific and Barbados (Watts 1970) in the Caribbean, indicating that if native plants are displaced due to direct or indirect human impact, they may regain their dominant positions when human-induced disturbance (such as phosphate mining on uplifted coral-limestone islands or browsing, grazing and rooting by introduced ungulates on volcanic islands) ceases. A general conclusion drawn from island comparisons herein is that alien plant communities (early successional) evidence no specific pattern of competitive advantage in situations where alien ungulates or other forms of human disturbance are absent.

Questions of competitive interactions involving "climax" or late successional aliens are less clearly understood. For example, can native insular plant communities dominated by endemic species that have undergone radical adaptive shifts (such as those found in the native *Scalesia* forests of the Galapagos Archipelago) ultimately compete against introduced "real" trees? The introduction of common guava to the Galapagos appears to have severely altered the present distribution of native vegetation (including the woody species of *Scalesia* and *Croton*) on at least three islands, San Cristobal, Floreana, and Isabela. Common guava "has grown out of control and covers vast areas of the humid highlands" (Eckhardt 1972). It has also been reported that the rank growth of alien elephant grass *(Pennisetum purpureum)* in some areas has prevented regeneration of native tree species such as *Scalesia* and *Eugenia* (Stone *et al.* 1988).

The role of feral ungulates in the relationships between native and alien plants in the Galapagos is quite instructive. Removal of feral cattle from anthropogenic grasslands in the highlands of San Cristobal about 50 years ago allowed previously cropped guava shoots to quickly develop into mature trees that dominate "the dense tangled and dwarf forests prevalent today" (Eckhardt 1972). It is possible that guava trees may now have a seed bank and other reproductive advantages (*e.g.*, allelopathy) that preclude the regeneration of native woody species. Although less radically affected than San Cristobal, parts of the highlands of Floreana and Isabela also appear to have experienced subsequent conversion of the native forests to grassland and then to guava-dominated scrub forest, through the initial presence and later removal of cattle (especially feral cattle).

The present paper suggests specific areas for further research to clarify the nature of competition between native and alien island plants with and without associated disturbance. Additional long-term studies of population dynamics in mixed native/alien ecosystems are needed. All the island groups described above offer opportunities for comparative studies under different disturbance regimes. For example, although common guava became a severe pest in significant areas of three of the four inhabited islands in the Galapagos, on the fourth of these islands, Santa Cruz, the species still formed only a minor part of the woody vegetation approximately 35 years after its introduction. In the 1980s, however, the tree apparently became more abundant. "What influences, man-made or natural, may induce its proliferation are still unknown" (Eckhardt 1972).

In summary, the comparisons made herein and studies cited from elsewhere suggest that competition of alien and native plants on islands with and without past and current disturbance by humans and livestock is a very complicated subject. Although severity and periodicity of natural and unnatural disturbances (including seabird colonies), proximity and kind of alien and native plant propagules, and site specificity can influence outcomes, it is likely that competitive advantages of native species are less important in early succession than late succession. In many cases, removal of disturbance caused by ungulates or humans results in dominance of native rather than alien species, particularly if enduring changes such as loss of pollinators, shading, allelopathy, or altered nutrient cycles have not been effected.

Literature Cited

Allan, H.H. 1936. Indigene versus alien in the New Zealand plant world. *Ecology* 17:187-193.

Allen, W.F. [1858] 1938. Unpublished letter dated May 31 to Messrs. Wyllie and Allen, describing Kahoolawe (original in Archives of Hawaii). Pub. 1938 in *Paradise Pac.* 50(5):22, 27.

Barrera, W., Jr. 1979. *Historic resources of Kahoolawe (partial inventory: prehistoric sites and historic properties).* National Register of Historic Places Inventory Nomination Form. Wash. D.C.: U.S. Dep. Interior Natl. Park Serv.

Bates, M. 1956. Man as an agent in the spread of organisms. In *Man's role in changing the face of the earth*, ed. W. Thomas, 788-804. Chicago: Univ. Chicago Pr.

Beck, 1923. The voyage of the *France. Nat. Hist.* 23:33-34.

Bourne, W.R., and A.C. David. 1983. Henderson Island, Central South Pacific, and its birds. *Notornis* 30 (3):233-252.

Brodie, W. 1851. *Pitcairn's Island, and the islanders in 1850.* London: Whittaker and Co.

Carlquist, S. 1974. *Island biology*. New York: Columbia Univ. Pr.

Carlquist, S. 1980. *Hawaii: a natural history*, 2nd ed. Lawai, Kauai, Hawaii: Pac. Trop. Botan. Gdn.

Carter, J. (ed.). 1981. *Pacific Island yearbook*. Sydney: Pac. Pub.

Conant, S., C. Christensen, P. Conant, W. Gagné, and M.L. Goff. 1984. The unique terrestrial biota of the Northwestern Hawaiian Islands. *Proc. Symp. Resour. Invent. Northwestern Hawn. Islands*, 25-27 May 1983, 77-94. Honolulu: Sea Grant Coll. Prog., Univ. Hawaii.

Conant, S. 1985. Recent observations on the plants of Nihoa Island, Northwestern Hawaiian Islands. *Pac. Sci.* 39(2):135-149.

Cooke, C.M., Jr., and Y. Kondo. 1960. *Revision of Tornatellinidae and Achatinellidae (Gastropoda, Pulmonata)*. B.P. Bishop Mus. Bull. 221. Honolulu: Bishop Mus. Pr.

Cuddihy, L.W., and C.P. Stone. 1990. *Alteration of native Hawaiian vegetation: effects of humans, their activities and introductions.* Univ. Hawaii Coop. Natl. Park Resour. Stud. Unit. Honolulu: Univ. Hawaii Pr.

Dasmann, R.F., J.P. Milton, and P.H. Freeman. 1973. *Ecological principles for economic development*. New York: John Wiley & Sons.

Decker, B.G. 1970. Plants, man and landscape in Marquesan valleys, French Polynesia. Berkeley, Calif.: PhD diss., Dept. Geogr., Univ. Calif.

Decker, B.G. 1973. Unique dry-island biota under official protection in northwestern Marquesas Islands (Iles Marqueses). *Biol. Conserv.* 5(1):66-67.

Duff, R. 1968. Stone adzes from Raoul, Kermadec Islands. *J. Polynesian Soc.* 77:386-401.

Eckhardt, R.C. 1972. Introduced plants and animals in the Galapagos Islands. *Bioscience* 22(10):587-590.

Egler, F.E. 1942. Indigene versus alien in the development of arid Hawaiian vegetation. *Ecology* 23:14-23.

Emory, K.P. 1928. *Archaeology of Nihoa and Necker Islands.* B.P. Bishop Mus. Bull. 53. Honolulu: Bishop Mus. Pr.

Environmental Impact Study Corporation. 1979. *Kahoolawe Training Area Environmental Impact Statement*. Honolulu.

Fosberg, F.R. 1965. The island ecosystem. In *Man's place in the island ecosystem: a symposium*, ed. F.R. Fosberg, 1-62. Honolulu: Bishop Mus. Pr.

Fosberg, F.R., M.-H. Sachet, and D.R. Stoddart. 1983. Henderson Island (southeastern Polynesia): summary of current knowledge. *Atoll Res. Bull.* 272.

Gade, D.W. 1985. Man and nature on Rodrigues: tragedy of an island common. *Environ. Conserv.* 12(3):207-216.

Harris, D.R. 1962. The invasion of oceanic islands by alien plants. *Trans. Pap. Inst. Brit. Geogr.* Pub. 31:67-82.

Hathaway, W.G. 1952. Composition of certain native dry forests: Mokuleia, Oahu, T.H. *Ecol. Monogr.* 22:153-168.

Hommon, R. 1980. *Kahoolawe: final report for the archaeological survey.* National Register of Historic Places Inventory Nomination Form. Wash. D.C.: Natl. Park Serv.

Kirch, P.V. 1982. The impact of the prehistoric Polynesians on the Hawaiian ecosystem. *Pac. Sci.* 36(1):1-14.

Lebarron, R., and R. Walker. 1970. Kahoolawe, a badly abused island -- how to rehabilitate it. Honolulu: Hawaii Div. Forestry. Typescript.

Mangenot, G. 1965. The effect of man on the plant world. In *Man's place in the island ecosystem: a symposium*, ed. F.R. Fosberg, 117-132. Honolulu: Bishop Mus. Pr.

Manner, H.I., R.R. Thaman, and D.C. Hassall. 1985. Plant succession after phosphate mining of Nauru. *Australian Geographer* 16(3):185-194.

Merlin, M.D. 1985. Woody vegetation in the upland region of Rarotonga. *Pac. Sci.* 39(1):81-99.

Merlin, M.D. 1991. Woody vegetation on the raised coral islands of Mangaia, southern Cook Islands. *Pac. Sci.* 45(2):131-151.

Montgomery, S.L., W.C. Gagné, and B.H. Gagné. 1980. Notes on birdlife and conservation in the Marquesas and Society Islands. *'Elepaio* 40(11):152-155.

Morton, E.K. 1964. *Crusoes of Sunday Island.* London: Bell.

Mueller-Dombois, D. 1975. Some aspects of island ecosystem analysis. In *Tropical ecological systems -- trends in aquatic and terrestrial research*, ed. F.B. Golley and E. Medina, 353-366. Vol. 11, Ecol. Stud. Ser., Analysis and synthesis. New York: Springer Verlag.

Mueller-Dombois, D., K.W. Bridges, and H.L. Carson, eds. 1981. *Island ecosystems: biological organization in selected Hawaiian communities.* Stroudsburg, Penn.: Hutchinson Ross Pub. Co.

Pacific Science Association. 1983. *Information Bull.* 35:17

Pacific Science Association. 1984. *Information Bull.* 36:3.

Parkes, J.P. 1984a. Feral goats on Raoul Island. I. Effect of control methods on their density, distribution, and productivity. *New Zealand J. Ecol.* 7:85-94.

Parkes, J.P. 1984b. Feral goats on Raoul Island. II. Diet and notes on the flora. *New Zealand J. Ecol.* 7:95-101.

Perrault, G.H. 1978. Peuplement Entomologique de Marquises. *Cahiers du Pacifique* 21(9):359-389.

St. John, H., and W.R. Philipson. 1962. An account of the flora of Henderson Island, South Pacific Ocean. *Trans. Royal Soc. New Zealand* 1:175-194.

Sinoto, Y.S. 1983. An analysis of Polynesian migrations based on the archaeological assessments. *J. Soc. Ocean.* 39:57-67.

Skottsberg, C. 1957. The vegetation of the Juan Fernandez and Desventuradas Islands. *Proc. 8th Pac. Sci. Congr.*, Quezon City, Philippines. Vol. 4:181-185.

Spriggs, M. 1986. Prehistoric man-induced landscape enhancement in the Pacific: examples and implications. In *Prehistoric intensive agriculture in the tropics,* ed. I.S. Farrington, Part I, 409-432. Oxford, England: Bar Internatl. Ser. 232.

Steadman, D.W., and S.L. Olson. 1985. Bird remains from the archaeological site on Henderson Island, South Pacific: man-caused extinctions on an "uninhabited" island. *Proc. Natl. Acad. Sci. USA* 82:6191-6195.

Stone, C.P., L.L. Loope, and C.W. Smith. 1988. Conservaton biology in the Galapagos Archipelago: perspectives from Hawaii. *'Elepaio* 48(1):1-8.

Straubel, C.R., ed. 1954. *The whaling journal of Captain W.B. Rhodes, 1836-1838.* Christchurch, N.Z.: Whitcombe and Tombs.

Sykes, W.R. 1969. The effect of goats on vegetation of the Kermadec Islands. *Proc. New Zealand Ecol. Soc.* 12:20-23.

Sykes, W.R. 1977. *Kermadec Islands flora: an annotated checklist.* Bull. 219. Wellington, New Zealand: Dep. Sci. Industr. Res.

Sykes, W.R. 1980. Botanical science. In *Bibliography of research on the Cook Islands*, ed. D.I. Kinloch, 9-68. New Zealand Man and Biosphere Rep. 4. Lower Hutt, New Zealand: Dep. Sci. Industr. Research.

Watts, D. 1970. Persistence and change in the vegetation of oceanic islands: an example from Barbados, West Indies. *Canadian Geographer* 14: 91-109.

Wirawan, N. 1978. Vegetation and soil-water regimes in the tropical rain forest valley on Oahu, Hawaiian Islands. PhD thesis, Univ. Hawaii, Honolulu.

Yen, D.E., P.V. Kirch, P. Rosendahl, and T. Riley. 1972. Prehistoric agriculture in the upper valley of Makaha, Oahu. *Makaha Valley Historical Project Interim Rep. No. 3, Pac. Anthropol. Rec. #18.* Anthropol. Dep., B.P. Bishop Mus., Honolulu.

ALIEN AND NATIVE PLANT RESPONSE TO RELEASE FROM FERAL SHEEP BROWSING ON MAUNA KEA

Paul G. Scowcroft and C. Eugene Conrad

ABSTRACT

Over 150 years of feral sheep damage to the subalpine woodlands on Mauna Kea, Hawai'i, ended when "eradication" of feral sheep *(Ovis aries)* was declared complete (even though a few sheep remained). Managers and others interested in conserving Mauna Kea's subalpine ecosystem are now concerned about the ability of native plants to reestablish dominance without the aid of humans in the absence of significant feral sheep grazing and browsing. Successful reestablishment will depend partly on the ability of native plants to compete with alien plants. Data from exclosures near tree line indicate that both herbaceous and woody native species establish in areas freed from sheep foraging pressure, in spite of alien plants, provided that the areas are not fully occupied by aliens. In areas that were densely populated by feral sheep before 1981, native and alien plants appear to be responding, as indicated in publications about earlier exclosure studies. Control of alien plants to favor native species does not now appear warranted. However, two noxious aliens, fountain grass *(Pennisetum setaceum)* and German ivy *(Senecio mikanioides)*, are established in the subalpine woodland and could become management problems in the future.

INTRODUCTION

Are native plants capable of successfully competing with introduced alien plants in insular terrestrial ecosystems? This question has been posed repeatedly in Hawai'i, where over 4,600 species of plants have been introduced over the last 200 years (St. John 1973). Of these species, 86 have been recognized as serious pests of native Hawaiian ecosystems (Smith 1985).

Answering the question is no simple task, because competitive interactions are not easily assessed in uncontrolled natural areas without experimentation. Furthermore, an answer based on native-alien plant interactions in one ecosystem cannot be generalized to others, especially in Hawai'i, where considerable terrestrial ecosystem diversity exists (*e.g.*, dry coastal lowlands, rain forests, and semiarid alpine tundra). The best one can hope for are answers on a case-by-case basis.

In this paper we examine alien and native plant responses to release from feral sheep grazing and browsing in the māmane *(Sophora chrysophylla)* forest on Mauna Kea, a subalpine woodland ecosystem that has been and remains the center of management controversy (Juvik and Juvik 1984). We also describe this woodland ecosystem, summarize the impacts of alien animals, and discuss potential interactions among native and alien plants and introduced ungulates. Specific objectives are: 1) to review responses of plants to release from sheep browsing in areas protected by exclosures prior to feral sheep eradication in 1981; 2) to describe plant responses inside and outside two exclosures five years after feral sheep "eradication;" and 3) to evaluate alien-native plant interactions in a portion of the Mauna Kea Forest Reserve that has been essentially sheep-free since about 1960 (R. Walker, pers. comm. 1986). We also address the question of whether plant recovery after "eradication" and outside exclosures follows a pattern similar to that observed inside exclosures before "eradication" and discuss suppression of native species by alien species.

The Subalpine Woodland Ecosystem

The subalpine ecosystem of Mauna Kea occurs between 5,900 and 9,840 ft (1,800-3,000 m) elevation and consists of two distinct woodland types: the māmane woodland and the naio *(Myoporum sandwicense)*-māmane woodland.

On eastern, northern, and western slopes of Mauna Kea, and also in a narrow band at tree line on the southern slope, māmane occurs in almost pure stands (Scott *et al.* 1984). Naio is absent in these areas, and other native tree species, such as pilo *(Coprosma montana)*, are scarce. The māmane trees average less than 26 ft (8 m) in height, and canopy cover ranges from very scattered (<5% cover) to partly closed (60% cover) (Scott *et al.* 1984), the latter being uncommon. Native woody species in the understory include pūkiawe (*Styphelia tameiameiae*), ʻōhelo (*Vaccinium* sp.), *Geranium cuneatum*, and *Dubautia arborea*. Herbaceous species are predominantly aliens and include several grasses: sweet vernalgrass, ripgut grass, hairy oatgrass, Kentucky bluegrass, needlegrass, and velvet grass *(Anthoxanthum odoratum, Bromus rigidus, Danthonia pilosa, Poa pratensis, Stipa cernua, Holcus lanatus)*. Native grasses are *Agrostis sandwicensis, Deschampsia nubigena*, and pili uka *(Trisetum glomeratum)*. Additional details about māmane woodland vegetation are given by Hartt and Neal (1940) and Mueller-Dombois and Krajina (1968).

The mixed naio-māmane woodland type is confined to the southwestern slopes of Mauna Kea (Scott *et al.* 1984). Naio accounts for 10 to over 85% of the trees, with māmane comprising most of the rest (van Riper 1980a). Other native tree species occurring as scattered individuals and colonies are *Dubautia arborea, Chamaesyce olowaluana*, and sandalwood (*Santalum ellipticum*). Tree heights average less than 26 ft (8 m), but tree canopy cover is generally higher than in the pure māmane forest and can exceed 85% (Scowcroft 1983). The shrubs in the mixed woodland are all native and include ʻāweoweo, ʻūlei, and ʻaʻaliʻi (*Chenopodium oahuense, Osteomeles anthyllidifolia, Dodonaea* sp.). Many of the herbaceous species found in the māmane woodland also occur in the naio-māmane woodland. Three noxious weeds are found in the mixed woodland: stinkweed, fountain grass, and German ivy (*Tagetes minuta, Pennisetum setaceum, Senecio mikanioides*) (Scott *et al.* 1984; Smith 1985).

The climate of the subalpine woodlands is a function of elevation and prevailing tradewinds. Temperatures range from nighttime lows near freezing to daytime highs above 68 F (20 C) (van Riper 1980a; Scowcroft 1981). The range of mean daily temperature over a year is smaller than the diurnal temperature range. The nighttime inversion layer, which generally lies below the subalpine woodland, breaks down by midday, and prevailing tradewinds bring fog or rain or both to northeastern to southeastern slopes. Median annual rainfall on windward slopes ranges from 28 in. (700 mm) at tree line (9,500 ft or 2,900 m elevation) to 63 in. (1,600 mm) at the lower boundary of the māmane woodland at 6,900 ft (2,100 m) elevation (Hawaii Division of Water and Land Development 1982). Rainfall is low at all elevations in the rain shadow of Mauna Kea, with an annual median of 16 in. (400 mm). Snow rarely falls in the subalpine woodlands, and if it does, it melts rapidly.

The soils are poorly developed, and in some cases plants are growing in undifferentiated parent material (cinders and volcanic sand). Where development has occurred, the soil is classified as a Typic Vitrandept, medial over cindery, isomesic (U.S. Soil Conservation Service 1973). Coarse structure and low amounts of organic matter result in low water-holding capacity. Steep slopes and coarse texture of cinders and ash result in unstable soils. In general, the edaphic conditions of the subalpine woodlands are harsh for plant growth (Hartt and Neal 1940).

Historical Impact of Ungulates

Domestic livestock were introduced to the Hawaiian Islands late in the 18th century (Kramer 1971). Feral populations of cattle, sheep, and goats (*Bos taurus, Ovis aries, Capra hircus*) soon became established in forests. By 1825, feral sheep had established in Mauna Kea's subalpine woodland. Lacking natural predators except for wild dogs (*Canis domesticus*), the sheep population reached about 40,000 animals by the early 1930s, one for every 5 a (2 ha) of habitat (Bryan 1937a). Sheep suppressed māmane and other tree reproduction over large areas, stripped bark from tree stems, and consumed herbaceous vegetation, thereby leaving the soil exposed to accelerated erosion (Warner 1960). Because damage to the ecosystem was severe and because feral sheep competed with commercial flocks (Judd 1936), Hawaiʻi Territorial foresters built a stock-proof fence around the Mauna Kea Forest Reserve (Bryan 1937b) and reduced the population through sheep drives and hunter-guide programs. Fewer than 500 feral sheep were left by 1950 (Bryan 1950). Control efforts were then relaxed, and populations increased.

Sustained yield management for public hunting was started in 1955, with the population kept below 5,000 animals. During the 1970s, the population averaged 1,500 animals. Even at this relatively low level, vegetation continued to deteriorate where sheep concentrated, especially at tree line (Scowcroft 1983; Scowcroft and Giffin 1983; Scowcroft and Sakai 1983).

Ecosystem damage has also been caused by mouflon sheep (*Ovis musimon*), which were released in the Mauna Kea Forest Reserve starting in 1962 (Giffin 1982). Food preferences, grazing and browsing behavior, and herding habits are similar to those of feral sheep, and native plants are particularly susceptible to damage by mouflon. In 1986, the largest concentrations of mouflon were on the southeastern and northwestern flanks of the mountain, and animals were moving into areas formerly occupied by feral sheep. The mouflon population was estimated at 500 animals (R.E. Bachman, pers. comm. 1986).

Continued degradation of the māmane woodland in the late 1970s posed a significant threat to the palila *(Loxioides bailleui)*, an endangered endemic Hawaiian bird now found only in the subalpine woodland of Mauna Kea (Berger 1972; van Riper *et al.* 1978; Scott *et al.* 1984). Palila depend on māmane and, to a lesser extent, on naio trees for food and nest sites (van Riper 1980b). Critical habitat for palila was designated in 1977 and included almost all State-owned māmane and naio-māmane woodlands on Mauna Kea (U.S. Fish and Wildlife Service 1977).

Because of continued habitat degradation and the attendant threat to the palila, a suit on behalf of the palila was brought by conservationists against the state of Hawai'i. Noncompliance with the U.S. Endangered Species Act of 1973 was charged. A detailed account of the initial palila lawsuit and events that preceded was given by Juvik and Juvik (1984). The suit was decided in favor of the bird, and the State was ordered to remove feral sheep and feral goats completely and permanently from those portions of the māmane forest designated as critical palila habitat. The status of mouflon sheep was not affected by the court order.

The feral sheep and goat "eradication" effort was completed in 1981. Five years later, it was evident that a few feral sheep had escaped death, as expected. The authors have seen a flock of 20 feral sheep at tree line on the western side of Mauna Kea. Sheep tracks and signs of recent browsing were common in the vicinity of that sighting.

Potential Plant-Animal Interactions

The interactions among native plants, alien plants, and alien ungulates are often manifest in composition of the plant communities and species densities. Relative species composition is the percent contribution of each species to total plant cover in a given area. Potential interactions among aliens and natives may be considered in terms of species composition and density of each group in a community.

Species Composition. Differential grazing may favor alien plants, which appear to be less palatable than some native plants. For instance, māmane is sometimes called an "ice cream plant" because it is preferentially browsed by sheep. The native climax perennial grasses in the Mauna Kea subalpine woodlands seem to be preferred by sheep over alien perennial grasses such as sweet vernalgrass.

The effect of sheep grazing on survival of native climax grasses may involve more than palatability. Williams (1968) and Williams and Roe (1975) reported that sheep-grazing decreased the probability of survival of climax grasses in semiarid portions of Australia. In contrast, the life spans of dominant disclimax grasses there were enhanced by grazing. On Mauna Kea the climax native grasses are tufted perennials, such as pili uka, which have their perennating buds above ground where they are susceptible to grazing damage. In contrast, several alien grasses (*e.g.*, Kentucky bluegrass, hairy oatgrass) have perennating buds below ground.

When released from grazing pressure, aliens may grow more rapidly and disperse more seeds over a larger area than native species. Thus, aliens may simply outcompete native species and dominate recovering plant communities. Kilcher and Looman (1983) concluded that most of the native grasses they studied in a temperate ecosystem did not have the potential to be competitive with adapted introduced and domesticated species and cultivars.

Conversely, although aliens may dominate a grazed plant community, they are not necessarily more competitive than native species. Rabinowitz *et al.* (1984) postulated that competitive inferiority may be a sufficient explanation for sparsity of certain herbaceous species in a plant community. But if abundance is regulated by another cause, such as differential herbivory, then competitive superiority may be a necessary conditon of persistence. In such a situation, we would expect the competitively superior but suppressed species to respond positively to release from grazing and become a biologically significant component of the community.

Plant Density. The reproductive capacity and population growth potential of alien plants may exceed that of native species. The greater potential of an alien may be genetically determined (*e.g.*, the growth potential for mullein (*Verbascum thapsus*), an alien species, is remarkable (Juvik and Juvik, this volume)). Lower population growth potential of native plants could also result from greater herbivory reducing the supply of photosynthate available for reproduction (Harper 1977). Ungulates may also preferentially eat germinants of native species even when native germinants are scarce.

Smith (1985) noted that some alien plant pests are known or suspected of allelopathy. Although true allelopathic effects are difficult to separate from other competitive interactions, allelopathy could be a mechanism controlling density of native competitors. Conversely, some native species, like ‘āweoweo, may be allelopathic.

Suitability of Alien and Native Species for Forage

We have assembled information from published sources about the forage value, relative palatability, and susceptibility to animal damage of many common plant species in the Mauna Kea Forest Reserve (Table 1). Four native species on the list are highly palatable and susceptible to grazing or browsing damage: pili uka, *Agrostis avenacea*, ‘āweoweo, and

Table 1. Relative forage value, relative palatability, and susceptibility to browsing or grazing damage of some common plant species in the Mauna Kea Forest Reserve, island of Hawai'i.

Species	Relative forage value	Relative palatability	Susceptibility to grazing/ browsing damage	References
GRASSES				
*Agrostis avenacea**	?	High	?	Whitney *et al.* 1939
Anthoxanthum odoratum	Low	Low	Low	Ripperton *et al.* 1933
*Bromus rigidus***	None	None	?	Whitney *et al.* 1939
Dactylis glomerata	High	High	Moderate	Ripperton *et al.* 1933
Danthonia pilosa	?	Moderate	Low	Whitney *et al.* 1939
*Deschampsia nubigena**	?	Low	?	Ripperton *et al.* 1933
Holcus lanatus	High	High	Moderate	Ripperton *et al.* 1933
Pennisetum setaceum	Low	Low	Low	Hosaka & Thistle 1954
Poa pratensis	High	High	Low	Ripperton *et al.* 1933
Trisetum glomeratum	?	High	High	Giffin 1982; Scowcroft & Giffin 1983; Ripperton *et al.* 1933
*Vulpia myuros***	?	Low-Moderate	?	Ripperton *et al.* 1933
FERN				
Pteridium aquilinum	Poisonous	Low	Low	Giffin 1982
HERBS AND FORBS				
*Bidens pilosa***	Low	Low	Low	McClelland 1915
Cirsium vulgare	Low	Low	Low	Giffin 1976
Epilobium billardierianum	High	High	High	Giffin 1976
*Erodium cicutarium***	Low	Moderate	Low	McClelland 1915; Giffin 1976
Hypochoeris radicata	Moderate	Moderate	Low	McClelland 1915; Giffin 1976; Giffin 1982; Ripperton *et al.* 1933

Species	Relative forage value	Relative palatability	Susceptibility to grazing/ browsing damage	References
Rumex acetosella	?	Moderate	Moderate	Giffin 1982
*Senecio sylvaticus***	?	Low	Moderate	Giffin 1976
*Sonchus oleraceus***	Moderate	High	Moderate	McClelland 1915; Giffin 1976
*Tagetes minuta***	None	None	Low	Hosaka & Thistle 1954
Taraxacum officinale	Moderate	Moderate	Low	McClelland 1915
*Urtica urens***	Low	Low	Moderate	Giffin 1976
*Verbascum thapsus***	Low	Low	Low	Giffin 1982
SHRUBS				
*Chenopodium oahuense**	?	High	High	Giffin 1976
Dodonaea sp.*	?	Low	Low	Giffin 1982
Dubautia sp.*	?	Low-Moderate	High	Giffin 1982
Geranium sp.*	?	Moderate	Moderate	Giffin 1982
*Styphelia tameiameiae**	Low	Low-Moderate	Low	Hosaka & Thistle 1954; Giffin 1976, 1982
Vaccinium sp.*	Low	Moderate	Moderate	McClelland 1915; Giffin 1982
TREES				
Coprosma sp.*	?	Low	Moderate	Giffin 1982
Euphorbia sp.*	?	Low	Moderate	McClelland 1915
*Myoporum sandwicense**	Low	Low	Moderate	Giffin 1976
*Sophora chrysophylla**	High	High	High	Giffin 1976, 1982

*Native species.
**Species with annual or biennial life spans. All others are perennials.

māmane. *Agrostis sandwicensis*, a native grass not listed in Table 1, is probably highly palatable and easily damaged also (Scowcroft and Giffin 1983). Several alien plant species are highly palatable: cocksfoot (*Dactylis glomerata*), velvet grass, Kentucky bluegrass, *Epilobium billardierianum*, and *Sonchus oleraceus*. However, susceptibility to damage by sheep for these and other alien plants is generally lower than for the native plants. Plant responses to release from grazing and browsing pressure in light of their forage characteristics will be discussed later.

RECOVERY OF ALIEN AND NATIVE PLANTS

Study Sites

Thirteen exclosures have been established within the Mauna Kea Forest Reserve (Loope and Scowcroft 1985). Quantitative data have been collected from six of these exclosures: Pu'u Nanahā, Pu'u Kole, Kaluamakani, Pu'u o Ka'uha, Wailuku, and Hale Pōhaku. All six exclosures are located at or near tree line (about 9,020 ft or 2,750 m elevation). Site attributes of the six exclosures are described by Scowcroft and Giffin (1983).

Methods

Methods used to study vegetation recovery at five of the exclosure sites before feral sheep "eradication" in 1981 were described by Scowcroft and Giffin (1983). At the sixth site, the Wailuku exclosure, data were collected four years after construction using stratified random sampling, but results were not published. Three strata were mapped and sampled inside Wailuku: a light grass stratum with eight sample quadrats, a cinders stratum with ten sample quadrats, and a rock stratum with three sample quadrats. Mature māmane trees and the vegetation under them constituted a fourth stratum, but it was not sampled. Quadrats measured 10 x 10 ft (3.05 x 3.05 m), and 3-6% of each stratum was sampled. The basal area of each plant was determined by species (Scowcroft and Giffin 1983). Mean basal area estimates were calculated, by species, for each stratum and for the whole exclosure.

Herbaceous vegetation was resampled in April and May 1986 at two of the six exclosures -- Pu'u o Ka'uha in the rain shadow on the western side of Mauna Kea, and Wailuku on the wetter, eastern side of the mountain. Methodology in 1986 differed from that used earlier. We reestablished the same sample quadrats used previously, but instead of measuring all plants in each quadrat, we subsampled as described by Cochran (1963). Each large quadrat (cluster) was divided into 36 cells measuring 0.8 x 0.8 ft (0.25 x 0.25 m). Twelve cells per cluster were randomly selected, and plants were measured for projected crown area, basal area, and height, by species. Overall means and standard errors were calculated according to Cochran (1963).

All pairwise comparisons in this paper were made using the Games and Howell (1976) procedure (see also Keselman and Rogan 1978). The Bonferroni "z" value (df = infinity) was used to calculate confidence limits for the difference between means (Bailey 1977).

We also estimated the percentage of each cell covered by rock, bare ground, litter, and live plants, regardless of species. We defined cover in this instance as vertical projected cover. Total cover for each cell had to equal 100%.

Woody species, which were exclusively native plants, were not as abundant as the herbaceous species, so we measured every individual inside the Wailuku exclosure and inside and outside the Pu'u o Ka'uha exclosure. Data included species, height, crown spread, and stem diameter 12 in. (30 cm) above the ground. For plants with multiple stems at 30 cm, we estimated the diameter of the "average"-sized stem. Mature individuals were tallied separately from immatures.

The māmane woodland was also sampled, along transects running along the contour and to either side of the Pu'u o Ka'uha exclosure, to determine if plant responses at the exclosure site were representative of the surrounding area. Samples were not taken outside the Wailuku exclosure, because grazing and browsing by mouflon sheep continued unabated there.

Using stopwatch, compass, and altimeter, we walked along the contour at the same elevation as the Pu'u o Ka'uha exclosure (9,020 ft or 2,750 m) and established five sample points 15 minutes apart. We estimated that the transects were about 1.24 mi (2 km) long. At each sample point we laid out two nested circular plots; the area of the inner one was 1,076 ft^2 (100 m^2), and the area of the outer one was 3,390 ft^2 (315 m^2). Vertical projected areas of rock, litter (including logs and branches), bare soil, and live herbaceous foliar cover, by species, were estimated within the smaller plots. Woody plant heights and crown diameters were measured in the larger plots, by species.

A similar pair of transects and set of nested plots were established near the Pu'u La'au cabin at an elevation of 7,710 ft (2,350 m). The woodland at this elevation has been essentially sheep-free since about 1960. The southern transect lies entirely within the mouflon-holding pen, which was built in 1960. Data collected along the lower-elevation transects were identical to those collected along the upper-elevation transects.

Results and Discussion

Recovery Before Feral Sheep Eradication. After 12 years of protection from sheep grazing, aerial cover of alien and native herbaceous species was generally greater inside than outside the Pu'u Nanahā, Pu'u Kole, and Kaluamakani exclosures. Total herbaceous cover was 24.5% inside and 9.2% outside at Pu'u Nanahā; 21.5% inside and 7.2% outside at Pu'u Kole; and 1.2% inside and <0.1% outside at Kaluamakani. Apparently, both alien and native species responded favorably to release from grazing.

Recompiled data from Scowcroft and Giffin (1983, Table 2) show that aliens comprised smaller proportions of total herbaceous aerial cover inside than outside the three old exclosures: alien aerial cover was 91% inside and 98% outside the Pu'u Nanahā exclosure; alien aerial cover was 79% inside and >99% outside the Pu'u Kole exclosure.

Native species became reestablished inside the exclosure in spite of alien species. The increased cover of native species was mainly due to reestablishment of the native grasses pili uka, *Agrostis sandwicensis*, and *Deschampsia nubigena*. These grasses were only found inside the exclosures.

Herbaceous species were taller inside than outside the three old exclosures (Scowcroft and Giffin 1983). Sample sizes for hairy oatgrass, velvet grass, gosmore (*Hypochoeris radicata*), sweet vernalgrass, and sheep sorrel (*Rumex acetosella*) were large enough to statistically compare mean height inside with that outside. All of these alien species were significantly taller inside than outside ($p < 0.05$).

Changes in basal area of alien and native herbaceous species after three years of protection from sheep browsing were reported for the Pu'u o Ka'uha and Hale Pōhaku exclosures, which were built in 1972 (Scowcroft and Giffin 1983).

Alien plant basal area increased significantly over a three-year period inside and outside the Pu'u o Ka'uha exclosure. No difference in mean alien basal area was detected between fenced and unfenced populations, which could indicate 1) that growth and reproduction of alien populations were not limited by feral sheep grazing, 2) that feral sheep avoided the areas in the immediate vicinity of the exclosure, or 3) that alien plants that established inside the exclosure served as a population center and source of propagules for colonization of adjacent sparsely occupied sites. We believe all three situations were operative, at least to some degree. As stated, feral sheep did affect aerial cover of aliens outside the older exclosures (Pu'u Nanahā, Pu'u Kole, and Kaluamakani), the effect being most pronounced at the last and harshest of the six sites.

Alien plant basal area did not change significantly inside or outside the Hale Pōhaku exclosure. Also, alien basal area after three years of protection was not significantly greater inside than outside the exclosure. This site, which was on the inside of a cinder cone and characterized by a steep slope and unstable, coarse sandy soil (like the Kaluamakani exclosure), is harsher than the Pu'u o Ka'uha site, so the lack of a significant increase in alien basal cover is not surprising. Hale Pōhaku was also the least impacted by sheep.

Māmane regeneration and establishment was observed only inside exclosures. The native shrub 'āweoweo became established inside but not outside the Pu'u o Ka'uha exclosure. Scowcroft and Giffin (1983) did not comment about possible inhibition of native woody plant regeneration by alien plants.

Recovery After Feral Sheep Eradication. In this section we detail the result of vegetation sampling done in April and May 1986 at Wailuku and Pu'u o Ka'uha exclosures, both erected in 1972. Data from Scowcroft and Giffin (1983) are presented to complement the 1986 data for Pu'u o Ka'uha.

The Wailuku Exclosure. Inside the Wailuku exclosure, basal area for alien herbaceous species in 1976 averaged 200 ± 25 cm^2/m^2, whereas in 1986 it averaged 232 ± 41 cm^2/m^2 (Fig. 1). The difference in basal area was not statistically different, however ($p > 0.05$). Likewise, changes in alien plant basal area within individual strata between years were not statistically significant (Table 2).

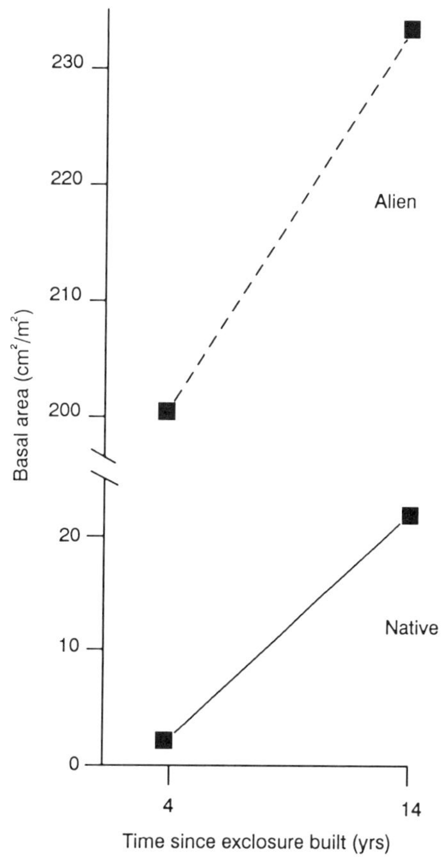

Figure 1. Basal area of alien and native herbaceous species inside the Wailuku exclosure at tree line on the east flank of Mauna Kea, by elapsed time since the exclosure was built.

We believe basal area changes in the cinders stratum were biologically significant, but because our sample size was small, we could not declare the difference statistically significant. In 1976, alien basal area averaged 47 ± 25 cm^2/m^2 (Table 2). By 1986, it had increased to 113 ± 62 cm^2/m^2. These data, together with our observations of unsampled portions of the cinders stratum during the period 1976 to 1986, indicate that alien species were colonizing the previously sparsely populated cinders stratum. Hairy oatgrass in particular was establishing in the coarse cinders (Table 2). We suspect that the decrease in alien plant basal area in the rock stratum may also have biological significance. Some alien species may be poorly equipped to cope with the harsh

Table 2. Mean basal area cover (cm^2/m^2) of herbaceous species in the Wailuku exclosure in 1976 and 1986 by major strata.

Species	Light grass stratum		Cinders stratum		Rock stratum		Entire exclosure	
	1976	1986	1976	1986	1976	1986	1976	1986
GRASSES AND SEDGES								
*Agrostis avenacea**	t**	--	--	--	--	--	t	--
*Agrostis sandwicensis**	--	0.1 (0.1)***	t	11.7 (11.4)	4.0 (2.3)	1.6 (1.6)	0.4 (0.2)	5.6 (5.3)
Anthoxanthum odoratum	78.2 (19.3)	123.5 (43.7)	31.6 (18.5)	55.2 (23.2)	26.5 (9.2)	15.7 (9.8)	52.3 (12.3)	82.9 (22.5)
Bromus carinatus	--	--	--	--	--	10.4 (10.2)	--	0.9 (0.8)
Bromus rigidus	--	--	0.2 (0.2)	--	--	--	0.1 (0.1)	--
Danthonia pilosa	263.9 (54.3)	188.1 (52.1)	8.8 (4.0)	54.0 (40.0)	74.7 (29.6)	10.7 (5.0)	129.7 (24.8)	111.1 (30.0)
*Luzula hawaiiensis**	0.8 (0.8)	--	--	--	1.8 (1.8)	t	0.5 (0.4)	t
Poa pratensis	--	t	--	--	--	--	--	--
Vulpia myuros	0.2 (0.2)	--	--	--	--	--	0.1 (0.1)	--

Species	Light grass stratum		Cinders stratum		Rock stratum		Entire exclosure	
	1976	1986	1976	1986	1976	1986	1976	1986
HERBS AND FORBS								
*Argyroxiphium sandwicense**	--	--	--	30.2 (29.9)	--	--	--	14.0 (13.9)
Hypochoeris radicata	6.5 (4.0)	73.0 (8.6)	6.8 (3.6)	4.0 (2.6)	89.5 (45.4)	24.3 (6.2)	13.5 (4.5)	36.9 (10.4)
Rumex acetosella	9.5 (9.4)	0.4 (0.4)	--	--	--	--	4.3 (4.2)	0.2 (0.2)
Senecio sylvaticus	t	--	t	--	t	--	t	--
*Tetramolopium humile**	3.0 (2.9)	0.7 (0.6)	--	t	3.2 (1.5)	1.4 (0.8)	1.6 (1.3)	0.4 (0.3)
FERNS								
*Asplenium trichomanes**	--	0.2 (0.2)	--	--	3.5 (3.0)	5.2 (3.7)	0.3 (0.3)	0.5 (0.3)
*Pellaea ternifolia**	t	1.5 (1.4)	--	--	2.0 (1.7)	6.1 (3.7)	0.2 (0.1)	1.2 (0.7)
Pteridium aquilinum	--	t	t	--	--	t	t	t

Table 2, continued.

Table 2, continued.

Species	Light grass stratum		Cinders stratum		Rock stratum		Entire exclosure	
	1976	1986	1976	1986	1976	1986	1976	1986
Alien species	358.3 (46.1)	385.0 (63.9)	47.4 (25.1)	113.2 (62.2)	190.7 (80.2)	61.1 (24.8)	200.0 (24.8)	232.0 (41.2)
Native species	3.8 (2.9)	2.5 (1.5)	t	41.9 (30.7)	14.5 (5.8)	14.3 (6.1)	3.0 (1.4)	21.7 (14.3)
Combined species	362.1 (48.0)	387.5 (64.0)	47.4 (25.1)	155.1 (67.3)	205.2 (75.0)	75.4 (23.9)	203.0 (25.5)	253.7 (42.7)

*Native species.
**Trace amount, $<0.1\ cm^2/m^2$.
***Values in parentheses are standard errors of the means.

conditions prevalent in the rock stratum, particularly the shallowness or complete absence of soil. About one-half of the decrease in alien basal area was due to hairy oatgrass (Table 2). Inside the Wailuku exclosure, total basal area for native herbaceous species, as shown in Figure 1, averaged 3 $\pm$ 1 cm^2/m^2 in 1976, and 22 $\pm$ 14 cm^2/m^2 in 1986. The difference between years was not statistically significant.

An examination of total native basal area data for individual strata shows that only the cinders community showed a change that could be interpreted as biologically significant (Table 2): 0.05 $\pm$ 0.05 cm^2/m^2 in 1976 and 42 $\pm$ 31 cm^2/m^2 in 1986. The primary native herbaceous colonizers of the cinders stratum were *Agrostis sandwicensis*, a perennial bunch grass that is conspicuously scarce in mouflon-browsed areas surrounding the Wailuku exclosure, and *Argyroxiphium sandwicense* var. *sandwicense*, the Mauna Kea silversword (Table 2). Data from a survey done in 1973, one year after the exclosure was built, showed that *Agrostis sandwicensis* was not found in the cinders stratum. In fact, we recorded only one individual at that time in the entire exclosure, and it was in the rock stratum. The silverswords in our sample and those found elsewhere in the cinders stratum are the result of natural regeneration by a small population planted by the Hawaii Division of Forestry and Wildlife in 1973 (Loope and Scowcroft 1985; C. Wakida, pers. comm. 1986; E. Powell, unpub. data). The data show that in the cinders stratum, the relative increase in basal cover was 9% per year for alien species, and 96% per year for native species.

Foliar cover inside the Wailuku exclosure was measured in 1986 for herbaceous species (Table 3). Aliens accounted for 99% of the foliar cover in the light grass community, 83% in the cinders community, and 92% in the rock community. Hairy oatgrass, sweet vernalgrass, and gosmore contributed most to alien foliar cover in all three communities.

Agrostis sandwicensis accounted for 40% of the native foliar cover in the cinders community. As we indicated earlier, this bunch grass seems to be an aggressive colonizer of barren cinder. The species comprised about 7% and 33% of the native herbaceous foliar cover in the light grass and rock communities.

The density of alien and native herbaceous species, each taken as a group, did not change significantly from 1976 to 1986, either in the separate strata or in the exclosure as a whole (Table 4). Generally, this result was also true for individual species. Hairy oatgrass in the rock stratum was an exception. Density of this alien grass decreased significantly from 26 plants/m^2 in 1976 to about 4 plants/m^2 in 1986. The decrease in basal area of hairy oatgrass noted earlier was apparently due to whole-plant mortality rather than death of portions of individuals. Gosmore in the light grass stratum was another individual species whose density changed significantly between measurements. It increased from less than 1 plant/m^2 in 1976 to over 16 plants/m^2 in 1986 (Table 4).

Table 3. Mean foliar cover (cm^2/m^2) for herbaceous species in the Wailuku exclosure in 1986, by major strata.

Species	Light grass stratum	Cinders stratum	Rock stratum	Entire exclosure
GRASSES AND SEDGES				
*Agrostis sandwicensis**	0.8 (0.8)**	30.8 (28.2)	3.3 (3.2)	14.9 (13.1)
Anthoxanthum odoratum	382.1 (131.8)	137.5 (49.9)	76.6 (55.5)	243.2 (64.2)
Bromus carinatus	--	--	71.2 (69.6)	5.9 (5.8)
Danthonia pilosa	469.0 (135.1)	140.3 (105.0)	45.9 (28.1)	281.3 (78.3)
*Luzula hawaiiensis**	--	--	0.4 (0.3)	t***
Poa pratensis	t	--	--	t
HERBS AND FORBS				
*Argyroxiphium sandwicense**	--	45.3 (44.8)	--	21.1 (20.8)
Hypochoeris radicata	349.5 (151.7)	91.9 (59.1)	347.8 (103.2)	229.7 (74.4)
Rumex acetosella	9.1 (8.1)	--	--	4.1 (3.7)
*tetramolopium humile**	6.2 (4.2)	0.4 (0.4)	8.6 (4.7)	3.7 (2.0)
FERNS				
*Asplenium trichomanes**	0.6 (0.6)	--	11.0 (8.8)	1.2 (0.8)
*Pellaea ternifolia**	3.9 (2.9)	--	26.1 (20.6)	3.9 (2.2)
Pteridium aquilinum	0.4 (0.3)	--	t	0.2 (0.2)
Alien species	1210.2 (292.0)	369.7 (207.4)	541.5 (219.9)	764.4 (164.6)
Native species	11.6 (5.0)	76.6 (50.0)	49.3 (17.1)	44.9 (23.4)
Combined species	1221.8 (291.9)	446.3 (205.9)	590.8 (228.6)	809.3 (164.2)

*Native species.
**Values in parentheses are standard errors of the means.
***Trace amount, < 0.1 cm^2/m^2.

Table 4. Mean plant density* (no./m^2) in the Wailuku exclosure in 1976 and 1986, by major strata.

Species	Light grass stratum		Cinders stratum		Rock stratum		Entire exclosure	
	1976	1986	1976	1986	1976	1986	1976	1986
GRASSES AND SEDGES								
*Agrostis avenacea***	0.2 (0.2)***	--	--	--	--	--	0.1 (0.1)	--
*Agrostis sandwicensis***	--	t#	t	0.3 (0.3)	0.2 (0.1)	0.2 (0.2)	t	0.2 (0.1)
Anthoxanthum odoratum	7.6 (1.9)	12.0 (3.1)	1.9 (1.1)	3.5 (1.0)	7.0 (2.7)	5.0 (3.0)	4.9 (1.0)	7.5 (1.5)
Bromus carinatus	--	--	--	--	--	0.7 (0.6)	--	t
Bromus rigidus	--	--	1.2 (1.2)	--	--	--	0.6 (0.5)	--
Danthonia pilosa	15.7 (3.5)	17.9 (3.9)	0.8 (0.4)	3.8 (2.1)	26.0 (7.2)	4.3 (1.2)	9.6 (1.7)	10.2 (2.0)
*Luzula hawaiiensis***	t	--	--	--	0.4 (0.4)	0.1 (0.1)	t	t
Poa pratensis##	--	t	--	--	--	--	--	t
Vulpia myuros	42.6 (40.7)	--	--	--	--	--	19.3 (18.4)	--

Table 4, continued.

Table 4, continued.

Species	Light grass stratum		Cinders stratum		Rock stratum		Entire exclosure	
	1976	1986	1976	1986	1976	1986	1976	1986
HERBS AND FORBS								
*Argyroxiphium sandwicense***	--	--	--	t	--	--	--	t
Hypochoeris radicata	0.6 (0.4)	16.3 (5.8)	0.5 (0.3)	6.7 (4.0)	17.4 (8.6)	18.6 (4.0)	1.9 (0.7)	12.0 (3.2)
Rumex acetosella	11.0 (10.8)	2.5 (2.2)	--	--	--	--	5.0 (4.9)	1.1 (1.0)
Senecio sylvaticus	0.4 (0.4)	--	t	--	3.2 (3.1)	--	0.5 (0.3)	--
*Tetramolopium humile***	0.3 (0.2)	0.5 (0.4)	--	t	1.5 (0.5)	0.4 (0.2)	0.2 (0.1)	0.3 (0.2)
FERNS								
*Asplenium trichomanes***	--	0.2 (0.2)	--	--	0.5 (0.4)	0.6 (0.4)	t	0.1 (0.1)
*Pellaea ternifolia***	t	0.2 (0.1)	--	--	0.6 (0.5)	0.3 (0.2)	t	0.1 (0.1)
Pteridium aquilinum	--	0.2 (0.1)	t	--	--	0.1 (0.1)	t	0.1 (0.1)

Species	Light grass stratum		Cinders stratum		Rock stratum		Entire exclosure	
	1976	1986	1976	1986	1976	1986	1976	1986
Alien species	77.8	48.9	4.4	14.0	53.6	28.7	41.8	30.9
	(54.0)	(9.4)	(2.9)	(6.8)	(12.1)	(6.0)	(24.5)	(5.3)
Native species	0.5	0.9	t	0.3	3.2	1.6	0.3	0.7
	(0.3)	(0.4)		(0.3)	(1.3)	(1.0)	(0.2)	(0.2)
Combined species	78.4	49.8	4.4	14.3	56.8	30.3	42.1	31.6
	(53.9)	(9.6)	(2.9)	(6.7)	(12.3)	(6.1)	(24.5)	(5.4)

*Densities for 1976 based on number of rooted plants/sampling unit; densities for 1986. based on number of plants either rooted in or overhanging each sampling unit. Consequently, estimates for 1986 are slightly inflated.

**Native species.

***Values in parentheses are standard errors of the means.

#Trace amount, <0.1 plant/m^2.

##Mean number of *Poa pratensis* patches, not individual plants.

The amount of ground covered by live herbaceous foliage, irrespective of species, inside the Wailuku exclosure was quite small -- only 843 cm^2/m^2 or about 8% (Table 5). Litter, including woody detritus, covered 8% of the ground, and rock covered 40%. The remaining 44% was bare soil and coarse cinders. The area of each stratum in each of the four ground cover categories is also shown in Table 5.

Without grazing, growth of alien grasses was lush under māmane trees inside the exclosure. Although the vegetation under māmane trees was not sampled, we made a few incidental measurements of grass heights inside as well as outside the exclosure. Kentucky bluegrass, which was a common grass beneath māmane, averaged about 18 in. (45 cm) tall inside the exclosure, but only 2.4 in. (6 cm) tall outside. Kentucky bluegrass litter formed a thick dense mat inside, but not outside. Sweet vernalgrass averaged 31 in. (79 cm) tall inside the exclosure and only 4 in. (11 cm) tall outside.

Characteristics of the woody plant population inside the Wailuku exclosure are listed in Table 6. All woody plants are native species. Of the five species represented, māmane, pūkiawe, and ‘ōhelo were regenerating, although ‘ōhelo was doing so slowly (*i.e.*, only one seedling after 14 years without sheep browsing). In comparison, māmane was regenerating rapidly, with 136 seedlings and 37 sprouts tallied in 1986. Most māmane seedlings established in the light grass stratum and under the canopy edge of seed trees. *Dubautia arborea*, an endangered species, was represented by a single flowering adult.

A survey of the woodlands adjacent to the Wailuku exclosure revealed no woody regeneration, even though alien plant cover generally appeared similar to that inside. We concluded that herbaceous vegetation at these densities in general, and alien herbaceous vegetation in particular, does not prevent reestablishment of native woody species.

<u>Pu‘u o Ka‘uha Exclosure</u>. Inside the Pu‘u o Ka‘uha exclosure, alien and native herbaceous basal areas increased at the same rate (Fig. 2). Initially, herbaceous basal area was very small for both alien and native species: basal area of each represented less than 0.03% of the exclosure surface area. After 14 years of no grazing, basal area was 184 $\pm$ 65 cm^2/m^2 for aliens, and 167 $\pm$ 57 cm^2/m^2 for natives, or the equivalent of 1.8% and 1.7% basal cover. Alien and native basal areas were both significantly greater in 1986 than in 1975 ($p < 0.05$).

Two alien species, Kentucky bluegrass and mullein, made up about 98% of the alien basal area inside the exclosure in 1986 (Table 7). However, the data indicated that only mullein had significantly more basal area in 1986 than in 1975. Juvik and Juvik (this volume) report on the phenomenal spread and adaptation of this alien species to Mauna Kea's subalpine woodland.

Table 5. Mean vertical projected area (cm^2/m^2) of live aerial herbaceous plant cover, litter, bare soil, and rock in the Wailuku exclosure in 1986.

	Projected area by ground cover category			
Stratum	Herbaceous cover	Litter	Bare soil	Rock
Light grass	1,303(317)*	1,431(329)	3,655(669)	3,611(679)
Cinders	422(227)	251(88)	5,602(775)	3,725(688)
Rock	689(313)	967(127)	1,269(359)	7,075(184)
Entire exclosure	843(116)	844(155)	4,363(471)	3,950(444)

*Values in parentheses are standard errors of the mean.

Table 6. Woody plant population characteristics in the Wailuku exclosure in 1986.

Species	Density (Plants/ha)	Height (m)	Plant crown (m^2/plant)	Total crown (m^2/ha)
Sophora chrysophylla				
mature trees	19	4.6(0.2)*	24.2(0.8)	460.4
sprouts**	37	1.2(0.5)	1.3(1.3)	47.4
seedlings	136	0.8(0.7)	0.4(1.8)	55.8
Styphelia tameiameiae				
mature shrubs	141	0.6(0.5)	1.3(2.2)	184.7
seedlings	11	0.2(0.4)	t***	0.4
Dubautia arborea	1	1.7(-)	4.5(-)	4.5
Dubautia ciliolata	1	0.5(-)	0.5(-)	0.5
Vaccinium sp.				
mature shrubs	14	0.9(0.3)	0.2(1.1)	2.2
seedlings	1	0.2(-)	t	t

*Values in parentheses are coefficients of variation.
**Statistics are for sprout clumps rather than individual stems.
***Trace amount, <0.1 m^2/plant or <0.1 m^2/ha.

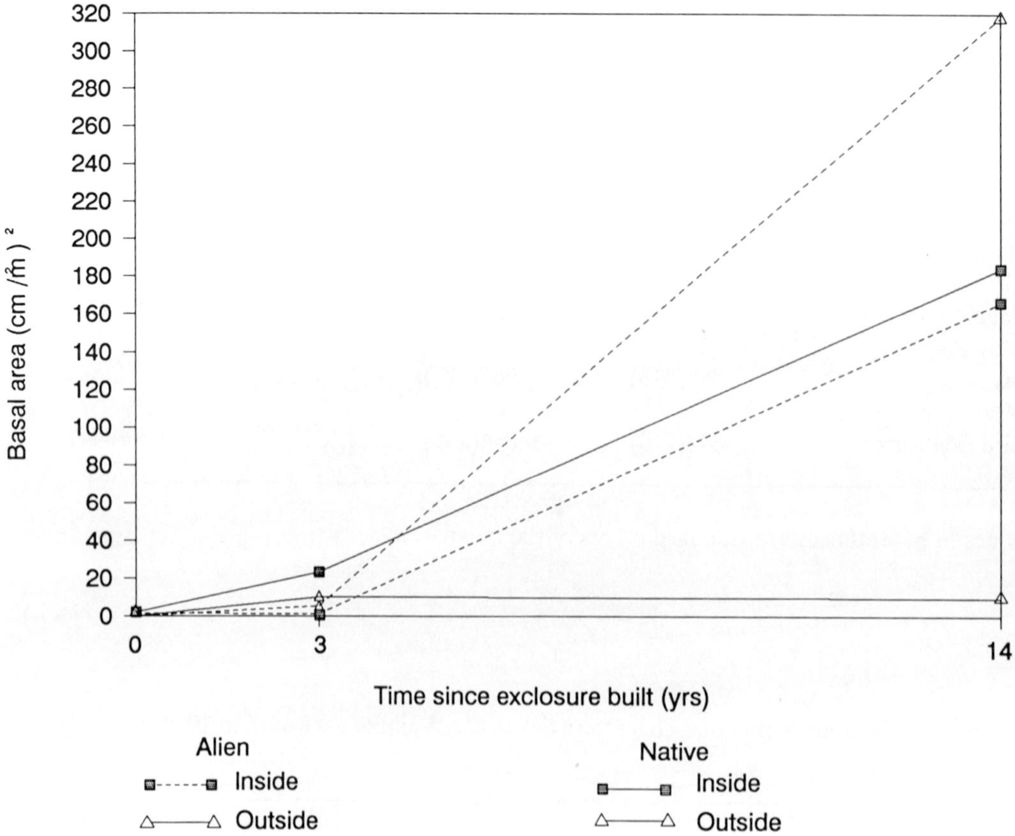

Figure 2. **Basal area of alien and native herbaceous species inside and outside the Pu'u o Ka'uha exclosure near tree line on the west flank of Mauna Kea, by elapsed time since the exclosure was built.**

Two native species, *Agrostis sandwicensis* and pili uka, comprised more than 99% of the native basal area in 1986. Only pili uka had significantly more basal area in 1986 than in 1975. Based on these data inside the exclosure, we conclude that native herbaceous species are reestablishing in spite of alien species. Probably interspecific competition is not widespread as yet inside the Pu'u o Ka'uha exclosure.

Plotted data for alien and native species just outside the exclosure appeared to indicate that basal area of aliens increased more rapidly than basal area of natives (Fig. 2). However, basal area variability for aliens was so large in 1986 that no significant difference was detected between years or between native and alien species in 1986 (Table 7). The large

Table 7. Average basal area (cm^2/m^2) of herbaceous plant species inside and outside Pu'u o Ka'uha sheep exclosure 2 months to almost 15 years after construction.*

Species	Inside			Outside		
	1972	1975	1986	1972	1975	1986
GRASSES AND SEDGES						
*Agrostis avenacea***	0.8(0.4)	8.8(3.2)	0.2(0.2)	0.1(0.1)	10.2(2.8)	1.1(0.4)
*Agrostis sandwicensis***	t#	0.1(0.1)	6.5(3.9)	--	t	--
Bromus rigidus	t	0.1(0.1)	t	--	--	0.4(0.1)
*Carex macloviana***	0.2(0.2)	4.8(4.5)	t	--	--	0.4(0.3)
Danthonia pilosa	--	--	--	--	--	0.1(0.1)
Hordeum murinum	--	t	--	--	--	t
Poa pratensis	--	t	95.5(57.3)	--	5.4(4.8)	255.2(222.3)
Stipa cernua	--	0.8(0.8)	2.9(1.7)	--	--	36.1(15.9)
*Trisetum glomeratum***	t	8.6(3.6)	177.2(66.1)	--	--	8.2(4.6)
Vulpia myuros	--	t	0.2(0.2)	--	t	t
FERN						
*Pellaea ternifolia***	0.2(0.2)	--	--	t	t	--
HERBS AND FORBS						
*Argemone glauca***	t	--	--	--	--	--
Cirsium vulgare	t	--	--	t	t	--
Epilobium billardierianum	--	--	--	--	--	2.8(2.7)
Erigeron sp.	--	--	--	--	0.1(0.1)	0.1(0.1)
Erodium cicutarium	t	t	--	--	t	t
*Gnaphalium purpureum***	0.9(0.1)	t	--	t	t	--
*Gnaphalium sandwicensium***	t	1.1(0.5)	0.2(0.1)	--	t	t
Heterotheca grandiflora	t	0.3(0.2)	0.5(0.2)	t	--	5.8(2.0)
Hypochoeris radicata	--	--	--	--	t	--

Table 7, continued.

Table 7, continued.

Species	Inside			Outside		
	1972	1975	1986	1972	1975	1986
Senecio sylvaticus	1.4(1.5)	t	--	0.1(0.0)	t	--
Sonchus oleraceus	t	0.1(0.1)	--	t	--	--
Taraxacum officinale	0.2(0.1)	t	--	t	t	t
Verbascum thapsus	t	0.2(0.0)	67.3(15.2)	t	t	17.3(3.6)
Alien species	1.6(0.6)	1.5(0.4)	166.4(56.9)	0.2(0.1)	5.6(5.2)	318.0(222.3)
Native species	2.1(n/a)##	23.4(n/a)	184.1(65.5)	0.2(0.1)	10.2	9.7(4.6)
Combined species	3.7(n/a)	24.9(n/a)	350.5(90.1)	0.4(n/a)	15.8(n/a)	327.7(222.0)

*1972 and 1975 based on Table 2 in Scowcroft and Giffin (1983).
**Native species.
#Trace amount, $<0.1\ cm^2/m^2$.
##Equivalent standard errors not given in Scowcroft and Giffin (1983).

alien mean and standard error were mainly due to Kentucky bluegrass, which was patchily distributed. Thus, our data indicated no significant increase in basal area of either alien or native species in the 11 years since 1975. Even without the inflationary effect of Kentucky bluegrass, we suspect that increases would not have been significant. We suggest that continued grazing outside the exclosure up to 1981 retarded plant establishment and growth, and that too little time had elapsed since 1981 for manifestation of basal area differences comparable to those inside the exclosure.

In 1986, mean foliar cover for all herbaceous species combined, like mean basal area for all species combined, was not statistically greater inside than outside the Pu'u o Ka'uha exclosure: 1,485 ± 230 cm^2/m^2 inside, versus 1,046 ± 236 cm^2/m^2 outside (Table 8). Similarly, mean foliar cover of aliens inside the exclosure was not significantly different from that outside. On the other hand, native foliar cover was significantly greater inside than outside the exclosure: 772 ± 175 cm^2/m^2 inside, versus 85 ± 56 cm^2/m^2 outside. Most of the native foliar cover was due to pili uka.

Plant density was significantly greater in 1986 than in 1972 just after the exclosure was built (Table 9). This finding was true for alien, native, and all species combined. The exceptionally large density values for aliens in 1975 and 1986 were due to two annual species, ripgut grass and rat tail fescue (*Vulpia myuros*).

Recovery trends for alien and native species outside Pu'u o Ka'uha in the years since sheep "eradication" appear to be following the general trend observed inside during the first few years after fencing. Total alien and total native plant densities in 1975 inside the exclosure were not significantly different from those in 1986 outside the exclosure. Only two individual species, both aliens, were significantly more abundant in 1986 outside than in 1975 inside -- *Heterotheca grandiflora* and mullein. Both are prolific seed producers, and mullein is aggressively colonizing the southern and western flanks of Mauna Kea, as we noted earlier. Furthermore, neither is an important forage species, so their populations probably were expanding even when sheep were grazing.

None of the native plant species were more abundant in 1986 outside than in 1975 inside. For example, density of pili uka, the more common native perennial bunch grass, was 0.35 ± 0.23 cm^2/m^2 outside the exclosure after about five years without sheep grazing. This density was not significantly different from that inside the exclosure in 1975, after three years of freedom from sheep grazing.

The amount of ground covered by live herbaceous vegetation inside and outside the Pu'u o Ka'uha exclosure, irrespective of species, was small, as it was at Wailuku -- 1,485 cm^2/m^2, or 14.8%, inside, and 1,046 cm^2/m^2, or about 10.5%, outside (Table 10). Bare soil comprised about 26% of ground cover inside and about 40% of the ground cover outside the exclosure. Rocks comprised 42% and 26% of the ground

Table 8. Mean aerial cover (cm^2/m^2) of herbaceous species inside and outside the Pu'u o Ka'uha exclosure and in surrounding woodland at the same elevation (2,750 m) and in woodland at a lower elevation (2,350 m) in 1986.

	Exclosure		2750 m		2350 m	
Species	Inside	Outside	South	North	South	North
GRASSES AND SEDGES						
*Agrostis avenacea**	t**	t	t	t	--	--
*Agrostis sandwicensis**	29(16)#	--	--	--	--	--
Anthoxanthum odoratum	--	--	--	t	250(189)	620(177)
Bromus rigidus	t	t	t	120(58)	t	--
Bromus sp.	--	--	--	--	--	t
*Carex macloviana**	t	t	--	--	--	t
Dactylis glomerata	--	--	--	t	1650(1007)	2200(921)
Danthonia pilosa	--	t	520(309)	1440(520)	--	--
Danthonia sp.	--	--	--	t	--	--
Holcus lanatus	--		--	--	25(1164)	1460(588)
Hordeum murinum	--	t	--	--	--	--
Lolium sp.	--		--	--	50(50)	t
Pennisetum setaceum	--	--	--	20(20)	--	--
Poa pratensis	96(58)	257(223)	--	620(571)	725(345)	940(492)
Stipa cernua	24(14)	176(70)	120(120)	120(80)	--	
*Trisetum glomeratum**	740(179)	82(56)	20(20)	--	300(191)	--
Vulpia myuros	t	t	--	--	--	--
FERNS						
Asplenium trichomanes	--	--	--	t	--	t
Pellaea ternifolia	--	--	--	t	--	--
Pteridium aquilinum	--	--	--	t	--	--
HERBS AND FORBS						
Cirsium vulgare	--	--	--	t	50(50)	t
Epilobium billardierianum	--	t	--	80(80)	t	t

Species	Exclosure		2750 m		2350 m	
	Inside	Outside	South	North	South	North
Erigeron sp.	--	t	t	t	t	--
Erodium cicutarium	--	t	t	t	--	--
Geranium homeanum	--	--	--	--	--	t
*Gnaphalium sandwicensium**	t	t	t	t	t	t
Heterotheca grandiflora	25(8)	173(34)	100(55)	120(80)	--	t
Hypochoeris radicata	--	--	--	--	--	20(20)
Medicago lupulina	--	--	--	--	50(50)	440(250)
Oxalis corniculata	--	--	--	--	t	--
Plantago sp.	--	--	--	--	--	t
Rumex acetosella	--	--	--	--	t	220(102)
Senecio sylvaticus	--	--	--	t	t	--
Sonchus oleraceus	--	--	t	--	t	--
Stenogyne sp.*	--	--	--	--	25(25)	--
Tagetes minuta	--	--	--	t	--	--
Taraxacum officinale	--	t	--	t	25(25)	2(20)
Urtica urens	--	--	--	--	t	--
Verbascum thapsus	564(136)	321(81)	460(214)	620(49)	100(71)	60(60)
Verbesina encelioides	--	--	--	t	--	--
Veronica serpyllifolia	--	--	--	--	--	t
Alien species##	713(141)	961(232)	1280(485)	3260(747)	4550(1805)	6120(306)
Native species##	772(175)	85(56)	40(24)	0(--)	300(192)	0(--)
All species combined##	1485(230)	1046(236)	1320(491)	3260(747)	4850(1618)	6120(306)

*Native species.
**Only trace amount found on sample plots.
#Values in parentheses are standard errors of the mean.
##Inclusion of trace amounts in summations results in subtotals that exceed the sum of tabled values for individual species.

Table 9. Average density* (no./m^2) of herbaceous plants inside and outside Pu'u o Ka'uha sheep exclosure 2 months to almost 15 years after construction.

Species	Inside 1972	Inside 1975	Inside 1986	Outside 1972	Outside 1975	Outside 1986
GRASSES AND SEDGES						
*Agrostis avenacea***	0.5(0.2)#	5.4(1.4)	0.2(0.1)	t##	6.7(1.9)	0.9(0.4)
*Agrostis sandwicensis***	t	t	0.2(0.1)	--	t	--
Bromus rigidus	t	17.8(16.9)	0.4(0.3)	--	t	56.5(17.7)
*Carex macloviana***	t	t	t	--	--	t
Danthonia pilosa	--	--	--	--	--	t
Hordeum murinum	--	0.2(0.1)	--	--	--	0.8(0.8)
Poa pratensis###	--	t	0.3(0.1)	--	0.2(0.2)	0.5(0.3)
Stipa cernua	t	t	0.3(0.2)	--	--	2.7(1.6)
*Trisetum glomeratum***	t	0.2(0.1)	2.8(0.6)	--	t	0.4(0.2)
Vulpia myuros	--	17.8(7.7)	t	--	1.0(0.8)	t
FERN						
*Pellaea ternifolia***	t	--	--	t	t	--
HERBS						
*Argemone glauca***	t	t	--	--	--	--
Cirsium vulgare	t	--	--	t	t	--
Epilobium billardierianum	--	--	--	--	--	t
Erigeron sp.	--	--	--	--	t	0.3(<0.1)
Erodium cicutarium	t	0.3(0.2)	--	--	t	8.0(3.6)
*Gnaphalium purpureum***	0.6(0.1)	1.4(0.4)	--	t	0.8(0.2)	--
*Gnaphalium sandwicensium***	t	1.1(0.6)	0.5(0.2)	--	t	0.2(0.1)
Heterotheca grandiflora	t	t	1.5(0.6)	--	t	15.6(3.4)
Hypochoeris radicata	--	--	--	--	t	--

Species	Inside 1972	Inside 1975	Inside 1986	Outside 1972	Outside 1975	Outside 1986
Senecio sylvaticus	0.3(0.1)	t	--	t	0.4 (0.2)	--
Sonchus oleraceus	t	t	--	t	--	--
Taraxacum officinale	t	t	--	t	t	t
Urtica urens	--	--	--	--	t	--
Verbascum thapsus	t	t	6.8 (2.1)	--	t	3.4 (0.8)
Alien Species	0.3(0.1)	36.3 (17.1)	9.3 (2.6)	0.2(0.1)	12.0 (0.5)	88.0 (18.4)
Native Species	1.2(0.2)	8.0 (1.1)	3.8 (0.5)	0.1(0.0)	7.5 (1.2)	1.5 (0.2)
Combined Species	1.5(0.3)	44.3 (18.1)	13.0 (2.7)	0.3(0.1)	9.1 (2.7)	89.5 (18.4)

*Densities for 1976 based on number of rooted plants in each sampling unit and densities for 1986 based on number of plants either rooted in or overhanging each sampling unit. Consequently, estimates for 1986 are slightly inflated.

**Native species.

#Values in parentheses are standard errors of the means.

##Trace amount, $<0.1/m^2$.

###Mean number of *Poa pratensis* patches, not individual plants.

Table 10. Mean vertical projected cover (cm^2/m^2) by cover class and data collection site at 2,750 m elevation in and around the Pu'u o Ka'uha exclosure in 1986.

		Pu'u o Ka'uha exclosure		
Cover class	North transect	Inside	Outside	South transect
Herbaceous cover	3260(747)*	1485(230)	1046(236)	1320(491)
Litter	2340(435)	1762(419)	2377(397)	1600(601)
Bare Ground	1140(808)	2580(289)	3977(396)	2140(730)
Rock	3260(568)	4173(504)	2600(337)	4940(816)

*Values in parentheses are standard errors of the mean.

cover inside and outside, with litter making up the rest of the ground cover. None of the differences between inside and outside were statistically significant ($p > 0.05$).

Māmane seedlings were more numerous outside than inside the Pu'u o Ka'uha exclosure (Table 11). A partial explanation for the higher seedling density outside is that the seed reserve outside was slightly larger than inside (Scowcroft and Giffin 1983). The ratio of seed density inside to outside was 0.59, and the ratio of seedling density inside to outside was 0.63.

Mean height and crown area were not significantly different for the two māmane seedling populations. Initially, we could not understand why the population means should be so similar when we knew that few seedlings survived outside before 1981 (Scowcroft and Giffin 1983). The explanation is that the distributions for seedling height and crown area (Fig. 3) were not the same inside and outside, and the variances were so large that statistical differences were not detectable. The distribution inside tended to be bimodal, thus indicating that some seedlings were older (larger) than the rest (*i.e.*, more advanced stand development inside). This result was expected because recovery inside had a nine-year head start on recovery outside.

More than twice as many adult 'āweoweo shrubs were found inside as outside Pu'u o Ka'uha. Mean plant height and canopy area for the two populations were not significantly different ($p > 0.05$). 'Āweoweo seedlings were over 13 times as abundant inside than outside. We believe that the greater abundance of both adult and seedling 'āweoweo inside the exclosure are due to heavier seed rain over a longer period. As Scowcroft and Giffin (1983) reported, 'āweoweo was among the first native woody species to establish inside Pu'u o Ka'uha, and its growth was rapid. Outside, sheep kept the species completely suppressed until about 1981 when the sheep were eliminated.

Table 11. Woody plant population characteristics in the Pu'u o Ka'uha exclosure and in a comparable area outside the exclosure in 1986.

Species	Density (Plants/ha)		Height (meters)		Plant Crown (sq. m/plant)		Total Crown (sq. m/ha)	
	Inside	Outside	Inside	Outside	Inside	Outside	Inside	Outside
Chenopodium oahuense								
Mature shrubs	391	183	1.2(0.4)*	1.4(0.3)	1.7(2.1)	1.6(0.9)	668.6	287.3
Seedlings	415	31	0.5(0.5)	0.6(0.4)	<0.1(1.7)	<0.1(0.9)	8.3	0.6
Myoporum sandwicense								
Mature trees	--	6	--	3.1(0.1)	--	17.8(0.6)	--	107.1
Seedlings	--	3	--	0.4(--)	--	0.3(--)	--	0.8
Sophora chrysophylla								
Mature trees	14	17	5.4(0.2)	5.8(0.3)	41.5(0.5)	36.1(0.6)	581.3	614.4
Sprouts**	34	23	2.5(0.4)	1.8(0.2)	12.0(2.2)	7.0(1.6)	409.4	160.8
Seedlings	204	326	1.7(0.7)	0.9(0.8)	2.0(1.2)	0.7(1.8)	406.0	224.9

*Value in parentheses is coefficient of variation.
**Statistics are for sprout clumps rather than individual stems.

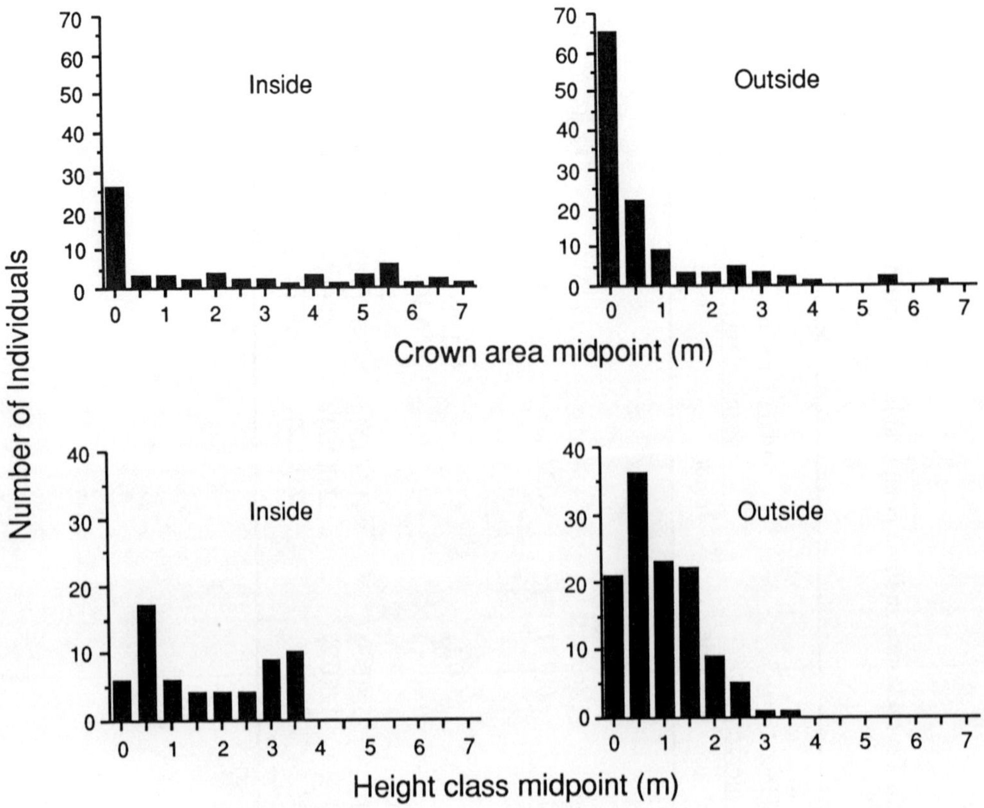

Figure 3. Height and crown area class distributions for seedling *Sophora chrysophylla* populations inside and outside the Pu'u o Ka'uha exclosure in 1986.

The more advanced recovery of vegetation inside the exclosure was not evident from examination of height and crown area data for 'āweoweo (Fig. 4). This species lives only about 10 years, and the original colonizers, although they had individual crown areas greater than 52 ft^2 (16 m^2), were all dead by 1986. The plants we measured were their progeny. Possibly, some of the adult 'āweoweo outside were also the offspring of the original colonizers inside the exclosure.

Naio was found outside the exclosure, but not inside (Table 11). When the exclosure was built, no naio trees were in the immediate vicinity. No seedlings or trees occurred within study boundaries in 1975 (Scowcroft and Giffin 1983). Based on mean height and canopy area data, we conclude that naio grows more rapidly than māmane in this tree line zone. The species is regenerating from seed.

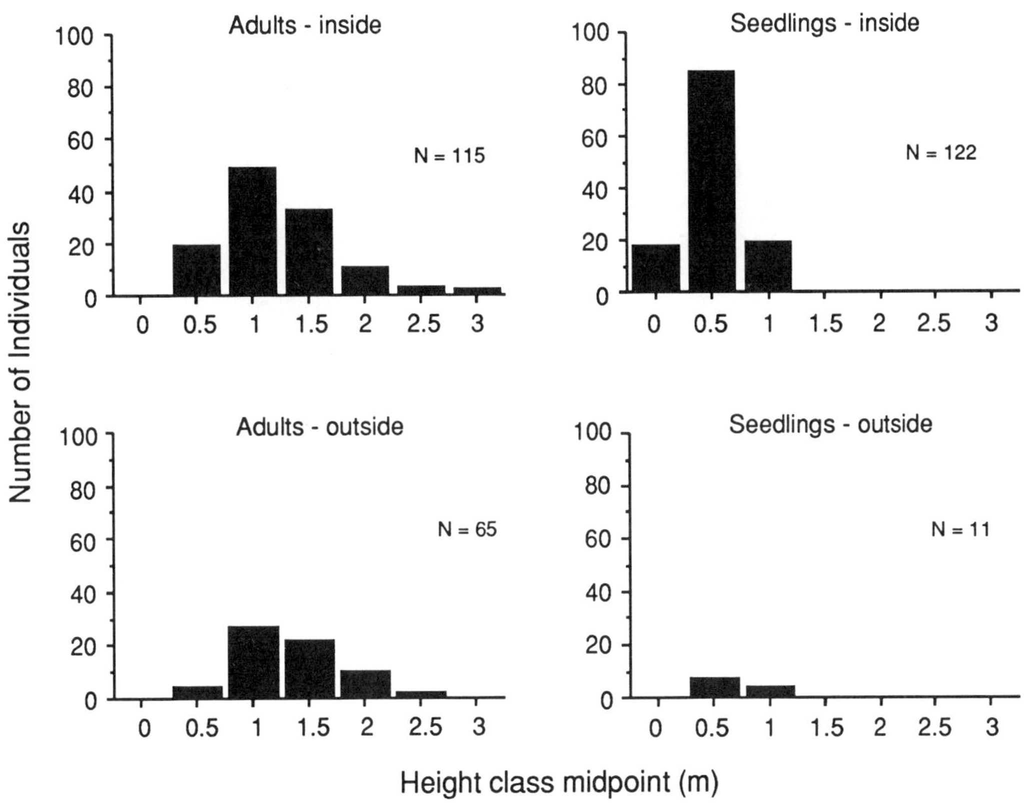

Figure 4. Height class distributions for adult and seedling *Chenopodium oahuense* populations inside and outside the Pu'u o Ka'uha exclosure in 1986.

Tree Line Zone Either Side of Pu'u o Ka'uha. We found no evidence that recovery of alien or native herbaceous species elsewhere in the tree line zone was different from that just outside the Pu'u o Ka'uha exclosure. Mean foliar cover estimates along transects at 9,020 ft (2,750 m) elevation were not significantly different from the estimate outside the exclosure (Table 8). The average proportions of ground covered by litter, bare soil, and rock along the two transects (Table 10) were not significantly different from outside the exclosure either. We conclude that the outside study area at Pu'u o Ka'uha is representative of the tree line zone on the western flank of the mountain.

Recovery of native herbaceous species inside the Pu'u o Ka'uha exclosure was more advanced than elsewhere in the tree line zone, *i.e.*, significantly more total native foliar cover occurred inside the exclosure than along the two transects at 9,020 ft (2,750 m) elevation (Table 8).

Recovery had been underway for 14 years inside, but only for five years elsewhere in the tree line zone. In contrast, judging from estimates of total alien foliar cover, recovery of alien herbaceous species inside the exclosure was not more advanced than elsewhere in the tree line zone (Table 8).

We call attention to the presence of fountain grass, along the north transect at 9,020 ft elevation (Table 8). This grass is a potential fire hazard and threatens native species, most of which are vulnerable to burning. We had not encountered it at this elevation in previous years, but a small population of scattered individuals has established in what we hope is an isolated situation. However, the seed source now exists near tree line to support expansion of this noxious species.

The only new woody species encountered along the 9,020 ft elevation transects was ʻaʻaliʻi, which was represented by one adult and one seedling in our sample (Table 12). No adult māmane trees occurred within sample plots, although scattered individuals grew all along the transects.

Naio was encountered only along the southern transect at 9,020 ft elevation and in the study area outside the Puʻu o Kaʻuha exclosure. Apparently the exclosure is located at an ecotone between māmane woodland to the north and naio-māmane woodland to the south. Characteristics of māmane seedling populations were not statistically different ($p > 0.05$) between the two woodland communities (Table 12). *Chenopodium* seedlings were found along the north transect in the māmane woodland, but not along the south transect in the naio-māmane woodland. Even adult ʻāweoweo shrubs were sparse in the mixed woodland.

Recovery of woody plant populations at the Puʻu o Kaʻuha exclosure (both inside and outside) was more advanced than elsewhere in the tree line zone (Tables 11 and 12). Crown area and average height of māmane seedlings were greater at the exclosure site than along the two transects established to either side of the exclosure site. Similarly, crown areas and density of ʻāweoweo adults and seedlings were greater inside and outside the exclosure than along transects to either side.

<u>Lower-elevation Woodland Around Puʻu Laʻau</u>. The woodland around Puʻu Laʻau has been relatively free of sheep for 20 to 30 years due to frequent human disturbance and the tendency of feral sheep to avoid areas with tall grass (Giffin 1976). Examination of vegetation recovery along a southern and northern transect at 7,710 ft (2,350 m) elevation showed that alien herbaceous species dominate the herbaceous community (Table 8). Total alien foliar cover averaged 4,550 $\pm$ 1,805 and 6,120 $\pm$ 306 cm^2/m^2 along the southern and northern transects. Total alien foliar cover was significantly greater along the north transect at 7,710 ft elevation than in the tree line zone. The only native herbaceous species with more than trace amounts of foliar cover were pili uka and *Stenogyne* sp., both of which were encountered only along the southern transect.

The proportion of ground covered by litter, rock, and moss or as exposed bare soil is shown in Figure 5. Significantly less rock cover existed at

Table 12. Woody plant population characteristics along transects at 2,350 and 2,750 m elevation on the west flank of Mauna Kea.

Transect	Elevation (m)	Species	Density* (no./ha)	Crown area* (m^2/ha)	Height** (m)
South	2350	*Sophora chrysophylla*			
		adults	151(78)	2346(1273)	4.7(2.6)
		seedlings	239(181)	38(36)	0.5(0.7)
		sprouts#	64(45)	46(45)	2.7(0.2)
		Chenopodium oahuense			
		seedlings	10(10)	< 1(< 1)	0.4(-)
North	2350	*Sophora chrysophylla*			
		adults	108(13)	2384(279)	4.9(2.0)
		seedlings	140(54)	17(8)	0.5(0.6)
		sprouts	38(38)	43(43)	1.8(1.1)
		Coprosma sp.			
		seedlings	13(8)	< 1(< 1)	0.6(0.3)
South	2750	*Sophora chrysophylla*			
		seedlings	471(308)	61(49)	0.5(0.6)
		Chenopodium oahuense			
		adults	13(13)	23(22)	1.2(0.8)
		Dodonaea sp.			
		adults	6(6)	5(5)	0.3(-)
		seedlings	6(6)	< 1(< 1)	0.3(-)
		Myoporum sandwicense			
		adults	102(78)	2059(914)	3.2(1.1)
		seedlings	153(102)	48(30)	0.7(0.8)
North	2750	*Sophora chrysophylla*			
		seedlings	414(334)	79(67)	0.5(0.5)
		sprouts	45(30)	272(143)	1.8(1.2)
		Chenopodium oahuense			
		adults	108(76)	243(168)	1.6(0.5)
		seedlings	57(42)	6(6)	0.5(0.3)

*Mean density and crown area per sample plot expanded to means per hectare and associated standard errors of the means.
**Mean height per individual and associated standard deviation in parentheses.
#Characteristics are for sprout clumps, not individual sprouts.

7,710 ft (2,350 m) elevation than at tree line. Moss-covered soil was found only along the lower transects and probably indicates moister conditions there owing to frequent afternoon fog and mist. Bare soil comprised less than 1% of the cover along the north transect at 7,710 ft elevation, but more than 29% of the cover along the south transect. Herbaceous plants occupied most of the available space along the north transect, whereas along the south transect, space was still available for alien and native plant establishment.

Recovery of two woody species in the low-elevation woodland was like that observed in the tree line woodland (Table 12). Māmane seedlings were just as numerous and about the same size at 7,710 ft (2,350 m) as at

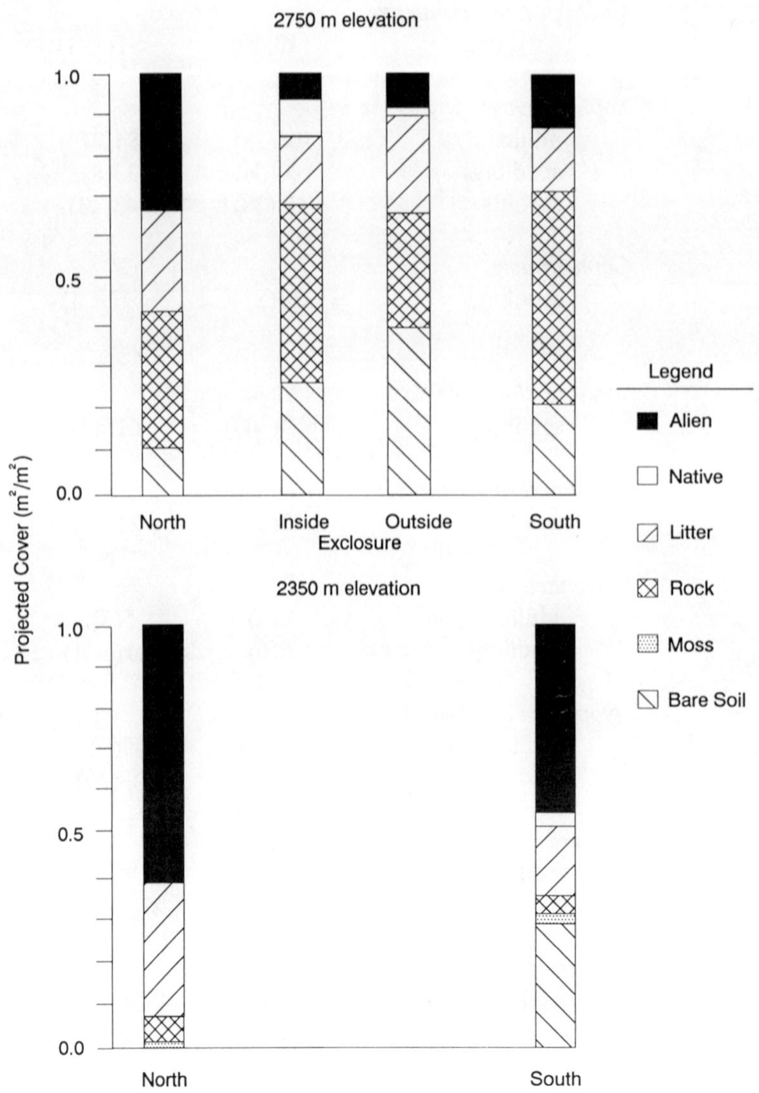

Figure 5. Projected cover of alien and native herbaceous plants, litter, rock, moss, and bare soil inside, outside, and in the vicinity of the Pu'u o Ka'uha exclosure and in the māmane woodland at 2,350 m elevation near Pu'u La'au, 1986.

9,020 ft (2,750 m) elevation. Density and size of 'āweoweo seedings were also alike at the two elevations.

If sheep foraging pressure has been low in the low elevation woodland for 20 to 30 years, why are populations of māmane and 'āweoweo regeneration not more advanced in terms of both density and size characteristics? Mature māmane trees appear more numerous and more vigorous in the lower stands than in the upper stands, and we would expect seed production and soil seed reserves to be greater as well. We doubt that lack of seed accounts for the lack of advanced regeneration. Pilo seedlings occurred at 7,710 ft (2,350 m) elevation, but not at the higher elevation. On the other hand, 'a'ali'i and naio reproduction were only found at higher elevation. Possibly, dense alien herbaceous cover inhibited woody native plant establishment at the low elevation. Our impression is that alien grass cover along the northern transect at 7,710 ft (2,350 m) elevation is dense enough now to interfere with establishment of native woody species. Whether it has been dense enough to do so for the past 20 to 30 years is unknown.

CONCLUSIONS

Release from grazing and browsing has not resulted in significant increases in basal area, density, or foliar cover of most alien plant species in the māmane woodland of Mauna Kea. Mullein increased significantly at the Pu'u o Ka'uha exclosure, but we saw no evidence that the increase was due to release from sheep grazing. Rather, the unpalatable, prolific seed-producing species appeared to aggressively colonize sites regardless of grazing pressure.

Release from grazing and browsing has benefitted native species in tree line communities, however, especially those species that are highly palatable and highly susceptible to damage (*e.g.*, māmane, 'āweoweo, pili uka, and *Agrostis sandwicensis*). The colonization and reestablishment of these species inside exclosures as well as outside exclosures after eradication of feral sheep indicates that these natives were able to persist in an ecosystem damaged by animals. As Rabinowitz *et al.* (1984) suggested, the persistence and ultimate resurgence of native grass species may be indicative of competitive superiority in the somewhat harsh environment of Mauna Kea's subalpine woodlands. Mueller-Dombois (1981) concluded that native Hawaiian plants cannot be considered weak competitors.

On the basis of our findings, we believe that control of most alien plant populations on upper Mauna Kea is unwarranted. Native species are recovering in spite of alien species. Even at lower elevation in the māmane woodland, recovery of native woody species is evident, although circumstantial evidence suggests that the rate of recovery is slower there than in the tree line zone.

Control of two alien species, fountain grass and possibly German ivy, may be desirable while their populations are still relatively small.

Fountain grass is not yet interfering with recovery of native plants in the māmane woodlands we examined, but it has the potential to do so. Furthermore, it is a fire-adapted species that is capable of replacing nearly all of the native ground layer species (Jacobi and Scott 1985). Repeated fires may result in the elimination of native woody plants as well. German ivy was not encountered in our sampling, but it is established nearby in the Kaohe Game Management Area just below Pu'u La'au. Both species have been declared noxious by the Hawaii Department of Agriculture (Hawaii Revised Statutes, Ch. 68, Noxious Weed Rules).

Data from Pu'u o Ka'uha and vicinity clearly show that native species will recover once sheep foraging pressure is eliminated. Data from Wailuku indicate that a similar native vegetation response can be expected on that side of the mountain if mouflon are eliminated. Both māmane and *Agrostis sandwicensis* reestablished at Wailuku in the absence of grazing and browsing. Neither species establishes where mouflon sheep can get to them. On the basis of these results and those of Scowcroft and Giffin (1983), we conclude that maintenance of mouflon sheep at 1986 population levels will prevent recovery of native species in those portions of the Mauna Kea Forest Reserve where the animals concentrate. Furthermore, should feral sheep reestablish, they too will interfere with native species recovery wherever the animals concentrate.

Because mouflon and feral sheep are problems only as defined by management objectives, managers must have a clear statement of land management policy for the land in question, especially if the land use is zoned Conservation (Riney 1964). With a clear statement of policy, managers can examine ecosystem trends and decide if they will continue to the detriment of the land. If the trends indicate that maintenance of large herbivores is in conflict with the policy statement, either in part or whole, then one of three decisions should be made: change the policy, get rid of the animals, or both. This is the prerogative of policy makers, not managers or researchers.

ACKNOWLEDGMENTS

We thank Don Goo, Alan Urakami, and Tats Kimoto for invaluable assistance in data collection; James Baldwin for statistical advice and review; and Charles Wakida, Libert Landgraf, and several anonymous reviewers of earlier drafts of this manuscript.

Literature Cited

Bailey, B.J.R. 1977. Tables of the Bonferroni 't' statistic. *J. Amer. Stat. Assn.* 72:469-478.

Berger, A.J. 1972. *Hawaiian birdlife*. Honolulu: Univ. Pr. Hawaii.

Bryan, L.W. 1937a. Wild sheep in Hawaii. *Paradise Pac.* 49(3):19, 31.

Bryan, L.W. 1937b. The big fence on the Big Island. *Paradise Pac.* 49(4):15, 30.

Bryan, L.W. 1950. Wild sheep on Mauna Kea Forest Reserve. *Paradise Pac.* 62:112-113, 122.

Cochran, W.G. 1963. *Sampling techniques*, 2nd ed. New York: John Wiley & Sons.

Games, P.A., and J.F. Howell. 1976. Pairwise multiple comparison procedures with unequal N's and/or variances: a Monte Carlo study. *J. Educ. Statistics* 1:113-125.

Giffin, J.G. 1976. Ecology of the feral sheep on Mauna Kea. Hawaii Dep. Land Nat. Resour., Pittman-Robertson Proj. W-15-5, Study XI, 1972-1975. Mimeo.

Giffin, J.G. 1982. Ecology of the mouflon sheep on Mauna Kea. Final Rep., Hawaii Dep. Land Nat. Resour., Pittman-Robertson Proj. W-17-R, Study No. R-III, 1975-1979. Mimeo.

Harper, J.L. 1977. *Population biology of plants*. London: Acad. Pr.

Hartt, C.E., and M.C. Neal. 1940. The plant ecology of Mauna Kea, Hawaii. *Ecology* 21:237-266.

Hawaii Division of Water and Land Development. 1982. *Median rainfall, state of Hawaii*. Circ. C88, Hawaii Dep. Land Nat. Resour.

Hosaka, E.Y., and A. Thistle. 1954. *Noxious plants of the Hawaiian ranges*. Univ. Hawaii Agric. Ext. Serv. Bull. 62. Honolulu: Univ. Hawaii.

Jacobi, J.D., and J.M. Scott. 1985. An assessment of the current status of native upland habitats and associated endangered species on the island of Hawaii. In *Hawai'i's terrestrial ecosystems: preservation and management*, ed. C.P. Stone and J.M. Scott, 3-22. Univ. Hawaii Coop. Natl. Park Resour. Stud. Unit. Honolulu: Univ. Hawaii Pr.

Judd, C.S. 1936. Hawaii's forests winning their battle with wild animals. *Honolulu Star Bull.*, March 21, section 3:1, 7.

Juvik, J.O., and S.P. Juvik. 1984. Mauna Kea and the myth of multiple use, endangered species, and mountain management in Hawaii. *Mountain Res. and Devel.* 4(3):191-202.

Juvik, J.O., and S.P. Juvik. [this volume] Mullein (*Verbascum thapsus*): the spread and adaptation of a temperate weed in the montane tropics.

Keselman, H.J., and J.C. Rogan. 1978. A comparison of the modified-Tukey and Scheffe methods of multiple comparisons for pairwise contrasts. *J. Amer. Stat. Assn.* 73:47-52.

Kilcher, M.R., and J. Looman. 1983. Comparative performance of some native and introduced grasses in southern Saskatchewan, Canada. *J. Range Manage.* 36:654-657.

Kramer, R.J. 1971. *Hawaiian land mammals.* Tokyo: Charles E. Tuttle Co.

Loope, L.L., and P.G. Scowcroft. 1985. Vegetation response within exclosures in Hawai'i: a review. In *Hawai'i's terrestrial ecosystems: preservation and management*, ed. C.P. Stone and J.M. Scott, 377-402. Univ. Hawaii Coop. Natl. Park Resour. Stud. Unit. Honolulu: Univ. Hawaii Pr.

McClelland, C.K. 1915. *Grasses and forage plants of Hawaii.* Hawaii Agric. Exp. Sta. Bull. 36. Honolulu: Univ. Hawaii.

Mueller-Dombois, D. 1981. Understanding Hawaiian forest ecosystems: the key to biological conservation. In *Island ecosystems: biological organization in selected Hawaiian communities*, ed. D. Mueller-Dombois, K.W. Bridges, H.L. Carson, 502-520. Stroudsburg, Penn.: Hutchinson Ross Pub. Co.

Mueller-Dombois, D., and V.J. Krajina. 1968. Comparison of east-flank vegetations on Mauna Loa and Mauna Kea, Hawaii. In *Proc. Symp. Recent Adv. Trop. Ecol.* Part II, ed. R. Misra and B. Gopal, 508-520. Internatl. Soc. Trop. Ecol.

Rabinowitz, D., J.K. Rapp, and P.M. Dixon. 1984. Competitive abilities of sparse grass species: means of persistence or cause of abundance. *Ecology* 65:1144-1154.

Riney, T. 1964. The impact of introductions of large herbivores on the tropical environment. *Internatl. Union Conserv. Nat. Pub.*, new ser. 4:261-273.

Ripperton, J.C., R.A. Goff, D.W. Edwards, and W.C. Davis. 1933. *Range grasses of Hawaii*. Hawaii Agric. Exp. Sta. Bull. 65. Honolulu: Univ. Hawaii.

St. John, H. 1973. *List and summary of the flowering plants in the Hawaiian Islands.* Mem. 1. Lawai, Kauai, Hawaii: Pac. Trop. Botan. Gdn.

Scott, J.M., S. Mountainspring, C. van Riper III, C.B. Kepler, J.D. Jacobi, T.A. Burr, and J.G. Giffin. 1984. Annual variation in the distribution, abundance, and habitat response of the palila *(Loxioides bailleui). Auk* 101:647-664.

Scowcroft, P.G. 1981. Regeneration of mamane: effects of seedcoat treatment and sowing depth. *For. Sci.* 27:771-779.

Scowcroft, P.G. 1983. Tree cover changes in mamane (*Sophora chrysophylla*) forests grazed by sheep and cattle. *Pac. Sci.* 37(2):109-119.

Scowcroft, P.G., and J.G. Giffin. 1983. Feral herbivores suppress mamane and other browse species on Mauna Kea, Hawaii. *J. Range Manage.* 36(5):638-645.

Scowcroft, P.G., and H.F. Sakai. 1983. Impact of feral herbivores on mamane forests of Mauna Kea, Hawaii: bark stripping and diameter class structure. *J. Range Manage.* 36:495-498.

Smith, C.W. 1985. Impact of alien plants on Hawai'i's native biota. In *Hawai'i's terrestrial ecosystems: preservation and management*, ed. C.P. Stone and J.M. Scott, 180-250. Univ. Hawaii Coop. Natl. Park Resour. Stud. Unit. Honolulu: Univ. Hawaii Pr.

U.S. Fish and Wildlife Service. 1977. Determination of critical habitat for six endangered species. *Fed. Regis.* 42(155):40685-40690.

U.S. Soil Conservation Service. 1973. *Soil survey of the island of Hawaii, State of Hawaii.* U.S. Dep. Agric. Soil Conserv. Serv. in coop. with Univ. Hawaii Agric. Exp. Sta. Washington, D.C.: U.S. Govt. Printing Office.

van Riper, C., III. 1980a. The phenology of the dryland forest of Mauna Kea, Hawaii, and the impact of recent environmental perturbations. *Biotropica* 12(4):282-291.

van Riper, C., III. 1980b. Observations on the breeding of the palila, *Psittirostra bailleui*, of Hawaii. *Ibis* 122:462-475.

van Riper, C., III, J.M. Scott, and D.M. Woodside. 1978. Distribution and abundance patterns of the palila on Mauna Kea, Hawaii. *Auk* 95:518-527.

Warner, R.E. 1960. A forest dies on Mauna Kea. *Pac. Discovery* 13:6-14.

Whitney, L.D., E.Y. Hosaka, and J.C. Ripperton. 1939. *Grasses of the Hawaiian ranges.* Hawaii Agric. Exp. Sta. Bull. 82. Honolulu: Univ. Hawaii.

Williams, O.B. 1968. The uneasy state between animal and plant in the manipulated situation. *Proc. Ecol. Soc. Australia* 3:167-174.

Williams, O.B., and R. Roe. 1975. Management of arid grassland for sheep: plant demography of six grasses in relation to climate and grazing. *Proc. Ecol. Soc. Australia* 9:140-146.

RESPONSES OF HAWAIIAN ECOSYSTEMS TO REMOVAL OF FERAL PIGS AND GOATS

Charles P. Stone,
Linda W. Cuddihy, and J. Timothy Tunison

ABSTRACT

Feral pigs (*Sus scrofa*) and goats (*Capra hircus*) damage a wide range of native ecosystems in Hawai'i, from dry to wet communities and from low to high elevations. Comparative studies suggest that once ungulate disturbance is curtailed, some communities recover better than others, depending on a number of interrelated variables. Remote and/or lightly disturbed rain forest, coastal strand, 'ōhi'a (*Metrosideros polymorpha*) and subalpine native shrub, and native bunchgrass/native shrub are among the least affected by long-term pig or goat disturbance, although most of these communities retain alien cover in the range of 10% or less. Lowland grasslands are less successfully reclaimed, retaining alien cover as high as 80% a decade or more after ungulate removal in some areas. Accessible (to humans, ungulates, and alien plant propagules) rain forests have a higher percentage of alien cover than remote rain forests after pigs are eliminated, but even remote and/or lightly disturbed areas can be invaded by aggressive alien species long after ungulate disturbance ceases. We conclude that because recovery depends on a number of variables and because some invasive plants can establish long after pig and goat removal, most disturbed sites require continual monitoring and specific alien plant management strategies after ungulates are eliminated. Even then, an enduring alien plant component will remain in near-natural areas.

INTRODUCTION

Removal of feral ungulates from areas managed for their values as native ecosystems in Hawai'i is a key step in reducing disturbance of native vegetation. Feral goats (*Capra hircus*) frequent dry to wet areas and consume native woody plants, although they also eat herbs and grasses (Yocum 1967; Baker and Reeser 1972; Spatz and Mueller-Dombois 1973; Mueller-Dombois and Spatz 1975). Goats may reduce many palatable native plant species to the point of extinction. Māmane (*Sophora chrysophylla*) is one example of a much-preferred or "ice cream" native species. Other native species such as pūkiawe (*Styphelia tameiameiae*) seem less sought out and are usually eaten after other plants become unavailable. Goat diets can readily change to predominantly alien species where native plants are absent or scarce, and some alien grasses such as African dropseed (*Sporobolus africanus*) are favored in some Hawaiian habitats (T.J. Ohashi, pers. comm.). Through removal of plant cover, trampling of soil, and repeated use of trails, feral goats may increase erosion and encourage degeneration of watersheds (Yocum 1967).

Feral pigs (*Sus scrofa*) are omnivores that consume both plant and animal materials in wet to dry ecosystems. In Hawai'i, pigs consume or damage native tree ferns and small ferns; fragile native lilies, mints, and bryophytes; lobelioids, koa (*Acacia koa*) seedlings, and other woody plants (Giffin 1978; Diong 1983; Stone 1985). Pigs also expose considerable ground by "rototilling" soil for earthworms, rhizomes, and tubers. Their presence in rain forests is associated with decrease of numbers and diversity of small, native soil arthropods, such as collembola (Vtorov, in press), and with greater nitrogenous resources (Vitousek *et al.* 1987). The role of feral pigs in dispersal of alien plants such as strawberry guava (*Psidium cattleianum*) and banana poka (*Passiflora mollissima*) has been well established (LaRosa, this volume; Diong 1983; Warshauer *et al.* 1983; Smith 1985).

What follows reduction or removal of pigs and goats in natural areas in Hawai'i is of increasing concern. The cost of ungulate management is high, and some pig-inhabited areas have been so heavily invaded by alien plants as to appear beyond the point of recovery. Deterioration and subsequent recovery of a given area depend on a multitude of factors. These include: duration and intensity of ungulate disturbance; presence of aggressive alien plants able to outcompete native species; soil and climatic conditions that influence erosion; and native plant propagules available for revegetation. Generalizations are difficult, and a better understanding of these influences is needed. Determining vegetation characteristics after ungulate removal or in areas with few or no ungulates, however, enables the development of some hypotheses.

Loope and Scowcroft (1985) reported on studies of 51 small exclosures established to demonstrate and evaluate vegetation response after ungulate removal in Hawai'i. They concluded that native vegetation usually holds its own or increases after ungulate removal. However, the vegetation response was highly variable and seemed related to extent of degradation. Alien grasses seemed to thrive after ungulate disturbance and removal in some highly impacted dry areas. Loope and Scowcroft noted that few exclosures had been studied in rain forest. Williams (1990) reviewed studies of ecological effects of long-term pressure of ungulates on recovery of vegetation. Included were increased insect attacks on injured native plants and increased suckering as a response to ungulates. Competitive disadvantages to native plants caused by opening up of forest by ungulates, soil compaction, and other effects have been discussed elsewhere (*e.g.*, Mueller-Dombois *et al.* 1981).

In this paper, vegetation response after removal or reduction of pig or goat disturbance is summarized for a number of areas where feral animals have been influences for long periods. Pig and goat exclosure and transect studies in large managed areas of rain forests and other vegetation types are also highlighted. A comparison of the vulnerability to invasion of different areas is considered to help provide better understandings of the recovery potential of different areas.

METHODS

Methods used in the studies summarized in this paper varied from site to site. Because of the number of uncontrolled and unmeasured variables affecting species composition and abundance in a given area, comparative methods were used. Vegetation differences were contrasted based on recent data and those collected several years before in the same area; from data taken inside and outside exclosures or from similar pig- and goat-inhabited and uninhabited areas; and from comparable areas of high and low animal density. Areas within communities of similar physiognomy (rain forests, grasslands, shrublands) are compared.

Point frequency measurements (with a 1-m tall point frame) and visual estimates were used to determine percent plant cover. Usually, 300 to 500 pin encounters comprised a point frequency sample in a given area. The Braun-Blanquet cover-abundance method was used to estimate percent cover for individual plant species and for other categories inside established relevés or plots. Plots were read by the same people to minimize observer error. Percent data from grouped 3 x 5 m Braun-Blanquet plots inside and outside exclosures were used for many rain forest comparisons. Cover was estimated (rather than directly measured) in one of seven or eight cover categories (Mueller-Dombois and Ellenberg 1974). Category means were transformed from percentage to arcsin values for analysis and reconverted to percentages for data presentation. Because mean category values were used, cover totals were not 100% in most cases, but those plots for which total cover did not total at least 70% were not analyzed. Braun-Blanquet plot data for the stratum from ground level to 0.5 m in height only are reported here because this stratum is most vulnerable to feral animals and most likely to respond quickly after their removal.

Native/alien species ratios and percent cover measurements for alien plants were used to estimate native species diversity and "intactness" of vegetation types in a number of areas. Responses of vegetation to ungulate removal or absence were partly a function of number of alien species that remained, although some invasive alien species were more important than others. Sampling strategy for, and sizes of, each of the areas discussed are briefly noted in the results and summary tables, for ease of association with data. Brief ungulate histories of study areas are presented in the results for the same reason.

RESULTS AND DISCUSSION

Feral Goats in Subalpine and Montane Parkland

Kahikinui Forest Reserve. Goats have likely been present in the subalpine shrub zone on the southern slope of Haleakalā, Maui, for 100-150 years, with populations similar to those present today. Year-round hunting and increased access since 1983 have resulted in reduction of goats on the southwestern part of the area, with most goats remaining in the southeastern portion, and intermediate densities between. A density of about 65 animals/km^2 was estimated in 1983 for the southeastern slopes. Two transects in each of eight equally spaced "bands" from west to east were used to measure percent cover by point frequency counts. Five hundred points were read in each band (T.J. Ohashi, pers. comm.).

Long-term goat foraging in the Forest Reserve has exposed soil (Table 1), especially in the southeastern portion, where goats have been abundant for long periods. Total percent cover of alien plants is also highest where goats are now abundant, but primarily because more alien grasses occur there with or without goats. Where goats are present, percent cover and diversity of native woody plants, especially the smaller (younger) individuals, is less. Pūkiawe masks this to some degree in the data set presented, since goats do not readily take this species, but the absence of kūkaenēnē (*Coprosma ernodeoides*), māmane, and ʻōhelo (*Vaccinium reticulatum*) in the higher goat density bands is apparent in tabulated data (Table 1).

Size class counts of kūkaenēnē and ʻōhelo showed little departure from J-shaped curves in the low goat density bands, but smaller size classes were missing in the high goat density bands (Ohashi 1984). However, the reduction in native woody species in the higher-density goat areas may, in fact, also be related to the influx of alien grasses in the eastern portion of the Reserve. Differences in microclimate between gulches and undissected slopes may have resulted in more woody plant habitat in some bands than in others in the low goat density areas.

Hawaii Volcanoes National Park. Goat populations below an exclosure fence at 6,600 ft (2,010 m) elevation were reduced considerably in the early 1970s when the fence was constructed, were controlled to remnant levels by 1982, and were eradicated by 1985. Above the fence, perhaps 20 goats remained in 1989. Forty paired 10 x 30 m plots were established in 1986, 20 above the fence where goats are still present and 20 below the fence (Tunison *et al.*, unpub. data). Plot elevations ranged from 5,740 to 6,740 ft (1,750-2,045 m), with 13 plot pairs in montane parkland and 7 in the subalpine zone. Point frequency counts were used to determine percent plant cover in two plant communities above and below the fence.

Only minor differences in cover of plants were found, although alien grasses and herbs comprised more cover in goat-inhabited areas in one plant community (Table 2) than where goats had been absent for six years. However, the big difference in the goat and goat-free areas in both plant communities was native tree abundance (Table 3), with māmane, the tree form of ʻaʻaliʻi (*Dodonaea viscosa*), and koa in goat-free areas showing

Table 1. Percent alien and native plant cover on southern slope of Haleakalā, 1983. Bands are subject to greater foraging pressure by goats from west to east (Band 1 to Band 8).*

Category and Species	Low Goat Densities (West)		Medium Goat Densities				High Goat Densities (East)	
	Band 1	Band 2	Band 3	Band 4	Band 5	Band 6	Band 7	Band 8
Alien Grasses								
Danthonia pilosa	2.03	0.58	4.54	1.54	5.56	0.62		0.47
Holcus lanatus				0.19		0.62		
Sporobolus africanus				0.19	2.68	5.19	6.49	3.58
Subtotals	2.03	0.58	4.54	1.93	8.24	6.43	6.49	4.04
Alien Herbs								
Hypochoeris radicata		1.56	2.17	3.08	1.34	4.77	0.81	2.95
Subtotals		1.56	2.17	3.08	1.34	4.77	0.81	2.95
Native Ferns								
Asplenium trichomanes			0.20					
Pteridium aquilinum	1.62	1.75	3.55	8.09	3.83	0.41	8.32	8.09
Subtotals	1.62	1.75	3.75	8.09	3.83	0.41	8.32	8.09
Native Grasses, Sedges								
Carex wahuensis	10.55	11.31	10.06	3.85	6.51	3.11	1.62	0.62
Deschampsia nubigena	5.27	4.29	7.30	8.09	8.05	14.11	9.13	13.06
Subtotals	15.82	15.59	17.36	11.95	14.56	17.22	10.75	13.69
Native Woody Plants								
Coprosma ernodeoides	1.83	4.29	4.54	2.12	1.72			
Coprosma montana			0.59					
Dodonaea viscosa	1.62		1.18					
Metrosideros polymorpha			1.18					
Osteomeles anthyllidifolia			0.39					

Category and Species	Low Goat Densities (West) Band 1	Band 2	Medium Goat Densities Band 3	Band 4	Band 5	Band 6	High Goat Densities Band 7	(East) Band 8
Sophora chrysophylla	1.01							
Styphelia tameiameiae	9.33	17.54	10.85	18.50	14.75	26.14	16.43	20.53
Vaccinium reticulatum	12.37	5.26	14.60	15.99	11.69			
Subtotals	26.16	27.09	33.33	36.61	28.16	26.14	16.43	20.53
Moss	2.43		0.59	0.77	0.96			
Litter	18.46	15.40	19.13	15.99	18.97	10.37	13.39	9.80
Dead Branches	7.51	3.31	4.54	3.47	1.53	3.73	1.22	3.11
Soil	3.85	6.04	3.94	6.94	6.70	21.99	15.62	17.73
Rock	22.11	28.65	10.65	11.18	15.71	8.92	26.77	19.75
Goat Droppings							0.20	0.31
Number of Native Species	8	6	11	6	6	4	4	4
Number of Alien Species	1	2	2	4	3	4	2	3
Percent Native Cover	43.61	44.44	54.45	56.65	46.55	43.77	35.50	42.31
Percent Alien Cover	2.03	2.14	6.71	5.01	9.58	11.20	7.30	6.99
Total Points/Band	493	513	507	519	522	482	493	643

*As determined by point frequency transects.

Table 2. Percent cover of alien and native plants in two communities above and below goat fence at 5,800 to 6,600 ft elevation on Mauna Loa.*

	Montane Parkland		Subalpine Scrub	
Category and Species	Goats	No Goats	Goats	No Goats
Native Ferns				
Asplenium trichomanes		00.08		
Pellaea ternifolia			00.33	00.08
Pteridium aquilinum				00.17
Native Grasses, Sedges				
Agrostis avenacea				00.50
Carex wahuensis	01.08	03.50	00.42	
Deschampsia nubigena	23.00	22.08	03.83	09.42
Gahnia gahniiformis	01.08		02.75	02.08
Luzula hawaiiensis			00.17	00.08
Panicum tenuifolium	03.83	01.92		
Alien Grasses, Sedges				
Anthoxanthum odoratum	00.08			
Danthonia pilosa	01.08			
Ehrharta stipoides	00.50			
Holcus lanatus	02.00	01.50	00.08	00.75
Alien Herbs				
Gnaphalium japonicum	00.08			
Hypochoeris radicata	02.00	00.67	00.42	00.33
Rumex acetosella	00.17	00.50		
Native Woody Plants				
Coprosma ernodeoides			00.20	04.42
Coprosma montana			00.08	
Dodonaea viscosa		00.33	02.92	04.83
Metrosideros polymorpha			00.33	00.50
Sophora chrysophylla	00.08	00.83		00.33
Styphelia tameiameiae	44.83	41.17	18.92	18.83
Vaccinium reticulatum			12.58	08.83
Bryophytes	01.00	06.67	00.58	00.83
Lichens			00.17	00.08
Litter	11.58	17.42	15.75	12.33
Soil	02.83	02.25	12.42	06.83
Rock	04.75	07.08	26.25	28.75
Number of Native Species	6	7	10	10
Number of Alien Species	7	3	3	4
Percent Native Cover	73.91	69.83	44.00	49.50
Percent Alien Cover	5.92	2.67	0.83	1.66

*As determined by point frequency samples; 4 transects with 300 sample points each (n = 1,200) for each of the 4 areas.

Table 3. Native tree abundance* in different size classes in two communities above and below goat fence (5,800-6,600 ft) on Mauna Loa, in 1986.**

	Montane Parkland				Subalpine Scrub			
	Goats		No Goats		Goats		No Goats	
Species and Size	#	%	#	%	#	%	#	%
Sophora chrysophylla (Māmane)								
<10 cm tall	13	21.31	16	10.06				
10 cm - 1 m	21	34.43	60	37.74	1	100	32	62.75
>1 m, <2cm dbh	18	29.51	57	35.85			16	31.37
2-5 cm dbh			21	13.21			3	5.88
6-10 cm dbh			2	1.26				
>10 cm dbh	9	14.75	3	1.89				
Subtotals	61		159		1		51	
Dodonaea viscosa ('A'ali'i)								
<10 cm tall	41	63.08	48	30.38				
10 cm - 1 m	12	18.46	55	34.81				
>1 m, <2 cm dbh	2	3.08	8	5.06				
2-5 cm dbh	4	6.15	7	4.43				
6-10 cm dbh			24	15.19			1	100
>10 cm dbh	6	9.23	16	10.13				
Subtotals	65		158				1	
Metrosideros polymorpha ('Ōhi'a)								
<10 cm tall								
10 cm - 1 m					1	6.67		

Table 3, continued.

Table 3, continued.

Species and Size	Montane Parkland				Subalpine Scrub			
	Goats		No Goats		Goats		No Goats	
	#	%	#	%	#	%	#	%
Metrosideros polymorpha ('Ōhi'a), cont.								
>1 m, <2 cm dbh					3	20.00	2	33.33
2-5 cm dbh					2	13.33		
6-10 cm dbh					2	13.33		
>10 cm dbh					7	46.67	4	66.67
Subtotals					15		6	
Acacia koa (Koa)								
<10 cm tall					5	83.33	18	12.77
10 cm - 1 m							80	56.74
>1 m, <2 cm dbh							38	26.95
2-5 cm dbh							4	2.84
6-10 cm dbh								
>10 cm dbh					1	16.67	1	0.71
Subtotals					6		141	
Total Native Trees	126		317		22		199	

*As determined by counts with 300 points per plot. Number of individuals and % of individuals of a species in particular size class.

**Area above or mauka with 100-200 goats at present; area below with few goats since early 1970s, no goats since 1985.

considerable regeneration. Significantly more (t test, $p = <0.05$) māmane, koa, and ‘a‘ali‘i seedlings, saplings, and sprouts were found below the fence in both 1986 and 1988. No comparable differences in seedling and sapling numbers of ‘ōhi‘a (*Metrosideros polymorpha*) were found in areas where goats were present and where they had been at remnant levels since 1982 and eradicated since 1985.

In the same subalpine area on Mauna Loa, a transect (utilizing point frequency with 300 points) established in the *Deschampsia*/native shrub community in the goat-free area showed appreciable reduction in alien plant cover from 1978 to 1984 (Table 4), although decreased pig digging may also explain some of the change. Not much change in alien cover occurred along one transect ($n = 300$ points) in the ‘ōhi‘a/native shrub type in the low goat density area above the fence (goat densities $<20/mi^2$ or $8/km^2$) (Table 4).

Table 4. Percent cover of native and alien plants in subalpine habitat (5,740-6,740 ft) on Mauna Loa in Hawaii Volcanoes National Park after goat removal.

Category and species	Open *Metrosideros* /Native Shrubs* 11/78	Open *Metrosideros* /Native Shrubs* 04/84	*Deschampsia* Grass /Native Shrubs** 11/78	*Deschampsia* Grass /Native Shrubs** 04/84
Native Ferns	0.7			
Alien Ferns			0.3	
Native Grasses, Sedges	3.6	6.0	33.9	37.3
Alien Grasses, Sedges			11.4	5.3
Alien Herbs	0.3		5.6	1.7
Native Woody Plants	61.4	49.4	24.6	21.0
Lichen	1.0	0.3		1.6
Moss	1.0	0.6	2.0	2.0
Litter	5.7	8.3	5.0	6.7
Rock	19.7	18.0	14.3	21.7
Soil	6.0	17.3	3.0	2.7
Number of Native Species	9	9	6	5
Number of Alien Species	1	0	6	3
Percent Native Cover	65.7	55.4	58.5	58.3
Percent Alien Cover	0.3	0.00	17.3	7.0

* Transect 118, 6,800 ft.
** Transect 119, 6,200 ft.

Feral Goats in Coastal Lowlands

Hawaii Volcanoes National Park Lowlands. Feral goats were probably present in the lowlands (area below fault scarp at <1,500 ft or 500 m elevation) of Hawaii Volcanoes National Park for 150 years at densities of about 170/mi^2 (60/km^2) (Stone and Keith 1986). Effective control of an estimated 15,000 animals began in 1971, with most goats gone (<200 left) by 1980, and populations in 1986 effectively zero. Lowland transects were established after initial goat reduction in the early 1970s. Results of point frequency studies (n = 300 points/transect) at that time in a number of lowland vegetation types (Tables 5a, b, and c) showed that current lowland vegetation was largely alien grasses. Percent alien cover values varied from 13 to 75% after six years at low goat densities for grasslands. Most of these areas had many more alien than native species, with only one native/alien species ratio reaching 1.0 (*Hyparrhenia* grassland in 1984). Changes in alien plant cover and native/alien ratios over a six-year period (1978-1984) were not great, and in many lowland grassland vegetation types, both alien and native cover declined somewhat, perhaps as a result of reduced precipitation levels and volcanic emissions from the current eruptive phase.

Lowland shrub systems (Table 5c) showed lower alien cover values and higher native/alien ratios than grasslands (Tables 5a and b), but change in these types after goat removal has also been slow. Alien cover values in lowland shrubland remained high (20.6-39.8%), but alien woody plants declined in cover over the six years. The native strand type showed a dramatic increase in percent native cover, largely as a result of increased naupaka kahakai (*Scaevola sericea*), a plant once browsed heavily by goats.

Native woody plant cover in a scrub ʻōhiʻa/native shrub community on ash substrate apparently decreased inside the Park and increased outside (where goats were not controlled) along one set of paired point frequency transects (n = 200 points/transect). However, native woody plants decreased outside the Park more than inside along another set of transects in a similar plant community on pāhoehoe substrate (Table 6). Variable goat browsing densities and distribution outside the Park, and weather, are again confounding factors. Small sample size and dramatic movement of ash with strong winds may also be important.

Two exclosures (Kaone and Kūkalauʻula) established in the lowlands of Hawaii Volcanoes National Park in 1969 and 1971 while goats were still abundant (Tables 7 and 8) showed low native/alien plant species ratios (0.28 to 0.50) after 13 to 15 years (1984) and high alien cover (over 70%). Alien plant cover (especially grasses) increased both inside and outside exclosures from 1971 to 1984. A previously undescribed jackbean (*Canavalia kauensis*, now included in *C. hawaiiensis*) appeared inside the Kūkalauʻula exclosure (St. John 1972; Mueller-Dombois 1981). A fire at the Kaone exclosure in 1980 probably encouraged increased cover of three alien perennial grasses, molasses grass (*Melinis minutiflora*), Natal red top (*Rhynchelytrum repens*), and *Hyparrhenia rufa* inside and outside exclosures. However, molasses grass and Natal red top also showed large increases inside and outside of the Kūkalauʻula exclosure, which did not burn. Mueller-Dombois (1981) noted the change over time from alien annual to alien perennial grasses. Native and alien cover may change gradually with time, or ratios may vary with succession, climatic trends, fires, or other variables.

Table 5a. Percent native and alien plant cover in lowland grasslands and shrublands, Hawaii Volcanoes National Park.*

Category and Species	*Rhynchelytrum repens* Grassland**		*Heteropogon contortus*/ *Rhynchelytrum repens* Grassland***		*Melinis minutiflora*/ *Rhynchelytrum repens* Grassland#		*Schizachyrium condensatum* Grassland##	
	12/78	03/84	10/78	01/84	12/78	03/84	12/79	01/84
Alien Ferns								0.7
Native Grasses, Sedges				15.3	17.0			
Alien Grasses, Sedges	41.7	31.4	9.6	11.0	71.7	72.7	62.7	59.7
Alien Herbs	8.0		1.0	1.0	0.3			
Native Woody Plants	5.7	4.3	10.0	7.3	0.7	0.7	14.7	5.0
Alien Woody Plants		5.3	1.0	1.0	11.7	1.0	2.0	4.0
Lichen				0.3			0.3	
Moss	0.7						0.3	0.3
Litter	11.3	25.3	8.7	9.0	9.3	16.3	8.7	11.7
Rock	29.7	32.7	53.7	52.0	6.3	9.3	10.7	18.7
Soil	3.0	1.0	1.3	0.7			0.3	
Dead wood							0.3	
Number of Native Species	1	1	3	2	1	1	2	2
Number of Alien Species	5	6	8	7	6	6	4	6
Percent Native Cover	5.7	4.3	25.3	24.3	0.7	0.7	14.7	5.0
Percent Alien Cover	49.7	36.7	11.6	13.0	83.7	73.7	64.7	64.4

*n = 300 for each sample, point frequency transects.
**Transect 113, 800 ft elevation.
***Transect 106, 120 ft elevation.
#Transect 110, 1,240 ft elevation.
##Transect 121 (burned in 1981), 1,250 ft elevation.

Table 5b. Percent native and alien plant cover in lowland grasslands and shrublands, Hawaii Volcanoes National Park.*

Category and Species	*Hyparrhenia rufa* Grassland**		*Rhynchelytrum repens* Grassland***		Mixed Grassland#	
	12/78	03/84	10/78	03/84	10/78	01/84
Native Ferns					1.0	
Alien Ferns						1.0
Native Grasses, Sedges	0.7					5.7
Alien Grasses, Sedges	80.3	75.0	48.3	39.3	27.7	25.6
Native Herbs	2.0	3.7				
Alien Herbs	0.7		0.7	0.6	1.7	1.0
Alien Vines				1.7	0.3	
Native Woody Plants	1.0		9.0	7.7	9.7	7.4
Alien Woody Plants	1.0		10.0	3.7	4.0	1.6
Litter	9.7	13.7	11.3	28.7	8.3	12.7
Rock			20.3	18.3	40.0	45.0
Soil	4.7	7.7	0.3			
Sand						
Number of Native Species	3	1	1	1	4	3
Number of Alien Species	5	1	6	8	9	8
Percent Native Cover	3.7	3.7	9.0	7.7	18.0	13.1
Percent Alien Cover	82.0	75.0	59.0	45.3	33.7	29.2

*n = 300 for each sample, point frequency transects.
**Transect 109, 1,020 ft elevation.
***Transect 115, 80 ft elevation.
#Transect 105, 60 ft elevation.

Table 5c. Percent native and alien plant cover in lowland grasslands and shrublands, Hawaii Volcanoes National Park.*

Category and Species	Lowland Scrub**		Lowland Scrub***		*Scaevola sericea* Strand#	
	09/78	01/84	10/78	01/84	12/78	03/84
Alien Ferns		0.6				
Native Grasses, Sedges	1.7	7.0	5.3	6.3	7.3	5.7
Alien Grasses, Sedges	35.7	37.0	18.3	15.3		
Native Herbs					18.0	4.0
Alien Herbs	1.0	0.3	2.0	1.3		
Native Woody Plants	17.7	23.0	47.0	43.4	47.9	94.0
Alien Woody Plants	6.7	1.7	7.9	4.0	0.7	
Moss		0.7				
Litter	20.0	17.7	9.7	20.7	7.7	2.0
Rock	17.3	12.7	9.0	9.0	3.7	
Sand					22.0	
Number of Native Species	5	5	7	7	5	4
Number of Alien Species	7	7	7	8	1	0
Percent Native Cover	19.4	30.0	52.3	49.7	65.9	98.0
Percent Alien Cover	43.4	39.6	28.2	20.6	0.7	0

*n = 300 for each sample, point frequency transects.
**Transect 103, 40 ft elevation.
***Transect 104, 70 ft elevation.
#Transect 116, 10 ft elevation.

Table 6. Percent cover of native and alien plants in *Metrosideros* shrubland inside and outside Hawaii Volcanoes National Park after goat removal program in the Park between 1971 and 1980 (for most animals).*

	Scrub *Metrosideros*/Native Shrubs**				Scrub *Metrosideros*/Native Shrubs ***			
	Inside		Outside		Inside		Outside	
Category and Species	01/79	01/84	01/79	01/84	01/79	01/84	01/79	01/84
Alien Grasses, Sedges	10.0	9.0	9.0	8.0	12.5	16.0	8.5	9.5
Native Woody Plants	4.0	1.0	4.5	6.0	6.0	5.0	4.5	2.0
Lichen	2.5	1.5		0.5	5.0	2.0	2.5	2.0
Moss				3.5	1.0	1.0		0.5
Litter	4.0	5.0	3.0	6.5	2.0	4.5	3.5	2.5
Rock	25.5	33.0	15.0	7.0	52.5	49.5	61.0	61.5
Ash	54.0	50.5	68.5	68.5	21.0	22.0	21.0	22.0
Number of Native Species	2	1	3	3	2	3	3	2
Number of Alien Species	4	3	4	4	5	4	4	3
Percent Native Cover	4.0	1.0	4.5	6.0	6.0	5.0	4.5	2.0
Percent Alien Cover	10.0	9.0	9.0	8.0	12.5	16.0	8.5	9.5

*n = 300 for each sample, point frequency transects.

**Transect 100 on ash substrate, 2,650 ft elevation.

***Transect 101 on pāhoehoe substrate, 2,650 ft elevation.

Table 7. Percent native and alien plant cover inside and outside Kaone Exclosure, Hawaii Volcanoes National Park.*

Category and Species	10/71		04/82		03/84	
	Inside	Outside	Inside	Outside	Inside	Outside
Alien Grasses, Sedges	15.8	9.6	42.4	51.2	61.0	68.0
Native Herbs				0.2	3.6	1.0
Alien Herbs	1.6	0.6	0.4	0.4		0.2
Alien Vines			0.4		1.3	
Native Woody Plants			1.4	1.0	0.6	0.6
Alien Woody Plants	11.2	10.8	38.2	33.4	15.1	13.4
Litter	50.4	58.2	2.2	4.0	6.3	6.6
Soil	21.0	20.4	15.0	9.2	12.2	10.0
Rock				0.6	12.2	0.2
Goat pellets		0.4				
Number of Native Species	0	0	1	2	2	2
Number of Alien Species	9	10	8	8	7	6
Percent Native Cover	0	0	1.4	1.2	4.2	1.6
Percent Alien Cover	28.6	21.0	81.4	85.0	77.4	81.6

*As determined by point frequency transects; n = 500 points/sample. The Kaone area burned in 1980.

Table 8. Percent native and alien plant cover inside and outside of Kūkalau'ula Exclosure, Hawaii Volcanoes National Park, 1971, 1982, and 1984.*

Category and Species	12/71		03/82		04/84	
	Inside	Outside	Inside	Outside	Inside	Outside
Native Grasses, Sedges	1.2		0.3		0.6	
Alien Grasses, Sedges	45.4	45.2	90.9	77.8	75.7	69.6
Native Herbs	45.0		6.0	13.2	8.9	17.2
Alien Herbs	1.2	0.4	2.3		1.7	
Native Woody Plants	2.4			1.4		1.0
Alien Woody Plants	2.4			5.2		1.4
Litter		14.4	0.6	1.6	13.1	9.6
Rock		38.0		0.8		1.2
Number of Native Species	3	0	2	2	2	2
Number of Alien Species	6	7	6	5	4	5
Percent Native Cover	47.4	0	6.3	14.6	9.5	18.2
Percent Alien Cover	49.0	45.6	93.2	83.0	77.4	71.0

*As determined by point frequency transects; n = 350 points inside, 500 outside. 1971 points were 270 and 140.

Feral Pigs in Subalpine and Montane Parkland Areas

Kalapawili Grasslands. Feral pigs have been present in the Kalapawili grasslands in Haleakala National Park since at least 1960 at low but unknown densities (probably <10/km^2). They were not effectively controlled until erection of fencing along the North Rim of the Crater and upper Kīpahulu Valley in 1985-1986. The native bunch grass *Deschampsia nubigena* is the dominant species, but velvet grass (*Holcus lanatus*) is a serious invader (Jacobi 1976, 1981).

Point frequency data (n = 500 points inside, 500 outside) from pig exclosures in the grasslands (J. Jacobi, pers. comm.) suggest that where disturbance by feral pigs continues, the percent cover of *Deschampsia* remains lower and the percent cover of velvet grass is somewhat higher than where disturbance by pigs ceases (Table 9). Estimates of plant cover from 100, 2 x 2 m plots along transects (Table 10), which cover a larger and more representative part of the grasslands, and where disturbance continued, show stable *Deschampsia* cover (57.3% in 1973 and 58.6% in 1986), but increasing *Holcus* (9.5% in 1973 to 15.3% in 1986) (J. Jacobi, pers. comm.). Pig control has been effective throughout the area since 1986. Blackberry (*Rubus argutus*) is gradually spreading at present.

Hawaii Volcanoes National Park. Three major intergrading plant communities were studied for recovery from feral pig damage in mountain parkland on Mauna Loa between elevations of 4,495 and 6,790 ft (1,370-2,070 m). Small stands of koa, shrublands dominated by pūkiawe and sometimes ‘a‘ali‘i, and grasslands dominated by native bunch grasses or alien grasses were sampled (Tunison *et al.*, in press). Twenty-seven 2 x 5-m plots were established in 1984 on sites recently (<30 days) disturbed by pigs. Plots in each of the three vegetation communities were each replicated nine times (clusters of three plots at three sites in each community). No disturbance by pigs occurred after plot establishment, since pig removal was nearly complete in 1984. Plots were sampled by the point intercept method from 1985 to 1988 with 10, 2-m long transects and 200 sample points per plot. Woody plants were counted in five height classes. Data from all 27 plots were pooled to give early successional trends.

Early results showed that bare soil and litter declined with increases in plant cover, as would be expected. Alien species cover increased at a faster rate than native species cover (Fig. 1), and alien grass cover accounted for most of the increase in total vegetative cover (Fig. 2). In lower-elevation parkland communities (4,400 ft or 1,340 m) dominated by alien grasses, especially sweet vernalgrass (*Anthoxanthum odoratum*) and velvet grass, these species increased in cover more rapidly than other native or alien grasses. At intermediate elevations (4,900 ft or 1,495 m) dominated by mixed alien and native grasses, dallis grass (*Paspalum dilatatum*) and sweet vernalgrass increased more rapidly than *Deschampsia.* At higher elevations (6,300 ft or 1,920 m) dominated largely by native grasses, velvet grass increased more rapidly than other species, but *Deschampsia* recovered nearly as rapidly. The total number of plant species increased slightly or remained the same for essentially all areas.

Table 9. Summary of percent cover of alien and native plants inside and outside Kalapawili Ridge Exclosure, Haleakala National Park, 1974 and 1985.*

	Transects 1-5				Transects 6-10			
	Inside		Outside		Inside		Outside	
Plant Species	'74	'85	'74	'85	'74	'85	'74	'85
Deschampsia nubigena	42	80	48	68	76	78	65	71
Holcus lanatus	1	16	2	22	7	16	5	21
Hypochoeris radicata	2	1	2	2	4		3	3
Rumex acetosella	9		4		1		4	
Other Species	3	3	5	3	4	4	4	3
Litter	36		32	4	8	2	17	2
Barren ground	7		7	1			2	

*As determined by point frequency sampling. Transects 1-5 received greater initial pig damage than 6-10 (from J. Jacobi, pers. comm.).

Table 10. Summary of percent cover of alien and native plants on Kalapawili grasslands transects, Haleakala National Park.*

	Transects	
Plant Species	1973	1986
Deschampsia nubigena	57.3	58.6
Holcus lanatus	9.5	15.3
Hypochoeris radicata	8.0	16.9
Rumex acetosella	0.9	0.1
Other Species	1.1	1.4
Rocks		0.8

*As determined by point frequency sampling (from J. Jacobi, pers. comm.).

Feral Pigs in Rain Forests

Feral pigs in rain forests in the Thurston and Puhimau areas of Hawaii Volcanoes National Park (Fig. 3) have probably been present for 60 years in densities of about $55/mi^2$ (20 animals/km^2). In this very accessible area, hunting has been moderate to heavy for at least 20 years. The vegetation in the Thurston-Puhimau area is open to closed ʻōhiʻa with a native tree subcanopy, an open tree fern or hapuʻu (*Cibotium glaucum*) layer beneath, and ground cover composed of uluhe (*Dicranopteris linearis*), other native ferns, and both native and alien shrubs, herbs, and grasses.

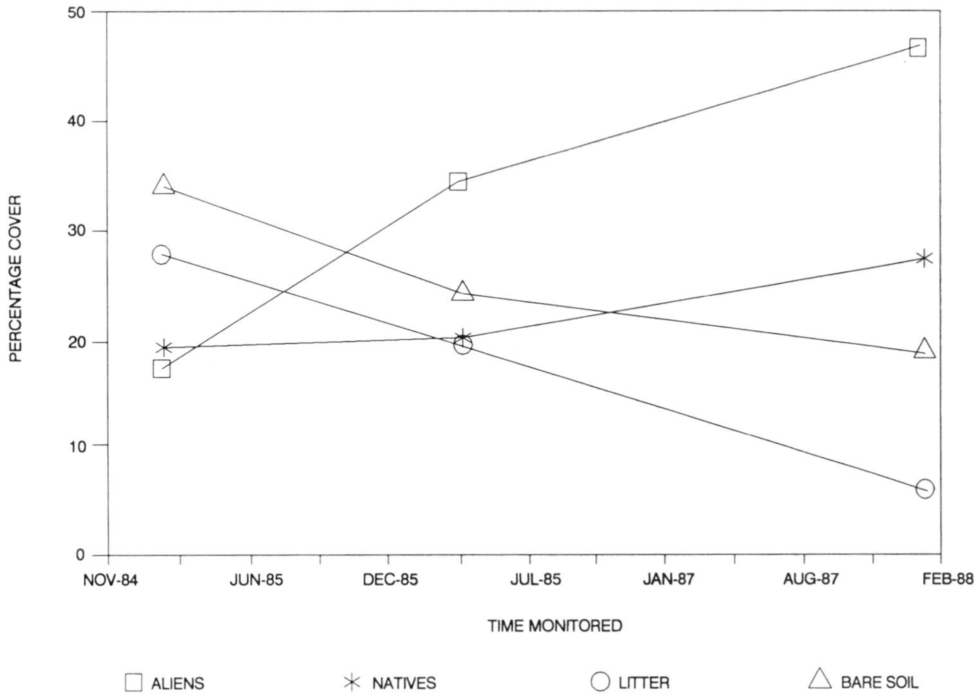

Figure 1. Changes in percentage cover (as determined by point intercept) in 27 plots in montane parkland area recently disturbed by pigs (elevation 4,495 to 6,790 ft) on Mauna Loa, Hawaii Volcanoes National Park, 1984-1985.

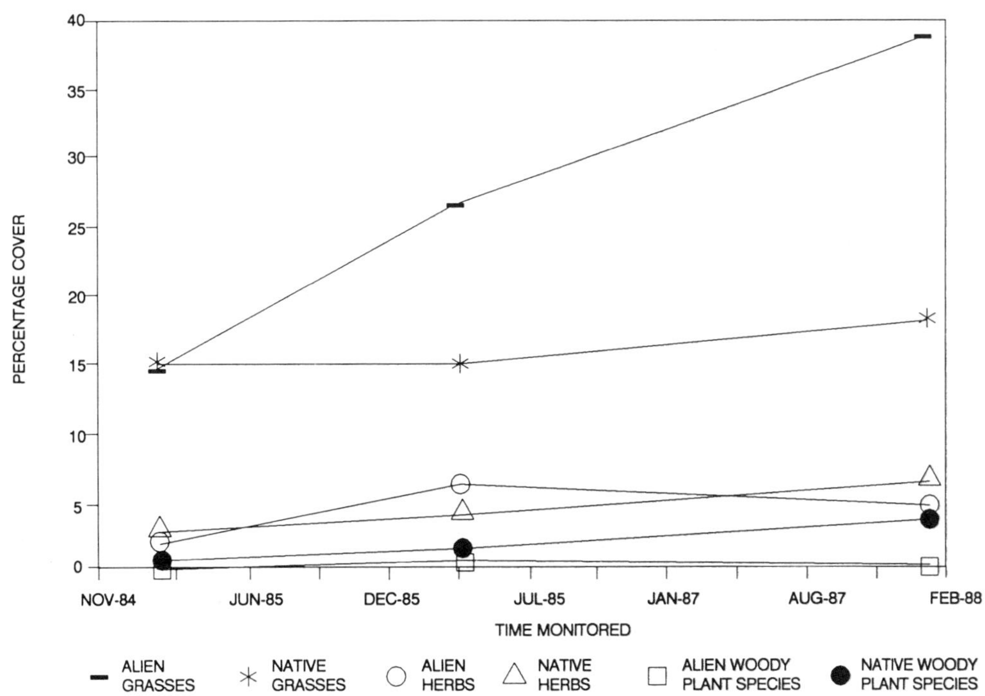

Figure 2. Changes in percentage cover of native and alien plants (as determined by point intercept) in 27 plots in montane parkland area recently disturbed by pigs (elevation 4,495 to 6,790 ft) on Mauna Loa, Hawaii Volcanoes National Park, 1984-1985.

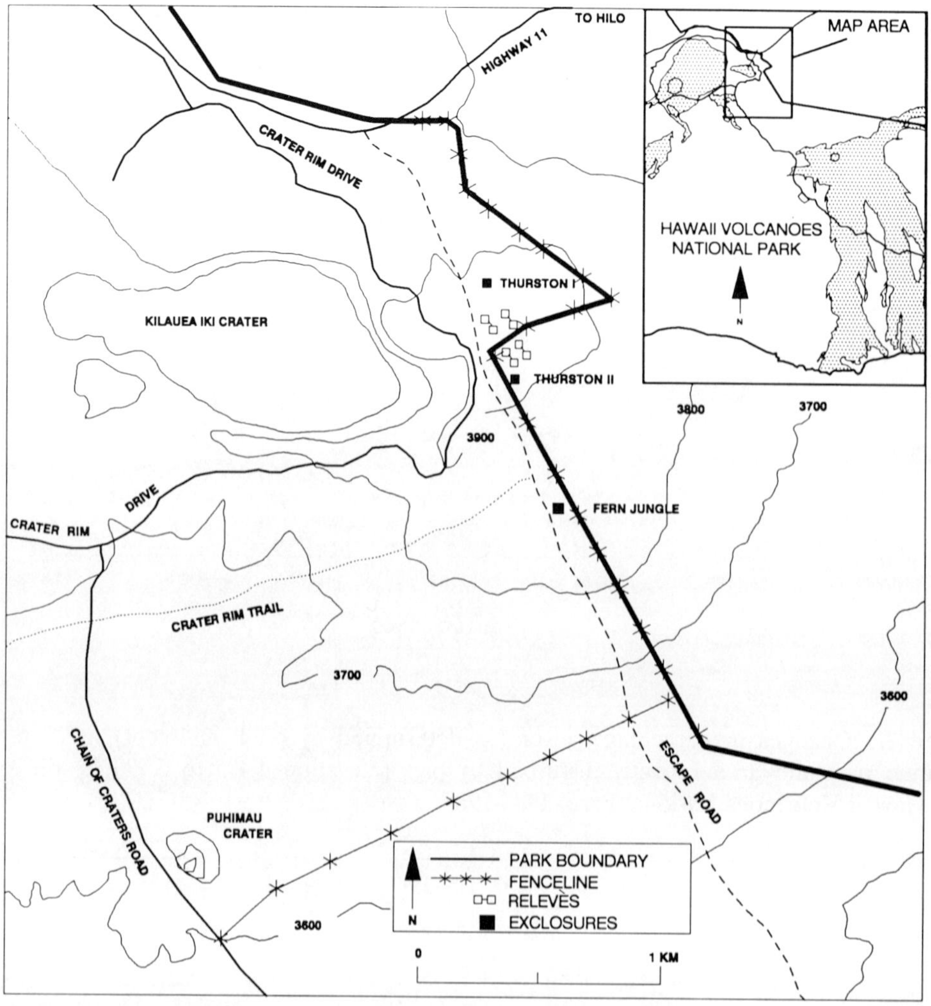

Figure 3. Locations of rain forest exclosures and relevés in Thurston area of Hawaii Volcanoes National Park.

One feral pig exclosure (Fern Jungle) was established in 1968 and two (Thurston I and II) in 1981. The enclosed 4.6 km^2 Thurston Lava Tube Management Area, which contains the Thurston I exclosure, has been pig-free since the fence was finished in December 1980. Another fence enclosing approximately 5.2 km^2, including the Fern Jungle Exclosure, Puhimau Crater, and part of the Crater Rim Trail, was finished in August 1984. Transects were established at that time to monitor alien plants and feral pig activity in the Thurston area and in an adjacent area on private (now State-owned) land still inhabited by pigs. Braun-Blanquet percentage cover estimates were obtained from 3- x 5-m plots. Unfortunately, data from many plots could not be included for the Thurston area analyses because cover totals did not exceed 70 percent. As indicated previously, sample sizes are given in tables for each area.

Fern Jungle Exclosure. Alien plant species cover inside the Fern Jungle Exclosure (0-0.5 m stratum) in 1981 was considerably less after 14 years without pigs than that outside, where disturbance continued (Table 11). Alien grasses, especially meadow ricegrass (*Ehrharta stipoides*), and the native ʻōhiʻa tree accounted for much more cover outside. In contrast, a native grass, ʻohe (*Isachne distichophylla*), and native bryophytes provided more cover inside the exclosure (Higashino and Stone 1982). Litter, especially that of uluhe, was greater inside the exclosure, and exposed ground was greater outside. Unfortunately, we were unable to directly compare 1968 data with 1981 data inside the exclosure, and no monitoring was done outside in 1968.

Table 11. Mean percent cover of alien and native plants in 3 x 5 m Braun-Blanquet relevés inside and outside Fern Jungle Exclosure, Hawaii Volcanoes National Park, 1981 (ground-0.5 m stratum).*

Category and Species	Inside**	Outside
Native Ferns, Clubmosses	0.5	0.3
Native Grasses, Sedges	5.0	2.8
Alien Grasses, Sedges	0.6	5.8
Alien Herbs	0.2	0.7
Native Woody Plants	1.3	4.8
Native Bryophytes	9.3	4.5
Litter	68.0	57.5
Exposed Ground		5.2
Total Cover***	84.9	81.6
Percent Native Cover	16.1	12.4
Percent Alien Cover	0.8	6.5
Number of Native Species	15	15
Number of Alien Species	9	12

*Sample size = 15 plots inside and 18 outside exclosure.
**Feral pigs excluded for 14 years.
***Because Braun-Blanquet cover estimates rely on category midpoints, percent cover does not necessarily add to 100.

Thurston I Exclosure. An increase in bryophyte cover outside the exclosure was the most significant change noted in the Thurston I exclosure study (0-0.5 m stratum) over four years (Table 12). Native grasses appeared to increase inside and outside the exclosure, and native woody plants appeared to decrease inside and outside the exclosure. Alien grasses, especially meadow ricegrass, showed some increase inside the exclosure. Litter appeared to decrease outside the exclosure, but the 1981 readings were taken in October, and the 1985, in July, perhaps accounting for some of the difference. Disturbed ground decreased inside and outside the exclosure. Since pigs have not been present outside Thurston I exclosure (established in 1981) since 1980, it was not surprising that similar vegetation responses occurred inside and outside.

Table 12. Mean percent cover of native and alien plants in 3 x 5 m Braun-Blanquet relevés inside and outside Thurston I Exclosure, Hawaii Volcanoes National Park, 1981 and 1985 (ground-0.5 m stratum).*

	1981		1985	
Category and Species	Inside	Outside	Inside	Outside
Native Ferns	0.2	1.4	0.5	2.4
Alien Ferns	**	**	**	
Native Grasses, Sedges	3.0	3.7	4.9	4.7
Alien Grasses	3.4	0.5	5.2	0.3
Native Herbs	**	**		0.1
Alien Herbs	0.8	0.3	0.1	0.1
Native Woody Plants	5.7	3.5	3.6	1.7
Alien Woody Plants	**	**	**	**
Native Bryophytes	8.1	1.9	8.1	15.9
Litter	47.6	69.6	47.6	50.0
Logs			2.5	
Exposed Ground	6.2	1.5		
Total Cover***	75.0	82.4	72.5	75.2
Percent Native Cover	17.0	10.5	17.1	24.8
Percent Alien Cover	4.2	0.8	5.3	0.4
Number of Native Species	14	13	17	15
Number of Alien Species	9	7	9	8

*Sample size = 8 plots outside, and 7 inside exclosure.

**Measurable cover but <0.1% mean.

***Because Braun-Blanquet cover estimates rely on category midpoints, percent cover does not necessarily add to 100.

Thurston II Exclosure. Insufficient plots were available with adequate cover estimates outside the exclosure, but obvious changes occurred inside over the four years (Table 13). Exposed ground decreased inside the exclosure, and the alien plant cover, especially the shade-tolerant meadow ricegrass, showed a great increase. Exposed ground available for ricegrass invasion was greater prior to construction of this exclosure, largely because pigs were not being controlled in the Thurston II area, as was the case in the Thurston I area. New alien plants in the plots in 1985 included strawberry guava (*Psidium cattleianum*), blackberry, broomsedge (*Andropogon virginicus*), and Hāmākua pāmakani (*Ageratina riparia*). Glenwood grass (*Sacciolepis indica*) was present in 1981 but not 1985.

Table 13. Mean percent cover of alien and native plants in 3 x 5 m Braun-Blanquet relevés inside Thurston II Exclosure, Hawaii Volcanoes National Park, 1981 and 1985. (ground - 0.5 m stratum).*

Category and Species	1981	1985
Native Ferns	0.4	0.4
Native Grasses, Sedges	2.1	2.0
Alien Grasses, Sedges	2.7	22.7
Alien Herbs	0.2	0.1
Native Woody Plants	1.3	2.4
Alien Woody Plants	**	**
Native Bryophytes	3.3	2.0
Exposed Ground	11.1	**
Litter	41.5	35.8
Total Cover***	62.6	65.4
Percent Native Cover	7.1	6.8
Percent Alien Cover	2.9	22.8
Number of Native Species	12	11
Number of Alien Species	3	6

*Sample size = 5 plots; sample size outside exclosure not adequate.

**Measurable cover, but <0.1% mean.

***Because Braun-Blanquet cover estimates rely on category midpoints, percent cover does not necessarily add to 100. Although all plots totalled >70%, means for the two years were somewhat lower after conversion from arcsin.

Thurston Relevés. Percent alien plant cover and number of alien species in plots in a pig-inhabited area near the Thurston II exclosure showed an increase in 1985, compared to 1981 (Table 14). Among the new invaders in the plots over the four-year period were kāhili ginger (*Hedychium gardnerianum*), strawberry guava, and firetree (*Myrica faya*), three of the most invasive species in the area. Meadow ricegrass continued to increase, although exposed ground from feral pig disturbance was higher in 1985 than 1981. A more extensive area was covered in this comparison than in plots limited to inside and immediately outside exclosures.

Table 14. Percent cover of alien and native plants in 3 x 5 m Braun-Blanquet relevés in pig-inhabited area of Thurston Lava Tube area, Hawaii Volcanoes National Park, 1981 and 1985 (ground-0.5 m stratum).*

Category and Species	1981	1985
Native Ferns	0.9	**
Native Grasses, Sedges	2.4	4.6
Alien Grasses, Sedges	11.5	18.9
Alien Herbs	0.1	0.3
Native Woody Plants	1.5	1.7
Alien Woody Plants	**	0.3
Native Bryophytes	2.8	4.9
Exposed Ground	0.4	4.2
Litter	58.5	36.6
Total Cover***	78.1	71.5
Percent Native Cover	7.6	11.2
Percent Alien cover	11.6	19.5
Number of Native Species	8	9
Number of Alien Species	3	11

*Sample size = 5 plots.
**Measurable cover, but <0.1% mean.
***Because Braun-Blanquet cover estimates rely on category midpoints, percent cover does not necessarily add to 100.

Preliminary population data for firetree and kāhili ginger in the Thurston area of Hawaii Volcanoes National Park five years after feral pig removal showed strong inverse J-curves indicative of stable or, more likely, expanding populations since removal (Figs. 4 and 5). Population expansion of these and probably other species subsequent to feral animal removal is likely to occur in some areas without aggressive alien plant

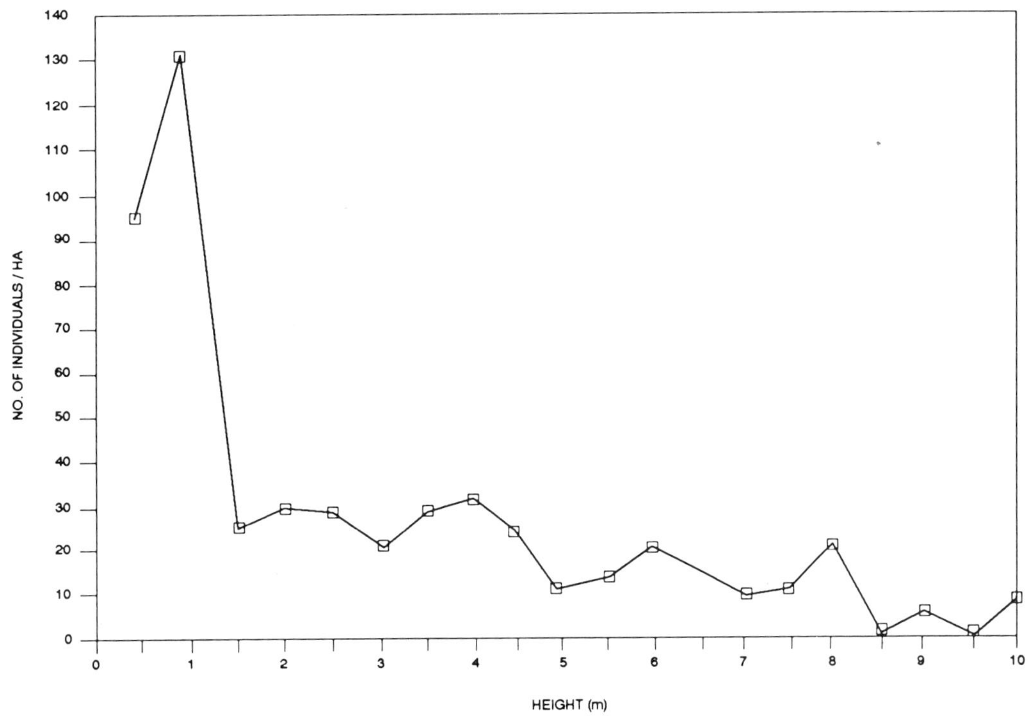

Figure 4. Population structure of firetree in Thurston area of Hawaii Volcanoes National Park, five years after feral pig removal.

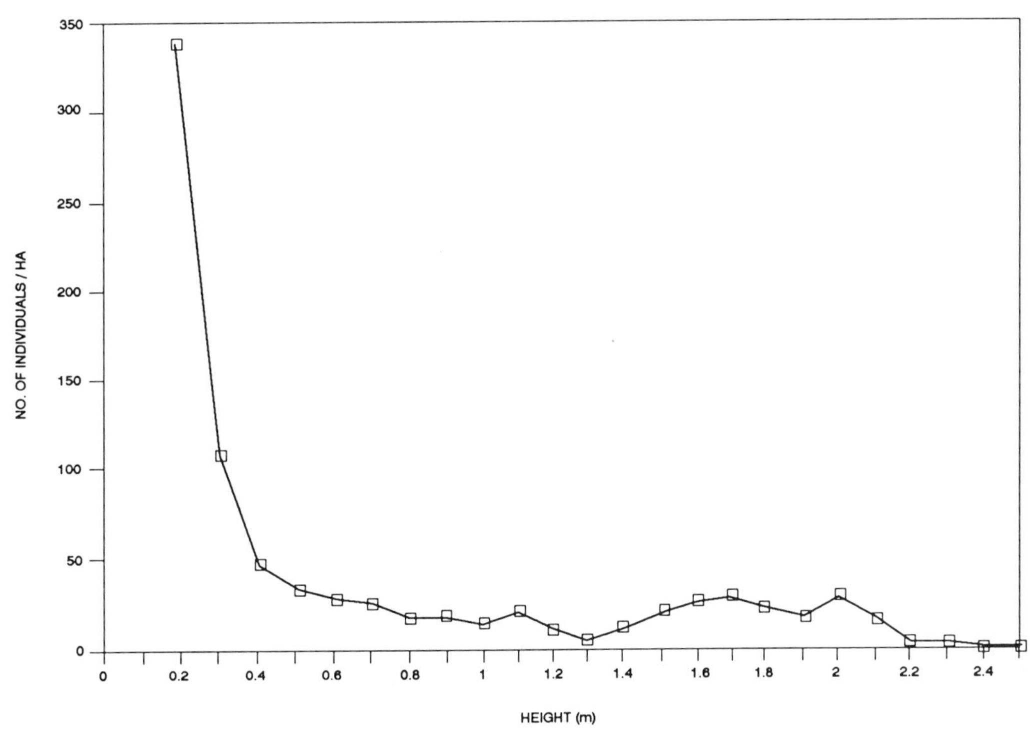

Figure 5. Population structure of kāhili ginger in Thurston area of Hawaii Volcanoes National Park, five years after feral pig removal.

management programs. Yellow Himalayan raspberry (*Rubus ellipticus*), Hilo grass (*Paspalum conjugatum*), meadow ricegrass, strawberry guava, kāhili ginger, firetree, banana poka (*Passiflora mollissima*), and palm grass (*Setaria palmifolia*) are the rain forest invaders of most concern in the Park, once feral pigs are gone.

Nāpau. Nāpau Exclosure, located a little over 0.3 mi (0.5 km) south-southeast of Nāpau Crater in Hawaii Volcanoes National Park, provided information on a remote rain forest area from which pigs had been excluded for 11 years when data were collected. The exclosure was built in 1975, and results to 1980 were reported by Katahira (1980) (based on 9 Braun-Blanquet plots established inside and 12 outside the exclosure). The rain forest in the area is closed-canopy ʻōhiʻa with tree fern (hapuʻu) understory. The exclosure is located in an area that is far removed from many of the plant propagules produced by aggressive alien species in Hawaii Volcanoes National Park.

Although measurable alien plant cover was not recorded inside or outside the exclosure in 1981, it was considerably higher by 1986, especially outside (Table 15). Exposed ground was much greater outside the exclosure in 1981 and 1983, but by 1986 tephra outfall from the ongoing volcanic eruption at Puʻu ʻŌʻō covered a large area inside and outside. Tephra probably was the overriding cause of vegetation changes both inside and outside the exclosure by 1986, including the invasion of Hilo grass, which accounted for most of the alien plant increase. Bryophytes that had begun to recover inside and outside the exclosure by 1983 decreased over the next three years, perhaps as a result of proximity to volcanic emissions. Alien species seemed to play a minor role in this remote area until the tephra outfall, and native species cover that had been slowly increasing seemed to decline over the same period.

Boundary Kīpuka. A natural feral pig exclosure on the East Rift of Hawaii Volcanoes National Park, located less than 0.3 mi (0.5 km) from the top of Puʻu Kamoamoa, was destroyed in 1983 by lava flow and cinder fall from the Puʻu ʻŌʻō eruption series. Unofficially called the Boundary Kīpuka, it probably had experienced only occasional pig use prior to that time because of isolation by rough lava flows and a pig fence that partially protected it. The Kīpuka was a 175-a (71-ha) closed canopy ʻōhiʻa rain forest. Percent cover values from 24 Braun-Blanquet releves in the Kīpuka and 24 in adjacent forest inhabited by pigs indicated that total native species cover was higher in the pig-inhabited forest than in the Kīpuka (Table 16), possibly because of a more closed canopy in the Kīpuka. Native woody plants accounted for a higher percent of cover in the East Rift forest with pigs than in the Kīpuka. Alien thimbleberry (*Rubus rosifolius*) shrubs, although not comprising a large percentage of cover, provided somewhat more cover in the pig-disturbed forest than in the Kīpuka. The Kīpuka contained fewer alien species (three) with measurable percent cover than the adjacent forest (six).

ʻŌlaʻa Tract. ʻŌlaʻa Tract in Hawaii Volcanoes National Park is primarily open canopy ʻōhiʻa rain forest with closed tree fern subcanopy and native-alien woody plant, vine, herb, and grass understory. It is

Table 15. Mean percent cover of alien and native plants in 3 x 5 m Braun-Blanquet relevés inside and outside Nāpau Exclosure, Hawaii Volcanoes National Park, 1981, 1983, and 1986 (ground - 0.5 m stratum).*

Category and Species	October 1981		November 1983		March 1986	
	Inside	Outside	Inside	Outside	Inside	Outside
Native Ferns	0.7	0.3	3.5	1.5	1.0	0.7
Native Grasses	0.2	**	1.6	0.9	0.2	**
Alien Grasses	**	**	1.1	0.5	2.7	9.5
Native Herbs		**		**		**
Alien Herbs		**		0.1		**
Native Woody Plants	0.5	0.1	1.3	2.0	0.4	0.1
Alien Woody Plants			0.1	0.1	0.3	0.4
Bryophytes	1.9	3.2	2.5	7.6	1.9	1.4
Exposed Ground	2.5	44.5	1.4	17.8		0.3
Litter	69.6	26.5	81.9	32.8	43.7	28.6
Tephra					23.8	31.1
Total Cover***	75.4	74.6	93.4	63.3	74.0	72.1
Percent Native Cover	3.3	3.6	8.9	12.0	3.5	2.2
Percent Alien Cover	0.0	0.0	1.2	0.7	3.0	9.9
Number of Native Species	15	25	12	25	10	19
Number of Alien Species	2	4	4	5	3	4

*Sample size = 4 plots inside exclosure, 10 plots outside exclosure.
**Measurable cover, but <0.1% mean.
***Because Braun-Blanquet cover estimates rely on category midpoints, percent cover does not necessarily add to 100.

Table 16. Percent cover of alien and native plants in 3 x 5 m Braun-Blanquet relevés on Boundary Kīpuka and in adjacent East Rift forest, Hawaii Volcanoes National Park, 1982 (ground to 0.5 m stratum).*

Category and Species	Kīpuka	Adjacent Forest
Native Ferns	1.6	1.4
Alien Grasses, Sedges		0.1
Native Herbs	3.3	1.2
Alien Herbs		0.2
Native Woody Plants	2.4	8.3
Alien Woody Plants	2.1	0.7
Bryophytes	15.7	20.4
Exposed Ground	0.2	2.9
Litter	65.9	66.0
Total Cover**	91.2	101.2
Percent Native Cover	23.0	31.3
Percent Alien Cover	2.1	1.0
Number of Native Species	33	34
Number of Alien Species	3	6

*Sample size = 24 plots in Kīpuka, 24 in adjacent forest.
**Because Braun-Blanquet cover estimates rely on category midpoints, percent cover does not necessarily add to 100.

underlain by fertile soils, and pig densities are probably <20/km^2 in the large, eastern (10,000 a or 40 km^2) section. The small, western section (320 a or 1.3 km^2) is an open canopy ʻōhiʻa rain forest with closed tree fern subcanopy and largely native fern understory. The small section was fenced in 1981, with fence upgrading in August of 1983 and 1984. The small size, disjunctness, and accessibility of the area to hunters have combined to limit pig ingress over the years. Pigs were finally removed from the small section in 1983.

Exposed ground was greater in the pig-inhabited large section in 1982 and 1985 than in the protected small section, but alien plant cover in both the large and small sections was low (Table 17). Native ferns began to recover in the small section but continued to decline in large section plots. The small section plots had more native species than the large section plots, but many more small section plots (28) than large (8) were sampled. The low alien plant cover in the small section is especially low with that in mind, but in 1985, three species of *Rubus (ellipticus, argutus,* and *rosifolius)* were found in plots there that had not been present in 1982. Banana poka, yellow Himalayan raspberry, and blackberry were present in large section plots in 1982; kāhili ginger appeared in

1985. Litter and bryophyte cover seemed to be lower in 1985 than 1981 in both sections.

A recent data set from extensive transects in small section ʻŌlaʻa shows that three species of alien plants continued to increase six years after feral pig removal (Table 18). Kāhili ginger, yellow Himalayan raspberry, and banana poka all showed increases in most size classes. The smallest class (<10 cm) is perhaps not as meaningful because survival is problematical.

Table 17. Percent cover (and standard deviations) of alien and native plants in large and small sections of ʻŌlaʻa Tract, Hawaii Volcanoes National Park, 1982 and 1985 (ground-0.5 m stratum).*

Category and species	ʻŌlaʻa Large Section 1982	ʻŌlaʻa Large Section 1985	ʻŌlaʻa Small Section 1982	ʻŌlaʻa Small Section 1985
Native Ferns	3.9	3.0	2.5	3.9
Alien Ferns	**	**	**	
Native Grasses, Sedges	0.2	0.1	0.3	0.1
Alien Grasses	**		**	
Native Herbs	0.8	0.3	1.4	1.1
Alien Herbs		0.1	**	**
Native Woody Plants	0.6	0.2	1.4	1.1
Alien Woody Plants	0.1	0.5		**
Bryophytes	6.9	3.4	9.4	5.4
Exposed Ground	3.8	2.3	0.7	0.3
Litter	82.2	55.8	77.9	60.9
Logfall		5.6		5.7
Total Cover***	98.5	71.3	93.6	78.5
Percent Native Cover	12.4	7.0	15.0	11.6
Percent Alien Cover	0.1	0.6	0.0	0.0
Number of Native Species	19	27	32	39
Number of Alien Species	5	5	3	4

*Sample size = 28 plots in small section, 8 plots in large section of ʻŌlaʻa Tract.

**Measurable cover, but <0.1% mean.

***Because Braun-Blanquet cover estimates rely on category midpoints, percent cover does not necessarily add to 100.

Table 18. Number of *Hedychium gardnerianum, Rubus ellipticus*, and *Passiflora mollissima* along 30-m wide belt transects in Small Tract 'Ōla'a Tract Special Ecological Area, Hawaii Volcanoes National Park, in 1986* and 1989.

Hedychium gardnerianum		Height (cm)			
(Kāhili ginger)	<10	10-50	50-100	100-155	
1986	820	372	54	20	
1989	324	281	75	47	
Rubus ellipticus			Height (cm)		
(Yellow Himalayan raspberry)	<10	10-50	50-200	200-300	>300
1986	20	82	35	5	2
1989	41	78	43	10	11
Passiflora mollissima		Height (cm)			
(Banana poka)	<10	10-200	>200		
1986	0	0	2		
1989	45	1	11		

*From J.T. Tunison, unpub. data.

Kīpahulu Valley. Anderson *et al.* (this volume) reported on distribution and spread of alien plants in Kīpahulu Valley rain forest above 2,300 ft (700 m) elevation during and subsequent to removal of feral pigs. Methodology will not be reiterated here, but some of the conclusions about alien plants and pigs bear repeating for purposes of this paper.

Based on frequency of occurrence data, the most rapid spread of weeds in the Valley during and after pig removal was a 17% increase in weeds in wet 'ōhi'a forest at 4,300 ft (1,310 m) elevation, probably resulting from a combination of factors, including increased use of newly-established trails by feral pigs, slow growth of uluhe (which had previously dominated the area), and wind and bird dispersal of the invading aliens Maui pāmakani (*Ageratina adenophora*) and thimbleberry.

Lowest weed frequencies were found in areas least disturbed by pigs, mostly at higher elevations. Removal or reduction of pigs did not result in decreased weed occurrence in koa forests at lower elevations that had been considerably disturbed, where sunlight needed for proliferation of alien plants was available, and where propagules of invasive weed species were numerous. Numbers of weed species, and percentage of weed cover in general, increased with decreasing elevation in the Valley.

Life forms of weeds seemed related to elevation, with alien grasses and sedges comprising most of the alien plant cover at lower elevations (3,600 ft or 1,100 m), but alien herbs (less invasive and lower percentage of cover than grasses) at higher elevations. The native matted fern, uluhe, inhibited alien plant invasions where undisturbed in the Valley, but the native fern hōʻiʻo (*Diplazium sandwichianum*) was preferred by pigs. Alien plants such as strawberry guava and Hilo grass often invaded pig-disturbed areas once dominated by this fern (Anderson *et al.*, this volume).

Feral pig disturbance was definitely not a prerequisite for invasions of kāhili ginger and strawberry guava in the Valley. Each was found in undisturbed areas, and all continue to spread at present (1989). That recovery of native plants after pig removal is not assured is demonstrated by alien grasses and sedges that still dominate in an exclosure at about 3,000 ft (915 m) elevation, which had been free of pigs for 10 years when last observed. Native shrubs and ground cover have not regenerated in the exclosure (Anderson *et al.*, this volume).

INTACTNESS AND MANAGEABILITY

"Intactness" of a native plant community in Hawaiʻi can potentially be indexed by the prevalence of invasive alien plants in it. Primary emphasis on alien plants rather than natives is useful, because aliens are the primary targets of management subsequent to feral ungulate removal. Native species richness is an additional piece of information important in setting management priorities. Although such indices are useful, other information is also used in decision-making. Knowledge of rare plant abundance and distribution and community rarity are valuable in making judgements; funding, personnel, and size of an area must be considered; and potential threats posed by particularly invasive alien plant species must also be evaluated and given considerable weight in decision-making.

Percent alien plant cover and native/alien species ratios of some areas considered in this paper are summarized in Table 19 (for grasslands and shrublands) and Table 20 (for rain forests). These intactness indices suggest that low elevation grasslands and at least some accessible (to humans and alien plant propagules) and long-damaged rain forests provide the greatest restoration challenges after ungulate removal. The Fern Jungle rain forest data (14 years without pigs inside the exclosure) suggest that simple protection from pigs may eventually result in near-native forest, but that native species richness may be low after considerable disturbance. Remote, subalpine grasslands and montane parklands may also show important alien components and/or decreased native plant richness with continual disturbance by feral ungulates, as well as higher alien plant cover indefinitely, especially at the lower elevational limits of these types. Adequate recovery of many Hawaiian ecosystems disturbed by ungulates will likely require protection, aggressive and continual alien plant management, considerable recovery time, and some reintroduction of missing native plant species. Long-term damage by ungulates can eliminate native plant species and deplete seed beds, change microclimates as well as water and nutrient systems, and slow growth of many natives. Tolerance of an alien species component in otherwise native communities will also be necessary.

Table 19. Grassland and shrubland intactness as measured by percent cover of alien plants and native/alien species ratios.

Vegetation Type (Locality)	Years w/out Goats*	% Alien Cover	Native/Alien Species Ratio
Scaevola strand (Hawaii Volcanoes Lowlands)	4	0.0	4.0+
Metrosideros/Native shrub (Mauna Loa)	2	0.3	6.0
Metrosideros/Native shrub (Mauna Loa)	2	0.3	9.0+
Subalpine shrub (Mauna Loa)		0.8-6	0.9-3.3
Subalpine shrub (Mauna Loa)	6	1.6-3.6	2.3-2.5
Subalpine shrub (Manawainui)	10	2.0-6.7	3.0-11.0
Deschampsia/Native shrub (Mauna Loa)	2	7.0	1.7
Metrosideros/Native shrub (Hawaii Volcanoes Lowlands)	4	9.0	0.3
Deschampsia grassland (Kalapawili Exclosure)	13	12-17	--
Heteropogon/Rhynchelytrum (Hawaii Volcanoes Lowlands)	4	13.0	0.3
Metrosideros/Native shrub (Hawaii Volcanoes Lowlands)	4	16.0	0.8
Lowland shrub (Hawaii Volcanoes Lowlands)	4	20.6	0.9
Mixed grass (Hawaii Volcanoes Lowlands)	4	29.2	0.4
Lowland shrub (Hawaii Volcanoes Lowlands)	4	39.6	0.7
Rhynchelytrum grassland (Hawaii Volcanoes Lowlands)	4	36.7	0.2
Rhynchelytrum grassland (Hawaii Volcanoes Lowlands)	4	45.3	0.1
Schizachyrium condensatum (Hawaii Volcanoes Lowlands)	4	64.4	0.3
Rhynchelytrum grassland (Hawaii Volcanoes, Kūkalau'ula)	13	71.0-77.4	0.4-0.5
Melinis/Rhynchelytrum (Hawaii Volcanoes Lowlands)	4	73.7	0.2
Hyparrhenia grassland (Hawaii Volcanoes Lowlands)	4	75.0	1.0
Hyparrhenia grassland (Hawaii Volcanoes, Kaone)	15	77.4-81.6	0.3

*See text for ungulate history.

Table 20. Rain forest intactness as measured by percent cover of alien plants and native/alien species ratios in pig-free plots in Hawaii Volcanoes National Park.

Vegetation Type (Locality)	Years w/out Pigs*	Mean % Alien Cover	No. of Native Species (n)	Native/Alien Species Ratio
ʻŌlaʻa (Small Tract)	3	0.0	39 (28)	9.8
Fern Jungle	14	0.8	15 (15)	1.7
Boundary Kīpuka	20?	2.1	33 (24)	11
Nāpau	11	3.0	10 (4)	3.3
Thurston I	4	5.3	17 (7)	1.9
Thurston II	4	22.8**	11 (5)	1.8

*See text for ungulate history.
**Largely meadow ricegrass invasion.

CONCLUSIONS

Feral pigs and goats are major stresses for many Hawaiian ecosystems, and their elimination usually produces beneficial responses, including recovery of native species. However, communities may be considerably simplified through years of ungulate damage, a significant alien component may remain, key alien species may continue to spread, and continued plant management may be required after ungulate removal. Many lowland grasslands subjected to over a century of stress from goats have responded to goat removal with an increase in percent cover of flammable alien grasses, including bush beardgrass (*Schizachyrium condensatum*), broomsedge, molasses grass, and *Hyparrhenia rufa*. Little hope exists that native plants will eventually dominate without more intensive management or biological control. Reintroduction of native species to these areas is a costly but perhaps necessary method to restore them in the future. Lowland shrublands seem less damaged by long-term goat use and retain reasonably low percentages of alien cover; however, native/alien plant species ratios may remain low (<1.0) even after up to 13 to 15 years of recovery from goat damage. Williams (1990) similarly found rapid recovery of native woody species within a few years after removal of goats in the eastern lowlands of Hawaii Volcanoes National Park.

In subalpine grassland and shrubland where native plant propagules are present and where the still-dominant native bunch grass *Deschampsia* can overtop and crowd out alien grasses and herbs, recovery progresses slowly, but stasis may prevail for the foreseeable future. Velvet grass is perhaps of primary concern after ungulate removal, although stabilization or even decline of this species is known to occur in some areas. Blackberry may also increase in some areas. Subalpine shrublands seem less affected by feral goats in some areas than in others, perhaps because goat densities vary greatly and because, where human access is good, hunting can provide effective protection for an area from this social ungulate. In montane woodlands, velvet grass and some other alien grasses (*e.g.*, meadow ricegrass) may continue to expand after pig or goal removal.

Responses of vegetation to release from goat damage can be rapid in terms of increased numbers of woody plant seedlings and vegetative reproduction (*e.g.*, koa, *Vaccinium*). However, exposed soil, continuing erosion, and alien grasses will undoubtedly remain in some subalpine and montane parkland areas long after goat removal. Restoring native components that are missing after long abuse may also be a problem.

In Hawaiian rain forests, responses of vegetation to pig removal are varied according to remoteness and presence of especially disruptive alien plants. The most intact (lowest alien cover and numbers of alien plants) rain forests are those rarely or never invaded by pigs and people. Lack of such disturbance may result from remoteness, rugged topography, restricted legal access by humans, or difficulty of penetration into the area. The small section of ʻŌlaʻa Tract and the East Rift zone in Hawaii Volcanoes National Park, together with Kīpahulu Valley in Haleakala National Park, were the most species-rich and least invaded of the rain forest areas considered in this paper. However, some aggressive alien plants remained and/or increased in these and other areas in the absence of feral pigs, over time intervals of 4 to 10 years.

Rain forest areas most accessible to humans and aliens in this study have been much disturbed by feral pigs and much invaded by alien plants. Recovery after four to six years without pigs is not marked. Aggressive alien plants can increase in such areas, especially near ranches or other weedy areas, and adequate forest recovery where weeds have become dominant will probably require active, enduring, and well-supported alien plant management programs.

Further studies of population structures, successional patterns, and competitive advantages of aggressive alien plants in different areas and different climates may eventually help prediction of long-term ecosystem recovery and establishment of effective management strategies for different areas. However, a number of variables influence the recovery of vegetation after feral ungulate removal from an area. These include: the duration, intensity, and frequency of previous ungulate disturbance; the intervals between disturbance events; the vulnerability and palatability of key native plant species in the community; the reproductive mode and dispersability of native plants in the area; the size of the affected area and the proximity to other similar areas; the sources and types of alien plant propagules available; the soil type and condition prior to and after ungulate disturbance; climatic cycles favorable and unfavorable to temporary alien and native plant establishment; ancillary disturbance; and alien and native plant propagule distribution by humans, birds, and other animals.

The variables are too numerous and too uncontrollable to allow definitive cause and effect statements about responses of vegetation due entirely to feral ungulate removal. They also defy predictive approaches, especially with currently limited databases. Not the least of the problems is lack of knowledge about what plant communities looked like in detail before a century or more of foraging by goats, sheep *(Ovis aries)*, and cattle (*Bos taurus*), confounded with more recent impacts caused by pigs, horses (*Equus caballus*), deer (*Odocoileus hemionus* and *Axis axis*), and other ungulates. We conclude that different areas are unique and will continue to require different monitoring and management strategies for alien plants long after ungulate removal.

RECOMMENDATIONS

Increasing our knowledge about the recovery and management potential of Hawaiian ecosystems after feral ungulate removal should receive more emphasis. We need to avoid costly efforts to remove ungulates from heavily invaded areas subject to further deterioration not readily amenable to management. In all vegetation types, losses of native components over time and gain of aliens through the usual colonization/extinction equilibria and through successional stages should be better documented and increasingly considered in management decisions. Size and proximity of vegetation types and dispersal and reproductive capabilities of alien and native plants are also relevant.

More emphasis should also be placed on determining alien plant distribution over wide areas within key ecosystems. We need to better understand the extensive dynamics of aliens for long periods. Exclosures and vegetation plots confined to small areas should be continued but enhanced with more extensive frequency of occurrence and percent cover data.

Continued research and management efforts on monitoring responses of vegetation to release from ungulate damage is needed. More areas should be studied to determine how released areas function with respect to alien plant invasion over time. Generalizations from a few areas are difficult because communities are variable and the complexity of determining factors are numerous and unpredictable. The fact that generalizations may not be possible should not preclude systematic data collection and long-term "management experiments."

Potentially generalizable measures such as "intactness" and "manageability" of invaded ecosystems should be considered further. Although they are no substitute for management judgments, such abstractions can serve as guides and be grasped by administrators. Increased predictability is needed for vegetation responses in relation to proximity of aggressive alien plants, sizes of areas, and other preserve design considerations (Franklin 1985). Cooperative and interactive efforts by natural resource managers and researchers, particularly in obtaining and interpreting monitoring data, will be necessary for a long period to understand the ecology of alien plants and their invasions subsequent to ungulate removal. The elimination of ungulates is but one of a number of important steps in active ecosystem restoration in Hawai'i.

ACKNOWLEDGMENTS

We thank the people who allowed us to use their data sets in this summary chapter, especially Jim Jacobi, Tim Ohashi, Larry Katahira, Dieter Mueller-Dombois, and Cliff Smith. Paul Higashino was the primary botanist in many of the rain forest studies, and Dina Kageler also contributed. We also thank Danielle Stone for typing and proofreading the manuscript.

Literature Cited

Anderson, S.J., C.P. Stone, and P.K. Higashino. [this volume] Distribution and spread of alien plants in Kīpahulu Valley, Haleakala National Park, above 2,300 ft elevation.

Baker, J.K., and D.W. Reeser. 1972. *Goat management problems in Hawaii Volcanoes National Park: a history, analysis, and management plan.* U.S. Dep. Interior, Natl. Park Serv. Nat. Resour. Rep. 2.

Diong, C.H. 1983. Population ecology and management of the feral pig (*Sus scrofa* L.) in Kipahulu Valley, Maui. Ph.D. Diss., Univ. Hawaii, Honolulu.

Franklin, J.F. 1985. Design of natural area preserves in Hawai'i. In *Hawai'i's terrestrial ecosystems: preservation and management*, ed. C.P. Stone and J.M. Scott, 459-474. Univ. Hawaii Coop. Natl. Park Resour. Stud. Unit. Honolulu: Univ. Hawaii Pr.

Giffin, J.G. 1978. Ecology of the feral pig on the island of Hawaii. Final Rep., Hawaii Dep. Land Nat. Resour., Pittman-Robertson Proj. W-15-3. Stud. 11. Mimeo.

Higashino, P.K., and C.P. Stone. 1982. The fern jungle exclosure in Hawaii Volcanoes National Park: 13 years without feral pigs in a rain forest. [Abstract] *Proc. 4th Conf. Nat. Sci., Hawaii Volcanoes Natl. Park,* 86. Honolulu: Univ. Hawaii Coop. Natl. Park Resour. Stud. Unit.

Jacobi, J.D. 1976. The influence of feral pigs on a native alpine grassland in Haleakala National Park. *Proc. 1st Conf. Nat. Sci., Hawaii Volcanoes Natl. Park*, 107-112. Honolulu: Univ. Hawaii Coop. Natl. Park Resour. Stud. Unit.

Jacobi, J.D. 1981. *Vegetation changes in a subalpine grassland in Hawaii following disturbance by feral pigs.* Tech. Rep. 41, Univ. Hawaii Coop. Natl. Park Resour. Stud. Unit. Honolulu.

Katahira, L. 1980. The effects of feral pigs on a montane rain forest in Hawaii Volcanoes National Park. *Proc. 3rd Conf. Nat. Sci., Hawaii Volcanoes Natl. Park*, 173-178. Honolulu: Univ. Hawaii Coop. Natl. Park Resour. Stud. Unit.

LaRosa, A.M. [this volume] The status of banana poka in Hawai'i.

Loope, L.L., and P.G. Scowcroft. 1985. Vegetation response within exclosures in Hawai'i: a review. In *Hawai'i's terrestrial ecosystems: preservation and management*, ed. C.P. Stone and J.M. Scott, 377-402. Univ. Hawaii Coop. Natl. Park Resour. Stud. Unit. Honolulu: Univ. Hawaii Pr.

Mueller-Dombois, D. 1981. Vegetation dynamics in a coastal grassland of Hawaii. *Vegetatio* 46:131-140.

Mueller-Dombois, D., and H. Ellenberg. 1974. *Aims and methods of vegetation ecology.* New York: John Wiley & Sons.

Mueller-Dombois, D., and G. Spatz. 1975. The influence of feral goats on the lowland vegetation in Hawaii Volcanoes National Park. *Phytocoenologia* 3(1):1-29.

Mueller-Dombois, D., K.W. Bridges, and H.L. Carson, eds. 1981. *Island ecosystems: biological organization in selected Hawaiian communities*. Stroudsburg, Penn.: Hutchinson Ross Pub. Co.

St. John, H. 1972. *Canavalia kauensis* (Leguminosae), a new species from the island of Hawaii. Hawn. Plant Stud. 39. *Pac. Sci.* 26(4):409-414.

Smith, C.W. 1985. Impact of alien plants on Hawaii's native biota. In *Hawai'i's terrestrial ecosystems: preservaton and management*, ed. C.P. Stone and J.M. Scott, 180-250. Univ. Hawaii Coop. Natl. Park Resour. Stud. Unit. Honolulu: Univ. Hawaii Pr.

Spatz, G., and D. Mueller-Dombois. 1973. The influence of feral goats on koa tree reproduction in Hawaii Volcanoes National Park. *Ecology* 54(4):870-876.

Stone, C.P. 1985. Alien animals in Hawai'i's native ecosystems: toward controlling the adverse effects of introduced vertebrates. In *Hawai'i's terrestrial ecosystems: preservaton and management*, ed. C.P. Stone and J.M. Scott, 251-297. Univ. Hawaii Coop. Natl. Park Resour. Stud. Unit. Honolulu: Univ. Hawaii Pr.

Stone, C.P., and J.O. Keith. 1986. Control of feral ungulates and small mammals in Hawaii's national parks: research and management strategies. In *Control of mammal pests*, ed. C.J. Richards and T.Y. Ku, 277-287. Trop. Pest Manage. 32, Suppl. 1. London: Taylor & Francis Ltd.

Tunison, J.T., L.W. Cuddihy, R. Loh-McMahon, and D. Kageler. In press. *Early succession in pig-disturbed mountain parkland, Hawaii Volcanoes National Park.* Tech. Rep., Univ. Hawaii Coop. Natl. Park Resour. Stud. Unit. Honolulu.

Vitousek, P.M., L.R. Walker, L.D. Whiteaker, D. Mueller-Dombois, and P.A. Matson. 1987. Biological invasion by *Myrica faya* alters ecosystem development in Hawaii. *Science* 238:802-804.

Vtorov, I. In press. Restoration of soil microarthropod population after feral pig removal in a Hawaiian rain forest ecosystem. *Pac. Sci.*

Warshauer, F.R., J.D. Jacobi, A.M. LaRosa, J.M. Scott, and C.W. Smith. 1983. *The distribution, impact and potential management of the introduced vine* Passiflora mollissima *(Passifloraceae) in Hawai'i.* Tech. Rep. 48, Univ. Hawaii Coop. Natl. Park Resour. Stud. Unit. Honolulu.

Williams, J. 1990. The coastal woodland of Hawaii Volcanoes National Park: vegetation recovery in a stressed ecosystem. Tech. Rept. 72, Univ. Hawaii Coop. Natl. Park Resour. Stud. Unit. Honolulu.

Yocum, C.F. 1967. Ecology of feral goats in Haleakala National Park, Maui, Hawaii. *Amer. Midland Nat.* 77(2):418-451.

CHOOSING WHICH PLANTS TO INTRODUCE

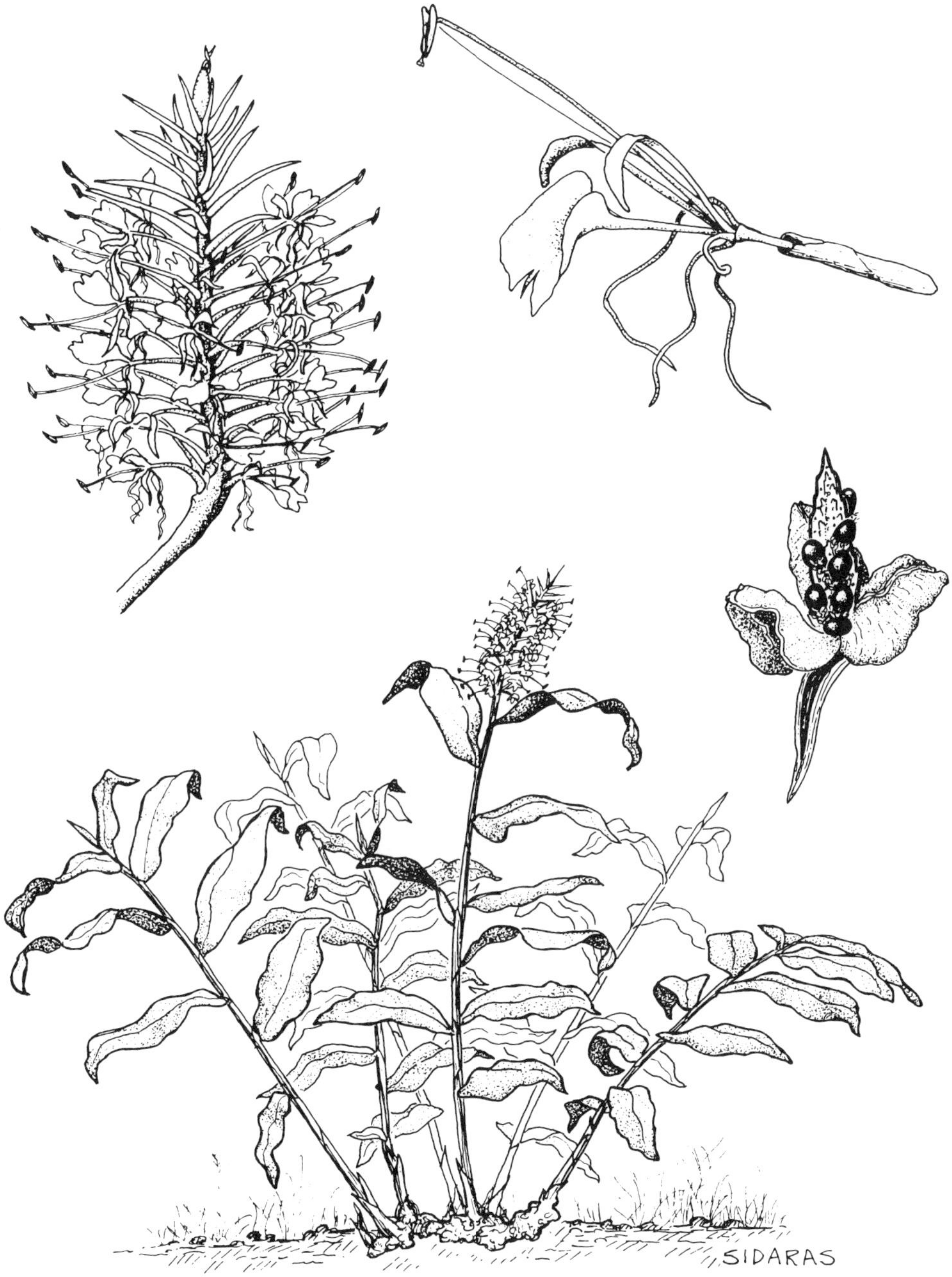

PLANT IMPORTATION RULES AND REGULATIONS

Stan Higa

ABSTRACT

The Plant Quarantine Branch of the Hawaii Department of Agriculture and the Plant Protection and Quarantine Service of the U.S. Department of Agriculture are complementary State and Federal agencies that deal with plants entering the State from foreign countries and the rest of the U.S. Certain plant groups from domestic sources are regulated under Hawaii Department of Agriculture Plant Import Rules, while foreign imports are regulated by the Code of Federal Regulations. People are free to transport most plants among islands with appropriate importation and certification procedures. Control of plants brought into the State depends upon voluntary declaration and is hampered by the inaccessibility of first class mail to inspection. Several possible improvements in alien plant importation monitoring are discussed.

INTRODUCTION

The Plant Quarantine Branch of the Hawaii Department of Agriculture and the Plant Protection and Quarantine Service of the U.S. Department of Agriculture are complementary State and Federal agencies that perform statewide inspections at airports, seaports, and post offices to regulate the importation of plants and plant products entering the State. Federal Plant Protection and Quarantine Service inspectors maintain jurisdiction over the importation of foreign commodities, while State Plant Quarantine Branch inspectors are responsible for monitoring articles arriving from domestic sources.

PROCEDURES FOR INTRODUCING PLANTS

Under State and Federal rules and regulations, many plant species are allowed entry by permit and/or inspection. Some are prohibited because they are considered noxious, while others are restricted, requiring (in addition to a permit) a period of isolated quarantine before being allowed into the State.

The importation of plants from foreign countries into Hawai'i is regulated under the Code of Federal Regulations (U.S. Code of Federal Regulations 1980). Importations generally require an "Application for Permit to Import Plants or Plant Products" (Plant Protection and Quarantine Service Form 587) approved in advance by the Plant Protection and Quarantine Service office in Honolulu. All plants imported from foreign countries require U.S. Customs declaration and must be inspected by Plant Protection and Quarantine Service inspectors upon arrival in Honolulu. Plants allowed entry by permit and inspection are subject to post-entry inspections at the premises of the importer for a period of up to one year. State Plant Quarantine inspectors, in cooperative agreement with the Plant Protection and Quarantine Service, conduct all post-entry inspections.

Plants sent or imported by cargo from foreign and domestic origins must be manifested as live plants or plant material for inspection. Not doing so constitutes false manifestation and subjects the importer to prosecution and penalty (Hawaii Department of Agriculture 1976a; U.S. Code of Federal Regulations 1980).

Importation of plants from domestic sources is generally allowed except for those plant groups such as bromeliads, grasses, orchids, coffee, and others related to crops representing viable industries in Hawai'i. These groups are regulated under the Hawaii Department of Agriculture Plant Import Rules (Hawaii Department of Agriculture 1981b). However, all plants, including cut flowers and foliage, are required to be declared or manifested for inspection upon arrival.

Plants and plant products sent or received by mail must be labeled as plant material on the shipping container for referral to agriculture inspectors for inspection. Plant material not properly designated is not made available for inspection and, as such, constitutes a violation of plant quarantine law.

NOXIOUS WEED AND SEED REGULATION

Although undesirable plants (weeds) occasionally enter Hawai'i inadvertently with nursery stock, as seed contaminants in livestock and poultry feed, or in agricultural or vegetable seed lots, the deliberate introduction of weed species is regulated under State and Federal laws and regulations (Hawaii Department of Agriculture 1976b; U.S. Code of Federal Regulations 1984; U.S. Code of Federal Regulations 1983). Under the Federal Seed Act, for example, the Hawaii Department of Agriculture Seed Laboratory cooperates with the U.S. Department of Agriculture to perform inspections of all bulk shipments of agricultural seeds into Hawai'i for noxious weed seeds. Under the Hawaii Seed Law (Hawaii Department of Agriculture 1976a) and Seed Rules (Hawaii Department of Agriculture 1981a), the Hawaii Department of Agriculture is also charged with the responsibility for testing agricultural and vegetable seeds entering from U.S. Mainland states for retail sales. Testing is conducted for seed viability and for noxious weed content. Small amounts of seeds brought in by individuals for personal or experimental use are generally not required to be inspected.

INTERISLAND MOVEMENT OF PLANTS

Although people are free to transport most plants among islands in the State, all plant materials are required to be inspected and certified (stamped and released) by a Plant Quarantine inspector before they can be taken to another Island. This is necessary because certain plants serve as hosts of insect pests and diseases not found on all islands; therefore, these plants require treatment before shipment. Movements may be prohibited altogether because of the absence of adequate treatment procedures (Hawaii Department of Agriculture 1976b, 1981c).

LOOPHOLES IN THE INSPECTION SYSTEM

Regulating the introduction of plants into the State from foreign and domestic sources depends on declaration statements by passengers and manifestation of cargo by shippers or importers. The system relies almost exclusively on the honesty and integrity of people. Since there is no formal inspection of passengers and personal baggage, individuals may be tempted to circumvent inspection by not declaring or manifesting plant materials and other regulated or restricted items.

Another major loophole in the inspection system lies in the inaccessibility of first class mail to agricultural inspection. The upgrading of all air mail parcels exceeding 13 oz to priority (first class) mail in the last 10 years has created a dilemma for Federal and State agriculture officials throughout the nation. The Privacy Act protects all forms of first class mail from inspection, even if certain parcels are suspected of containing contraband articles. Individuals knowingly take advantage of this loophole to import/export plant materials and products to circumvent inspection requirements.

POSSIBLE SOLUTIONS TO LOOPHOLES

The Plant Quarantine law and Hawaii Department of Agriculture administrative rules are being revised to facilitate enforcement. One change being proposed is for the airlines to be fully responsible for distributing, collecting, and delivering, immediately upon arrival, all passenger declaration forms to the Plant Quarantine inspector(s) on duty at the first port of entry. Presently, the law requires only that the airlines distribute the passenger declaration forms.

The Plant Quarantine Branch is also implementing a citations and summons procedure for prosecuting and fining carriers and individuals who violate the quarantine law and administrative rules. The Branch's Federal counterpart, the Plant Protection and Quarantine Service, has already implemented a civil penalties procedure for assessing on-the-spot fines of violators of their regulations. Since the inception of this procedure, the number of violations has declined.

The Plant Quarantine Branch has proposed a substantial increase in staffing to allow for around-the-clock coverage of Honolulu International Airport and more extensive coverage of nearby military bases. Funding has also been requested to develop resources for educating visitors and local residents on plant importation requirements. Public awareness and education programs entail production of videos for viewing on aircraft, at conventions, theatres, schools, and on TV; printing and disseminating brochures, leaflets, and posters; promoting advertisements through the various news media; and making appearances at schools, community meetings, and radio talk shows.

The U.S. Department of Agriculture and Hawaii Department of Agriculture, through affiliated organizations such as the National Plant Board and the National Association of State Departments of Agriculture, and through the Hawai'i Congressional delegation, are working toward a solution to the first class mail problem. A solution to alien species introductions by mail will not be found easily, as this is a problem of national concern and has been the subject of numerous discussions among state, U.S. Department of Agriculture, and U.S. Postal Service officials over the years. Thus far, the rights of the individual under the Privacy Act have prevailed.

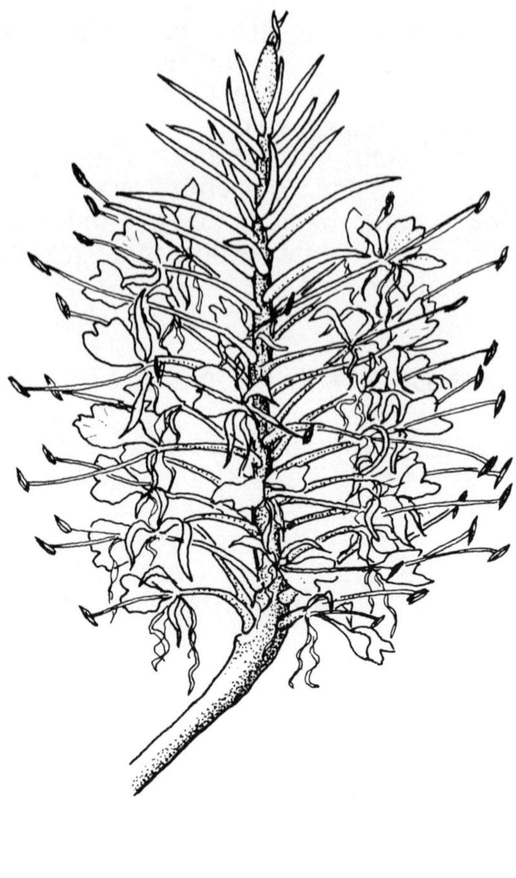

Literature Cited

Hawaii Department of Agriculture. 1976a. *Hawaii revised statutes*. Chapter 150, Plant and animal life, seeds, and soils. Vol. 3, 188-192. Honolulu.

Hawaii Department of Agriculture. 1976b. *Hawaii revised statutes*. Chapter 150A, Plant and non-domestic animal quarantine. Vol. 3, 192-201. Honolulu.

Hawaii Department of Agriculture. 1981a. Chapter 67, Seed rules. In *Administrative rules*. Honolulu.

Hawaii Department of Agriculture. 1981b. Chapter 70, Plant import rules. In *Administrative rules*. Honolulu.

Hawaii Department of Agriculture. 1981c. Chapter 72, Plant intrastate rules. In *Administrative rules*. Honolulu.

U.S. Code of Federal Regulations. 1980. *Nursery stock, plants, roots, bulbs, seeds, and other plant products*. Title 7, Subpart 319.37. Washington, D.C.

U.S. Code of Federal Regulations. 1983. Title 7, 1151. *Federal Seed Act.* Washington, D.C.

U.S. Code of Federal Regulations. 1984. *Noxious weeds*. Title 7, Part 160. Washington, D.C.

ACTIVITIES AND NEEDS OF THE HORTICULTURE INDUSTRY IN RELATION TO ALIEN PLANT PROBLEMS IN HAWAI'I

Rylan S.N. Yee and Wayne C. Gagné

ABSTRACT

The horticulture industry in Hawai'i can be characterized by product types. They are, in order of economic importance, cut flowers, potted foliage plants, and potted ornamental-landscape material. Other products such as plant rentals are also of some importance. Most of the farms involved in growing these products are small, the majority being single-family or small-partnership operations of less than one acre. Although more than 8,000 taxa have been introduced or developed in Hawai'i, only about 15% are commonly cultivated, while the rest, seldom or rarely cultivated, are found in collections. Most plants marketed are those commonly or occasionally cultivated; however, the species composition of this group changes because plant preferences are trendy. Trends are often predicated by new introductions or newly developed cultivars that have been propagated and marketed by one source. The return from investment is often the highest for the first release and decreases almost logarithmically for each following release. The market encourages secretiveness and speed of propagation and culture, and the importation of many different taxa is required to achieve economic success. Of at least 8,000 species or cultivars introduced, about 10% have escaped and naturalized; of these, about one third were intentionally introduced by the industry. As a result, the alien plant problem when viewed from the horticulture industry's viewpoint is insignificant compared with its economic benefit. Plant material may be imported by collectors, which include professionals such as those at botanical gardens, as well as amateurs. Amateur collectors are difficult to regulate because plant material is easily shipped into Hawai'i illegally, especially if it originates on the U.S. Mainland. Many undesired pests are probably introduced in this way. The horticulture industry needs a steady influx of new plant material, some of which is collected from the wild. The industry cannot afford to screen out potential weeds. Regulation can only be minimal, since overregulation will force noncompliance because of economics. Education is necessary to promote understanding and change philosophies and may be the best long-range solution to alien plant problems. In the short term, strict enforcement is required.

INTRODUCTION

Agriculture ranks seventh in sales and income in Hawai'i's economy, behind tourism, national defense, retail and wholesale sales, construction, and manufacturing (Table 1). Within agriculture, the horticulture industry ranks fifth behind sugar, fruit and nuts, pineapple, and cattle (Table 2). Like any business, agriculture is market driven, with industry perceptions and practices reflecting the need to satisfy demand while maximizing profit.

Table 1. Sales and income of major Hawaiian industries.*

Industry	Value ($ in millions)
Tourism	2,875
Retail sales	1,900
National defense	1,850
Wholesale sales	1,600
Manufacturing (including processing of agricultural products)	1,249
Construction	767
Agriculture	559
Fishing	10
Forestry	4

*From Schmitt *et al.* 1984.

While these perceptions and practices are similar in many respects to those outside Hawai'i, the island environment and isolation modify them. Limited land areas, increasing population sizes, small farm sizes, high land values, and high labor costs make agriculture expensive. Isolation increases the cost of marketing outside Hawai'i and, coupled with a lack of resources, requires importation of many raw materials required by industries, thereby increasing cost.

Alien plants are essential to the horticulture industry. They are introduced intentionally for sale and often dictate trends. However, some also arrive unintentionally with essential nursery products vital to the industry or with incoming plant material. Some plants introduced in this way have escaped and have become serious problems in native Hawaiian ecosystems. Plants are not only imported by commercial growers but by botanical gardens and amateur collectors. These collectors are believed to be a major source of alien plant introductions.

Table 2. Value of major Hawaiian agricultural industries.*

Industry	Value ($ in millions)
Sugar	266
Fruits and nuts	113
Pineapple	100
Cattle	45
Horticulture	36
Milk	24
Poultry	21
Vegetables	18
Eggs	14
Hogs and pigs	9
Grains	3
Prawns	1.3
Sheep	0.8

*From U.S. Bureau of the Census 1984b.

HORTICULTURE INDUSTRY COMPONENTS

Three types of products dominate the horticulture nursery industry in Hawai'i: cut flowers, potted foliage plants, and potted ornamental plants and landscape material. Each product type is composed of various crop types, and each has its own requirement for new plant material.

Cut Flowers

Cut and lei flowers are the horticultural industry's largest segment in terms of sales and number of farms (Table 3). Major crops consist of anthuriums (*Anthurium andreanum*), dendrobium orchids (*Dendrobium* spp.), and roses (*Rosa* spp.), which collectively represent about 50% of the total. Most flowers are marketed intrastate, with some exported to Asia and the U.S. Mainland. Most are purchased by commercial users such as florists, lei-makers, and decorators affiliated with hotels, restaurants, and other public areas. Buyers are primarily concerned with availability, cost, and quality; they adapt to what flowers are available, and specific types, species, or cultivars are not important considerations. Buyers use or sell any flower that will maximize profits.

Most growers specialize in a limited number of crops, often no more than a few selected cultivars of a given species that are industry standards. Many are selected for specific characteristics desired by industry. Examples include *Anthurium andreanum* cv. 'Ozaki,' cultivated for its color, high yield, and sucker formation; *A. andreanum* cv. 'Marion Seefurth' for its color and flower form, which facilitates packing;

Dendrobium x *Louis Bleriot* for its availability during the winter months when other flowers are not available; and *D.* x *Jac-Hawaii* for its color and continual flowering. Specialization allows predictability, simplicity of culture, and steady turnover of valuable growing space, which in turn maximizes profits.

Similar crop types with similar cultural requirements allow larger farms that take advantage of economy of scale; less complexity lowers labor costs, further lowering unit costs. Since high volume and quality production are major considerations, cultivars selected for production are often restricted to those that have been carefully studied and proven.

Potted Foliage Plants

Foliage plants for indoor use rank second to cut flowers in sales (Table 3). Most are sold as potted plants for indoor decoration, with the rest grown as nursery stock and for landscaping. Like cut flowers, most foliage plants are sold intrastate, although some are exported to the U.S. Mainland. Specific species and cultivars are not primary considerations to the majority of consumers; availability and ease of cultivation in indoor situations are more important. Only a small range of plants such as aroids, palms, and ferns are commonly grown. However, unlike cut flowers, there is a demand for different species and cultivars in these groups to satisfy a wide range of situations and consumers, including many nonprofessional buyers. Single farms may cultivate plants ranging from palms and aroids such as *Dieffenbachia* for floor displays and indoor planters, *Philodendron* for totems and hanging baskets, and peperomias (*Peperomia* spp.) and ferns for table decorations. Like cut flowers, foliage plant cultivars are selected for ease of production and rapid turnover. Still, a broad market requires a wide range of crops to satisfy demand; thus, numbers of introductions of new species or cultivars of foliage plants are greater than for cut flowers.

Potted Ornamental and Landscape Plants

Businesses that sell ornamental potted plants and trees rank third in sales (Table 3). The market for these plants is broadest of all categories, ranging from designers and landscapers to homeowners and collectors, and from large estates and public plantings to small homes. The ornamental and landscape plant market also reflects changing tastes and preferences. Almost all plants in this group are sold intrastate. Unlike the market for cut flowers and potted foliage plants, there are definite demands for specific kinds of plants (often for specific cultivars), especially by hobbyists and collectors. This is particularly true for orchids, succulents, and bromeliads, in which there is a continuing interest in Hawai'i.

TRENDS AND ORIGINS OF HORTICULTURAL PLANTS

Interest in specific plants or groups of plants is trendy and often cyclic. The history of horticulture is one of waxing and waning interest in various groups of plants. In Hawai'i, interest in croton (*Codiaeum*

Table 3. Value of various components of ornamental plant industries in Hawai'i.*

Industry	Value ($ in millions)	Number of Farms
Cut flowers and lei flowers	17.098	723
Anthuriums	7.351	205
Orchids	1.667	138
Roses	1.556	8
Chrysanthemums	0.710	415
Proteas	0.470	46
Carnations	0.219	12
Gingers	0.267	60
Birds of paradise	0.186	50
Other lei flowers	3.880	123
Other	1.052	75
Foliage plants	16.820	150
Indoor	14.050	85
Stock	1.525	20
Landscape	1.245	45
Potted ornamental plants	5.544	291
Potted orchids	2.69	155
Other flowers	1.939	94
Ornamental trees	0.915	42
Others (bedding, rental, etc.)	2.311	102
TOTAL	41.773	1,266

*From Hawaii Agriculture Reporting System 1985.

variegatum var. *pictum*) and ti (*Cordyline fruticosa*) started after World War II and continued into the 1950s, followed by bougainvillea (*Bougainvillea* spp.) in the 1960s, with ixora (*Ixora* spp.) and a resurgence of ti in the 1970s and heliconia (*Heliconia* spp.) in the 1980s (K. Wooliams, pers. comm.). As with any broad consumer market, there is a constant demand for new introductions that may or may not dictate new trends. These new introductions are acquired in a number of ways. They may be selected from mutations or sports discovered in existing stock, purposely developed from systematic breeding programs, or collected in the wild.

Many new introductions come from serendipitous discoveries in existing stock such as *Dieffenbachia amoena* cv. 'Mary Ann' from *D. amoena* cv. 'compacta' and the "double" *Bougainvillea* cv. 'Bridal Bouquet,' cv.

'Purple Prince' from single cultivars. Others, such as orchid hybrids and roses, are developed from breeding programs. Marketing such discoveries is usually financially safe and in many cases cheap, as cultural requirements are known and a ready market exists. Most "new" introductions are obtained through these methods.

Plants collected in the wild are the most sought after. New material from existing stock and breeding programs are permutations of a similar theme, but new wild-collected material has the excitement of being completely different. Growers capitalize on this excitement, which is the quickest way of turning large profits. The combination of publicity and product scarcity produce the largest return per unit cost and may start a trend that may provide an edge over the competition, a goal that many growers seek to achieve. Heliconias, gingers, and orchids are examples of ornamentals subject to such marketing.

ECONOMICS

Economic success for most ornamental farms is based on a wide range of currently popular species and cultivars, the composition of which varies as new plants are introduced and old ones are phased out. Timely introduction of newly acquired plant materials with quick propagation and release is a necessity. As a result, a large number of plants must be obtained for every one that is successfully marketed. To maximize profits, new plant materials are often imported and propagated in secrecy to avoid competition, as the amount of return per unit is highest at first release and diminishes rapidly thereafter. Many ornamental farms thus have a wide range of crop types, requiring different cultural conditions that are in some cases unknown. Operations of this kind are the most complex and are very capital- and labor-intensive; costs are further exacerbated by high land values and lack of local raw materials. These farms are also expensive to operate, and because of limited market size, many are only marginally profitable.

Most Hawaiian farms are small in size (Tables 4, 5) and generate little revenue (Table 6). Farms growing potted ornamentals are usually the smallest and are either operated by families or are one- to two-owner cooperatives (inferred from Tables 3, 4, and 7, assuming a narrow range of unit prices). Because of limited market size, high labor costs, small amounts of cash flow, and small staffs, these farms have limited resources to spend for non-production activities. This type of farm has the largest interest in the introduction of alien plants.

OTHER IMPORTATION SOURCES

Our preliminary findings indicate that professional and nonprofessional collectors may be responsible for more alien plant introductions than commercial enterprises. Preliminary examination of checklists and herbarium specimens at the B.P. Bishop Museum in Honolulu indicates large-scale importation of alien plants by Hawaiian botanical gardens. Of

Table 4. Sizes of farms by farm type.*

Type	Greenhouse**	Saran**	Natural***	Open Field***
Anthuriums	--	16,446	81	--
Dendrobiums	400	1,981	--	2
Other orchids	436	420	3	66
Potted ornamentals (including foliage)	528	5,295	1	149
Proteas	--	--	--	119
Others	1,050	2,675	7	623
TOTALS	2,414	26,817	92	959

*From Schmitt *et al.* 1984.
**Area is in 10^3 square feet.
***Area is in acres.

Table 5. Number of farms by size.*

Size in acres	Number of farms
1 - 9	2,743
10 - 49	1,261
50 - 179	320
180 - 499	126
500 - 999	33
1,000 - 1,999	33
2,000 and over	79
Total farms	4,595

*From Schmitt *et al.* 1984. Data summarized from U.S. Bureau of the Census, *1982 census of agriculture.* Geogr. Area Ser., Hawaii, state and co. data, AC82-A-11, Sept. 1982, chap. 1, tables 1, 4.

Table 6. All horticultural farms in Hawai'i by value of sales.*

Value of Sales	Number of Farms
250,000 -- +	161
100,000 -- 249,999	192
40,000 -- 99,999	343
20,000 -- 39,999	469
10,000 -- 19,000	610
5,000 -- 9,999	653
0 -- 4,999	2,160
Total number of farms	4,588

*U.S. Bureau of the Census 1984b.

Table 7. Number of farms and farm size.*

Farm Type	Number of Farms	Size (Acres)
Individual - family	3,852	481,212
Partnership	351	130,761
Family-held corporation	272	507,280
Corporations other than family	80	492,941
Other -- corporations, estates, trusts, institutions	40	345,307
Total	4,595	1,957,501

*Derived from Schmitt *et al.* 1984; data from U.S. Bureau of the Census, *1982 census of agriculture*. Geogr. Area Ser., Hawaii, state and co. data, AC82-A-11, Sept. 1982, chap. 1, tables 5, 46.

more than 8,000 names of introduced plant species or cultivars found in Hawai'i, about 15% are commonly or occasionally cultivated (H. Clay, pers. comm.). The remainder are rarely cultivated or are single-specimen plants in various botanical gardens, parks, universities, or other collections. Many botanical gardens have exchange programs with other botanical gardens to increase their collections. Many of the larger botanical gardens have propagation and distribution programs for the most promising plants. This is also one mechanism by which commercial nurseries obtain new plants.

Amateur collectors appear to be another major source of alien plant introductions. Although it is more difficult to determine conclusively, preliminary examinations of herbarium specimens collected as part of the Bishop Museum's "In Gardens of Hawaii II" project and surveys of botanical gardens, private gardens, and popular literature have indicated that a significant number of alien plants are introduced by amateur collectors. This is especially true for theme collections such as bonsai, orchids, succulents, and bromeliads. Both amateur collectors and botanical gardens have considerable motivation and, in most cases, the resources to import plant material, perhaps much more so than commercial enterprises, where profits are the overriding consideration and importation of new cultivars of known crop types is much more economical.

CONTRIBUTIONS OF THE HORTICULTURE INDUSTRY TO THE ALIEN PLANT PROBLEM

About 820 species of persisting and naturalized plants can be found in Hawai'i today (W.L. Wagner, pers. comm.) -- about 10% of the 8,000 species found in the "In Gardens of Hawaii II" project checklist. Many cultivars that can be found in Hawai'i are not included on the list of 8,000, nor are plants that did not survive to be included on checklists, nor those in private or public herbariums. Of the naturalized plants, only one-third of the 820 (3% of the 8,000 introductions) were intentionally introduced (W.L. Wagner, pers. comm.). The industry considers these consequences very small when compared with the economic benefits of introducing new plants.

In addition to intentionally introduced alien plants escaping into and persisting in the native environment, importation of related materials for the horticultural industry also may result in accidental introductions. Non-growing plant materials that are imported in large quantities, such as Christmas trees, cut flowers, and greens, may be hosts for insect pests and plant pathogens as well as alien plant propagules. Soil additives and other non-sterile organic materials are substrates for fungi and pathogenic microorganisms as well as another vehicle for the entry of alien plants.

NEEDS OF THE HORTICULTURAL INDUSTRY

The horticultural industry believes that it is necessary to continually import new species and cultivars to remain economically competitive. Although most new plants are cultivars of existing species, some importation of new alien plants of sufficient diversity is also considered necessary to maintain a successful program. Most of the members of this industry have limited incomes and small staffs and therefore have little desire to fund research and development programs related to importation impacts. Reluctance is supported by the perceptions that the number of intentionally introduced alien plants that have become established in Hawai'i is very small and that the economic benefits far exceed the problems.

POSSIBLE SOLUTIONS

The need to control indiscriminate introductions of alien plants and to produce enforceable regulations is apparent. The horticultural industry's need to import plants and raw materials vital to success is also apparent. The industry must recognize the consequences of uncontrolled importation of alien plants, and the following are suggestions to address these needs:

1. A thorough assessment of the present impact of alien plants on the environment, and the role the horticultural industry has had in inducing these changes, should be made;

2. An assessment of the threats of various alien plant groups to the native biota should be made, and regulations and mechanisms should be developed for prohibiting entry of these high-risk groups;

3. Strict enforcement of existing and formulated regulations is needed;

4. Educational programs explaining findings and regulations should be made a part of formal horticultural curricula, and participation in these programs should be a requirement for obtaining an import permit. Hawaiian natural history should be a part of school curricula to foster an appreciation of Hawaiian ecosystems.

Horticultural Industry's Contribution to Impacts

It is important to know what plants have had significant impacts on Hawai'i's environment, how they arrived, and for what reasons. This information is important for formulating necessary regulations, as well as for modifying the industry's role if necessary. The industry needs to understand both short- and long-range consequences of irresponsible alien plant importation.

Plants arriving through accidental introductions from horticulturally related goods must also be determined. Non-botanical material such as insects and diseases that arrive in conjunction with importation of alien plants or non-growing plant material such as Christmas trees and imported flowers should also be included. Only recently has the magnitude of this problem been appreciated. Many details need to be examined, and a detailed compilation as well as a summary of assessments need to be provided.

Potential Threats of Various Alien Plant Groups

A thorough assessment of threats posed by various plant groups to different Hawaiian environments is critically needed. The analysis by Smith (1985) is a good start for plants already present in Hawai'i. Assessment of threats posed by proposed introductions must be performed in a timely manner; it will not work if nurseries have to wait for results, as they need to have plants when markets are receptive. Information is vital for evaluating potentially undesirable plants and for formulating regulations for their control. Risk assessments will be difficult and expensive, and the horticultural industry alone cannot afford them. The most efficient solution may be funding by government, by pooled resources paid in the form of importation fees, or a combination of both.

A database on the present status of each cultivated taxon needs to be established. This is being accomplished in part by the B.P. Bishop Museum in the "In Gardens of Hawai'i II" project in a collection of cultivated plant specimens currently being made. The project will also make an assessment of each plant's abundance and distribution. However, the database will need periodic updating if it is to remain useful.

Assessment of risk should take place in the plant's original habitat to eliminate as many obvious undesirable plants as possible. Observations of plants in various botanical gardens can provide some information on potential naturalization, and further screening could be performed at botanical gardens located near native habitats. Using botanical gardens has significant advantages: 1) Plants can be cultivated under controlled conditions simulating horticultural environments; 2) Most botanical gardens have skilled horticultural staffs that can evaluate reproductive biology under cultivated conditions, and 3) most have extensive collections, providing a good perspective on various plant groups.

An existing *ad hoc* organization of Hawaiian botanical gardens could be used for screening. Collectively, the members represent a range of horticultural environments and have some expertise in evaluating plants. Commercial importers may accept regulation more willingly under these conditions. Botanical gardens could serve as a conduit for importation of all alien plants. However, this function should be made separate from the rest of their operations. Separate guidelines should be formulated for plant introductions initiated by industry importers that will remove the profit motive, minimize possible abuse, and prevent any shortcuts. Confidentiality between importer and competitors must be maintained, and evaluations must be made in a timely manner. Distribution should remain with the initiating agency or grower. The botanical garden should also be responsible for control of plants, and in return, gardens should be funded for this service.

A final level of assessment can be at the nursery. Importing nurseries should evaluate potential plant growth problems in addition to insect pests and diseases. Most nurseries perform these evaluations as part of their planning programs. Hopefully, by the time any plant reaches this level of evaluation, the probability of it being undesirable will not be great, the number rejected will be small, and importers will be more willing to accept controls. In general, a multi-level program would distribute the cost of evaluation among all who have interests in alien plant importation problems. It would also minimize the cost to the horticultural industry, which should encourage compliance.

Strict Enforcement

One of the major problems of enforcement is the perception that, unlike animals, plants are non-mobile, take time to spread, and are therefore easily controlled. Present plant importation regulations reflect this philosophy, allowing all plants to be imported, with only undesirables prohibited (Higa, this volume). Strict enforcement of import regulations is one of the most effective methods of regulating alien plant importation. Inspection at ports of departure is desirable but nearly impossible because of cost and time. Routine airport and harbor inspections of all imported materials may be more feasible and would help control much noncommercial material from entering. While it may also be too expensive to inspect all incoming material, spot checks and penalties must be more frequent and consistent than at present.

Importation through the mail is more difficult to enforce. To encourage compliance with mailing import regulations, import loopholes must be closed (Higa, this volume). Inspections of wholesale and retail nurseries coupled with economic penalties such as confiscation and destruction of illegal plants seem the most efficient deterrents. Many nurseries are currently inspected for certification for export, and this mechanism could be expanded to include imports. Non-commercial sources should also be inspected. The removal of any significant economic benefit should illegally imported plants be sold, and severe penalties such as confiscation even if imported for personal use regardless of economic value, would encourage compliance.

Regulations need to be resonable. It is very difficult to control importation even under the best of conditions. Unreasonable regulations force non-compliance, particularly smuggling, and will amplify the problem. Reasonable regulations will minimize secretiveness and facilitate control.

Education Programs

Educational programs may be the only way of ultimately resolving the alien plant importation problem in Hawai'i. The public perception of the values of Hawai'i's natural environments must change. As long as present opinions remain, it will be very difficult to effectively regulate alien plant introductions. Education should play a key role in providing understanding of the need for controlling introductions of alien plants, and it is probably the most effective long-term strategy in encouraging compliance.

Horticulturists rarely appreciate the magnitude of the environmental impact of their industry. It is difficult to reconcile profits with conservation of the native biota, as they are at times mutually exclusive. Because of marginal profitability, there is a strong reluctance to add to farm operations the additional burden of the cost of environmental regulations. Lack of appreciation of the problems is often reflected by disregard for conservation and, in many instances, for regulations. Programs that provide information and encourage understanding might alleviate this problem. Courses in ecology and conservation should be made part of formal agricultural curricula and short courses for non-students. Courses should also be made part of the import permitting process.

Society will have to bear the economic cost for controls. Willingness will be a function of appreciation of all facets of the alien plant problem, including an appreciation for Hawai'i's natural environment. This should start with children and continue throughout their adult lives. A curriculum needs to be developed and adopted for all of Hawai'i's schools, starting in lower grade levels and continuing through the entire educational system. Adult education materials should also be produced. Understanding of Hawai'i's environmental problems could encourage creative solutions.

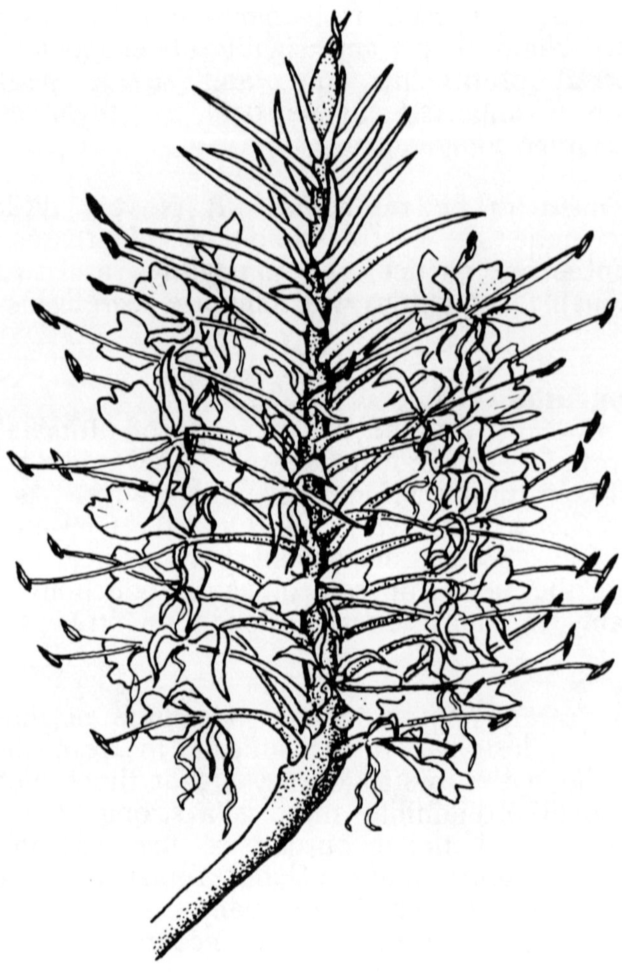

Literature Cited

Hawaii Agricultural Reporting System. 1985. *Statistics of Hawaiian agriculture 1984*. Honolulu: Dep. Agric.

Higa, S.J. [this volume] Plant importation rules and regulations.

Schmitt, R.C., S. Nishi, B. Stanfield, and K. Yamashita. 1984. *The state of Hawai'i data book: 1985 statistical abstract*. Honolulu: Dep. Plan. Econ. Devel.

Smith, C.W. 1985. Impact of alien plants on Hawai'i's native biota. In *Hawai'i's terrestrial ecosystems: preservation and management*, ed. C.P. Stone and J.M. Scott, 180-250. Univ. Hawaii Coop. Natl. Park Resour. Stud. Unit. Honolulu: Univ. Hawaii Pr.

U.S. Bureau of the Census. 1984a. *1982 census of agriculture, preliminary report, Hawaii*. January. AC82-A-15-000(P)-0009(P). Washington D.C.: U.S. Bur. Census.

U.S. Bureau of the Census. 1984b. *1982 census of agriculture, Hawaii*. September. AC82-A-11. Washington D.C.: U.S. Bur. Census.

PLANT INTRODUCTION NEEDS OF THE HAWAIIAN SUGAR INDUSTRY

Robert V. Osgood and Robert D. Wiemer

ABSTRACT

The Hawaiian sugar industry has introduced plants for three major purposes: 1) expansion of the genetic base from which new sugar cane (*Saccharum officinarum*) cultivars are developed, 2) protection of watersheds from erosion, and 3) development of new crops to supplement sugar plantation income. New, higher-yielding and disease-resistant cultivars of sugar cane were initially developed as a result of the introduction of closely related wild species of *Saccharum*. More recently, related genera have also been introduced to increase genetic diversity. It was determined early in the 1900s that, for several reasons, native plant species would not readily cover terrain devastated by the indiscriminate grazing of cattle, goats and sheep (*Bos taurus, Capra hircus, Ovis aries*). A large plant introduction program, initiated by the Experiment Station of the Association under the direction of H.L. Lyon, was aimed at the reforestation of the watersheds. The introduction of plants has also followed from programs to develop new agricultural industries in the State. These crop introductions are especially important today, with the continuing decline in sugar cane acreage and the need to supplement the income generated by the sugar industry.

INTRODUCTION

This paper addresses the historical and current plant introduction needs of the sugar industry rather than agriculture in general. We have identified three areas that have required the introduction of alien plants: 1) increasing the genetic diversity of sugar cane (*Saccharum officinarum*), 2) protecting watersheds, and 3) developing new crops.

INCREASING GENETIC DIVERSITY OF SUGARCANE

The earliest introductions of sugar cane to Hawai'i were made by the Polynesians. Captain James Cook recorded sugar cane on the island of Kaua'i in 1778 (Cook 1785):

> We saw no wood, but what was up in the interior part of the island, except a few trees about the villages; near which, also we could observe several plantations of plantains and sugar canes, and spots that seemed cultivated for roots.

Little is known about early sugar cane introductions by the Polynesians; however, some canes were classified by Moir (1932). Early Polynesian introductions were all that were available when the first attempts were made to develop a sugar industry in Hawai'i in the mid-1800s.

Problems with diseases quickly developed in sugar cane propagated by cloning, and agriculturists attempted to hybridize sugar cane to produce disease-resistant cultivars (Mangelsdorf 1946). These attempts failed, and the planters were forced to introduce cultivars from other countries. The first recorded successful importation of sugar cane was in 1854, when the cultivar "Bourbon (Otaheite)" was introduced from Tahiti. This cane was first planted in Lahaina, Maui, and became known as "Lahaina." Many introductions followed when the "Lahaina" cultivar failed. Introductions were interrupted in 1904 for a 19-year period due to a self-imposed quarantine on sugar cane importation owing to the (probable) accidental importation of the sugar cane leafhopper (*Perkinsiella saccharicida*) from Australia on cane shipments.

In the early 1900s, sugar cane seed was found to be fertile, and a breeding program was initiated. Even with the ability to make genetic improvements with locally available material, introduction of additional sugar cane cultivars and other *Saccharum* species were required, primarily to increase disease resistance. Many additional cultivars and related *Saccharum* species were introduced in 1923 when the importation embargo was lifted. A quarantine station was set up on the island of Moloka'i to protect commercial sugar cane from accidental introduction of insects and diseases. The introduced species included *Saccharum robustum, S. sinense, S. spontaneum, S. officinarum,* and *S. barberi.* The Java (POJ) varieties were imported and formed the basis for the development of the modern Hawaiian varieties. Two hundred introductions had been made as of 1953, and many more cultivars have been introduced since (Mangelsdorf 1953). At present, to introduce sugar cane to Hawai'i, a three-year quarantine is required at the U.S. Department of Agriculture in Maryland, followed by a one-year quarantine at Waimānalo, O'ahu. Genera allied to *Saccharum* have also been introduced for the purpose of making wide crosses to insure genetic diversity.

WATERSHED PROTECTION

In the late 1800s, sugar planters became increasingly aware and concerned that watershed sources of irrigation water on all the Islands were in a state of decline. The destruction of large areas of forest by cattle (*Bos taurus*) resulted in extensive runoff; heavy erosion was evident in the muddiness of streams during storms, and extensive areas of topsoil had eroded to bedrock in many instances. Dr. Harold L. Lyon, pathologist at Hawaiian Sugar Planters' Association, was asked to visit

Maui in 1909, to determine the reason for the decline of the native forests. Lyon formed an hypothesis (to be summarized below) for the decline of native forests (Lyon 1909). He became so concerned that he recommended extensive reforestation of the damaged watershed areas. The Association's Experiment Station began introducing plants for reforestation in 1909; however, no field plantings were made. Instead, the introduced seeds were turned over to the Territory Bureau of Agriculture and Forestry. According to Lyon (1918), not much was accomplished, and it was his recommendation that the Experiment Station become directly involved in planting trees for watershed protection. Lyon was also very much concerned with the conservation of the existing native forests (Lyon 1909):

> What ever planting is attempted, the native trees and shrubs still clothing the slopes of the gulches should receive special consideration. Their preservation should be the main feature of any plan adopted.

In 1918, under the direction of Lyon, a forestry research program for the purpose of watershed protection commenced. This work was considered essential to the survival of the sugar plantations and the domestic water supply, especially on O'ahu.

Lyon concluded that the native species of plants in the forests could not be depended upon to protect the valuable water resources (Lyon 1918). For this reason, he recommended the planting of introduced plants. Over 8,000 species of trees, shrubs, and economic plants were introduced between 1918 and 1934. Lyon was convinced that a forest should be planted as a community of plants, including trees, shrubs, and vines, for the purpose of maximum water preservation. Associated plants would then act as a natural forest for water preservation. A forest planted in such a manner would not be used for commercial logging. Lyon was especially interested in the use of the genus *Ficus* in the watersheds due to its ability to prevent erosion and because of easy establishment and natural regeneration (with the presence of pollinating wasps). Hundreds of *Ficus* introductions were made, but only a few pollinating wasps were successfully introduced.

By 1918, Lyon was convinced that those recommending the use of native species for watershed reforestation were wrong. He was fully convinced that Hawai'i's native forests were doomed and would fall into a condition that would not afford protection of watersheds. He proposed two reasons for the forest decline:

1. The natural makai (seaward) forest frontier was destroyed by humans and uncontrolled grazing, thus exposing the interior forest to invasion by weeds and the devastating effects of the weather. Interior forests were also encroached upon from the mauka (landward) side, primarily by ranching interests, and became exposed, narrow strips deprived of natural protective barriers.

2. Native trees in the watersheds were not well adapted to poorly drained soil conditions present in the forest as a result of disturbances. Without a secondary invasion of native trees suited to the evolving soils

and poor drainage, the native forests were doomed. Establishment of native trees on denuded watershed areas was difficult to accomplish.

Lyon's recommendation was to establish conservation areas, remove cattle from watershed areas, and plant introduced trees. He was quite sure that commercial timber should not be planted in the watershed areas. For this reason, he proposed the planting of dense forest formations for one purpose only -- preservation of water.

In 1919, Lyon elaborated further on his original hypothesis and proposed again that pure culture forests were not acceptable for water conservation (Lyon 1919). He wrote:

> Now, the demands on the local water supply are going to increase rather than decrease, but the available local water supply [for O'ahu] is most certainly going to decrease if the denudation of our watersheds is allowed to continue at the present rate. Oahu will suffer from periodic water famine, which will become more and more frequent until eventually drought will be a permanent condition.

Under the direction of Lyon, thousands of denuded forest acres were reforested. He continued to write about the successes and failures of the reforestation program until 1951 (Lyon 1922, 1924, 1926, 1927, 1929, 1934, 1940).

Without the pioneering reforestation work of the sugar planters, including the removal of cattle from watersheds, water would not have become available to the leeward sugar plantations and expanding cities such as Honolulu.

INTRODUCTION OF SUPPLEMENTAL CROPS

A third reason for the continued introduction of plant material to Hawai'i arose from the need to establish new agricultural crops. Of course, the earliest introductions were brought to Hawai'i by the Polynesians. In addition to sugar cane, these included taro (*Colocasia esculenta*), sweet potato (*Ipomoea batatas*), yam (*Dioscorea alata*), arrowroot (*Tacca leontopetaloides*), mountain apple (*Syzygium malaccense*), breadfruit (*Artocarpus altilis*), banana (*Musa x paradisiaca*), wauke (*Broussonetia papyrifera*), kukui (*Aleurites moluccana*), 'ōlena (*Curcuma longa*), hau (*Hibiscus tiliaceus*, considered by some to be indigenous (Wagner *et al.* 1990)), kou (*Cordia subcordata*), kamani (*Calophyllum inophyllum*), and possibly coconut (*Cocos nucifera*). These early introductions for food, fiber, and energy, along with the clearing of large areas and the introduction of the domestic pig (*Sus scrofa*), which foraged around settlements, certainly took a toll of lowland native plants (Cuddihy and Stone 1990).

A second wave of economic plant and animal introductions came with the Europeans. These included many fruit, fiber, medicinal, forage, grain, vegetable, and ornamental crops. Well-established crops of considerable economic importance include macadamia nut (*Macadamia integrifolia*), papaya (*Carica papaya*), coffee (*Coffea arabica*), and pineapple (*Ananas comosus*). Introductions that have become weeds in addition to crops include common guava (*Psidium guajava*) and passion fruit (*Passiflora* spp.).

Sugar planters have been especially interested in plant introduction under three circumstances: during periods of war or anticipation of war; during periods of forced agricultural acreage reduction; and during periods of poor sugar returns. During both World Wars, the Association and the sugar plantations encouraged local production of subsistence food crops and optional food plants. Many vegetable introductions were made, and some superior types were bred and selected. Most notable was the development of "Hawaiian Sugar" sweet corn (*Zea mays*) by A.J. Mangelsdorf. Many vegetable seeds were distributed to the plantations, but no new plant industry developed.

Probably the most active period for the introduction of food crops was during the period just after the enactment of the Jones-Costigan Act of 1934, which limited the tonnage of sugar that could be produced in Hawai‘i. This act created 17,000 surplus acres (6,800 ha) for which crops were needed. An Hawaiian Sugar Planters' Association Diversified Crops Committee was founded in 1935 to address the new crops issue.

Today, because of low sugar prices and competition from other sweeteners, thousands of acres have been taken out of sugar cane production. Of a potential 230,000 a (93,115 ha), only 180,000 a (72,875 ha) are now in production. The Association and its member companies are now engaged in a search for alternate uses for sugar as well as new crops that can supplement low sugar returns.

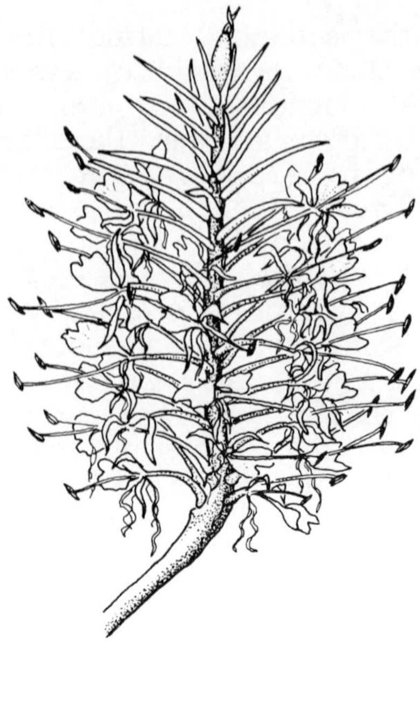

Literature Cited

Cook, James. 1785. *Cook's voyages*, 3rd ed. Order of the Lord's Commissioners of the Admiralty, London. Vol. 2, p. 193. Printed by H. Hughes.

Cuddihy, L.W., and C.P. Stone. 1990. *Alteration of native Hawaiian vegetation: effects of humans, their activities and introductions.* Univ. Hawaii Coop. Natl. Park Resour. Stud. Unit. Honolulu: Univ. Hawaii Pr.

Lyon, H.L. 1909. The forest disease on Maui. *Hawn. Planters' Rec.* 1:151-159.

Lyon, H.L. 1918. The forests of Hawaii. *Hawn. Planters' Rec.* 18:276-280.

Lyon, H.L. 1919. Some observations on the forest problems of Hawaii. *Hawn. Planters' Rec.* 19:3-12.

Lyon, H.L. 1922. Fig trees for Hawaiian forests. *Hawn. Planters' Rec.* 26:78-87.

Lyon, H.L. 1924. Remarkable growth of fig trees in a forest planting. *Hawn. Planters' Rec.* 28:188-191.

Lyon, H.L. 1926. Exotic trees in Hawaii. *Hawn. Planters' Rec.* 30:255-258.

Lyon, H.L. 1927. Forestry on Oahu. *Hawn. Planters' Rec.* 31:307-311.

Lyon, H.L. 1929. Ten years of Hawaiian forestry. *Hawn. Planters' Rec.* 33:55.

Lyon, H.L. 1934. Current forestry projects. *Annual Rep., Exp. Sta., Hawn. Sugar Planters' Assoc.*, 100-112.

Lyon, H.L. 1940. Polymorphic species in Hawaii. *Proc. Pac. Sci. Congr. of the Pac. Sci. Assoc.* VI, Vol. 4, p. 657.

Mangelsdorf, A.J. 1946. Sugarcane breeding in Hawaii, Part I, 1778-1920. *Hawn. Planters' Rec.* 50:141-160.

Mangelsdorf, A.J. 1953. Sugarcane breeding in Hawaii, Part II, 1921-1952. *Hawn. Planters' Rec.* 54:101-137.

Moir, W.W.G. 1932. The native Hawaiian canes. *Proc. Internatl. Soc. Sugar Cane Technol.*, Bull. 7:1-8.

CHOOSING PLANT INTRODUCTIONS: NEEDS OF GRAZIERS

Gordon Cran and Ken Dillingham

ABSTRACT

Forage plants introduced to Hawai'i have come from both temperate and tropical areas of the world. Although introduced species have changed the face of the landscape, they have often failed to provide long-term benefits to the livestock industry. Proper management of rangelands can result in increased livestock productivity and reduce the need for further introductions of forage plants.

Almost from the start of the cattle grazing industry, graziers in Hawai'i have been introducing new forage plants. Native Hawaiian grasses evolved without grazing pressure, so when subjected to ruminants and overgrazing, they soon made room for introduced species better adapted to grazing. Today it is difficult to find indigenous and endemic grasses in a sward or grassland. Alien plants have been introduced both intentionally and unintentionally, most flourishing for a number of years then decreasing and giving way to less palatable, and in a few cases hardier, plants.

As the early grazing industry needed improved varieties, it initially drew upon proven forage plants that were available from temperate areas. In general, these plants did not do well at lower elevations but improved the forage in the cooler regions of Hawai'i. It would be difficult to imagine what Mauna Kea would look like without sweet vernalgrass (*Anthoxanthum odoratum*), or what Ka'ū would be like without meadow ricegrass (*Ehrharta stipoides*) or kikuyu grass (*Pennisetum clandestinum*). They have become part of the landscape along with many other introduced plants.

In more recent times some of the forage plant introductions have come from tropical areas and are thus well adapted to the conditions in the lower and drier areas in Hawai'i. Some of the best examples of these highly adapted grasses are buffelgrass (*Cenchrus ciliaris*) in dry areas, and green panic (*Panicum maximum* var. *trichylum*) and Guinea grass (*Panicum maximum* var. *maximum*) in slightly wetter areas.

Many legumes have also found a new home in Hawai'i in the warmer areas. Koa haole *(Leucaena leucocephala)*, greenleaf desmodium (*Desmodium* sp.) and glycine (*Glycine wightii*) are examples of newer introductions. As a young boy G.C. recalls the bleak arid makai (seaward) areas in Ka'ū before koa haole and Guinea grass were introduced. However, Christmas berry (*Schinus terebinthifolius*) and common guava (*Psidium guajava*) had not spread as they have now.

THE NEED FOR PLANT INTRODUCTIONS

The present selection of forage plants is adequate for current grazing needs. However, the door should be kept open, with reasonable quarantine requirements, for new introductions. If a new introduction produces more forage than another plant, it usually requires more moisture, fertility or energy than a less productive plant. Some areas may still exist where a highly adapted introduction may prove superior in soil protection and forage by being perfectly adapted to the environment.

As desirable as it would be to have a new wonder forage available, we already have many available forage plants that will yield many times the present production through better grazing management. Any new introduction will require the same careful grazing management.

RANGE MANAGEMENT AND MISMANAGEMENT

"Overgrazing" is a term that needs to be better understood. When forage is grazed for more than a short period, the individual plant may be cropped more than once without a rest period to recover or recuperate. This is overgrazing. If this situation continues indefinitely, the desirable forage will be eaten off as it grows, reducing plant reserves and eventually killing the plant.

The more palatable the plant to livestock, the faster it will be eliminated from the sward with overgrazing. Overgrazing, then, allows the desirable palatable plant to be replaced by one less palatable. As time goes by, woody plants invade the sward, and plant succession proceeds into brushland, which in turn has a greater moisture requirement. If moisture is the limiting factor, the grasses and legumes will give way to brush. If moisture is not lacking, succession will continue into woodland. Change is very rapid in the tropics, where the seasons of the year are not well defined.

Humans have spent much time and money over the years to improve forage and ranges but have been failing miserably. Repeated investments have been made in new forage plants and land clearing, only to have swards degenerate to less desirable plants. Many crawling grasses (with stolons or rhizomes) are popular in Hawai'i today, as they can withstand longer periods of overgrazing before they give way to less palatable plants. Kikuyu grass is an excellent example of this; but with overgrazing such grasses also will regress. We have seen areas go from grassland to brush, and tree and forest areas to savanna and grass, in our short lifetimes on Hawai'i Island. Rocky, dry areas may become very productive grasslands, and the perennial grasses may give way to annuals (or in wet areas to rank, unpalatable grasses). Examples of dramatic change are the spread of fountain grass (*Pennisetum setaceum*) on the western side of Hawai'i Island, and bush beardgrass (*Schizachyrium condensatum*) and broomsedge (*Andropogon virginicus*) on the eastern side. It is obvious that introductions of new species without adequate study and proper grazing management waste time and money; they can also be detrimental to the grazing industry. With adequate grazing management, new introductions become less important, as there are many plants already available from which to choose.

The intent of the grazier is to obtain food and fiber from an area with suitable forage. Pasture management, as with other businesses, is expected to be profitable. Land managers must establish goals for the landscape and expectations of profit, and then determine whether they are economically feasible. Environmental and human influences on successful pasture management can be viewed at several levels (Fig. 1).

RECYCLING AND GRAZING MANAGEMENT

Each plant and animal is adapted to a particular environment. Any change, even the slightest, will have an impact on all the rest of the environment. Nature controls recycling of nutrients in an area through selection of adapted plant and animal life. The goals of the grazier are achieved by taking advantage of and managing the process of recycling. Under a grazing program to maximize returns (profit), forage is utilized at its peak of nutrition as much as possible. At the same time, plant succession can be largely controlled by increasing or decreasing grazing pressure. Grazing pressure in any livestock management system is controlled by stock density (area and numbers), plus time of exposure. The rest period between grazing periods will favor certain plants if the interval is short, and others, if longer. Control of grazing can include many paddocks to prevent overgrazing (time control) yet maximize the animal impact through foraging, hoof action, and excrement on the soil. The New Zealand system of "flogging," which involves high densities of livestock in an area for a short time, is an extreme example of intensive livestock management used in Hawai'i. Use of different species of livestock to enhance productivity and plant control have rarely been attempted in Hawai'i because of logistical difficulties.

The use of browsing animals will control woody plants and will recycle minerals and energy that have been locked up in ungrazed grass or brush for a long time. Recycling can also be promoted with machinery, chemicals, or fire, but at a greater expense and less profit. The more rapid recycling of minerals and energy in an area, the better the quality of forage that results. Plants of higher nutritional value will increase in the sward as recycling increases. However, supplemental minerals must eventually be provided grazing stock to replace soil minerals removed over years of grazing. All the supplements are not assimilated by the animals, and much is passed on to the ecosystem in the excrement.

In summary, through proper range management, livestock production can be more effective, thus eliminating the need to introduce new plants to Hawai'i. In extreme conditions, use of fire, chemical, and mechanical controls may be necessary to enhance management of grazing animals.

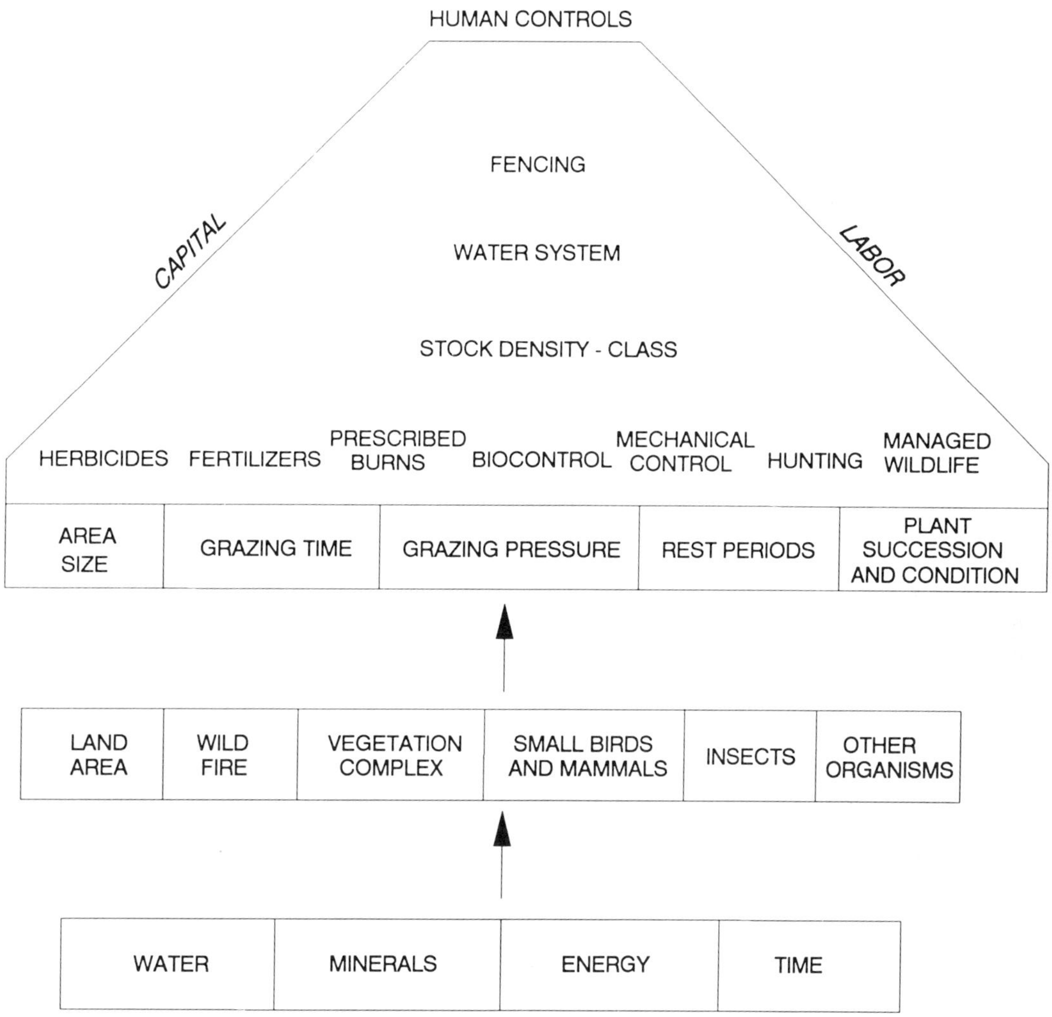

Figure 1. Interrelationships of environmental variables controlled and uncontrolled by humans in a grazing system.

CHOOSING WHICH PLANTS TO CONTROL IN NATIVE ECOSYSTEMS

MODELLING TO PREDICT THE FATE OF INVADING PLANTS

Richard N. Mack and Michael E. Moody

Despite the enormous worldwide effort expended in contending with plant invasions, control -- much less eradication -- is infrequent. We hypothesized that more effective control (and even eradication) of alien plants would be achieved by recognizing the importance of new foci in an invasion. We based our hypothesis on two observations: 1) The time course for invasions by terrestrial plants follows a logistic curve: a phase in which the new range is invaded slowly is followed by a phase of rapid range expansion. 2) Length of the initial phase of slow spread may be inversely dependent on the number of initial foci. Dividing a focus into many isolated foci will enormously increase the rate of new range occupation, assuming the same expansion rate for all foci (Mack 1985).

We formalized our hypothesis in a simple model in which each expanding focus is represented as an expanding circle on a two-dimensional plane. We compared the efficacy of a one-time control of a large or primary focus while neglecting small "satellite" foci, with control of small satellites while neglecting the large focus. Let a population initially consist of a primary focus (area $= \pi R_0^2$) and $N \geq 1$ non-overlapping satellite foci (the area of each satelite $= \pi r_0^2$). We assume that $\beta = Nr_0^2/R_0^2 << 1$, *i.e.*, the total area occupied by the satellites is initially negligible compared to the area of the primary focus. Allow the radii of all foci to increase with time (t) at the constant rate k. In the first strategy, satellites are ignored and the outermost annulus of the primary focus with area $\propto \pi R_0^2$ is removed. The second strategy prescribes elimination of the satellites, and the primary focus is ignored. If the population is allowed to expand, then the ratio (ρ) at t of the total area infested after implementation of the first strategy versus the area infested after implementation of the second strategy is:

$$\rho(t) = \frac{(R_0\sqrt{1-\propto} + kt)^2 + N(r_0 + kt)^2}{(R_0 + kt)^2}$$

A simple calculation shows that $d\rho/dt > 0$ (if $R_o > r_o$), so that the relative area is strictly increasing. Because $\rho(t) \sim N + 1$ as $t \to \infty$, the ultimate gain or advantage in eradicating the satellites is in proportion to their number. The gain depends on the ratio of the initial area of the primary focus to the growth rate k. In Figure 1, $R_o k = 10$, and in Figure 2 $R_o k = 2$. In both Figures 1 and 2, $\alpha = 0.2$ (20% reduction of the primary focus), and $Nr_o^2/R_o^2 = 0.1$ (satellites are initially 10% of the area of the primary focus). The model assumes that the density of the invader is uniform and that no extinction occurs through stochastic fluctuations in the environment. (For a more detailed account of this and other models of spread through the growth of foci, see Moody and Mack 1988.)

Our model shows that any gain achieved by ignoring small foci is short term and is disadvantageous in proportion to the number of foci. Continual removal of new foci (and monitoring the sites of old foci for reemergence) would enhance control.

A successful control effort for a plant invasion with spatial conditions strikingly similar to those prescribed in our model substantiates our recommendations. Witchweed (*Striga asiatica*), an aggressive hemi-parasite on corn (*Zea mays*) and sorghum (*Sorghum vulgare*), was accidentally introduced 30 years ago in the southeastern United States. The alien has nonetheless been confined to the eastern Carolinas by prohibiting the plant's further transport and concentrating eradication efforts on satellite infestations in counties peripheral to the primary focus (Eplee 1979, 1981).

Our model incorporates no properties restricted to organisms, but the importance of the foci's circumferences to plant spread is nonetheless clear. For large foci, most disseminules land inside the boundary of the focus. For smaller foci, proportionally more parents occur on or near the circumference; their seeds are more likely to fall into the adjacent uninfested area, thus affecting spread.

Most current practice in alien plant control appears to emphasize the tactics ("the art of handling forces in battle or in the immediate presence of the enemy") of controlling the invasion, *i.e.*, the specific procedures used locally to remove alien plants by burning, excavation, herbicide application, etc. Based on the predictions emergent from this model, we believe much more effort should be exerted in developing an effective strategy ("the art of projecting and directing the larger military movements and operations of a campaign") (Anonymous 1971) for controlling the spread of alien plants.

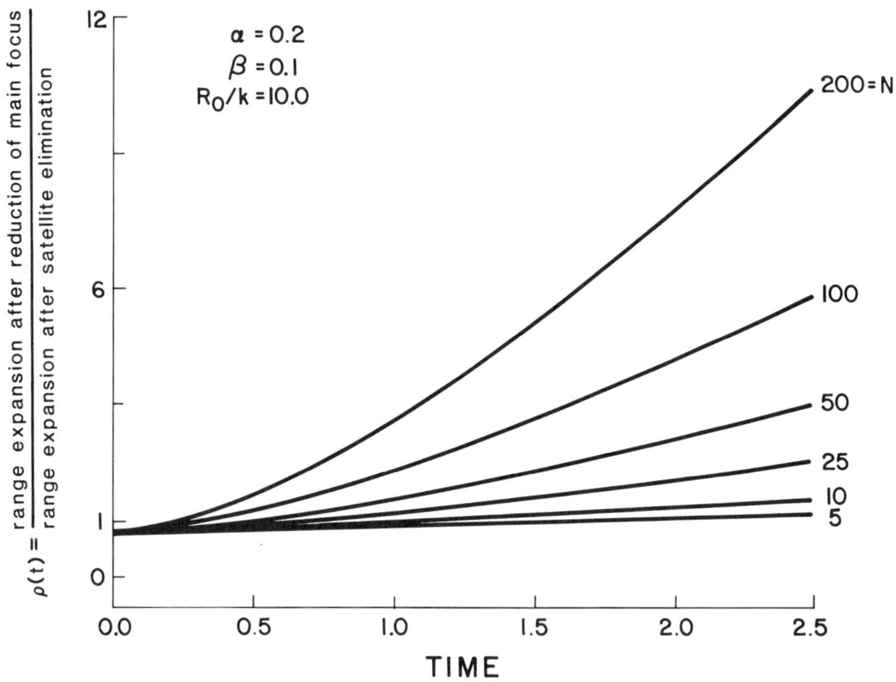

Figure 1. Comparison of the efficacy of two alternative approaches for halting the spread of an alien population of terrestrial plants. R_0 k = 10, = 0.2 (20% reduction of the primary focus) and ß = 0.1 (satellites are 10% of the area of the primary focus).

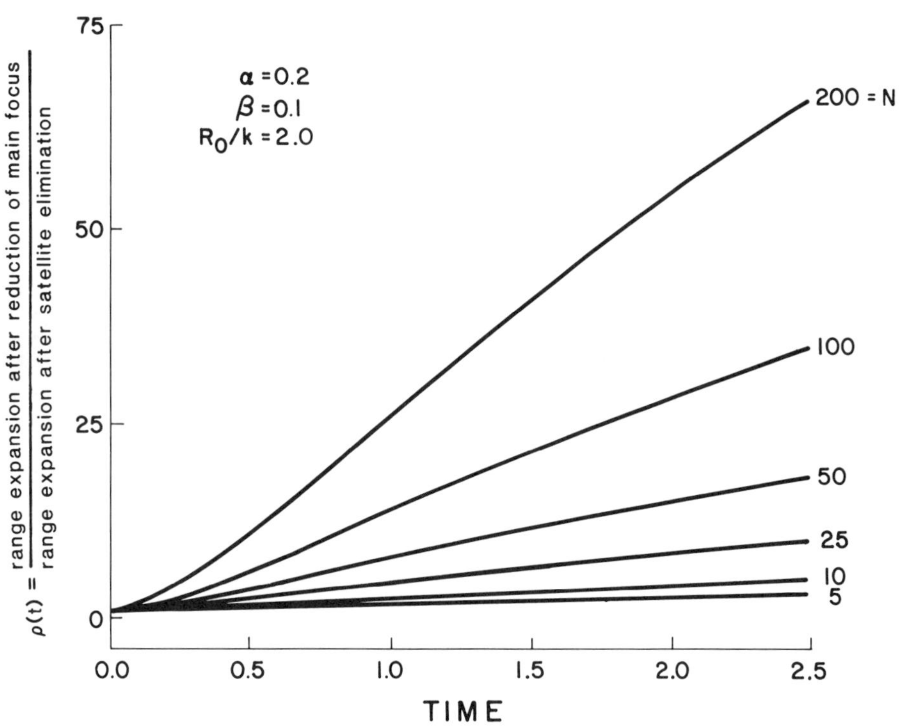

Figure 2. Comparison of the efficacy of two alternative approaches for halting the spread of an alien population of terrestrial plants. R_0k = 2, = 0.2 (20% reduction of the primary focus) and ß = 0.1 (satellites are 10% of the area of the primary focus).

Literature Cited

Anonymous. 1971. *The compact edition of the Oxford English Dictionary*. Oxford, England: Oxford Univ. Pr.

Eplee, R.E. 1979. The *Striga* eradication program in the United States of America. In *Proc. Second Symp. Parasitic Weeds*, ed. L.J. Musselman, A.D. Worsham, and R.E. Eplee, 269-272. Raleigh: No. Carolina State Univ.

Eplee, R.E. 1981. Striga's status as a plant parasite in the United States. *Plant Disease* 65:951-954.

Mack, R.N. 1985. Invading plants; their potential contribution to population biology. In *Studies on plant demography: a Festschrift for John L. Harper*, ed. J. White, 127-142. London: Acad. Pr.

Moody, M.E., and R.N. Mack. 1988. Controlling the spread of plant invasions: the importance of nascent foci. *J. Applied Ecol.* 25:1009-1021.

RULE-BASED DECISION MAKING: A WAY TO DETERMINE WHICH ALIEN SPECIES TO CONTROL

Kent W. Bridges

ABSTRACT

Most decisions that managers make are based on information. Presumably the more relevant the data, the better the decision. Managers rarely feel that they have enough information, but it is not always possible to easily obtain more. What is required are decision-making techniques that make the best use of incomplete information. In addition, a good decision-support system should point to what data might best be collected in future studies. We are entering a period of rapid development of sophisticated decision-making tools. It is appropriate to look at how these tools can help managers decide whether alien species are likely to become problems. While it is too early to expect the new technologies to provide direct assistance, an understanding of their operations should help guide future data collection efforts.

HOW DECISIONS ARE MADE

We generally work under the premise that more relevant information enables us to make better decisions. If this is true, we are then faced with obtaining enough relevant data. For most management problems, some information will always be at hand; but the amount is rarely considered sufficient. Sometimes we know that data exist that will assist us with a particular decision-making problem, but we are unable to actually get the data. Reasons for unavailability range from location inaccessibility to inability to take the time to find data. Collection of new data is frequently unacceptable because of time constraints, or resources are not available. Sometimes it is not entirely clear what new information is needed. Biological problems cannot always be solved with the collection of new data anyway. For example, when historical patterns are being examined, "old" data have great value. More often than not, however, such data are incomplete and ambiguous.

How do we make the most effective use of data in the decision-making process? At the least, it would be reassuring to know that data are actually being used to make decisions in an unbiased, objective manner. We can be even more demanding and pose the question of whether it is possible to have a system in which decision making is so well organized that it is effectively "reproducible," that is, other people who use the same data would be led to the same conclusion. We refer to such formal decision making as an <u>inference process</u> and will explain how we can use specific inference techniques to help us answer questions about weed control below.

WEED CONTROL: WHICH ALIEN SPECIES ARE LIKELY TO BECOME PESTS?

Two aspects of alien species control are examined here. The first concerns the problem of determining the existing distributions of species. This involves evaluating information on the geographic range of the weed in its native habitat and the extent of its spread in its new habitat. The second aspect involves making predictions about the likely changes in distributions. How rapidly will the weed spread, and what will be the extent of the invasion? Species distribution questions are too broad to be answered directly. Instead, they need to be divided into a series of related "operational questions."

Many operational questions are relevant to the evaluation of species distributions. For example:

1. What is the highest elevation at which each species is found?
2. What is the usual range of each species?
3. What species have broad ranges?
4. What other species are associated with a particular species?
5. What is the life-form spectrum at a particular location?
6. What species might be found at a particular location?

These questions can be analyzed in several different ways, ranging from the direct use of information to the need to do complex, indirect inferences. A naturalist who is intimately familiar with species distributions in a region will perform these analyses mentally. But the concern here is how such questions can be answered by someone relatively unfamiliar with the region, who has access to distributional data, such as herbarium records and field notes. The point is not to remove experienced people from the decision-making process, but to examine whether there are ways that this process can be assisted so that all managers will be able to make the best possible decisions.

A useful place to start examining the existing decision-making process is by evaluating how each of the operational species-distribution questions listed above can be answered. The first question shows the direct use of information: the existing species distribution records, or new collections, are searched for the highest elevation value. The value provides the answer to the question. The second and third questions involve the use of statistical concepts. The term "usual range" implies that an answer includes an interpretation of the modal (most common) occurrence of the species as well as something like its interquartile range. Questions of this type are answered by producing some sort of statistical distribution for all of the recorded distribution information.

The fourth and fifth questions are examples of questions that require the use of complex definitions. The answer to question four involves the definition of the term "associated with," while the fifth question requires the definition of "life-form spectrum." These terms make use of relatively complex categorization processes that must be made based on analyses of large amounts of data. As a result, considerable work must be done on information relating to all the species in the geographic region of concern before specific questions can be answered.

The most complex type of question is one that requires a series of individual decisions. This is illustrated by the sixth question. The sort of individual decisions that make up the question-answering strategy will be examined after a few comments on the sources of data and how these can be organized and searched.

SOURCES OF INFORMATION AND THEIR ORGANIZATION

Information on species distributions comes in many forms. Although historical records are of considerable importance, the way in which information is recorded often reflects the period when it was produced. Some types of records, such as herbarium sheets, show relatively little change in the last hundred years in how annotations have been made. In contrast, field notes are particularly variable and strongly reflect the skills and purposes of their author.

Some information can be found in the published literature. Notes on species distributions are part of almost every ecological paper, for example, even though they are not the primary focus of the study being reported. Published papers are a diffuse source of information. They are also a relatively incomplete resource since some critical information can be lacking. The increasing trend to edit journal articles so that they do not stray from their specific topic makes incompleteness likely.

One of the most valuable sources of information on species distributions is the "common" or "general" knowledge held by people familiar with the species in a region. This sort of information is known as "heuristics." It is rarely recorded. Instead, it is either acquired independently through experience or passed along in informal discussion.

When a large amount of information has a consistent structure, it is possible to record it in a formal database. This is the case with herbarium information relating to species distributions, where species names, collection dates, and collection locations can be recorded in a simple table. It is also possible to record data on the sites on which species have been found (*e.g.*, elevation, soil type, rainfall, and temperature). Tables of such information can be used to answer specific questions.

Many tools can be applied to formal databases. Personal computers are increasingly being used both for the storage and use of databases. Complex searches, even when they involve information from several tables, are made quite easily and are usually done to produce a short list of species (or sites) that meet some set of criteria (Jacobi, this volume). For example, a search may be made of all records of species that are now listed as threatened or endangered, collected at a specific elevation on a particular island during a particular time period. Questions of this complexity can be handled routinely, provided an appropriately structured database exists. Designing such a database is a relatively simple matter, given the sophistication of the relational database tools that are now available.

Not all information can be stored in a formal database. Information such as heuristics (*i.e.*, "common knowledge") can only be stored in what can be called an "*ad hoc* organization" database. Such a database has two kinds of information, referred to as "facts" and "rules." The facts record simple observations and appear in the database like a simple English sentence. For example, "Pu'u Maka'ala has an elevation of 5,610 ft (1,700 m)." Rules are more complicated and correspond to a definition. The form of the definition is that something will be true if a set of conditions is met. For example: "A particular site will be considered a bog if it has high rainfall and poor soil drainage and topography that restricts runoff." More complete and stylistically accurate examples of facts and rules are given below.

Relatively few database tools in widespread use handle *ad hoc* database structures. The work reported here examines database organization using PROLOG, a programing language based on predicate calculus (Clocksin and Mellish 1984). PROLOG stands for "programing in logic" and is one of the logic programing languages. It provides a very natural way to do the sort of species distribution investigations that are described here, in large part because of the way information is stored, and because the statements used to pose search questions in PROLOG closely mimic English descriptions. However, the match is not exact. Since PROLOG has its own formalities, like any other computer language, and the purpose here is not to teach this language, all the following examples will be paraphrased compromises between English and PROLOG.

EXAMPLES OF AN AD HOC DATABASE SYSTEM

Examples of facts and rules that could be used in an *ad hoc* database are given below. Facts and rules use a notation in which some compound words end with the letters x, y or z. These words indicate that something appropriate has to be substituted. For example,

species-x found-on site-y on-date-z

requires three substitutions: x, y, and z. Some of the substitutions are made when the database is constructed. This would be the case in the above example. There would be many listings of species distributions in the form shown in the example, such as passiflora-mollissima found-on puu-makaala on-date-11-24-1974.

This sort of substitution is most characteristic of the facts that are stored in the database.

Rules, in contrast, generally do not have all the substitutions made when they are stored. A rule could be stored that states:

```
site-y is-a-bog if
      site-y has-an-annual-rainfall-very-wet and
      site-y has-drainage-poor and
      site-y has-topography-flat or
      site-y has-topography-basin.
```

Note that site-y has not been identified in the rule. When PROLOG uses rules, it will substitute sites in the database, one at a time, to determine which ones satisfy the criteria listed in the body of the rule.

When PROLOG structure is applied to species distribution problems, the following sorts of facts, with the appropriate values substituted, would probably be included in the database:

```
species-x found-on site-y on-date-z
site-y has-soil-type-z
site-y has-elevation-z
site-y has-annual-rainfall-z
site-y has-slope-z
site-y is-on-island-z
botanist-w collected-at site-y on-date-z.
```

These sorts of facts are likely to come from herbarium records and published site descriptions or field notes.

PROLOG-like rules that relate to a database with such facts would be like:

```
species-x may-be-found-on site-y if
      species-x found-on site-y on-date-z and
      z = this-year and
      confidence = 1.0

species-x may-be-found-on site-y if
      species-x found-on site-y and
      site-y has-elevation elevation-similar-to site-z and
      site-y has-annual-rainfall rainfall-similar-to site-z and
      site-y is-on-island-z and
      site-y is-on-island-z and
      confidence = 0.75

site-y elevation-similar-to site-z if
      site-y has-elevation-x and
      site-z has-elevation-y and
      abs (x-y) less-than-200
```

A few observations should be made about these rules because they show so much of the power of this type of database. Several rules can define any particular situation. In the example above, there were two definitions of the "may-be-found-on" rule. The first alternative implies that one would expect to find a particular species on a site if it had been recorded on that site this year. The second alternative means that one would expect to find a particular species on a site if it had been found sometime on a site on the same island and the sites had similar elevations and amounts of annual rainfall. The alternatives are used one at a time. If the first "may-be-found-on" rule is discovered to be true, then the remaining ones are not used. If the first one is not found to be true, then the second one is tried, and so forth, until all the alternatives are exhausted.

Confidence statements were given at the end of the "may-be-found-on" rules. These are not conditions that are being tested, but values that are being saved when an alternative rule is found to be true. This provides a way of storing the likelihood of the results. In this example, it means that a recent sighting of a plant on a site is stronger evidence that it is still likely to occur there, than finding that it was only found once on a site with a similar climate. Whether these values (100% versus 75% certain) are correct is a matter of debate and is not the issue here.

Sometimes it is useful to use indirect rules. That is true in the case of the "elevation-similar-to" rule, where a simple test is performed to see if two sites are within 200 m of each other. This is useful because it makes it possible to simplify the statement of rules. It also makes it possible to change a definition such as "elevation-similar-to" in a single place, rather than having to modify every place that it has been used in other rules.

The purpose of a species distribution database is to obtain answers to the two basic, specific questions that can be asked. The first question tests whether there is evidence in the database to show that a statement is true. For the example used here **passiflora-mollissima found-on puu-makaala**: PROLOG automatically searches the database, looking at facts and testing rules, until it finds a "found-on" fact that is true (in which case it gives the response YES), or until it gets through all the rules with none of them being true (and it responds NO).

The second basic question that can be used with PROLOG provides lists of solutions, not just YES or NO answers. In particular, the responses list the values for items in statements for which specific information has not been supplied. This is seen with three questions and typical PROLOG responses (shown here as indented lines below the question):

x found-on puu-makaala on-date-11-24-1974
x = passiflora-mollissima
x = rubus-ellipticus

passiflora-mollissima found-on puu-makaala on-date-x
x = 11-24-1974
x = 9-17-82

x found-on puu-makaala on-date-y

x = passiflora-mollissima	y = 11-14-1974
x = rubus-ellipticus	y = 11-24-1974
x = passiflora-mollissima	y = 9-17-1982

IMPLICATIONS FOR MANAGEMENT

The sort of *ad hoc* database shown here has a number of implications for managers who are faced with alien species control problems. This new database structure is well adapted for storing species distribution and site information. The *ad hoc* database differs from the formal database in its ability to include complex rules as part of the overall database. This means that it should be possible to use information that was once considered to be too informal to be stored or used in a database system.

The flexibility provided managers can be seen in another example. Suppose that a database has been constructed that includes all the species distribution information from a large set of herbarium records. This information has been transcribed as faithfully as possible, but when it comes time to use it, the manager discovers that a particular plant collector was relatively inaccurate in the identification of a particular species. Rather than going back into the files and changing the species identification made by this collector, it would be possible to add a rule to the database that had the effect of saying "so-and-so wasn't good at identifying x so ignore these records." This maintains the integrity of the database and clearly identifies a constraint that will be applied when the database is used. It is obvious that it would be relatively easy to remove such a rule-based restriction, while any changes in the database would be particularly difficult to reverse.

Since rules in the database provide a way of recording "common sense," it is possible that a manager could build a database that has "institutional memory." As researchers and field workers come to understand some of the subtle relationships that they feel occur in the species distribution patterns, these can be coded as rules and entered into the database. This information can remain in the database, regardless of turnover of personnel. Eventually, the database could be an extremely rich source of very practical information that would otherwise have to be rediscovered by new personnel before it could be used. The recording of "common sense" may also have a valuable side benefit. New personnel should be able to gain considerable insight by examining the database, which could likely shorten the apprenticeship necessary to become familiar with a new environment.

It may be possible for managers to undertake analyses of alien species problems more objectively by using a database of the sort described here. If a management question is posed as a rule, then it is possible to examine the way that the answer is obtained. At the least, this forces questions to be clearly stated.

No species distribution databases exist that have the structure described here. The potential benefits that have been shown provide motivation for the storing of information from herbarium records, field notes, and the thoughts of field workers in a readily accessible, usable form. Exciting possibilities await the manager who has an opportunity to search such a database.

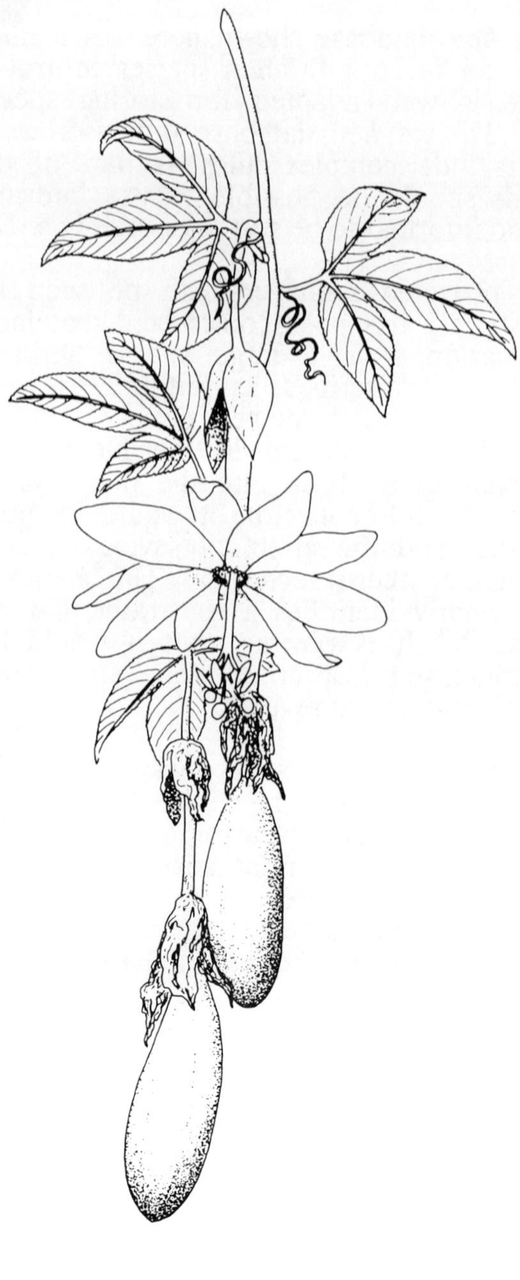

Literature Cited

Clocksin, W.F., and C.S. Mellish. 1984. *Programming in Prolog*, 2nd ed. Berlin: Springer-Verlag.

Jacobi, J.D., and F.R. Warshauer. [this volume] Distribution of six alien plants in upland habitats on the island of Hawai'i.

CONTROLLING WEEDS IN NATURAL AREAS IN HAWAI'I: A MANAGER'S PERSPECTIVE

Dan Taylor

ABSTRACT

Managers of most natural areas in Hawaii have a simple mandate: to know, restore, protect, and maintain native ecological communities and natural processes. Applying this to management of nonnative plants requires skillful evaluation of ecological principles and consideration of manpower and financial limits. Professional opinions, lay public demands, and even administrative preference must be considered. Hawaii Volcanoes National Park integrates most of these considerations in order to delineate priority categories for alien plant control work. The Park prioritization guide is a useful model for other lands, even those with more complicated land use mandates. Nonnative species that are disruptive (because they replace native species and communities) and those that are benign (because they do not dominate) are distinguished, and disruptive species are categorized according to distribution. Localized species are prioritized and placed on a control schedule, whereas widely distributed species are targeted for control only in designated areas, called Special Ecological Areas. Funding and staffing levels are inadequate for parkwide control of widely distributed disruptive plant species.

MANDATES AND OBJECTIVES

In Hawaii, the National Park Service manages Haleakala and Hawaii Volcanoes National Parks, as well as smaller historical parks and sites on Hawai'i Island, Maui, Moloka'i, and O'ahu. The Service operates under a land management mandate to maintain native plant and animal communities and the natural processes that shape them, and to minimize unnatural disturbances. "Native" is defined as endemic or indigenous. Species introduced by Polynesian settlers prior to the arrival of Europeans in the late 1770s are usually not disruptive of native communities or have cultural values. "Natural" is defined as unaffected to a great degree by humans, in the past or present, including subsequent restoration (removal of aliens and/or replacement or some encouragement of natives.) The

Service mandate may be contrasted with that of ranching or agriculture, which have the objective of producing food or money. It also contrasts with mandates of multiple use agencies such as the U.S. Forest Service or the Hawaii Department of Land and Natural Resources, which include resource conservation, commercial resource exploitation, and recreation. The mandates of the U.S. Fish and Wildlife Service, the Hawaii Natural Area Reserves System, and the privately endowed Nature Conservancy most nearly resemble that of the National Park Service, in that these three entities manage lands primarily to conserve natural resources. The mandates of all the interests listed are valid and serve the collective needs of Hawai'i's people.

The intent of National Park Service management is to restore and maintain the ecological diversity of plant and animal communities, to maintain native ecosystems, and to promote natural dynamics with minimal human interference. Native plant and animal communities in the areas we manage have been disrupted by ungulates, predators, alien plants, and other disturbances. Managers, therefore, must remove introduced species in order to restore natural communities and processes. We recognize that ecosystems will change, in community composition, structure, and distribution. We know that some species will naturally disappear, that some will evolve, and that these changes will occur over a very long period of time, as they did prior to the arrival of technological human societies in Hawai'i in the 18th century. We also realize that some changes brought about by humans are irreversible. National Park Service management objectives might be viewed as attempting to neutralize, where practical, some of the changes imposed on native Hawaiian ecosystems after the arrival of James Cook's expedition in 1778. The National Park Service is implementing a Congressional mandate that presumes inherent value in preserving for human society remnants of wild, natural America. In attempting to manage toward this end, we acknowledge our rudimentary understanding of Hawaiian ecological processes.

The magnitude of disruption of Hawaiian ecosystems sometimes seems overwhelming. Many of the animals and plants brought to the Islands during the past two centuries have become established, especially in lowland areas, and have thoroughly changed the landscape. In some instances the changes have been deliberate (*e.g.*, conversion of native forests to plantation agricultural systems and ranchland pastures, introduction of wild ungulates to forests for recreational hunting, and conversion of lowland beach strands for tourist recreation). In other instances, changes have not been deliberate, for example, conversion of native dry forest to flammable fountain grass (*Pennisetum setaceum*) lands in western Hawai'i, statewide degradation of rain forests and montane forests by feral ungulates, and the early degradation of forest bird habitat caused by mosquitoes. As a consequence of both deliberate and accidental changes, however, only remnants of native plant and animal communities remain in Hawai'i, some in Hawaii Volcanoes and Haleakala National Parks. Even native plant and animal communities in these protected areas are deteriorating because of aggressive species of introduced plants and animals. In the remainder of the paper, I will consider how the National Park Service makes difficult choices in selecting target weeds for control.

SELECTING TARGET SPECIES FOR MANAGEMENT

The plant checklist for Hawaii Volcanoes National Park includes approximately 725 vascular species (Higashino *et al.* 1988). Approximately 475 (66%) of these are introduced or alien. The list of approximately 250 natives is missing some species that have become extinct or are no longer able to survive in the Park. The nonnative species in the checklist include a variety of life forms and threats. However, because many introduced species are not invasive, it is clearly unnecessary to control all of them, even if it were possible to do so. The manager's job is to identify which nonnative plants are relatively harmless and which are disruptive; then a decision must be made about which disruptive species should first be targeted for control.

We have identified 53 species from the list of 475 aliens as actually or potentially disruptive. Disruptive species are defined as those threatening the integrity of native plant communities because they displace native species, hinder native plant reproduction and growth, or alter community dynamics (such as changing the fire regime, soil structure, nutrient cycling, or soil chemistry, etc.). Good examples of disruptive alien species include banana poka (*Passiflora mollissima*), which smothers native vegetation in ʻŌlaʻa Tract in the Park; fountain grass, which alters fire regimes in the Kaʻū Desert; firetree (*Myrica faya*), which alters nutrient cycling in submontane ʻōhiʻa (*Metrosideros polymorpha*) forests; kāhili ginger (*Hedychium gardnerianum*), which displaces native vegetation in ʻōhiʻa rain forests; and strawberry guava (*Psidium cattleianum*), which changes forest composition by shading out native species in ʻōhiʻa forests from 500 to 5,000 ft (150-1,525 m) elevation. These species are discussed by others in this volume. Aesthetic factors usually are not seriously considered in evaluating an alien plant's disruptive potential. For example, the aesthetically out-of-place eucalypts (*Eucalyptus* spp.) at Nāmakani Paio Campground, the pines at ʻĀinahou Ranch, and the fuchsias (*Fuchsia magellanica*) at Thurston Lava Tube are relegated to low priority for control because they are localized, non-disruptive species.

The Division of Resources Management program at Hawaii Volcanoes is under strict funding and manpower constraints and therefore must prioritize the list of 53 target species to bring the workload into line with financial means, and the chances of success. Target species have been categorized into two general groups: widely distributed (16 species) and localized (37 species). Species within each category have been prioritized according to feasibility for control, with an intent to promote ecological resources management as well as demonstrate positive results. Widely distributed alien plant species cannot be controlled throughout their range because of funding and staffing limits, so they are controlled only in areas where native vegetation is relatively intact and where we are able to maximize success. Management of these Special Ecological Areas is described in Tunison and Stone (this volume). Localized alien plant species are confined to small enough areas to enable workers to control them and thus prevent their spread and possible attainment of uncontrollable status. This aspect of the vegetation management program is described by Tunison and Zimmer (this volume).

APPLICATION

The selection and prioritization process practiced at Hawaii Volcanoes National Park has several applications. **First**, it gives an otherwise overwhelming alien plant control problem a systematic framework within which certain decisions about weed control work can be made. The framework also identifies needs for future work and funds, regardless of any probabilities we will ever receive adequate funds and manpower. **Second**, the selection and prioritization process also articulates procedures and objectives for others, namely administrators, other agencies, and the lay public. **Finally**, the selection and prioritization process presents an opportunity for regional cooperation. The system helps solve shared problems by fostering communication among State, Federal, and private land managers. The degree to which responsible authorities outside of the Park share concerns about fountain grass, banana poka, firetree, clidemia (*Clidemia hirta*), and Christmas berry (*Schinus terebinthifolius*) is affirmation of this. I anticipate that we will eventually conduct regional control programs for some alien plant species, as we have done already on a limited scale for firetree.

Literature Cited

Higashino, P.K., L.W. Cuddihy, S.J. Anderson, and C.P. Stone. 1988. *Checklist of vascular plants of Hawaii Volcanoes National Park.* Tech. Rep. 64, Univ. Hawaii Coop. Natl. Park Resour. Stud. Unit. Honolulu.

Tunison, J.T., and C.P. Stone. [this volume] Special Ecological Areas: an approach to alien plant control in Hawaii Volcanoes National Park.

Tunison, J.T., and N.G. Zimmer. [this volume] Success in controlling localized alien plants in Hawaii Volcanoes National Park.

EVALUATING PROPOSED BIOLOGICAL CONTROL PROGRAMS FOR INTRODUCED PLANTS

George P. Markin and Ernest Yoshioka

ABSTRACT

Introduced plants in native ecosystems can be managed with a variety of techniques, including biological control, the introduction and establishment of natural enemies of the weed. While generally accepted as a permanent and ecologically accepted method of management, biocontrol has drawbacks, including high initial costs, long delays before results are visible, and a variable success record. It is therefore recommended that before a new biological control program is initiated, a full evaluation be made of the weed and its role in native ecosystems, local agriculture, and other areas in an effort to identify potential problems. A 20-point checklist with a subjective scoring system is proposed for making this evaluation. The scoring system provides a numerical total, which should indicate the potential for success of the program. The system may be more valuable as a method of ranking similar programs (to indicate which may be the best investment of funds and time) than as a predictive tool. To demonstrate the scoring system, three weeds considered major threats to native ecosystems in Hawai'i are evaluated.

INTRODUCTION

Other authors have discussed methods for evaluating introduced plants in native ecosystems to identify those that affect the systems severely enough to justify control efforts. Others have also discussed various control or management techniques available to resource managers to achieve control. One management technique is biological control, the introduction and establishment of the natural enemies of the introduced plants. Biological control is particularly attractive because, if successful, it can bring about a permanent solution to the weed problem; environmentally, it may be the least damaging control technique available for use in native ecosystems.

The biological control of weeds in Hawai'i and elsewhere has historically been almost entirely restricted to agriculture, usually focusing on weeds that are major problems over large areas of rangeland. These weeds have caused easily recognizable losses for which costs can readily be identified and quantified. Obtaining support for biocontrol programs in such cases and justifying costs have therefore been relatively easy.

Securing support for biological control as a management option in native ecosystems may be more difficult to justify on economic grounds. Many remaining native ecosystems consist of small scattered blocks of public or private lands, increasing the chances of conflicts of interest and the potential for adverse impacts on interspersed private lands. Other problems, including low success rates, long lapses in initiating new programs, and initial high costs of biological control, may become more critical on natural areas than for programs aimed at weeds on agricultural lands with better funding. We have already discussed several possible problems with biological control and potential impacts on new programs (Markin *et al.*, this volume). The need for more public education about biological control in natural areas is another problem to consider.

The potential difficulties are great enough to justify developing criteria for evaluating proposed biological control programs. Ideally, such evaluation would allow predictions of program success and would provide indications of the difficulties in, and expenses of, a new program. Such a process would also provide a checklist for planning and organizing a proposed program. This paper presents a checklist of 20 factors that could influence the success of a proposed program and a scoring system for evaluating each factor.

PROCEDURES IN CONDUCTING BIOLOGICAL CONTROL

The history of biological control of weeds dates to 1902 (Julien 1982). Since that date, this science has undergone many changes, particularly in our understanding of the underlying natural processes that we are trying to control (population dynamics) and the approach we are using for selecting and establishing new agents. Biological control today follows a fairly standard procedure for selecting, testing, and establishing control agents (Huffaker 1957, 1959; DeBach 1964, 1974; Frick 1976). In initiating a program to control a weed, we generally follow a fairly prescribed sequence. For planning purposes alone, almost 100 steps or decision-making points have been identified in flow diagrams attempting to systematize the entire process of conducting a biological control program (Grabau and Spencer 1976). We have summarized these in 10 sequential steps, each incorporating a separate part of the process (Table 1).

To identify where problems are most likely to occur in a biological control program, we have reviewed past programs in Hawai'i and elsewhere in the world, including those that were not successful. Based on this review and on personal experience in planning and implementing weed control programs in Hawai'i, we have identified 20 critical factors that have contributed to the successes, failures, or difficulties of earlier programs

Table 1. Major steps in conducting a program aimed at biological control of an introduced plant.

1. Determine range, economic, and ecological impact of the target plant.
2. Conduct literature review to determine what is known of the target plant.
3. Identify the original natural range of the target plant and the probable genetic epicenter for the genus.
4. Explore natural range of the target plant and of related species to inventory associated natural enemies.
5. Select the most promising natural enemies for potential biological control agents.
6. Determine biology and ecological requirements of candidate agents.
7. Conduct host testing to determine plants other than the target weed that each candidate agent can survive on, or temporarily feed on.
8. Select the most promising agents and obtain approval for their release.
9. Begin releasing agents.
10. Conduct an evaluation of the program to determine degree of success or reason for failure.

(Table 2). These factors apply to programs that use either insects or plant pathogens as the controlling agent. However, we restrict our discussion to programs using plant-eating (phytophagous) insects, since Gardner (this volume) has already discussed plant pathogens.

EVALUATION OF CRITICAL FACTORS

To evaluate the importance of each factor in a proposed program, we have used a scoring system on a scale of 0 to 5. A factor that could hinder or deter a program's success is scored 0. A factor that should simplify a program or contribute significantly to its success is scored 5. One factor has been used to eliminate a biocontrol program entirely, based on its impact on past programs. A program should not be started if this factor is rated zero. Another factor is weighted much more heavily (25) than other factors to encourage biological control if this condition is met.

Scoring of each critical factor is the responsibility of the resource manager and/or researcher, based on knowledge of the weed's biology, history, and role in local agriculture and society. (The role of the weed

Table 2. Checklist of critical factors that can influence the success of a proposed biological control program against an introduced plant, with a scoring system for indicating the relative importance of each factor to the overall program.

Critical Factor	Score
1. <u>Availability of Quarantine Facilities</u>	
--Quarantine facilities do not exist.	0
--Quarantine facilities and support laboratories available.	5
2. <u>Administrative Support</u>	
--Program administrators unfamiliar with or negative towards biological control.	0
--Program administrators personally familiar with biological control and committed to its support.	5
3. <u>Long-term Funding</u>	
--Uncertain: appropriated yearly or dependent upon outside sources.	0
--Funding totalling over $1 million committed for 5 years.	5
4. <u>Available Scientific Expertise</u>	
--No local scientists available or familiar with biological control of weeds.	0
--Trained and experienced scientists and technicians available to be assigned to or assist the project.	5
5. <u>Geographic Area Affected</u>	
--Weed established on a large geographic area such as a continent.	0
--Weed established in a very isolated geographic area, e.g., an island system.	5
6. <u>Ecological Range of Weed</u>	
--Weed well established in a number of ecological zones.	0
--Weed limited to a very narrow ecological zone.	5
7. <u>Life Cycle of Weed</u>	
--Weed an herbaceous annual.	0
--Weed a perennial shrub.	5

Critical Factor	Score
8. Density and Local Distribution of Weed	
--Weed occurs as widely scattered individuals.	0
--Weed forms dense stands covering large areas.	5
9. Associated Insects and Pathogens	
--Many insects or pathogens, both native or introduced, attack weed and support numerous parasites and predators.	0
--Few insects or pathogens have become established on the weed.	5
10. Economic Importance of Target Plant	
--In parts of its range, weed is a major agricultural plant.	Don't begin
--The weed has no recognized agricultural uses.	5
11. Redeeming Values	
--Weed is recognized for some socially accepted purpose, e.g., esthetic values, cover for wildlife, watershed protection, etc.	0
--Weed has no redeeming values, now or predicted for the immediate future.	5
12. Related Agriculturally Important Plant Species	
--A related species of the plant is a major economic crop within same ecological range.	0
--A related species of plants is of minor economic importance, with different ecological requirements.	5
13. Use by Beekeepers	
--Weed is widely used by beekeepers as a major source of nectar and pollen.	0
--Weed has not been recognized, or used, as a source of honey and pollen.	5

Table 2, continued.

Table 2, continued.

Critical Factor	Score
14. Related Endangered Species of Plants	
--A related species of plants (within the same genus) is an endangered species and shares the same ecological range as the weed.	0
--No closely related species are endangered or threatened, at least in the same geographic area.	5
15. Status of Target Weed Worldwide	
--Plant is recognized as a problem only in a very limited area.	0
--Plant has been introduced to many other areas of the world and is recognized in most as a major weed.	10
16. Previous Biological Control Programs	
--No programs attempted in other parts of the world.	0
--Major efforts made at biological control in other parts of the world, including release of one or more successful agents.	25+
17. Accessibility of the Plant for Study in its Natural Range	
--Natural range covers underdeveloped, unfriendly, or politically unstable countries.	0
--Natural range includes at least one developed country that is friendly towards the United States.	5
18. Availability of Research Facilities in its Natural Range	
--Agricultural research facilities staffed by trained entomologist lacking, or at least not available.	0
--Well-staffed agricultural research facilities engaged in biological control available.	5
19. Availability of Wild Genotypes	
--Wild forms of the plant in undisturbed native ecosystems unknown or represented by limited isolated populations.	0
--Plant still growing wild over large areas of the original geographic range.	5

Critical Factor	Score
20. Availability of Closely Related Species	
--No closely related species within the same genus.	0
--Weed has many closely related species within the same genus.	5

in native ecosystems is considered serious if a biocontrol program for it is being considered, and thus is not scored here.) Each critical factor is generally self explanatory, but a narrative for each extreme is provided for demonstration purposes (Table 2).

Availability of Quarantine Facilities

The introduction of any live insect into the United States requires special permits issued by the U.S. Department of Agriculture Animal Plant Health Inspection Service. Permitting is totally dependent upon the availability of an approved quarantine facility in which the insects can be received, screened to remove diseased or parasitized individuals, and studied. There are at present two insect quarantine facilities in Hawai'i and nine on the U.S. Mainland (Klingman and Coulson 1982). Information on facilities and workers in biological control of weeds in the United States has been compiled by Coulson and Hagan (1985).

It was formerly possible to make introductions at a certified facility where preliminary screening and testing could be done, then conduct less-critical study at an approved state or university laboratory or insectary. This approach was recently used for introducing a native plant pathogen into Hawai'i. The screening was accomplished at the Federal Plant Pathogen Quarantine Facility at Frederick, Maryland, then the pathogen was introduced to Hawai'i and studied at the Department of Plant Pathology at the Mānoa campus of the University of Hawai'i (Trujillo *et al.* 1986). New regulations now make this procedure impossible in Hawai'i.

If a quarantine facility does not exist or is unavailable for use, an alternative temporary facility can be constructed and equipped. This, however, can cost $150,000 to $250,000 and take one to two years before it is fully approved. For a permanent, certified facility, the expected expense could exceed $1 million.

Administrative Support

This criterion, although obvious, warrants serious consideration. Administrators define long-term goals and program directions and allocate space, funding, and manpower. Their attitudes and continual support, and the infrastructure of the agency, are critical in approving a new biological control program and in obtaining the necessary support. A legislative mandate for control acts to encourage administrative support.

A new biological control program usually requires 7-10 years before the first agents are released. Several years of monitoring should follow release. Therefore, an organization with primarily short-term objectives and priorities that often change, or that are revised yearly, probably does not provide suitable continuity in which to initiate a biological control program.

Long-term Funding

Biological control of weeds is expensive. Even if a quarantine facility is available, the cost of a full-scale program (Table 1) aimed at a single weed has been estimated at $1-2 million (Andres 1977; Harris 1979). Will the supporting agency commit itself for this amount over a 7-10 year period? If sufficient long-term funding is guaranteed, a major factor has been overcome. If the program must compete for available funding or find outside support, it may be dropped before it is completed. Of course, the funding cycles of most Federal and state agencies create a certain amount of risk that must be accepted for progress to occur.

Available Scientific Expertise

Biological control is now a recognized science taught in entomological departments throughout the U.S. According to Coulson and Hagan (1985), 219 investigators at present directly or indirectly conduct research related to biological control in the U.S. Department of Agriculture. Most of these researchers are presently engaged in the control of insects, with only 35 listed as actively working on the control of weeds. We estimate that an additional 20 to 30 qualified persons are located in at least six state plant and insect laboratories. If an experienced investigator can be found to head or at least participate in a new program, initiation and operations are greatly simplified. If experienced personnel are not available, inexperienced people must be hired and trained or existing entomologists transferred and trained. This increases the time before the new program is fully operational by an additional year or two.

Geographic Area Affected

The immediate goal of any biological control program is the introduction and release of an insect (occasionally a nematode, mite, or pathogen) into the ecosystem in which the weed is present. Once established, agents do not usually remain at their point of release, but eventually spread over part or most of the range of the weed. The total range of the weed must be considered in evaluating the impact of a biological control program, not just the natural area that is to be protected. The chance of conflict of interest (Andres 1980) increases in proportion to the size of the area and the number of people familiar with the weed in that area. If the weed is newly established and localized in a natural area, opposition to biological control should be minimal. (Interest in biocontrol at this point might also be slight until a larger area is occupied.) In contrast, a weed that is found well beyond the boundaries of a natural area is more likely to be considered beneficial in another area with different land use priorities.

The geographic range of a weed is not limited by political boundaries, and the effect of a biological control program on neighboring countries must also be considered. On the U.S. Mainland, when a researcher applies for permission to release a new biological control agent, the release request is submitted to a review committee representing various U.S. government agencies and institutions. Applications for agent release are also routinely forwarded by this review committee to Canada and Mexico if the weed exists in these countries (Klingman and Coulson 1982).

Ecological Range of Weed

The range of ecological areas over which an introduced weed has become established can affect the degree of success of a biological control program. In many past programs, the climatic range of introduced agents was not as broad as that of their host weed, resulting in pockets of unaffected plants. In Hawai'i, a successful program against prickly pear cactus (*Opuntia* spp.) has left much of this species unaffected at elevations above 3,300 ft (1,000 m). Lantana (*Lantana camara*), despite 50 years of effort and the introduction of over 35 different insects, still remains a serious weed in parts of several of the Islands (Davis *et al.* this volume).

A similar problem with extensive range has occurred with Klamath weed (*Hypericum perforatum*) in North America. While three introduced agents were successful in controlling this weed in the United States and most of Canada, the northernmost extreme of the weed's range in British Columbia was beyond the climatic tolerance of released insects; there the weed remains unaffected (Williams 1984). Surviving pockets can be ignored, handled by conventional methods of management, or efforts can be made to find additional biological control agents adapted to additional areas.

Life Cycle of Weed

In the past, most successes in biological control of weeds have been against shrubs and other perennial plants. This suggests that annual weeds, with their short life cycles and long periods of dormancy while in the seed stage, are difficult for biological control agents to attack (Huffaker 1957). In Hawai'i and other tropical and subtropical areas where many annuals can grow throughout the year, this may not be as critical as elsewhere. Two very successful weed programs in Hawai'i have been against two annuals, the puncture vine (*Tribulus terrestris*) and the spiny emex (*Emex spinosa*) (Markin *et al.*, this volume).

Density and Local Distribution of Weed

Generally, biological control programs have been conducted against weeds that formed extensive and dense stands, replacing large areas of desirable vegetation (usually range plants). To protect undisturbed native ecosystems, biological control may be considered against plants that are scattered, particularly early in the process of invasion. Biological control may not be as effective against scattered target plants as in dense stands of the weed because of insect survival and dispersal difficulties. For example, biological control of the prickly pear cactus *Opuntia aurantiaca* in Australia with the use of *Dactylopius austrina* (a wingless sucking insect with immature stages dependent upon wind dispersal) was very effective in dense stands of cactus, but was ineffective in controlling isolated cactus patches. The effective but expensive solution was to disperse the insect manually to isolated clusters of plants (Hosking and Deighton 1979).

Associated Insects and Pathogens

The success rate for establishing new biocontrol agents in an area to date is about 50% (Markin *et al.*, this volume). Unfortunately, many established insects fail to build sufficient populations to overwhelm or significantly damage their host due to biologic interference by local parasites and predators that attack the introduced agents (Goeden and Louda 1976). A survey of naturally occurring predators (particularly ants) and general parasites attacking local Lepidoptera (butterflies and moths) and Coleoptera (beetles) in native ecosystems may indicate whether or not such problems might occur. No guidelines can be given for interpreting the results of such a survey. Presumably, however, the more abundant and diverse the parasites and predators, the greater the chance of interference with biological control introductions.

A preliminary survey of the insects already associated with the target plant is also useful to identify existing phytophagous insects. The survey may indicate that some potential biological control agents are already present. In North America, the introduced weed Scotch broom (*Cytisus scoparius*) once was a target for biological control (Goeden 1978; Julien 1982). A preliminary survey, however, showed that 10 European species of phytophagous insects had been accidentally introduced and were already attacking the plant (Waloff 1966).

Economic Importance of Target Weed

Although it may be difficult for resource managers to visualize any redeeming value for a weed invading a native ecosystem, they should realize that many weeds were introduced as potential crops and may still be used in agriculture. Of 20 major introduced plants in national parks in Hawai'i (Gardner and Davis 1982), ginger (*Hedychium* spp.) and common guava (*Psidium guajava*) are grown commercially, and koa haole (*Leucaena leucocephala*) and at least four species of grasses are recognized as livestock forage plants. Proposing biological control programs for common guava, ginger, koa haole, or any of these grasses may not be well supported because of opposition from agricultural interests. The original biological control program against cactus in Hawai'i was opposed by some ranchers, since they had learned to use cactus as an emergency food and moisture source for cattle during drought (Fullaway 1954). However, the cactus control program eventually was started anyway (Davis *et al.*, this volume).

Related Agriculturally Important Plants

Initiation of a biological control program for a weed closely related to a plant of recognized agricultural value is difficult but has been successfully accomplished. Biological control was undertaken and insects released successfully against thistles and knapweeds (*Cirsium*, *Carduus* spp., and *Centaurea* spp.) (Frick 1976; Goeden 1978; Julien 1982) in the U.S. despite their close relationships (all are in the same tribe, Cynareae) to the safflower *Carthamus tinctorius* and the artichoke *Cynara scolymus*, both important agricultural plants. The ultimate decision to release biological control agents against this group of weeds was made by a review committee, which probably paid considerable attention to the benefits to be derived from a successful program against these major agricultural weeds, as opposed to possible losses if the agents proved capable of attacking an agricultural crop. Such a crossover is not likely and would have been demonstrated by extensive studies of the host range of the insect in its native homeland and by quarantine studies in the United States. However, the extra time required for these studies no doubt increased the expense of screening agents for thistles and knapweeds.

Holm *et al.* (1977) identified the 18 worst weeds of agriculture on a worldwide basis, and 10 of these are grasses. Yet not a single biological control program has ever been attempted for the group (Pemberton 1980), due to their relationships to major crop plants. Some programs are now being given serious consideration (C.W. Smith, pers. comm.). In Hawai'i, the sugar cane industry has been an important influence on proposals to control grasses in the past. As a rule, biocontrol programs should generally not be sought where target plants are closely related to crops of major economic importance (Table 2, item 10).

In Hawai'i, we are presently facing the problem of commercial species closely related to the forest weed banana poka (*Passiflora mollissima*). This plant, which grows at elevations of 1,000-2,000 ft (300-600 m) (LaRosa 1984, this volume), is related to the edible passion fruit *P. edulis*, which was grown commercially on 40 acres (16 ha) in 1984 (Anonymous 1985), but always at an elevation of less than 900 ft (300 m). Although it will be difficult to find a control agent that will not attempt to feed on passion fruit when starving, we hope to find one that not only does not prefer it but cannot successfully complete its life cycle on passion fruit or cannot survive in the warmer, drier conditions below 900 ft (300 m) elevation. To date, we have identified several insects that fit these criteria, but it will entail considerable extra time and cost in screening potential agents.

Use by Beekeepers

Beekeepers regularly oppose biological control programs because even the most noxious weeds may be important sources of nectar and pollen. The subject has been discussed earlier (Markin *et al.*, this volume), so it is sufficient to note that if the target weed is regularly used by beekeepers over any part of its range, opposition can be expected.

Related Endangered Species

Besides economically important plants related to the target weed, threatened or endangered species must be considered when releasing a new biological control agent. If one or more endangered species is even distantly related to the weed to be controlled, much more extensive host testing will be necessary to demonstrate the safety of potential biological control agents. The problem of related endangered species has recently complicated efforts to control leafy spurge (*Euphorbia esula*) in North America (Pemberton 1985).

Redeeming Values of Target Weed

In Hawai'i, some of our weeds were accidentally introduced, but many were deliberately brought in as potential agricultural crops. Others were introduced for landscaping, medicinal purposes, or ornamentals. Many of these plants are still valued for these purposes, or new purposes that have been since discovered, and conflicts of interest can be expected with their user groups (Andres 1980). Learning the history of the weed and its present intrinsic value to the local population are important parts of planning a new biological control program; identifying potential conflicts of interest before they occur is critical.

Status of Target Weed Worldwide

There is considerable value in working on weeds that are recognized as problems in other parts of the world. Other countries may be willing to cooperate in a new biological control program by participating in the actual research, providing financial support, or exchanging information if they already have a program of their own under way.

Previous Biological Control Programs

Emphasizing weeds that have already been successfully controlled by biological control programs in other countries is one of the surest guarantees of success. This factor is given great weight in Table 2 (25 points). Usually, the control agents are readily available and need only a minor amount of additional host testing against related native plants before they can be released. In Hawai'i, at least three of our most successful programs followed this pattern -- prickly pear cactus, Klamath weed, and puncture vine. However, even selecting weeds which have been successfully controlled in other countries does not guarantee success in a new area. We have been successful in controlling lantana in many areas of Hawai'i, but introduction of the insects effective in Hawai'i into 18 other countries where lantana is also a problem gave only partial success in five of the cases (Julien 1982).

Accessibility of Plant in Its Natural Range

An essential early step in conducting a biological control program is sending an explorer to the weed's country of origin to study and select potential agents. Easy and continued accessibility to the natural range of the weed is essential. If a weed's natural range is in undeveloped countries, the lack of maps, transportation systems, and general accommodations can hamper exploration. There are also countries where, for political reasons, an American would not be welcomed or want to venture.

Availability of Research Facilities in Plant's Natural Range

An additional way to simplify a proposed study and increase its chance of success is to conduct as much of the biological work and screening as possible within the natural range of the target weed. This can often be done cheaply and easily by using foreign facilities where a visiting scientist can be stationed, or a local scientist hired. In Italy, France, Greece, Argentina, and Korea, the U.S. Department of Agriculture maintains facilities where this type of work can be done (Coulson and Hagan 1985). Other facilities are maintained by the Commonwealth Institute of Biological Control in countries such as Trinidad, where the Hawaii Department of

Agriculture is presently doing most of its work on clidemia (Nakahara *et al.*, this volume).

Availability of Wild Genotypes of Target Weed

Another problem in biological control is the possibility that in its natural range the target weed may have become extinct or its range greatly reduced due to deforestation or replacement by agriculture. This problem has been encountered in our effort to control banana poka. To date, no wild plants have been discovered in three different surveys of the Andes, and the few insects collected have come from domestic plants growing around homes or in commercial plantings. At this time we are beginning to suspect that the original natural range of this plant, the forested areas of the Andes, has been almost totally eliminated by agricultural development (G. Taniguchi and R. Pemberton, pers. comm.). The situation limits our selection of races and genotypes of the few insects we have been able to find. Good genetic variation is an important characteristic of a successfully introduced species (Wilson 1963; Mackauer 1976).

Availability of Closely Related Species

Successful introduction programs for biological control were reviewed by Hokkanen and Pimental (1984), who pointed out that almost half of all successful biological control agents were not collected on the target hosts (both plants or insects), but on species closely related to the target. The implication is that insects associated directly with the target weed may have evolved a balance with their host, which prevents them from overwhelming and destroying it and therefore destroying themselves. Insects not directly associated with target weeds may be more likely to reach population levels needed to eventually overwhelm targets. Work with a weed in a genus with few species could therefore be a disadvantage, while working with a weed with many closely related species may provide a much larger pool of candidate biological control agents.

USE OF SCORING SYSTEM

To use the scoring system for evaluating chances of success with biological control of a particular weed, a resource manager or researcher, based on familiarity with his or her agency or organization, the native ecosystem at risk, and the target weed, can assign a value from 0 to 5 (more than 5 in two cases) to 18 of the 20 critical factors outlined above. The more values other than 0 or 5 are used, the more subjective the system becomes. The same people should rate all possible programs. When totalled, the final score can be used as an indication of probable success. A score of 100 or more would indicate that a program for a particular weed has a very good chance of success. Realistically, the scores are not a true indication of success of a proposed program; rather, scores provide an indication of the difficulty or ease with which a program can be completed. Difficulties with any one of the 20 factors mentioned may make a successful program very difficult or even impossible. Generally, however, no factor alone will totally negate a successful program. Most difficulties can be overcome with sufficient commitment of manpower and resources.

RATINGS FOR THREE IMPORTANT WEEDS IN NATIVE ECOSYSTEMS OF HAWAI'I

To demonstrate the proposed scoring system, three weeds that are major threats to native Hawaiian ecosystems were selected and scored. All three weeds have been discussed by others in this volume -- banana poka (LaRosa), firetree (*Myrica faya*) (Whiteaker and Gardner), and gorse (*Ulex europaeus*) (Tulang). Scores applied to each of the 20 critical factors and the totals are shown in Table 3. Scores for factors 1 through 5 are the same for all weeds, since it is presumed the work would be done by the same research organization.

Banana poka. A biological control program against this weed has begun, and we have already referred to some of the problems encountered. Several preliminary studies have also been completed on the plant's local biology (LaRosa 1984) and distribution (Warshauer *et al.* 1983). Its present status in Hawai'i has also been discussed (LaRosa, this volume). In summary, the weed is presently established on over 100,000 acres (40,000 ha) but is restricted to forests between 3,000 and 6,500 ft (1,000-2,000 m). It is a perennial vine and forms very dense, continuous stands, usually worse where the forest canopy has been opened.

Our surveys have indicated the presence of few general insect predators associated with this plant, and our study of insects on blackberry (*Rubus argutus*), another introduced weed with similar distribution, has led us to believe there may not be a large population of general larval parasites in the rain forest ecosystem where banana poka is found (Nagata and Markin 1986).

The major disadvantage of biological control for this species in Hawai'i is that banana poka is related to the edible passion fruit *P. edulis*, which is commercially raised on a very limited scale (Anonymous 1985). Fortunately, passion fruit is grown only at lower elevations (below 900 ft or 300 m), so we hope to find an agent with a sufficiently narrow climatic range to exclude it from lower elevations. Banana poka is not used by beekeepers, and there are no native species in the same genus, or family, in Hawai'i. There are several possible areas of conflict, including the fact that native birds now utilize its nectar and that the feral pig (*Sus scrofa*) and the kalij pheasant (*Lophura leucomelana*), both game species, eat its fruit.

Banana poka is established in Australia and New Zealand but is not considered a weed in either country. A survey in South America has been conducted to identify potential biological control agents (Pemberton 1982), so some information exists on its natural enemies. This plant's natural range is in the high Andes (9,000 ft or 3,000 m and above) of northern South America, and problems encountered in trying to work in some parts of this area include political instability, few research facilities, and few trained entomologists with whom to collaborate. Also, the plant apparently has been exterminated by agricultural development over much of its natural range, although a large number of domestic varieties and closely related species exist in South America.

Table 3. Evaluation of three proposed biological control programs of introduced plants of Hawaii.

Critical Factor	Assigned Score: Banana poka program	Assigned Score: Firetree program	Assigned Score: Gorse program
1. Availability of quarantine facilities	5	5	5
2. Administrative support	5	5	5
3. Long-term funding	3	3	3
4. Available scientific expertise	5	5	5
5. Geographic area affected	5	5	5
6. Ecological range of weed	5	3	5
7. Life cycle of weed	5	5	5
8. Density and local distribution of weed	3	5	5
9. Associated insects and pathogens	5	5	4
10. Economic importance of target plant	5	5	5
11. Redeeming values	3	5	5
12. Related agriculturally important plants	0	5	5
13. Use by beekeepers	5	5	5
14. Related endangered species of plants	5	5	1
15. Status of target weed worldwide	0	0	5
16. Previous biological control programs	0	0	5
17. Accessibility of the plant for study in its natural range	0	4	5
18. Availability of research facilities in its natural range	1	4	5
19. Availability of wild genotypes	0	3	5
20. Availability of closely related species	5	3	2
Total	65	80	90

Banana poka appears to be a difficult biological control subject to undertake, based on the final total score of 65. The major problems are conflict of interest due to its close relationship to the edible passion fruit, and the difficulty in working in its natural range, either to study it or to obtain a usable supply of insects for evaluation.

Firetree. One survey of the natural range of firetree for associated natural enemies has been conducted in the Azores, Madeira, and Canary Islands in the Atlantic (Hodges and Gardner 1985). In Hawai‘i, the plant is distributed on all major islands except Moloka‘i and infests an estimated 85,000 a (35,000 ha) ranging from 1,900 to 6,600 ft (500-2,000 m). It has invaded many different ecosystems, which suggests that we may have difficulty finding biological control agents that can cover the extremes of its range. It is a perennial small to large tree, and in many areas it forms very dense stands.

Firetree has never been considered an agricultural plant in Hawai'i or in its natural range, nor are any of its close relatives used in agriculture. However, wax myrtle (*M. cerifera*) is grown as an ornamental in Hawai'i (Neal 1965). While the weed has a small flower, it is presently not recognized by beekeepers as a source of nectar or pollen. Similarly, the weed has no relatives in the same family in Hawai'i, so it is not related to any rare plants. While *M. faya* at one time was used for reforestation and ground cover, since 1944 its noxious nature has been recognized, and several mechanical and herbicidal programs have been initiated for its control (Kim 1969; Gardner and Kageler 1982).

Firetree is recognized as a weed only in Hawai'i; therefore, no biological control programs have been attempted except in Hawai'i, where one insect, *Strepsicrates smithiana*, was established in 1955. The insect has only been found to attack the closely related wax myrtle (Krauss 1964; Julien 1982). The original natural range of the plant is restricted to islands in the Atlantic Ocean, and political stability is not a problem. Most of these islands contain limited agriculture research facilities which could be used by visiting scientists. Unfortunately, wild stands of this plant in undisturbed native ecosystems are apparently limited (Hodges and Gardner 1985). Finally, firetree belongs to a fairly small family with only 45 species in two genera (Neal 1965).

In conclusion, the extensive distribution and ecological range of firetree in Hawai'i may make it a difficult plant to control with a single biological control agent, although its native range is well suited for conducting foreign exploration. Unfortunately, the limited wild populations of this plant in its natural range, lack of closely related species, and few reported associated insects probably indicate that there will be some difficulty finding natural agents with which to work. The species does not show potential for agricultural conflicts, but being a problem only in Hawai'i, and belonging to a very small family of plants, means that little work has been undertaken to provide background information. However, when compared to banana poka (with a score of 65), firetree, with a score of 80, may be more amenable to biological control.

Gorse. Gorse is a perennial shrub with spines, which, once established, rapidly forms very thick, impenetrable stands. It is a potentially significant problem in native ecosystems in Hawai'i, although many resource managers are unfamiliar with it. It is presently restricted to two infestations, totalling about 35,000 a (14,000 ha), on Maui and Hawai'i; dense stands cover less than 7,500 a (3,000 ha) (Markin *et al.* 1988). In Hawai'i its range is generally limited to between 3,000 and 6,600 ft (1,000-2,000 m) in open pasture, although it has begun to invade forests dominated by both native and alien species. Insects associated with gorse in Hawai'i have been surveyed (Markin 1984), and no native insects were found attacking it, although several introduced Lepidoptera already feed on its flowers. One biological control agent, a small seed weevil (one of three insects introduced in the 1950s) is now well established.

Gorse was once used for food by sheep (*Ovis aries*) and to form hedgerows in Hawai'i. However, since the 1940s, it has been declared a noxious weed. It has no close relatives in Hawai'i that are presently used in agriculture. However, as a legume, it is distantly related to many agricultural and native species. Fortunately, gorse is in a different tribe (Genisteae) than all native and important introduced species. In other parts of the world, gorse is widely used by beekeepers as a major source of nectar and pollen, but not in Hawai'i.

To date, we have been unable to find any recognized uses for gorse, although in other parts of the world it is still used for livestock foraging, hedgerows, and firewood. Because of its status as a major weed in many different parts of the world, a considerable amount of literature on it exists, including several annotated bibliographies (Anonymous 1975, 1978; Gaynor and MacCarter 1981).

From its native range in western Europe, gorse has been introduced to many parts of the world, including India, South America, New Zealand, Australia, the west coast of the United States, as well as Hawai'i. In all of these areas it is considered a major weed problem.

The natural range of this plant is western Europe, and extensive surveys have been conducted on its natural enemies in England, France, Spain, and Portugal. In Europe, at least 20 closely related species of *Ulex* are known (although many are considered synonymous by some authorities), and over 100 insects have been found attacking them. The British Commonwealth Institute for Biological Control, headquartered in England, has, at the request of New Zealand, conducted extensive studies of this plant in Europe.

In general, we believe that the past efforts devoted to the study of this weed, both by the State of Hawai'i and foreign countries; the earlier efforts at biological control; the extensive facilities and availability of experienced scientists familiar with gorse in Europe; and the respectable list of natural enemies known to attack it and its close relatives indicate that there is a good chance that a biological control program against this weed in Hawai'i would be successful.

CONCLUSION

This checklist provides resource managers or researchers with a decision-making tool that can be used to predict the potential success of biological control of an introduced plant in native ecosystems. The scoring system is subjective, and we undoubtedly have omitted many critical factors that other investigators would include. Values we have assigned to each critical factor may not be considered by others to be realistic or representative of the importance of a factor under the local conditions and environment in which they have to work. However, if a resource manager concludes that biological control of a weed in a natural area is a viable management option, the scoring system may be useful as a method for comparing the chances of success for a number of weed species.

ACKNOWLEDGMENTS

We thank G. Funasaki, P.-Y. Lai, R.W. Pemberton, C.J. Davis, and D. Taylor for reviewing an early draft of the manuscript.

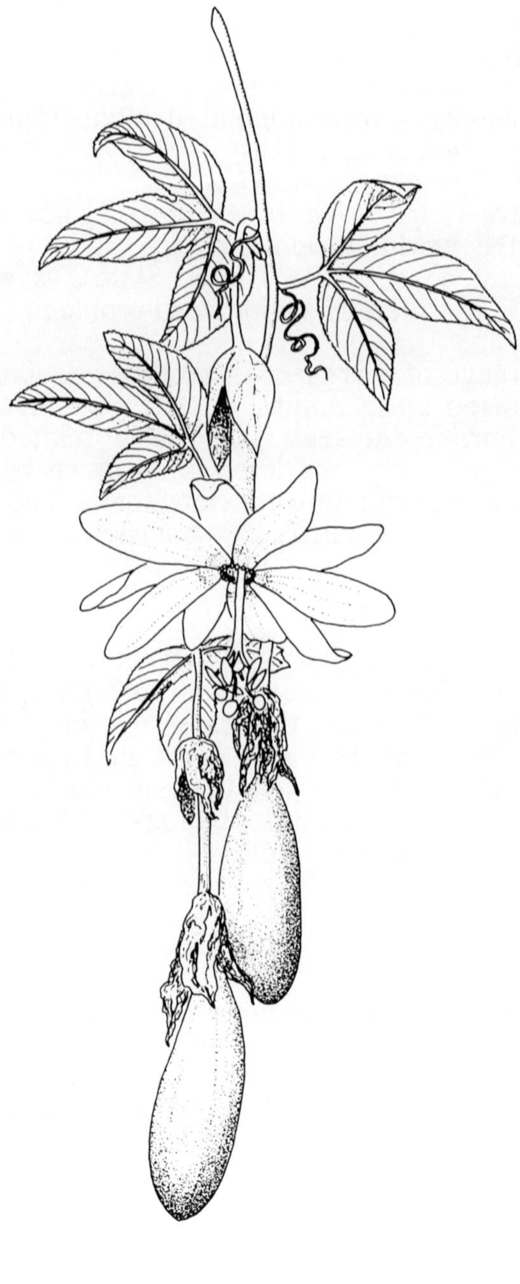

Literature Cited

Andres, L.A. 1977. The economics of biological control of weeds. *Aquatic Botany* 3:111-123.

Andres, L.A. 1980. Conflicting interests and the biological control of weeds. *Proc. V. Internatl. Symp. Biol. Contr. Weeds*, 11-20. Brisbane, Australia.

Anonymous. 1975. *Selected references to the control of gorse (*Ulex europaeus*) 1940-1975.* Annot. Bibliogr. 89. Begbroke Hill, Yarnton, Oxford, U.K.: Agric. Res. Council, Weed Research Organization.

Anonymous. 1978. *Selected references to the control of gorse (*Ulex europaeus*) 1968-1978.* Annot. Bibliogr. 122. Begbroke Hill, Yarnton, Oxford, U.K.: Agric. Res. Council, Weed Research Organization.

Anonymous. 1985. *Statistics of Hawaiian Agriculture 1984.* Hawaii Agric. Reporting Service, Honolulu.

Coulson, J.R., and J.H. Hagan. 1985. *Information on U.S. and Canadian biological control programs using natural enemies and other beneficial organisms.* Doc. No. 0006I. Beltsville, Maryland: USDA Agric. Res. Serv., ARS Biol. Control Documentation Center.

Davis, C.J., E. Yoshioka, and D. Kageler. [this volume] Biological control of lantana, prickly pear, and Hāmākua pāmakani in Hawai'i: a review and update.

DeBach, P. 1964. *Biological control of insect pests and weeds.* New York: Reinhold Pub. Corp.

DeBach, P. 1974. *Biological control by natural enemies.* London: Cambridge Univ. Pr.

Frick, K.E. 1976. Biological control of weeds: Introduction, history, theoretical and practical applications. In *Proc. Summer Institute on Biological Control of Plants, Insects and Disease*, ed. F.G. Maxwell and F.A. Harris, 204-223. Jackson, Mississippi: Univ. Pr. of Mississippi.

Fullaway, D.T. 1954. Biological control of cactus in Hawaii. *J. Econ. Entomol.* 47(4):696-700.

Gardner, D.E. [this volume] Plant pathogens as biocontrol agents in native Hawaiian ecosystems.

Gardner, D.E., and C.J. Davis. 1982. *The prospects for biological control of nonnative plants in Hawaiian national parks.* Tech. Rep. 45, Univ. Hawaii Coop. Natl. Park Resour. Stud. Unit. Honolulu.

Gardner, D.E., and V.A.D. Kageler. 1982. *Herbicidal control of firetree in Hawaii Volcanoes National Park: a new approach.* Natl. Park Serv. Ecol. Serv. Bull. 7. Washington, D.C.: U.S. Dep. Interior.

Gaynor, D.L., and L.E. MacCarter. 1981. Biology, ecology and control of gorse (*Ulex europaeus* L.): a bibliography. *New Zealand J. Agric. Res.* 24:123-137.

Goeden, R.D. 1978. Biological control of weeds. In *Introduced parasites and predators of arthropod pests and weeds: a world review*, ed. C.P. Clausen, 357-545. U.S. Dep. Agric. Handb. No. 480. Washington, D.C.: U.S. Govt. Print. Office.

Goeden, R.D., and S.M. Louda. 1976. Biotic interference with insects imported for weed control. *Ann. Rev. Entomol.* 21:325-342.

Grabau, W.E., and N.R. Spencer. 1976. A management procedure for the introduction of biological agents for control of weeds. *Proc. IV Internatl. Symp. Biol. Contr. Weeds*, Gainesville, Florida, 1976, 13-34.

Harris, P. 1979. Cost of biological control of weeds by insects in Canada. *Weed Science* 27:242-250.

Hodges, C.S., Jr., and D.E. Gardner. 1985. *Myrica faya: potential biological control agents*. Tech. Rep. 54, Univ. Hawaii Coop. Natl. Park Resour. Stud. Unit. Honolulu.

Hokkanen, H., and D. Pimentel. 1984. New approach for selecting biological control agents. *Can. Entomol.* 116:1109-1121.

Holm, L.G., D.L. Plucknett, J.V. Pancho, and J.P. Herberger. 1977. *The world's worst weeds, distribution and biology*. East-West Center. Honolulu: Univ. Pr. Hawaii.

Hosking, J.R., and P.J. Deighton. 1979. The distribution and control of *Opuntia aurantiace*, New South Wales. *Proc. Seventh Asian-Pacific Weed Sci. Soc. Conf.* Sidney, Australia, 195-200.

Huffaker, C.B. 1957. Fundamentals of biological control of weeds. *Hilgardia* 27:101-157.

Huffaker, C.B. 1959. Biological control of weeds with insects. *Ann. Review Entomol.* 4:251-276.

Julien, M.H. 1982. *Biological control of weeds, a world catalogue of agents and their target weeds.* Commonwealth Inst. Biol. Contr. Farnham Royal, Slough, U.K.: Commonwealth Agric. Bur.

Kim, J.Y. 1969. *Myrica faya* control in Hawai'i. *Down to earth* 25:23-25.

Klingman, D.L., and J.K. Coulson. 1982. Guidelines for introducing foreign organisms into the United States for biological control of weeds. *Weed Sci.* 30:661-667.

Krauss, N.L.H. 1964. Insects associated with firebush (*Myrica faya* Aiton). *Proc. Hawn. Entomol. Soc.* 18(3):405-411.

LaRosa, A.M. 1984. *The biology and ecology of* Passiflora mollissima *in Hawaii*. Tech. Rep. 50, Univ. Hawaii Coop. Natl. Park Resour. Stud. Unit. Honolulu.

LaRosa, A.M. [this volume] The status of banana poka in Hawai'i.

Mackauer, M. 1976. Genetic problems in the production of biological control agents. *Ann. Review Entomol.* 21:369-385.

Markin, G.P. 1984. Biological control of the noxious weed gorse *Ulex europaeus* L.: a status report. [abstract] *Proc. Fifth Conf. Nat. Sci., Hawaii Volcanoes Natl. Park*, 77. Honolulu: Univ. Hawaii Coop. Natl. Park Resour. Stud. Unit.

Markin, G.P., P.-Y. Lai, and G. Funasaki. [this volume] Status of biological control of weeds in Hawai'i and implications for managing native ecosystems.

Markin, G.P., L.A. Decker, L.A. Lapp, and R.F. Nagata. 1988. Distribution of gorse (*Ulex europaeus* L.), a noxious weed in Hawaii. *Newsl. Hawn. Bot. Soc.* 27:110-117.

Nagata, R.F., and G.P. Markin. 1986. Status of insects introduced into Hawaii for biological control of wild blackberry, *Rubus argutus*. *Proc. Fifth Conf. Nat. Sci., Hawaii Volcanoes Natl. Park*, 53-64. Honolulu: Univ. Hawaii Coop. Natl. Park Resour. Stud. Unit.

Nakahara, L.M., R.M. Burkhart, and G.Y. Funasaki. [this volume] Review and status of biological control of clidemia in Hawai'i.

Neal, M.C. 1965. *In gardens of Hawaii*. B.P. Bishop Mus. Spec. Pub. 50. Honolulu: Bishop Mus. Pr.

Pemberton, R.W. 1980. International activity in biological control of weeds: patterns, limitations and needs. *Proc. V Internatl. Symp. Biol. Contr. Weeds*, Brisbane, Australia, 1980, 57-71.

Pemberton, R.W. 1982. Exploration for natural enemies of *Passiflora mollissima* in the Andes. U.S. Dep. Agric. Biol. Contr. Weeds Lab, Albany, Calif. Unpubl. Rep.

Pemberton, R.W. 1985. Native plant consideration in the biological control of leafy spurge. *Proc. VI Internatl. Symp. Biol. Contr. Weeds*, Vancouver, Canada, 1984, 365-390.

Trujillo, E.E., F.M. Latterell, and A.E. Rossi. 1986. *Colletotrichum gloeosporioides*, a possible biological control agent for *Clidemia hirta* in Hawaiian forests. *Plant Disease* 70:974-976.

Tulang, M. [this volume] The U.S. Department of Agriculture's rural development approach to alien plant control in Hawai'i.

Waloff, N. 1966. Scotch broom and its insect fauna introduced into the Pacific Northwest of America. *J. Applied Ecol.* 3:293-311.

Warshauer, F.R., J.D. Jacobi, A.M. LaRosa, J.M. Scott, and C.W. Smith. 1983. *The distribution, impact, and potential management of the introduced vine* Passiflora mollissima *(Passifloraceae) in Hawai'i.* Tech. Rep. 48, Univ. Hawaii Coop. Natl. Park Resour. Stud. Unit. Honolulu.

Whiteaker, L.D., and D.E. Gardner. [this volume] Firetree (*Myrica faya*) distribution in Hawai'i.

Williams, K.S. 1984. Climatic influences on weeds and their herbivores: biological control of St. John's Wort in British Columbia. *Proc. VI Internatl. Symp. Biol. Contr. Weeds*, Vancouver, Canada, 1984, 127-132.

Wilson, F. 1963. Biological control and the genetics of colonizing species. In *The genetics of colonizing species*, ed. F. Wilson, 307-329. New York: Academic Pr., Inc.

ECOSYSTEM APPROACHES, EDUCATION, AND COMMUNITY INVOLVEMENT

SPECIAL ECOLOGICAL AREAS: AN APPROACH TO ALIEN PLANT CONTROL IN HAWAII VOLCANOES NATIONAL PARK

J. Timothy Tunison and Charles P. Stone

ABSTRACT

A Special Ecological Areas approach to alien plant management was adopted at Hawaii Volcanoes National Park in 1985. The approach involves control of widespread, disruptive alien plant species in Special Ecological Areas, which are intensive management and research units in the Park. Special Ecological Areas management developed from these perceptions: 1) unmanaged alien species were affecting many of the most valuable biological sites in the Park; 2) some widespread alien plants had increased their ranges dramatically in recent years; 3) alien plant control efforts were focused on species in degraded parts of the Park; and 4) biological control was not a panacea for widespread weed problems. Areas were selected according to the following criteria: Representativeness/rarity of vegetation type, vegetation intactness, plant species diversity/richness, manageability, presence of rare flora, preserve design considerations, immediacy of threat from alien plants, and research and interpretive values. Eleven species were targeted for control in Special Ecological Areas, were usually monitored by counting individuals encountered, and were controlled on an annual basis. All target species have declined sharply since initiation of treatment. For example, density or biomass of the three most important alien species, firetree (*Myrica faya*), strawberry guava (*Psidium cattleianum*), and kāhili ginger (*Hedychium gardnerianum*), declined by 86%, 85%, and 96% respectively in Special Ecological Areas treated from 1985 to 1988. The number of worker days required for control efforts declined 73% over the same interval. Tentative conclusions about alien plant control in Special Ecological Areas are: 1) populations of alien plant species can be significantly reduced and maintained at very low levels after three to four years of control work; 2) recruitment of alien plant species is usually very low; 3) continued follow-up treatments are required in all areas and may be needed indefinitely; 4) workloads drop significantly after initial control efforts; and 5) Special Ecological Areas serve as effective models for management of other areas. Special Ecological Areas also provide a focus for research on and monitoring of native and alien species as related to management. Permanent monitoring systems will soon be established for birds, Hawaiian bats (*Lasiurus cinereus semotus*), invertebrates, and rare plants, including candidate endangered species. Habitat correlates for key species and groups will be determined so that Park managers can better protect essential areas in the Park. Baseline information on the ecology and responses of Special Ecological Areas to management may lead to translocation and reestablishment of birds and plants previously found in the Park.

INTRODUCTION

A Special Ecological Areas approach to alien plant management was adopted in Hawaii Volcanoes National Park in 1985. The approach involves control of widespread, disruptive alien plant species in the Areas, intensive management and research units in the Park selected for their biological and management values. Eight Special Ecological Areas totalling 12,954 a (5,266 ha) have been established to date (1988) (Fig. 1). The purpose of this paper is to:

1. Describe the current alien plant control program in Special Ecological Areas and how the Special Ecological Areas strategy developed.

2. Document changes in alien plant populations under management.

3. Summarize long-term monitoring and research supporting alien plant control programs in Special Ecological Areas.

4. Discuss strengths and weaknesses of a Special Ecological Areas approach to alien plant management.

5. Outline long-term plans for alien plant management in Special Ecological Areas.

REASONS FOR SPECIAL ECOLOGICAL AREAS

The Special Ecological Areas approach markedly changed existing alien plant control strategies at Hawaii Volcanoes National Park. Prior to this strategy, alien plant management emphasized control of localized alien plant species and control of firetree (*Myrica faya*), fountain grass (*Pennisetum setaceum*), koa haole (*Leucaena leucocephala*), and other widespread weeds found mostly in the coastal lowland or submontane seasonal zones, the two ecological zones in the Park with the highest proportion of alien plant species. Control was not begun on a number of target widespread weeds found in the more intact plant communities (*e.g.*, banana poka (*Passiflora mollissima*), kāhili ginger (*Hedychium gardnerianum*), and strawberry guava (*Psidium cattleianum*)) because these species could not be controlled parkwide. Biological control was perceived as the only solution to the control of most widespread alien plant species.

New alien plant distribution information and uncertainties about the effectiveness and timeliness of biological control provoked a reevaluation of our approach to alien plant management. Distribution mapping completed

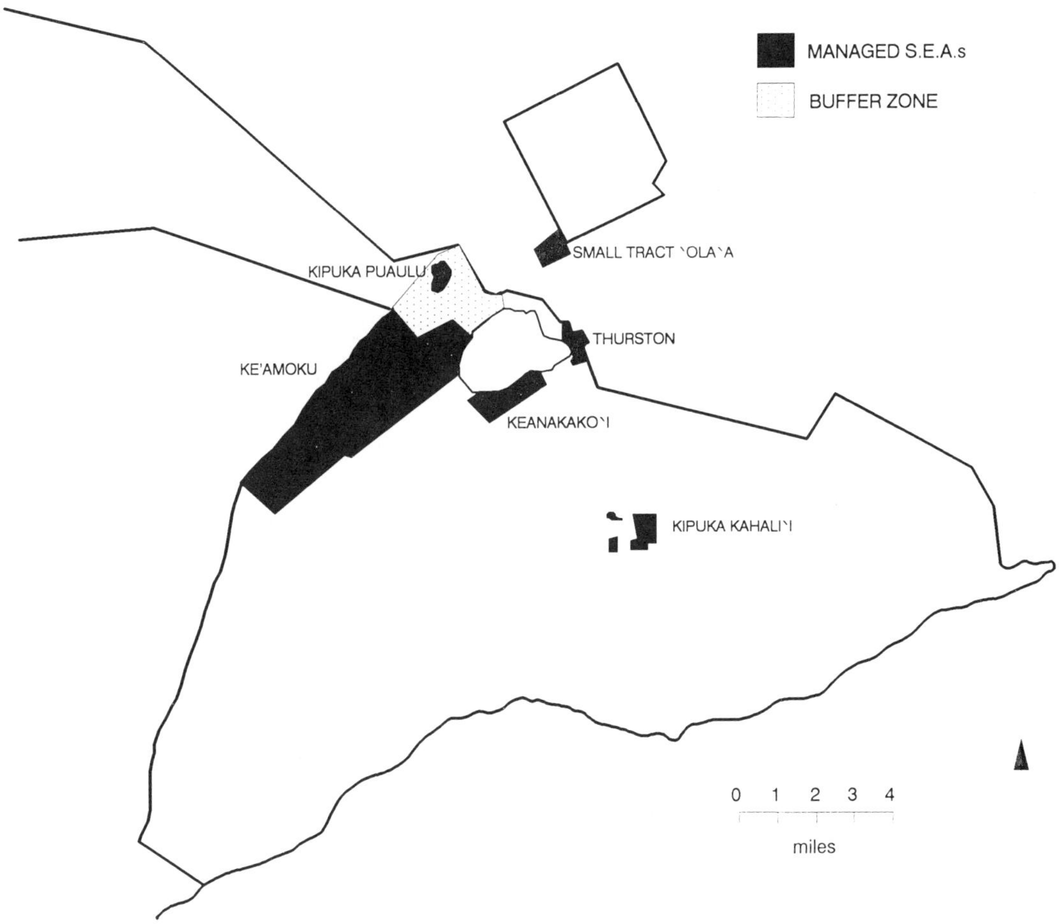

Figure 1. Special Ecological Areas in Hawaii Volcanoes National Park as of 1988.

in 1985 indicated that many target species, such as fountain grass, koa haole, and silk oak (*Grevillea robusta*), were too widely distributed for parkwide control without marked funding increases (Tunison *et al.*, in press; Tunison, this volume). Distribution mapping also indicated that firetree and banana poka had spread rapidly into extensive, mostly intact areas of the Park (Warshauer *et al.* 1982; Whiteaker and Gardner 1985). As the biocontrol program developed, it became clear that effective biocontrol agents might not be found or that irreversible ecological damage might occur before biological controls for such widespread species became effective (Gardner and Davis 1982; Markin *et al.*, this volume).

Additional considerations important in adopting and using a Special Ecological Areas approach are:

1. The most intact, diverse, unique, and valuable research and interpretive sites in the Park can be protected from the impacts of alien plant species with alien plant control in Special Ecological Areas. Uprooting of plants or chemical control can protect such sites until biological controls become effective or if biological controls are not found.

2. Alien plant control is more feasible in Special Ecological Areas than in more degraded parts of Hawaii Volcanoes National Park. Native vegetation in these areas is mostly intact, and alien plant control is more feasible because target alien populations are localized or at low densities.

3. Special Ecological Areas allows an incremental approach to management, a strategy that has proven effective in control of alien species in Hawai'i. When adequate control of an alien plant species in a Special Ecological Area is assured, the area can be expanded or additional areas established.

4. Special Ecological Areas can be models for management of other areas. Treatment methods and management strategies can be developed in Special Ecological Areas and applied to other areas. Prior to this approach, alien plant management was not conducted in many plant communities, and many of the most important alien plant species were not controlled.

5. Special Ecological Areas are sites for integrating feral animal and alien plant management. Feral goats (*Capra hircus*) have been controlled in all but the subalpine and alpine zones of the Park (Katahira and Stone 1982). Feral pigs (*Sus scrofa*) have been controlled in approximately 30% of their habitat in the Park (Katahira *et al.*, in press). Special Ecological Areas represent a way to organize the necessary follow-up management of alien plant species in those portions of the Park where ungulate management has been successful. For example, it is probably not feasible or worthwhile to control feral pigs and alien plants in all rain forest areas. Special Ecological Areas provide a means of selecting the best areas to manage for both pigs and alien plants.

6. Special Ecological Areas serve as a focus for research and interpretation. Biological understanding can be gained through long-term monitoring and research in the most diverse, intact, representative, and unique areas. Special Ecological Areas can be used to illustrate intact native ecosystems to the public and interpret the need for intensive management of alien species in Hawai'i.

SELECTION OF SPECIAL ECOLOGICAL AREAS

Special Ecological Areas were selected on the basis of the following criteria (Tunison and Stone 1986):

1. **Representativeness and/or rarity of vegetation.** These features were considered on parkwide, regional, and statewide bases. Special Ecological Areas were proposed in all six major ecological zones of the Park to assure representativeness. No systematic attempt was made to locate Special Ecological Areas in all vegetation types in these zones.

2. **Vegetation intactness.** Special Ecological Areas have low densities of alien plant species, usually resist most alien plant invasions, are

often less vulnerable to fire, and are typically more manageable than more degraded areas.

3. **Plant species diversity/richness.** Plants rather than animals were considered initially because we lack information about invertebrates, native avifauna are scarce, and because bird species have broad habitat correlates or ranges.

4. **Manageability.** Manageability is a function of size, accessibility, intactness, and kinds of alien plant problems.

5. **Presence of rare flora and, to the extent known, rare fauna.**

6. **Preserve design considerations.** Proximity to other Special Ecological Areas, presence of similar but less intact habitat as a buffer against threats or as a source of genetic stock, and the size of the Special Ecological Area were factors. Preference was given to proposed areas that are contiguous with or near other intact areas to enhance chances of dispersal or gene flow among areas. Larger areas were preferred because they have greater interior to edge ratios, which reduces the probability of dispersal of alien species into Special Ecological Areas and decreases adverse edge effects. Larger areas also were more capable of supporting larger populations of native species (species/area relationships), which minimizes population extinctions from stochastic demographic events and genetic losses.

7. **Degree of immediacy of threats from alien biota.** Some areas contain small populations of disruptive species, and successful management is feasible if control efforts are begun immediately. These areas were designated Special Ecological Areas before other areas with fewer alien species problems.

8. **Research and interpretive values.** Intact and diverse sites are ideal for understanding native Hawaiian ecosystems and obtaining baseline information on them. Managed and unmanaged communities can be studied in Special Ecological Areas and adjacent areas. Areas accessible to the public along roads and trails are preferred for educational purposes because they represent natural communities. Guided walks can be conducted or nature trails established in these areas to interpret not only natural history but also management of Hawaiian systems.

The selection criteria for Special Ecological Areas were generally ranked in the above order. However, lower-priority criteria, particularly manageability, immediacy of threat, and potential for interpretation, were overriding criteria in designating some dry ʻōhiʻa (*Metrosideros polymorpha*) scrub and woodland areas along Chain of Craters and Hilina Pali roads. Some rain forest and mesic forest and scrub had considerable past research data available and were considered especially important.

The designation of Special Ecological Areas to date has been based on existing biological information about potential areas, rather than waiting for a systematic program of data collection. Specific boundaries were refined through ground surveys by Park resource managers, who prepared management plans. Alien plant management commenced after a consensus of interested managers and researchers developed. Alien plant control efforts were started by a large volunteer group in residence in the Park in 1985. Work was begun in six Special Ecological Areas requiring immediate management: Thurston, Ke'āmoku, North 'Āinahou, South 'Āinahou, Small Tract 'Ōla'a, and Kīpuka Puaulu (and buffer zone). Alien plant control work began in Kīpuka Kahali'i, Keanakāko'i, and Hilina Pali Road Special Ecological Areas in 1986-1988. Kīpuka Puaulu and Thurston Special Ecological Areas were expanded as control efforts became successful. Selection criteria, ecological zones, major management problems, and research for managed and proposed Special Ecological Areas are presented in Table 1.

MONITORING ALIEN PLANT POPULATIONS

Populations of target alien plants were monitored in each managed Special Ecological Area to determine the effectiveness of control efforts. Usually, numbers of individual alien plants treated (*i.e.*, found or occurring in the Special Ecological Area), as reported by field workers, were used to monitor populations. In one case, the area infested by a target species was mapped at yearly intervals; in another case, the weight of rhizomes removed was estimated by the volume removed, and treated individuals were categorized according to those uprooted and those treated with herbicide. Belt transects were also established in Kīpuka Puaulu, Thurston, and Small Tract 'Ōla'a Special Ecological Areas to facilitate annual determination of the densities of target species by size classes. Large permanent plots were established in Keamoku and Keanakāko'i Special Ecological Areas to assess recruitment of firetree in these areas. Monitoring and inventory of biota other than alien plants is being explored by researchers in a project discussed later in this paper.

POPULATION CHANGES OF TARGET ALIEN PLANT SPECIES

Eleven alien plant species were targeted for control in Special Ecological Areas: firetree, kāhili ginger, strawberry guava, banana poka, Jerusalem cherry (*Solanum pseudocapsicum*), palm grass (*Setaria palmifolia*), silk oak, olive (*Olea europaea* subsp. *africana*), fountain grass (*Pennisetum setaceum*), yellow Himalayan raspberry (*Rubus ellipticus*), and nasturtium (*Tropaeolum majus*). The three target species requir- ing the greatest expenditure of worker resources and found in most Special Ecological Areas were firetree, strawberry guava, and kāhili ginger; all declined greatly after four years of control work. Firetree density de- clined by 86%, strawberry guava density by 95%, and kāhili ginger rhizome biomass declined by 96% (Fig. 2). Population levels after control were very low: firetree occurred at approximately 0.8 plant/a (2 plants/ha), strawberry guava at less than 0.5 plant/a (1 plant/ha), and kāhili ginger rhizome biomass at less than 4.5 lb/a (5 kg/ha) in infested areas.

Table 1. Ecological and management features of Special Ecological Areas in Hawaii Volcanoes National Park.

Special Ecological Area	Proposed/ Managed	Area (ha)	Ecological Zone	Plant Community/ Vegetation Type	Selection Criteria	Management/ Threats
Hōlei	Proposed	286	Coastal lowlands	Pili grassland/ savanna	Representativeness Intactness Interpretation Manageability Rarity in State	Thatching grass, fountain grass; goats controlled
Kamoamoa Makai	Proposed	77	Coastal lowlands	Native scrub, dry forest	Uniqueness Intactness Rare plant species	Strawberry guava, Christmas berry, fountain grass; goats controlled
Kamoamoa Mauka	Proposed	758	Submontane seasonal/ montane rain forest	Mesic forest, wet forest	As above	Strawberry guava; goats controlled
Poliokeawe	Proposed	257	Coastal lowlands	Dry ʻōhiʼa woodland	Intactness Manageability Rarity in State	Located in fountain grass control unit; goats controlled
Kaʻaha	Proposed	391	Coastal lowlands	Pili grassland	Intactness Representativeness Rarity in State	Thatching grass in site; located in fountain grass control unit; goats controlled
ʻĀpua Point	Proposed	23	Coastal lowlands	Coastal strand	Intactness Rare plants Uniqueness in Park	Threats not known; goats controlled

Table 1, continued.

Table 1, continued.

Special Ecological Area	Proposed/ Managed	Area (ha)	Ecological Zone	Plant Community/ Vegetation Type	Selection Criteria	Management/ Threats
ʻĀinahou North	Managed	17	Submontane seasonal	Dry ʻōhiʻa woodland and scrub	Plant species richness Representativeness Rare plants	Firetree, olive, strawberry guava? Goats controlled; pigs not controlled
ʻĀinahou South	Managed	25	Submontane seasonal	Dry ʻōhiʻa woodland	Rare plants Representativeness	Same as ʻĀinahou North
Keʻāmoku	Managed	3,380	Submontane seasonal	Dry ʻōhiʻa scrub	Intactness Representativeness Potential for Interp.	Firetree, silk oak. Fountain grass controlled, goats controlled; pigs not a problem
Kīpuka Kahaliʻi	Managed	272	Submontane seasonal	Dry ʻōhiʻa scrub/ woodland; lava flows	Manageability Intactness Representativeness Potential for Interp. Rare plants	Firetree, olive controlled; goats controlled; pigs not a problem
Keanakākoʻi	Managed	366	Submontane seasonal	ʻŌhiʻa forest/ woodland/scrub/ lava flows	Intactness Rare plants Potential for Interp. Manageability	Firetree controlled; goats and pigs controlled

Special Ecological Area	Proposed/ Managed	Area (ha)	Ecological Zone	Plant Community/ Vegetation Type	Selection Criteria	Management/ Threats
Hilina Pali Road	Proposed	401	Submontane seasonal	Dry ʻōhiʻa scrub/ woodland	Potential for Interp. Manageability Representativeness Rare plants	Firetree controlled; goats controlled; pigs not a problem
Kamakaiʻa Hills	Proposed	2,535	Submontane seasonal	Dry ʻōhiʻa scrub	Manageability Representativeness	Fountain grass controlled; goats controlled; pigs not a problem
Nāulu	Proposed	76	Submontane seasonal	Lama forest lava flows	Species richness Rare plants	Fountain grass
Puhimau Hot Spot	Proposed	10	Submontane seasonal	Hydrothermal	Uniqueness Rare plants	No alien species threats
ʻŌlaʻa Large Tract	Proposed	2,279	Montane rain forest	Hapuʻu/ʻōhiʻa	Intactness Representativeness Rare plants	Several Areas may be selected in this area; Areas will be less than total proposed, which is fenced for pig control. Threats are kāhili ginger, yellow Himalayan raspberry, palm grass, and strawberry guava

Table 1, continued.

Table 1, continued.

Special Ecological Area	Proposed/ Managed	Area (ha)	Ecological Zone	Plant Community/ Vegetation Type	Selection Criteria	Management/ Threats
ʻŌlaʻa Small Tract	Managed	133	Montane rain forest	Hapuʻu/ʻōhiʻa	Intactness Representativeness Rare plants	Alien plants/rare plants monitored; experimental alien plant control plot established; pigs controlled
Thurston	Managed	48	Montane rain forest	ʻŌhiʻa/hapuʻu, ʻōhiʻa/uluhe	Intactness Representativeness Potential for Interp.	Firetree, strawberry guava, kāhili ginger, palm grass, yellow Himalayan raspberry controlled; pigs controlled
East Rift	Proposed	2,704	Montane rain forest	ʻŌhiʻa/hapuʻu ʻōhiʻa/uluhe lava flows	Intactness Representativeness Rare plants	Strawberry guava; pigs need to be controlled
Kīpuka Puaulu	Managed	73	Montane seasonal	Mesic forest	Uniqueness Plant species richness Potential for Interp. Rare plants	Kāhili ginger, strawberry guava, Jerusalem cherry, nasturtium controlled; goats and pigs controlled; alien grasses and fire are problems

Special Ecological Area	Proposed/ Managed	Area (ha)	Ecological Zone	Plant Community/ Vegetation Type	Selection Criteria	Management/ Threats
Kīpuka Puaulu Buffer Zone	Managed	952	Montane seasonal	ʻŌhiʻa or koa woodlands/forest	N/A	Firetree controlled; pigs and goats controlled
Kīpuka Kī	Proposed	87	Montane seasonal	Mesic forest	Plant species richness Potential for Interp. Uniqueness	Jerusalem cherry, blackberry, alien grasses are major problems; goats and pigs controlled
Mauna Loa	Proposed	4,142	Montane seasonal Subalpine	Mountain parkland; Subalpine scrub	Intactness Representativeness	No major alien plant problems; firetree and fountain grass are potential threats; goats and pigs controlled

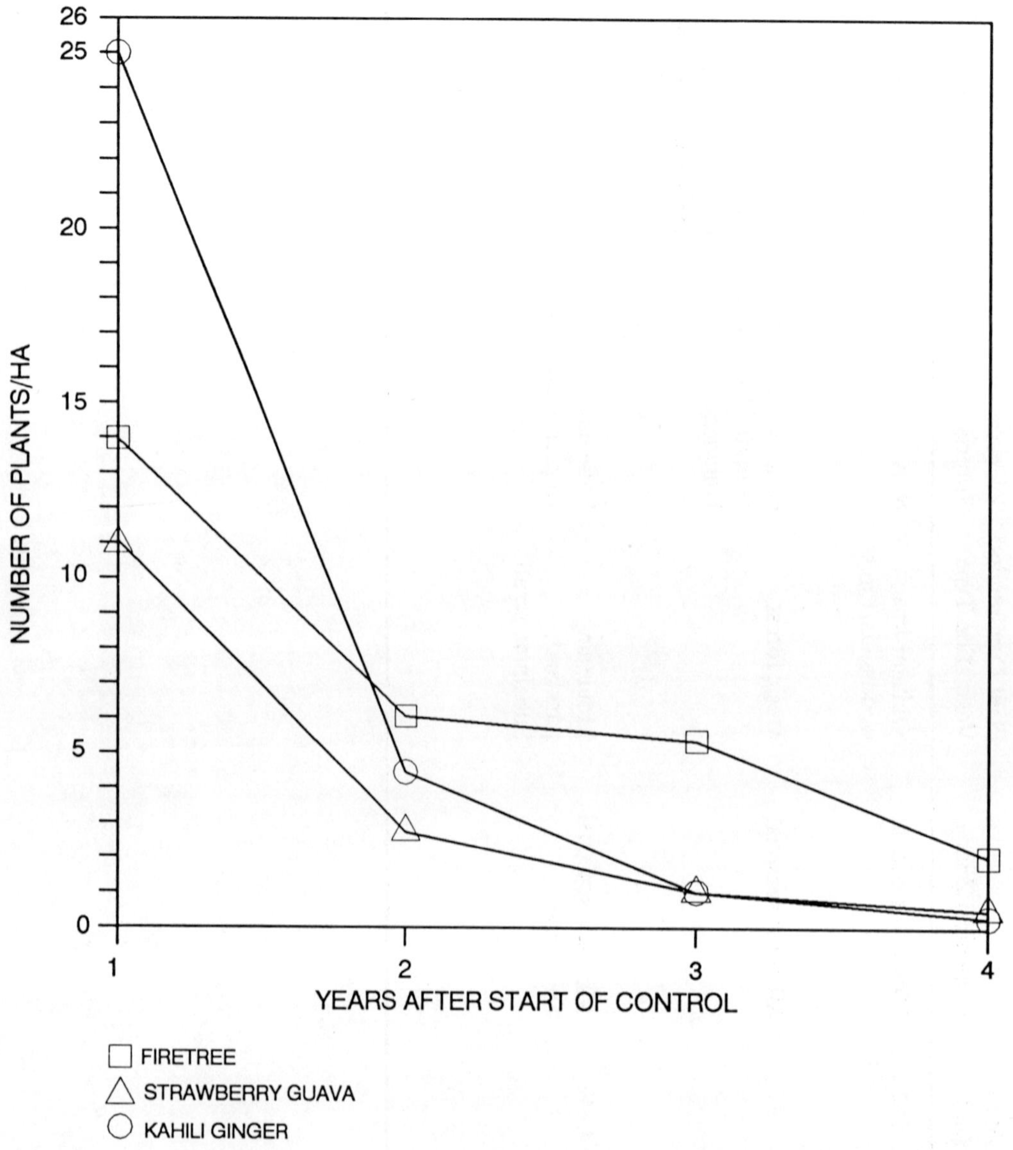

Figure 2. Densities of firetree (*Myrica faya*), strawberry guava (*Psidium cattleianum*), and kāhili ginger (*Hedychium gardnerianum*) in all Special Ecological Areas in Hawaii Volcanoes National Park. Density (number of plants/ha) was used to indicate population levels because size of Special Ecological Areas was increased over time.

Populations of less-widespread target alien plant species also declined sharply (Table 2). The area infested by Jerusalem cherry decreased by 96%, from 65,070 ft^2 to 2,690 ft^2 (6,045 m^2 to 250 m^2) in Kīpuka Puaulu Special Ecological Area. The frequency of nasturtium along five belt transects in Kīpuka Puaulu Special Ecological Area declined from 65% to 2%. Palm grass decreased 54%, from 1,984 ft^2 of infested area to 872 ft^2 (175 m^2 to 81 m^2) in Thurston Special Ecological Area. Silk oak declined 97%, from 110 to 3 plants in Ke'āmoku

Table 2. Population levels of minor target alien plant species in Special Ecological Areas in Hawaii Volcanoes National Park.

Target Species	Location (Special Ecological Area)	Monitoring Units	Population Levels after Treatment			
			0-12 mo	13-24 mo	25-36 mo	37-48 mo
Jerusalem cherry (*Solanum pseudocapsicum*)	Kīpuka Puaulu	m^2 infested	N/A	6,045	325	250
Fountain grass (*Pennisetum setaceum*)	Ke'āmoku	No. of plants	27	71	8	34
Nasturtium (*Tropaeolum majus*)	Kīpuka Puaulu	% cover in 29 monitoring plots	65%	44%	14%	2%
Palm grass (*Setaria palmifolia*)	Thurston	m^2 infested	175	90	81	N/A
Olive (*Olea europaea* subsp. *africana*)	Kīpuka Kahali'i	No. of plants	10	N/A	N/A	N/A
Silk oak (*Grevillea robusta*)	Ke'āmoku	No. of plants	110	1	10	3
Yellow Himalayan raspberry (*Rubus ellipticus*)	Thurston	No. of plants	287	33	40	0

Special Ecological Area. Yellow Himalayan raspberry declined 100%, from 287 to 0 plants in Thurston Special Ecological Area. Olive decreased 80%, from 665 plants to 134 plants in ʻĀinahou North, ʻĀinahou South, and Kīpuka Kahaliʻi Special Ecological Areas. Population changes of alien plant species are described in greater detail by Tunison *et al.* (1992). Fountain grass populations increased slightly, from 27 to 34 plants in Keʻāmoku Special Ecological Area.

The number of worker days required for alien plant control efforts in all Special Ecological Areas declined 73% (Fig. 3). In nearly all cases, as many worker days were required in the first 12 months of control as in the subsequent 36 months. The work load requirement dropped sharply in the second 12 months, with the rate of decrease declining over the next three years. Approximately 280 worker field days were expended on all alien plant control work in all Special Ecological Areas in 1988.

DISCUSSION

Need for Long-term Management

Follow-up alien plant treatments are required in all Special Ecological Areas. No target species have been eliminated in any Special Ecological Area, even though some species have been reduced to very low levels.

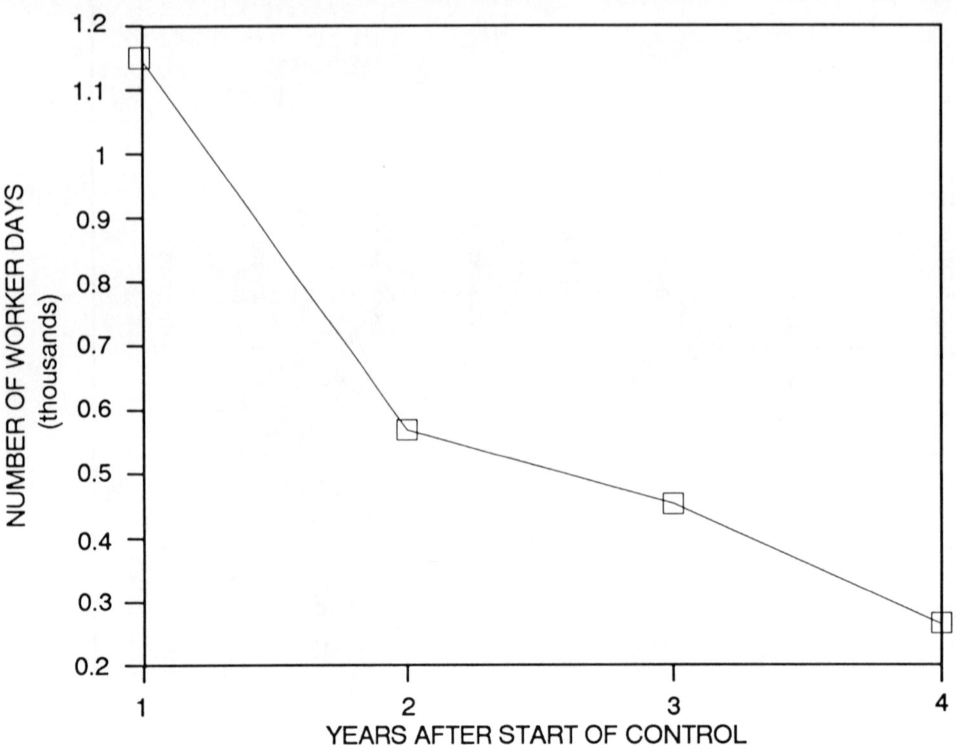

Figure 3. Number of worker days expended in controlling alien plants in Special Ecological Areas in Hawaii Volcanoes National Park.

Population levels of firetree are still noticeably dropping, but populations of kāhili ginger and strawberry guava have plateaued at low levels (Fig. 2). Results from localized alien plant control records indicate that even small populations of localized species can be very persistent (Tunison and Zimmer, this volume), mainly because of recruitment from long-lived seed banks. In addition, managed species are also established in surrounding areas and will continue to be dispersed into Special Ecological Areas. Since most alien plants probably cannot be eliminated in Special Ecological Areas, follow-up maintenance treatments may be required indefinitely.

Decreases in Work Load

Work loads decrease over time because alien plant population densities decline and because uprooting replaces cut-stump treatments. Treatments also become more efficient as workers become more familiar with the location of alien plant populations. The work load in Keamoku and Thurston Special Ecological Areas is expected to drop dramatically because low population levels now permit follow-up every other year rather than annually. Biennial treatments may eventually be applied to other areas. However, a minimum number of worker days will be required because of the need to systematically resurvey each Special Ecological Area, even if no plants are treated. As the work load in Special Ecological Areas decreases, new areas can be established or current areas expanded.

Weaknesses of the Special Ecological Areas Approach

A potential shortcoming of the Special Ecological Areas approach is that areas will become surrounded by increasing densities of alien plant species, dispersal of aliens into the area will increase greatly, and recruitment will reach unmanageable levels. Three strategies are designed to reduce alien plant species populations in areas between Special Ecological Areas. First, many of the plant species controlled in Special Ecological Areas are targeted for biological control, which may provide effective long-term control. Second, new alien plants are prevented from entering the Park by controlling localized alien plant species (Tunison and Zimmer, this volume). Third, buffer zones serve to keep population levels of selected alien plants low in surrounding areas. Kīpuka Puaulu is surrounded by a 2,350-a (952-ha) buffer zone. Finally, feral ungulate control occurs in larger management units containing Special Ecological Areas and serves to reduce disturbance and alien plant dispersal.

Another shortcoming of Special Ecological Areas may be the small size of some areas. Special Ecological Areas range from 42 to 8,350 a (17-3,380 ha) and average 1,576 a (638 ha). Native plant and animal populations in smaller areas may be too low to maintain fitness or evolutionary potential. Small populations are also vulnerable to stochastic demographic erosion.

Special Ecological Areas as Models for Managing Other Areas

Managed Special Ecological Areas provide information leading to expansion or establishment of new Special Ecological Areas. The effectiveness of alien plant control treatment methods in Special Ecological Areas as well as the effects of treatment methods on nontarget species can be tested. Work load requirements and recruitment rates of target alien plants can be determined, allowing efficient planning for additional efforts.

Additional Special Ecological Areas are continually being added to the system in Hawaii Volcanoes National Park. Management of alien plants was started in Special Ecological Areas where problems required immediate attention, especially in submontane seasonal environments. About 79% of the area of managed Special Ecological Areas is in this ecological zone, although nearly half of the alien plant control field days are now allocated to rain forest and mesic forest areas, and more than half if monitoring activities are considered. Thirteen additional areas totalling 31,825 a (12,884 ha) have been proposed as Special Ecological Areas (Table 1). Additions to the Special Ecological Areas system will emphasize rain forest, mesic forest, mountain parkland, subalpine, alpine, and native grasslands and forest in the coastal lowlands. Most proposed Special Ecological Areas appear to have low densities of disruptive but manageable alien plant species. However, monitoring will be required to determine management needs. A few Special Ecological Areas (Large Tract 'Ōla'a, East Rift, and Kīpuka Kī) have high densities of alien plants requiring management.

Research and Other Monitoring Emphases

Special Ecological Areas are attractive foci for research efforts because of their intactness, richness, representativeness, and the presence of rare species. They are also important areas in which to study responses to management actions, including alien plant control, and to establish long-term monitoring programs for native and alien species. In Hawaii Volcanoes National Park, where most effort has been spent in controlling alien species of plants and animals, the efficacy of management in restoring native biota is especially important. Probably no better place exists to study dynamics of invasions, native and alien interactions, and the response to management of aliens. Major long-term research projects funded through Stanford University and the University of Hawaii, as well as National Park Service efforts, are leading to better understandings of plant and animal invasions and their management.

The necessary emphasis of management and research on alien species to date has resulted in lack of knowledge about the native flora and fauna. No Resources Base Inventory has ever been conducted in the Park, although some information is available from research conducted during the International Biological Program in parts of Hawaii Volcanoes. Abundance and distribution of important groups, including birds, various invertebrates, and higher and lower plants, have not received systematic study, even in the most valuable ecological areas of the Park. Population sizes, dynamics, limiting factors, and long-term natural variability are unknown for most native species, especially in relation to variation caused by humans and their management programs. Undisturbed behaviors of many important native animals have not received thorough study, and nests of some avian species have not even been described.

Statistically adequate, long-term monitoring systems have not been established in the Park, even for such visible groups as birds. Lack of avian population data hampers our understandings of long-term trends, responses to disease and predators, and response to management practices, including removal of ungulates such as pigs and goats and removal of alien plants from native bird habitat. Indicator species have not been determined for any groups. Voucher specimens of animals have not been systematically and adequately collected for most species to allow detection of changes with time. Native plants, including candidate endangered species, are not monitored adequately, and a database for the future is not available. Similarly, native invertebrates have not been systematically surveyed for distribution, abundance, variability, and their value as indicators of environmental change or management effectiveness. Since Hawaii Volcanoes National Park is an International Biosphere Reserve and a World Heritage Site, collection of good baseline data is an extremely important function. The Park should serve as a well-documented, near-native resource for comparison with more-rapidly changing unmanaged areas elsewhere in Hawai'i. The purpose of management programs for alien plants and animals is ultimately to return communities to near-native states that can be maintained for future generations. Adequate methods of monitoring progress toward that goal are essential.

Beginning in 1991, a three-year research project will attempt to remedy some deficiencies in our understanding of native species abundance, distribution, and interactions with other native and alien species. It is beyond the scope of this paper to do more than summarize the approach, but a major objective is to establish permanent monitoring transects, plots, and protocols for Special Ecological Areas throughout the Park. Native and alien plants, birds, invertebrates, and mammals will be sampled in Special Ecological Areas to determine abundance, distribution, and variability, at least on a preliminary basis. Special research projects will emphasize impacts of important alien species and their management on native flora and fauna; recovery of vegetation after ungulate and alien plant removal; population structures and dynamics of alien and native plants and animals; and other key topics. Systematic studies of insect, bird, and Hawaiian bat distribution and habitat correlates should help us to further identify the most important and responsive areas in the Park to target for protection and management of native species. Adequate information about native species will help refine the ranking of areas in which to expend efforts on control of aliens, the primary driving force for Special Ecological Areas management to date.

ACKNOWLEDGMENTS

The authors would like to thank Linda Cuddihy, who was an early advocate and supporter of Special Ecological Areas; and Chris Zimmer, who, once he became convinced of their value, has been the main manager in the field implementing the concept.

Literature Cited

Gardner, D.E., and C.J. Davis. 1982. *The prospects for biological control of nonnative plants in Hawaiian national parks.* Tech. Rep. 45, Univ. Hawaii Coop. Natl. Park Resour. Stud. Unit. Honolulu.

Katahira, L.K., and C.P. Stone. 1982. Status of management of feral goats in Hawaii Volcanoes National Park. *Proc. 4th Conf. Nat. Sci., Hawaii Volcanoes Natl. Park,* 102-108. Honolulu: Univ Hawaii Coop. Natl. Park Resour. Stud. Unit.

Katahira, L.K., P.M. Finnegan, and C.P. Stone. In press. Eradicating feral pigs in montane seasonal habitat at Hawaii Volcanoes National Park. *Wildlife Soc. Bull.*

Markin, G.P., P.-Y. Lai, and G.Y. Funasaki. [this volume] Status of biological control of weeds in Hawai'i and implications for managing native ecosystems.

Mueller-Dombois, D., and F.R. Fosberg. 1974. *Vegetation map of Hawaii Volcanoes National Park.* Tech. Rep. 4, Univ. Hawaii Coop. Natl. Park Resour. Stud. Unit. Honolulu.

Tunison, J.T., and C.P. Stone. 1986. SEAs provide ecosystem focus for management and research. *Park Sci.* 6(3):10-12.

Tunison, J.T., and N.G. Zimmer. [this volume] Success in controlling localized alien plants in Hawaii Volcanoes National Park.

Tunison, J.T., M.R. Gates, and N.G. Zimmer. In press. *Alien plant control in Special Ecological Areas: a progress report, 1985-1991.* Tech. Rep., Univ. Hawaii Coop. Natl. Park Resour. Stud. Unit. Honolulu.

Tunison, J.T., L.D. Whiteaker, L.W. Cuddihy, A.M. LaRosa, D.V. Kageler, M.R. Gates, N.G. Zimmer, and L. Stemmermann. 1992. *The distribution of selected localized alien plant species in Hawaii Volcanoes National Park.* Tech. Rep. 84, Univ. Hawaii Coop. Natl. Park Resour. Stud. Unit. Honolulu.

Warshauer, F.R., J.D. Jacobi, A.M. La Rosa, J.M. Scott, and C.W. Smith. 1983. *The distribution, impact, and potential management of the introduced vine* Passiflora mollissima *(Passifloraceae) in Hawai'i.* Tech. Rep. 48, Univ. Hawaii Coop. Natl. Park Resour. Stud. Unit. Honolulu.

Whiteaker, L.D., and D.E. Gardner. 1985. *The distribution of* Myrica faya *Ait. in the state of Hawaii.* Tech. Rep. 55, Univ. Hawaii Coop. Natl. Park Resour. Stud. Unit. Honolulu.

ENVIRONMENTAL EDUCATION IN HAWAI'I: HISTORY AND OVERVIEW

Charles P. Stone

ABSTRACT

Environmental education in Hawai'i has been pragmatic throughout its history but insufficient to the task of stimulating enough concern about irreplaceable native ecosystems. Developments since 1970 that have set the stage for a more positive future include: the continuing activities of the Moanalua Gardens Foundation and other private conservation groups; an infusion of money from the John D. and Catherine T. MacArthur Foundation, which led to production of curriculum materials by the 'Ōhi'a Project and to stimulation of other agencies and organizations; increased efforts by the Hawaii Department of Education and the Department of Land and Natural Resources; and formation of a statewide environmental education organization to enhance communication and competence among environmental educators. More specific targeting of decision makers affecting the rapidly changing face of Hawai'i and increased emphasis on teaching environmental ethics and critical thinking skills to adults and children are needed in the future. The complexities, urgencies, and scale of many of Hawai'i's environmental problems make environmental education a critical component of informed problem solving.

IN THE BEGINNING

Environmental education in Hawai'i began with the first Polynesian colonizers, who, like most aboriginal peoples, had numerous natural gods. The power of the gods was expressed in divine force called *mana*, which was manifested in various forms of nature as well as in people. Some things and places in nature and society had greater mana than others and were treated with much respect (Burrows 1989). The people valued nature but did not revere all life. Before the arrival of continental humans, Hawaiians cleared large areas under 1,500 ft (455 m) elevation of native vegetation. Agriculture and villages needed for expanding populations replaced native species. A number of flightless birds, other animals, and plants were driven to extinction in the process (Olson and James 1982; Cuddihy and Stone 1990). Natural resources, including those of the sea (probably less affected), were needed for subsistence by the Polynesians (Dye and Steadman 1990). Environmental education was less than perfect if judged in terms of effective stewardship of natural resources (Stone and Machlis 1989) and preservation of biological diversity.

EARLY EUROPEAN EMPHASES

With the arrival of the European ships, trade in agricultural products (such as pigs, bananas, and taro; firewood for whaling ships; sandalwood for the Orient; *pulu* or tree fern "hairs" for use in mattresses and pillows; and wild cattle, goats, and sheep) began to produce widespread effects on large expanses of native forest. The "Great Mahele" of 1848 and later land reform laws gave foreigners the right to own land and allowed Hawaiians to sell their interests. Clearing of forests continued to move upslope, and land ownership was soon concentrated in the hands of non-Hawaiian planters and business people. Agricultural crops such as sugar cane, pineapple, and coffee (*Saccharum officinarum, Ananas comosus, Coffea arabica)* began to become important, and commercial logging and ranching became firmly established after land ownership changed in the mid-19th century (Cuddihy and Stone 1990). Economic exploitation (rather than mere subsistence) of the land was foremost in the minds of the new immigrants; native Hawaiians were caught in the momentous changes of the times. Technology for affecting the environment improved. Little, if any, thought was given to conservation, much less to environmental education (Gill 1989).

WATERSHED WORRIES

In the last few decades of the 19th century, hundreds of thousands of dollars were spent on water projects in Hawaiʻi. Effects of logging, clearing, ranching, and foraging by ungulates became apparent to the sugar industry, and forest loss was assumed to be responsible for reduced runoff. Hawaiian forests were viewed as watersheds rather than commercial forests by the late 1920s, and water was believed to be the most important forest product (Judd 1918, 1927). A government forest reserve system was considered essential to preserving forests and watersheds as early as 1904 (Hall 1904). Big business, at least and at last, saw the value of conservation when loss of water resources threatened the economy; people drinking salty and brackish water also became more environmentally aware for a time. Watershed degradation was alleviated temporarily by establishment of forest reserves and a professional forestry program, construction of ditches, planting of alien trees, removal of ungulates from many areas, and development of new technologies. The approach to environmental education was strongly utilitarian.

DON'T WORRY, BE HAPPY

In the 1940s and 50s, conservation organizations began to focus on native wildlife and natural resource conservation (Gagné and Gill 1989).

Introductions of additional game species to Hawai'i and the change in emphasis from watershed protection to multiple use and sustained yield hunting were points of conflict among conservation organizations, hunters, and public and private landowners. In 1970, a group of outdoor educators formed the Citizens Committee for Environmental Education, and this soon became the Environmental Education Association. The organization dissolved about a year later, partly because people seemed to feel that Hawai'i was "Paradise" and lacked real U.S. Mainland problems such as air and water pollution. Attempts by community groups to integrate environmental education into the Hawaii Department of Education curriculum about this time were unsuccessful, probably because there was not overriding concern about the environment, because mainland models did not work (Gill 1989), and because of other priorities for existing dollars.

Several initiatives of the 1970s bore fruit, including the High School Hikers Program of the Hawaii Chapter of the Sierra Club, a Compendium of Critical Field Sites for teachers, and the Hawaii Nature Study Program developed by the University of Hawaii (Gagné and Gill 1989). Some concerned teachers also began to teach environmental education in the classroom and field. An Outdoor Education Center for 6th graders was established on Hawai'i Island by the Department of Education in 1976. The National Park Service interpretation programs at Hawaii Volcanoes National Park on Hawaii Island and Haleakala National Park on Maui educated visitors in biological, geological, and cultural themes.

LATE-BREAKING DEVELOPMENTS

In 1978, a State constitutional requirement was passed to teach environmental education in the public schools. By 1981, the Office of Instructional Services of the Hawaii Department of Education had developed *Environmental Education: K-12 Curriculum Guide*, including annotated lists of journals and magazines, agencies and organizations, and definitions of terms concerned with environmental education. Despite the legal requirement and the guide, little occurred in the schools, probably for the same reasons noted for previous inaction. Establishment of an Environmental Information Office in the Department of Education, opening of the Hawaii Nature Center on O'ahu, an increased emphasis on environmental education at Moanalua Gardens on O'ahu, and preparation of environmental education materials by more government agencies, including the Hawaii Department of Land and Natural Resources, were other landmarks in the late 70s and 80s.

Most recent emphases on environmental education can be traced to several initiatives based on a growing realization of real threats to the Hawaiian environment. Probably the most important of these is the gift of more than $3.5 million to Hawai'i from the John D. and Catherine T. MacArthur Foundation in Chicago. The grant allows a diverse approach to preserving biological diversity in Hawai'i, including environmental education (Tangley 1988). Grants for land acquisition, research, and advocacy group support have also helped make people aware of conservation problems in Hawaii.

The ‘Ōhi‘a Project

Money given to the Bernice P. Bishop Museum ($765,500) specifically for environmental education has allowed curriculum materials to be produced for Hawai‘i's children in elementary grades K through 8. Suggested activities in natural areas, audiovisual tools, training, testing of materials by teachers, and review of materials by authorities are part of the curriculum development program, named the *‘Ōhi‘a Project* for Hawai‘i's most common native tree. Key concepts within five major topics (geology, geography, plants and animals, humans and the environment, global interactions) are designed to integrate with existing elementary school curricula. In the first phase of the project, about 1,300 (25%) of Hawai‘i's teachers were targeted (D. King, pers. comm.). Final copies of materials for grades K through 8 have now been produced.

Accelerated Activities

Many agencies and organizations have increased their efforts in environmental education in the past few years. Reasons for this undoubtedly include the stimulation of MacArthur money, the growing number of conflicts between development and environmental needs, the realization of the importance of environmental education in addressing Hawai‘i's environmental problems wisely, mutual stimulation and synergistic efforts, and the high profile of global problems. Earth Day 1990 also increased awareness.

Examples of educational initiatives in the past few years are numerous and worth mentioning to increase awareness and publicize positive approaches. They include: annual Hawaii Department of Education training camps in environmental education; *Greenprint*, the National Audubon Society's bimonthly newsletter; *A Teacher's Guide to the Endangered Birds of Hawai‘i* and a video on Hawai‘i's birds, prepared by the Hawaii Department of Land and Natural Resources; *Resource*, a quarterly newsletter produced by the Hawaii Department of Land and Natural Resources; *Preserving Hawai‘i's Natural Treasures: Hawai‘i's Natural Area Reserves System*, produced by the Department of Land and Natural Resources; *Hiking Softly in Hawai‘i*, by the Hawai‘i Chapter of the Sierra Club; *Kilauea Pointers*, by the U.S. Fish and Wildlife Service; *Recycle*, the newsletter of the Recycling Association of Hawaii; "Interpret Hawai‘i" programs at community colleges on several islands; the Visitor Industry Training and Education Center (VITEC) on Maui; a conservation education workshop held at Hawaii Volcanoes National Park in 1988; "Science in Hawai‘i" and "It's Science, Isn't It?" videos produced by the Hawaii Department of Education; *Conservation Biology in Hawai‘i,* edited by National Park Service personnel; *Islands in a Far Sea*, published by the Sierra Club; *Extinctions in Paradise: Protecting Our Hawaiian Species*, by the Natural Resources Defense Council; *Hawai‘i: the Islands of Life*, by The Nature Conservancy of Hawaii; "Puppets on the Path," an environmental education group; "We All Need the Forest," an ‘Ōhi‘a Project video; Conservation Council of Hawaii and Sierra Club newsletters; 1990 and 1991 environmental education workshops on Hawai‘i Island, Kaua‘i, and O‘ahu; educator training sponsored by the National Park Service for Thailand and South Pacific Islands personnel; recruitment of an Environmental Education Specialist by the Hawaii Division of Forestry and

Wildlife; plans to establish Department of Education outdoor education centers on Kaua'i, windward O'ahu, and Maui; attempts to make the Hawaii Department of Education Environmental Education Specialist in Honolulu a permanent position; and the excellent video produced by National Geographic Society entitled "Hawaii: Strangers in Paradise."

Professional People

Large numbers of people are involved in educating others about the environment in some way, shape, or form in Hawai'i. Some may not even realize that they are involved. Environmental education at its best involves stimulating the appreciation and concern of others and instilling in them an environmental ethic. Hawai'i's environmental problems are often urgent and usually complex, so the job of environmental education has to be done effectively. Time and money for education are limited, and environmental educators often need help. Networking, training, and materials are essential.

The Hawaii Environmental Education Association was formed in 1989 to help improve communication and effectiveness. Consisting of a statewide (four islands at present) Board of Directors, officers, and island chapters, the Association seeks to improve environmental education. It does not take positions on environmental issues. The emphasis is on accomplishing effective environmental education in Hawai'i and helping environmental educators toward that end. Hawaii Environmental Education Association is incorporated, has nonprofit status, and held its first annual statewide conference on O'ahu in June 1991. Projects include a statewide catalog of programs available to educators (including information on subject, target audience, contacts, cost, location); a growing island-by-island mailing list of those who consider themselves environmental educators; planned workshops entitled "The Art of Environmental Education through Art" (Hawai'i Island) and "Earth-wise -- the Art of Environmental Education" (on Kaua'i). Island chapters of the Hawaii Environmental Education Association meet at one- or two-month intervals to share information, often at different sites. Some chapters plan educational projects such as training, others create and participate in programs, and others just share information about educational activities. The emphasis is on meeting the needs of the educators on each island -- as they define them. The Board of Directors meets every two months to share information on progress on each island, discuss organizational progress, problems, and direction, and plan for statewide meetings. Some island groups meet before work or on weekends to more readily accommodate school teachers as well as other educators. As the accomplishments of the organization have increased and been publicized, participation has also increased; for example, over 250 people attended the 1991 statewide conference.

CONCLUSIONS

An estimated $25 million was spent in conservation in Hawai'i in 1988, of which about $1 million was spent on environmental education (Gagné 1988). Although environmental education in Hawai'i has a long way to go in influencing the apathetic, the prognosis is hopeful. However, strategies are needed to reach urban and rural residents, visitors, and adults as well as children. Insufficient environmental education in Hawai'i's schools leads to lack of concern of adult decision-makers, but that catch-22 situation is changing with the influence of the 'Ōhi'a Project and other recent educational efforts (Hawkins, this volume). Environmental ethics, critical thinking, and practice in dealing with complex problems with many "right" answers are needed in educating young and old alike in good citizenship.

Visitors and the visitor industry seem especially difficult to educate about Hawai'i's natural values and threats to them. "Ecotourism" does not seem to be the emerging strategy in Hawai'i that it is in other areas of the world. All decision-makers also need to be targeted specifically -- politicians, business persons, realtors, and others who are actively and rapidly changing the face of Hawai'i without full understandings of environmental losses. Media representatives influence great numbers of people, yet often they lack in-depth information, continuity, and real involvement. To be effective in the conservation of Hawai'i's rapidly diminishing natural resources, environmental educators (and that includes many of us) must do a better job of educating others about the environment, significant threats to it, and about the values of environment to all of us. Environmental education in Hawai'i today is needed more than ever to provide appreciation and concern about the world in which we live and the decisions we make about it.

ACKNOWLEDGMENTS

I would like to thank some of the dedicated and effective environmental educators who have inspired me recently and through the years. They include: Paul Banko, Dave Boynton, Mike Buck, Hamp Carson, Tamar Chotzen, Sheila Conant, Linda Cuddihy, Marsha Erickson, Betsy and Wayne Gagné, Lorin Gill, Sam Gon III, Paul Higashino, Jim Jacobi, Dina Kageler and the other Puppets on the Path, Larry Katahira, Kathy Kawaguchi, Diana King, Dana Kokubun, Mardie Lane, Lloyd Loope, Art Medeiros, Dick Mortemore, Bill and Mae Mull, Colleen Murakami, Maura O'Connor, Lenny Paik and his cadre of teachers, Kim Sikoryak, Cliff Smith, Dan Taylor, Tim Tunison, Ron Walker, and Julie Williams. Some may not think of themselves as environmental educators, but they (and others) really are. May their numbers increase! I'm grateful to Marsha Erickson, without whom the Hawaii Environmental Education Association would not have begun, and to Danielle Stone, who word processed and otherwise helped me finish this manuscript.

Literature Cited

Burrows, C.K.P.M. 1989. Hawaiian conservation values and practices. In *Conservation biology in Hawai'i*, ed. C.P. Stone and D.B. Stone, 203-213. Univ. Hawaii Coop. Natl. Park Resour. Stud. Unit. Honolulu: Univ. Hawaii Pr.

Cuddihy, L.W., and C.P. Stone. 1990. *Alteration of native Hawaiian vegetation: effects of humans, their activities and introductions*. Univ. Hawaii Coop. Natl. Park Resour. Stud. Unit. Honolulu: Univ. Hawaii Pr.

Culliney, J.L. 1988. *Islands in a far sea: nature and man in Hawaii*. San Francisco, Calif.: Sierra Club Books.

Daws, G. 1988. *Hawaii: the islands of life*. The Nature Conservancy of Hawaii. Honolulu: Signature Pub.

Dye, T., and D.W. Steadman. 1990. Polynesian ancestors and their animal world. *Amer. Scientist* 78:207-215.

Gagné, W.C. 1988. Conservation priorities in Hawaiian natural systems. *Bioscience* 38(4):264-271.

Gagné, W.C., and L.T. Gill. 1989. Conservation education in Hawai'i: history, problems, and needs. In *Conservation biology in Hawai'i*, ed. C.P. Stone and D.B. Stone, 179-183. Univ. Hawaii Coop. Natl. Park Resour. Stud. Unit. Honolulu: Univ. Hawaii Pr.

Gill, L.T. 1989. Perspectives on environmental education in Hawai'i. In *Conservation biology in Hawai'i*, ed. C.P. Stone and D.B. Stone, 177-178. Univ. Hawaii Coop. Natl. Park Resour. Stud. Unit. Honolulu: Univ. Hawaii Pr.

Hall, W.L. 1904. *The forests of the Hawaiian Islands*. Bull. 48. U.S. Dep. Agric., Bureau Forestry. Washington, D.C.: U.S. Govt. Printing Ofc.

Hawaii Department of Education. 1981. *Environmental education K-12 curriculum guide*. Honolulu: Ofc. Instruc. Svcs., Gen. Educ. Br.

Hawaii Department of Land and Natural Resources. 1986. *A teacher's guide to endangered birds of Hawai'i*. Honolulu: Div. Forestry and Wildlife.

Hawaii Department of Land and Natural Resources. n.d. *Preserving Hawaii's natural treasures: Hawaii's Natural Area Reserves System*. Honolulu: Nat. Area Reserves System.

Hawkins, J.W. [this volume] Environmental education in the schools.

Judd, C.S. 1918. Forestry as applied in Hawaii. *Hawn. Forester and Agriculturist* 15(5):117-133.

Judd, C.S. 1927. The natural resources of the Hawaiian forest regions and their conservation. *Hawn. Forester and Agriculturist* 24:40-47.

King, L.B., F. Campbell, D. Edelson, and S. Miller. 1989. *Extinction in Paradise: protecting our Hawaiian species*. Washington, D.C.: Nat. Resour. Def. Coun.

National Audubon Society. *Greenprint: Audubon Hawai'i conservation news*.

Olson, S.L., and H.F. James. 1982. Prodromus of the fossil avifauna of the Hawaiian Islands. *Smithsonian Contrib. Zool.* 365:1-59.

Sierra Club, Hawaii Chapter. n.d. *Hiking softly in Hawai'i. A guide to the enjoyment of Hawaiian wilderness*. Honolulu.

Stone, C.P., and G.E. Machlis. 1989. Conservation education: techniques and targets. In *Conservation biology in Hawai'i*, ed. C.P. Stone and D.B. Stone, 184-189. Univ. Hawaii Coop. Natl. Park Resour. Stud. Unit. Honolulu: Univ. Hawaii Pr.

Stone, C.P., and D.B. Stone, eds. 1989. *Conservation biology in Hawai'i*. Univ. Hawaii Coop. Natl. Park Resour. Stud. Unit. Honolulu: Univ. Hawaii Pr.

Tangley, L. 1988. Fighting deforestation at home. *Bioscience* 38(4):220-224.

ENVIRONMENTAL EDUCATION IN THE SCHOOLS

John W. Hawkins

ABSTRACT

Problems stemming from uncontrolled technological impact on global ecology include: 1) the need for resource management, 2) pollution of the global environment, 3) increasing human population, and 4) inadequacy of knowledge about the nature and interactions of ecological relationships. Environmental problems are perennial; already many have become critical and many others must at least be characterized as pressing. Concern has brought about national as well as local calls for educational programs that will prepare students to cope rationally and effectively with current problems and those that may loom in the future. Environmental education in the schools is thematic; it is both interdisciplinary and multidisciplinary, integrating content from such subject areas as science, social studies, mathematics, and language arts, and incorporating such process skills as problem-solving and decision-making. Rarely does anyone apply skills and knowledge from a single academic discipline in solving environmental problems. Environmental education can be used as a vehicle to equip students with skills for continued lifelong learning, including those associated with the collection of information; the organization, manipulation, and expression of ideas; the evaluation of information and ideas, including competent analysis and critique; and the production of new plans and proposals for action.

INTRODUCTION

An environmentally literate population must be developed and nurtured to promote knowledge and responsible action dealing with wise natural resource management decisions. Education for environmental literacy must begin in early childhood and continue throughout the formal school experiences of students. To do less is to continue to add graduates to an adult population that in large measure has become callous in its considerations and interactions with Hawai'i's environment.

The island state of Hawai'i includes ecosystems that are unique and heavily impacted by humans. Many native organisms have become rare or endangered, and many ecosystems have been seriously altered. It is important that students learn to value the unique environment of Hawai'i and understand its ecosystems, and learn how to protect and make wise decisions about these valuable resources.

To meet the need for an informed citizenry, the Hawaii Department of Education is making a concerted effort to develop a comprehensive and well-balanced environmental education program for all levels in the school system. Environmental education is a thematic area; it is both interdisciplinary and multidisciplinary, integrating such subject areas as science, mathematics, social studies, and language arts, and incorporating such process skills as problem-solving and decision-making. Schools and districts are currently seeking ways to develop optimum instructional programs to ensure delivery of these necessary educational experiences to all students. Environmental education offers the opportunity for significant contributions to cognitive and affective development and can be used as a vehicle to equip students with skills for continued life-long learning, including those associated with the collection of information; the organization, manipulation, and expression of ideas; the evaluation of information and ideas, including competent analysis and critique; and the production of plans and proposals for action. Rarely does anyone apply skills and knowledge from a single academic discipline in solving environmental problems. Similarly, several areas must usually be integrated and applied holistically to meet life's demands.

Interest in environmental education carries over into the community in which agencies, organizations, and interest groups provide resources and environmental education activities for schools (Stone, this volume). Coordination of activities and development of viable resources that strengthen the academic, decision-making, and problem-solving skills of students necessitates the delivery of a comprehensive program from kindergarten through 12th grade.

OPERATIONAL DEFINITIONS

The term "environment" is variously used in educational, popular, and technical literature. Each use has particular meaning to a specific group, and understanding of such particularities is a function of education. However, some broad common characteristics of "environment" can be identified as common to most uses. These are:

1. An environment is a place. It is a region, setting, or context, real or imagined.

2. An environment is the contents of a place, including biotic or a-biotic matter and energy. Humans and their artifacts may or may not be involved.

3. An environment includes dynamic processes and interactions and is bounded by time.

4. An environment is a totality. It is a holistic, time-bound reality including substance, energy, and processes of interaction, physical, biological, societal, or psychological.

The following principles were developed and adopted by the World Intergovernmental Conference on Environmental Education, Tbilisi, USSR, October 14-26, 1977, to provide guidance for environmental education programs worldwide. Environmental education should:

1. Consider the environment in its totality -- natural and built, technological and social, economic, political, moral, cultural, historical, and aesthetic aspects;

2. Be a continuous lifelong process; it should begin at the preschool level and continue through all formal and non-formal stages;

3. Be interdisciplinary in its approach, drawing on the specific content of each discipline in making possible a holistic and balanced perspective;

4. Emphasize active participation in preventing environmental problems and working toward their solution;

5. Examine major environmental issues from local, national, regional, and international points of view, so that students receive insights into environmental conditions in other geographical areas;

6. Focus on current and potential environmental situations;

7. Emphasize the complexity of environmental problems and thus the need to develop critical thinking and problem-solving skills;

8. Utilize diverse learning environments and a broad array of educational approaches to teaching and learning about and from the environment, with due emphasis on practical activities and first-hand experiences;

9. Focus on the student's own community and relate topics being discussed to state, regional, national, and international issues and perspectives.

PROBLEM STATEMENT

The problem being addressed in education is that instructional services regarding resource management are not being provided at the desired program level. This deficiency is being aggravated by: 1) the impact humans have on terrestrial and aquatic ecosystems; 2) the need for cooperation and coordination of total community efforts in delivering services to schools and students; 3) the need to provide teachers access to resources that can be systematically supplied; 4) the need to enhance academic and problem-solving skill development in students; and 5) the need to develop and implement a staff development plan to promote improvements in school programs related to developing environmentally literate students.

Formal recognition of the environment, the emerging problems caused by human impacts on it, and the limited carrying capacity of the Earth have occurred rather recently. Suddenly and rather dramatically, humans have become aware that unless attention is given and safeguards are established toward protecting environmental quality and maintaining ecological balance, irreversible damage will result. Not only will choices in lifestyle and quality of life be reduced by lack of attention to environmental needs, but survival of both humans and other life is at stake.

METHODS

Major educational activities focus on providing a systematic instructional support system to include: 1) instructional modules and activities related to conservation efforts and wise-use decisions about environmental resources; 2) an instructional coordination network; 3) an in-service training program for integrating the topic of resource management into existing subject areas; 4) an experiential component of direct classroom-related services to schools; and 5) coordination of services by governmental and other community agencies involved in environmental and resource management education.

Instructional Resources

The development of instructional materials should be designed to integrate environmental education into existing curricula while specifically addressing environmental matters, thereby providing students with a broader, deeper knowledge of environmental concepts, issues, and problems. Materials should be developed in conjunction with various subject area programs as appropriate, with the involvement of local community groups or governmental agencies knowledgeable about relevant environmental matters.

Instructional Coordination

Implementing an effective instructional support system should involve the cooperative efforts of the total educational community, including kindergarten through university levels, and interdisciplinary efforts across content areas.

In-service Training

In-service training for teachers and administrators should concentrate on raising teachers' and administrators' awareness of environmental issues and problems, increasing their knowledge of various environments, and helping them to provide optimum environmental education experiences for students through the use of the viable new teaching strategies and approaches.

Experiential Activities

Direct student services should provide studies with environmental education experiences not normally attainable in the regular classroom situation. These types of services and activities in the state of Hawai'i include: the Blue-Water Marine Laboratory Program, the Student Marine Affairs Symposium, Keakealani Outdoor Education Center in Volcano on the island of Hawaii, Hawaii Nature Center on O'ahu, Moanalua Gardens Foundation on O'ahu, private residential outdoor camps, Scout camps, county zoos, Waikīkī Aquarium instructional programs, programs at U.S. Fish and Wildlife Service refuges (*e.g.*, Ki'i Wildlife Refuge on O'ahu and Kilauea Refuge on the island of Kaua'i), visits to national parks such as Hawaii Volcanoes National Park on Hawai'i Island and Haleakala National Park on Maui, teacher-initiated field trips, and camping experiences for students.

Community Services

A critical component in development and implementation of an effective instructional support system is the interaction of pre-college educational agencies with resource expertise at universities and in governmental and community agencies involved with environmental education. A strong need for better coordination and communication efforts exists in this area.

A comprehensive education program that can be integrated with existing curricula, maximizing usage of available resource materials and implemented in a cost-effective manner, is necessary. This need can be met by a program in which the interdisciplinary-multidisciplinary nature is realized and the environmental concepts, issues, and problems are taught within established disciplines. To fully implement such a program necessitates development of supplemental instructional materials in gap areas. The 'Ōhi'a Project is meeting some of these needs (Stone, this volume). In addition, in-service training must be supplied for teachers and administrators to raise their awareness of environmental issues and problems, increase their knowledge of various environments, and help them provide optimum environmental experiences for students through the use of viable new teaching strategies and approaches. Department of Education environmental education cadres, in which teachers with a strong interest in environmental education develop skills, help meet these needs, as does networking of environmental educators through organizations such as the Hawaii Environmental Education Association (Stone, this volume). Besides optimizing utilization of community, national, and international resources, coordination ensures that gap areas where materials are now lacking will be identified and materials promoting awareness and concern about Hawai'i's unique environments and resources will be incorporated into the curriculum.

Literature Cited

Stone, C.P. [this volume] Environmental education in Hawai'i: history and overview.

THE USE OF VOLUNTEERS FOR ALIEN PLANT CONTROL AT HAWAII VOLCANOES NATIONAL PARK AND THE NATURE CONSERVANCY'S KAMAKOU PRESERVE

J. Timothy Tunison and Edwin T. Misaki

ABSTRACT

Hawaii Volcanoes National Park on Hawai'i Island and The Nature Conservancy of Hawai'i Kamakou Preserve on Moloka'i require labor-intensive management of alien plant species beyond the level available from paid staff. Volunteers are used to make up part of the worker shortfall and have made significant contributions in both areas. Kamakou Preserve has focused on developing a program that depends primarily on volunteers. Hawaii Volcanoes has a more extensive program relying mainly on paid staff. The level of volunteer participation differs because of differences in size of the alien plant control programs, remoteness of the areas, and organizational objectives. Volunteers are most productive when simple control treatment methods are used for readily recognizable alien plants located in localized areas. Motivation is highest in situations in which progress can be observed, and where work on alien plant projects is followed by hiking and other kinds of outdoor enjoyment.

INTRODUCTION

The maintenance and restoration of native ecosystems in preserves of The Nature Conservancy of Hawaii and in the national parks require intensive management of alien plants. Protected areas often contain some of the most undisturbed native vegetation in Hawai'i, but preserves and parks also support increasing populations of alien plant species. Introduced species degrade native plant communities, and some alien plant species threaten to completely alter native ecosystems (Jacobi and Warshauer, this volume). Removal of feral ungulates appears to be insufficient to control some introduced plants (Stone *et al.*, this volume; Tunison and Stone, this volume). Biological control is a possible solution only for some widespread species (Gardner, this volume; Markin and Yoshioka, this volume). New plant species continue to be introduced into Hawai'i, and disruptive species continue to expand into protected areas.

HAWAII VOLCANOES NATIONAL PARK

At Hawaii Volcanoes, the level of staffing is inadequate to manage alien plants, and volunteers are used to supplement the efforts of paid employees. The introduced plant program consists of control efforts on 39 localized species on a parkwide basis (Tunison and Zimmer, this volume), and nine widespread species in eight localized areas called Special Ecological Areas (the least disturbed, most diverse or representative areas in the Park) (Tunison and Stone, this volume). Extending parkwide control programs to additional widespread species (*e.g.*, fountain grass, *Pennisetum setaceum*, and kāhili ginger, *Hedychium gardnerianum*) identified by Park staff would require approximately 20 alien plant workers per year (National Park Service 1986) for initial control stages.

Although volunteers have contributed significantly to the alien plant control program at Hawaii Volcanoes, the level of volunteer contribution is significantly lower than that of paid staff (Table 1). In five years (1981-1985) volunteers contributed 430 worker-days (11.4% of total effort) to introduced plant control projects. Paid staff accounted for 3,345 worker-days (88.6% of total effort) during this period (Table 1). Volunteer contributions have increased somewhat each year, with especially high levels of support in 1985 because of a single British group, Operation Raleigh. Most contributions were made by volunteer groups working in Special Ecological Areas or by individual volunteers monitoring vegetation. Conservation groups such as Friends of the Forest and Sierra Club have adopted a Special Ecological Area or portions of one, in which they conduct some of the needed management under the direct supervision of a National Park Service employee. Volunteers working as individuals are typically students who map weeds or monitor plots and transects. Most volunteers are from out of State, but numbers of local volunteers are increasing. Volunteers cost approximately $7 per day in stipends for food and lodging.

Volunteer support at Hawaii Volcanoes National Park is limited by four factors:

1. Alien plant control programs require timely follow-up treatments. Volunteer contributions tend to be unpredictable in level and timing. Therefore, most programs are carried out by paid staff.

2. It is difficult to motivate volunteers for many control programs because field activities are strenuous and monotonous, target plants are difficult to locate and identify, and progress in controlling plants cannot be perceived by volunteers after one or two treatments or in the short period they are involved.

3. The population base of Hawai'i Island is low, and interest in conservation is lower than many other areas of the nation. This is made worse by the fact that costs for airfare, housing, and supplies discourage the use of volunteers from off-Island; therefore, off-island volunteers are no longer recruited at Hawaii Volcanoes.

Table 1. Volunteer contributions (worker days) to alien plant control programs at Hawaii Volcanoes National Park and The Nature Conservancy of Hawaii Kamakou Preserve.

	Worker Status			Volunteers				
				Work Unit		Origin		
Year	Volunteers	Paid Staff	Total	Individuals	Groups	Local	Off-Island	Out-of-State
HAWAII VOLCANOES								
1981	5	350	355	5		5		
1982	20	350	370	10	10	10	10	
1983	40	630	670		20		35	5
1984	60	810	870	60				60
1985	305	1,205	1,510	60	235	45		250
Subtotals	430	3,345	3,775	135	265	60	45	315
KAMAKOU								
1983	81	15	96		81		81	
1984	73	15	88	10	63	24	39	10
1985	36	8	44		36	7	29	
Subtotals	190	38	228	10	180	31	149	10
Totals	620	3,383	4,003	145	445	91	194	325

4. Park managers believe that the time required to recruit, motivate, care for, and supervise volunteers is often more productively spent on other projects.

KAMAKOU PRESERVE

In contrast to Hawaii Volcanoes, 83% (190 worker-days) of the alien plant control work done at Kamakou Preserve from 1983 through 1985 was accomplished by volunteers (Table 1). Kamakou had only one paid staff member during that time. This individual coordinated or conducted all alien plant and animal control and interpretive and fund-raising activities at the Preserve. At least 50% of the volunteer worker-days were spent on reducing population levels of the higher-priority target alien plant species such as karakanut (*Corynocarpus laevigata*), black wattle (*Acacia mearnsii*), paperbark (*Melaleuca quinquenervia*), and strawberry guava or waiawī *(Psidium cattleianum)*. Volunteers are successful and maintain high motivation levels with these particular plant species. Target plants occur in highly localized populations, are easy to locate and identify, and are readily eradicable by simple techniques such as cut stump or frill (notching) treatments. Progress in controlling these species is noticeable, and several areas have been cleared except for seedling stages.

A steady flow of volunteers at Kamakou issues from two main sources: the Honolulu chapter of the Sierra Club and local Moloka'i groups. The impetus for Sierra Club involvement started during a two-week summer weed control project in 1983. All expenses were provided by The Nature Conservancy of Hawaii. Eight volunteers contributed a total of 80 worker-days at a total cost of approximately $2,000 ($25/worker-day).

By working with key members of Sierra Club work parties, the Preserve Manager facilitated the establishment of a long-term volunteer group, known as the "Longweekenders." Longweekenders work in Kamakou on alien plant control projects eight to ten long weekends or holidays each year. Veterans of previous outings lead, with minimal supervision from the Preserve Manager, work parties of eight to ten people, approximately half of whom are new to the Preserve. Motivation and productivity are high, and the groups are essentially self-supervised. The Preserve provides supplies and materials for plant control, transportation from the airport, and lodging. These volunteers cost the Preserve $10/day/person.

The Longweekender trips have become very popular. One day of work is followed by two days of exploring the Preserve or Moloka'i, which appeals to many O'ahu residents, the majority of Longweekenders. An important factor motivating volunteers to come to Kamakou is the intact nature of Preserve ecosystems and the rural nature of Moloka'i.

Service trips by Moloka'i groups are not as cost-effective as Longweekender trips, but they expose local citizens to the Preserve and elicit support for Preserve programs on Moloka'i. Most local groups -- Teachers for Kamakou, 4-H groups, and Moloka'i Boy Scouts -- participate on a semi-annual basis, and trips are conducted in two parts. The morning session is devoted to weed control work, while the afternoon session is spent on an interpretive hike guided by the Preserve Manager. The Preserve also provides transportation and refreshments.

Management plans for Kamakou mandate increasing use of volunteers. However, achieving this objective depends largely on enhanced staffing and improved vehicle support.

VOLUNTEER MOTIVATION

A number of lessons about volunteer programs have been learned at Kamakou and Hawaii Volcanoes:

1. The most cost-effective volunteer groups are those with trained leaders who can organize, motivate, and supervise themselves and their groups. The Longweekenders at Kamakou are a good example of a productive and self-directed volunteer group.

2. Volunteers are most productive and motivated when working with mechanical, cut-stump, or frill treatments, with readily recognizable plants, and with target species with localized geographical ranges.

3. To motivate volunteers, managers need to provide projects in which progress is demonstrable and where recognizable work units can be completed by the end of each work session. Localized introduced plants on Kamakou and in Special Ecological Areas in Hawaii Volcanoes provide projects that meet these criteria.

4. Recognizing volunteer contributions through media exposure increases volunteer support as well as public education. (However, efforts of paid staff should also be recognized to prevent conflicts.)

5. Integrating natural history interpretation, recreation, and work projects helps increase volunteer understanding and motivation.

Literature Cited

Gardner, D.E. [this volume] Plant pathogens as biocontrol agents in native Hawaiian ecosystems.

Jacobi, J.D., and F.R. Warshauer. [this volume] Distribution of six alien plants in upland habitats on the island of Hawai'i.

Markin, G.P., and E. Yoshioka. [this volume] Evaluating proposed biological control programs for introduced plants.

National Park Service. 1986. Natural Resources Management Plan and Environmental Assessment, revised January 1986. Hawaii Volcanoes National Park.

Stone, C.P., L.W. Cuddihy, and J.T. Tunison. [this volume] Responses of Hawaiian ecosystems to removal of feral pigs and goats.

Tunison, J.T., and C.P. Stone. [this volume] Special Ecological Areas: an approach to alien plant control in Hawaii Volcanoes National Park.

Tunison, J.T., and N.G. Zimmer. [this volume] Success in controlling localized alien plants in Hawaii Volcanoes National Park.

IN CONCLUSION . . .

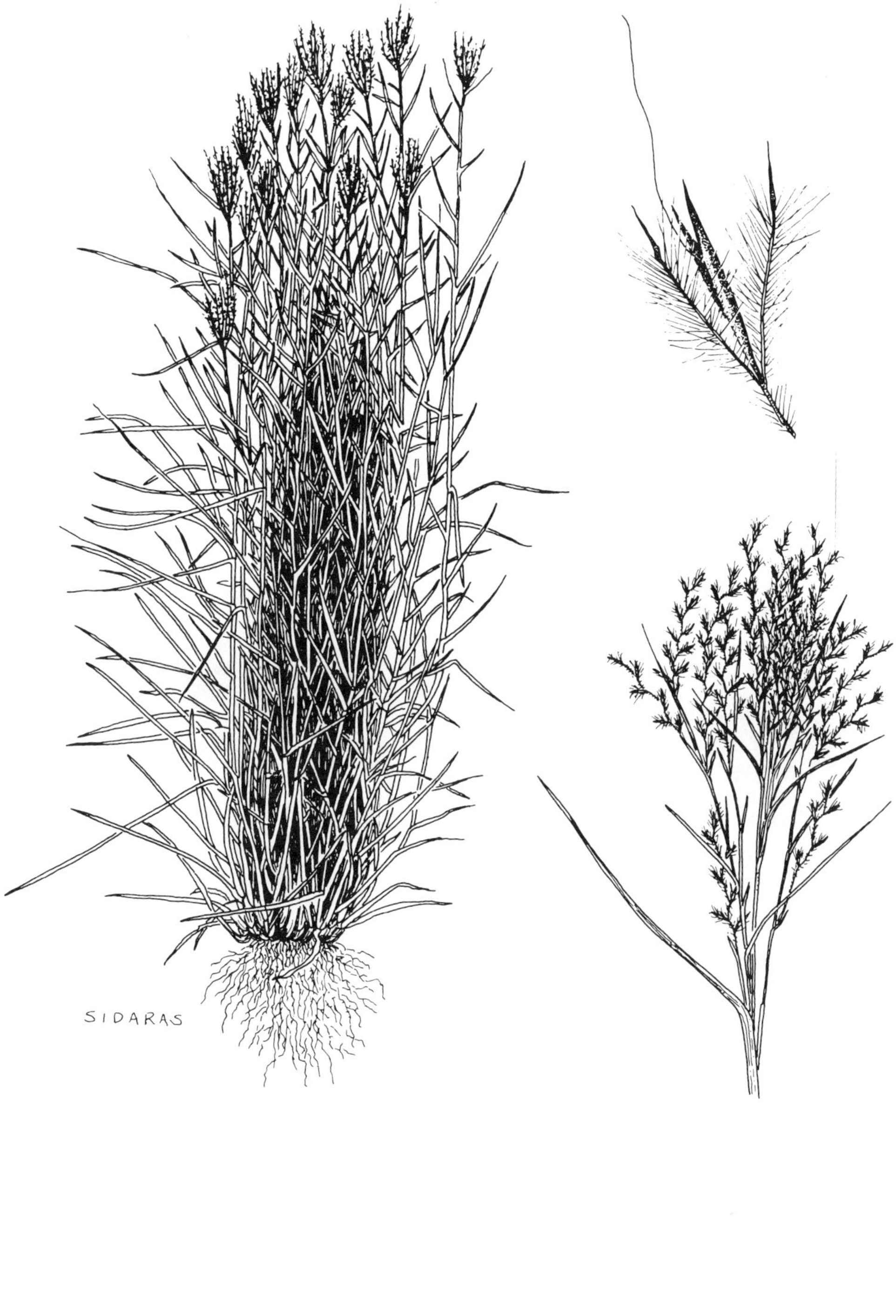

ALIEN PLANT MANAGEMENT IN HAWAI'I: CONCLUSIONS

J. Timothy Tunison,
Clifford W. Smith, and Charles P. Stone

This volume has clearly shown the importance of the weed problem in the Hawaiian Islands. Since arrival of the first human settlers and especially with establishment of continental humans and commerce, introduced plants have colonized "Paradise" at an increasingly rapid rate (Wester, this volume). All ecosystems have been affected, including those in the remote Northwestern Hawaiian Islands of the archipelago (Herbst and Wagner, this volume).

Approximately 90 of the 861 naturalized alien plant species in Hawai'i are significant threats to native ecosystems (Smith 1985). A number of these species continue to expand occupied ranges. Among those are firetree (*Myrica faya*) (Whiteaker and Gardner, this volume); strawberry guava (*Psidium cattleianum*), fountain grass (*Pennisetum setaceum*), yellow Himalayan raspberry (*Rubus ellipticus*) (Jacobi and Warshauer, this volume); and clidemia (*Clidemia hirta*), (Smith, this volume). These species and others have not yet attained potential distributions (or, probably, densities), and species widespread on some islands are now becoming established on others. For example, *Miconia calvescens* (Melastomataceae) is established on Hawai'i Island and is now a concern on Maui (Gagné *et al.*, in press). In addition, new potentially invasive species are still being introduced to Hawai'i. Some of the most disruptive weeds are such efficient competitors that they can form monospecific stands (Smith 1985), particularly when accompanied by disturbance. Others can alter ecosystem level processes such as nutrient cycling, hydrological, and fire regimes (Vitousek, this volume; Smith and Tunison, this volume). These changes can result in widespread replacement of native vegetation by alien-dominated plant communities and the loss of native ecosystems. However, few studies have actually documented the replacement of native plants and alteration of native plant communities.

The severity of the alien plant problem and the fact that it is so widespread in the Islands make a rigorously organized approach based on relevant information especially necessary. Moreover, development of a variety of approaches to weed control to deal with different situations

(Tunison, this volume a; Stone *et al.*, this volume), use of cooperative efforts in management (Taylor, this volume; Tunison and Misaki, this volume), the implementation and enforcement of legislation and rules to control entry and spread of aliens (Higa, this volume), and emphasis on environmental education (Hawkins, this volume; Stone, this volume) are necessary components of weed management programs.

Weed management approaches in Hawai'i can be summarized in several ways, but in our opinion major emphasis must occur in three areas: 1) protecting parks and preserves from, and managing for, invasive species; 2) reducing entry and establishment of weeds in Hawai'i; and 3) developing a regional approach to accomplish 1) and 2). Minimizing disturbance that allows alien plants to establish and spread is an important part of all emphases.

EFFECTIVE WEED MANAGEMENT IN PARKS AND PRESERVES

Managers of parks and preserves in Hawai'i all have developed strategies for dealing with important introduced plants (Herbst and Wagner, this volume; Holt, this volume; Jacobi and Warshauer, this volume; Loope *et al.*, this volume; Spoehr, this volume; Stone *et al.* 1991; Tanimoto and Char, this volume; Tunison, this volume a). The weed management program in Hawaii Volcanoes National Park probably provides the most advanced model of weed management for Hawai'i's natural areas, primarily because of comparatively good support to deal with severe invasions. The ten management strategies in the Park are discussed thoroughly elsewhere (Tunison, this volume a). The following major emphases can be recommended for natural areas in Hawai'i (Fig. 1).

Removal of Artificial Disturbance

Disturbance by ungulates is one of the primary factors accounting for spread of alien plants (Loope and Scowcroft 1985; Loope and Mueller-Dombois 1989; Merlin and Juvik, this volume; Scowcroft and Conrad, this volume; Stone *et al.*, this volume). Ungulate control has traditionally been an initial and often major step in alien plant management in Hawai'i. Native vegetation often recovers well, and sometimes dramatically, once ungulates are removed (Loope and Scowcroft 1985; Merlin and Juvik, this volume; Scowcroft and Conrad, this volume; Stone *et al.*, this volume), although an introduced plant component may remain in the community. However, certain alien plants may persist and even spread in spite or because of ungulate removal or following other disturbances (Stone *et al.*, this volume). Thus, ungulate control is a necessary but not always sufficient measure in management of alien plants.

Establishment of roads and trails and transport of weed seeds by humans and their animals can also encourage establishment of invasive weeds (Tunison, this volume b; Anderson *et al.*, this volume). The widespread presence of invasive alien plant propagules now favors weed establishment and spread even after natural perturbations in many areas. Thus, physical disturbance of the land should be minimized in parks and preserves.

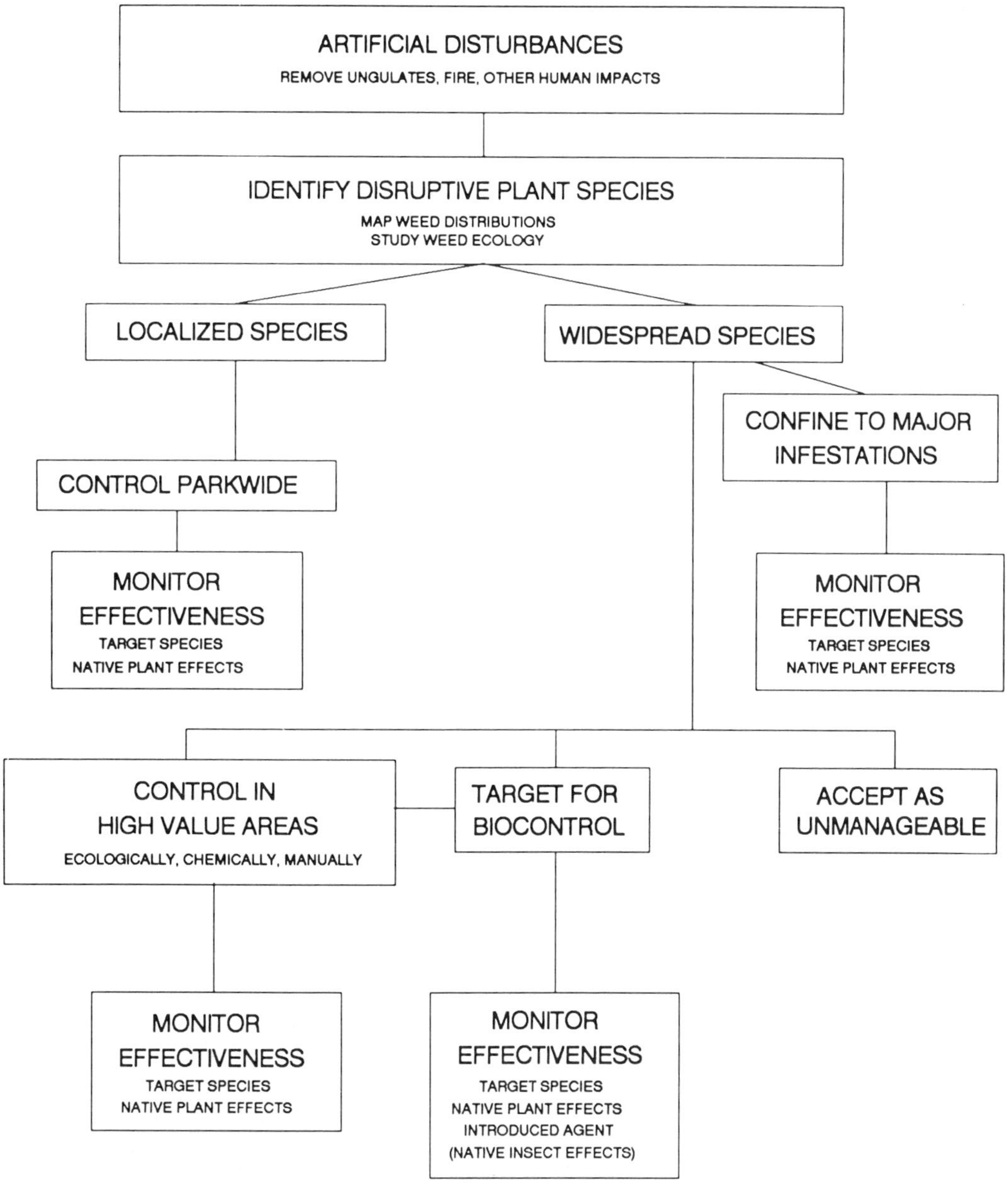

Figure 1. Alien plant control strategies or actions in the alien plant control program at Hawaii Volcanoes National Park.

Disturbance by fire often encourages introduced plants and is as important in some vegetation communities as ungulates are in others (Smith and Tunison, this volume). Uncontrolled natural fire and prescribed burning are not effective tools for managing native ecosystems in Hawai'i. Fire management policy in natural areas should mandate prevention, immediate suppression of all fires, and strategies for reducing fire starts and sizes, including fuel breaks and hazardous fuel reduction.

Biological, Chemical, Manual, and Ecological Control

Biological control is a possible strategy for control of some widespread invasive alien plants that continually reinvade parks and preserves. As with other control methods, strengths and weaknesses are inherent in this approach (Gardner and Davis 1982; Howarth 1983; Gardner, this volume; Markin *et al.*, this volume; Tunison, this volume a). However, for some widespread species no other alternative seems feasible, and economic costs and risks (in carefully managed programs) would seem to be warranted. Successes in alien plant management through biocontrol in natural areas are eagerly anticipated in the near future. It is likely that only a few major weeds will be controlled in the next decade, and perhaps only in a portion of their geographic ranges or in certain situations (moist or dry, etc.). Inadequate long-term monitoring of the results of past releases has been a weakness in the biocontrol approach, a deficiency that must be corrected in the future.

Herbicides have been successfully used to control a number of woody and non-woody plants in Hawai'i's natural areas (Santos *et al.*, this volume; Cuddihy *et al.* 1988; Santos *et al.* 1981a, b; Cuddihy *et al.* 1991). Herbicides have numerous advantages and disadvantages but despite labor requirements and some risk would appear to be a necessary component of weed management in most parks and preserves for the long term. Control of new invasions, of target species not amenable to biological control, and control to supplement biocontrol are legitimate uses for herbicides. A major disadvantage is the need for repeated treatments of widespread species that reinvade. Very little chemical is needed in careful control of selected weeds (mostly woody species) in natural areas (<16 liters in 1988 in 12,000 a (4,860 ha) in Hawaii Volcanoes National Park). As with biocontrol, little long-term monitoring of effects on vegetation recovery has been accomplished in the past to determine whether target weeds reinvade or more serious weeds establish in controlled areas in time. Native plant species diversity should also be monitored to detect changes after herbicide control. Beyond these basic questions about program benefits, more knowledge about persistence, mobility, and effects on invertebrates and other components of native species communities is desirable. Transferring knowledge about the most effective new chemicals and techniques from researchers to managers is important as new and safer chemicals are found. Adequate long-term weed control research programs are necessary to meet many of these needs.

Manual control is even more labor intensive than herbicidal control but is equally likely to be used indefinitely for some species in small areas of high ecological value in Hawai'i. Manual methods can open areas to reinvasion by the target weeds or even more invasive species, but the risk is often necessary. Manual methods are amenable to use of volunteers as a means of overcoming some of the costs of control (Tunison and Misaki, this volume). As with other control methods, long-term monitoring of results is recommended.

Ecological control, by encouraging the spread of native woody species and closed-canopy forest to shade out invasive species, has been experimentally attempted on the edges of high-value areas. Removal of alien grass cover through herbicide treatment resulted in recruitment of native trees (G.L. Santos and L.W. Cuddihy, pers. comm.). However, direct planting of native plants to shade out alien species and prevent their establishment has not been widely used as yet. Planting may become more necessary in the future as more weed control problems are solved and management of less-intact (more highly invaded) natural areas is emphasized.

High-Value Areas

Some areas within legally protected parks and preserves in Hawai'i are less invaded and contain more native elements than others (Tunison and Stone, this volume). Such areas may contain high native biological diversity or important rare species. These areas may also be representative of widespread or rare communities, contain very few invasive alien species, or be readily manageable. Protecting such areas from alien plant invasion is a realistic strategy, especially if the areas are sizable (variously defined for the taxa and communities at risk) and can be buffered against future invasions by less intensive management of aliens in the surrounding area. Widespread as well as localized invasive species should be controlled in such areas where feasible (Tunison and Zimmer, this volume; Tunison, this volume a; Tunison and Stone, this volume). Focusing management effort on alien plants in a few important areas at a time is an incremental approach to management that has proven very successful where problems are almost overwhelming and human resources are comparatively limited. As alien plant management needs in one area are reduced, workers are freed for managing new and/or larger areas.

Confining Invasive Species to Major Sites of Infestation

Some invasive species that show great potential for invading significant parts of a protected area can be confined to major infestation sites (Tunison, this volume b), albeit with considerable effort. This strategy involves knowing enough about the species to recognize invasion corridors, major sites of concentration, and ecology to accomplish adequate confinement. Control of outlying colonization sites is more important than attacking the major infestation itself (Mack, this volume; Tunison, this volume b). A long-term monitoring program is also necessary to evaluate the effectiveness of this strategy.

Research and Monitoring

Managers and researchers need to work together in alien plant management programs in Hawai'i. The need for research in biocontrol and herbicide programs has been mentioned above, and research is also necessary to objectively evaluate long-term plant community recovery and diversity after operational management programs. Studies of alien plant ecology to determine best means and applications for control are also desirable. Monitoring of the effects of management for introduced plants on other community components, including invertebrates, birds, rare native plants, and the interactions among these, increases the level of knowledge that can be applied to accomplish effective stewardship of natural resources.

Education in Parks and Preserves

Active management of introduced plants in Hawai'i involves such controversial practices as introduction of biological control agents, use of herbicides, and establishment of marked monitoring and research plots and transects in natural areas. The public is generally interested and supportive once it understands that the reasons for action are logical and necessary to perpetuate the native biota. Interest, concern, and action, including much-needed support for expensive programs of alien plant control, can be generated through educational programs that share knowledge and concern (Stone, this volume; Hawkins, this volume). Strategies for environmental education are an important part of alien plant management programs in natural areas, whether managers, researchers, or interpreters deliver the message.

REDUCING ALIEN PLANT ENTRY AND ESTABLISHMENT IN HAWAI'I

Among the ten actions needed in "reversing the extinction trend" in Hawai'i, according to the Hawaii Department of Land and Natural Resources, the U.S. Fish and Wildlife Service, and The Nature Conservancy of Hawaii (1991) is "Halt the flow of new foreign pest species into Hawai'i."

> Congress and the state legislature must recognize Hawai'i's exceptional vulnerability to destructive, foreign plant and animal pests and enact special measures to prevent the establishment of additional problem species. Hawai'i cannot keep pace with its extinction problem if new, destructive pest species continue arriving in commercial cargo, personal baggage, or packages mailed to the islands.

Unwanted alien plants arrive accidently in cargo, on travellers, or in their belongings. Seeds or spores are especially likely to arrive this way. Alien plants also arrive legally as feed for livestock or poultry, in seed lots for agriculture (Higa, this volume), as biological control agents, for horticulture (Yee and Gagné, this volume), and for research. Alien species will be brought in intentionally and illegally despite all efforts to intercept them. Wildflower seeds sent by mail as promotional material from the U.S. Mainland provide a good example of a deliberate entry that is difficult to counteract. Most inspection systems cannot possibly check all items, nor can thorough checks for a variety of threats be made.

Several groups in Hawai'i in addition to the regulatory agencies are now working to strengthen the process of preventing alien species entry into the State. These include The Nature Conservancy of Hawaii and the Natural Resources Defense Council, which are preparing a background study and have organized and sponsored working groups on the subject; the Alien Species Alert Program of the National Audubon Society, which is emphasizing educational materials and legislation; and the Noxious Plants Task Force of the Conservation Council of Hawaii, which is working on legislation, rulings, and education. The Office of Technology Assessment of the U.S. Congress is doing a nationwide study on the problem of introduced species, with some emphasis on Hawai'i. Increased congressional funding for introduced species programs should follow this study. The World Wide Fund for Nature is also preparing a guide entitled "Invasive species: the threat to natural ecosystems worldwide."

Problems identified by The Nature Conservancy of Hawaii and the Natural Resources Defense Council (1992) include those under the general headings of jurisdictional gaps, overlaps, and conflicts; uncoordinated decision-making; lack of resources; and technical limitations. A number of difficulties and inefficiencies arise because: 1) Hawai'i's quarantine and inspection systems have been developed piecemeal over the past century, with each piece designed to address only part of the problem; and 2) Hawai'i has been treated the same as any other state in international traffic of biological material, despite much greater vulnerability to biological invasions. Sources of entry for alien species into Hawai'i and regulating pathways are shown in Figure 2. Five Federal agencies (the U.S. Treasury Department, the U.S. Department of the Interior, the U.S. Department of Agriculture, the U.S. Department of Defense, and the U.S. Postal Service) and two branches of State government (the Plant Industry Division and the Plant Pest Control Branch of the Hawaii Department of Agriculture) are involved in preventing alien plant entry for different programs (The Nature Conservancy of Hawaii and Natural Resources Defense Council 1992). Some current activities and concerns of the regulatory agencies are discussed by Higa (this volume).

Prevention of Establishment

Quarantine and inspection mechanisms, no matter how good, will never be able to prevent entry of all alien plants into Hawai'i. However, if detected early enough, some invasions can be stopped. A combination of awareness, expertise, monitoring, cooperation, and the capability and will to take rapid action are necessary to prevent establishment once an invasive species has arrived. Movements of introduced species from place to place on the same island and from island to island are difficult to prevent, but success is possible. The State Noxious Plant list specifies islands on which at least some plants are considered noxious. Unfortunately, unless a plant is on the noxious species list, no governmental entity is responsible for preventing establishment. A species can only become officially noxious by meeting all of several criteria. When a weed is no longer considered manageable, it is removed from the list. Currently, plant species listed on the State Noxious Plant list and the Federal Noxious Weed Seed list show little overlap.

Rapid detection of alien plant invasions is a serious problem, and action must be taken soon after discovery. The emphasis in legislation, rules, and enforcement is on threats to agriculture, forestry, and public health rather than native ecosystems. The Hawaii Department of Land and Natural Resources lacks statutory authority to list noxious plants that are problems in parks and reserves. Controlling, eradicating, or preventing the entry of noxious plant species that have not yet been listed is also a problem, since hearings must be conducted to add new species to the list, and a number of criteria must be met. An example is *Miconia calvescens,* apparently introduced by nurseries to Maui and Hawai'i islands, but not placed on the noxious list for several years because of delays in hearings required to revise the list. The Hawaii Department of Agriculture must

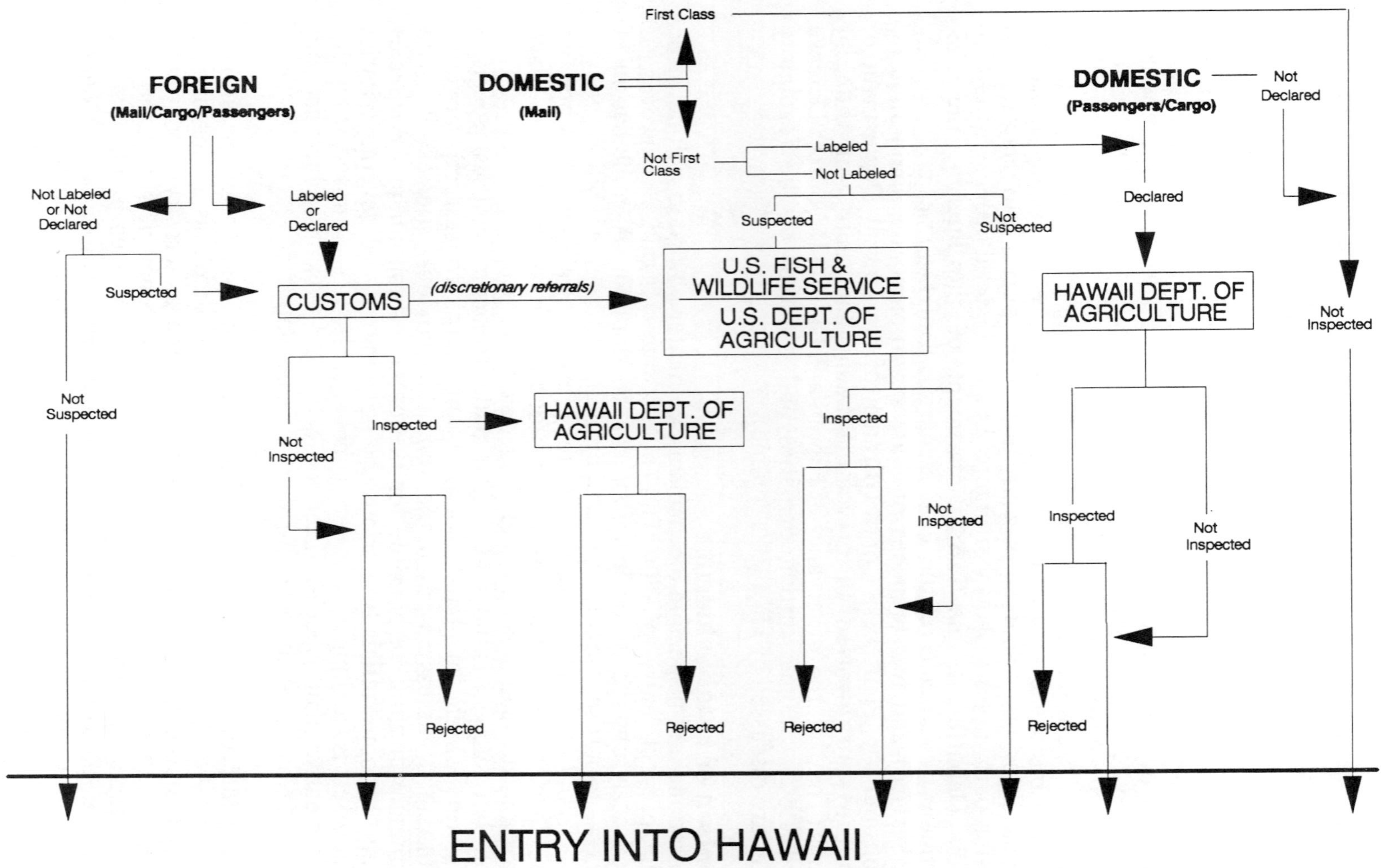

Figure 2. Prevention process for alien species entering Hawai'i. (From The Nature Conservancy of Hawaii and Natural Resources Defense Council 1992.)

remove listed noxious weeds from private property if requested but cannot enter private property to remove plants if opposed by the landowner. A rapid response to *Miconia calvescens* invasion on Maui by National Park Service, The Nature Conservancy, and Sierra Club volunteers appears to have removed 90% of the individuals in seven populations (Gagné *et al.*, in press).

Prevention of establishment of noxious plants after entry by chemical, mechanical, and biological control is a function of the Hawaii Department of Agriculture Plant Pest Control Branch. This jurisdiction extends to forest and conservation lands. The Hawaii Department of Land and Natural Resources Division of Forestry and Wildlife forestry personnel also work with private landowners to control noxious plants on private lands, as well as in state-owned conservation areas. Control programs for alien plants on lands managed by the U.S. Fish and Wildlife Service, the National Park Service, The Nature Conservancy of Hawaii, and various private landowners have already been mentioned. The U.S. Forest Service, the National Park Service, and other state and federal research entities participate in biological control programs for introduced plants (Gardner, this volume; Markin *et al.*, this volume).

REGIONAL AND STATEWIDE EDUCATION, LEGISLATION, AND COOPERATION

Public awareness and involvement are important in preventing entry of introduced plants and in detecting early establishment, as well as in supporting active management programs in natural areas. Attractive pamphlets and brochures, continual emphasis in print and electronic media, and local slide shows and talks can raise public consciousness statewide. The curriculum materials for grades K-8 produced by the ʻŌhiʻa Project will do much to educate youth and their parents in public and private schools about alien species and the problems they cause in Hawaiʻi. A video on introduced species for use on arriving commercial airlines, prepared by the Alien Species Alert Program of the National Audubon Society, a flier on miconia prepared by the Conservation Council of Hawaii, posters on banana poka (*Passiflora mollissima*) and firetree prepared by the Hawaii Department of Land and Natural Resources, a poster/pamphlet on firetree prepared by the Big Island Resource Conservation and Development Council, and the excellent National Geographic Society video entitled "Strangers in Paradise" are but a few examples of recent educational activities related to alien plant threats. The Nature Conservancy of Hawaii, the Hawaii Department of Land and Natural Resources, the National Audubon Society, the U.S. Fish and Wildlife Service, and the Conservation Council of Hawaii have been especially active in both educational and legislative initiatives recently, often working together on these efforts. They are all to be congratulated for their success in overcoming jurisdictional and organizational barriers in a concerted effort to deal with alien species problems and the extinction crisis in Hawaiʻi.

Regional and statewide cooperation in weed management is difficult to address because of the number of entities and management missions involved. Nevertheless, some examples of regional approaches such as the multi-agency biological control committee, the gorse (*Ulex europaeus*) and firetree committees, and the Hawaiian Home Lands procedures (Markin *et al.*, this volume; Tulang, this volume; Gardner, this volume; Spoehr, this volume) show promise. Information-sharing among federal and state agencies and private landowners has occurred through publications, reports, and the loan of knowledgeable personnel on occasion. However, much more could be done in regional and statewide management of alien plant problems. We think that the following statement about government and private conservation efforts could equally apply to alien plant management needs and have inserted appropriate wording in brackets (Hawaii Department of Land and Natural Resources, U.S. Fish and Wildlife Service, and The Nature Conservancy of Hawaii 1991):

> A task force of federal, state, and private conservation organizations is necessary to coordinate land acquisitions, habitat management, research, enforcement, and other conservation measures in the most cost-effective manner possible. Shared databases on the status and biology of native [and alien] species need reliable funding. The existing network of environmental organizations should consider their legislative [and educational] advocacy and together prepare regular reports on progress in the fight to reduce extinction [or alien plant invasions] in the islands.

The primary difference in emphasis is that alien species are managed for reduction, whereas the native biota are managed for stability or increase. However, managers need shared databases, reliable funding, coordinated management and research, legislative advocacy, and reporting for both management efforts. It is particularly important that education, legislation, and management related to alien plant entry and invasions achieve statewide cooperation by all concerned. Not only are the resources of any one agency not equal to the task in Hawai'i, but combined efforts will be better able to reach beyond state boundaries to congressional delegations in Washington as well as to commerce and visitors from all over the world. A concerted effort will also be needed to effectively reach the local populace. Reducing alien plant ingress and establishment necessitate increasing monitoring, control, and research, and stepping up educational, enforcement, and legislative efforts all at the same time. Major enduring commitments from many quarters are necessary. It has been said that the race between developing environmental consciousness and deterioration of the environment will be decided in the next human generation. Increased effectiveness in controlling introduced plants during that period will certainly be an essential determinant of Hawai'i's future environmental quality.

ACKNOWLEDGMENTS

We thank Alan Holt of The Nature Conservancy of Hawaii for helpful comments on this paper, and Danielle B. Stone for typing and helping to edit it. Michelle Fulton produced the figures, and we appreciate her help.

Literature Cited

Anderson, S.J., C.P. Stone, and P.K. Higashino. [this volume] Distribution and spread of alien plants in Kīpahulu Valley, Haleakala National Park, above 2,300 ft elevation.

Cuddihy, L.W., G.L. Santos, and C.P. Stone. 1991. *Control of firetree (*Myrica faya *Aiton) with herbicides in Hawaii Volcanoes National Park.* Tech. Rep. 82, Univ. Hawaii Coop. Natl. Park Resour. Stud. Unit. Honolulu.

Cuddihy, L.W., C.P. Stone, and J.T. Tunison. 1988. Alien plants and their management in Hawaii Volcanoes National Park. *Trans. West. Sect. Wildl. Soc.* 24:42-46.

Gagné, B.H., L.L. Loope, A.C. Medeiros, and S.J. Anderson. In press. *Miconia calvescens*: a threat to native forests of the Hawaiian Islands. *Pac. Sci.*

Gardner, D.E. [this volume] Plant pathogens as biocontrol agents in native Hawaiian ecosystems.

Gardner, D.E., and C.J. Davis. 1982. *The prospects for biological control of nonnative plants in Hawaiian national parks.* Tech. Rep. 45, Univ. Hawaii Coop. Natl. Park Resour. Stud. Unit. Honolulu.

Hawaii Department of Land and Natural Resources, U.S. Fish and Wildlife Service, and The Nature Conservancy of Hawaii. 1991. *Hawaii's extinction crisis: a call to action.* Honolulu.

Hawkins, J.W. [this volume] Environmental education in the schools.

Herbst, D.R., and W.L. Wagner. [this volume] Alien plants on the Northwestern Hawaiian Islands.

Higa, S.J. [this volume] Plant importation rules and regulations.

Holt, R.A. [this volume] Control of alien plants on Nature Conservancy preserves.

Howarth, F.G. 1983. Classical biocontrol: panacea or Pandora's box. *Proc. Hawn. Entomol. Soc.* 24(2,3):239-244.

Jacobi, J.D., and F.R. Warshauer. [this volume] Distribution of six alien plants in upland habitats on the island of Hawai'i.

Loope, L.L., and D. Mueller-Dombois. 1989. Characteristics of invaded islands, with special reference to Hawaii. In *Biological invasions, a global perspective*, ed. J.A. Drake, H.A. Mooney, F. di Castri, R.H. Groves, F.J. Kruger, M. Rejmànek, and M. Williamson, 257-280. SCOPE 37. New York: John Wiley & Sons.

Loope, L.L., and P.G. Scowcroft. 1985. Vegetation response within exclosures in Hawai'i: a review. In *Hawai'i's terrestrial ecosystems: preservation and management*, ed. C.P. Stone and J.M. Scott, 377-402. Univ. Hawaii Coop. Natl. Park Resour. Stud. Unit. Honolulu: Univ. Hawaii Pr.

Loope, L.L., R.J. Nagata, and A.C. Medeiros. [this volume] Alien plants in Haleakala National Park.

Markin, G.P., P.-Y. Lai, and G.Y. Funasaki. [this volume] Status of biological control of weeds in Hawai'i and implications for managing native ecosystems

Merlin, M.D., and J.O. Juvik. [this volume] Relationships among native and alien plants on Pacific islands with and without significant human disturbance and feral ungulates.

Santos, G.L., L.W. Cuddihy, and C.P. Stone. 1991a. *A test of two herbicides for use on banana poka (*Passiflora mollissima *(Kunth) L.H. Bailey) in Hawaii Volcanoes National Park.* Tech. Rep. 79, Univ. Hawaii Coop. Natl. Park Resour. Stud. Unit. Honolulu.

Santos, G.L, L.W. Cuddihy, and C.P. Stone. 1991b. *Control of yellow Himalayan raspberry (*Rubus ellipticus *Sm.) with cut stump herbicide treatments in Hawaii Volcanoes National Park.* Tech. Rep. 80, Univ. Hawaii Coop. Natl. Park Resour. Stud. Unit. Honolulu.

Santos, G.L., D. Kageler, D.E. Gardner, L.W. Cuddihy, and C.P. Stone. [this volume] Herbicidal control of selected alien plant species in Hawaii Volcanoes National Park.

Scowcroft, P.G., and C.E. Conrad. [this volume] Alien and native plant response to release from feral sheep browsing on Mauna Kea.

Smith, C.W. 1985. Impact of alien plants on Hawai'i's native biota. In *Hawai'i's terrestrial ecosystems: preservation and management*, ed. C.P. Stone and J.M. Scott, 180-250. Univ. Hawaii Coop. Natl. Park Resour. Stud. Unit. Honolulu: Univ. Hawaii Pr.

Smith, C.W. [this volume] Distribution, status, phenology, rate of spread, and management of clidemia in Hawai'i.

Smith, C.W., and J.T. Tunison. [this volume] Fire and alien plants in Hawai'i: research and management implications for native ecosystems.

Spoehr, H. [this volume] Public/private sector partnerships to preserve native ecosystems: the Department of Hawaiian Home Lands experience.

Stone, C.P. [this volume] Environmental education in Hawai'i: history and overview.

Stone, C.P., L.W. Cuddihy, and J.T. Tunison. [this volume] Responses of Hawaiian ecosystems to removal of feral pigs and goats.

Stone, C.P., P.K. Higashino, L.W. Cuddihy, and S.J. Anderson. 1991. *Preliminary survey of feral ungulate and alien and rare plant occurrence on Hakalau Forest National Wildlife Refuge.* Tech. Rep. 81, Univ. Hawaii Coop. Natl. Park Resour. Stud. Unit. Honolulu.

Tanimoto, V.M., and W.P. Char. [this volume] Alien plant control on state lands including Natural Area Reserves.

The Nature Conservancy of Hawaii and Natural Resources Defense Council. 1992. The alien pest species invasion in Hawai'i: background study and recommendations for interagency planning. Review draft 3. Honolulu.

Tulang, M. [this volume] The U.S. Department of Agriculture's rural development approach to alien plant control in Hawai'i.

Tunison, J.T. [this volume a] Alien plant control strategies in Hawaii Volcanoes National Park.

Tunison, J.T. [this volume b] Fountain grass control in Hawaii Volcanoes National Park: management considerations and strategies.

Tunison, J.T., and E. Misaki. [this volume] The use of volunteers for alien plant control at Hawaii Volcanoes National Park and The Nature Conservancy's Kamakou Preserve.

Tunison, J.T., and C.P. Stone [this volume] Special Ecological Areas: an approach to alien plant control in Hawaii Volcanoes National Park.

Tunison, J.T., and N.G. Zimmer. [this volume] Success in controlling localized alien plants in Hawaii Volcanoes National Park.

Vitousek, P.M. [this volume] Effects of alien plants on native ecosystems.

Wester, L. [this volume] Origins and introductions of adventive alien flowering plants in Hawai'i.

Whiteaker, L.D., and D.E. Gardner. [this volume] Firetree (*Myrica faya*) distribution in Hawai'i.

Yee, R.S.N., and W.C. Gagné. [this volume] Activities and needs of the horticulture industry in relation to alien plant problems in Hawai'i.

INDEX

INDEX

'a'ali'i (*Dodonaea, D. viscosa*) 351, 352, 354, 355, 359, 394, 399, 401, 402, 403, 559, 564, 585, 627, 658, 661, 669, 673, 675, 683; see also *Dodonaea, D. viscosa*
Abelmoschus moschatus 142
Abronia 88
Abrus precatorius 137
Abutilon grandifolium (hairy abutilon) 142, 217
Acacia 304, 309-310
Acacia confusa (Formosan koa) 137, 487, 506, 507, 508, 514, 521, 538
Acacia farnesiana (klu) 137, 215, 538
Acacia koa (koa) 155, 168, 186, 271, 274, 304, 363, 394, 399, 532, 558, 580, 667, 674; see also koa
Acacia mearnsii (black wattle) 137, 289, 487, 506, 508, 513, 514, 528, 538, 547, 568, 816; see also black wattle
Acacia melanoxylon (blackwood acacia) 487, 508, 513, 514
Acalypha indica 136
Acalypha wilkesiana (beefsteak plant) 214
Acanthaceae (acanthus) 125, 209
acanthus (Acanthaceae) 125, 209
Acanthospermum 127
Achillea millefolium 70, 71, 72, 81, 127
Achyranthes aspera 125, 210, 613-614
Achyranthes atollensis 199, 210
Acridotheres tristis (common myna, Indian myna) 227, 281, 412
Adelges piceae (balsam wooly aphid) 20
Adiantum jordanii 90
'ae'ae (*Bacopa monnieri*) 220
Aerenicopsis championi 415
Aeschynomene virginica 433
African dropseed (*Sporobolus africanus*) 209, 667; see also *Sporobolus africanus*
African tulip tree (*Spathodea campanulata*) 212, 316, 318, 547, 559, 566, 618; see also *Spathodea campanulata*
Agavaceae 116, 205; see also agave
agave 116, 487; see also Agavaceae, *Agave americana*
Agave americana (century plant) 487, 508, 514, 521; see also agave
Agave sisalana (sisal) 116, 205, 487, 507, 508, 514, 521, 538
Ageratina adenophora (Maui pāmakani) 127, 304, 315, 326-327, 425, 468, 538, 544, 547, 558, 563, 696; see also Maui pāmakani
Ageratina riparia (Hāmākua pāmakani) 127, 327, 411, 425, 426, 432, 434, 468, 475, 538, 544, 547, 585, 689; see also Hāmākua pāmakani
Ageratum conyzoides 127, 327
Ageratum houstonianum 127
Agonopterix ulicitella 564, 580
Agraulis vanillae 291
agriculture 3, 4, 7, 8, 13, 16, 56, 70, 73, 76-77, 79, 93, 99, 100, 108, 113, 114, 115ff, 189, 249, 271, 317, 318, 397, 432, 434, 436, 439, 441, 442, 446, 455, 456, 466, 467, 472, 474, 476, 478, 479, 502, 533, 546, 554, 559, 577, 591, 602, 618, 708, 709, 713, 714, 727, 729, 753, 757, 758, 759, 761, 762, 766, 767, 768, 769, 770, 771, 772, 773, 799, 800, 826, 827
agromyzid 413; see also *Ophiomyia lantanae*
Agropyron cristatum (bunchgrass) 15; see also bunchgrass
Agropyron desertorum (bunchgrass) 35; see also bunchgrass
Agropyron repens 88
Agropyron spicatum (bunchgrass) 35; see also bunchgrass
Agroseris grandiflora 81

Agrostis avenacea 119, 629, 630, 636, 641, 647, 650, 652, 672
Agrostis sandwicensis 626, 632, 634, 636, 639, 640, 641, 646, 647, 650, 652, 661, 662
Agrostis semiverticillata 88, 119
Agrostis stolonifera 17, 119
'ahu'awa (*Mariscus javanicus*) 101, 207; see also *Mariscus javanicus*
Ailanthus altissima 20
Aira caryophyllea 119
air plant (*Kalanchoe pinnata*) 214; see also *Kalanchoe pinnata*
Aizoaceae 81, 125, 209; see also carpetweed, fig-marigold
'ākala (*Rubus hawaiiensis*) 476; see also *Rubus hawaiiensis*
'ākepa (*Loxops coccineus*) 300
'ākia (*Wikstroemia, W. sandwicensis*) 359, 400, 402; see also *Wikstroemia*
'akoko (*Chamaesyce celastroides* var. *celastroides*) 190, 215
'ākulikuli (*Sesuvium portulacastrum*) 209
'alalā (*Corvus hawaiiensis*, Hawaiian crow) 282, 288
albatross 201
albizia (*Albizia*) 487, 508, 513, 514
Albizia julibrissin 20
Albizia lebbeck (siris tree) 137, 215
alena (*Boerhavia repens*) 218
Aleurites moluccana (candlenut, kukui) 102, 136, 214, 607, 729; see also candlenut
Aleurodicus dispersus (spiraling whitefly) 456
alfalfa (*Medicago sativa*) 216; see also *Medicago sativa*
algae 21, 33
Alismataceae (water plantain) 116
allamanda (*Allamanda cathartica*) 210
allelopathy 77, 227, 313, 315, 620, 629
Allium cepa (onion) 207
Allium fistulosum (bunch onion) 207
Allium praecox 81
Alocasia cucullata (Chinese taro) 205
Alocasia macrorrhiza 102, 116, 610
aloe (*Aloe*) 205
Aloeaceae 205
Alopex lagopus (arctic fox) 21
alpine zone 7, 268, 396, 541, 542, 551, 556, 626, 784, 786, 796
Alpinia 124
Alternanthera (joyweed) 125; see also joyweed
Alternanthera tenella (joyweed) 125, 210
Alternaria 291
Alternaria passiflora 278
Alysicarpus vaginalis 123, 137
'amakihi (*Hemignathus virens*) 278
amaranth (Amaranthaceae) 125, 210
Amaranthaceae (amaranth) 125, 210
Amaranthus brownii 210
Amaranthus dubius (pakai) 125, 210, 555
Amaranthus hybridus (green amaranth) 125, 210
Amaranthus lividus 126
Amaranthus spinosus (spiny pigweed) 126, 210, 555
Amaranthus viridis (slender amaranth) 126, 210, 555
Amaryllidaceae 81, 487, 508
'ama'u 307, 367, 400, 401, 562; see also *Sadleria cyatheoides, S. pallida*
'ama'uma'u (*Sadleria cyatheoides, S. pallida*) 348, 363; see also *Sadleria cyatheoides*
Amblyopappus pusillus 70, 81
Ambrosia artemisiifolia 127
Ambrosia chamissonis 81
American crocodile (*Crocodylus acutus*) 50, 53, 55
Ammannia 142
Ammophila arenaria (European beach grass) 8, 16, 119, 208
Amsinckia 72
Amsinckia intermedia 83, 131
Amsinckia spectabilis 83
Anacardiaceae 81, 126, 210, 493; see also mango
Anacardium occidentale 126
Anagallis arvensis (scarlet pimpernel) 90, 147; see also scarlet pimpernel
Ananas comosus (pineapple) 420, 444, 453, 554, 730, 800; see also pineapple industry

'ānaunau (*Lepidium bidentatum* var. *o-waihiense)* 190, 212
anchialine pools 542; see also aquatic habitat
Andropogon 34, 35, 399, 401; see also bunchgrass
Andropogon glomeratus. See *Schizachyrium condensatum*
Andropogon virginicus (broomsedge) 119, 250, 314, 322, 394, 398, 434, 486, 487, 538, 547, 552, 563, 689, 734; see also broomsedge
Anenome hupehensis 147
Anethum graveolens 126
angiosperms 116; see also species listings
Annonaceae 126
Annona cherimola 126
annual bluegrass (*Poa annua*) 209, 566; see also *Poa annua*
Anous minutus melanogenys (black noddy) 197
Anredera cordifolia 131
Anthemis cotula 127
Anthonomus grandis (cotton boll weevil) 52
Anthoxanthum odoratum (sweet vernalgrass) 119, 314, 322, 487, 557, 626, 630, 636, 640, 641, 650, 672, 683, 732; see also sweet vernalgrass
anthropogenic effects 31, 64, 292, 394, 395, 397, 398, 400, 403, 404, 597, 598, 600, 604, 606, 609, 611, 612, 614, 618, 620, 796; see also human disturbance
anthurium *(Anthurium andraeanum)* 205, 714, 716, 718
Anthurium andraeanum (anthurium) 205, 714; see also anthurium
Anthurium andreanum cv. 'Marion Seefurth' 714
Anthurium andreanum cv. 'Ozaki' 714
Antiblemma acclinalis 452, 459, 460, 461
Antidesma 560
Antigonon leptopus 146
Antirrhinum nuttallianum 91
Antirrhinum orontium 148
ants 11, 12, 766
'ape (*Xanthosoma*) 205
Aphanisma blitoides 84
Aphelandra aurantiaca 125
Apiaceae 81, 126, 210; see also carrot, parsley
Apiastrum angustifolium 81
Apienulicis 580
Apion ulicis 475
Apis mellifera (honeybee) 278
Apocynaceae 126, 210; see also dogbane, periwinkle
aquatic habitat 6, 9, 17, 18, 20, 21, 32, 52, 59; see also anchialine pools, riparian habitat
aquatic plants 20, 30, 32, 42, 52, 59
Aquifoliaceae (holly) 127
Araceae 81, 116, 205; see also taro
Araliaceae (ginseng) 127, 211, 489, 510
aramina (*Urena lobata)* 104
Araucariaceae (araucaria) 204
Araucaria heterophylla (Norfolk Island pine) 204
Archlagocheirus funestus 20, 422
arctic fox (*Alopex lagopus*) 21
arctic ground squirrel *(Citellus undulatus)* 21
Arctium lappa (gobo) 127, 211
ardisia (*Ardisia elliptica*) 538; see also *Ardisia elliptica*
Ardisia crenata (Hilo holly) 143, 487, 508, 514, 521
Ardisia elliptica (ardisia) 143, 538
Ardisia humilis 618
Ardisia solanacea (shoebutton ardisia) 52, 58
Arecaceae 81, 205; see also palm
Arenaria serpyllifolia 133
Argemone glauca 647, 652
Argemone mexicana 145
Argyroxiphium sandwicense (silversword) 267, 637, 640, 642; see also silversword
Argyroxiphium sandwicense subsp. *macrocephalum* (Haleakalā silversword) 556
Argyroxiphium sandwicense var. *sandwicense* (Mauna Kea silversword) 639; see also Mauna Kea silversword
Argyroxiphium virescens (greensword) 6
armyworm 455
arnotto (Bixaceae) 131

aroids 715
arrowroot (*Tacca leontopetaloides*) 729; see also *Tacca leontopetaloides*
Artemisia californica (coastal sage) 70, 72, 82
Artemisia pontica 127
Artemisia vulgaris 127
Arthraxon hispidus 119
arthropods 291, 499, 601, 667
Arthrostema ciliatum (arthrostema) 142, 487, 508, 515, 521
artichoke (*Cynara scolymus*) 767
artillery plant (*Pilea microphylla*) 21; see also *Pilea microphylla*
Artocarpus altilis (breadfruit) 102, 143, 729
Arundina graminifolia (bamboo orchid) 119, 402
Asclepiadaceae (milkweed) 127
Asclepias curassavica 127
Asclepias physocarpa (balloon plant) 127, 538
ash 532; see also tropical ash
Asiatic butterfly bush (*Buddleia asiatica*) 547; see also *Buddleia asiatica*
Asiatic pennywort (*Centella asiatica*) 558; see also *Centella asiatica*
asparagus fern (*Asparagus setaceus*) 207
Asparagus setaceus (asparagus fern) 207
aspen (*Populus tremuloides*) 7
Aspidiaceae 81
Asplenium 363
Asplenium trichomanes 637, 640, 642, 650, 670, 672
Astelia menziesiana (pa'iniu) 345-346, 347, 349
Asteraceae (daisy, sunflower) 81, 82, 83, 127, 157, 211
Aster subulatus 127
Astragalus (locoweed) 7, 86
Astragalus miguelensis 71, 72, 86
asystasia (*Asystasia gangetica*) 125, 209
Asystasia gangetica (asystasia) 125, 209
Ategumia matutinalis 452, 454, 455, 459, 461
Athyriopsis japonica 321
Athyrium 308-310
Athyrium microphyllum 558
Atriplex 85
Atriplex californica 71, 84
Atriplex eardleyae 134
Atriplex lentiformis 85, 134
Atriplex semibaccata (Australian saltbush) 72, 74, 75, 76, 85, 134
Atriplex suberecta (salt bush) 134, 213, 555
Australian fern weevil (*Syagrius fulvitarsus*) 562
Australian fireweed (*Erechtites glomerata*) 76; see also *Erechtites glomerata*, fireweed
Australian pine (*Casuarina*) 47, 49, 52, 57, 59; see also *Casuarina*
Australian saltbush (*Atriplex semibaccata*) 75, 76, 78; see also *Atriplex semibaccata*
Australian tree fern (*Cyathea cooperi*) 314, 318, 319, 551, 559, 561, 562, 571; see also *Cyathea cooperi*
Avena 21
Avena barbata (slender oats) 73, 77, 88, 119
Avena fatua (wild oats) 72, 73, 74, 77, 89, 119; see also wild oats
avian pox 201
avocado (*Persea americana*) 441, 455; see also *Persea americana*
'āweoweo (*Chenopodium oahuense*) 213, 627, 629, 634, 654, 656, 658, 661; see also *Chenopodium oahuense*
Axis axis (axis deer) 701
Axonopus fissifolius (narrow-leaved carpetgrass) 119, 314, 323; see also narrow-leaved carpetgrass

Baccharis 82
Baccharis pilularis 71, 72, 82
Bacopa monnieri ('ae'ae) 220
bacteria 8, 33
bagnit (*Tristellateia australasiae*) 217
balloon plant (*Asclepias physocarpa*) 538; see also *Asclepias physocarpa*
Balsaminaceae 131; see also touch-me-not
balsam wooly aphid (*Adelges piceae*) 20

Baltimora recta 127
bamboo (*Phyllostachys*) 511, 518, 521, 563, 572
bamboo orchid (*Arundina graminifolia*) 402; see also *Arundina graminifolia*
Bambusa vulgaris (feathery bamboo) 538
banana 119, 207, 729, 800; see also *Musa*, Musaceae, *Musa* x *paradisiaca*
banana poka (*Passiflora mollissima*) 13, 41, 155-158, 163, 168, 169, 170, 172, 175, 176, 177, 178, 249, 271-293, 372, 392, 433, 438, 442, 443, 444, 445, 472, 474, 476, 477, 486, 491, 498, 501, 502, 536, 537, 538, 539, 540, 551, 567-568, 570, 571, 572, 593, 667, 692, 694, 695, 696, 754, 755, 767, 769, 770-771, 772, 782, 783, 786, 829; see also *Passiflora mollissima*, *Passiflora*
Barleria cristata 125
barley 72, 93
barringtonia (*Barringtonia asiatica*, Lecythidaceae) 217
Barringtonia asiatica (barringtonia) 217
Basella (Basellaceae) 131
Bassia hyssopifolia 134
bat, Hawaiian (*Lasiurus cinereus semotus*) 781, 797
Bataceae (saltwort) 131
Batis maritima 131
bayberry 143; see also Myricaceae
beach dropseed (*Sporobolus virginicus*) 209
beaches 50, 52, 55, 66, 70, 75, 101, 197; see also coastal zone
beach morning glory (*Ipomoea pes-caprae* subsp. *brasiliensis*) 214
beach naupaka (*Scaevola sericea*) 197, 201; see also naupaka kahakai, *Scaevola sericea*
beefsteak plant (*Acalypha wilkesiana*) 214
beekeepers 477, 761, 767, 770, 771, 772, 773
beetles 766; see also Coleoptera
beggar's tick (*Bidens pilosa*) 42; see also *Bidens pilosa*
begonia (*Begonia*, Begoniaceae) 131
Belamcanda chinensis 118
Belgian hare 93; see also rabbits
bellflower (Campanulaceae) 133
bell pepper. See *Capsicum annuum*
Benicasa hispida (Chinese melon) 487, 509, 513, 515
Berberidaceae 83
Berberis pinnata insularis 83
Berberis thunbergii 20
Bermuda grass (*Cynodon dactylon*) 199, 208, 364; see also *Cynodon dactylon*
berries 45; see also entries for *Rubus* species
Berula erecta 81
be-still tree (*Cascabela thevetia*) 210
Beta vulgaris 85
Beta vulgaris var. *cicla* (Swiss chard) 213
Bidens alba 127
Bidens alba var. *radiata* (Spanish needles) 211
Bidens cynapiifolia 127
Bidens gardneri 128
Bidens laevis 112, 128
Bidens lanceolata 128
Bidens pilosa (beggar's tick) 42, 128, 211, 630; see also Spanish needles
big-headed ant (*Pheidole megacephala*) 454
Bignoniaceae (bignonia) 131, 212
biocontrol 14, 16, 236, 242, 387, 411-428, 432-447, 466, 467, 472, 473, 474, 475, 491, 492, 493, 498, 500, 502, 513, 563, 564, 568, 571, 572, 757-773, 783, 823, 824, 825; see also biological control
bioherbicides. See mycoherbicides
biological control 7, 9, 13, 16, 19, 57, 58, 79, 158, 178, 226, 228, 236, 242, 250, 267, 274, 291, 292, 315, 377, 386, 387, 318, 392, 411-428, 452-462, 466-479, 485, 486, 487, 489, 490, 491, 495, 498, 499, 503, 507, 518, 522, 533, 537, 539, 540, 544, 545, 567, 572, 578, 580, 581, 582, 585, 699, 757-773, 781, 782, 783, 795, 813, 824, 825, 826, 829, 830; see also biocontrol

bird of paradise (*Strelitzia reginae*) 207, 716
birds 51, 75, 93, 104, 105, 156, 157, 158, 189, 190, 193, 195, 197, 200, 201, 227, 236, 241, 249, 271, 272, 278, 281, 287, 290, 292, 304, 308, 311, 313, 316, 319, 343, 373, 412, 477, 536, 540, 548, 553, 557, 561, 562, 567, 569, 601, 604, 612, 614, 628, 696, 700, 770, 781, 782, 785, 796, 797, 799, 825; see also dispersal by birds, seabirds, species listings, waterbirds
bison (*Bison bison*) 18
bitter herb (*Centaurium erythraea* subsp. *erythraea*) 217
Bixaceae (arnotto, *Bixa*) 131
black bamboo (*Phyllostachys nigra*) 559, 563, 572
blackberry (*Rubus argutus*) 316, 341, 359-366, 467, 470, 476, 492, 527, 529, 531, 532, 538, 544, 545, 551, 557, 558, 561, 569, 570, 571, 572, 683, 689, 699, 770, 791; see also Florida blackberry, *Rubus argutus*
black-footed albatross (*Diomedea nigripes*) 197
black medic (*Medicago lupulina*) 216; see also *Medicago lupulina*
black mustard (*Brassica nigra*) 212; see also *Brassica nigra*
black noddy (*Anous minutus melanogenys*) 197
black wattle (*Acacia mearnsii*) 289, 487, 506, 508, 513, 514, 528, 538, 547, 568, 569, 816; see also *Acacia mearnsii*
blackwood acacia (*Acacia melanoxylon*) 487, 508, 513, 514
blechnum fern (*Blechnum occidentale*) 42; see also *Blechnum occidentale*
Blechnum occidentale (blechnum fern) 42, 321
Blepharomastix acutangulalis 415
Blepharomastix ebulealis 454; see also *Ategumia matutinalis*
blowflies (*Calliphora*) 278
blue gum (*Eucalyptus globulus*) 75, 76, 554; see also eucalyptus, *Eucalyptus globulus*
blue ginger (*Dichorisandra thyrsiflora*) 206
Blumea laciniata 112, 128
Blumea sessiliflora 112, 128
Bocchoris fatualis 454
Bocconia frutescens 145
Boehm australis var. *dealbata* 610
Boehmeria nivea 151
Boerhavia coccinea 144
Boerhavia herbstii 218
Boerhavia repens (alena) 218
bogs 34, 304, 314, 395, 396, 541, 542, 551, 554, 557, 558, 564, 565, 571, 746, 747
boobies (*Sula*) 604
borage 131; see also Boraginaceae
Boraginaceae (borage, heliotrope) 83, 131, 212
boreal vegetation 7
Bos taurus (cattle) 35, 235, 281, 342, 553, 577, 586, 602, 627, 701, 726, 727; see also cattle
botanical gardens 712, 713, 717, 719, 720, 722
Bothriochloa 119
Bothriochloa intermedia 112, 119
Bothriospermum tenellum 131
bougainvillea (*Bougainvillea*) 218, 716
Bougainvillea cv. 'Bridal Bouquet,' cv. 'Purple Prince' 716-717
Bougainvillea spectabilis (bougainvillea) 218; see also bougainvillea
bowstring hemp (*Sanseviera*) 205
boxthorn (*Lycium californicum*) 71, 94; see also *Lycium californicum*
Brachiaria mutica (paragrass) 119, 208
Brachymeria obscurata 455
bracken fern (*Pteridium aquilinum*) 400, 401, 557; see also *Pteridium aquilinum*
bracket plant (*Chlorophytum capense*) 207
braconid, Braconidae 414, 416, 455
Bracon mellitor 414
brambles 433, 435; see also *Rubus* listings
Brassica campestris (field mustard) 132, 212
Brassicaceae (mustard) 84, 132, 212
Brassica geniculata 84
Brassica juncea 132

Brassica nigra (black mustard, wild mustard) 73, 74, 84, 132, 212
Brassica rapa (wild mustard) 4, 73, 74
Brazilian peppertree (*Schinus terebinthifolius*) 6, 10, 31, 32, 43, 47, 49, 51, 52, 53, 54, 57, 58, 59, 476; see also Christmas berry, *Schinus terebinthifolius*
breadfruit (*Artocarpus altilis*) 729; see also *Artocarpus altilis*
Brevipalpus phoenicis 425
Breynia disticha var. *rosi-picta* (snow bush) 215
Brickellia californica 82
bristly foxtail (*Setaria verticillata*) 209; see also *Setaria verticillata*
Briza 119
Brodiaea jolonensis 81
brome fescue (*Vulpia bromoides*) 566; see also *Vulpia bromoides*
brome grass (*Bromus*) 72; see also *Bromus*
bromeliads 708, 715, 720
Bromus (brome grass) 72, 650
Bromus arizonicus 89
Bromus carinatus 89, 636, 640, 641
Bromus catharticus 610
Bromus diandrus 74, 89
Bromus inermis (smooth brome) 7, 15
Bromus madritensis 89
Bromus maritimus 89
Bromus mollis (soft chess) 77, 89, 120
Bromus pseudolaevipes 89
Bromus rigidus (ripgut grass) 120, 626, 630, 636, 641, 647, 650, 652; see also ripgut grass
Bromus rubens (red brome) 70, 77, 89, 120
Bromus tectorum (cheat grass) 8, 15, 18, 43, 120; see also cheat grass
Bromus trinii 89
Bromus willdenowii (prairie grass) 120, 208
broomsedge (*Andropogon virginicus*) 250, 314, 318, 394, 398, 399, 401, 402, 403, 405, 486, 487, 538, 547, 563, 689, 699, 734; see also *Andropogon virginicus*
Broussaisia 308
Broussaisia arguta (pū'aha-nui) 313
Broussonetia papyrifera (wauke) 102, 143, 729
browsing 21, 496, 560, 611, 619, 625, 626, 628, 629, 630ff, 632, 633, 634, 639, 644, 661, 662, 676, 734; see also grazing
Brugmansia candida 149
Bruguiera gymnorrhiza 147
bryophytes 307, 396, 558, 667, 672, 687, 688, 689, 690, 692, 693, 694, 695
buckwheat 146, 219; see also Polygonaceae
Buddleiaceae (butterfly bush) 133
Buddleia asiatica (Asiatic butterfly bush) 133, 547
Buddleia madagascariensis 133
buffalo grass 209; see also *Stenotaphrum secundatum*
buffelgrass (*Cenchrus ciliaris*) 402, 547, 732; see also *Cenchrus ciliaris*
Bulbostylis capillaris 117
bull thistle (*Cirsium vulgare*) 6, 528; see also *Cirsium vulgare*, thistle
bumble-mustard (*Sisymbrium altissimum*) 8; see also *Sisymbrium altissimum*
bunchgrass 16, 34, 35, 72, 73, 376, 398, 400, 563, 568, 649; see also *Agropyron*, *Andropogon*, *Stipa pulchra*
bunch onion (*Allium fistulosum*) 207
buprestid, Buprestidae 459, 461
bur clover (*Medicago polymorpha*) 77, 216; see also *Medicago polymorpha*
burro (*Equus asinus*) 10, 14, 17
bush beardgrass (*Schizachyrium condensatum*) 394, 398, 399, 402, 403, 405, 486, 493, 495, 699, 734; see also *Schizachyrium condensatum*
buttercup (Ranunculaceae) 147; see also Ranunculaceae
butterflies 414, 416, 423, 766; see also Lepidoptera
butterfly bush (Buddleiaceae) 133
button sedge (*Fimbristylis cymosa*) 206
buttonwood (*Conocarpus erectus*) 51, 213

cabbage palm (*Livistona australis*) 205
cacao (Sterculiaceae) 150
Cactaceae 133, 212, 422, 490, 511; see also cactus
Cactoblastis 420, 423, 424
Cactoblastis cactorum 420, 421, 422, 520
cactus 34, 133, 212, 411, 419, 420, 421, 422, 423, 424, 425, 426, 473, 475, 477, 585, 766; see also Cactaceae, *Opuntia*
cactus scale 420
Caesalpinia 137
Caesalpinia bonduc (kākalaioa, kinikini) 137, 215
Caesalpinia decapetala (cat's claw) 137, 538
Cajanus cajan 138
cajeput (*Melaleuca quinquenervia*) 47, 49, 50, 51, 52, 53, 54, 58, 59; see also *Melaleuca quinquenervia*
Cakile (sea-rocket) 8, 84
caladium (*Caladium bicolor*) 205
Calandrinia 90
Calliphora (blowflies) 278
Callisia fragrans 116
Calophyllum antillanum 135
Calophyllum inophyllum (kamani) 102, 104, 135, 213, 616, 729
Calyptocarpus vialis 128
Calystegia macrostegia 70, 72, 85
Calystegia soldanella 85
Camellia sinensis 150
Camissonia 88
Campanulaceae (bellflower) 133
Canada thistle (*Cirsium arvense*) 6, 9; see also *Cirsium arvense*, thistle
Canavalia 138
Canavalia ensiformis 215; see also jackbean
Canavalia hawaiiensis 676; see also jackbean
Canavalia kauensis 676; see also jackbean
candlenut (*Aleurites moluccana*) 607, 612, 613, 618; see also *Aleurites moluccana*, kukui
Canis domesticus (dog) 553, 627; see also dog
canna (*Canna*, Cannaceae) 116, 206
Cannabis sativa (marijuana) 249
caper (Capparaceae) 133, 212
Capparis sandwichiana (puapilo) 212
Capra hircus (goat) 35, 73, 235, 301, 377, 398, 496, 532, 551, 553, 581, 586, 602, 627, 666, 726, 784; see also goat
Caprifoliaceae (honeysuckle) 133
Capsella bursa-pastoris 84
Capsella rubella (shepherd's purse) 132, 212
Capsicum annuum (nīoi) 149, 220, 461, 555
Cardamine 132
Cardamine californica 84
Cardionema ramosissimum 84
Cardiospermum 148
Carduus (thistle) 766
Carduus nutans (musk thistle) 6, 19; see also thistle
Caretta caretta caretta (loggerhead turtle) 50; see also sea turtle
Carex 308, 314, 558, 565
Carex alligata 304; see also sedge
Carex echinata 557, 564
Carex macloviana 647, 650, 652
Carex wahuensis 355, 670, 672
Caribbean pine (*Pinus caribaea*) 42; see also *Pinus caribaea*
caribou (*Rangifer arcticus*) 21
Caricaceae (papaya) 133, 213
Carica papaya (papaya) 133, 213, 730
Carissa macrocarpa (Natal plum) 210
carnations 716
carpetweed 143, 209; see also Aizoceae, Molluginaceae
Carpobrotus 72; see also Hottentot fig, sea fig
Carpobrotus aequilaterus (sea fig) 71, 75, 78, 81; see also sea fig, ice plant
Carpobrotus edulis (Hottentot fig) 76, 81
Carposina atlanticella 445
Carposina bullata 452, 458, 459, 461
Carposinidae 459
Carthamus tinctorius (safflower) 767
carrot 104, 210; see also Apiaceae, *Daucus pusillus*
Caryophyllaceae (pink) 84, 133, 213
Cascabela thevetia (be-still tree) 210

Casinaria infesta 455
Cassia 215
Cassia floribunda 610
Castilleja affinis 72, 91
Castilleja arvensis 149
Castilleja hololeuca 91
castor bean (*Ricinus communis*) 215, 492, 506, 512, 519, 521, 522, 566, 569; see also *Ricinus communis*
Casuarina (casuarina) 10, 33, 47, 49, 50, 52, 53, 55, 57, 58; see also Australian pine, *Casuarina equisetifolia*
Casuarinaceae (she-oak) 134, 193, 213, 488, 509
Casuarina equisetifolia (ironwood) 10, 21, 49, 50, 134, 189, 193, 213, 488, 509, 513, 515, 528; see also *Casuarina*, ironwood
Casuarina glauca 49, 50, 134; see also *Casuarina*
Casuarina littoralis 616
Catabena esula 415, 416, 417
Catalina cherry (*Prunus ilicifolia* subsp. *lyonii*) 71
Catharanthus roseus (periwinkle) 126, 210
Catimbium speciosum (shell ginger) 209
cats 93, 94
cat's claw (*Caesalpinia decapetala*) 538; see also *Caesalpinia decapetala*
cattail 124; see also Typhaceae
cattle (*Bos taurus*) 6, 7, 18, 21, 73, 77, 235, 259, 281, 287, 288, 342, 419, 497, 553, 554, 557, 577, 586, 602, 603, 619, 620, 701, 726, 727, 729, 766, 800; see also *Bos taurus*, cattle industry
cattle industry 412, 419, 584, 713, 714, 732; see also *Bos taurus*, cattle
cayenne vervain (*Stachytarpheta dichotoma*) 538; see also *Stachytarpheta dichotoma*
Ceanothus megocarpus subsp. *insularis* (island bigpod ceanothus) 71, 91
cecidomyiid (Cecidomyiidae) 452, 459
Cecropia 134
Cecropiaceae 134
Cecropia obtusifolia 134
Cecropia palmata 618
Celastrus orbiculatus 20
Cenchrus agrimonioides var. *laysanensis* (kāmanomano) 199, 208
Cenchrus ciliaris 120, 547, 732; see also buffelgrass
Cenchrus echinatus (sandbur) 120, 194, 195, 200, 208
Centaurea 9, 19, 766; see also *Centaurea maculosa*, knapweed
Centaurea maculosa (spotted knapweed) 7, 15, 16, 128; see also *Centaurea*
Centaurea melitensis (tocalote) 74, 77, 79, 82; see also yellow star thistle
Centaurium erythraea 140
Centaurium erythraea subsp. *erythraea* (bitter herb) 217
Centella asiatica (Asiatic pennywort) 126, 558
Centranthus ruber 151
century plant (*Agave americana*) 508, 514, 521
cerambycid, Cerambycidae 415, 416, 452, 459, 461
Cerastium fontanum 133
Cerastium fontanum var. *triviale* (common mouse ear) 213
Ceratophyllaceae (hornwort) 134
Ceratophyllum demersum 17, 134
Ceratophyllum glomeratum 84
Cercosporella. See *Entyloma*
Cereus uruguayanus 133
Ceroplastis cerripidiformis 291
Cestrum 149
chalcid 455
Chamaecrista nictitans (partridge pea) 138, 399
Chamaesyce albomarginata 136
Chamaesyce celastroides var. *celastroides* ('akoko) 190, 215
Chamaesyce degeneri 402
Chamaesyce hirta (hairy spurge) 136, 215, 604
Chamaesyce hypericifolia 136, 215
Chamaesyce hyssopifolia 137
Chamaesyce maculata 137, 215
Chamaesyce olowaluana 626
Chamaesyce prostrata (small ground fig) 137, 215

Chamaesyce thymifolia 137
Chasmanthe aethiopica 87
chandelier plant (*Kalanchoe tubiflora*) 214; see also *Kalanchoe tubiflora*
chaparral 8, 17, 67, 71
cheat grass (*Bromus tectorum*) 8, 43, 44; see also *Bromus tectorum*
cheese weed (*Malva parviflora*) 77, 218; see also *Malva parviflora*
Cheirodendron 308-309
Cheirodendron trigynum (ʻōlapa) 290, 347; see also ʻōlapa
Chelinidea tabulata 422
Chelinidea vittiger 420
Chelonia mydas mydas (green sea turtle, sea turtle) 50
chemical control 7, 10, 13, 15, 19, 20, 47, 54, 55, 58, 59, 79, 100, 178, 228, 236, 250, 274, 291, 292, 293, 318, 319, 320, 341-373, 425, 440, 442, 445, 454, 455, 479, 488, 490, 491, 499, 500-501, 502, 526, 527, 532, 546, 548, 561, 562, 566, 570, 572, 580, 581, 586, 734, 735, 740, 772, 783, 795, 816, 817, 823, 824, 825, 829; see also herbicides
Chenopodiaceae (goosefoot) 84, 134, 213
Chenopodium 658
Chenopodium ambrosioides 134
Chenopodium californicum 85
Chenopodium carinatum 134, 555
Chenopodium murale (nettle-leaved goosefoot) 74, 77, 85, 134, 213; see also pigweed
Chenopodium oahuense (ʻāweoweo) 213, 627, 631, 655, 657, 659; see also ʻāweoweo
Chinese banyan (*Ficus microcarpa*) 218, 547, 564; see also *Ficus microcarpa*
Chinese melon (*Benicasa hispida*) 487, 509, 513, 515
Chinese tallow (*Sapium sebiferum*) 20
Chinese taro (*Alocasia cucullata*) 205
Chionanthus ligustrina 144
Chloris (feather fingergrass, swollen fingergrass) 120, 208
Chlorophytum capense (bracket plant) 207
cholla (*Opuntia prolifera*) 70; see also *Opuntia prolifera*
Chondrilla juncea (rush skeleton weed) 435
Christella dentata 321
Christella parasitica 322
Christmas berry (*Schinus terebinthifolius*) 6, 31, 43, 49, 51, 210, 470, 476, 493, 522, 530, 538, 545, 547, 548, 559, 564, 569, 571, 584, 585, 586, 593, 733, 755, 787; see also Brazilian peppertree, *Schinus terebinthifolius*
Christmas trees 720, 721
chrycoleus (*Plectranthus scutellarioides*) 217; see also *Plectranthus scutellarioides*
chrysanthemum (*Chrysanthemum*) 128, 716
Chrysodeixis chalcites 412; see also noctuid
chrysomelid (Chrysomelidae) 414, 415, 416, 417, 452, 459; see also leaf beetle
Chrysopogon aciculatus 120
Cibotium (tree fern) 168, 204, 283, 314, 366, 561, 562; see also hāpuʻu, tree fern
Cibotium glaucum (tree fern) 347, 367, 397, 400, 684; see also hāpuʻu, hāpuʻu pulu, tree fern
Ciclospermum leptophyllum (fine-leaved celery) 126, 210
Cinnamomum burmannii 141
Cirsium (thistle) 766; see also thistle
Cirsium arvense (thistle) 6, 15, 17, 19, 20; see also Canada thistle, thistle
Cirsium occidentale 82
Cirsium vulgare (thistle) 128, 528, 566, 630, 647, 650, 652; see also bull thistle, Eurasian thistle
Cissus nodosa (thistle) 6, 152; see also bull thistle, thistle
Cistaceae 85
Citellus undulatus (arctic ground squirrel 21
Citharexylum 151
citrus (Rutaceae) 220
Claoxylon sandwicense 560
Clarkia epilobioides 88

Claytonia perfoliata 90
Cleome gynandra 133
Clermontia parviflora ('ōhā) 347, 349
Clerodendrum philippinum 151
clidemia (*Clidemia*) 241, 242, 243, 244, 246, 248, 249, 250, 251, 320, 435, 452-457, 458, 460, 461, 467, 468, 472, 474, 528, 532, 536, 537, 538, 539, 540, 544, 551, 567, 570, 572, 593, 755, 769, 821; see also *Clidemia hirta*, Koster's curse
Clidemia debilis 460
Clidemia hirta (clidemia) 142, 241, 242, 243, 245, 247, 248, 314, 399, 432, 435, 452, 456, 457, 458, 459, 460, 461, 462, 467, 468, 525, 528, 532, 536, 537, 538, 547, 551, 567, 593, 755, 821; see also clidemia, Koster's curse
climbing fig (*Ficus pumila*) 509, 516, 521
Clitoria ternata 138
cloud forest 542
clubmosses 687
Clusiaceae (mangosteen) 135, 213
Clusia rosea (copey) 135, 213
cluster pine (*Pinus pinaster*) 42; see also *Pinus pinaster*
coastal dune community 542
coastal lowlands 386, 396, 398, 400, 402, 513, 617, 626
coastal sage (*Artemisia californica*) 70; see also *Artemisia californica*
coastal scrub 8, 67, 68, 70
coastal zone 8, 16, 17, 50, 101, 103, 104, 189, 190, 402-403, 419, 496, 541, 559, 593, 601, 617, 618, 666, 675, 676, 753, 782, 787, 796; see also beaches
coat buttons (*Tridax procumbens*) 211; see also *Tridax procumbens*
Coccinellidae 424
Coccinia grandis 136, 569
Coccoloba uvifera (sea grape) 219
cochineal cactus 419; see also cactus
cocklebur (*Xanthium strumarium*) 43, 211; see also *Xanthium strumarium*
cocksfoot (*Dactylis glomerata*) 632; see also *Dactylis glomerata*
cocoa 221; see also Sterculiaceae
coconut (*Cocos nucifera*) 205, 613, 729; see also *Cocos nucifera*
coconut leaf roller (*Hedylepta*) 455
Cocos nucifera (coconut) 102, 205, 613, 729
Codiaeum variegatum var. *pictum* (croton) 215, 715-716
Coffea arabica (coffee) 148, 455, 730, 800; see also coffee
coffee (*Coffea arabica*) 148, 220, 455, 730, 800; see also Rubiaceae
coffee senna (*Senna occidentalis*) 566, 604, 606, 607; see also *Senna occidentalis*
Coix lachryma-jobi 120
Coleoptera (beetles) 424, 459, 766
collembola 667
Colletotrichum 278
Colletotrichum gloeosporioides 250, 432, 435
Colletotrichum gloeosporioides f.sp. *aeschynomene* 433
Colletotrichum gloeosporioides f.s. *clidemiae* 457, 461
Colocasia esculenta (taro) 101, 102, 116, 120, 205, 729; see also taro
Columbia grass (*Paspalum fimbriatum*) 208; see also *Paspalum fimbriatum*
Columbrina asiatica (lather leaf) 52, 58; see also lather leaf
Comarostaphylos diversifolia 85
Combretaceae (Indian almond) 135, 213
Commelina benghalensis 116
Commelinaceae (spiderwort) 116, 206
Commelina diffusa (honohono) 116, 206, 327, 363
Commelina erecta 117
common chickweed (*Stellaria media*) 213; see also *Stellaria media*
common guava 52, 311, 315, 319, 399, 492, 529, 538, 545, 547, 559, 565, 570, 584, 586, 616, 618, 620, 730, 733, 766; see also guava, *Psidium*, *Psidium guajava*
common mouse ear (*Cerastium fontanum* var. *triviale*) 213
common myna (*Acridotheres tristis*) 227, 281-282; see also *Acridotheres tristis*
common plantain (*Plantago major*) 219; see also *Plantago major*

common purslane (*Portulaca oleracea*) 219; see also *Portulaca oleracea*
common speedwell (*Veronica plebeia*) 364; see also *Veronica plebeia*
competition 16, 19, 21, 33, 34, 35, 44, 45, 49, 51, 54, 64, 76, 100, 110, 112, 113, 193, 281, 301, 313, 314, 315, 394, 397, 398, 435, 440, 544, 556, 597, 598, 611, 616, 619, 620, 625, 629, 646, 661, 667, 700, 821
Compositae 493, 512
Compsolechia seductella 452, 459, 460
congeners 42, 44, 387, 565
coniferous forest 7, 8, 9, 16, 19, 554; see also pine forest
conifers 533, 554, 564
Conium maculatum 16
Conocarpus erectus (buttonwood) 213; see also buttonwood
Convolvulaceae (morning glory) 85, 135, 214; see also morning glory
Convolvulus 214
Convolvulus arvensis 82, 135
Conyza 611
Conyza bonariensis (ilioha) 128, 195, 211, 610
Conyza canadensis 128
Copestylum (=*Volucella*) *tamaulipana* 420
copey (*Clusia rosea*) 213; see also *Clusia rosea*
Coprosma (pilo) 308, 313, 631, 659; see also pilo
Coprosma acutifolia 609
Coprosma ernodioides (kūkaenēnē) 401, 669, 670, 672
Coprosma montana (pilo) 626, 670, 672; see also pilo
Coprosma ochracea (pilo) 345, 347, 355, 363; see also pilo
Coptotermes formosanus (ground termite) 198
Cordia lutea 606
Cordia sebestena (kou-haole) 212
Cordia subcordata (kou) 102, 104, 132, 212, 729
Cordyline 205
Cordyline fruticosa (ti) 102, 116, 205, 607, 716
coreid bugs 420
Coreopsis 93, 94
Coreopsis gigantea (giant coreopsis) 70, 72, 82
Coreopsis lanceolata 128
Corethrogyne filaginifolia 82
Coriandrum sativum 126
corn (*Zea mays*) 209, 730, 740
Coronopus didymus (swine cress) 132, 212
Cortaderia jubata 16
Cortaderia selloana 89
Corvus hawaiiensis (Hawaiian crow) 282; see also ʻalalā
Corynocarpaceae 136; see also karakanut
Corynocarpus laevigatus (karakanut) 136, 528, 532, 816
Cosmos caudatus 128
cossid (Cossidae) 414, 415
cotton boll weevil (*Anthonomus grandis*) 52
cotton, wild (*Gossypium hirsutum*) 52; see also *Gossypium hirsutum*
Cotula australis 128
Cotula coronopifolia 82
crape jasmine (*Ervatamia*) 210
Crassocephalum crepidioides 128
Crassula 214
Crassulaceae (orpine, stonecrop) 85, 136, 214
Crassula erecta 82
Cremastobombycia lantanella (gracillariid) 413, 417
creosote bush 152; see also Zygophyllaceae
Crepis capillaris 128
Cressa truxillensis vallicola 85
crested honeycreeper (*Palmeria dolei*) 300
Crinum asiaticum (spider lily) 207
Crisicoccus 455
Crocodylus acutus (American crocodile) 50; see also American crocodile
Crocosmia crocosmiiflora (montbretia) 118, 363; see also montbretia
Croesia zimmermanii 476
Crotalaria 138
Crotalaria incana (fuzzy rattle-pod) 138, 215, 555
Crotalaria pallida (rattle-pod) 138, 215
croton (*Codiaeum variegatum* var. *pictum*) 215, 715

Croton 620
Cryphonectria 433, 445
Cryptantha 83
Cryptoblabes aliena (phycitid) 412
Cryptolaemus montrouzieri 424
Cryptomeria japonica (tsugi pine) 284
Cryptostegia grandiflora (Palay rubber vine) 56
Cryptotaenia canadensis 126
crystalline iceplant (*Mesembryanthemum crystallinum, M. nodiflorum*) 72, 76, 78, 93; see also *Mesembryanthemum crystallinum, M. nodiflorum*
cucumber (*Cucumis sativus*) 214, 454, 461
Cucumis dipsaceus 136
Cucumis sativus (cucumber) 214, 454; see also cucumber
Cucurbitaceae (gourd) 85, 136, 214, 487, 509
Cucurbita pepo (squash, zucchini) 214
cudweed (*Gnaphalium purpureum*) 211; see also *Gnaphalium purpureum*
cultivated plants 4, 75, 189, 190, 192, 193, 197, 199, 203ff, 272, 412, 433, 453, 617, 708, 712, 719, 722, 730; see also agriculture, horticulture, ornamentals
cultivation 100, 263, 278, 284, 616; see also agriculture, horticulture, ranching
Cuphea 142
Cuphea carthagenensis 142, 328
Cupressaceae 204; see also cypress
Cupressus 204
Curculionidae 459
Curcuma longa ('ōlena) 102, 729
curly dock (*Rumex crispus*) 436; see also *Rumex crispus*
currant tomato (*Lycopersicon pimpinellifolium*) 105; see also *Lycopersicon pimpinellifolium*
Cuscuta campestris 136
Cuscutaceae (dodder) 85, 136
Cuscuta salina 85
cyanea (*Cyanea superba*) 545
Cyanea grimesiana 560
Cyanea superba (cyanea) 6, 545
Cyanotricha necyria 568
Cyathea 562
Cyathea cooperi (Australian tree fern) 314, 322, 551, 559, 561; see also Australian tree fern
Cycadaceae (*Cycas*, sago palm) 204
Cymbopogon citratus (lemon grass) 208
Cymbopogon refractus 120
Cynara scolymus (artichoke) 767
Cynareae 767
Cynodon dactylon (Bermuda grass) 89, 120, 199, 208, 364
Cynoglossum amabile 132
Cyperaceae 85, 117, 206; see also sedge
Cyperus 117, 311
Cyperus alternifolius subsp. *flabelliformis* (umbrella plant) 206
Cyperus brevifolia 610
Cyperus esculentus (yellow nutsedge) 117, 436
Cyperus halpan (sedge) 117, 314, 319, 323, 558; see also sedge
Cyperus laevigatus (makaloa) 206
Cyperus papyrus (papyrus) 206
Cyperus rotundatus (purple nutsedge) 468
Cyperus rotundus (nutgrass) 117, 206
cypress (Cupressaceae) 51, 204
Cytisus monspessulanus 16, 138
Cytisus palmensis 138
Cytisus scoparius (Scotch broom) 16, 17, 33, 138, 766

Dactylis glomerata (orchard grass) 120, 557, 630, 632, 650; see also orchard grass
Dactyloctenium aegyptium 120
Dactylopius 420, 422, 424
Dactylopius austrina 765
Dactylopius coccus 419, 424
Dactylopius opuntiae 420, 421, 422, 520
daisy (Asteraceae) 211; see also Asteraceae
dallis grass (*Paspalum dilatatum*) 315, 318, 364, 491, 683; see also *Paspalum dilatatum*
dandelion (*Taraxacum officinale*) 7; see also *Taraxacum officinale*
Danthonia 650

Danthonia pilosa (hairy oatgrass) 120, 566, 626, 630, 636, 640, 641, 647, 650, 652, 670, 672; see also hairy oatgrass
Danthonia semiannularis 120
date palm (*Phoenix dactylifera*) 206, 511, 513, 518
Datura stramonium 149, 555
Daucus pusillus (carrot) 81, 104, 126; see also carrot
Davalliaceae 204; see also sword fern
day lily (*Hemerocallis*) 207
deciduous forest/plants 9, 19, 20, 31, 49, 50, 51, 54
deer (*Axis axis, Odocoileus hemionus*) 701
Delonix regia (royal poinciana) 215
Delphinium (larkspur) 7
Delphinium parryi 90
Dendrobium (dendrobium orchid) 714, 718
Dendrobium x *Jac-Hawaii* 715
Dendrobium x *Louis Bleriot* 715
Deschampsia 315, 316, 317, 675, 683, 699
Deschampsia grassland community 303-304, 308, 698
Deschampsia nubigena 307, 400, 401, 557, 626, 630, 634, 670, 672, 683, 684
Descurainia pinnata menziesii 84
Descurainia sophia 18, 555, 556
desert 9, 14, 17, 18, 32, 551, 556
desert bighorn sheep (*Ovis canadensis nelsoni*) 6, 10
Desmanthus virgatus (virgate mimosa) 138, 215
Desmodium 138, 733
Desmodium cajanifolium 138, 509, 513, 515
Desmodium sandwicense (Spanish clover) 138, 215
Dianthus armeria 133
Diaspis echinocacti 420
Diastema tigris 415
Dichanthelium cynodon 557, 565
Dichanthium 120
Dichelostemma 72
Dichelostemma pulchellum 72, 81
Dichondra occidentalis 85
Dichondra repens 135
Dichorisandra thyrsiflora (blue ginger) 206
Dicksoniaceae 204
Dicliptera chinensis 125
Dicotyledonae (dicotyledons) 125, 209
Dicranopteris 308-309
Dicranopteris emarginata (uluhe) 372; see also uluhe
Dicranopteris linearis (uluhe) 246, 302, 396, 538, 684; see also matted fern, uluhe
Dicrostonyx groenlandicus (Greenland collared lemming) 21
dieback 201, 283, 416, 441, 461; see also 'ōhi'a dieback
Dieffenbachia (dumb cane) 205, 715
Dieffenbachia amoena cv. 'compacta' 716
Dieffenbachia amoena cv. 'Mary Ann' 716
Digitaria 121
Digitaria ciliaris (kūkaepua'a) 120, 208; see also kūkaepua'a
Digitaria insularis (sourgrass) 121, 208
Digitaria longiflora 112, 121
Digitaria setigera (kūkaepua'a) 101, 102, 121; see also kūkaepua'a
Dioclea wilsonii (sea bean) 138, 216
Diomedea nigripes (black-footed albatross) 197
Dioscorea alata (yam) 102, 729; see also yam
Dioscorea bulbifera 102, 118
Dioscoreaceae 118; see also yam
Dioscorea pentaphylla 102, 118
Diospyros sandwicensis (lama) 397; see also lama
Diplazium sandwichianum (hō'i'o, pahole) 311, 558, 697
Diptera (flies) 413, 459
diseases 35, 228, 284, 411, 412, 420, 425, 426, 432-447, 537, 539, 709, 721, 722, 727, 763, 797; see also pathogens
dispersal 4, 9, 11, 35, 44, 49, 56, 77, 99, 100, 103, 104, 105, 108, 110, 112, 113, 158, 175, 177, 189, 192, 193, 200, 203ff, 227, 241, 243, 246, 248, 249, 251, 255, 267, 271, 272, 281-284, 287, 288, 292, 300, 302, 304, 311, 315, 319, 320, 343, 376, 382, 386, 387, 388, 412, 432, 433ff, 435, 436ff, 498, 500, 501,

dispersal [cont.] 540, 547, 548, 557, 559, 560, 561, 567, 579, 597, 609, 614, 619, 626, 701, 785, 795, 822
dispersal by animals 45, 73, 156, 157, 282, 313, 700
dispersal by birds 51, 101, 104, 105, 156, 157, 158, 193, 227, 241, 249, 271, 272, 281, 287, 288, 304, 311, 313, 316, 319, 343, 376, 412, 510, 540, 548, 557, 561, 562, 567, 569, 614, 696, 700
dispersal by humans 73, 74, 101, 103, 113, 158, 159, 189, 192, 193, 198, 199, 200, 203, 225, 226, 241, 249, 251, 255, 256, 272, 281, 285, 289, 293, 313, 314, 315, 316, 319, 320, 376, 532, 612, 613-614, 616, 618, 700, 766; see also agriculture, horticulture, Polynesian introduction, ranching, reforestation
dispersal by mammals 35-36, 158, 227, 241, 249, 271, 272, 281, 287, 288, 315, 316, 496, 540, 548, 557, 558-559, 561, 567, 570, 667
dispersal by water 9, 17, 103, 104, 105, 193, 282, 288, 289, 319, 376, 532
dispersal by wind 9, 52, 56, 73, 74, 157, 158, 159, 193, 199, 278, 304, 314, 315, 316, 319, 376, 388, 435, 437, 696, 765
Dissotis hensii 454
Dissotis rotundifolia 142
Distichlis spicata (saltgrass) 72, 89
ditch polypogon (*Polypogon interruptus*) 209; see also *Polypogon interruptus*
Dithyrea californica maritima 84
dodder 136; see also Cuscutaceae
Dodecatheon clevelandii insulare 90
Dodonaea ('a'ali'i) 396, 627, 631, 657; see also 'a'ali'i
Dodonaea viscosa ('a'ali'i) 351, 355, 394, 399, 559, 585, 615, 616, 669, 670, 672, 673; see also 'a'ali'i
dog (*Canis domesticus*) 93, 541, 553, 627
dogbane 126; see also Apocynaceae
donkey (*Equus asinus*) 21; see also *Equus asinus*
Dopatrium junceum 149
Doryctes palliatus 416
Dothiorella 441
Dothiorella berengeriana 228
Douglas fir 35
downy rose myrtle (*Rhodomyrtus tomentosa*) 538; see also *Rhodomyrtus tomentosa*
Dracaena 205
Drepanidinae 477
Drosophila, drosophilids 12, 278
Druentia 460
Druentia inscita 452, 459, 460, 461
dry forest 225, 297, 342, 533, 541, 542, 545, 559, 560, 593, 602, 753, 787
dry habitat 177, 180-182, 350, 357, 378, 398, 412, 416, 417, 418, 419, 468, 469, 533, 593, 600, 601, 602, 603, 606, 666, 667, 732, 733, 787, 788, 789
Drymaria cordata 134, 328
Dryopteris 308
Dryopteris arguta 81
Dubautia 396, 631
Dubautia arborea 626, 644, 645
Dubautia ciliolata 645
duckweed (Lemnaceae) 118
Ducula (pigeon) 612
Dudleya caespitosa 70, 72, 85
Dudleya greenei 71, 85
Dudleya traskiae 67, 70, 85
dumb cane (*Dieffenbachia*) 205; see also *Dieffenbachia*

eagle 93
earthworms 667
Echinochloa 121
Echinochloa phyllopogon 112, 121
Echinodorus rostratus 116
Echium 265
Echium pininana 265
Echium plantagineum 477
Eclipta alba 128
Ecthromorpha fuscator 416
education, environmental 15, 57, 251, 293, 362, 404, 478, 485, 495, 502, 540, 541, 543, 568, 571, 582, 590, 710, 712, 721, 723-724, 758, 784, 785, 799-804, 807-811, 817, 822, 825-826, 829, 830
Egeria densa 118
eggplant (*Solanum melongena*) 221, 461

Ehrharta stipoides (meadow ricegrass) 121, 315, 364, 372, 488, 528, 672, 687, 732; see also meadow ricegrass
Eichhornia crassipes (water hyacinth) 9, 20, 43, 52, 124
Elaeagnaceae (oleaster) 136, 488, 509
eleagnus 513
Eleagnus umbellata (oleaster) 136, 488, 509, 513, 515
Eleocharis 117
elephant grass *(Pennisetum purpureum)* 620; see also *Pennisetum purpureum*
Elephantopus mollis (elephant's foot) 128, 468, 618
Elephantopus scaber 468
Elephantopus spicatus 128
elephant's foot *(Elephantopus mollis)* 468; see also *Elephantopus mollis*
Eleusine indica (goosegrass) 121, 208, 610
elm 151; see also Ulmaceae
Elymus caput-medusae 476
Elymus condensatus 89
Elymus pacificus 89
emex 469, 475
Emex australis (emex) 469, 475
Emex spinosa (spiny emex) 146, 469, 765; see also emex
Emilia coccinea 128
Emilia fosbergii (red pualele) 129, 211
Emilia sonchifolia 129
'emoloa (*Eragrostis variabilis*) 208, 400, 401; see also *Eragrostis variabilis*
'ena'ena (*Gnaphalium sandwicensium* var. *sandwicensium*) 211
Encelia californica 70
Encelia farinosa 129
endangered species 300, 545, 644, 797, 807-808; see also specific taxa
endemism 189, 192, 193, 199, 203ff, 597, 598, 601, 604, 609, 611, 613, 619, 628, 752
English ivy (*Hedera helix*) 489, 510, 517, 521; see also *Hedera helix*
English plantain (*Plantago lanceolata*) 219
Entada scandens 216
Entyloma 426, 427
Entyloma ageratinae 432, 434
Entyloma compositarum 432, 434
Epidendrum x obrianianum 119
Epilobium adenocaulon 88
Epilobium angustifolium (fireweed) 35; see also fireweed
Epilobium billardierianum 144, 630, 632, 647, 650, 652
Epilobium billardierianum subsp. *cinerium* 328-329
Epilobium ciliatum 144
Epinotia lantana 413, 414, 417; see also tortricid
Epiphyllum oxypetalum (gooseneck cactus) 212
epiphytes 272, 313, 562
Equus asinus (burro) 10, 17; see also burro, donkey
Equus caballus (horse) 21, 287, 554, 602, 701; see also horse
Equus caballus x *asinus* (mule) 602
Eragrostis 121
Eragrostis brownei (sheepgrass) 121, 558
Eragrostis grandis ('emoloa) 400; see also 'emoloa
Eragrostis paupera 208
Eragrostis tenella (Japanese lovegrass) 121, 208, 616
Eragrostis variabilis ('emoloa) 199, 208; see also 'emoloa
Eragrostis xerophila (tussock grass) 604, 606
Erechtites glomerata (Australian fireweed) 74, 76, 79, 82; see also fireweed
Erechtites hieracifolia 129
Erechtites valerianifolia 129, 329, 558
Ericaceae 85
Erigeron 647, 651, 656
Erigeron annuus 129
Erigeron bellioides 129
Erigeron foliosus 82
Erigeron glaucus 70, 82
Erigeron karvinskianus 129
Erigeron sanctarum 82
Eriobotrya japonica (loquat) 147, 488, 509, 513, 515
Eriogonum arborescens 70, 90
Eriogonum giganteum var. *compactum* 67, 90

Eriogonum grande 70, 71, 90
Eriophyllum 70
Eriophyllum confertiflorum 82; see also *Eriophyllum*
Eriophyllum nevinii 67, 82
Eriophyllum staechadifolium 82
Erodium 72
Erodium cicutarium (red-stemmed filaree) 72, 73, 74, 77, 87, 140, 630, 647, 651
Erodium moschatum 74, 77, 87
erosion 29, 30, 73, 76, 78, 667; see also soil disturbance
Ervatamia (crape jasmine) 210
Eryngium foetidum 126
Erysimum insulare 71, 84
Erythrina 126
Erythrina sandwicensis (wiliwili) 397, 603; see also wiliwili
Erythrina variegata var. *orientalis* (tiger's claw) 216
Eschscholzia 88
eucalypts 33, 557, 754; see also *Eucalyptus*
Eucalyptus (eucalyptus) 78, 289, 488, 509, 516, 521, 554, 569, 571, 754; see also eucalypts, *Eucalyptus globulus*
Eucalyptus globulus (blue gum) 75, 76, 88, 488, 551, 554; see also *Eucalyptus*
Eucalyptus robusta (swamp mahogany) 528
Eucalyptus saligna (Sydney blue gum) 528
Euchylaena tomentosa 134
Eucrypta chrysanthemifolia 87
Eugenia 620
Eugenia uniflora (Surinam cherry) 144, 218
eumenid wasp 414
Euphorbia 631
Euphorbiaceae (spurge) 136, 214, 492, 512
Euphorbia cotinifolia 137
Euphorbia cyathophora (wild poinsettia) 137, 215
Euphorbia esula (leafy spurge) 9, 15, 19, 767
Euphorbia graminea 137
Euphorbia heterophylla 137, 215
Euphorbia peplus (petty spurge) 137, 215
Euphorbia pulcherrima (poinsettia) 215
Eurasian thistle (*Cirsium vulgare*) 566; see also *Cirsium vulgare*
Eurasian water milfoil (*Myriophyllum spicatum*) 20
European beach grass (*Ammophila arenaria*) 8, 208; see also *Ammophila arenaria*
European rabbit (*Oryctolagus cuniculus*) 76; see also rabbits
Eurytoma 452, 458, 459
Eurytomidae 459
Eustachys petraea 121, 209
Eutreta xanthochaeta 413, 417; see also tephritid flies
evening primrose 144, 219; see also Onagraceae
evolution 4, 11, 12, 21, 43, 397, 495
exclosures 398, 496, 526, 556, 557, 559, 560, 625, 626, 632, 633, 634, 635, 636-638, 639, 640-643, 644, 645, 646, 647-648, 649, 650-651, 652-653, 654, 655, 656, 657, 658, 660, 661, 667, 668, 669, 676, 681, 682, 683, 684, 686, 687, 688, 689, 690, 692, 693, 697, 701; see also fencing
extinction 6, 12, 110, 112, 199, 241, 598, 666, 701, 754, 769, 785, 799, 826, 829, 830

Fagaceae (peas) 6, 137, 215
Fagopyrum esculentum 146
Falco peregrinus (peregrine falcon) 50
false kamani (*Terminalia catappa*) 213; see also *Terminalia catappa*
false mallow (*Malvastrum coromandelianum* subsp. *coromandelianum*) 218
false staghorn fern (*Dicranopteris linearis*) 538; see also *Dicranopteris linearis*
fan palm (*Washingtonia filifera*) 75, 77; see also *Washingtonia filifera*
faya tree (*Myrica faya*) 225, 399; see also *Myrica faya*, firetree
feather fingergrass (*Chloris virgata*) 208; see also *Chloris*
feathery bamboo (*Bambusa vulgaris*) 538

Feijoa sellowiana (guavasteen) 488, 509, 513, 516

fence, fence line, fencing 301, 302, 303, 311, 317, 318, 556, 558, 570, 627, 634, 669, 673-674, 675, 683, 686, 692; see also exclosures

feral animals 14, 67, 234-235, 301, 317, 391, 394, 551, 556, 559, 571, 572, 602, 614, 668, 690, 784; see also feral ungulates

feral goat. See goat

feral pig. See pig

feral ungulates 267, 398, 496, 499, 526, 601, 603, 666, 697, 700, 701, 795, 813; see also cattle, deer, goat, mouflon, pig, sheep

ferns 32, 42, 199, 204, 272, 301, 303, 307, 308, 309, 321-322, 400, 544, 553, 558, 561, 613, 616, 630, 637, 640, 642, 650, 652, 667, 670, 672, 675, 677, 678, 679, 684, 687, 688, 689, 690, 693, 694, 695, 697, 715; see also Polypodiaceae, species listings

Festuca rubra 121

Ficus 218, 728

Ficus benghalensis (Indian banyan) 218

Ficus carica 87

Ficus elastica (Indian rubber tree) 218

Ficus microcarpa (Chinese banyan) 143, 218, 547, 564

Ficus prolixa 615, 616

Ficus pumila (climbing fig) 509, 516, 521

field mustard (*Brassica campestris*) 212; see also *Brassica campestris*

fig-marigold (Aizoaceae) 125

figwort 148, 220; see also Scrophulariaceae

Filago californica 82

Filago gallica 129

Fimbristylis aestivalis 117

Fimbristylis cymosa (button sedge) 206, 616; see also button sedge

Fimbristylis schoenoides 117

fine-leaved celery (*Ciclospermum leptophyllum*) 210; see also *Ciclospermum leptophyllum*

fire 3, 6, 8, 11, 17, 18, 21, 31, 35, 49, 50, 51, 70, 71, 76, 77, 79, 93,

fire [cont.] 94, 100, 158, 194, 235, 241, 242, 249, 250, 376, 386, 394-405, 452, 476, 485, 486, 495, 496-497, 500, 502, 533, 551, 552, 553, 559, 560, 563, 565, 568, 570, 577, 581, 584, 586-587, 600, 602, 658, 662, 676, 681, 699, 734, 735, 740, 754, 785, 790, 821, 823

firethorn (*Pyracantha angustifolia*) 538; see also *Pyracantha angustifolia*

firetree (*Myrica faya*) 13, 33, 42, 225-240, 387, 388, 391, 392, 411, 433, 441, 445, 467, 470, 476, 486, 490, 498, 501, 502, 538, 568-569, 571, 577, 690, 691, 692, 754, 755, 770, 771-772, 781, 782, 783, 786, 788, 789, 790, 791, 792, 795, 821, 829, 830; see also *Myrica faya*

fireweed (*Epilobium angustifolium*) 35, 77; see also Australian fireweed, *Erechtites glomerata*

fishes 6, 9

fish geranium (*Pelargonium hortorum*) 217

Flaveria trinervia 129

flax (Linaceae) 141

Flemingia strobilifera 139

flies 413; see also Diptera

Flindersia brayleyana (silkwood) 547

Florida blackberry (*Rubus argutus*) 316, 318, 319, 343, 359, 445; see also blackberry, *Rubus argutus*

flower flies (Syrphidae) 278

flowers 712, 714, 715, 716, 720, 721; see also horticulture, ornamentals

Flueggea neowawraea (mehamehame) 545

Foeniculum vulgare (sweet fennel) 16, 126, 566

foraging 235, 436, 511ff, 584, 609, 661, 662, 670ff, 701, 732, 733, 734, 766, 773, 800; see also grazing, browsing

forbs 8, 15, 72, 77, 78, 402, 557, 571, 630-631, 637, 640, 642, 647-648, 650-651, 652-653; see also herbaceous plants, herbs

forest birds 159, 290, 300, 752; see also birds, species listings

Formosan koa (*Acacia confusa*) 487, 506, 507, 508, 514, 521, 538; see also *Acacia confusa*

fountain grass *(Pennisetum setaceum)* 42, 155, 156, 157, 158, 161, 164, 168, 169, 170, 171, 173, 175, 176, 178, 376-392, 394, 399, 401, 402, 485, 491, 495, 499-500, 501, 507, 513, 522, 538, 551, 568, 571, 587, 625, 627, 658, 661, 662, 734, 753, 754, 755, 782, 783, 786, 787, 788, 789, 791, 793, 794, 814, 821; see also *Pennisetum setaceum*
four o'clock *(Mirabilis jalapa)* 144, 218; see also Nyctaginaceae
foxtail (*Hordeum glaucum*) 77, 93; see also *Hordeum glaucum*
foxtail fescue (*Vulpia myuros*) 209; see also *Vulpia myuros*
foxtail grass *(Setaria gracilis)* 64; see also *Setaria gracilis*
fox-tongue melastoma (*Melastoma sanguineum*) 538; see also *Melastoma sanguineum*
Fragaria chiloensis (strawberry) 401
Fragaria vesca (wild strawberry) 148, 364
Frankeniaceae *(Frankenia)* 87, 140, 216
Frankenia grandifolia 71, 86
Frankenia salina 140, 216
Fraxinus americana (white ash) 488, 509, 513, 516
Fraxinus uhdei (Mexican ash, tropical ash) 144, 284, 488, 510, 513, 516, 528, 531, 532, 547
Fregata minor (frigate bird) 604
frog's-bits (Hydrocharitaceae) 118
fruit industry 156, 158, 713, 714, 770, 771
fuchsia (*Fuchsia magellanica, F. paniculata*) 510, 513, 516, 754; see also *Fuchsia magellanica, F. paniculata*
Fuchsia magellanica (fuchsia) 144, 754; see also fuchsia
Fuchsia paniculata (fuchsia) 145, 510, 513, 516; see also fuchsia
fungus, fungi 45, 51, 228, 250, 275, 291, 411, 417, 420, 426, 432-435, 436, 437-438, 441, 443-446, 457, 466, 720
Furcraea foetida (Mauritius hemp) 43, 116
Fusarium oxysporum 420, 432, 433, 434, 443, 444
fuzzy rattle-pod *(Crotalaria incana)* 215; see also *Crotalaria incana*

Gahnia gahniiformis 672
Gaillardia pulchella (waikāhuli) 129, 211
Galinsoga 129
Galium angustifolium foliosum 91
Galium aparine 91
Galium buxifolium 70, 91
Galium californicum miguelense 91
Galium divaricatum 148
gall flies 544; see also tephritid flies
Gallicolumba rubescens (Marquesan ground dove) 604
gallinaceous birds 287
Gallus gallus (jungle fowl) 288
gardenia (*Gardenia*) 220, 455
Gardenia jasminoides (gardenia) 455
Garnotia acutigluma 121
Gastridium ventricosum 121
Gelechiidae 459
genetic attributes, diversity, traits, variation 4, 12, 36, 42, 43, 44, 45, 110, 112, 268, 442, 447, 727, 785
Genisteae 773
gentian (Gentianaceae) 140, 217
Gentianaceae (gentian) 140, 217
geometrid (looper, *Scotorythra*) 412
Geraniaceae (geranium) 87, 140, 217
geranium (*Geranium)* 631
Geranium cuneatum 626
Geranium dissectum 112, 140
Geranium homeanum 140, 651
Geranium pusillum 140
Geranium retrorsum 140
German ivy (*Senecio mikanioides*) 155-157, 159, 161, 167, 169-172, 175, 176, 178, 538, 625, 627, 661, 662; see also *Senecio mikanioides*
giant coreopsis (*Coreopsis gigantea)* 70; see also *Coreopsis gigantea*
Gilia angelensis 89
Gilia capitata 146
Gilia clivorum 90
Gilia nevinii 90
ginger 45, 124, 209, 345, 349, 716, 717, 766; see also *Hedychium*, kāhili ginger, Zingiberaceae
ginseng (Araliaceae) 127, 211; see also Araliaceae

Glenwood grass (*Sacciolepis indica*) 42, 316, 319, 558, 565, 570, 689; see also *Sacciolepis indica*
glorybush (lasiandra) 341, 343, 370-372, 467, 470, 472, 493, 512, 519, 521, 538; see also *Tibouchina urvilleana*
glossy nightshade (pōpolo, *Solanum americanum*) 105; see also *Solanum americanum*, pōpolo
Glycine soja (soy bean) 216
Glycine wightii (glycine) 733
Gnaphalium 82
Gnaphalium japonicum 129, 672
Gnaphalium purpureum (cudweed) 82, 129, 211, 647, 652
Gnaphalium sandwicensium 647, 651, 652
Gnaphalium sandwicensium var. *sandwicensium* ('ena'ena) 211
goat (*Capra hircus*) 14, 35, 79, 93, 235, 301, 307, 314, 316, 317, 319, 377, 391, 398, 477, 485, 495, 496, 532, 551, 553, 557, 559, 560, 563, 565, 566, 570, 571, 581, 586, 602, 603, 606, 608, 609, 610, 611, 613, 619, 627, 628, 666, 667, 668, 669-682, 698, 699, 700, 701, 708, 726, 732, 733, 734, 784, 787, 788, 789, 790, 791, 797, 800; see also *Capra hircus*
gobo (*Arctium lappa*) 211; see also *Arctium lappa*
golden crown-beard (*Verbesina encelioides*) 199, 211; see also *Verbesina encelioides*
Gomphrena globosa 126
Gonocarpus chinensis 141
goodenia (Goodeniaceae) 217
Goodeniaceae (goodenia) 217
goosefoot 134, 213; see also Chenopodiaceae
goosegrass (*Eleusine indica*) 208; see also *Eleusine indica*
gooseneck cactus (*Epiphyllum oxypetalum*) 212
gorse (*Ulex europaeus*) 255, 470, 472, 475, 477, 502, 527, 530, 531, 538, 540, 551, 557, 563-564, 569, 570, 571, 577-582, 593, 770, 771, 772-773, 830; see also *Ulex europaeus*
gorse mite (*Teranycus lintearious*) 580
gorse thrips (*Sericothrips staphylinum*) 580
gosmore (*Hypochoeris radicata*) 304, 307, 315, 317, 319, 554, 556, 558, 565, 634, 639; see also *Hypochoeris radicata*
Gossypium barbardense 142
Gossypium hirsutum (wild cotton) 52, 142
gourd 136, 214; see also Cucurbitaceae
gracillariid 413; see also *Cremastobobycia lantanella*
Gramineae 487, 488, 489, 490, 491, 493, 511
grape 152, 222; see also Vitaceae, *Vitis*
grasses 7, 8, 9, 10, 11, 15, 16, 17, 21, 34, 42, 44, 56, 70, 71, 72, 73, 76, 77, 78, 79, 93, 119, 156, 158, 199, 208, 250, 259, 283, 303, 308, 309-310, 312, 313, 314, 316, 322-326, 343, 382, 386, 394, 396, 398, 399, 400, 401, 403, 404, 405, 412, 434, 476, 496, 511, 531, 533, 552, 553, 557, 559, 560, 562, 565, 566, 571, 586, 602, 604, 606, 626, 628, 629, 630, 632, 634, 636, 639, 640, 641, 642, 647, 650, 652, 658, 661, 666, 667, 669, 670, 672, 675, 676, 677, 678, 679, 680, 681, 682, 684, 685, 687, 688, 689, 690, 692, 693, 694, 695, 696, 697, 699, 700, 766, 767; see also Poaceae, species listings
grasslands 8, 9, 17, 18, 19, 21, 67, 68, 69, 70, 71, 72, 73, 76, 77, 78, 79, 170, 180ff, 300, 301, 302, 304, 307, 309, 315, 316, 317, 319, 321ff, 395, 396, 397, 398, 401, 403, 422, 541, 542, 544, 551, 553, 557, 561, 565, 571, 586, 602, 620, 666, 668, 676, 677, 678, 679, 683, 697, 698, 699, 732, 787, 796; see also prairie
grass webworm (*Herpetogramma licarsisalis*) 585
grazing 4, 6, 7, 8, 10, 11, 15, 17, 18, 19, 21, 35, 45, 70, 71, 73, 74, 77, 79, 93, 94, 259, 267, 287, 288, 389, 477, 496, 502, 526, 536, 553, 554, 566, 571, 577, 578, 581, 586,

grazing [cont.] 598, 603, 610, 619, 625, 626, 628, 629, 630ff, 632, 633, 634, 644, 649, 661, 662, 726, 728, 732-735; see also browsing
green amaranth (*Amaranthus hybridus*) 210; see also *Amaranthus hybridus*
green bean (*Phaseolus vulgaris*) 216
greenhouse orthezia (*Orthezia insignis*) 412
Greenland collared lemming (*Dicrostonyx groenlandicus*) 21
green sea turtle (*Chelonia mydas mydas*) 50; see also sea turtle
greensword (*Argyroxiphium virescens*) 6
Grevillea 342, 359
Grevillea banksii (kāhili flower, silk oak) 147, 341, 342, 356-359, 528; see also silk oak
Grevillea robusta (silk oak) 147, 341, 342, 356-359, 488, 528, 547, 783, 793; see also silk oak
Grindelia latifolia 72, 82
groundsel (*Senecio*) 45
ground termite (*Coptotermes formosanus*) 198
guava 45, 218, 318, 548; see also common guava, *Psidium, P. cattleianum, P. guajava, strawberry guava, waiwī*
guavasteen (Feijoa sellowiana) 488, 509, 513, 516
Guinea grass (*Panicum maximum* var. *maximum*) 732, 733
gunpowder tree (*Melochia umbellata*) 490, 512, 520, 521, 522; see also *Melochia umbellata*
Gygis alba (white tern) 197, 604
Gymnetron tetrum 267
Gymnoconia nitens 433, 446
Gymnospermae (gymnosperms) 116, 204
Gynaikothrips ficorum 454

Haematoxylum campechianum (logwood tree) 139, 216
hairy abutilon (*Abutilon grandifolium*) 217; see also *Abutilon grandifolium*
hairy merremia (kuahulu) 105; see also *Merremia aegyptica*
hairy oatgrass (*Danthonia pilosa*) 566, 626, 629, 634, 635, 639; see also *Danthonia pilosa*
hairy spurge (*Chamaesyce hirta*) 215, 604; see also *Chamaesyce hirta*
hala (*Pandanus odoratissimus, P. tectorius*) 207, 402, 612
Haleakalā silversword (*Argyroxiphium sandwicense* subsp. *macrocephalum*) 556
Halogeton glomeratus (halogeton) 8
Haloragaceae 141; see also water milfoil
Hāmākua pāmakani (*Ageratina riparia*) 411, 425-428, 432, 434, 446, 468, 475, 538, 544, 547, 585, 689; see also *Ageratina riparia*
Haplopappus 68, 70
Haplopappus canus 82
Haplopappus detonsus 70, 82
Haplopappus ericoides 82
Haplopappus squarrosus 82
Haplopappus venetus 70, 71, 82
hāpu'u (tree fern) 366, 400, 401, 684, 692, 789, 790; see also *Cibotium, C. glaucum*
hāpu'u pulu (*Cibotium glaucum*) 347, 367; see also *Cibotium glaucum*
hare, Belgian 93
Harrisia martinii 133
hau (*Hibiscus tiliaceus*) 103, 218, 538, 618, 729
ha'uōwī (*Verbena litoralis*) 221; see also *Verbena litoralis*
Hawaiian bat (*Lasiurus cinereus semotus*) 781, 797
Hawaiian crow ('alalā, *Corvus hawaiiensis*) 282
Hawaiian goose (nēnē, *Nesochen sandwicensis*) 403; see also nēnē
Hawaiian monk seal (*Monachus schauinslandi*) 189
Hawaiian nama (*Nama sandwicensis*) 217
Hawaiian stilt (*Himantopus mexicanus knudseni*) 6
Hawaiian tree fern (*Cibotium*) 204
Hebe 266
Hebe breviracemosa 610
Hedera helix (English ivy) 20, 489, 510, 517, 521

Hedychium (ginger) 45, 766
Hedychium coronarium (white ginger) 124, 529
Hedychium flavescens (yellow ginger) 124, 529
Hedychium gardnerianum (kāhili ginger) 124, 209, 311, 341, 342, 344-350, 387, 489, 498, 529, 551, 562, 584, 696, 754, 781, 782, 790, 792, 814; see also kāhili ginger
Hedylepta 455
Hedyotis centranthoides 347
Hedyotis corymbosa 148
Helianthemum greenei 85
Helianthus annuus var. *macrocarpus* (sunflower) 211; see also sunflower
Helichrysum foetidum 129
Heliconia (heliconia) 716
Heliconia psittacorum 207
Heliocarpus popayanensis 150
heliotrope 198, 212; see also Boraginaceae, *Heliotropium procumbens*
Heliotropium amplexicaule 132
Heliotropium curassivicum (nena) 83, 198, 212
Heliotropium procumbens (heliotrope) 132, 189, 198-199, 212
Hemerocallis (day lily) 207
Hemignathus lucidus affinis (Maui nukupu'u) 300
Hemignathus virens ('amakihi) 278
Hemigraphis reptans 125
Hemiptera (lace bugs) 413, 475
Hemizonia clementina 82
Hemizonia fasciculata 83
herbaceous plants 645, 646, 647ff, 649, 650ff, 652ff, 654, 657, 658, 660, 661, 760; see also forbs, herbs
herbariums 100, 101, 103, 106-110, 112, 113, 155ff, 193, 198, 199, 203ff, 243, 251, 256, 258, 717, 720, 744, 745, 747, 749, 750
herbicides 9, 13, 47, 53, 54, 79, 198, 228, 250, 291, 292, 318, 320, 343-373, 432, 433, 434, 435, 439, 441, 442, 474, 485, 486, 495, 498, 500, 501, 502, 513, 514-520, 526, 532, 545, 546, 548, 562, 571, 581, 584, 586, 786, 824, 826; see also chemical control
herbs 17, 31, 42, 64, 70, 75, 76, 77, 78, 198, 266, 303, 307, 311, 312, 326-335, 498, 508ff, 513, 558, 565, 610, 611, 618, 619, 626, 627, 629, 630, 632, 633, 634, 635, 636-638, 639, 640-643, 644, 647-648, 650-651, 652-653, 666, 669, 670, 672, 675, 677, 678, 679, 681, 682, 684, 685, 687, 688, 689, 690, 692, 693, 694, 695, 696, 699; see also forbs
Herpestes auropunctatus (small Indian mongoose) 241, 249, 403, 540, 567
herpetofauna 612; see also lizard, turtle
Herpetogramma licarsisalis (grass webworm) 585
Hesperocnide tenella 92
Heterocentron subtriplinervium (pearl flower) 142, 372, 510, 517, 521; see also pearl flower
Heteromeles arbutifolia (toyon) 71, 91
Heteropogon contortus (pili grass) 380, 396, 602, 677; see also pili grass
Heteropogon/Rhynchelytrum community 698
Heterotheca grandiflora 129, 647, 649, 651, 652
Heterotrichum cymosum 454
Heuchera maxima 91
hi'aloa (*Waltheria americana*) 402
hibiscus (*Hibiscus*) 218, 454
Hibiscus esculentus (okra) 217
Hibiscus macrophyllus 142
Hibiscus rosa-sinensis (red hibiscus) 218
Hibiscus tiliaceus (hau) 103, 218, 538, 618, 729
Hilaria jamesii 18
Hilo grass (*Paspalum conjugatum*) 4, 6, 42, 250, 304, 311, 313, 314, 317, 318, 320, 490, 558, 561, 570, 571, 618, 692, 697; see also *Paspalum conjugatum*
Hilo holly (*Ardisia crenata*) 487, 508, 514, 521; see also *Ardisia crenata*
Himatione sanguinea freethii (Laysan honeycreeper) 199
Himantopus mexicanus knudseni (Hawaiian stilt) 6

Hippeastrum 207
Hippeastrum puniceum 118
Hippobroma longiflora 133
hog (*Sus scrofa*) 20, 93; see also pig, *Sus scrofa*
hō'i'o (*Diplazium sandwichianum*, pahole) 311; see also *Diplazium sandwichianum*
Holcus 683; see also *Holcus lanatus*
Holcus lanatus (velvet grass) 6, 20, 122, 304, 315, 323-324, 364, 401, 489, 557, 562, 626, 630, 650, 670, 672, 683, 684; see also velvet grass
holly (Aquifoliaceae) 127
Homalanthus polyandrus 610
honeybee (*Apis mellifera*) 278
honeycreeper. See Drepanidinae
honeysuckle (Caprifoliaceae) 133
honohono (*Commelina diffusa*) 206, 363; see also *Commelina diffusa*
Hordeum 21
Hordeum californicum 89
Hordeum glaucum (foxtail) 70, 72, 77, 89; see also foxtail
Hordeum leporinum 73, 89, 122, 208
Hordeum murinum 647, 650, 652
Hordeum pusillum 89
Hordeum vulgare 122
horehound (*Marrubium vulgare*) 75; see also *Marrubium vulgare*
hornwort (Ceratophyllaceae) 134
horse (*Equus caballus*) 21, 93, 287, 554, 602, 603, 701
horticulture 44, 56, 115ff, 158, 197, 342, 343, 419, 432, 436, 439, 495, 546, 569, 708, 712-724, 768, 826; see also ornamentals
Hottentot fig (*Carpobrotus edulis*) 76; see also *Carpobrotus edulis*
hu'ahu'akō (*Rumex albescens*) 190; see also *Rumex albescens*
human disturbance, impact, intervention 4, 7, 11, 12, 32, 34, 44, 45, 49, 50, 67, 78, 93-94, 99, 100, 103, 104, 112, 178, 193-198, 199, 235, 283, 343, 526, 543, 544, 553, 555, 616, 619, 620, 658, 700, 728, 752, 753, 799, 807, 809, 810, 822, 823; see also agriculture, anthropogenic effects, cultivation, dispersal, logging, Polynesian introduction, ranching
Hunnemannia fumariifolia 145
hunting 249, 477, 536, 541, 543, 609, 627, 669, 684, 694, 699, 753, 801
Hutchinsia procumbens 84
hydric forest 169; see also wet forest, hydric habitat
hydric habitat 170, 184-186; see also wet forest, wet shrub community
hydric scrub 169
Hydrilla verticillata (hydrilla, milfoil) 9, 52
Hydrocharitaceae (frog's-bits) 118
Hydrocleys nymphoides 118
Hydrocotyle 126, 329-330
Hydrophyllaceae (water-leaf) 87, 217
hydrothermal plant community 789
Hylocereus undatus (night blooming cereus) 133, 212
Hymenoptera 459, 461
Hyparrhenia 676
Hyparrhenia grassland 698
Hyparrhenia hirta 122
Hyparrhenia rufa (thatching grass) 122, 398, 489, 676, 678, 699; see also thatching grass
Hypena strigata 414, 415, 416, 417
Hypericum 135
Hypericum degeneri 476
Hypericum perforatum (Klamath weed) 7, 135, 469, 473, 765; see also Klamath weed
Hypochoeris 83
Hypochoeris glabra 129
Hypochoeris radicata (gosmore) 129, 304, 315, 330, 554, 565, 630, 634, 637, 640, 642, 647, 651, 652, 670, 672, 684; see also gosmore
Hyptis pectinata 141

ice plant (*Mesembryanthemum* spp.) 8, 21, 34, 72, 73, 76, 77, 93; see also *Mesembryanthemum crystallium*, *M. nodiflorum*
iceplant community 67, 68, 71
ichneumonid (Ichneumonidae) 415, 416, 455
'ihi (*Portulaca lutea*) 219; see also *Portulaca lutea*
'ihi'ai (*Oxalis corniculata*) 103; see also yellow wood sorrel, *Oxalis corniculata*
'i'iwi (*Vestiaria coccinea*) 278, 477

Ilex 127
Ilex anomala (kāwa'u) 347, 397; see also kāwa'u
'iliahialo'e *(Santalum ellipticum)* 220; see also *Santalum ellipticum*
'ilima (*Sida fallax*) 218
ilioha (*Conyza bonariensis*) 211; see also *Conyza bonariensis*
Impatiens 131
importation 707-710, 712, 713, 717, 720, 721, 722, 723, 727, 822, 826-829, 830
Indian almond (Combretaceae) 135, 213
Indian banyan (*Ficus benghalensis*) 218
Indian mallow 104
Indian mulberry (*Morinda citrifolia,* noni) 616; see also *Morinda citrifolia,* noni
Indian myna (*Acridotheres tristis*) 412; see also *Acridotheres tristis*
Indian pluchea (*Pluchea indica*) 211; see also *Pluchea indica*
Indian rhododendron (*Melastoma candidum*) 538; see also *Melastoma candidum*
Indian rubber tree (*Ficus elastica*) 218
indigo (*Indigofera suffruticosa*) 103, 399; see also *Indigofera suffruticosa*
Indigofera spicata 139
Indigofera suffruticosa (indigo) 102, 103, 139, 399
'inikō (*Indigo suffruticosa*) 103; see also *Indigo suffruticosa*
insecticides 344
insects 14, 16, 19, 51, 75, 201, 228, 242, 267, 291, 385, 411, 412, 413, 414, 415, 416, 417, 418, 420, 424, 425, 426, 433, 436, 437, 438, 440, 441, 445, 452-462, 466, 467, 468, 469, 470, 472, 473, 474, 475, 476, 479, 537, 539, 562, 563, 564, 567, 578, 580, 585, 612, 667, 709, 720, 721, 722, 727, 728, 759, 761, 763, 764, 765, 766, 767, 768, 769, 770, 771, 772, 773, 797, 823
invertebrates 290, 291, 349, 373, 557, 560, 601, 602, 604, 612, 781, 785, 796, 797, 824, 825; see also nematodes, species listings
Ipomoea (morning glory) 214, 365, 538; see also morning glory
Ipomoea alba 135
Ipomoea aquatica 135
Ipomoea batatas (sweet potato) 102, 135, 214, 729
Ipomoea cairica var. *hederacea* 135
Ipomoea congesta (morning glory) 363; see also morning glory
Ipomoea hederifolia 135
Ipomoea indica f. albiflora 214
Ipomoea indica f. indica (koali-'awahia) 214; see also koali'awahia
Ipomoea obscura 135
Ipomoea ochracea 135
Ipomoea pes-caprae subsp. *brasiliensis* (beach morning glory) 214
Ipomoea purpurea 135
Ipomoea triloba (little bell) 135, 214
Ipomoea violacea 136
Iridaceae (iris) 87, 118
iris (Iridaceae) 118; see also Iridaceae
ironwood (*Casuarina equisetifolia*) 189, 193, 197-198, 199, 200, 201, 203, 213, 488, 509, 513, 515, 528; see also *Casuarina equisetifolia*
Isachne distichophylla ('ohe) 347, 687
island bigpod ceanothus (*Ceanothus megacarpus* subsp. *insularis*) 71; see also *Ceanothus megacarpus* subsp. *insularis*
island night lizard (*Xantusia riversiana*) 71; see also herpetofauna
island oak (*Quercus tomentella*) 71; see also *Quercus tomentella*
island sagebrush 74
Ixora (ixora) 716

jackbean (*Canavalia*) 215, 676
jade tree (*Portulacaria afra*) 220
Japanese honeysuckle *(Lonicera japonica)* 9; see also *Lonicera japonica*
Japanese lovegrass (*Eragrostis tenella*) 208; see also *Eragrostis tenella*
Japanese white-eye (*Zosterops japonicus*) 227, 278, 281-282, 548, 567

Jasminum sambac (pikake) 219
Jatropha 137
Jaumea carnosa 83
Java plum (*Syzygium cumini*) 399, 493, 538, 545, 547, 559; see also *Syzygium cumini*
Jerusalem cherry (*Solanum pseudocapsicum*) 364, 372, 493, 786, 790, 791, 792, 793; see also *Solanum pseudocapsicum*
Joinvillea gaudichaudiana 560
joyweed (*Alternanthera tenella*) 210; see also *Alternanthera tenella*
Juncaceae (rushes) 87, 118
Juncus 72
Juncus acuminatus 118
Juncus balticus 87
Juncus bufonius 87, 118
Juncus effusus 118
Juncus ensifolius 118
Juncus mexicanus 87
Juncus planifolius 118, 558, 564, 565, 570
Juncus polyanthemos 118
Juncus tenuis 118
jungle fowl (*Gallus gallus*) 288; see also gallinaceous birds
Justicia 125

kā'e'e (*Mucuna gigantea*) 216
kāhili flower (*Grevillea banksii*) 528; see also *Grevillea banksii*, silk oak
kāhili ginger (*Hedychium gardnerianum*) 209, 311, 318, 341, 342, 344-350, 372, 387, 489, 498, 529, 551, 562, 570, 571, 572, 584, 691, 692, 694, 695, 696, 697, 754, 781, 782, 786, 789, 790, 792, 795, 814; see also *Hedychium gardnerianum*
kākalaioa (*Caesalpinia bonduc*) 215; see also *Caesalpinia bonduc*
kākonakona (*Panicum torridum*) 208
Kalanchoe daigremontiana x tubiflora 214
Kalanchoe pinnata (air plant) 136, 214
Kalanchoe tubiflora (chandelier plant) 136, 214
kalij pheasant *(Lophura leucomelana)* 281, 288, 477, 770
Kalimeris pinnatifida 129
kamani (*Calophyllum inophyllum*) 104, 213, 729; see also *Calophyllum inophyllum*
kāmanomano (*Cenchrus agrimonioides* var. *laysanensis*) 199; see also *Cenchrus agrimonioides* var. *laysanensis*
Kamehameha butterfly (*Vanessa tameamea*) 278
kāmole (primrose willow) 104; see also *Ludwigia octovalvis*
karakanut (Corynocarpaceae, *Corynocarpus laevigatus*) 136, 528, 532, 816
kau-la'ai ('ōhelo, *Vaccinium reticulatum*) 372; see also 'ōhelo, *Vaccinium reticulatum*
kāwa'u (*Ilex anomala*) 347, 397, 401
Keckiella cordifolia 91
Kentucky bluegrass (*Poa pratensis*) 566, 626, 629, 632, 644, 646, 649; see also *Poa pratensis*
kiawe (*Prosopis pallida*) 216, 402, 403, 492, 512, 513, 519, 603; see also *Prosopis pallida*
kikuyu grass *(Pennisetum clandestinum)* 6, 315, 320, 364, 387, 491, 529, 544, 559, 562, 563, 571, 732, 733; see also *Pennisetum clandestinum*
kīlau (*Pteridium aquilinum* var. *decompositum*) 363
kili'o'opu (*Kyllinga nemoralis*) 101; see also *Kyllinga nemoralis*
kinikini (kākalaioa) 215; see also *Caesalpinia bonduc*
Klamath weed (*Hypericum perforatum)* 7, 469, 473, 475, 476, 765, 768; see also *Hypericum perforatum*
klu (*Acacia farnesiana*) 215, 538; see also *Acacia farnesiana*
knapweed (*Centaurea*) 766, 767; see also *Centaurea*
koa (*Acacia koa*) 155, 168, 170, 177, 180, 181, 182, 183, 184, 185, 271, 274, 284, 285, 287, 288, 290, 304, 307, 311, 313, 363, 394, 401, 403, 532, 558, 560, 561, 571, 580, 581, 593, 667, 669, 675, 683, 696, 700, 791
koa forest 168, 273, 290, 304, 309,

koa forest [cont.] 311, 403, 532, 542, 558, 560, 561, 571, 593
koa haole (*Leucaena leucocephala*) 42, 112, 216, 399, 402, 489, 544, 547, 552, 585, 733, 766, 782, 783; see also *Leucaena leucocephala*
koali'awahia (*Ipomoea indica* f. *indica*) 214
koa-māmane forest 288
koa-'ōhi'a forest 168, 170, 177, 271, 285, 287, 288, 290, 314, 315, 532
kōlea (*Myrsine lessertiana*) 347, 349; see also *Myrsine lessertiana*
kolomona (*Senna surattensis*) 216; see also *Senna surattensis*
Koster's curse (clidemia, *Clidemia hirta*) 241, 314, 320, 399, 432, 435, 452, 468, 532, 538, 544, 547, 567; see also *Clidemia hirta*, clidemia
kou (*Cordia subcordata*) 104, 212, 729; see also *Cordia subcordata*
kou-haole (*Cordia sebestena*) 212
kuahulu (hairy merremia) 105; see also *Merremia aegyptica*
kudzu (*Pueraria lobata*) 9, 20, 492, 512, 519, 521
Kuehneola uredinis 433, 445, 446
kūkaenēnē (*Coprosma ernodioides*) 401, 669; see also *Coprosma ernodioides*
kūkaepua'a (*Digitaria ciliaris, D. setigera*) 101, 208; see also *Digitaria ciliaris, D. setigera*
kukui (*Aleurites moluccana*, candlenut) 214, 729; see also *Aleurites moluccana*, candlenut
Kyllinga brevifolia 117, 324
Kyllinga nemoralis (kili'o'opu) 102, 117; see kili'o'opu

Lablab purpureus 139
lace bugs 413; see also Hemiptera
lace-necked dove (*Streptopelia chinensis*) 412
Lactuca sativa (lettuce) 211
Lactuca serriola 83, 129
Lactuca serriola integrata 83
Lagenaria siceraria 102, 136
lama (*Diospyros sandwicensis*) 397, 400, 402, 789
Lamarckia aurea 74, 89
Lamiaceae 87, 141, 217; see also mint
Lamium amplexicaule 555, 556
Langsdorfia franckii 414, 415
lantana (*Lantana camara*) 42, 221, 399, 411, 412, 413, 414, 415, 416, 417, 418, 425, 426, 439, 467, 473, 475, 476, 489, 538, 545, 547, 584, 585, 593, 618, 765, 768; see also *Lantana camara*
Lantana camara (lantana) 42, 151, 221, 399, 411, 412, 440, 467, 469, 489, 538, 545, 547, 584, 593, 618, 765; see also lantana
Lantanophaga (=*Platyptilia*) *pussillidactyla* (pterophorid) 413, 417
Laportea interrupta 112, 151
Lappula redowskii 18
Lapsana communis 129, 330-331
Lascosomidae 459
lasiandra (glorybush) 343; see also *Tibouchina urvilleana*
Lasiurus cinereus semotus (Hawaiian bat) 781; see also bat
larkspur (*Delphinium barkeyi, D. occidentale*) 7
Lasthenia chrysostoma 72, 83
lather leaf (*Colubrina asiatica*) 52, 57, 58
Lathyrus laetiflorus barbarae 86
Lathyrus latifolius 139
Lathyrus tingitanus 139
laua'e (*Phymatodes scolopendria*) 204
Lauraceae (laurel) 141
laurel (Lauraceae) 141
lava flows 288, 376, 378, 380, 394, 396, 398, 403, 500, 542, 692, 788, 789, 790
Lavatera assurgentiflora 87
Layia platyllossa 83
Laysan finch (*Telespyza cantans*) 200
Laysan honeycreeper (*Himatione sanguinea freethii*) 199
Laysan rail (*Porzana palmeri*) 199
leadwort (Plumbaginaceae) 219
leaf beetle 452, 459, 461; see also chrysomelid
leaf roller 412, 455, 461, 476; see also tortricid
leafy spurge (*Euphorbia esula*) 9, 767; see also *Euphorbia esula*

Lecythidaceae (barringtonia) 217
legumes (Leguminosae) 586, 733, 773; see also Leguminosae
Leguminosae (legumes) 487, 489, 492, 508, 509, 510, 512; see also legumes
Leiothrix lutea (red-billed leiothrix) 227
Lemnaceae (duckweed) 118
Lemna perpusilla 118
lemonade berry (*Rhus integrifolia*) 71; see also *Rhus integrifolia*
lemon grass (*Cymbopogon citratus*) 208
Leonotis 141
Lepidium 132
Lepidium bidentatum var. *o-waihiense* ('ānaunau) 190, 212
Lepidium nitidum nitidum 84
Lepidium oblongum 84, 132
Lepidium strictum 84
Lepidium virginicum (pepper grass) 132, 212
Lepidoptera 413, 416, 455, 459, 461, 766, 772; see also butterflies, moths, plume moths, species or group listings
Leptobyrsa decora 414, 415, 416, 417, 418
Leptospermum 144
Leptospermum scoparium (tea tree) 144, 544
Lepturus repens 208
lettuce (*Lactuca sativa*) 211
Leucaena 33
Leucaena leucocephala (koa haole) 42, 112, 139, 216, 399, 489, 544, 547, 552, 555, 585, 733, 766, 782; see also koa haole
Leucobryum gracile 243
lianas 271, 290; see also vines
lichens 33, 396, 672, 675, 677, 680
Ligustrum 20
Liliaceae 87, 118, 207, 491, 511; see also lily
liliko'i (*Passiflora edulis*) 219, 443, 529; see also *Passiflora edulis*
Lilium humboldtii 71, 87
lily 118, 207, 667; see also Liliaceae
Limnocharitaceae 118
Linaceae (flax) 141
Linanthus androcaceus micranthus 90
Linaria canadensis var. *texana* 149
Linaria texana 91
linden (Tiliaceae) 150
Lindernia crustacea 149
Linum 141
Liothrips urichi (thrips) 242, 453, 454, 567; see also thrips
Lipochaeta integrifolia (nehe) 211
little bell (*Ipomoea triloba*) 214; see also *Ipomoea triloba*
Lius poseidon 452, 459, 460, 461
liverworts 33, 243, 544
Livistona australis (cabbage palm) 205
lizard 71
Lobelia erinus 133
lobelioids 301, 667
Lobularia maritima (sweet alyssum) 132, 199, 212
locoweed (*Astragalus*) 7
loggerhead turtle (*Caretta caretta caretta*) 50; see also sea turtle
logging 20, 21, 34-35, 49, 175, 274, 283, 287, 288, 290, 476, 532, 533, 536, 729, 800
logwood tree (*Haematoxylum campechianum*) 216; see also *Haematoxylum campechianum*
Lolium 650
Lolium multiflorum 122
Lolium perenne 73, 74, 89
Lomatium caruifolium 81
long-horned beetle 452, 459, 461
Lonicera japonica (Japanese honeysuckle) 9, 20, 133
looper (geometrid, *Scotorythra*) 412
loosestrife (Lythraceae) 142
Lophospermum erubescens 149
Lophostemon confertus (vinegar tree) 529
Lophura leucomelana (kalij pheasant) 281, 477, 770; see also kalij pheasant
loquat (*Eriobotrya japonica*) 488, 509, 513, 515; see also *Eriobotrya japonica*
Lotus 86
Lotus dendroideus 70, 86
loulu (*Pritchardia martii, P. remota*) 206; see also *Pritchardia remota*

lowland community 17, 255, 378, 386, 388, 389, 390, 392, 397, 402, 419, 420, 486, 508, 510, 511, 512, 522, 533, 561, 617, 675-679, 698, 699, 753
Loxioides bailleui (palila) 628
Loxops coccineus ochraceus (Maui 'ākepa) 300
Luculia gratissima 510, 513, 517
Ludwigia octovalvis (kāmole, primrose willow) 102, 104, 145, 331
Ludwigia palustris 145
lupine (*Lupinus hybridus*) 489, 510, 517, 521
Lupinus 510, 517, 521
Lupinus albifrons (shrub lupine) 71, 86
Lupinus arboreus (shrub lupine) 71, 86
Lupinus bicolor 86
Lupinus chamissonis 86
Lupinus hybridus (lupine) 489; see also lupine
Lupinus succulentus 86
Lupinus truncatus 86
Luzula hawaiiensis 636, 640, 641, 672
lycaenid 413; see also *Strymon*
Lycium californicum (boxthorn) 71, 72, 92; see also boxthorn
Lycopersicon esculentum (tomato) 92, 149, 220; see also tomato
Lycopersicon pimpinellifolium (currant tomato) 102, 105, 149, 555
Lycopodium cernuum (wawae'iole) 345, 347; see also wawae'iole
Lysimachia nummularia 20
Lytanthus 266
Lythraceae (loosestrife) 142
Lythrum maritimum (pūkāmole) 102, 104, 142, 331-332, 363
Lythrum salicaria (purple loosestrife) 9

macadamia (*Macadamia integrifolia)* 730
Macaranga mappa 137
Machaerina angustifolia ('uki) 347, 372
Macroptilium lathyroides 139
Madagascar olive (*Noronhia emarginata*) 219
Madia sativa 129
mahogany (Meliaceae) 143
maile pilau (*Paederia scandens*) 490, 511, 513, 518, 538; see also *Paederia scandens*
makaloa (*Cyperus laevigatus*) 206
Malabar melastome (*Melastoma candidum, M. malabathricum*) 155, 156, 157, 161, 162, 168, 169, 171, 174, 175, 176, 178, 402; see also *Melastoma candidum, M. malabathricum*
Malachra alceifolia 142
Malacothrix foliosa 83
Malacothrix implicata 70
Malacothrix incana 71, 72, 83
Malacothrix indecora 83
Malacothrix saxatilis 70, 83
Malacothrix squalida 83
Malacothrix succulenta 72, 83
Malephora (sea fig) 72; see also sea fig, ice plant, sea fig vine
Malephora crocea (sea fig vine) 76, 81
mallow 142, 217; see also Malvaceae
malphigia 217
Malpighiaceae (malphigia) 217
Malvaceae (mallow) 87, 142, 217
Malva parviflora (cheese weed) 74, 77, 87, 142, 218
Malvastrum coromandelianum 142, 555
Malvastrum coromandelianum subsp. *coromandelianum* (false mallow) 218
Malvaviscus arboreus (turks cap) 218
māmaki (*Pipturus albidus*) 363
māmane (*Sophora chrysophylla*) 180, 181, 182, 183, 288, 351, 355, 363, 399, 562, 580, 593, 626, 627, 628, 632, 633, 634, 644, 654, 656, 658, 660, 661, 662, 666, 669, 673, 675; see also *Sophora chrysophylla*
māmane forest/habitat 593, 626, 627, 628, 633, 658, 660, 661, 662
māmane/naio forest 400
mammals 11, 21, 272, 281, 537, 541, 598, 602, 612, 614, 616, 619, 797; see also species listings
mānele (*Sapindus saponaria*) 363
Mangifera indica (mango) 210, 441, 559; see also mango
mango (*Mangifera indica*) 126, 210, 441, 559; see also Anacardiaceae

mangosteen (Clusiaceae) 135, 213
mangrove (Rhizophoraceae) 6, 147; see also mangrove habitat, red mangrove
mangrove habitat 50, 51, 53, 58
manual control 16, 17, 19, 53, 198, 246, 248, 250, 267, 268, 314, 315, 318, 319, 320, 356, 377, 382, 385, 391, 441, 479, 513, 514ff, 526, 532, 545, 546, 783, 823, 824; see also mechanical control
Marah macrocarpus 70, 72, 85
marijuana (*Cannabis sativa*) 249
marine habitat 21, 47, 66, 104, 105, 541, 542
Mariscus 611
Mariscus javanicus ('ahuawa) 101, 102, 207
Mariscus pennatiformis subsp. *bryanii* 207
maritime cactus scrub 71
Marquesan ground dove (*Gallicolumba rubescens*) 604
Marrubium vulgare (horehound) 75, 87, 141
marshes 9, 32, 48, 50, 51, 101
Matricaria chamomilla 130
matted fern (*Dicranopteris linearis*, uluhe) 30, 311, 320, 697; see also *Dicranopteris linearis*, uluhe
Maui 'ākepa (*Loxops coccineus ochraceus*) 300
Maui blight 412
Maui nukupu'u (*Hemignathus lucidus affinis*) 300
Maui pāmakani (*Ageratina adenophora*) 304, 315, 319, 425, 468, 538, 544, 547, 558, 559, 563, 696; see also *Ageratina adenophora*
Maui parrotbill (*Pseudonestor xanthophrys*) 300
Mauna Kea silversword (*Argyroxiphium sandwicense* var. *sandwicense*) 593, 639
Maurandya antirrhiniflora 149
Mauritius hemp (*Furcraea foetida*) 43; see also *Furcraea foetida*
meadow community 17
meadow ricegrass (*Ehrharta stipoides*) 315, 318, 320, 364, 372, 488, 528, 687, 688, 689, 690, 692, 699, 700, 732; see also *Ehrharta stipoides*
mealybug 455
mechanical control 7, 9, 10, 13, 15, 18, 19, 47, 53, 54, 55, 58, 78, 79, 80, 100, 178, 198, 199, 201, 250, 274, 291, 292, 293, 318, 420, 432, 433, 440, 442, 445, 454, 479, 486, 495, 499, 500, 501, 502, 519-520, 526, 545, 546, 566, 568, 584, 585, 734, 735, 740, 772, 795, 829; see also manual control
Medicago intertexta 139
Medicago lupulina (black medic) 139, 216, 651
Medicago polymorpha (bur clover) 72, 73, 74, 77, 86, 139, 216
Medicago sativa (alfalfa) 86, 216
Medinilla venosa 143
mehamehame (*Flueggea neowawraea*) 545
melaleuca (*Melaleuca quinquenervia*, cajeput) 10, 50, 51, 52, 53, 55, 57; see also cajeput, *Melaleuca quinquenervia*
Melaleuca quinquenervia (cajeput, melaleuca, paperbark) 4, 21, 32, 47, 49, 50, 144, 489, 510, 513, 517, 529, 547, 816; see also cajeput, melaleuca
Melastoma 156, 168
Melastoma candidum (Indian rhododendron) 143, 155, 156, 157, 169, 175, 186, 402, 454, 538; see also Malabar melastome
Melastoma sanguineum (fox-tongue melastoma) 143, 538
Melastomataceae 142, 157, 241, 452, 454, 458, 460, 461, 487, 493, 508, 510, 512, 821; see also melastome
melastome 142, 242, 343, 403, 452, 454, 456, 460, 461; see also Melastomataceae
Meleagris gallopavo (turkey) 281; see also turkey
Melia azedarach (pride of India) 20, 143, 547
Meliaceae (mahogany) 143
Melia imperfecta 89
Melilotus 19, 77; see also *Melilotus indica*
Melilotus alba (white sweet clover) 216, 555
Melilotus indica (sweet clover) 77, 86, 139, 216; see also *Melilotus*

Melilotus officinalis 139
Melinis 34; see also *Melinis minutiflora*
Melinis minutiflora (molasses grass) 35, 122, 208, 315, 324, 372, 394, 398, 490, 533, 538, 547, 551, 559, 560, 676, 677; see also molasses grass
Melinis/Rhynchelytrum community 698
Melitara 422
melochia (*Melochia umbellata*) 402, 510, 517, 521, 522, 538; see also *Melochia umbellata*
Melochia umbellata (gunpowder tree) 150, 402, 490, 510, 517, 521; see also melochia, gunpowder tree
Mentha 20
Mentha pulegium 141
Mentha x *villosa* 93, 141
Merremia 136
Merremia aegyptica (hairy merremia, kuahulu) 102, 105
Merremia tuberosa (wood rose) 136, 214
Mesembryanthemum 93
Mesembryanthemum chilense (ice plant) 8; see also ice plant
Mesembryanthemum crystallinum (ice plant) 21, 34, 71, 76, 78, 81; see also ice plant
Mesembryanthemum edule (ice plant) 8; see also ice plant
Mesembryanthemum nodiflorum (ice plant) 21, 71, 73, 74, 76, 81; see also ice plant
mesic forest/habitat 167, 168, 169, 170, 174, 177, 178, 182-184, 190, 226, 231, 234, 241, 242, 271, 284, 285, 287, 288, 290, 376, 396, 398, 532, 533, 541, 542, 545, 559, 562, 785, 787, 790, 791, 796
mesic grassland 169
mesic scrub 169, 785
mesic shrublands 533
Meteorus laphygmae (braconid) 455; see also braconid
Metrosideros 308-309, 680
Metrosideros/Acacia/Sapindus forest (ʻōhiʻa/koa/mānele forest) 359
Metrosideros/native shrub community 698
Metrosideros polymorpha (ʻōhiʻa) 33, 155, 168, 186, 227, 271, 283, 304, 344, 347, 394, 400, 532, 544,
Metrosideros polymorpha (ʻōhiʻa) [cont.] 558, 666, 670, 672, 673, 674, 675, 754, 785; see also ʻōhiʻa
Mexican ash (*Fraxinus uhdei*) 528; see also *Fraxinus uhdei*
miconia (*Miconia calvescens*) 569, 828
Miconia 458, 460, 461
Miconia acinodendrum 460
Miconia calvescens (miconia) 143, 569, 821, 827, 829; see also miconia
Miconia mirabilis 461
microfauna. See invertebrates
Microlepia strigosa (palapalai) 363; see also palapalai
Microseris 83
Microsorium (Polypodium) scolopendria 616
Microstegium vimineum 20
Microtus oeconomus (tundra vole) 21
midge 452, 458, 459
Mikania micrantha 618
milfoil (*Hydrilla verticillata*) 52; see also *Hydrilla verticillata*
military activities 73, 76, 93, 189, 190, 194-196, 197, 198, 199, 200, 244, 249, 554
milkweed (Asclepiadaceae) 127
milkwort (Polygalaceae) 146
milo (*Thespesia populnea*) 103, 218; see also *Thespesia populnea*
mimiʻilio (*Spergularia marina*) 213; see also *Spergularia marina*
Mimosa pudica 139
Mimulus flemingii 91
mint 141, 217 667; see also Lamiaceae
Mirabilis californica 88
Mirabilis jalapa (four o'clock) 144, 218
mites 425, 467, 764
Mitracarpus hirtus 148
mixed grass community 698
moa (*Psilotum nudum*) 204
mock orange (*Murraya paniculata* 220
Modiola caroliniana 142
mokihana (*Pelea anisata*) 289
molasses grass *(Melinis minutiflora)* 35, 208, 315, 318, 372, 394, 398, 399, 402, 405, 490, 533, 538, 547, 551, 559, 560, 561, 565, 570, 572, 676, 699; see also *Melinis minutiflora*
Molluginaceae 143; see also carpetweed

Mollugo cerviana 143
Momordica charantia 136
Mompha trithalama 452, 458, 459, 461
Momphidae 459
Monachus schauinslandi (Hawaiian monk seal) 189
Moneilema 422
mongoose, small Indian (*Herpestes auropunctatus*) 241, 249, 403, 540, 567
monkeypod (*Samanea saman*) 216, 547; see also *Samanea saman*
monk seal, Hawaiian (*Monachus schauinslandi*) 189
Monochoria vaginalis 124
Monocotyledonae, monocotyledons 116, 205
Monolepis nuttalliana 85
monstera (*Monstera deliciosa*) 205
montane bog 542, 554, 557, 558, 562, 563, 564, 572
montane forest 343, 787
montane habitat 254, 255, 256, 262, 267, 268, 272, 400, 525, 618, 619
montane parkland 401, 669, 672, 673-674, 683, 685, 697, 700; see also mountain parkland
montane rain forest 225, 229, 231, 234, 284, 401, 486, 753, 789, 790
montane seasonal forest 229, 512
montane seasonal vegetation zone 343, 386, 401, 403, 405, 486, 496, 790-791
Montandoniola moraguesi 454
montanoa (*Montanoa hibiscifolia*) 547; see also *Montanoa hibiscifolia*
Montanoa hibiscifolia (montanoa) 130, 547
montbretia (*Crocosmia crocosmiiflora*) 363, 544, 545; see also *Tritonia crocosmiflora*
Monterey pine (*Pinus radiata*) 33; see also *Pinus radiata*
Moraceae (mulberry) 87, 143, 218, 509
Morinda citrifolia (Indian mulberry, noni) 102, 105, 148, 616
morning glory 135, 363, 538; see also Convolvulaceae, *Ipomoea*
Morus alba (white mulberry) 143, 218
mosquito 201, 753
mosses 243, 272, 544, 660, 671, 675, 677, 679, 680
moths 413, 423, 454, 455, 564, 568, 580, 766; see also Lepidoptera
mouflon (*Ovis musimon*) 628, 633, 639, 662
mountain apple (*Syzygium malaccense*) 455, 729; see also *Syzygium malaccense*
mountain goat (*Oreamnos americanus*) 10
mountain parkland 791, 796; see also montane parkland
mountain pili (*Panicum tenuifolium*) 401; see also *Panicum tenuifolium*
Mucuna gigantea (kā‘e‘e) 216
Mucuna urens (sea bean) 139, 216; see also sea bean
Muehlenbeckia axillaris (wire vine) 146, 507, 511, 517, 521
Muhlenbergia microsperma 89, 122
mulberry 143, 218; see also Moraceae
mule (*Equus caballus* x *asinus*) 602
mullein (*Verbascum thapsus*) 254-256, 258-259, 261-263, 265-268, 494, 557, 629, 644, 649, 661; see also *Verbascum thapsus*
Murraya paniculata (mock orange) 220
Musa 102, 207, 332
Musaceae 119, 207; see also banana
Musa troglodytarum 119
Musa x *paradisiaca* (banana) 119; see also banana
musk thistle 19
mustard (Brassicaceae) 132, 212
Myadestes obscurus obscurus (‘ōma‘o) 278; see also ‘ōma‘o
mycoherbicides 432, 433, 435
myna, common (*Acridotheres tristis*) 227, 281-282; see also *Acridotheres tristis*
Myoporum (naio) 284, 402, 611; see also naio
Myoporum sandwicense (naio) 186, 363, 399, 626, 631, 655, 659; see also naio
Myrica 228, 445
Myricaceae (bayberry) 143, 225, 445, 490
Myrica cerifera (wax myrtle) 227, 772
Myrica faya (faya tree, firetree) 13, 33, 42, 143, 225, 387, 399, 411, 433, 441, 467, 469, 486, 490, 538,

Myrica faya (faya tree, firetree) [cont.] 568-569, 577, 690, 754, 770, 772, 781, 782, 792, 821; see also firetree
Myriophyllum (water milfoil) 9, 20; see also water milfoil
Myriophyllum aquaticum 141
Myriophyllum spicatum (Eurasian water milfoil) 20; see also *Myriophyllum*
Myrsinaceae (myrsine) 143, 487
Myrsine lessertiana (kōlea) 45, 347, 349
Myrsine saliciana 45
Myrtaceae (myrtle) 88, 144, 157, 218, 488, 489, 492, 493, 508, 509, 510, 512
myrtle 144, 218; see also Myrtaceae
Mytosis discolor 132

naio (*Myoporum sandwicense*) 181, 182, 183, 284, 363, 399, 626, 627, 628, 656, 658, 661; see also *Myoporum sandwicense*
naio-māmane forest/woodland 626, 627, 628, 658
Nama sandwicensis (Hawaiian nama) 217
narrow-leaved carpetgrass (*Axonopus fissifolius*) 314, 319; see also *Axonopus fissifolius*
nasturtium (*Tropaeolum majus*) 151, 221, 494, 506, 512, 520, 521, 786, 790, 792, 793; see also *Tropaeolum majus*
Nasturtium microphyllum 132
Nasturtium sarmentosum 102, 132
Natal plum (*Carissa macrocarpa*) 210
Natal redtop (*Rhynchelytrum repens*) 209, 380, 398, 565, 676; see also *Rhynchelytrum repens*
naupaka kahakai (*Scaevola sericea*) 217, 402, 676; see also beach naupaka, *Scaevola sericea*
Nectria galligena 433
needlegrass (*Stipa cernua*) 626; see also *Stipa cernua*
nehe (*Lipochaeta integrifolia*) 211
nematodes 284, 433, 467, 764
Nemophila pedunculata 87
nena (*Heliotropium curassavicum*) 198, 212; see also *Heliotropium curassavicum*
nēnē (Hawaiian goose, *Nesochen sandwicensis*) 403, 502
Nephrolepis biserrata 616
Nephrolepis multiflora (sword fern) 199, 204
Nerium oleander (oleander) 210
Nesochen sandwicensis (Hawaiian goose) 403; see also nēnē
Nesofregetta fuliginosa (storm-petrel) 612
nettle 151, 221
nettle-leaved goosefoot (*Chenopodium murale*) 213; see also *Chenopodium murale*
New Zealand flax (*Phormium tenax*) 491, 506, 511, 513, 518, 529, 532; see also *Phormium tenax*
New Zealand red rabbit 93; see also rabbits
New Zealand spinach (*Tetragonia expansa, T. tetragonioides*) 76, 209; see also *Tetragonia tetragonioides*
Nicandra physalodes 149
Nicotiana glauca (tree tobacco) 17, 150, 220, 555, 603
Nicotiana tabacum (tobacco) 150, 220
night blooming cereus (*Hylocereus undatus*) 212; see also *Hylocereus undatus*
nightshade (Solanaceae) 75, 149, 220; see also Solanaceae, *Solanum nodiflorum*
nīoi (*Capsicum annuum*) 220; see also *Capsicum annuum*
nitrogen cycle/fixation 33, 78, 193, 235, 501, 568, 598, 667
nitrogen-fixing bacteria 8; see also bacteria
noctuid, Noctuidae 412, 415, 416, 459
noddy 200
nohu (*Tribulus cistoides*) 222; see also *Tribulus cistoides*
noni (Indian mulberry) 105, 616; see also *Morinda citrifolia*
Nopalea 419
Norfolk Island pine (*Araucaria heterophylla*) 204
Noronhia emarginata (Madagascar olive) 219
northern jointvetch 433, 435
Nothocestrum 560
noxious plants 7, 19, 156, 157, 158-159, 178, 226, 236, 241, 271, 314, 320, 342, 343, 411, 412, 425, 477,

noxious plants [cont.] 547, 625, 627, 658, 662, 707, 708, 767, 772, 773, 826, 827, 829
nutgrass (*Cyperus rotundus*) 206; see also *Cyperus rotundus*
nutrient cycle 6, 33, 486, 598, 620, 667, 697, 734, 754, 821; see also nitrogen cycle
nutrients 3, 29, 30, 31, 32, 34, 35, 36, 76, 193
Nyctaginaceae (four o'clock) 88, 144, 218

Ocimum 141
octopus tree (*Schefflera actinophylla*) 211, 547; see also *Schefflera actinophylla*
Octotoma 414, 415, 416, 417
Odocoileus hemionus (deer) 701
odontonema *(Odontonema stictum)* 209
Odontonema cuspidatum 125
Odontonema stictum (odontonema) 209
Oenothera 145
Oenothera laciniata 145, 219
'ōhā (*Clermontia parviflora*) 347; see also *Clermontia parviflora*
'ohai (*Sesbania tomentosa*) 190, 216
'ohe (*Isachne distichophylla*) 347, 687
'ōhelo (*Vaccinium, V. reticulatum*) 307, 348, 372, 400, 401, 626, 644, 669; see also *Vaccinium, V. reticulatum*
'ōhelo/pukiawe scrub 401
'ōhi'a (*Metrosideros polymorpha*) 33, 155, 168, 169, 177, 189, 181, 182, 183, 184, 185, 186, 227, 229, 271, 283, 284, 285, 287, 288, 290, 306, 307, 345, 347, 349, 351, 359, 367, 369, 372, 394, 400, 401, 402, 532, 544, 548, 558, 561, 666, 673, 674, 675, 676, 687, 692, 694, 696, 754, 785, 787, 788, 789, 790, 791; see also *Metrosideros polymorpha*
'ōhi'a dieback 545, 548; see also dieback
'ōhi'a forest 168, 170, 177, 285, 287, 288, 290, 304, 308, 310, 344, 366, 370, 532, 542, 548, 558, 561, 754, 788
'ōhi'a-koa forest 155, 168, 285, 314
'ōhi'a/koa/mānele forest (Metrosideros/ Acacia/Sapindus forest) 359
'ōhi'a rain forest 155, 177, 284, 306, 344, 350, 357
'ōhi'a-tree fern forest 284, 285
'ōhi'a vegetation type 666, 675, 684
oī (*Stachytarpheta dichotoma*) 221; see also *Stachytarpheta dichotoma*
Oidaematophorus beneficus (plume moth) 425; see also plume moth
okra (*Hibiscus esculentus*) 217
'ōlapa (*Cheirodendron trigynum*) 290, 347, 367
Oleaceae (olive) 144, 219, 488, 490, 509, 510; see also olive
Olea europaea subsp. *africana* (olive) 144, 341, 342, 351, 353, 354, 490, 523, 786, 793; see also olive
oleander (*Nerium oleander*) 210
oleaster 136, 488; see also Eleagnaceae, *Eleagnus umbellata*
'ōlena (*Curcuma longa*) 729; see also *Curcuma longa*
Oligomeris linifolia 91
olive 144, 219, 341, 342, 350-356, 490, 523, 786, 788, 793, 794; see also Oleaceae, *Olea europaea* subsp. *africana*
olomea (*Perrottetia sandwicensis*) 313
'ōma'o *(Myadestes obscurus obscurus)* 227, 278
Onagraceae (evening primrose) 88, 144, 219, 510
onion (*Allium cepa*) 207
Ophiomyia lantanae (seed fly, agromyzid) 413, 414, 417
Opius lantanae 414
Oplismenus compositus 122
Oplismenus hirtellus 122, 325
ōpuhe (*Urera sandwicensis*) 104
Opuntia (prickly pear cactus) 34, 411, 419, 420, 422, 585, 765; see also cactus, pānini cactus, prickly pear cactus
Opuntia aurantiaca (prickly pear cactus) 765; see also prickly pear cactus
Opuntia cochenillifera 133
Opuntia cordobensis 419, 469
Opuntia ficus-indica (prickly pear cactus) 84, 133, 419, 420, 432, 434, 467, 469, 490, 511, 518, 521, 529;

Opuntia ficus-indica (prickly pear cactus) [cont.] see also pānini cactus, prickly pear cactus
Opuntia littoralis (prickly pear cactus) 70, 84; see also prickly pear cactus
Opuntia megacantha. See *O. ficus-indica*
Opuntia oricola 70, 84
Opuntia prolifera (cholla) 70, 84
Opuntia tuna 419; see also *Opuntia ficus-indica*
Opuntia vulgaris 133
orchard grass (*Dactylus glomerata*) 557, 566; see also *Dactylus glomerata*
Orchidaceae (orchids) 119, 207, 708, 715, 716, 717, 718, 720
Oreamnos americanus (mountain goat) 10
Oreobolus 558
Oreobolus furcatus 557
ornamentals 44, 45, 49, 50, 51, 100, 108, 110, 113, 114, 115ff, 156, 158, 159, 190, 193, 196, 197, 199, 200, 225, 289, 342, 343, 412, 436, 441, 443, 454, 497, 507, 508ff, 567, 568, 569, 584, 616, 617, 712-724, 729, 768, 772; see also horticulture
Orobanchaceae (*Orobanche*) 88
orpine 136; see also Crassulaceae
Orthezia insignis (greenhouse orthezia) 412
Orthocarpus densiflorus 91
Orthocarpus purpurascens 91, 149
Oryctolagus cuniculus (European rabbit) 76; see also rabbits
Osteomeles anthyllidifolia (ʻūlei) 352, 355, 394, 402, 559, 627, 670; see also ʻūlei
Osteospermum calendulaceum 130
Ovis aries (sheep) 21, 577, 586, 602, 625, 627, 701, 726, 773; see also sheep
Ovis canadensis nelsoni (desert bighorn sheep) 6; see also desert bighorn sheep
Ovis musimon (mouflon) 628; see also mouflon
ōwī (*Stachytarpheta jamaicensis*) 221; see also *Stachytarpheta jamaicensis*
Oxalidaceae (wood sorrel) 145, 219
Oxalis corniculata (ʻihi, yellow wood sorrel) 102, 103, 145, 219, 610, 651
Oxalis corniculata var. *repens* 612
Oxalis corymbosa (pink wood sorrel) 145, 219
Oxyspora paniculata 100, 143, 569
oyster plant (*Tradescantia spathacea*) 206; see also *Tradescantia spathacea*

Pachodynerus nasidens 414
Paederia scandens (maile pilau) 148, 490, 511, 513, 518, 538
pahole (*Diplazium sandwichianum*, hōʻiʻo) 311; see also *Diplazium sandwichianum*
paʻiniu (*Astelia menziesiana*) 345, 347, 349
pakai (*Amaranthus dubius*) 210; see also *Amaranthus dubius*
Palafoxia callosa 130
palapalai (*Microlepia strigosa*) 363
Palay rubber vine (*Cryptostegia grandiflora*) 56
palila (*Loxioides bailleui*) 628
palm 205, 715; see also Arecaceae
Palmaceae 511
Palmeria dolei (crested honeycreeper) 300
palmgrass (*Setaria palmifolia*) 493, 538, 547, 692, 786, 789, 790, 792, 793; see also *Setaria palmifolia*
panax (*Polyscias guilfoylei*) 211
Pancratium littorale (spider lily) 207
Pandanaceae (screwpine) 207
Pandanus 613
Pandanus odoratissimus (hala) 402; see also hala
Pandanus tectorius (hala) 207, 612; see also hala
panic grass (*Panicum maximum* var. *trichylum*) 732
Panicum maximum 122
Panicum maximum var. *maximum* (Guinea grass) 732; see also Guinea grass
Panicum maximum var. *trichylum* (panic grass) 732
Panicum miliaceum 122

Panicum repens (torpedo grass) 122, 538
Panicum tenuifolium 672
Panicum torridum (kākonakona) 208
pānini cactus 419, 422, 469, 518, 521, 522; see also *Opuntia*, prickly pear cactus, *Opuntia ficus-indica*
Papaveraceae 88; see also poppy
papaya (*Carica papaya*) 133, 213, 730; see also Caricaceae, *Carica papaya*
paperbark (*Melaleuca quinquenervia*) 489, 510, 513, 517, 529, 547, 816; see also *Melaleuca quinquenervia*
papyrus (*Cyperus papyrus*) 206
paragrass (*Brachiaria mutica*) 208; see also *Brachiaria mutica*
Parapholis incurva 89
Paraserianthes falcataria 139
parasites 45, 290, 412, 414, 416, 418, 435, 437, 455, 461, 472, 766, 770
Parentucellia viscosa 149
Parietaria hespera 92
Parkinsonia aculeata 139
parsley 126; see also Apiaceae
Parthenium hysterophorus 130
partridge pea (*Chamaecrista nictitans*) 399; see also *Chamaecrista nictitans*
Partulina porcellana 560; see also invertebrates
Paspalidium radiatum 122
Paspalum 122, 194, 208, 309-310, 365, 372, 401
Paspalum conjugatum (Hilo grass) 4, 42, 122, 250, 304, 313, 315, 490, 491, 558, 561, 618, 692; see also Hilo grass
Paspalum dilatatum (dallis grass) 122, 315, 364, 491, 610, 683; see also dallis grass
Paspalum fimbriatum (Columbia grass) 122, 208
Paspalum urvillei (vasey grass) 122, 208, 315, 491, 538; see also vasey grass
Passiflora 35, 272, 278, 280, 443, 444, 730; see also banana poka, *Passiflora mollissima*, passion flower, passion fruit
Passifloraceae 145, 157, 219, 443, 444, 491; see also passion flower
Passiflora edulis (liliko'i, passion fruit) 145, 219, 271, 443, 444, 454, 529, 767, 770; see also liliko'i, passion fruit
Passiflora foetida (water lemon) 145, 443, 444, 604
Passiflora laurifolia 145
Passiflora ligularis (sweet granadilla) 145, 272, 443, 444
Passiflora mixta 274
Passiflora mollissima (banana poka) 13, 41, 145, 155-157, 169, 175, 186, 249, 271-272, 274, 277, 293, 372, 392, 433, 438, 442, 444, 472, 486, 491, 536, 537, 538, 551, 567-568, 593, 667, 692, 696, 754, 767, 782, 829; see also banana poka
Passiflora pulchella 145
Passiflora quadrangularis 145
Passiflora suberosa 145, 443, 444
Passiflora subpeltata 145
Passiflora tripartita 274, 442
Passiflora x *caerulea* 145
passion flower 35, 145, 219, 567; see also *Passiflora*, Passifloraceae
passion fruit 271, 443, 454, 730, 767, 770, 771; see also *Passiflora*, *P. edulis*
passion vine butterfly (*Agraulis vanillae*) 291
pasture 228, 229, 235, 259, 317, 412, 420, 436, 453, 468, 548, 553, 554, 559, 562, 563, 564, 565, 584, 585, 586, 587, 591, 734, 753; see also cattle industry, grazing, rangelands
pathogens 36, 44, 45, 228, 272, 278, 284, 290, 387, 411, 425, 426, 432-447, 457, 462, 432-447, 457, 461, 462, 466, 467, 468, 472, 475, 498, 580, 720, 759, 761, 763, 764, 771; see also diseases
Paulownia tomentosa 20
pea 137, 215; see also Fabaceae
pearl flower (*Heterocentron subtriplinervium*) 372, 510, 517, 521, 522; see also *Heterocentron subtriplinervium*
Pedilanthus tithymaloides (slipper flower) 215
Pelargonium hortorum (fish geranium) 217
Pelargonium zonale 140

Pelea anisata (mokihana) 289
Pelea hawaiensis 560
Pellaea 90
Pellaea ternifolia 637, 640, 642, 647, 650, 652, 672
Penestes 452, 459, 461
Pennisetum 34; see also *Pennisetum clandestinum, P. setaceum*
Pennisetum clandestinum (kikuyu grass) 6, 89, 122, 315, 364, 387, 491, 529, 544, 559, 562, 732; see also kikuyu grass
Pennisetum macrostachyum 122
Pennisetum polystachion 122
Pennisetum purpureum (elephant grass) 122, 620
Pennisetum setaceum (fountain grass) 18, 42, 123, 155, 156, 157, 169, 175, 186, 376, 394, 485, 491, 495, 507, 538, 551, 568, 587, 625, 627, 630, 650, 734, 753, 782, 786, 793, 814, 821; see also fountain grass
Penstemon cordifolia 71
Peperomia 715
Peperomia pellucida 146
pepper (Piperaceae) 146
pepper grass (*Lepidium virginicum*) 212; see also *Lepidium virginicum*
peregrine falcon (*Falco peregrinus*) 50
Perisierola emigrata 414
Perityle emoryi 70, 83
periwinkle (*Catharanthus roseus*) 210; see also *Catharanthus roseus*
Perkinsiella saccharicida (sugar cane leafhopper) 727
Perrottetia sandwicensis (olomea) 313
Persea americana (avocado) 141, 441-442, 455
pesticides 13, 440, 441, 442, 501, 525, 526; see also herbicides
Petrorhagia velutina 134
Petroselinum crispum 126
petty spurge (*Euphorbia peplus*) 215; see also *Euphorbia peplus*
Phacelia 87
Phaeornis obscurus 227
Phaius tankarvilleae 119
Phalaris minor 89
Phalaris tuberosa 16
Phaseolus vulgaris (green bean) 216
Pheidole megacephala (big-headed ant) 454
Philaenus spumarius (spittle bug) 425
Philodendron (philodendron) 15, 205
Phlaeothripidae 242
Phleum pratense (timothy) 7, 15
phlox 146; see also Polemoniaceae
Phoenix dactylifera (date palm) 206, 511, 513, 518
Pholistoma 87
Phormium tenax (New Zealand flax) 116, 491, 506, 511, 513, 518, 529, 532
photosynthesis 31, 34, 290, 344
Phragmidium violaceum 435, 436, 445, 446
phycitid (*Cryptoblabes aliena*) 412
Phyla nodiflora 221
Phyllanthus 137
Phyllanthus amarus 616
Phyllospadix 92
Phyllostachys (bamboo) 511, 518, 521; see also bamboo
Phyllostachys nigra (black bamboo) 559, 563; see also black bamboo
Phyllostegia variabilis 199, 217
Phymatodes scolopendria (laua'e) 204
Physalis 150
Phytolaccaceae (pokeweed) 145
Phytolacca octandra 145
phytophagous insects 433; see also insects
pickerel weed (Pontederiaceae) 124
Picris hieracioides 130
Piesmopoda 452, 459, 460
pig 10, 11, 14, 16, 21, 35, 101, 158, 227, 235, 241, 242, 244, 249, 271, 275, 281, 283, 287, 288, 292, 300, 301, 302, 303, 305, 306, 307, 311, 313, 314, 315, 316, 317, 318, 319, 320, 343, 391, 477, 485, 495, 496, 513, 532, 540, 544, 545, 546, 548, 551, 554, 557, 558, 559, 560, 561, 562, 565, 567, 570, 571, 600, 602, 606, 608, 613, 619, 666, 667, 668, 675, 683-697, 699, 700, 701, 714, 729, 770, 784, 788, 789, 790, 791, 797, 800; see also *Sus scrofa*
pigeon (*Ducula*) 612
pigweed (*Chenopodium murale, Portulaca oleracea*) 77, 601, 604, 607
pikake (*Jasminum sambac*) 219
Pilea microphylla (artillery plant) 151, 221

pili grass (*Heteropogon contortus*) 380, 396, 397, 398, 400, 402, 602; see also *Heteropogon contortus*
pili uka (*Trisetum glomeratum*) 626, 629, 634, 646, 649, 658, 661; see also *Trisetum glomeratum*
pilo (*Coprosma, C. montana, C. ochracea*) 313, 345, 347, 349, 355, 363, 626, 661; see also *Coprosma, C. montana*
Pinaceae 116, 491, 511; see also pine
pine 32, 116, 511, 518, 521, 529, 533, 551, 554, 557, 564, 569, 571, 754; see also Pinaceae, *Pinus*
pineapple (*Ananas comosus*) 420, 444, 453, 554, 730, 800; see also pineapple industry
pineapple industry 453, 713, 714; see also pineapple
pine forest/trees/pinelands 6, 42, 49, 50, 51, 53, 54, 344, 554; see also coniferous forest
pink 133, 213; see also Caryophyllaceae
pink wood sorrel (*Oxalis corymbosa*) 219; see also *Oxalis corymbosa*
Pinus 511, 518, 521, 529, 533, 551, 565; see also pine
Pinus caribaea (Caribbean pine) 42, 491; see also slash pine
Pinus elliottii (slash pine) 20, 54, 116, 580; see also slash pine
Pinus patula 116, 554, 564
Pinus pinaster (cluster pine) 42, 116, 554, 564
Pinus radiata (Monterey pine) 16, 33, 554, 564
Piperaceae (pepper) 146
Piper methysticum 102, 146
Pipturus albidus (māmaki) 363
Pisonia 604
Pisonia grandis 218, 604, 606
Pistia stratiotes 116
Pithecellobium dulce 13
Pittosporaceae (*Pittosporum*) 146, 492, 511
pittosporum (*Pittosporum*) 146, 492; see also Pittosporaceae, *Pittosporum undulatum*
Pittosporum undulatum (pittosporum) 146, 492, 511, 513, 518
Pityrogramma triangularis 90
Plagiobothrys californicus 83, 84
Plagiohammus spinipennis 415, 416, 417
Plantaginaceae (plantain) 88, 146, 219
Plantago 651
Plantago aristata 146
Plantago australis 146
Plantago bigelovii 88
Plantago debilis 146
Plantago erecta erecta 88
Plantago hirtella 88
Plantago insularis 88
Plantago lanceolata (English plantain) 146, 219
Plantago major (common plantain) 146, 219
Plantago ovata 88
plantain 146, 219; see also Plantaginaceae
plantations 284
plant succession 477, 500, 676, 701
Platystemon californicus 88
Plectranthus scutellarioides (coleus) 141, 217
Pleomele auwahiensis 561
pluchea 538; see also *Pluchea symphytifolia*
Pluchea indica (Indian pluchea) 130, 211
Pluchea symphytifolia (pluchea) 199, 130, 211, 399, 469, 538; see also sourbush
Pluchea x *fosbergii* 130, 211
Plumbaginaceae (leadwort) 219
plumbago (*Plumbago auriculata*) 219
plume moth 413, 425, 426; see also Lepidoptera, moths
plumeria (*Plumeria*) 210
Plumeria obtusa (Singapore plumeria) 210
Poa annua (annual bluegrass) 89, 123, 209, 566
Poaceae 88, 119, 157, 208; see also grass
Poa douglasii 72, 89
Poa pratensis (Kentucky bluegrass) 18, 123, 566, 626, 630, 636, 640, 641, 643, 747, 650, 652, 653; see also Kentucky bluegrass
Poa scabrella 89
poinsettia (*Euphorbia pulcherrima*) 215
pokeweed (Phytolaccaceae) 145

Polemoniaceae (phlox) 89, 90, 146
pōlinalina (*Vitex trifolia*) 221
pollination, pollinators 35, 44, 51, 278, 284, 620, 728
Polycarpon tetraphyllum 134, 610
Polygalaceae (milkwort) 146
Polygala paniculata 146
Polygonaceae (buckwheat) 90, 146, 219, 511
Polygonum 146
Polygonum aviculare 146, 555
Polygonum convolvulus 147, 555
Polygonum glabrum 147
Polygonum punctatum 147, 332
Polynesian effects, impacts, influences, introductions 12, 100, 101, 102, 103, 104, 105, 396, 397, 400, 401, 553, 566, 602, 604, 607, 611, 612, 613, 616, 726, 727, 729, 752, 799-800; see also anthropogenic effects, human disturbance
Polynesian rat (*Rattus exulans*) 612; see also rat
Polypodiaceae 90, 204; see also ferns
Polypodium californicum 90
Polypogon interruptus (ditch polypogon) 89, 123, 209
Polypogon monspeliensis (rabbitfoot grass) 74, 89, 123, 209
Polyscias guilfoylei (panax) 211
pomace flies (*Drosophila*) 278; see also *Drosophila*
pondweed (Potamogetonaceae) 124
Pontederiaceae (pickerel weed) 124
pōpolo (nightshade, *Solanum americanum, S. nelsonii*) 105, 190, 221; see also *Solanum americanum*
poppy 145; see also Papaveraceae
Populus alba 20
Populus tremuloides (aspen) 7
Portulacaceae (purslane) 90, 147, 219
Portulaca lutea (ʻihi) 194, 219
Portulaca oleracea (common purslane) 147, 194, 219, 555, 601, 604, 610; see also pigweed
Portulacaria afra (jade tree) 220
Portulaca villosa 220
Porzana palmeri (Laysan rail) 199
Potamogeton 112, 124
Potamogetonaceae (pondweed) 124
potato (*Solanum tuberosum*) 93, 221; see also *Solanum tuberosum*
Potentilla agedii grandis 91
Pouteria sandwicensis 560-561
prairie 9, 18, 49, 50, 51, 53; see also grasslands
prairie grass (*Bromus willdenowii*) 208; see also *Bromus willdenowii*
predators, predation 44, 45, 267, 272, 275, 403, 418, 437, 451, 627, 753, 766, 770, 797
prescribed burning 10, 16, 18, 25, 49, 53, 54, 55, 526; see also fire
preserve design 20, 701, 785-786
prickly pear cactus (*Opuntia* spp.) 70, 93, 419, 432, 434, 440, 467, 469, 477, 490, 511, 529, 765, 768; see also *Opuntia, O. ficus-indica, O. littoralis*, pānini cactus
pride of India (*Melia azedarach*) 547; see also *Melia azedarach*
primrose 147, 220; see also Primulaceae
primrose willow (kāmole) 104; see also *Ludwigia octovalvis*
Primulaceae (primrose) 90, 147, 220
princess flower. See glorybush, *Tibouchina urvilleana*
Pristomenes hawaiiensis 414
Pritchardia 199, 206, 601
Pritchardia martii (loulu) 206
Pritchardia pacifica (vin) 206
Pritchardia remota (loulu) 190, 206
Procecidochares alani (stem gall fly) 426
Procecidochares utilis (stem gall fly) 563
Prosopis pallida (kiawe) 139, 216, 402, 492, 512, 513, 519, 603; see also kiawe
protea 147, 716, 718; see also Proteaceae
Proteaceae 147, 488; see also protea
Prunella vulgaris (selfheal) 141, 332-333, 558
Prunus ilicifolia subsp. *lyonii* (Catalina cherry) 71
Prunus lyonii 91
Prunus x *cerasifera* 148
pseudococcid 455
Pseudonestor xanthophrys (Maui parrotbill) 300
Psidium 45, 158, 310; see also common guava, guava

Psidium cattleianum (strawberry guava) 35, 42, 144, 155-157, 169, 175, 186, 250, 303, 313, 335, 387, 437, 476, 492, 501, 529, 531, 544, 547, 551, 558, 560, 618, 667, 689, 754, 781, 782, 792, 816, 821; see also *Psidium*, strawberry guava
Psidium cattleianum var. *cattleianum* (strawberry guava) 538; see also strawberry guava
Psidium cattleianum var. *littorale* (waiawī) 547; see also waiawī
Psidium cattleianum var. *lucidum* (waiawī) 538; see also waiawī
Psidium guajava (common guava) 52, 144, 218, 311, 315, 335-336, 399, 492, 521, 538, 545, 547, 559, 565, 584, 616, 730, 733, 766; see also common guava, guava, *Psidium*
Psilotum nudum (moa) 204
Pteridaceae 90
Pteridium aquilinum (bracken fern) 400, 630, 637, 640, 642, 650, 670, 672; see also bracken fern
Pteridium aquilinum var. *decompositum* (bracken fern, kīlau) 363, 557; see also bracken fern, kīlau
pterophorid 413; see also *Lantanophaga (=Platyptilia) pussillidactyla*
Pteridophyta 204
Pterolepis glomerata 143
Pterostegia drymarioides 70, 90
pū'aha-nui (*Broussaisia*) 313
puapilo (*Capparis sandwichiana*) 212
Puccinia canaliculata 436
Puccinia chondrillina 435
Pueraria lobata (kudzu) 9, 20, 492, 512, 519, 521
Puffinus pacificus chlorohynchus (wedge-tailed shearwater) 200
pūkāmole (*Lythrum maritimum*) 104, 363; see also *Lythrum maritimum*
pūkiawe (*Styphelia tameiameiae*) 35, 307, 352, 355, 372, 394, 400, 401, 402, 559, 564, 626, 644, 666, 669, 683; see also *Styphelia tameiameiae*
pūkiawe/'a'ali'i shrubland 564
puncture vine (*Tribulus cistoides, T. terrestris*) 470, 475, 476, 765, 768; see also *Tribulus cistoides, T. terrestris*
purple loosestrife (*Lythrum salicaria*) 9
purple nutsedge (*Cyperus rotundatus*) 468
purslane 147, 219; see also Portulacaceae
Pycnotus cafer (red-vented bulbul) 548
Pycreus polystachyos 207
Pyracantha angustifolia (firethorn) 148, 538
Pyracantha crenati-serrata 148
pyralid, Pyralidae 459, 461
Pyrausta perelegans 568
pyraustid, Pyraustidae 414, 415

quarantine 236, 388, 391, 414, 426, 438, 439, 442, 454, 455, 456, 457, 461, 467, 472, 473, 474, 495, 498, 523, 707, 708, 709, 727, 733, 760, 763, 764, 767, 771, 827
Quercus tomentella (island oak) 71, 86

rabbitfoot grass (*Polypogon monspeliensis*) 209; see also *Polypogon monspeliensis*
rabbits 21, 76, 93, 94, 195
radish (*Raphanus sativus*) 212; see also *Raphanus sativus*
Rafinesquia californica 83
rainbow trout (*Salmo gairdneri*) 20
rain forest 34, 170, 225, 242, 300, 302, 315, 318, 320, 342, 343, 366, 377, 386, 388, 396, 397, 398, 401, 486, 497, 498, 500, 501, 551, 558, 560, 563, 565, 567, 571, 626, 666, 667, 668, 684, 686, 692, 694, 696, 697, 699, 700, 702, 753, 754, 770, 784, 785, 796
Ramularia destructiva 433
ranching 7, 9, 17, 19, 34, 67, 73, 75, 76, 77, 78, 93, 158, 175, 283, 388, 412, 420, 425, 426, 432, 467, 473, 476, 477, 497, 502, 546, 577, 578, 580, 581, 584, 585, 586, 602, 616, 700, 728, 753, 758, 772, 800
rangelands/pastures 7, 9, 17, 19, 34, 158, 229, 236, 250, 343, 357, 387, 411, 414, 425, 435, 436, 440, 467, 476, 477, 502, 507, 548, 554, 562, 568, 580, 584, 585, 587, 732, 753,

rangelands/pastures [cont.] 758, 772; see also grazing, ranching
Rangifer arcticus (caribou) 21
Rangifer tarandus (reindeer) 21
Ranunculaceae (buttercup) 90, 147
Ranunculus 147
Ranunculus californicus 90
Ranunculus repens 17, 147
Raphanus raphanistrum 132
Raphanus sativus (radish) 132, 212
raspberry (*Rubus glaucus*) 492; see also *Rubus glaucus*
rat 201, 604; see also *Rattus, R. exulans*
rat tail fescue (*Vulpia myuros*) 649; see also *Vulpia myuros*
rattail grass (*Sporobolus africanus*) 559, 565; see also *Sporobolus africanus*
rattle-pod (*Crotalaria pallida*) 215
Rattus 201; see also rat
Rattus exulans (Polynesian rat) 604, 612; see also rat
red-billed leiothrix (*Leiothrix lutea*) 227
red brome (*Bromus rubens*) 77; see also *Bromus rubens*
red-fruited cactus 420
red hibiscus (*Hibiscus rosa-sinensis*) 218
red mangrove (*Rhizophora mangle*) 6; see also *Rhizophora mangle*
red pualele (*Emilia fosbergii*) 211; see also *Emilia fosbergii*
red-stemmed filaree (*Erodium cicutarium*) 77; see also *Erodium cicutarium*
red-vented bulbul (*Pycnotus cafer*) 548
reforestation 107, 113, 226, 231, 235, 342, 399ff, 497, 502, 507, 508ff, 578, 726, 728, 729, 772; see also plant succession, revegetation
Reichardia picroides 130
reindeer (*Rangifer tarandus*) 21
Resedaceae 91
restoration 8, 15, 394, 403, 526, 598, 625, 697
revegetation 7, 78, 496, 497, 525, 557, 563, 667, 697, 699-701
Rhamnaceae (*Rhamnus*) 91
Rhaphidophora aurea (taro vine) 205
Rhizogonium spiniforme 243
Rhizophoraceae (mangrove) 147; see also mangrove, mangrove habitat
Rhizophora mangle (red mangrove) 6, 147
Rhodomyrtus tomentosa (downy rose myrtle) 144, 538
Rhus integrifolia (lemonade berry) 71, 81
Rhynchelytrum 698
Rhynchelytrum grassland 698
Rhynchelytrum repens (Natal redtop) 123, 209, 380, 398, 565, 676, 677, 678
Rhynchosia minima 139
Rhynchospora 566
Rhynchospora caduca 325-326
Ribes malvaceum 91
Richardia brasiliensis 148
Ricinis communis (castor bean) 137, 215, 492, 506, 512, 519, 521, 566; see also castor bean
riparian habitat 9, 14, 18, 31, 32, 72, 314, 315, 316, 541, 542, 558; see also aquatic habitat
ripgut grass (*Bromus rigidus*) 626, 649; see also *Bromus rigidus*
Rivina humilis 146
robin (*Turdus migratorius*) 51
roadside habitat 20, 22, 110, 260, 262, 268, 377, 378, 388, 389, 390, 391, 453, 499, 507, 510, 557, 578, 580, 581, 587; see also ruderal habitat
rodent 11, 50; see also rat
Rosa 148, 220, 454, 714; see also rose
Rosaceae (rose) 91, 147, 157, 220, 488, 492, 509
Rosa multiflora 20
Rosa wichuriana 20
rose 147, 220, 714, 716, 717; see also *Rosa*, Rosaceae
rose apple (*Syzygium jambos*) 493, 512, 513, 519, 530, 558, 559, 564, 570; see also *Syzygium jambos*
royal poinciana (*Delonix regia*) 215
Roystonea 206
Rubiaceae 91, 148, 220, 490, 510, 511; see also coffee, *Rubus* spp.
Rubus 13, 45, 158, 308, 362, 367, 392, 433, 435, 445, 446, 476, 477

Rubus argutus (blackberry, Florida blackberry) 148, 316, 341, 343, 355-366, 367, 445, 446, 467, 470, 492, 525, 527, 529, 532, 538, 544, 551, 557, 561, 683, 694, 770; see also blackberry, Florida blackberry
Rubus constrictus 436
Rubus discolor 148
Rubus ellipticus (yellow Himalayan raspberry) 148, 155, 156, 157, 169, 175, 186, 341, 343, 366-370, 399, 446, 472, 492, 692, 694, 696, 786, 793, 821; see also yellow Himalayan raspberry
Rubus glaucus (raspberry) 446, 492
Rubus hawaiiensis ('ākala) 445, 446, 476
Rubus lucidus. See *Rubus argutus*
Rubus macraei 445, 446
Rubus niveus 149
Rubus penetrans. See blackberry, *Rubus argutus*
Rubus rosifolius (thimbleberry) 148, 304, 333, 446, 558, 692, 694; see also thimbleberry
Rubus sieboldii 148
Rubus ulmifolius 436
Rubus ursinus 91
Rudbeckia laciniata 130
ruderal habitat 4, 7, 17, 18, 50, 51, 52, 53, 56, 77, 100, 105, 110, 190, 235, 242, 249, 254, 255, 256, 258, 259, 261, 267, 320, 376, 386, 392, 452, 498, 506, 507, 509-512, 513, 514ff, 523, 544, 554, 555-556, 565, 566, 567, 571, 822; see also roadside habitat
Ruellia 125
Ruellia brittoniana 125, 209
Rumex acetosella (sorrel) 147, 307, 333-334, 556, 631, 634, 637, 640, 642, 651, 672, 684; see also sheep sorrel
Rumex albescens (hu'ahu'akō) 190, 219
Rumex brownei 147, 610
Rumex crispus (curly dock) 90, 147, 436
Rumex fueginus 90
Rumex obtusifolius 147
Rumex salicifolius 90
rush 118; see also Juncaceae
rush skeleton weed (*Chondrilla juncea*) 435
Russian thistle (*Salsola kali*, thistle) 8, 411, 470; see also *Salsola kali*, thistle
rust (fungi) 435, 436, 445, 446
Rutaceae (citrus) 220

Saccharum 727
Saccharum officinarum (sugar cane) 102, 444, 524, 726, 727, 800; see also sugar cane
Sacciolepis indica (Glenwood grass) 42, 123, 316, 326, 558, 565, 689; see also Glenwood grass
Sadleria cyatheoides ('ama'u, 'ama'uma'u) 307, 363, 400, 562
Sadleria pallida ('ama'uma'u) 348
safflower (*Carthamus tinctorius*) 767
sage 70
sagebrush desert 8
Sagina decumbens 84, 134
Sagina japonica 134
Sagittaria latifolia 116
sago palm 204; see also Cycadaceae, *Cycas*
St. Augustine grass (*Stenotaphrum secundatum*) 364; see also *Stenotaphrum secundatum*
Salicaceae 91
Salicornia subterminalis 85
Salicornia virginica (samphire) 85, 135, 213
Salix lasiolepis (willow) 72, 91
Salmo gairdneri (rainbow trout) 20
Salpichroa rhomboidea 150
Salsola kali (Russian thistle) 8, 17, 18, 135, 411, 470; see also thistle
saltbush (*Atriplex suberecta*) 213; see also *Atriplex suberecta*
salt cedar (tamarisk, *Tamarix*) 17, 31
salt concentrations 21, 34, 49, 50, 72-73, 76
saltgrass (*Distichlis spicata*) 72
saltwort (Bataceae) 131
Salvia coccinea 141
Salvia mellifera 70
Salvia occidentalis 141
Salvinia molesta 32, 43
Samanea saman (monkeypod) 139, 216, 547
Sambucus mexicana 133

samphire (*Salicornia virginica*) 213; see also *Salicornia virginica*
sandalwood (Santalaceae, *Santalum, S. ellipticum*) 199, 220, 400, 626, 800
sandbur (*Cenchrus echinatus*) 200, 208; see also *Cenchrus echinatus*
Sanicula arguta 81
Sanicula sandwicensis 561
Sanseviera (bowstring hemp) 205
Santalaceae 220; see also sandalwood
Santalum (sandalwood) 400; see also sandalwood
Santalum ellipticum ('iliahialo'e) 199, 220, 626; see also sandalwood
Sapindaceae (soapberry) 148
Sapindus 365
Sapindus saponaria (mānele) 363
Sapium sebiferum (Chinese tallow) 20
Sargasso muticum 21
savanna 51, 395, 397, 398, 526, 734, 787
sawgrass 51
Saxifragaceae 91
Scaevola sericea (naupaka kahakai) 197, 217, 402, 676, 679; see also beach naupaka
Scaevola strand 698
Scaevola taccada 616
Scalesia 620
scarlet pimpernel (*Anagallis arvensis*) 220; see also *Anagallis arvensis*
Schefflera actinophylla (octopus tree) 127, 211, 547
Schiedea verticillata 190, 213
Schinus 564
Schinus molle 44
Schinus terebinthifolius (Christmas berry) 6, 21, 31, 43, 44, 47, 49, 51, 52, 53, 54, 58, 126, 210, 470, 476, 493, 522, 530, 538, 545, 547, 559, 564, 584, 593, 733, 755; see also Brazilian peppertree, Christmas berry
Schizachyrium 401
Schizachyrium condensatum (bush beardgrass) 123, 394, 398, 434, 486, 487, 493, 552, 677, 698, 699, 734; see also bush beardgrass
Schizostachyum glaucifolium 102, 123
Schoenoplectus californicus 117
Scirpus 611
Scirpus cernuus californicus 85
Scotch broom (*Cytisus scoparius*) 33, 766; see also *Cytisus scoparius*
Scotorythra (geometrid, looper) 412
screwpine (Pandanaceae) 207
Scrophulariaceae (figwort) 91, 148, 220, 255, 266, 494
Scrophularia villosa 91
scrub 225, 229, 234, 401, 604, 787, 788, 789
scrub communities 67, 68, 70-71, 78, 170, 180-186, 544, 613, 615, 620, 666, 675, 669, 676, 677, 678, 679, 680, 683
Scythris grandipennis (spine-feeding moth) 580
sea bean (*Dioclea wilsonii, Mucuna urens*) 216; see also *Dioclea wilsonii, Mucuna urens*
seabirds 94, 197, 199, 200, 601, 604, 609, 610, 612, 620; see also birds
sea blite (*Suaeda californica*) 77, 94; see also *Suaeda californica*
sea cliff community 67, 70, 73, 75, 564
sea fig (*Carpobrotus aequilaterus*) 72, 75; see also ice plant, *Carpobrotus, Malephora*
sea fig vine (*Malephora crocea*) 76; see also sea fig, ice plant
sea grape (*Coccoloba uvifera*) 219
sea-rocket (*Cakile*) 8; see also *Cakile*
sea turtle 55, 189; see also green sea turtle, loggerhead turtle
Sechium edule 136
sedges 117, 206, 303, 304, 311, 313, 314, 322-326, 396, 557, 565, 566, 616, 636, 640, 641, 650, 670, 672, 675, 677, 678, 679, 680, 681, 682, 687, 688, 689, 690, 694, 695, 696, 697; see also *Carex* spp., Cyperaceae, *Cyperus* spp.
sedge-rush-native shrub community 168
seed fly (*Ophiomyia lantanae*) 413; see also *Ophiomyia lantanae*
Selaginella bigelovii 92
Selaginellaceae 92
Selaginella deflera 565
Selca brunella 454

selfheal (*Prunella vulgaris*) 558; see also *Prunella vulgaris*
Senecio (groundsel) 45
Senecio jacobea (tansy ragwort) 7, 16, 17; see also tansy ragwort
Senecio mikanioides (German ivy) 130, 155, 156, 157, 169, 175, 186, 538, 625, 627; see also German ivy
Senecio sylvaticus (wood groundsel) 16, 34, 35, 130, 631, 637, 642, 648, 651, 653
Senecio vulgaris 83, 130
Senna didymobotrya 139
Senna obtusifolia 139
Senna occidentalis (coffee senna) 140, 566, 571, 604; see also coffee senna
Senna pendula var. *advena* 140
Senna septemtrionalis 140
Senna surattensis (kolomona) 140, 216
Sericothrips staphylinum (gorse thrips) 580
sesame (*Sesamum indicum*) 476
Sesbania sesban 140
Sesbania tomentosa ('ohai) 190, 216
Sesuvium portulacastrum ('ākulikuli) 209
Setaria glauca 123
Setaria gracilis (foxtail grass) 123, 364
Setaria palmifolia (palmgrass) 123, 493, 538, 547, 692, 786, 793; see also palmgrass
Setaria verticillata (bristly foxtail) 123, 209
sheep (*Ovis aries*) 21, 93, 497, 577, 581, 586, 602, 603, 605, 606, 607, 625, 626, 627, 628, 629, 630, 633, 634, 644, 649, 652, 656, 658, 661, 662, 701, 714, 726, 773, 800
sheepgrass (*Eragrostis brownei*) 558; see also *Eragrostis brownei*
sheep ranching 577
sheep sorrel (*Rumex acetosella*) 307, 317, 634; see also *Rumex acetosella*
shell ginger (*Catimbium speciosum*) 209
she-oak 134; see also Casuarinaceae
shepherd's purse (*Capsella rubella*) 212; see also *Capsella rubella*
Sherardia arvensis 148
shoebutton ardisia (*Ardisia solanacea*) 52, 58
shrub community 68, 69, 77, 168, 170, 302, 304, 307, 308, 314, 316, 342, 343, 401, 403, 525, 532, 551, 554, 557, 559, 562, 564, 566, 571, 593, 668, 679, 697, 698, 699
shrub lupine (*Lupinus albifrons, L. arboreus*) 71; see also *Lupinus albifrons, L. arboreus*
shrubs 8, 31, 42, 64, 70, 72, 73, 75, 76, 78, 93, 193, 156, 193, 309, 313, 316, 396, 399, 400, 401, 477, 509ff, 544, 558, 560, 562, 565, 566, 567, 585, 606, 616, 626, 631, 684, 697, 728, 760, 772
Sicyos 214
Sida acuta 142
Sida fallax ('ilima) 218
Sida rhombifolia 142, 610
Sida spinosa 142
Sidalcea malvaeflora 87
Sigesbeckia orientalis 130, 134
Silene 84
Silene gallica 74, 84
Silene laciniata 71, 84
silk oak (*Grevillea banksii, G. robusta*, kāhili flower) 341, 342, 356-359, 488, 528, 547, 783, 786, 788, 792, 793; see also *Grevillea banksii, G. robusta*
silkwood (*Flindersia brayleyana*) 547
silversword (*Argyroxiphium sandwicense*) 267, 639; see also *Argyroxiphium sandwicense*
silviculture 32, 33, 290
Silybum marianum 16, 74, 83
Singapore plumeria (*Plumeria obtusa*) 210
siris tree (*Albizia lebbeck*) 215; see also *Albizia lebbeck*
sisal (*Agave sisalana*) 205, 487, 507, 508, 514, 521, 538; see also *Agave sisalana*
Sisymbrium altissimum (bumblemustard) 8, 132, 556
Sisymbrium irio 132
Sisymbrium officinale 132
Sisyrinchium bellum 87
Sisyrinchium rosulatum 118
Sisyrinchium strictum 118

slash pine (*Pinus caribaea, P. elliottii*) 20, 51, 54, 491, 580; see also *Pinus caribaea, P. elliottii*
slender amaranth (*Amaranthus viridis*) 210; see also *Amaranthus viridis*
slender oats (*Avena barbata*) 77; see also *Avena barbata*
slipper flower (*Pedilanthus tithymaloides*) 215
small ground fig (*Chamaesyce prostrata*) 215; see also *Chamaesyce prostrata*
small mammals 158, 499
smooth brome (*Bromus inermis*) 7; see also *Bromus inermis*
snails 560; see also invertebrates
snow bush (*Breynia disticha* var. *rosi-picta*) 215
soapberry (Sapindaceae) 148
soft chess (*Bromus mullis*) 77; see also *Bromus mullis*
soil disturbance 11, 29, 35, 476, 496, 565, 598, 602-603, 604, 611, 614-615, 616, 619, 667, 669, 688, 700, 727; see also cattle, erosion, goats, pigs, sheep, ungulates
soil fertility, nutrients 32, 33, 34, 35, 290, 477
soil properties 35, 349, 728, 754
Solanaceae (nightshade) 92, 149, 220, 493; see also nightshade
Solanum 150, 611
Solanum americanum (glossy nightshade) 102, 105, 221; see also pōpolo
Solanum douglasii 92
Solanum mauritianum 150, 618
Solanum melongena (eggplant) 221, 461
Solanum nelsonii (pōpolo) 190, 221; see also pōpolo
Solanum nigrum 610
Solanum nodiflorum (nightshade) 75, 610; see also nightshade
Solanum pseudocapsicum (Jerusalem cherry) 150, 364, 372, 493, 786, 793; see also Jerusalem cherry
Solanum tuberosum (potato) 150, 221; see also potato
Solidago canadensis 130
Soliva sessilis 493, 512, 519, 521
Sonchus asper 73, 83, 130
Sonchus oleraceus (sow thistle) 72, 74, 78, 83, 130, 211, 610, 631, 632, 648, 651, 653
Sonchus tenerrimus 83
sooty tern (*Sterna fuscata*) 604
Sophora chrysophylla (māmane) 186, 288, 351, 355, 363, 365, 399, 562, 580, 626, 631, 645, 655, 656, 659, 666, 671, 672, 673; see also māmane
sorghum (*Sorghum vulgare*) 740
Sorghum halpense 20, 123
Sorghum sudanense 123
Sorghum vulgare (sorghum) 740
sorrel (*Rumex acetosella*) 556; see also *Rumex acetosella*
sourbush (*Pluchea symphytifolia*) 199, 211, 399, 469; see also *Pluchea symphytifolia*
sourgrass (*Digitaria insularis*) 208; see also *Digitaria insularis*
sow thistle (*Sonchus oleraceus*) 78, 211; see also *Sonchus oleraceus*
soy bean (*Glycine soja*) 216
Spanish clover (*Desmodium sandwicense*) 215; see also *Desmodium sandwicense*
Spanish needles (*Bidens alba* var. *radiata, B. pilosa*) 211; see also *Bidens pilosa*
sparrow, Santa Barbara song 94
Spartium junceum 140
Spathodea campanulata (African tulip tree) 131, 212, 316, 547, 559, 566, 618; see also African tulip tree
Spathoglottis plicata 119
Special Ecological Areas 22, 292, 372, 377, 387, 388, 391, 392, 445, 485, 488, 489, 490, 491, 492, 493, 494, 495, 497, 498-499, 500, 503, 506, 507, 513, 522, 696, 752, 754, 781-797, 814, 817
Spergula arvensis 112, 134
Spergularia macrotheca 71, 84
Spergularia marina (mimi‘ilio) 134, 213
Spermacoce 148
spider lily (*Crinum asiaticum, Pancratium littorale*) 207
spiderwort (Commelinaceae) 116, 206
spine-feeding moth (*Scythris grandipennis*) 580

spineless cactus 419, 420, 426

spiny emex (*Emex spinosa*) 765; see also *Emex spinosa*

spiny pigweed (*Amaranthus spinosus*) 210; see also *Amaranthus spinosus*

spiraling whitefly (*Aleurodicus dispersus*) 456

Spirodela polyrhiza 118

spittle bug (*Philaenus spumarius*) 425

Sporobolus africanus (African dropseed, rattail grass) 123, 209, 559, 565, 610, 667, 670; see also African dropseed, rattail grass

Sporobolus diander 123

Sporobolus elongatus 123

Sporobolus indicus (West Indian dropseed) 123, 209

Sporobolus pyramidatus 123, 209

Sporobolus virginicus (beach dropseed) 209

spotted dove (*Streptopelia chinensis*) 412

spotted knapweed (*Centaurea, C. maculosa*) 7; see also *Centaurea, Centaurea maculosa*

spreading mistflower 425

spurge 136, 214; see also Euphorbiaceae

squash (*Cucurbita pepo,* zucchini) 214

Stachys arvensis (staggerweed) 141, 217

Stachytarpheta 112, 151

Stachytarpheta dichotoma (cayenne vervain) 151, 221, 538; see also oī

Stachytarpheta jamaicensis (ōwī) 151, 221

Stachytarpheta mutabilis 151

Stachytarpheta urticifolia 151, 334

staggerweed (*Stachys arvensis*) 217; see also *Stachys arvensis*

Stapelia gigantea 127

Stellaria media (common chickweed) 74, 84, 134, 213

stem gall fly (*Procecidochares* spp.) 426, 563

Stenogyne 651, 658

Stenotaphrum secundatum (buffalo grass, St. Augustine grass) 123, 209, 364, 610

Stephanomeria virgata 83

Sterculiaceae 150, 221, 490, 510; see also cacao, cocoa

Stereocaulon vulcani 33

Sterna fuscata (sooty tern) 604

Stictocardia tiliifolia 136

stinkweed (*Tagetes minuta*) 627; see also *Tagetes minuta*

Stipa cernua (needlegrass) 123, 626, 647, 650, 652

Stipa lepida 89

Stipa pulchra (bunchgrass) 72, 73, 89; see also bunchgrass

stonecrop 214; see also Crassulaceae

storm-petrel (*Nesofregetta fuliginosa*) 612

strawberry (*Fragaria chiloensis*) 401

strawberry guava (*Psidium cattleianum*) 35, 42, 155-158, 165, 168-171, 173, 175, 176, 178, 250, 303, 311, 313, 315, 317, 318, 319, 387, 437, 476, 492, 501, 529, 531, 538, 544, 545, 547, 551, 558, 559, 560, 561, 564, 565, 566, 570, 572, 618, 667, 689, 690, 692, 697, 754, 781, 782, 786, 787, 788, 789, 790, 792, 795, 816, 821; see also *Psidium cattleianum*

Strelitzia reginae (bird of paradise) 207; see also bird of paradise

Strepsicrates smithiana 772

Streptopelia chinensis (spotted dove) 412

Streptosolen jamesoni 150

Striga asiatica (witchweed) 740

Strongylodon ruber 560

Strymon (lycaenid) 413, 414, 416, 417

Stylomecon heterophylla 88

Styphelia 308

Styphelia tameiameiae (pūkiawe) 35, 307, 352, 355, 394, 400, 559, 626, 631, 645, 666, 671, 672; see also pūkiawe

Suaeda californica (sea blite) 70, 73, 77, 78, 85; see also sea blite

subalpine community 7, 263, 268, 386, 396, 400-401, 486, 532, 533, 557, 562, 564, 571, 625, 626, 627, 628, 644, 661, 666, 669, 672, 673-674, 675, 683, 697, 698, 699, 700, 784, 791, 796

submontane forest 343, 754

submontane seasonal forest 229, 234, 386, 396, 398, 400, 401, 402, 486, 496, 513, 782, 787, 788, 789, 796

succession 8, 9, 15, 30, 31, 33, 34, 51, 54, 235, 254, 255, 267, 285, 288, 290, 316, 376, 397, 399, 401, 477, 500, 534, 539, 565, 607, 609, 615, 616, 620, 676, 683, 700, 701, 733, 734
succulents 34, 67, 70, 402, 565, 715, 720
sugar cane (*Saccharum officinarum*) 318, 444, 554, 726, 727, 729, 730, 767, 800; see also *Saccharum officinarum*
sugar cane leafhopper (*Perkinsiella saccharicida*) 727
sugar cane leaf roller (*Hedylepta*) 455
sugar industry 453, 713, 714, 726-730; see also sugar cane
Sula (boobies) 604
sunflower (*Helianthus annuus* var. *macrocarpus*) 127, 211; see also Asteraceae
Surinam cherry (*Eugenia uniflora*) 218; see also *Eugenia uniflora*
Sus scrofa (pig) 10, 20, 35, 158, 227, 241, 242, 271, 275, 301, 343, 477, 496, 532, 540, 551, 554, 600, 666, 667, 729, 770, 784; see also hog, pig, swine
swamp mahogany (*Eucalyptus robusta*) 528
sweet alyssum (*Lobularia maritima*) 199, 212; see also *Lobularia maritima*
sweet clover (*Melilotus indica*) 216; see also *Melilotus indica*
sweet fennel (*Foeniculum vulgare*) 566; see also *Foeniculum vulgare*
sweet granadilla (*Passiflora ligularis*) 272; see also *Passiflora ligularis*
sweet pepper. See *Capiscum annuum*
sweet potato (*Ipomoea batatas*) 214, 729; see also *Ipomoea batatas*
sweet vernalgrass (*Anthoxanthum odoratum*) 314, 317, 319, 401, 487, 557, 566, 626, 628, 634, 639, 644, 683, 732; see also *Anthoxanthum odoratum*
swine 605; see also pig, *Sus scrofa*
swine cress (*Coronopus didymus*) 212; see also *Coronopus didymus*
Swiss chard (*Beta vulgaris* var. *cicla*) 213
swollen fingergrass (*Chloris barbata*) 208; see also *Chloris*
sword fern (*Nephrolepis multiflora*) 199, 204
Syagrius fulvitarsus (Australian fern weevil) 562
Sydney blue gum (*Eucalyptus saligna*) 528
Syncarpia glomulifera 530
Synedrella nodifolia 130
Syngamia 414
Syngamia haemorrhoidalis 414, 415, 417
syngonium (*Syngonium podophyllum*) 205
Syrphidae (flower flies) 278
Syzygium 618
Syzygium cumini (Java plum) 144, 399, 493, 538, 545, 547, 559
Syzygium jambos (rose apple) 144, 493, 512, 513, 519, 530, 558, 564; see also rose apple
Syzygium malaccense (mountain apple) 102, 144, 455, 729

Tabebuia 212
tacca (Taccaceae) 124
Taccaceae (tacca) 124
Tacca leontopetaloides (arrowroot) 102, 124, 729
Tacsonia 274, 443
Tagetes minuta (stinkweed) 130, 627, 631, 651
Talinum 147
Tamarindus indica 140
tamarisk (salt cedar, *Tamarix*) 6, 9, 10, 14, 17, 18, 31
Tamarix 17
Tanacetum vulgare 130
tansy ragwort (*Senecio jacobea*) 7, 13, 16; see also *Senecio jacobea*
Taraxacum laevigatum 130
Taraxacum officinale (dandelion) 7, 15, 130, 631, 648, 651, 653
taro (*Colocasia esculenta*) 101, 104, 116, 205, 729, 800; see also Araceae, *Colocasia esculenta*
taro patches 101, 104
taro vine (*Rhaphidophora aurea*) 205
Taxodium distichum (dwarf cypress) 32

tea (Theaceae) 150
tea tree (*Leptospermum scoparium*) 544; see also *Leptospermum scoparium*
Teleonemia scrupulosa (tingid) 413, 414, 417, 475, 476; see also tingid
Teleonemia vanduzii 415
Telespyza cantans (Laysan finch) 200
tephritid fly (*Eutreta xanthochaeta*) 413, 544; see also gall flies
Tephrosia purpurea 102, 140
Teranycus lintearious (gorse mite) 580
Terminalia catappa (false kamani) 135, 213
Terminalia myriocarpa 213
termite (*Coptotermes formosanus*) 198
Tetragonia expensa (New Zealand spinach) 76
Tetragonia tetragonioides (New Zealand spinach) 81, 125, 209; see also New Zealand spinach
Tetramolopium humile 637, 640, 642
Tetrazygia bicolor 143
thatching grass (*Hyparrhenia rufa*) 398, 402, 489, 787; see also *Hyparrhenia rufa*
Theaceae (tea) 150
Thelypodium lasiophyllum 84
Themeda villosa 123
Thespesia populnea (milo) 102, 103, 142, 218
Tibouchina urvilleana (lasiandra) 341, 343, 370-372; see also glorybush
thimbleberry (*Rubus rosifolius*) 304, 558, 692, 696; see also *Rubus rosifolius*
thistle 6, 566, 569, 570, 766, 767; see also bull thistle, Canada thistle, *Carduus nutans, Cirsium arvense, Cirsium vulgare,* musk thistle, Russian thistle, *Salsola kali*
thrips (*Liothrips urichi*) 242, 250, 278, 453, 454, 567
Thunbergia 125
Thysanoptera 242, 278; see also thrips
ti (*Cordyline fruticosa*) 205, 607, 612, 613, 618, 716; see also *Cordyline fruticosa*
Tibouchina herbacea 143
Tibouchina urvilleana (glorybush) 143, 454, 470, 493, 512, 519, 521, 538; see also glorybush
tiger's claw (*Erythrina variegata* var. *orientalis*) 216
Tiliaceae (linden) 150
timothy (*Phleum pratense*) 7; see also *Phleum pratense*
tingid, Tingidae 413, 414, 415, 416, 475; see also *Teleonemia scrupulosa*
Tithonia diversifolia 130
tobacco (*Nicotiana tabacum*) 220; see also *Nicotiana tabacum*
tocalote (yellow star thistle) 77; see also *Centaurea melitensis*
tomato (*Lycopersicon pimpinellifolium*) 105, 220; see also *Lycopersicon pimpinellifolium*
Toona ciliata 143
Torenia asiatica 149
Torilis nodosa 74, 81, 126
torpedo grass (*Panicum repens*) 538; see also *Panicum repens*
tortricid 412, 413; see also leafroller, *Epinotia lantana*
Torulinium odoratum 117
touch-me-not (Balsaminaceae) 131
Tournefortia argentea (tree heliotrope) 132, 200, 212; see also tree heliotrope
Toxicodendron diversilobum 81
toxic plants 7, 8, 16, 597, 603, 606
toyon (*Heteromeles arbutifolia*) 71; see also *Heteromeles arbutifolia*
Tradescantia geniculata 117
Tradescantia spathacea (oyster plant) 117, 206
Tradescantia zebrina (wandering Jew) 117, 206
Tragopogon porrifolius 130
Tragus berteronianus 123
Trathala flavoorbitalis 455
tree cactus. See *Opuntia ficus-indica*
tree fern 168, 170, 184, 185, 283, 284, 285, 287, 314, 397, 561, 562, 667, 684, 692, 694, 800; see also cibotium, *Cibotium, C. glaucum, Cyathea cooperi*, hāpu'u
tree fern forest 170, 283

tree heliotrope (*Tournefortia argentea*) 199, 200, 212; see also *Tournefortia argentea*

trees 10, 13, 20, 42, 64, 67, 75, 76, 78, 101, 156, 158, 193, 200, 225-237, 303, 311, 312, 313, 316, 335-336, 396, 397, 399, 477, 508ff, 521, 527, 531, 532, 537, 552, 559, 560, 562, 566, 569, 616, 620, 631, 632, 715, 727, 728, 729, 733, 771; see also species listings

tree tobacco (*Nicotiana glauca*) 220, 603; see also *Nicotiana glauca*

trema (*Trema orientalis*) 151, 493, 512, 520, 521

Trembleya phylogiformis 143

tribulus 222; see also Zygophyllaceae

Tribulus cistoides (nohu, puncture vine) 222, 470, 475; see also nohu, puncture vine

Tribulus terrestris (puncture vine) 152, 470, 475, 765; see also puncture vine

Tricachne insularis 123

Trichogramma 414, 416

Tridax procumbens (coat buttons) 130, 211, 616

Trifolium 15, 86, 140

Trimezia martinicensis 118

Triodanis biflora 133

Trisetum glomeratum (pili uka) 626, 630, 647, 650, 652; see also pili uka

Tristellateia australasiae (bagnit) 217

Tritonia crocosmiflora (montbretia) 544; see also montbretia

Triumfetta 150

Tropaeolaceae 151, 221, 493, 512; see also nasturtium

Tropaeolum majus (nasturtium) 151, 221, 494, 506, 512, 520, 521, 786, 793; see also nasturtium

tropical ash (*Fraxinus uhdei*) 284, 488, 510, 513, 516, 531, 532, 547; see also *Fraxinus uhdei*

tsugi pine (*Cryptomeria japonica*) 284

tundra 7

tundra vole (*Microtus oeconomus*) 21

Turdus migratorius (robin) 51

turkey, wild (*Meleagris gallopavo*) 281, 287

turks cap (*Malvaviscus arboreus*) 218

Turnera (Turneraceae) 151

Turnera ulmifolia 151

turtle 189, 502; see also green sea turtle, herpetofauna, loggerhead turtle, sea turtle

tussock grass (*Eragrostis xerophila*) 604; see also *Eragrostis xerophila*

Typhaceae (cattail) 92, 124

Typha domingensis 72, 92, 112, 124

Typha latifolia 124

'uhaloa (*Waltheria indica*) 104, 221; see also *Waltheria indica*

'uki (*Machaerina angustifolia*) 347, 372

'ūlei (*Osteomeles anthyllidifolia*) 352, 354, 355, 394, 402, 559, 627; see also *Osteomeles anthyllidifolia*

Ulex 33, 773

Ulex europaeus (gorse) 140, 255, 470, 472, 502, 527, 530, 538, 540, 551, 557, 563-564, 577, 593, 770, 830; see also gorse

Ulmaceae (elm) 151, 494, 512

uluhe 246, 302, 304, 311, 313, 320, 396, 400, 401, 402, 684, 687, 696, 697, 790; see also matted fern, *Dicranopteris* spp.

umbrella plant (*Cyperus alternifolius* subsp. *flabelliformis*) 206

ungulates 10, 11, 14, 22, 175, 315, 317, 497, 501, 503, 536, 543, 597, 598, 600, 601, 602, 603, 604, 606, 608, 609, 611, 619, 620, 626, 628, 629, 666, 667, 668, 697, 698, 699, 700, 701, 753, 784, 797, 800, 822, 823; see also cattle, feral ungulates, goat, pig, sheep

upland forest 619

Urena lobata (aramina) 102, 104

Urera sandwicensis (ōpuhe) 104

Uromyces rumicis 436

Uroplata girardi 415, 416, 417

Urticaceae (nettle) 92, 104, 151, 221

Urtica dioica holosericea 92

Urtica urens 92, 151, 631, 651, 653

Vaccinium ('ōhelo) 307, 308, 348, 400, 626, 631, 645, 700; see also 'ōhelo
Vaccinium reticulatum (kau-la'ai, 'ōhelo) 372, 669, 671, 672; see also 'ōhelo
valerian (Valerianaceae) 151
vanda orchid 207
Vanessa tameamea (Kamehameha butterfly) 278
vasey grass (*Paspalum urvillei*) 208, 315, 318, 491, 538; see also *Paspalum urvillei*
velvet grass (*Holcus lanatus*) 6, 304, 307, 315, 317, 319, 364, 401, 489, 557, 558, 562-563, 565, 570, 626, 632, 634, 683, 699, 700; see also *Holcus lanatus*
Verbascum blattaria 149
Verbascum lasiostachys 92
Verbascum thapsus (mullein) 15, 16, 149, 254-257, 259-260, 262-268, 494, 557, 629, 631, 648, 651, 653; see also mullein
Verbascum virgatum 149
verbena 151, 221; see also Verbenaceae
Verbena bonariensis 151
Verbenaceae (verbena) 92, 151, 221, 489
Verbena litoralis (ha'uōwī, weed verbena) 152, 221, 364
Verbena rigida 152
Verbesina encelioides (golden crownbeard) 131, 199, 211, 556, 651
Vernonia cinerea 131, 616
Veronica arvensis 149
Veronica plebeia (common speedwell) 149, 364
Veronica serpyllifolia 149, 651
Vestiaria coccinea ('i'iwi) 278, 477
Vicia americana 86
Vicia angustifolia 140
Vicia exigua 86
Vicia sativa 555, 556
Vigna adenantha 102, 103
vin (*Pritchardia pacifica*) 206
Vinca minor 20
vinegar tree (*Lophostemon confertus*) 529
vines 20, 42, 72, 75, 76, 156, 158, 159, 271-293, 316, 442, 498, 506, 507, 509ff, 521, 567, 568, 678, 681, 692, 728, 770; see also lianas
Violaceae (violet) 221
Viola chamissoniana 561
Viola maviensis 565
Viola odorata (violet) 221
violet 221
virgate mimosa (*Desmanthus virgatus*) 215; see also *Desmanthus virgatus*
viruses 45, 433, 453
Vitaceae (grape, *Vitis*) 152, 222
Vitex trifolia (pōlinalina) 221
Vitis (grape) 222; see also Vitaceae
volcanism 11, 235, 236, 395, 396, 398, 607, 610, 676, 692
Vulpia 89
Vulpia bromoides (brome fescue) 124, 566
Vulpia myuros (foxtail fescue, rat tail fescue) 124, 209, 630, 636, 641, 647, 649, 650, 652
Vulpia octoflora 89, 124

Wahlenbergia gracilis 133
waiawī (strawberry guava, *Psidium cattleianum*) 538, 547; see also strawberry guava, *Psidium cattleianum*
waikāhuli (*Gaillardia pulchella*) 211; see also *Gaillardia pulchella*
Waltheria americana (hi'aloa) 402
Waltheria indica ('uhaloa) 102, 104, 221
Waltheria lophanthus 606
wandering Jew (*Tradescantia zebrina*) 206; see also *Tradescantia zebrina*
Washingtonia filifera (fan palm) 17, 75; see also fan palm
wasp 414
waterbirds 101, 104
watercourses 17, 554, 562, 563, 564, 565, 567
water hyacinth (*Eichhornia crassipes*) 9, 20, 43, 52; see also *Eichhornia crassipes*
water-leaf (Hydrophyllaceae) 217; see also Hydrophyllaceae
water lemon (*Passiflora foetida*) 604; see also *Passiflora foetida*
water milfoil (Haloragaceae) 9, 141; see also *Myriophyllum*
water plantain (Alismataceae) 116
water regimes 3, 6, 18, 29, 30, 32, 35, 36, 49, 51, 76, 93, 241, 697, 821

watersheds 107, 290, 477, 502, 537, 539, 554, 564, 577, 580, 593, 667, 726, 727, 728, 729, 761, 800, 801
water table 9, 14
wauke (*Broussonetia papyrifera*) 729; see also *Broussonetia papyrifera*
wawae'iole (*Lycopodium cernuum*) 345, 347, 349, 372
wax myrtle (*Myrica cerifera*) 772; see also *Myrica cerifera*
wedelia 131, 211
Wedelia trilobata (wedelia) 211
wedge-tailed shearwater (*Puffinus pacificus chlorophynchus*) 200
weed verbena (*Verbena litoralis*) 364; see also *Verbena litoralis*
weevil 267, 475
West Indian dropseed (*Sporobolus indicus*) 209; see also *Sporobolus indicus*
wet forest 226, 271, 285, 287, 288, 290, 304, 343, 344, 532, 541, 544, 696, 787
wet habitat 168, 170, 171, 177, 178, 184-186, 241, 242, 284, 313, 317, 343, 396, 416, 417, 468, 614, 666, 667
wetlands 6, 9, 21, 47, 49, 51, 54, 101, 104, 541
wet shrub community 542
white ash (*Fraxinus americana*) 488, 509, 513, 516
white field daisy 7; see also chrysanthemum
whitefly (*Aleurodicus dispersus*) 456
white ginger (*Hedychium coronarium*) 529; see also *Hedychium coronarium*
white mulberry (*Morus alba*) 218; see also *Morus alba*
white sweet clover (*Mililotus alba*) 216; see also *Mililotus alba*
white tern (*Gygis alba*) 197, 604
white vine (*Muehlenbeckia axillaris*) 507, 511, 517, 521; see also *Muehlenbeckia axillaris*
Wikstroemia ('ākia) 359, 403; see also 'ākia
Wikstroemia sandwicensis ('ākia) 400; see also 'ākia
wildfires. See fires
wild mustard (*Brassica nigra, B. rapa*) 73; see also *Brassica nigra, B. rapa*
wild oats (*Avena fatua*) 72, 77, 79; see also *Avena fatua*
wild poinsettia (*Euphorbia cyathopora*) 215; see also *Euphorbia cyathopora*
wild strawberry (*Fragaria vesca*) 364; see also *Fragaria vesca*
wild turkey (*Meleagris gallopavo*) 281; see also turkey
wiliwili (*Erythrina sandwicensis*) 397, 402, 603
willow (*Salix lasiolepis*) 72; see also *Salix lasiolepis*
wilt disease 443, 444, 445
wilt virus 453
wind disturbance 11, 52, 53, 66, 110, 198, 202, 241, 242, 244, 249, 282, 283, 287, 288, 289
witchweed (*Striga asiatica*) 740
wood groundsel (*Senecio sylvaticus*) 34; see also *Senecio sylvaticus*
woodlands 17, 67, 71, 396, 402, 403, 625, 626, 644, 646, 650ff, 658, 661, 733
wood rose (*Merremia tuberosa*) 214; see also *Merremia tuberosa*
wood sorrel (Oxalidaceae) 145, 219
woody plants 13, 17, 31, 303, 313, 343, 349, 376, 397, 399, 434, 584, 585, 611, 619, 626, 633, 634, 644, 645, 654, 655, 658, 659, 660, 661, 662, 666, 667, 669, 670, 672, 675, 676, 677, 678, 679, 680, 681, 682, 683, 685, 687, 688, 689, 690, 692, 693, 694, 695, 699, 700, 733, 734, 824

Xanthium spinosum 74, 83
Xanthium strumarium (cocklebur) 43, 131, 211, 556
Xanthosoma ('ape) 205
Xantusia riversiana (island night lizard) 71

xeric habitats 155, 169, 180-182, 376, 604
Xyridaceae (yellow-eyed grass) 124
Xyris 124

yam (Dioscoreaceae, *Dioscorea alata)* 118, 729; see also Dioscoreaceae, *Dioscorea alata*
yellow-eyed grass (Xyridaceae) 124
yellow ginger *(Hedychium flavescens)* 529; see also *Hedychium flavescens*
yellow Himalayan raspberry (*Rubus ellipticus*) 155-158, 161, 166, 169-171, 174-176, 178, 341, 343, 366-370, 372, 399, 401, 472, 492, 692, 694, 695, 696, 786, 789, 790, 793, 794, 819; see also *Rubus ellipticus*
yellow nutsedge *(Cyperus esculentus)* 436; see also *Cyperus esculentus*
yellow star thistle (tocalote) 77; see also *Centaurea melitensis*
Yellowstone elk 15
yellow sweet clover (*Melilotus, Melilotus indica*) 77; see also *Melilotus, Melilotus indica*
yellow wood sorrel (*Oxalis corniculata*) 103, 219; see also *Oxalis corniculata*
Youngia japonica 131, 334-335

Zantedeschia aethiopica 81
Zanthoxylum kauaense 561
Zauschneria 88
Zea mays (corn) 209, 730, 740
Zigadenus fremontii 87
Zingiberaceae 124, 209, 489; see also ginger
Zingiber zerumbet 102, 124
Zinnia 131
Zosteraceae 92
Zostera marina 92
Zosterops japonicus (Japanese white-eye) 227, 278, 548, 567; see also Japanese white-eye
zucchini (*Cucurbita pepo,* squash) 214
Zygophyllaceae (creosote bush, *Tribulus*) 152, 222